T0327376

Digital Electronics

Digital Electronics

Principles, Devices and Applications

Anil K. Maini
Defence Research and Development Organization (DRDO), India

John Wiley & Sons, Ltd

Telephone (+44) 1243 779777

Email (for orders and customer service enquiries): cs-books@wiley.co.uk
Visit our Home Page on www.wiley.com

Other Wiley Editorial Offices

John Wiley & Sons Inc., 111 River Street, Hoboken, NJ 07030, USA

Jossey-Bass, 989 Market Street, San Francisco, CA 94103-1741, USA

Wiley-VCH Verlag GmbH, Boschstr. 12, D-69469 Weinheim, Germany

John Wiley & Sons Australia Ltd, 42 McDougall Street, Milton, Queensland 4064, Australia

John Wiley & Sons (Asia) Pte Ltd, 2 Clementi Loop #02-01, Jin Xing Distripark, Singapore 129809

John Wiley & Sons Canada Ltd, 6045 Freemont Blvd, Mississauga, ONT, Canada L5R 4J3

Wiley also publishes its books in a variety of electronic formats. Some content that appears in print may not be available in electronic books.

Anniversary Logo Design: Richard J. Pacifico

Library of Congress Cataloging in Publication Data

Maini, Anil Kumar.
Digital electronics : principles, devices, and applications / Anil Kumar Maini.
p. cm.
Includes bibliographical references and index.
ISBN 978-0-470-03214-5 (Cloth)
1. Digital electronics. I. Title.
TK7868.D5M275 2007
621.381—dc22 2007020666

British Library Cataloguing in Publication Data

A catalogue record for this book is available from the British Library

ISBN 978-0-470-03214-5 (HB)

Typeset in 9/11pt Times by Integra Software Services Pvt. Ltd, Pondicherry, India
Printed and bound in Great Britain by Antony Rowe Ltd, Chippenham, Wiltshire
This book is printed on acid-free paper responsibly manufactured from sustainable forestry in which at least two trees are planted for each one used for paper production.

In the loving memory of my father, Shri Sukhdev Raj Maini, *who has been a source of inspiration, courage and strength to me to face all challenges in life, and above all instilled in me the value of helping people to make this world a better place.*

Anil K. Maini

Contents

Preface

Digital electronics is essential to understanding the design and working of a wide range of applications, from consumer and industrial electronics to communications; from embedded systems, and computers to security and military equipment. As the devices used in these applications decrease in size and employ more complex technology, it is essential for engineers and students to fully understand both the fundamentals and also the implementation and application principles of digital electronics, devices and integrated circuits, thus enabling them to use the most appropriate and effective technique to suit their technical needs.

Digital Electronics: Principles, Devices and Applications is a comprehensive book covering, in one volume, both the fundamentals of digital electronics and the applications of digital devices and integrated circuits. It is different from similar books on the subject in more than one way. Each chapter in the book, whether it is related to operational fundamentals or applications, is amply illustrated with diagrams and design examples. In addition, the book covers several new topics, which are of relevance to any one having an interest in digital electronics and not covered in the books already in print on the subject. These include digital troubleshooting, digital instrumentation, programmable logic devices, microprocessors and microcontrollers. While the book covers in entirety what is required by undergraduate and graduate level students of engineering in electrical, electronics, computer science and information technology disciplines, it is intended to be a very useful reference book for professionals, R&D scientists and students at post graduate level.

The book is divided into sixteen chapters covering seven major topics. These are: *digital electronics fundamentals* (chapters 1 to 6), *combinational logic circuits* (chapters 7 and 8), *programmable logic devices* (chapter 9), *sequential logic circuits* (chapters 10 and 11), *data conversion devices and circuits* (chapter 12), *microprocessors, microcontrollers and microcomputers* (chapters 13 to 15) and *digital troubleshooting and instrumentation* (chapter 16). The contents of each of the sixteen chapters are briefly described in the following paragraphs.

The first six chapters deal with the fundamental topics of digital electronics. These include different number systems that can be used to represent data and binary codes used for representing numeric and alphanumeric data. Conversion from one number system to another and similarly conversion from one code to another is discussed at length in these chapters. Binary arithmetic, covering different methods of performing arithmetic operations on binary numbers is discussed next. Chapters four and five cover logic gates and logic families. The main topics covered in these two chapters are various logic gates and related devices, different logic families used to hardware implement digital integrated circuits, the interface between digital ICs belonging to different logic families and application information such as guidelines for using logic devices of different families. Boolean algebra and its various postulates and theorems and minimization techniques, providing exhaustive coverage of both Karnaugh mapping and Quine-McCluskey techniques, are discussed in chapter six. The discussion includes application of these minimization techniques for multi-output Boolean functions and Boolean functions with larger number of variables. The concepts underlying different fundamental topics of digital electronics and discussed in first six chapters have been amply illustrated with solved examples.

As a follow-up to logic gates – the most basic building block of combinational logic – chapters 7 and 8 are devoted to more complex combinational logic circuits. While chapter seven covers arithmetic circuits, including different types of adders and subtractors, such as half and full adder and subtractor, adder-subtractor, larger bit adders and subtractors, multipliers, look ahead carry generator, magnitude comparator, and arithmetic logic unit, chapter eight covers multiplexers, de-multiplexers, encoders and decoders. This is followed by a detailed account of programmable logic devices in chapter nine. Simple programmable logic devices (SPLDs) such as PAL, PLA, GAL and HAL devices, complex programmable logic devices (CPLDs) and field programmable gate arrays (FPGAs) have been exhaustively treated in terms of their architecture, features and applications. Popular devices, from various international manufacturers, in the three above-mentioned categories of programmable logic devices are also covered with regard to their architecture, features and facilities.

The next two chapters, 10 and 11, cover the sequential logic circuits. Discussion begins with the most fundamental building block of sequential logic, that is, *flip flop*. Different types of flip flops are covered in detail with regard to their operational fundamentals, different varieties in each of the categories of flip flops and their applications. Multivibrator circuits, being operationally similar to flip flops, are also covered at length in this chapter. Counters and registers are the other very important building blocks of sequential logic with enormous application potential. These are covered in chapter 11. Particular emphasis is given to timing requirements and design of counters with varying count sequence requirements. The chapter also includes a detailed description of the design principles of counters with arbitrary count sequences. Different types of shift registers and some special counters that have evolved out of shift registers have been covered in detail.

Chapter 12 covers data conversion circuits including digital-to-analogue and analogue-to-digital converters. Topics covered in this chapter include operational basics, characteristic parameters, types and applications. Emphasis is given to definition and interpretation of the terminology and the performance parameters that characterize these devices. Different types of digital-to-analogue and analogue-to-digital converters, together with their merits and drawbacks are also addressed. Particular attention is given to their applications. Towards the end of the chapter, application oriented information in the form of popular type numbers along with their major performance specifications, pin connection diagrams etc. is presented. Another highlight of the chapter is the inclusion of detailed descriptions of newer types of converters, such as quad slope and sigma-delta types of analogue-to-digital converters.

Chapters 13 and 14 discuss microprocessors and microcontrollers – the two versatile devices that have revolutionized the application potential of digital devices and integrated circuits. The entire range of microprocessors and microcontrollers along with their salient features, operational aspects and application guidelines are covered in detail. As a natural follow-up to these, microcomputer fundamentals, with regard to their architecture, input/output devices and memory devices, are discussed in chapter 15.

The last chapter covers digital troubleshooting techniques and digital instrumentation. Troubleshooting guidelines for various categories of digital electronics circuits are discussed. These will particularly benefit practising engineers and electronics enthusiasts. The concepts are illustrated with the help of a large number of troubleshooting case studies pertaining to combinational, sequential and memory devices. A wide range of digital instruments is covered after a discussion on troubleshooting guidelines. The instruments covered include digital multimeters, digital oscilloscopes, logic probes, logic analysers, frequency synthesizers, and synthesized function generators. Computer-instrument interface standards and the concept of virtual instrumentation are also discussed at length towards the end of the chapter.

As an extra resource, a companion website for my book contains lot of additional application relevant information on digital devices and integrated circuits. The information on this website includes numerical and functional indices of digital integrated circuits belonging to different logic families, pin connection diagrams and functional tables of different categories of general purpose digital integrated circuits and application relevant information on microprocessors, peripheral devices and microcontrollers. Please go to URL http://www.wiley.com/go/maini_digital.

The motivation to write this book and the selection of topics to be covered were driven mainly by the absence a book, which, in one volume, covers all the important aspects of digital technology. A large number of books in print on the subject cover all the routine topics of digital electronics in a conventional way with total disregard to the needs of application engineers and professionals. As the author, I have made an honest attempt to cover the subject in entirety by including comprehensive treatment of newer topics that are either ignored or inadequately covered in the available books on the subject of digital electronics. This is done keeping in view the changed requirements of my intended audience, which includes undergraduate and graduate level students, R&D scientists, professionals and application engineers.

Anil K. Maini

1

Number Systems

The study of *number systems* is important from the viewpoint of understanding how data are represented before they can be processed by any digital system including a digital computer. It is one of the most basic topics in digital electronics. In this chapter we will discuss different number systems commonly used to represent data. We will begin the discussion with the decimal number system. Although it is not important from the viewpoint of digital electronics, a brief outline of this will be given to explain some of the underlying concepts used in other number systems. This will then be followed by the more commonly used number systems such as the binary, octal and hexadecimal number systems.

1.1 Analogue Versus Digital

There are two basic ways of representing the numerical values of the various physical quantities with which we constantly deal in our day-to-day lives. One of the ways, referred to as *analogue*, is to express the numerical value of the quantity as a continuous range of values between the two expected extreme values. For example, the temperature of an oven settable anywhere from 0 to 100 °C may be measured to be 65 °C or 64.96 °C or 64.958 °C or even 64.9579 °C and so on, depending upon the accuracy of the measuring instrument. Similarly, voltage across a certain component in an electronic circuit may be measured as 6.5 V or 6.49 V or 6.487 V or 6.4869 V. The underlying concept in this mode of representation is that variation in the numerical value of the quantity is continuous and could have any of the infinite theoretically possible values between the two extremes.

The other possible way, referred to as *digital*, represents the numerical value of the quantity in steps of discrete values. The numerical values are mostly represented using binary numbers. For example, the temperature of the oven may be represented in steps of 1 °C as 64 °C, 65 °C, 66 °C and so on. To summarize, while an analogue representation gives a continuous output, a digital representation produces a discrete output. Analogue systems contain devices that process or work on various physical quantities represented in analogue form. Digital systems contain devices that process the physical quantities represented in digital form.

Digital Electronics: Principles, Devices and Applications Anil Kumar Maini

Digital techniques and systems have the advantages of being relatively much easier to design and having higher accuracy, programmability, noise immunity, easier storage of data and ease of fabrication in integrated circuit form, leading to availability of more complex functions in a smaller size. The real world, however, is analogue. Most physical quantities – position, velocity, acceleration, force, pressure, temperature and flowrate, for example – are analogue in nature. That is why analogue variables representing these quantities need to be digitized or discretized at the input if we want to benefit from the features and facilities that come with the use of digital techniques. In a typical system dealing with analogue inputs and outputs, analogue variables are digitized at the input with the help of an analogue-to-digital converter block and reconverted back to analogue form at the output using a digital-to-analogue converter block. Analogue-to-digital and digital-to-analogue converter circuits are discussed at length in the latter part of the book. In the following sections we will discuss various number systems commonly used for digital representation of data.

1.2 Introduction to Number Systems

We will begin our discussion on various number systems by briefly describing the parameters that are common to all number systems. An understanding of these parameters and their relevance to number systems is fundamental to the understanding of how various systems operate. Different characteristics that define a number system include the number of independent digits used in the number system, the place values of the different digits constituting the number and the maximum numbers that can be written with the given number of digits. Among the three characteristic parameters, the most fundamental is the number of independent digits or symbols used in the number system. It is known as the *radix* or *base* of the number system. The decimal number system with which we are all so familiar can be said to have a radix of 10 as it has 10 independent digits, i.e. 0, 1, 2, 3, 4, 5, 6, 7, 8 and 9. Similarly, the binary number system with only two independent digits, 0 and 1, is a radix-2 number system. The octal and hexadecimal number systems have a radix (or base) of 8 and 16 respectively. We will see in the following sections that the radix of the number system also determines the other two characteristics. The place values of different digits in the integer part of the number are given by r^0, r^1, r^2, r^3 and so on, starting with the digit adjacent to the radix point. For the fractional part, these are r^{-1}, r^{-2}, r^{-3} and so on, again starting with the digit next to the radix point. Here, r is the radix of the number system. Also, maximum numbers that can be written with n digits in a given number system are equal to r^n.

1.3 Decimal Number System

The decimal number system is a radix-10 number system and therefore has 10 different digits or symbols. These are 0, 1, 2, 3, 4, 5, 6, 7, 8 and 9. All higher numbers after '9' are represented in terms of these 10 digits only. The process of writing higher-order numbers after '9' consists in writing the second digit (i.e. '1') first, followed by the other digits, one by one, to obtain the next 10 numbers from '10' to '19'. The next 10 numbers from '20' to '29' are obtained by writing the third digit (i.e. '2') first, followed by digits '0' to '9', one by one. The process continues until we have exhausted all possible two-digit combinations and reached '99'. Then we begin with three-digit combinations. The first three-digit number consists of the lowest two-digit number followed by '0' (i.e. 100), and the process goes on endlessly.

The place values of different digits in a mixed decimal number, starting from the decimal point, are 10^0, 10^1, 10^2 and so on (for the integer part) and 10^{-1}, 10^{-2}, 10^{-3} and so on (for the fractional part).

The value or magnitude of a given decimal number can be expressed as the sum of the various digits multiplied by their place values or weights.

As an illustration, in the case of the decimal number 3586.265, the integer part (i.e. 3586) can be expressed as

$$3586 = 6 \times 10^0 + 8 \times 10^1 + 5 \times 10^2 + 3 \times 10^3 = 6 + 80 + 500 + 3000 = 3586$$

and the fractional part can be expressed as

$$265 = 2 \times 10^{-1} + 6 \times 10^{-2} + 5 \times 10^{-3} = 0.2 + 0.06 + 0.005 = 0.265$$

We have seen that the place values are a function of the radix of the concerned number system and the position of the digits. We will also discover in subsequent sections that the concept of each digit having a place value depending upon the position of the digit and the radix of the number system is equally valid for the other more relevant number systems.

1.4 Binary Number System

The binary number system is a radix-2 number system with '0' and '1' as the two independent digits. All larger binary numbers are represented in terms of '0' and '1'. The procedure for writing higher-order binary numbers after '1' is similar to the one explained in the case of the decimal number system. For example, the first 16 numbers in the binary number system would be 0, 1, 10, 11, 100, 101, 110, 111, 1000, 1001, 1010, 1011, 1100, 1101, 1110 and 1111. The next number after 1111 is 10000, which is the lowest binary number with five digits. This also proves the point made earlier that a maximum of only 16 ($= 2^4$) numbers could be written with four digits. Starting from the binary point, the place values of different digits in a mixed binary number are 2^0, 2^1, 2^2 and so on (for the integer part) and 2^{-1}, 2^{-2}, 2^{-3} and so on (for the fractional part).

Example 1.1

Consider an arbitrary number system with the independent digits as 0, 1 and X. What is the radix of this number system? List the first 10 numbers in this number system.

Solution

- The radix of the proposed number system is 3.
- The first 10 numbers in this number system would be 0, 1, X, 10, 11, 1X, X0, X1, XX and 100.

1.4.1 Advantages

Logic operations are the backbone of any digital computer, although solving a problem on computer could involve an arithmetic operation too. The introduction of the mathematics of logic by George Boole laid the foundation for the modern digital computer. He reduced the mathematics of logic to a binary notation of '0' and '1'. As the mathematics of logic was well established and had proved itself to be quite useful in solving all kinds of logical problem, and also as the mathematics of logic (also known as Boolean algebra) had been reduced to a binary notation, the binary number system had a clear edge over other number systems for use in computer systems.

Yet another significant advantage of this number system was that all kinds of data could be conveniently represented in terms of 0s and 1s. Also, basic electronic devices used for hardware implementation could be conveniently and efficiently operated in two distinctly different modes. For example, a bipolar transistor could be operated either in cut-off or in saturation very efficiently.

Lastly, the circuits required for performing arithmetic operations such as addition, subtraction, multiplication, division, etc., become a simple affair when the data involved are represented in the form of 0s and 1s.

1.5 Octal Number System

The octal number system has a radix of 8 and therefore has eight distinct digits. All higher-order numbers are expressed as a combination of these on the same pattern as the one followed in the case of the binary and decimal number systems described in Sections 1.3 and 1.4. The independent digits are 0, 1, 2, 3, 4, 5, 6 and 7. The next 10 numbers that follow '7', for example, would be 10, 11, 12, 13, 14, 15, 16, 17, 20 and 21. In fact, if we omit all the numbers containing the digits 8 or 9, or both, from the decimal number system, we end up with an octal number system. The place values for the different digits in the octal number system are 8^0, 8^1, 8^2 and so on (for the integer part) and 8^{-1}, 8^{-2}, 8^{-3} and so on (for the fractional part).

1.6 Hexadecimal Number System

The hexadecimal number system is a radix-16 number system and its 16 basic digits are 0, 1, 2, 3, 4, 5, 6, 7, 8, 9, A, B, C, D, E and F. The place values or weights of different digits in a mixed hexadecimal number are 16^0, 16^1, 16^2 and so on (for the integer part) and 16^{-1}, 16^{-2}, 16^{-3} and so on (for the fractional part). The decimal equivalent of A, B, C, D, E and F are 10, 11, 12, 13, 14 and 15 respectively, for obvious reasons.

The hexadecimal number system provides a condensed way of representing large binary numbers stored and processed inside the computer. One such example is in representing addresses of different memory locations. Let us assume that a machine has 64K of memory. Such a memory has 64K ($= 2^{16} = 65\,536$) memory locations and needs 65 536 different addresses. These addresses can be designated as 0 to 65 535 in the decimal number system and 00000000 00000000 to 11111111 11111111 in the binary number system. The decimal number system is not used in computers and the binary notation here appears too cumbersome and inconvenient to handle. In the hexadecimal number system, 65 536 different addresses can be expressed with four digits from 0000 to FFFF. Similarly, the contents of the memory when represented in hexadecimal form are very convenient to handle.

1.7 Number Systems – Some Common Terms

In this section we will describe some commonly used terms with reference to different number systems.

1.7.1 Binary Number System

Bit is an abbreviation of the term 'binary digit' and is the smallest unit of information. It is either '0' or '1'. A *byte* is a string of eight bits. The byte is the basic unit of data operated upon as a single unit in computers. A *computer word* is again a string of bits whose size, called the 'word length' or 'word size', is fixed for a specified computer, although it may vary from computer to computer. The word length may equal one byte, two bytes, four bytes or be even larger.

The *1's complement* of a binary number is obtained by complementing all its bits, i.e. by replacing 0s with 1s and 1s with 0s. For example, the 1's complement of $(10010110)_2$ is $(01101001)_2$. The *2's complement* of a binary number is obtained by adding '1' to its 1's complement. The 2's complement of $(10010110)_2$ is $(01101010)_2$.

1.7.2 Decimal Number System

Corresponding to the 1's and 2's complements in the binary system, in the decimal number system we have the 9's and 10's complements. The *9's complement* of a given decimal number is obtained by subtracting each digit from 9. For example, the 9's complement of $(2496)_{10}$ would be $(7503)_{10}$. The *10's complement* is obtained by adding '1' to the 9's complement. The 10's complement of $(2496)_{10}$ is $(7504)_{10}$.

1.7.3 Octal Number System

In the octal number system, we have the 7's and 8's complements. The *7's complement* of a given octal number is obtained by subtracting each octal digit from 7. For example, the 7's complement of $(562)_8$ would be $(215)_8$. The *8's complement* is obtained by adding '1' to the 7's complement. The 8's complement of $(562)_8$ would be $(216)_8$.

1.7.4 Hexadecimal Number System

The 15's and 16's complements are defined with respect to the hexadecimal number system. The *15's complement* is obtained by subtracting each hex digit from 15. For example, the 15's complement of $(3BF)_{16}$ would be $(C40)_{16}$. The *16's complement* is obtained by adding '1' to the 15's complement. The 16's complement of $(2AE)_{16}$ would be $(D52)_{16}$.

1.8 Number Representation in Binary

Different formats used for binary representation of both positive and negative decimal numbers include the sign-bit magnitude method, the 1's complement method and the 2's complement method.

1.8.1 Sign-Bit Magnitude

In the sign-bit magnitude representation of positive and negative decimal numbers, the MSB represents the 'sign', with a '0' denoting a plus sign and a '1' denoting a minus sign. The remaining bits represent the magnitude. In eight-bit representation, while MSB represents the sign, the remaining seven bits represent the magnitude. For example, the eight-bit representation of +9 would be 00001001, and that for −9 would be 10001001. An n−bit binary representation can be used to represent decimal numbers in the range of $-(2^{n-1}-1)$ to $+(2^{n-1}-1)$. That is, eight-bit representation can be used to represent decimal numbers in the range from −127 to +127 using the sign-bit magnitude format.

1.8.2 1's Complement

In the 1's complement format, the positive numbers remain unchanged. The negative numbers are obtained by taking the 1's complement of the positive counterparts. For example, +9 will be represented as 00001001 in eight-bit notation, and −9 will be represented as 11110110, which is the 1's complement of 00001001. Again, n-bit notation can be used to represent numbers in the range from $-(2^{n-1}-1)$ to $+(2^{n-1}-1)$ using the 1's complement format. The eight-bit representation of the 1's complement format can be used to represent decimal numbers in the range from −127 to +127.

1.8.3 2's Complement

In the 2's complement representation of binary numbers, the MSB represents the sign, with a '0' used for a plus sign and a '1' used for a minus sign. The remaining bits are used for representing magnitude. Positive magnitudes are represented in the same way as in the case of sign-bit or 1's complement representation. Negative magnitudes are represented by the 2's complement of their positive counterparts. For example, +9 would be represented as 00001001, and −9 would be written as 11110111. Please note that, if the 2's complement of the magnitude of +9 gives a magnitude of −9, then the reverse process will also be true, i.e. the 2's complement of the magnitude of −9 will give a magnitude of +9. The n-bit notation of the 2's complement format can be used to represent all decimal numbers in the range from $+(2^{n-1}-1)$ to $-(2^{n-1})$. The 2's complement format is very popular as it is very easy to generate the 2's complement of a binary number and also because arithmetic operations are relatively easier to perform when the numbers are represented in the 2's complement format.

1.9 Finding the Decimal Equivalent

The decimal equivalent of a given number in another number system is given by the sum of all the digits multiplied by their respective place values. The integer and fractional parts of the given number should be treated separately. Binary-to-decimal, octal-to-decimal and hexadecimal-to-decimal conversions are illustrated below with the help of examples.

1.9.1 Binary-to-Decimal Conversion

The decimal equivalent of the binary number $(1001.0101)_2$ is determined as follows:

- The integer part = 1001
- The decimal equivalent $= 1 \times 2^0 + 0 \times 2^1 + 0 \times 2^2 + 1 \times 2^3 = 1 + 0 + 0 + 8 = 9$
- The fractional part = .0101
- Therefore, the decimal equivalent $= 0 \times 2^{-1} + 1 \times 2^{-2} + 0 \times 2^{-3} + 1 \times 2^{-4} = 0 + 0.25 + 0$
  $+ 0.0625 = 0.3125$
- Therefore, the decimal equivalent of $(1001.0101)_2 = 9.3125$

1.9.2 Octal-to-Decimal Conversion

The decimal equivalent of the octal number $(137.21)_8$ is determined as follows:

- The integer part = 137
- The decimal equivalent $= 7 \times 8^0 + 3 \times 8^1 + 1 \times 8^2 = 7 + 24 + 64 = 95$

- The fractional part = .21
- The decimal equivalent = $2 \times 8^{-1} + 1 \times 8^{-2} = 0.265$
- Therefore, the decimal equivalent of $(137.21)_8 = (95.265)_{10}$

1.9.3 Hexadecimal-to-Decimal Conversion

The decimal equivalent of the hexadecimal number $(1E0.2A)_{16}$ is determined as follows:

- The integer part = 1E0
- The decimal equivalent = $0 \times 16^0 + 14 \times 16^1 + 1 \times 16^2 = 0 + 224 + 256 = 480$
- The fractional part = 2A
- The decimal equivalent = $2 \times 16^{-1} + 10 \times 16^{-2} = 0.164$
- Therefore, the decimal equivalent of $(1E0.2A)_{16} = (480.164)_{10}$

Example 1.2

Find the decimal equivalent of the following binary numbers expressed in the 2's complement format:

(a) 00001110;
(b) 10001110.

Solution

(a) The MSB bit is '0', which indicates a plus sign.
The magnitude bits are 0001110.

$$\begin{aligned}\text{The decimal equivalent} &= 0\times2^0+1\times2^1+1\times2^2+1\times2^3+0\times2^4+0\times2^5+0\times2^6\\ &= 0+2+4+8+0+0+0=14\end{aligned}$$

Therefore, 00001110 represents +14

(b) The MSB bit is '1', which indicates a minus sign
The magnitude bits are therefore given by the 2's complement of 0001110, i.e. 1110010

$$\begin{aligned}\text{The decimal equivalent} &= 0\times2^0+1\times2^1+0\times2^2+0\times2^3+1\times2^4+1\times2^5\\ &\quad +1\times2^6\\ &= 0+2+0+0+16+32+64=114\end{aligned}$$

Therefore, 10001110 represents −114

1.10 Decimal-to-Binary Conversion

As outlined earlier, the integer and fractional parts are worked on separately. For the integer part, the binary equivalent can be found by successively dividing the integer part of the number by 2 and recording the remainders until the quotient becomes '0'. The remainders written in reverse order constitute the binary equivalent. For the fractional part, it is found by successively multiplying the fractional part of the decimal number by 2 and recording the carry until the result of multiplication is '0'. The carry sequence written in forward order constitutes the binary equivalent of the fractional part of the decimal number. If the result of multiplication does not seem to be heading towards zero in the case of the fractional part, the process may be continued only until the requisite number of equivalent bits has been obtained. This method of decimal–binary conversion is popularly known as the double-dabble method. The process can be best illustrated with the help of an example.

Example 1.3

We will find the binary equivalent of $(13.375)_{10}$.

Solution

- The integer part = 13

Divisor	Dividend	Remainder
2	13	—
2	6	1
2	3	0
2	1	1
—	0	1

- The binary equivalent of $(13)_{10}$ is therefore $(1101)_2$
- The fractional part = .375
- $0.375 \times 2 = 0.75$ with a carry of 0
- $0.75 \times 2 = 0.5$ with a carry of 1
- $0.5 \times 2 = 0$ with a carry of 1
- The binary equivalent of $(0.375)_{10} = (.011)_2$
- Therefore, the binary equivalent of $(13.375)_{10} = (1101.011)_2$

1.11 Decimal-to-Octal Conversion

The process of decimal-to-octal conversion is similar to that of decimal-to-binary conversion. The progressive division in the case of the integer part and the progressive multiplication while working on the fractional part here are by '8' which is the radix of the octal number system. Again, the integer and fractional parts of the decimal number are treated separately. The process can be best illustrated with the help of an example.

Example 1.4

We will find the octal equivalent of $(73.75)_{10}$.

Solution

- The integer part = 73

Divisor	Dividend	Remainder
8	73	—
8	9	1
8	1	1
—	0	1

- The octal equivalent of $(73)_{10} = (111)_8$
- The fractional part = 0.75
- $0.75 \times 8 = 0$ with a carry of 6
- The octal equivalent of $(0.75)_{10} = (.6)_8$
- Therefore, the octal equivalent of $(73.75)_{10} = (111.6)_8$

1.12 Decimal-to-Hexadecimal Conversion

The process of decimal-to-hexadecimal conversion is also similar. Since the hexadecimal number system has a base of 16, the progressive division and multiplication factor in this case is 16. The process is illustrated further with the help of an example.

Example 1.5

Let us determine the hexadecimal equivalent of $(82.25)_{10}$.

Solution

- The integer part = 82

Divisor	Dividend	Remainder
16	82	—
16	5	2
—	0	5

- The hexadecimal equivalent of $(82)_{10} = (52)_{16}$
- The fractional part = 0.25
- $0.25 \times 16 = 0$ with a carry of 4
- Therefore, the hexadecimal equivalent of $(82.25)_{10} = (52.4)_{16}$

1.13 Binary–Octal and Octal–Binary Conversions

An octal number can be converted into its binary equivalent by replacing each octal digit with its three-bit binary equivalent. We take the three-bit equivalent because the base of the octal number system is 8 and it is the third power of the base of the binary number system, i.e. 2. All we have then to remember is the three-bit binary equivalents of the basic digits of the octal number system. A binary number can be converted into an equivalent octal number by splitting the integer and fractional parts into groups of three bits, starting from the binary point on both sides. The 0s can be added to complete the outside groups if needed.

Example 1.6

Let us find the binary equivalent of $(374.26)_8$ and the octal equivalent of $(1110100.0100111)_2$.

Solution

- The given octal number = $(374.26)_8$
- The binary equivalent = $(011\ 111\ 100.010\ 110)_2 = (011111100.010110)_2$

- Any 0s on the extreme left of the integer part and extreme right of the fractional part of the equivalent binary number should be omitted. Therefore, $(011111100.010110)_2 = (11111100.01011)_2$
- The given binary number $= (1110100.0100111)_2$
- $(1110100.0100111)_2 = (1\ 110\ 100.010\ 011\ 1)_2$
  $= (001\ 110\ 100.010\ 011\ 100)_2 = (164.234)_8$

1.14 Hex–Binary and Binary–Hex Conversions

A hexadecimal number can be converted into its binary equivalent by replacing each hex digit with its four-bit binary equivalent. We take the four-bit equivalent because the base of the hexadecimal number system is 16 and it is the fourth power of the base of the binary number system. All we have then to remember is the four-bit binary equivalents of the basic digits of the hexadecimal number system. A given binary number can be converted into an equivalent hexadecimal number by splitting the integer and fractional parts into groups of four bits, starting from the binary point on both sides. The 0s can be added to complete the outside groups if needed.

Example 1.7

Let us find the binary equivalent of $(17E.F6)_{16}$ and the hex equivalent of $(1011001110.011011101)_2$.

Solution

- The given hex number $= (17E.F6)_{16}$
- The binary equivalent $= (0001\ 0111\ 1110.1111\ 0110)_2$
  $= (000101111110.11110110)_2$
  $= (101111110.1111011)_2$
- The 0s on the extreme left of the integer part and on the extreme right of the fractional part have been omitted.
- The given binary number $= (1011001110.011011101)_2$
  $= (10\ 1100\ 1110.0110\ 1110\ 1)_2$
- The hex equivalent $= (0010\ 1100\ 1110.0110\ 1110\ 1000)_2 = (2CE.6E8)_{16}$

1.15 Hex–Octal and Octal–Hex Conversions

For hexadecimal–octal conversion, the given hex number is firstly converted into its binary equivalent which is further converted into its octal equivalent. An alternative approach is firstly to convert the given hexadecimal number into its decimal equivalent and then convert the decimal number into an equivalent octal number. The former method is definitely more convenient and straightforward. For octal–hexadecimal conversion, the octal number may first be converted into an equivalent binary number and then the binary number transformed into its hex equivalent. The other option is firstly to convert the given octal number into its decimal equivalent and then convert the decimal number into its hex equivalent. The former approach is definitely the preferred one. Two types of conversion are illustrated in the following example.

Example 1.8

Let us find the octal equivalent of $(2F.C4)_{16}$ and the hex equivalent of $(762.013)_8$.

Solution

- The given hex number = $(2F.C4)_{16}$.
- The binary equivalent = $(0010\ 1111.1100\ 0100)_2 = (00101111.11000100)_2$
  $= (101111.110001)_2 = (101\ 111.110\ 001)_2 = (57.61)_8$.
- The given octal number = $(762.013)_8$.
- The octal number = $(762.013)_8 = (111\ 110\ 010.000\ 001\ 011)_2$
  $= (111110010.000001011)_2$
  $= (0001\ 1111\ 0010.0000\ 0101\ 1000)_2 = (1F2.058)_{16}$.

1.16 The Four Axioms

Conversion of a given number in one number system to its equivalent in another system has been discussed at length in the preceding sections. The methodology has been illustrated with solved examples. The complete methodology can be summarized as four axioms or principles, which, if understood properly, would make it possible to solve any problem related to conversion of a given number in one number system to its equivalent in another number system. These principles are as follows:

1. Whenever it is desired to find the decimal equivalent of a given number in another number system, it is given by the sum of all the digits multiplied by their weights or place values. The integer and fractional parts should be handled separately. Starting from the radix point, the weights of different digits are r^0, r^1, r^2 for the integer part and r^{-1}, r^{-2}, r^{-3} for the fractional part, where r is the radix of the number system whose decimal equivalent needs to be determined.
2. To convert a given mixed decimal number into an equivalent in another number system, the integer part is progressively divided by r and the remainders noted until the result of division yields a zero quotient. The remainders written in reverse order constitute the equivalent. r is the radix of the transformed number system. The fractional part is progressively multiplied by r and the carry recorded until the result of multiplication yields a zero or when the desired number of bits has been obtained. The carrys written in forward order constitute the equivalent of the fractional part.
3. The octal–binary conversion and the reverse process are straightforward. For octal–binary conversion, replace each digit in the octal number with its three-bit binary equivalent. For hexadecimal–binary conversion, replace each hex digit with its four-bit binary equivalent. For binary–octal conversion, split the binary number into groups of three bits, starting from the binary point, and, if needed, complete the outside groups by adding 0s, and then write the octal equivalent of these three-bit groups. For binary–hex conversion, split the binary number into groups of four bits, starting from the binary point, and, if needed, complete the outside groups by adding 0s, and then write the hex equivalent of the four-bit groups.
4. For octal–hexadecimal conversion, we can go from the given octal number to its binary equivalent and then from the binary equivalent to its hex counterpart. For hexadecimal–octal conversion, we can go from the hex to its binary equivalent and then from the binary number to its octal equivalent.

Example 1.9

Assume an arbitrary number system having a radix of 5 and 0, 1, 2, L and M as its independent digits. Determine:

(a) the decimal equivalent of (12LM.L1);
(b) the total number of possible four-digit combinations in this arbitrary number system.

Solution

(a) The decimal equivalent of (12LM) is given by

$$\begin{aligned} \mathrm{M} \times 5^0 + \mathrm{L} \times 5^1 + 2 \times 5^2 + 1 \times 5^3 &= 4 \times 5^0 + 3 \times 5^1 + 2 \times 5^2 + 1 \times 5^3 (\mathrm{L} = 3, \mathrm{M} = 4) \\ &= 4 + 15 + 50 + 125 = 194 \end{aligned}$$

The decimal equivalent of (L1) is given by

$$\mathrm{L} \times 5^{-1} + 1 \times 5^{-2} = 3 \times 5^{-1} + 5^{-2} = 0.64$$

Combining the results, $(12\mathrm{LM.L1})_5 = (194.64)_{10}$.

(b) The total number of possible four-digit combinations $= 5^4 = 625$.

Example 1.10

The 7's complement of a certain octal number is 5264. Determine the binary and hexadecimal equivalents of that octal number.

Solution

- The 7's complement = 5264.
- Therefore, the octal number = $(2513)_8$.
- The binary equivalent = $(010\ 101\ 001\ 011)_2 = (10101001011)_2$.
- Also, $(10101001011)_2 = (101\ 0100\ 1011)_2 = (0101\ 0100\ 1011)_2 = (54\mathrm{B})_{16}$.
- Therefore, the hex equivalent of $(2513)_8 = (54\mathrm{B})_{16}$ and the binary equivalent of $(2513)_8 = (10101001011)_2$.

1.17 Floating-Point Numbers

Floating-point notation can be used conveniently to represent both large as well as small fractional or mixed numbers. This makes the process of arithmetic operations on these numbers relatively much easier. Floating-point representation greatly increases the range of numbers, from the smallest to the largest, that can be represented using a given number of digits. Floating-point numbers are in general expressed in the form

$$N = m \times b^e \tag{1.1}$$

where m is the fractional part, called the *significand* or *mantissa*, e is the integer part, called the *exponent*, and b is the *base* of the number system or numeration. Fractional part m is a p-digit number of the form ($\pm d.dddd \ldots dd$), with each digit d being an integer between 0 and $b - 1$ inclusive. If the leading digit of m is nonzero, then the number is said to be normalized.

Equation (1.1) in the case of decimal, hexadecimal and binary number systems will be written as follows:

Decimal system

$$N = m \times 10^e \tag{1.2}$$

Hexadecimal system

$$N = m \times 16^e \tag{1.3}$$

Binary system

$$N = m \times 2^e \tag{1.4}$$

For example, decimal numbers 0.0003754 and 3754 will be represented in floating-point notation as 3.754×10^{-4} and 3.754×10^3 respectively. A hex number 257.ABF will be represented as $2.57\text{ABF} \times 16^2$. In the case of normalized binary numbers, the leading digit, which is the most significant bit, is always '1' and thus does not need to be stored explicitly.

Also, while expressing a given mixed binary number as a floating-point number, the radix point is so shifted as to have the most significant bit immediately to the right of the radix point as a '1'. Both the mantissa and the exponent can have a positive or a negative value.

The mixed binary number $(110.1011)_2$ will be represented in floating-point notation as $.1101011 \times 2^3 = .1101011e+0011$. Here, .1101011 is the mantissa and $e+0011$ implies that the exponent is +3. As another example, $(0.000111)_2$ will be written as $.111e-0011$, with .111 being the mantissa and $e-0011$ implying an exponent of −3. Also, $(-0.00000101)_2$ may be written as $-.101 \times 2^{-5} = -.101e-0101$, where −.101 is the mantissa and $e-0101$ indicates an exponent of −5. If we wanted to represent the mantissas using eight bits, then .1101011 and .111 would be represented as .11010110 and .11100000.

1.17.1 Range of Numbers and Precision

The range of numbers that can be represented in any machine depends upon the number of bits in the exponent, while the fractional accuracy or precision is ultimately determined by the number of bits in the mantissa. The higher the number of bits in the exponent, the larger is the range of numbers that can be represented. For example, the range of numbers possible in a floating-point binary number format using six bits to represent the magnitude of the exponent would be from 2^{-64} to 2^{+64}, which is equivalent to a range of $10^{-19}$to 10^{+19}. The precision is determined by the number of bits used to represent the mantissa. It is usually represented as decimal digits of precision. The concept of precision as defined with respect to floating-point notation can be explained in simple terms as follows. If the mantissa is stored in n number of bits, it can represent a decimal number between 0 and $2^n - 1$ as the mantissa is stored as an unsigned integer. If M is the largest number such that $10^M - 1$ is less than or equal to $2^n - 1$, then M is the precision expressed as decimal digits of precision. For example, if the mantissa is expressed in 20 bits, then decimal digits of precision can be found to be about 6, as $2^{20} - 1$ equals 1 048 575, which is a little over $10^6 - 1$. We will briefly describe the commonly used formats for binary floating-point number representation.

1.17.2 Floating-Point Number Formats

The most commonly used format for representing floating-point numbers is the IEEE-754 standard. The full title of the standard is IEEE Standard for Binary Floating-point Arithmetic (ANSI/IEEE STD 754-1985). It is also known as Binary Floating-point Arithmetic for Microprocessor Systems, IEC

60559:1989. An ongoing revision to IEEE-754 is IEEE-754r. Another related standard IEEE 854-1987 generalizes IEEE-754 to cover both binary and decimal arithmetic. A brief description of salient features of the IEEE-754 standard, along with an introduction to other related standards, is given below.

ANSI/IEEE-754 Format

The IEEE-754 floating point is the most commonly used representation for real numbers on computers including Intel-based personal computers, Macintoshes and most of the UNIX platforms. It specifies four formats for representing floating-point numbers. These include single-precision, double-precision, single-extended precision and double-extended precision formats. Table 1.1 lists characteristic parameters of the four formats contained in the IEEE-754 standard. Of the four formats mentioned, the single-precision and double-precision formats are the most commonly used ones. The single-extended and double-extended precision formats are not common.

Figure 1.1 shows the basic constituent parts of the single- and double-precision formats. As shown in the figure, the floating-point numbers, as represented using these formats, have three basic components including the sign, the exponent and the mantissa. A '0' denotes a positive number and a '1' denotes a negative number. The n-bit exponent field needs to represent both positive and negative exponent values. To achieve this, a bias equal to $2^{n-1} - 1$ is added to the actual exponent in order to obtain the stored exponent. This equals 127 for an eight-bit exponent of the single-precision format and 1023 for an 11-bit exponent of the double-precision format. The addition of bias allows the use of an exponent in the range from -127 to $+128$, corresponding to a range of 0–255 in the first case, and in the range from -1023 to $+1024$, corresponding to a range of 0–2047 in the second case. A negative exponent is always represented in 2's complement form. The single-precision format offers a range from 2^{-127} to 2^{+127}, which is equivalent to 10^{-38} to 10^{+38}. The figures are 2^{-1023} to 2^{+1023}, which is equivalent to 10^{-308} to 10^{+308} in the case of the double-precision format.

The extreme exponent values are reserved for representing special values. For example, in the case of the single-precision format, for an exponent value of -127, the biased exponent value is zero, represented by an all 0s exponent field. In the case of a biased exponent of zero, if the mantissa is zero as well, the value of the floating-point number is exactly zero. If the mantissa is nonzero, it represents a denormalized number that does not have an assumed leading bit of '1'. A biased exponent of $+255$, corresponding to an actual exponent of $+128$, is represented by an all 1s exponent field. If the mantissa is zero, the number represents infinity. The sign bit is used to distinguish between positive and negative infinity. If the mantissa is nonzero, the number represents a 'NaN' (Not a Number). The value NaN is used to represent a value that does not represent a real number. This means that an eight-bit exponent can represent exponent values between -126 and $+127$. Referring to Fig. 1.1(a), the MSB of byte 1 indicates the sign of the mantissa. The remaining seven bits of byte 1 and the MSB of byte 2 represent an eight-bit exponent. The remaining seven bits of byte 2 and the 16 bits of byte 3 and byte 4 give a 23-bit mantissa. The mantissa m is normalized. The left-hand bit of the normalized mantissa is always

Table 1.1 Characteristic parameters of IEEE-754 formats.

Precision	Sign (bits)	Exponent (bits)	Mantissa (bits)	Total length (bits)	Decimal digits of precision
Single	1	8	23	32	>6
Single-extended	1	≥ 11	≥ 32	≥ 44	>9
Double	1	11	52	64	>15
Double-extended	1	≥ 15	≥ 64	≥ 80	>19

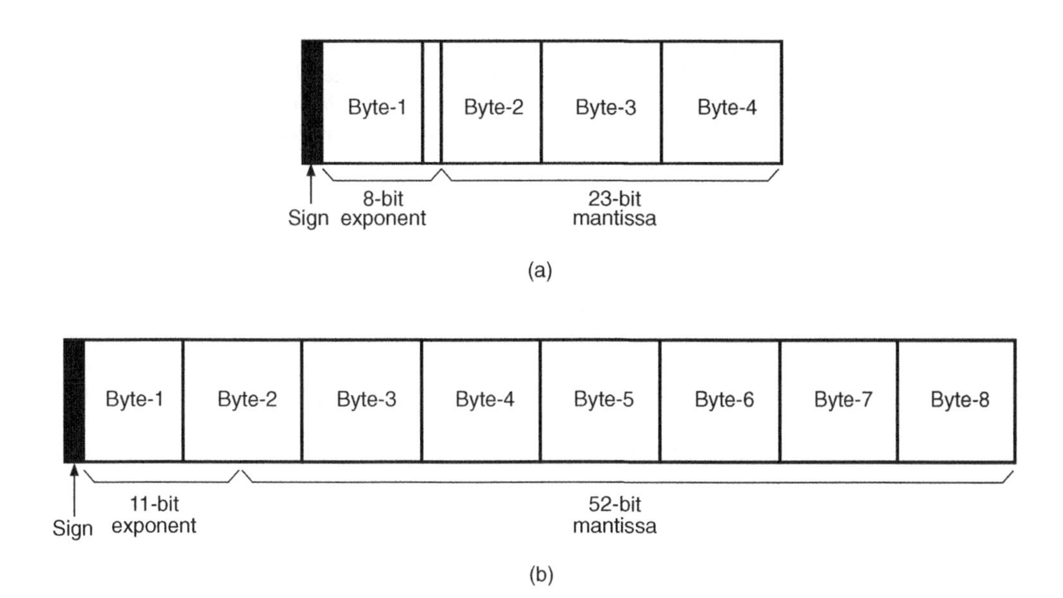

Figure 1.1 Single-precision and double-precision formats.

'1'. This '1' is not included but is always implied. A similar explanation can be given in the case of the double-precision format shown in Fig. 1.1(b).

Step-by-step transformation of $(23)_{10}$ into an equivalent floating-point number in single-precision IEEE format is as follows:

- $(23)_{10} = (10111)_2 = 1.0111e + 0100$.
- The mantissa = 0111000 00000000 00000000.
- The exponent = 00000100.
- The biased exponent = 00000100 + 01111111 = 10000011.
- The sign of the mantissa = 0.
- $(+23)_{10}$ = 01000001 10111000 00000000 00000000.
- Also, $(-23)_{10}$= 11000001 10111000 00000000 00000000.

IEEE-754r Format

As mentioned earlier, IEEE-754r is an ongoing revision to the IEEE-754 standard. The main objective of the revision is to extend the standard wherever it has become necessary, the most obvious enhancement to the standard being the addition of the 128-bit format and decimal format. Extension of the standard to include decimal floating-point representation has become necessary as most commercial data are held in decimal form and the binary floating point cannot represent decimal fractions exactly. If the binary floating point is used to represent decimal data, it is likely that the results will not be the same as those obtained by using decimal arithmetic.

In the revision process, many of the definitions have been rewritten for clarification and consistency. In terms of the addition of new formats, a new addition to the existing binary formats is the 128-bit 'quad-precision' format. Also, three new decimal formats, matching the lengths of binary formats, have been described. These include decimal formats with a seven-, 16- and 34-digit mantissa, which may be normalized or denormalized. In order to achieve maximum range (decided by the number of exponent bits) and precision (decided by the number of mantissa bits), the formats merge part of the exponent and mantissa into a combination field and compress the remainder of the mantissa using densely packed decimal encoding. Detailed description of the revision, however, is beyond the scope of this book.

IEEE-854 Standard

The main objective of the IEEE-854 standard was to define a standard for floating-point arithmetic without the radix and word length dependencies of the better-known IEEE-754 standard. That is why IEEE-854 is called the IEEE standard for radix-independent floating-point arithmetic. Although the standard specifies only the binary and decimal floating-point arithmetic, it provides sufficient guidelines for those contemplating the implementation of the floating point using any other radix value such as 16 of the hexadecimal number system. This standard, too, specifies four formats including single, single-extended, double and double-extended precision formats.

Example 1.11

Determine the floating-point representation of $(-142)_{10}$ using the IEEE single-precision format.

Solution

- As a first step, we will determine the binary equivalent of $(142)_{10}$. Following the procedure outlined in an earlier part of the chapter, the binary equivalent can be written as $(142)_{10} = (10001110)_2$.
- $(10001110)_2 = 1.000\ 1110 \times 2^7 = 1.0001110e + 0111$.
- The mantissa = 0001110 00000000 00000000.
- The exponent = 00000111.
- The biased exponent = 00000111 + 01111111 = 10000110.
- The sign of the mantissa = 1.
- Therefore, $(-142)_{10}$ = 11000011 00001110 00000000 00000000.

Example 1.12

Determine the equivalent decimal numbers for the following floating-point numbers:

(a) 00111111 01000000 00000000 00000000 (IEEE-754 single-precision format);
(b) 11000000 00101001 01100 ... 45 0s (IEEE-754 double-precision format).

Solution

(a) From an examination of the given number:

The sign of the mantissa is positive, as indicated by the '0' bit in the designated position.

The biased exponent = 01111110.

The unbiased exponent = 01111110 − 01111111 = 11111111.

It is clear from the eight bits of unbiased exponent that the exponent is negative, as the 2's complement representation of a number gives '1' in place of MSB.

The magnitude of the exponent is given by the 2's complement of $(11111111)_2$, which is $(00000001)_2 = 1$.

Therefore, the exponent = −1.
The mantissa bits = 11000000 00000000 00000000 ('1' in MSB is implied).
The normalized mantissa = 1.1000000 00000000 00000000.
The magnitude of the mantissa can be determined by shifting the mantissa bits one position to the left. That is, the mantissa = $(.11)_2 = (0.75)_{10}$.

(b) The sign of the mantissa is negative, indicated by the '1' bit in the designated position.
The biased exponent = 10000000010.
The unbiased exponent = 10000000010 − 01111111111 = 00000000011.
It is clear from the 11 bits of unbiased exponent that the exponent is positive owing to the '0' in place of MSB. The magnitude of the exponent is 3. Therefore, the exponent = +3.
The mantissa bits = 1100101100 . . . 45 0s ('1' in MSB is implied).
The normalized mantissa = 1.100101100 . . . 45 0s.
The magnitude of the mantissa can be determined by shifting the mantissa bits three positions to the right.
That is, the mantissa = $(1100.101)_2 = (12.625)_{10}$.
Therefore, the equivalent decimal number = −12.625.

Review Questions

1. What is meant by the radix or base of a number system? Briefly describe why hex representation is used for the addresses and the contents of the memory locations in the main memory of a computer.
2. What do you understand by the l's and 2's complements of a binary number? What will be the range of decimal numbers that can be represented using a 16-bit 2's complement format?
3. Briefly describe the salient features of the IEEE-754 standard for representing floating-point numbers.
4. Why was it considered necessary to carry out a revision of the IEEE-754 standard? What are the main features of IEEE-754r (the notation for IEEE-754 under revision)?
5. In a number system, what decides (a) the place value or weight of a given digit and (b) the maximum numbers representable with a given number of digits?
6. In a floating-point representation, what represents (a) the range of representable numbers and (b) the precision with which a given number can be represented?
7. Why is there a need to have floating-point standards that can take care of decimal data and decimal arithmetic in addition to binary data and arithmetic?

Problems

1. Do the following conversions:

   (a) eight-bit 2's complement representation of $(-23)_{10}$;
   (b) The decimal equivalent of $(00010111)_2$ represented in 2's complement form.

   (a) 11101001; (b) +23

2. Two possible binary representations of $(-1)_{10}$ are $(10000001)_2$ and $(11111111)_2$. One of them belongs to the sign-bit magnitude format and the other to the 2's complement format. Identify.

   $(10000001)_2$ = sign-bit magnitude and $(11111111)_2$ = 2's complement form

3. Represent the following in the IEEE-754 floating-point standard using the single-precision format:

   (a) 32-bit binary number 11110000 11001100 10101010 00001111;
   (b) $(-118.625)_{10}$.

(a) 01001111 01110000 11001100 10101010;
(b) 11000010 11101101 01000000 00000000

4. Give the next three numbers in each of the following hex sequences:

   (a) 4A5, 4A6, 4A7, 4A8, . . . ;
   (b) B998, B999, . . .

(a) 4A9, 4AA, 4AB; (b) B99A, B99B, B99C

5. Show that:

   (a) $(13A7)_{16} = (5031)_{10}$;
   (b) $(3F2)_{16} = (1111110010)_2$.

6. Assume a radix-32 arbitrary number system with 0–9 and A–V as its basic digits. Express the mixed binary number $(110101.001)_2$ in this arbitrary number system.

1L.4

Further Reading

1. Tokheim, R. L. (1994) *Schaum's Outline Series of Digital Principles*, McGraw-Hill Companies Inc., USA.
2. Atiyah, S. K. (2005) *A Survey of Arithmetic*, Trafford Publishing, Victoria, BC, Canada.
3. Langholz, G., Mott, J. L. and Kandel, A. (1998) *Foundations of Digital Logic Design*, World Scientific Publ. Co. Inc., Singapore.
4. Cook, N. P. (2003) *Practical Digital Electronics*, Prentice-Hall, NJ, USA.
5. Lu, M. (2004) *Arithmetic and Logic in Computer Systems*, John Wiley & Sons, Inc., NJ, USA.

2

Binary Codes

The present chapter is an extension of the previous chapter on *number systems*. In the previous chapter, beginning with some of the basic concepts common to all number systems and an outline on the familiar decimal number system, we went on to discuss the binary, the hexadecimal and the octal number systems. While the binary system of representation is the most extensively used one in digital systems, including computers, octal and hexadecimal number systems are commonly used for representing groups of binary digits. The binary coding system, called the straight binary code and discussed in the previous chapter, becomes very cumbersome to handle when used to represent larger decimal numbers. To overcome this shortcoming, and also to perform many other special functions, several binary codes have evolved over the years. Some of the better-known binary codes, including those used efficiently to represent numeric and alphanumeric data, and the codes used to perform special functions, such as detection and correction of errors, will be detailed in this chapter.

2.1 Binary Coded Decimal

The binary coded decimal (BCD) is a type of binary code used to represent a given decimal number in an equivalent binary form. BCD-to-decimal and decimal-to-BCD conversions are very easy and straightforward. It is also far less cumbersome an exercise to represent a given decimal number in an equivalent BCD code than to represent it in the equivalent straight binary form discussed in the previous chapter.

The BCD equivalent of a decimal number is written by replacing each decimal digit in the integer and fractional parts with its four-bit binary equivalent. As an example, the BCD equivalent of $(23.15)_{10}$ is written as $(0010\ 0011.0001\ 0101)_{BCD}$. The BCD code described above is more precisely known as the 8421 BCD code, with 8, 4, 2 and 1 representing the weights of different bits in the four-bit groups, starting from MSB and proceeding towards LSB. This feature makes it a weighted code, which means that each bit in the four-bit group representing a given decimal digit has an assigned

Table 2.1 BCD codes.

Decimal	8421 BCD code	4221 BCD code	5421 BCD code
0	0000	0000	0000
1	0001	0001	0001
2	0010	0010	0010
3	0011	0011	0011
4	0100	1000	0100
5	0101	0111	1000
6	0110	1100	1001
7	0111	1101	1010
8	1000	1110	1011
9	1001	1111	1100

weight. Other weighted BCD codes include the 4221 BCD and 5421 BCD codes. Again, 4, 2, 2 and 1 in the 4221 BCD code and 5, 4, 2 and 1 in the 5421 BCD code represent weights of the relevant bits. Table 2.1 shows a comparison of 8421, 4221 and 5421 BCD codes. As an example, $(98.16)_{10}$ will be written as 1111 1110.0001 1100 in 4221 BCD code and 1100 1011.0001 1001 in 5421 BCD code. Since the 8421 code is the most popular of all the BCD codes, it is simply referred to as the BCD code.

2.1.1 BCD-to-Binary Conversion

A given BCD number can be converted into an equivalent binary number by first writing its decimal equivalent and then converting it into its binary equivalent. The first step is straightforward, and the second step was explained in the previous chapter. As an example, we will find the binary equivalent of the BCD number 0010 1001.0111 0101:

- BCD number: 0010 1001.0111 0101.
- Corresponding decimal number: 29.75.
- The binary equivalent of 29.75 can be determined to be 11101 for the integer part and .11 for the fractional part.
- Therefore, $(0010\ 1001.0111\ 0101)_{BCD} = (11101.11)_2$.

2.1.2 Binary-to-BCD Conversion

The process of binary-to-BCD conversion is the same as the process of BCD-to-binary conversion executed in reverse order. A given binary number can be converted into an equivalent BCD number by first determining its decimal equivalent and then writing the corresponding BCD equivalent. As an example, we will find the BCD equivalent of the binary number 10101011.101:

- The decimal equivalent of this binary number can be determined to be 171.625.
- The BCD equivalent can then be written as 0001 0111 0001.0110 0010 0101.

2.1.3 Higher-Density BCD Encoding

In the regular BCD encoding of decimal numbers, the number of bits needed to represent a given decimal number is always greater than the number of bits required for straight binary encoding of the same. For example, a three-digit decimal number requires 12 bits for representation in conventional BCD format. However, since $2^{10} > 10^3$, if these three decimal digits are encoded together, only 10 bits would be needed to do that. Two such encoding schemes are *Chen-Ho encoding* and the *densely packed decimal*. The latter has the advantage that subsets of the encoding encode two digits in the optimal seven bits and one digit in four bits like regular BCD.

2.1.4 Packed and Unpacked BCD Numbers

In the case of unpacked BCD numbers, each four-bit BCD group corresponding to a decimal digit is stored in a separate register inside the machine. In such a case, if the registers are eight bits or wider, the register space is wasted.

In the case of packed BCD numbers, two BCD digits are stored in a single eight-bit register. The process of combining two BCD digits so that they are stored in one eight-bit register involves shifting the number in the upper register to the left 4 times and then adding the numbers in the upper and lower registers. The process is illustrated by showing the storage of decimal digits '5' and '7':

- Decimal digit 5 is initially stored in the eight-bit register as: 0000 0101.
- Decimal digit 7 is initially stored in the eight-bit register as: 0000 0111.
- After shifting to the left 4 times, the digit 5 register reads: 0101 0000.
- The addition of the contents of the digit 5 and digit 7 registers now reads: 0101 0111.

Example 2.1

How many bits would be required to encode decimal numbers 0 to 9999 in straight binary and BCD codes? What would be the BCD equivalent of decimal 27 in 16-bit representation?

Solution

- Total number of decimals to be represented $= 10\,000 = 10^4 = 2^{13.29}$.
- Therefore, the number of bits required for straight binary encoding = 14.
- The number of bits required for BCD encoding = 16.
- The BCD equivalent of 27 in 16-bit representation = 0000000000100111.

2.2 Excess-3 Code

The excess-3 code is another important BCD code. It is particularly significant for arithmetic operations as it overcomes the shortcomings encountered while using the 8421 BCD code to add two decimal digits whose sum exceeds 9. The excess-3 code has no such limitation, and it considerably simplifies arithmetic operations. Table 2.2 lists the excess-3 code for the decimal numbers 0–9.

The excess-3 code for a given decimal number is determined by adding '3' to each decimal digit in the given number and then replacing each digit of the newly found decimal number by

Table 2.2 Excess-3 code equivalent of decimal numbers.

Decimal number	Excess-3 code	Decimal number	Excess-3 code
0	0011	5	1000
1	0100	6	1001
2	0101	7	1010
3	0110	8	1011
4	0111	9	1100

its four-bit binary equivalent. It may be mentioned here that, if the addition of '3' to a digit produces a carry, as is the case with the digits 7, 8 and 9, that carry should not be taken forward. The result of addition should be taken as a single entity and subsequently replaced with its excess-3 code equivalent. As an example, let us find the excess-3 code for the decimal number 597:

- The addition of '3' to each digit yields the three new digits/numbers '8', '12' and '10'.
- The corresponding four-bit binary equivalents are 1000, 1100 and 1010 respectively.
- The excess-3 code for 597 is therefore given by: 1000 1100 1010 = 100011001010.

Also, it is normal practice to represent a given decimal digit or number using the maximum number of digits that the digital system is capable of handling. For example, in four-digit decimal arithmetic, 5 and 37 would be written as 0005 and 0037 respectively. The corresponding 8421 BCD equivalents would be 0000000000000101 and 0000000000110111 and the excess-3 code equivalents would be 0011001100111000 and 0011001101101010.

Corresponding to a given excess-3 code, the equivalent decimal number can be determined by first splitting the number into four-bit groups, starting from the radix point, and then subtracting 0011 from each four-bit group. The new number is the 8421 BCD equivalent of the given excess-3 code, which can subsequently be converted into the equivalent decimal number. As an example, following these steps, the decimal equivalent of excess-3 number 01010110.10001010 would be 23.57.

Another significant feature that makes this code attractive for performing arithmetic operations is that the complement of the excess-3 code of a given decimal number yields the excess-3 code for 9's complement of the decimal number. As adding 9's complement of a decimal number B to a decimal number A achieves A – B, the excess-3 code can be used effectively for both addition and subtraction of decimal numbers.

Example 2.3

Find (a) the excess-3 equivalent of $(237.75)_{10}$ *and (b) the decimal equivalent of the excess-3 number 110010100011.01110101.*

Solution

(a) Integer part = 237. The excess-3 code for $(237)_{10}$ is obtained by replacing 2, 3 and 7 with the four-bit binary equivalents of 5, 6 and 10 respectively. This gives the excess-3 code for $(237)_{10}$ as: 0101 0110 1010 = 010101101010.

Fractional part = .75. The excess-3 code for $(.75)_{10}$ is obtained by replacing 7 and 5 with the four-bit binary equivalents of 10 and 8 respectively. That is, the excess-3 code for $(.75)_{10} = .10101000$.
Combining the results of the integral and fractional parts, the excess-3 code for $(237.75)_{10} = 010101101010.10101000$.

(b) The excess-3 code = 110010100011.01110101 = 1100 1010 0011.0111 0101.
Subtracting 0011 from each four-bit group, we obtain the new number as: 1001 0111 0000.0100 0010.
Therefore, the decimal equivalent = $(970.42)_{10}$.

2.3 Gray Code

The Gray code was designed by Frank Gray at Bell Labs and patented in 1953. It is an unweighted binary code in which two successive values differ only by 1 bit. Owing to this feature, the maximum error that can creep into a system using the binary Gray code to encode data is much less than the worst-case error encountered in the case of straight binary encoding. Table 2.3 lists the binary and Gray code equivalents of decimal numbers 0–15. An examination of the four-bit Gray code numbers, as listed in Table 2.3, shows that the last entry rolls over to the first entry. That is, the last and the first entry also differ by only 1 bit. This is known as the *cyclic property* of the Gray code. Although there can be more than one Gray code for a given word length, the term was first applied to a specific binary code for non-negative integers and called the *binary-reflected Gray code* or simply the Gray code.

There are various ways by which Gray codes with a given number of bits can be remembered. One such way is to remember that the least significant bit follows a repetitive pattern of '2' (11, 00, 11, . . .), the next higher adjacent bit follows a pattern of '4' (1111, 0000, 1111, . . .) and so on. We can also generate the n-bit Gray code recursively by prefixing a '0' to the Gray code for $n-1$ bits to obtain the first 2^{n-1} numbers, and then prefixing '1' to the reflected Gray code for $n-1$ bits to obtain the remaining 2^{n-1} numbers. The reflected Gray code is nothing but the code written in reverse order. The process of generation of higher-bit Gray codes using the reflect-and-prefix method is illustrated in Table 2.4. The columns of bits between those representing the Gray codes give the intermediate step of writing the code followed by the same written in reverse order.

Table 2.3 Gray code.

Decimal	Binary	Gray	Decimal	Binary	Gray
0	0000	0000	8	1000	1100
1	0001	0001	9	1001	1101
2	0010	0011	10	1010	1111
3	0011	0010	11	1011	1110
4	0100	0110	12	1100	1010
5	0101	0111	13	1101	1011
6	0110	0101	14	1110	1001
7	0111	0100	15	1111	1000

Table 2.4 Generation of higher-bit Gray code numbers.

One-bit Gray code		Two-bit Gray code		Three-bit Gray code		Four-bit Gray code
0	0	00	00	000	000	0000
1	1	01	01	001	001	0001
	1	11	11	011	011	0011
	0	10	10	010	010	0010
			10	110	110	0110
			11	111	111	0111
			01	101	101	0101
			00	100	100	0100
					100	1100
					101	1101
					111	1111
					110	1110
					010	1010
					011	1011
					001	1001
					000	1000

2.3.1 Binary–Gray Code Conversion

A given binary number can be converted into its Gray code equivalent by going through the following steps:

1. Begin with the most significant bit (MSB) of the binary number. The MSB of the Gray code equivalent is the same as the MSB of the given binary number.
2. The second most significant bit, adjacent to the MSB, in the Gray code number is obtained by adding the MSB and the second MSB of the binary number and ignoring the carry, if any. That is, if the MSB and the bit adjacent to it are both '1', then the corresponding Gray code bit would be a '0'.
3. The third most significant bit, adjacent to the second MSB, in the Gray code number is obtained by adding the second MSB and the third MSB in the binary number and ignoring the carry, if any.
4. The process continues until we obtain the LSB of the Gray code number by the addition of the LSB and the next higher adjacent bit of the binary number.

The conversion process is further illustrated with the help of an example showing step-by-step conversion of $(1011)_2$ into its Gray code equivalent:

Binary 1011
Gray code 1- - -
Binary 1011
Gray code 11- -
Binary 1011
Gray code 111-
Binary 1011
Gray code 1110

2.3.2 Gray Code–Binary Conversion

A given Gray code number can be converted into its binary equivalent by going through the following steps:

1. Begin with the most significant bit (MSB). The MSB of the binary number is the same as the MSB of the Gray code number.
2. The bit next to the MSB (the second MSB) in the binary number is obtained by adding the MSB in the binary number to the second MSB in the Gray code number and disregarding the carry, if any.
3. The third MSB in the binary number is obtained by adding the second MSB in the binary number to the third MSB in the Gray code number. Again, carry, if any, is to be ignored.
4. The process continues until we obtain the LSB of the binary number.

The conversion process is further illustrated with the help of an example showing step-by-step conversion of the Gray code number 1110 into its binary equivalent:

Gray code 1110
Binary 1- - -
Gray code 1110
Binary 10 - -
Gray code 1110
Binary 101
Gray code 1110
Binary 1011

2.3.3 n-ary Gray Code

The binary-reflected Gray code described above is invariably referred to as the 'Gray code'. However, over the years, mathematicians have discovered other types of Gray code. One such code is the n-ary Gray code, also called the non-Boolean Gray code owing to the use of non-Boolean symbols for encoding. The generalized representation of the code is the (n, k)-Gray code, where n is the number of independent digits used and k is the word length. A ternary Gray code ($n=3$) uses the values 0, 1 and 2, and the sequence of numbers in the two-digit word length would be (00, 01, 02, 12, 11, 10, 20, 21, 22). In the quaternary ($n=4$) code, using 0, 1, 2 and 3 as independent digits and a two-digit word length, the sequence of numbers would be (00, 01, 02, 03, 13, 12, 11, 10, 20, 21, 22, 23, 33, 32, 31, 30). It is important to note here that an (n, k)-Gray code with an odd n does not exhibit the cyclic property of the binary Gray code, while in case of an even n it does have the cyclic property.

The (n, k)-Gray code may be constructed recursively, like the binary-reflected Gray code, or may be constructed iteratively. The process of generating larger word-length ternary Gray codes is illustrated in Table 2.5. The columns between those representing the ternary Gray codes give the intermediate steps.

2.3.4 Applications

1. The Gray code is used in the transmission of digital signals as it minimizes the occurrence of errors.
2. The Gray code is preferred over the straight binary code in angle-measuring devices. Use of the Gray code almost eliminates the possibility of an angle misread, which is likely if the

Table 2.5 Generation of a larger word-length ternary Gray code.

One-digit ternary code	Two-digit ternary code		Three-digit ternary code	
0	0	00	00	000
1	1	01	01	001
2	2	02	02	002
	2	12	12	012
	1	11	11	011
	0	10	10	010
	0	20	20	020
	1	21	21	021
	2	22	22	022
			22	122
			21	121
			20	120
			10	110
			11	111
			12	112
			02	102
			01	101
			00	100
			00	200
			01	201
			02	202
			12	212
			11	211
			10	210
			20	220
			21	221
			22	222

angle is represented in straight binary. The cyclic property of the Gray code is a plus in this application.

3. The Gray code is used for labelling the axes of Karnaugh maps, a graphical technique used for minimization of Boolean expressions.
4. The use of Gray codes to address program memory in computers minimizes power consumption. This is due to fewer address lines changing state with advances in the program counter.
5. Gray codes are also very useful in genetic algorithms since mutations in the code allow for mostly incremental changes. However, occasionally a one-bit change can result in a big leap, thus leading to new properties.

Example 2.4

Find (a) the Gray code equivalent of decimal 13 and (b) the binary equivalent of Gray code number 1111.

Solution

(a) The binary equivalent of decimal 13 is 1101.
Binary–Gray conversion

Binary 1101
Gray 1- - -
Binary 1101
Gray 10 - -
Binary 1101
Gray 101 –
Binary 1101
Gray 1011

(b) *Gray–binary conversion*

Gray 1111
Binary 1- - -
Gray 1111
Binary 10- -
Gray 1111
Binary 101-
Gray 1111
Binary 1010

Example 2.5

Given the sequence of three-bit Gray code as (000, 001, 011, 010, 110, 111, 101, 100), write the next three numbers in the four-bit Gray code sequence after 0101.

Solution

The first eight of the 16 Gray code numbers of the four-bit Gray code can be written by appending '0' to the eight three-bit Gray code numbers. The remaining eight can be determined by appending '1' to the eight three-bit numbers written in reverse order. Following this procedure, we can write the next three numbers after 0101 as 0100, 1100 and 1101.

2.4 Alphanumeric Codes

Alphanumeric codes, also called character codes, are binary codes used to represent alphanumeric data. The codes write alphanumeric data, including letters of the alphabet, numbers, mathematical symbols and punctuation marks, in a form that is understandable and processable by a computer. These codes enable us to interface input–output devices such as keyboards, printers, VDUs, etc., with the computer. One of the better-known alphanumeric codes in the early days of evolution of computers, when punched cards used to be the medium of inputting and outputting data, is the 12-bit Hollerith code. The Hollerith code was used in those days to encode alphanumeric data on punched cards. The code has, however, been rendered obsolete, with the punched card medium having completely vanished from the scene. Two widely used alphanumeric codes include the ASCII and the EBCDIC codes. While the former is popular with microcomputers and is used on nearly all personal computers and workstations, the latter is mainly used with larger systems.

Traditional character encodings such as ASCII, EBCDIC and their variants have a limitation in terms of the number of characters they can encode. In fact, no single encoding contains enough characters so as to cover all the languages of the European Union. As a result, these encodings do not permit multilingual computer processing. Unicode, developed jointly by the Unicode Consortium and the International Standards Organization (ISO), is the most complete character encoding scheme that allows text of all forms and languages to be encoded for use by computers. Different codes are described in the following.

2.4.1 ASCII code

The ASCII (American Standard Code for Information Interchange), pronounced 'ask-ee', is strictly a seven-bit code based on the English alphabet. ASCII codes are used to represent alphanumeric data in computers, communications equipment and other related devices. The code was first published as a standard in 1967. It was subsequently updated and published as ANSI X3.4-1968, then as ANSI X3.4-1977 and finally as ANSI X3.4-1986. Since it is a seven-bit code, it can at the most represent 128 characters. It currently defines 95 printable characters including 26 upper-case letters (A to Z), 26 lower-case letters (a to z), 10 numerals (0 to 9) and 33 special characters including mathematical symbols, punctuation marks and space character. In addition, it defines codes for 33 nonprinting, mostly obsolete control characters that affect how text is processed. With the exception of 'carriage return' and/or 'line feed', all other characters have been rendered obsolete by modern mark-up languages and communication protocols, the shift from text-based devices to graphical devices and the elimination of teleprinters, punch cards and paper tapes. An eight-bit version of the ASCII code, known as US ASCII-8 or ASCII-8, has also been developed. The eight-bit version can represent a maximum of 256 characters.

Table 2.6 lists the ASCII codes for all 128 characters. When the ASCII code was introduced, many computers dealt with eight-bit groups (or bytes) as the smallest unit of information. The eighth bit was commonly used as a parity bit for error detection on communication lines and other device-specific functions. Machines that did not use the parity bit typically set the eighth bit to '0'.

Table 2.6 ASCII code.

Decimal	Hex	Binary	Code	Code description
0	00	0000 0000	NUL	Null character
1	01	0000 0001	SOH	Start of header
2	02	0000 0010	STX	Start of text
3	03	0000 0011	ETX	End of text
4	04	0000 0100	EOT	End of transmission
5	05	0000 0101	ENQ	Enquiry
6	06	0000 0110	ACK	Acknowledgement
7	07	0000 0111	BEL	Bell
8	08	0000 1000	BS	Backspace
9	09	0000 1001	HT	Horizontal tab
10	0A	0000 1010	LF	Line feed
11	0B	0000 1011	VT	Vertical tab
12	0C	0000 1100	FF	Form feed
13	0D	0000 1101	CR	Carriage return
14	0E	0000 1110	SO	Shift out
15	0F	0000 1111	SI	Shift in
16	10	0001 0000	DLE	Data link escape
17	11	0001 0001	DC1	Device control 1 (XON)

Table 2.6 (*continued*).

Decimal	Hex	Binary	Code	Code description
18	12	0001 0010	DC2	Device control 2
19	13	0001 0011	DC3	Device control 3 (XOFF)
20	14	0001 0100	DC4	Device control 4
21	15	0001 0101	NAK	Negative acknowledgement
22	16	0001 0110	SYN	Synchronous idle
23	17	0001 0111	ETB	End of transmission block
24	18	0001 1000	CAN	Cancel
25	19	0001 1001	EM	End of medium
26	1A	0001 1010	SUB	Substitute
27	1B	0001 1011	ESC	Escape
28	1C	0001 1100	FS	File separator
29	1D	0001 1101	GS	Group separator
30	1E	0001 1110	RS	Record separator
31	1F	0001 1111	US	Unit separator
32	20	0010 0000	SP	Space
33	21	0010 0001	!	Exclamation point
34	22	0010 0010	"	Quotation mark
35	23	0010 0011	#	Number sign, octothorp, pound
36	24	0010 0100	$	Dollar sign
37	25	0010 0101	%	Percent
38	26	0010 0110	&	Ampersand
39	27	0010 0111	’	Apostrophe, prime
40	28	0010 1000	(	Left parenthesis
41	29	0010 1001	)	Right parenthesis
42	2A	0010 1010	*	Asterisk, ‘star’
43	2B	0010 1011	+	Plus sign
44	2C	0010 1100	,	Comma
45	2D	0010 1101	-	Hyphen, minus sign
46	2E	0010 1110	.	Period, decimal Point, ‘dot’
47	2F	0010 1111	/	Slash, virgule
48	30	0011 0000	0	0
49	31	0011 0001	1	1
50	32	0011 0010	2	2
51	33	0011 0011	3	3
52	34	0011 0100	4	4
53	35	0011 0101	5	5
54	36	0011 0110	6	6
55	37	0011 0111	7	7
56	38	0011 1000	8	8
57	39	0011 1001	9	9
58	3A	0011 1010	:	Colon
59	3B	0011 1011	;	Semicolon
60	3C	0011 1100	<	Less-than sign
61	3D	0011 1101	=	Equals sign
62	3E	0011 1110	>	Greater-than sign
63	3F	0011 1111	?	Question mark
64	40	0100 0000	@	At sign
65	41	0100 0001	A	A

(*continued overleaf*)

Table 2.6 (*continued*).

Decimal	Hex	Binary	Code	Code description
66	42	0100 0010	B	B
67	43	0100 0011	C	C
68	44	0100 0100	D	D
69	45	0100 0101	E	E
70	46	0100 0110	F	F
71	47	0100 0111	G	G
72	48	0100 1000	H	H
73	49	0100 1001	I	I
74	4A	0100 1010	J	J
75	4B	0100 1011	K	K
76	4C	0100 1100	L	L
77	4D	0100 1101	M	M
78	4E	0100 1110	N	N
79	4F	0100 1111	O	O
80	50	0101 0000	P	P
81	51	0101 0001	Q	Q
82	52	0101 0010	R	R
83	53	0101 0011	S	S
84	54	0101 0100	T	T
85	55	0101 0101	U	U
86	56	0101 0110	V	V
87	57	0101 0111	W	W
88	58	0101 1000	X	X
89	59	0101 1001	Y	Y
90	5A	0101 1010	Z	Z
91	5B	0101 1011	[	Opening bracket
92	5C	0101 1100	\	Reverse slash
93	5D	0101 1101	]	Closing bracket
94	5E	0101 1110	^	Circumflex, caret
95	5F	0101 1111	_	Underline, underscore
96	60	0110 0000	`	Grave accent
97	61	0110 0001	a	a
98	62	0110 0010	b	b
99	63	0110 0011	c	c
100	64	0110 0100	d	d
101	65	0110 0101	e	e
102	66	0110 0110	f	f
103	67	0110 0111	g	g
104	68	0110 1000	h	h
105	69	0110 1001	i	i
106	6A	0110 1010	j	j
107	6B	0110 1011	k	k
108	6C	0110 1100	l	l
109	6D	0110 1101	m	m
110	6E	0110 1110	n	n
111	6F	0110 1111	o	o
112	70	0111 0000	p	p
113	71	0111 0001	q	q
114	72	0111 0010	r	r

Table 2.6 (*continued*).

Decimal	Hex	Binary	Code	Code description
115	73	0111 0011	s	s
116	74	0111 0100	t	t
117	75	0111 0101	u	u
118	76	0111 0110	v	v
119	77	0111 0111	w	w
120	78	0111 1000	x	x
121	79	0111 1001	y	y
122	7A	0111 1010	z	z
123	7B	0111 1011	{	Opening brace
124	7C	0111 1100	|	Vertical line
125	7D	0111 1101	}	Closing brace
126	7E	0111 1110	~	Tilde
127	7F	0111 1111	DEL	Delete

Looking at the structural features of the code as reflected in Table 2.6, we can see that the digits 0 to 9 are represented with their binary values prefixed with 0011. That is, numerals 0 to 9 are represented by binary sequences from 0011 0000 to 0011 1001 respectively. Also, lower-case and upper-case letters differ in bit pattern by a single bit. While upper-case letters 'A' to 'O' are represented by 0100 0001 to 0100 1111, lower-case letters 'a' to 'o' are represented by 0110 0001 to 0110 1111. Similarly, while upper-case letters 'P' to 'Z' are represented by 0101 0000 to 0101 1010, lower-case letters 'p' to 'z' are represented by 0111 0000 to 0111 1010.

With widespread use of computer technology, many variants of the ASCII code have evolved over the years to facilitate the expression of non-English languages that use a Roman-based alphabet. In some of these variants, all ASCII printable characters are identical to their seven-bit ASCII code representations. For example, the eight-bit standard ISO/IEC 8859 was developed as a true extension of ASCII, leaving the original character mapping intact in the process of inclusion of additional values. This made possible representation of a broader range of languages. In spite of the standard suffering from incompatibilities and limitations, ISO-8859-1, its variant Windows-1252 and the original seven-bit ASCII continue to be the most common character encodings in use today.

2.4.2 EBCDIC code

The EBCDIC (Extended Binary Coded Decimal Interchange Code), pronounced 'eb-si-dik', is another widely used alphanumeric code, mainly popular with larger systems. The code was created by IBM to extend the binary coded decimal that existed at that time. All IBM mainframe computer peripherals and operating systems use EBCDIC code, and their operating systems provide ASCII and Unicode modes to allow translation between different encodings. It may be mentioned here that EBCDIC offers no technical advantage over the ASCII code and its variant ISO-8859 or Unicode. Its importance in the earlier days lay in the fact that it made it relatively easier to enter data into larger machines with punch cards. Since, punch cards are not used on mainframes any more, the code is used in contemporary mainframe machines solely for backwards compatibility.

It is an eight-bit code and thus can accommodate up to 256 characters. Table 2.7 gives the listing of characters in binary as well as hex form in EBCDIC. The arrangement is similar to the one adopted for Table 2.6 for the ASCII code. A single byte in EBCDIC is divided into two four-bit groups called

Table 2.7 EBCDIC code.

Decimal	Hex	Binary	Code	Code description
0	00	0000 0000	NUL	Null character
1	01	0000 0001	SOH	Start of header
2	02	0000 0010	STX	Start of text
3	03	0000 0011	ETX	End of text
4	04	0000 0100	PF	Punch off
5	05	0000 0101	HT	Horizontal tab
6	06	0000 0110	LC	Lower case
7	07	0000 0111	DEL	Delete
8	08	0000 1000		
9	09	0000 1001		
10	0A	0000 1010	SMM	Start of manual message
11	0B	0000 1011	VT	Vertical tab
12	0C	0000 1100	FF	Form feed
13	0D	0000 1101	CR	Carriage return
14	0E	0000 1110	SO	Shift out
15	0F	0000 1111	SI	Shift in
16	10	0001 0000	DLE	Data link escape
17	11	0001 0001	DC1	Device control 1
18	12	0001 0010	DC2	Device control 2
19	13	0001 0011	TM	Tape mark
20	14	0001 0100	RES	Restore
21	15	0001 0101	NL	New line
22	16	0001 0110	BS	Backspace
23	17	0001 0111	IL	Idle
24	18	0001 1000	CAN	Cancel
25	19	0001 1001	EM	End of medium
26	1A	0001 1010	CC	Cursor control
27	1B	0001 1011	CU1	Customer use 1
28	1C	0001 1100	IFS	Interchange file separator
29	1D	0001 1101	IGS	Interchange group separator
30	1E	0001 1110	IRS	Interchange record separator
31	1F	0001 1111	IUS	Interchange unit separator
32	20	0010 0000	DS	Digit select
33	21	0010 0001	SOS	Start of significance
34	22	0010 0010	FS	Field separator
35	23	0010 0011		
36	24	0010 0100	BYP	Bypass
37	25	0010 0101	LF	Line feed
38	26	0010 0110	ETB	End of transmission block
39	27	0010 0111	ESC	Escape
40	28	0010 1000		
41	29	0010 1001		
42	2A	0010 1010	SM	Set mode
43	2B	0010 1011	CU2	Customer use 2
44	2C	0010 1100		
45	2D	0010 1101	ENQ	Enquiry
46	2E	0010 1110	ACK	Acknowledge
47	2F	0010 1111	BEL	Bell
48	30	0011 0000		

Table 2.7 (*continued*).

Decimal	Hex	Binary	Code	Code description
49	31	0011 0001		
50	32	0011 0010	SYN	Synchronous idle
51	33	0011 0011		
52	34	0011 0100	PN	Punch on
53	35	0011 0101	RS	Reader stop
54	36	0011 0110	UC	Upper case
55	37	0011 0111	EOT	End of transmission
56	38	0011 1000		
57	39	0011 1001		
58	3A	0011 1010		
59	3B	0011 1011	CU3	Customer use 3
60	3C	0011 1100	DC4	Device control 4
61	3D	0011 1101	NAK	Negative acknowledge
62	3E	0011 1110		
63	3F	0011 1111	SUB	Substitute
64	40	0100 0000	SP	Space
65	41	0100 0001		
66	42	0100 0010		
67	43	0100 0011		
68	44	0100 0100		
69	45	0100 0101		
70	46	0100 0110		
71	47	0100 0111		
72	48	0100 1000		
73	49	0100 1001		
74	4A	0100 1010	¢	Cent sign
75	4B	0100 1011	.	Period, decimal point
76	4C	0100 1100	<	Less-than sign
77	4D	0100 1101	(	Left parenthesis
78	4E	0100 1110	+	Plus sign
79	4F	0100 1111	|	Logical OR
80	50	0101 0000	&	Ampersand
81	51	0101 0001		
82	52	0101 0010		
83	53	0101 0011		
84	54	0101 0100		
85	55	0101 0101		
86	56	0101 0110		
87	57	0101 0111		
88	58	0101 1000		
89	59	0101 1001		
90	5A	0101 1010	!	Exclamation point
91	5B	0101 1011	$	Dollar sign
92	5C	0101 1100	*	Asterisk
93	5D	0101 1101	)	Right parenthesis
94	5E	0101 1110	;	Semicolon
95	5F	0101 1111	∧	Logical NOT
96	60	0110 0000	-	Hyphen, minus sign

(*continued overleaf*)

Table 2.7 *(continued)*.

Decimal	Hex	Binary	Code	Code description
97	61	0110 0001	/	Slash, virgule
98	62	0110 0010		
99	63	0110 0011		
100	64	0110 0100		
101	65	0110 0101		
102	66	0110 0110		
103	67	0110 0111		
104	68	0110 1000		
105	69	0110 1001		
106	6A	0110 1010		
107	6B	0110 1011	,	Comma
108	6C	0110 1100	%	Percent
109	6D	0110 1101	_	Underline, underscore
110	6E	0110 1110	>	Greater-than sign
111	6F	0110 1111	?	Question mark
112	70	0111 0000		
113	71	0111 0001		
114	72	0111 0010		
115	73	0111 0011		
116	74	0111 0100		
117	75	0111 0101		
118	76	0111 0110		
119	77	0111 0111		
120	78	0111 1000		
121	79	0111 1001	‘	Grave accent
122	7A	0111 1010	:	Colon
123	7B	0111 1011	#	Number sign, octothorp, pound
124	7C	0111 1100	@	At sign
125	7D	0111 1101	’	Apostrophe, prime
126	7E	0111 1110	=	Equals sign
127	7F	0111 1111	“	Quotation mark
128	80	1000 0000		
129	81	1000 1001	a	a
130	82	1000 1010	b	b
131	83	1000 1011	c	c
132	84	1000 1100	d	d
133	85	1000 0101	e	e
134	86	1000 0110	f	f
135	87	1000 0111	g	g
136	88	1000 1000	h	h
137	89	1000 1001	i	i
138	8A	1000 1010		
139	8B	1000 1011		
140	8C	1000 1100		
141	8D	1000 1101		
142	8E	1000 1110		
143	8F	1000 1111		
144	90	1001 0000		
145	91	1001 0001	j	j

Table 2.7 (*continued*).

Decimal	Hex	Binary	Code	Code description
146	92	1001 0010	k	k
147	93	1001 0011	l	l
148	94	1001 0100	m	m
149	95	1001 0101	n	n
150	96	1001 0110	o	o
151	97	1001 0111	p	p
152	98	1001 1000	q	q
153	99	1001 1001	r	r
154	9A	1001 1010		
155	9B	1001 1011		
156	9C	1001 1100		
157	9D	1001 1101		
158	9E	1001 1110		
159	9F	1001 1111		
160	A0	1010 0000		
161	A1	1010 0001	~	Tilde
162	A2	1010 0010	s	s
163	A3	1010 0011	t	t
164	A4	1010 0100	u	u
165	A5	1010 0101	v	v
166	A6	1010 0110	w	w
167	A7	1010 0111	x	x
168	A8	1010 1000	y	y
169	A9	1010 1001	z	z
170	AA	1010 1010		
171	AB	1010 1011		
172	AC	1010 1100		
173	AD	1010 1101		
174	AE	1010 1110		
175	AF	1010 1111		
176	B0	1011 0000		
177	B1	1011 0001		
178	B2	1011 0010		
179	B3	1011 0011		
180	B4	1011 0100		
181	B5	1011 0101		
182	B6	1011 0110		
183	B7	1011 0111		
184	B8	1011 1000		
185	B9	1011 1001		
186	BA	1011 1010		
187	BB	1011 1011		
188	BC	1011 1100		
189	BD	1011 1101		
190	BE	1011 1110		
191	BF	1011 1111		
192	C0	1100 0000	{	Opening brace
193	C1	1100 0001	A	A

(*continued overleaf*)

Table 2.7 (*continued*).

Decimal	Hex	Binary	Code	Code description
194	C2	1100 0010	B	B
195	C3	1100 0011	C	C
196	C4	1100 0100	D	D
197	C5	1100 0101	E	E
198	C6	1100 0110	F	F
199	C7	1100 0111	G	G
200	C8	1100 1000	H	H
201	C9	1100 1001	I	I
202	CA	1100 1010		
203	CB	1100 1011		
204	CC	1100 1100		
205	CD	1100 1101		
206	CE	1100 1110		
207	CF	1100 1111		
208	D0	1101 0000	}	Closing brace
209	D1	1101 0001	J	J
210	D2	1101 0010	K	K
211	D3	1101 0011	L	L
212	D4	1101 0100	M	M
213	D5	1101 0101	N	N
214	D6	1101 0110	O	O
215	D7	1101 0111	P	P
216	D8	1101 1000	Q	Q
217	D9	1101 1001	R	R
218	DA	1101 1010		
219	DB	1101 1011		
220	DC	1101 1100		
221	DD	1101 1101		
222	DE	1101 1110		
223	DF	1101 1111		
224	E0	1110 0000	\	Reverse slant
225	E1	1110 0001		
226	E2	1110 0010	S	S
227	E3	1110 0011	T	T
228	E4	1110 0100	U	U
229	E5	1110 0101	V	V
230	E6	1110 0110	W	W
231	E7	1110 0111	X	X
232	E8	1110 1000	Y	Y
233	E9	1110 1001	Z	Z
234	EA	1110 1010		
235	EB	1110 1011		
236	EC	1110 1100		
237	ED	1110 1101		
238	EE	1110 1110		
239	EF	1110 1111		
240	F0	1111 0000	0	0
241	F1	1111 0001	1	1

Table 2.7 (*continued*).

Decimal	Hex	Binary	Code	Code description
242	F2	1111 0010	2	2
243	F3	1111 0011	3	3
244	F4	1111 0100	4	4
245	F5	1111 0101	5	5
246	F6	1111 0110	6	6
247	F7	1111 0111	7	7
248	F8	1111 1000	8	8
249	F9	1111 1001	9	9
250	FA	1111 1010	|	
251	FB	1111 1011		
252	FC	1111 1100		
253	FD	1111 1101		
254	FE	1111 1110		
255	FF	1111 1111	eo	

nibbles. The first four-bit group, called the 'zone', represents the category of the character, while the second group, called the 'digit', identifies the specific character.

2.4.3 Unicode

As briefly mentioned in the earlier sections, encodings such as ASCII, EBCDIC and their variants do not have a sufficient number of characters to be able to encode alphanumeric data of all forms, scripts and languages. As a result, these encodings do not permit multilingual computer processing. In addition, these encodings suffer from incompatibility. Two different encodings may use the same number for two different characters or different numbers for the same characters. For example, code 4E (in hex) represents the upper-case letter 'N' in ASCII code and the plus sign '+' in the EBCDIC code. Unicode, developed jointly by the Unicode Consortium and the International Organization for Standardization (ISO), is the most complete character encoding scheme that allows text of all forms and languages to be encoded for use by computers. It not only enables the users to handle practically any language and script but also supports a comprehensive set of mathematical and technical symbols, greatly simplifying any scientific information exchange. The Unicode standard has been adopted by such industry leaders as HP, IBM, Microsoft, Apple, Oracle, Unisys, Sun, Sybase, SAP and many more.

Unicode and ISO-10646 Standards

Before we get on to describe salient features of Unicode, it may be mentioned that another standard similar in intent and implementation to Unicode is the ISO-10646. While Unicode is the brainchild of the Unicode Consortium, a consortium of manufacturers (initially mostly US based) of multilingual software, ISO-10646 is the project of the International Organization for Standardization. Although both organizations publish their respective standards independently, they have agreed to maintain compatibility between the code tables of Unicode and ISO-10646 and closely coordinate any further extensions.

The Code Table

The code table defined by both Unicode and ISO-10646 provides a unique number for every character, irrespective of the platform, program and language used. The table contains characters required to represent practically all known languages and scripts. The list includes not only the Greek, Latin, Cyrillic, Arabic, Arabian and Georgian scripts but also Japanese, Chinese and Korean scripts. In addition, the list also includes scripts such as Devanagari, Bengali, Gurmukhi, Gujarati, Oriya, Telugu, Tamil, Kannada, Thai, Tibetan, Ethiopic, Sinhala, Canadian Syllabics, Mongolian, Myanmar and others. Scripts not yet covered will eventually be added. The code table also covers a large number of graphical, typographical, mathematical and scientific symbols.

In the 32-bit version, which is the most recent version, the code table is divided into 2^{16} subsets, with each subset having 2^{16} characters. In the 32-bit representation, elements of different subsets therefore differ only in the 16 least significant bits. Each of these subsets is known as a plane. Plane 0, called the basic multilingual plane (BMP), defined by 00000000 to 0000FFFF, contains all most commonly used characters including all those found in major older encoding standards. Another subset of 2^{16} characters could be defined by 00010000 to 0001FFFF. Further, there are different slots allocated within the BMP to different scripts. For example, the basic Latin character set is encoded in the range 0000 to 007F. Characters added to the code table outside the 16-bit BMP are mostly for specialist applications such as historic scripts and scientific notation. There are indications that there may never be characters assigned outside the code space defined by 00000000 to 0010FFFF, which provides space for a little over 1 million additional characters.

Different characters in Unicode are represented by a hexadecimal number preceded by 'U+'. For example, 'A' and 'e' in basic Latin are respectively represented by U+0041 and U+0065. The first 256 code numbers in Unicode are compatible with the seven-bit ASCII-code and its eight-bit variant ISO-8859-1. Unicode characters U+0000 to U+007F (128 characters) are identical to those in the ASCII code, and the Unicode characters in the range U+0000 to U+00FF (256 characters) are identical to ISO-8859-1.

Use of Combining Characters

Unicode assigns code numbers to combining characters, which are not full characters by themselves but accents or other diacritical marks added to the previous character. This makes it possible to place any accent on any character. Although Unicode allows the use of combining characters, it also assigns separate codes to commonly used accented characters known as precomposed characters. This is done to ensure backwards compatibility with older encodings. As an example, the character 'ä' can be represented as the precomposed character U+00E4. It can also be represented in Unicode as U+0061 (Latin lower-case letter 'a') followed by U+00A8 (combining character '..').

Unicode and ISO-10646 Comparison

Although Unicode and ISO-10646 have identical code tables, Unicode offers many more features not available with ISO-10646. While the ISO-10646 standard is not much more than a comprehensive character set, the Unicode standard includes a number of other related features such as character properties and algorithms for text normalization and handling of bidirectional text to ensure correct display of mixed texts containing both right-to-left and left-to-right scripts.

2.5 Seven-segment Display Code

Seven-segment displays [Fig. 2.1(a)] are very common and are found almost everywhere, from pocket calculators, digital clocks and electronic test equipment to petrol pumps. A single seven-segment display or a stack of such displays invariably meets our display requirement. There are both LED and

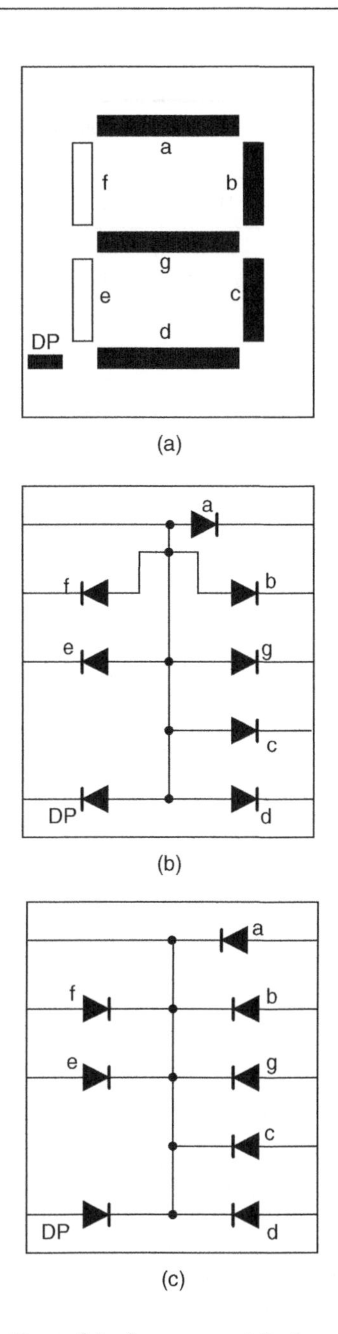

Figure 2.1 Seven-segment displays.

Table 2.8 Seven-segment display code.

Common cathode type '1' means ON									Common anode type '0' means ON								
	a	b	c	d	e	f	g	DP		a	b	c	d	e	f	g	DP
0	1	1	1	1	1	1	0		0	0	0	0	0	0	0	1	
1	0	1	1	0	0	0	0		1	1	0	0	1	1	1	1	
2	1	1	0	1	1	0	1		2	0	0	1	0	0	1	0	
3	1	1	1	1	0	0	1		3	0	0	0	0	1	1	0	
4	0	1	1	0	0	1	1		4	1	0	0	1	1	0	0	
5	1	0	1	1	0	1	1		5	0	1	0	0	1	0	0	
6	0	0	1	1	1	1	1		6	1	1	0	0	0	0	0	
7	1	1	1	0	0	0	0		7	0	0	0	1	1	1	1	
8	1	1	1	1	1	1	1		8	0	0	0	0	0	0	0	
9	1	1	1	0	0	1	1		9	0	0	0	1	1	0	0	
a	1	1	1	1	1	0	1		a	0	0	0	0	0	1	0	
b	0	0	1	1	1	1	1		b	1	1	0	0	0	0	0	
c	0	0	0	1	1	0	1		c	1	1	1	0	0	1	0	
d	0	1	1	1	1	0	1		d	1	0	0	0	0	1	0	
e	1	1	0	1	1	1	1		e	0	0	1	0	0	0	0	
f	1	0	0	0	1	1	1		f	0	1	1	1	0	0	0	

LCD types of seven-segment display. Furthermore, there are common anode-type LED displays where the arrangement of different diodes, designated *a, b, c, d, e, f* and *g*, is as shown in Fig. 2.1(b), and common cathode-type displays where the individual diodes are interconnected as shown in Fig. 2.1(c). Each display unit usually has a dot point (DP).

The DP could be located either towards the left (as shown) or towards the right of the figure '8' display pattern. This type of display can be used to display numerals from 0 to 9 and letters from A to F. Table 2.8 gives the binary code for displaying different numeric and alphabetic characters for both the common cathode and the common anode type displays. A '1' lights a segment in the common cathode type display, and a '0' lights a segment in the common anode type display.

2.6 Error Detection and Correction Codes

When we talk about digital systems, be it a digital computer or a digital communication set-up, the issue of error detection and correction is of great practical significance. Errors creep into the bit stream owing to noise or other impairments during the course of its transmission from the transmitter to the receiver. Any such error, if not detected and subsequently corrected, can be disastrous, as digital systems are sensitive to errors and tend to malfunction if the bit error rate is more than a certain threshold level. Error detection and correction, as we will see below, involves the addition of extra bits, called check bits, to the information-carrying bit stream to give the resulting bit sequence a unique characteristic that helps in detection and localization of errors. These additional bits are also called redundant bits as they do not carry any information. While the addition of redundant bits helps in achieving the goal of making transmission of information from one place to another error free or reliable, it also makes it inefficient. In this section, we will examine some common error detection and correction codes.

2.6.1 Parity Code

A parity bit is an extra bit added to a string of data bits in order to detect any error that might have crept into it while it was being stored or processed and moved from one place to another in a digital system.

We have an *even parity*, where the added bit is such that the total number of ls in the data bit string becomes even, and an *odd parity*, where the added bit makes the total number of ls in the data bit string odd. This added bit could be a '0' or a '1'. As an example, if we have to add an even parity bit to 01000001 (the eight-bit ASCII code for 'A'), it will be a '0' and the number will become 001000001. If we have to add an odd parity bit to the same number, it will be a 'l' and the number will become 101000001. The odd parity bit is a complement of the even parity bit. The most common convention is to use even parity, that is, the total number of 1s in the bit stream, including the parity bit, is even.

The parity check can be made at different points to look for any possible single-bit error, as it would disturb the parity. This simple parity code suffers from two limitations. Firstly, it cannot detect the error if the number of bits having undergone a change is even. Although the number of bits in error being equal to or greater than 4 is a very rare occurrence, the addition of a single parity cannot be used to detect two-bit errors, which is a distinct possibility in data storage media such as magnetic tapes. Secondly, the single-bit parity code cannot be used to localize or identify the error bit even if one bit is in error. There are several codes that provide self-single-bit error detection and correction mechanisms, and these are discussed below.

2.6.2 Repetition Code

The repetition code makes use of repetitive transmission of each data bit in the bit stream. In the case of threefold repetition, '1' and '0' would be transmitted as '111' and '000' respectively. If, in the received data bit stream, bits are examined in groups of three bits, the occurrence of an error can be detected. In the case of single-bit errors, '1' would be received as 011 or 101 or 110 instead of 111, and a '0' would be received as 100 or 010 or 001 instead of 000. In both cases, the code becomes self-correcting if the bit in the majority is taken as the correct bit. There are various forms in which the data are sent using the repetition code. Usually, the data bit stream is broken into blocks of bits, and then each block of data is sent some predetermined number of times. For example, if we want to send eight-bit data given by 11011001, it may be broken into two blocks of four bits each. In the case of threefold repetition, the transmitted data bit stream would be 110111011101100110011001. However, such a repetition code where the bit or block of bits is repeated 3 times is not capable of correcting two-bit errors, although it can detect the occurrence of error. For this, we have to increase the number of times each bit in the bit stream needs to be repeated. For example, by repeating each data bit 5 times, we can detect and correct all two-bit errors. The repetition code is highly inefficient and the information throughput drops rapidly as we increase the number of times each data bit needs to be repeated to build error detection and correction capability.

2.6.3 Cyclic Redundancy Check Code

Cyclic redundancy check (CRC) codes provide a reasonably high level of protection at low redundancy level. The cycle code for a given data word is generated as follows. The data word is first appended by a number of 0s equal to the number of check bits to be added. This new data bit sequence is then divided by a special binary word whose length equals $n+1$, n being the number of check bits to be added. The remainder obtained as a result of modulo-2 division is then added to the dividend bit sequence to get the cyclic code. The code word so generated is completely divisible by the divisor used in the generation of the code. Thus, when the received code word is again divided by the same divisor, an error-free reception should lead to an all '0' remainder. A nonzero remainder is indicative of the presence of errors.

The probability of error detection depends upon the number of check bits, n, used to construct the cyclic code. It is 100 % for single-bit and two-bit errors. It is also 100 % when an odd number of bits are in error and the error bursts have a length less than $n + 1$. The probability of detection reduces to $1 - (1/2)^{n-1}$ for an error burst length equal to $n + 1$, and to $1 - (1/2)^n$ for an error burst length greater than $n + 1$.

2.6.4 Hamming Code

We have seen, in the case of the error detection and correction codes described above, how an increase in the number of redundant bits added to message bits can enhance the capability of the code to detect and correct errors. If we have a sufficient number of redundant bits, and if these bits can be arranged such that different error bits produce different error results, then it should be possible not only to detect the error bit but also to identify its location. In fact, the addition of redundant bits alters the 'distance' code parameter, which has come to be known as the Hamming distance. The Hamming distance is nothing but the number of bit disagreements between two code words. For example, the addition of single-bit parity results in a code with a Hamming distance of at least 2. The smallest Hamming distance in the case of a threefold repetition code would be 3. Hamming noticed that an increase in distance enhanced the code's ability to detect and correct errors. Hamming's code was therefore an attempt at increasing the Hamming distance and at the same time having as high an information throughput rate as possible.

The algorithm for writing the generalized Hamming code is as follows:

1. The generalized form of code is $P_1P_2D_1P_3D_2D_3D_4P_4D_5D_6D_7D_8D_9D_{10}D_{11}P_5 \ldots$, where P and D respectively represent parity and data bits.
2. We can see from the generalized form of the code that all bit positions that are powers of 2 (positions 1, 2, 4, 8, 16, . . .) are used as parity bits.
3. All other bit positions (positions 3, 5, 6, 7, 9, 10, 11, . . .) are used to encode data.
4. Each parity bit is allotted a group of bits from the data bits in the code word, and the value of the parity bit (0 or 1) is used to give it certain parity.
5. Groups are formed by first checking $N-1$ bits and then alternately skipping and checking N bits following the parity bit. Here, N is the position of the parity bit; 1 for P_1, 2 for P_2, 4 for P_3, 8 for P_4 and so on. For example, for the generalized form of code given above, various groups of bits formed with different parity bits would be $P_1D_1D_2D_4D_5 \ldots$, $P_2D_1D_3D_4D_6D_7 \ldots$, $P_3D_2D_3D_4D_8D_9 \ldots$, $P_4D_5D_6D_7D_8D_9D_{10}D_{11} \ldots$ and so on. To illustrate the formation of groups further, let us examine the group corresponding to parity bit P_3. Now, the position of P_3 is at number 4. In order to form the group, we check the first three bits ($N-1=3$) and then follow it up by alternately skipping and checking four bits ($N=4$).

The Hamming code is capable of correcting single-bit errors on messages of any length. Although the Hamming code can detect two-bit errors, it cannot give the error locations. The number of parity bits required to be transmitted along with the message, however, depends upon the message length, as shown above. The number of parity bits n required to encode m message bits is the smallest integer that satisfies the condition $(2^n - n) > m$.

Table 2.9 Generation of Hamming code.

	P_1	P_2	D_1	P_3	D_2	D_3	D_4
Data bits (without parity)			0		1	1	0
Data bits with parity bit P_1	1		0		1		0
Data bits with parity bit P_2		1	0			1	0
Data bits with parity bit P_3				0	1	1	0
Data bits with parity	1	1	0	0	1	1	0

The most commonly used Hamming code is the one that has a code word length of seven bits with four message bits and three parity bits. It is also referred to as the Hamming (7, 4) code. The code word sequence for this code is written as $P_1P_2D_1P_3D_2D_3D_4$, with P_1, P_2 and P_3 being the parity bits and D_1, D_2, D_3 and D_4 being the data bits. We will illustrate step by step the process of writing the Hamming code for a certain group of message bits and then the process of detection and identification of error bits with the help of an example. We will write the Hamming code for the four-bit message 0110 representing numeral '6'. The process of writing the code is illustrated in Table 2.9, with even parity.

Thus, the Hamming code for 0110 is 1100110. Let us assume that the data bit D_1 gets corrupted in the transmission channel. The received code in that case is 1110110. In order to detect the error, the parity is checked for the three parity relations mentioned above. During the parity check operation at the receiving end, three additional bits X, Y and Z are generated by checking the parity status of $P_1D_1D_2D_4$, $P_2D_1D_3D_4$ and $P_3D_2D_3D_4$ respectively. These bits are a '0' if the parity status is okay, and a '1' if it is disturbed. In that case, ZYX gives the position of the bit that needs correction. The process can be best explained with the help of an example.

Examination of the first parity relation gives $X = 1$ as the even parity is disturbed. The second parity relation yields $Y = 1$ as the even parity is disturbed here too. Examination of the third relation gives $Z = 0$ as the even parity is maintained. Thus, the bit that is in error is positioned at 011 which is the binary equivalent of '3'. This implies that the third bit from the MSB needs to be corrected. After correcting the third bit, the received message becomes 1100110 which is the correct code.

Example 2.6

By writing the parity code (even) and threefold repetition code for all possible four-bit straight binary numbers, prove that the Hamming distance in the two cases is at least 2 in the case of the parity code and 3 in the case of the repetition code.

Solution

The generation of codes is shown in Table 2.10. An examination of the parity code numbers reveals that the number of bit disagreements between any pair of code words is not less than 2. It is either 2 or 4. It is 4, for example, between 00000 and 10111, 00000 and 11011, 00000 and 11101, 00000 and 11110 and 00000 and 01111. In the case of the threefold repetition code, it is either 3, 6, 9 or 12 and therefore not less than 3 under any circumstances.

Example 2.7

It is required to transmit letter 'A' expressed in the seven-bit ASCII code with the help of the Hamming (11, 7) code. Given that the seven-bit ASCII notation for 'A' is 1000001 and that the data word gets

Table 2.10 Example 2.6.

Binary number	Parity code	Three-time repetition Code	Binary number	Parity code	Three-time repetition code
0000	00000	000000000000	1000	11000	100010001000
0001	10001	000100010001	1001	01001	100110011001
0010	10010	001000100010	1010	01010	101010101010
0011	00011	001100110011	1011	11011	101110111011
0100	10100	010001000100	1100	01100	110011001100
0101	00101	010101010101	1101	11101	110111011101
0110	00110	011001100110	1110	11110	111011101110
0111	10111	011101110111	1111	01111	111111111111

corrupted to 1010001 in the transmission channel, show how the Hamming code can be used to identify the error. Use even parity.

Solution

- The generalized form of the Hamming code in this case is $P_1P_2D_1P_3D_2D_3D_4P_4D_5D_6D_7 = P_1P_21P_3000P_4001$.
- The four groups of bits using different parity bits are $P_1D_1D_2D_4D_5D_7$, $P_2D_1D_3D_4D_6D_7$, $P_3D_2D_3D_4$ and $P_4D_5D_6D_7$.
- This gives $P_1 = 0$, $P_2 = 0$, $P_3 = 0$ and $P_4 = 1$.
- Therefore, the transmitted Hamming code for 'A' is 00100001001.
- The received Hamming code is 00100101001.
- Checking the parity for the P_1 group gives '0' as it passes the test.
- Checking the parity for the P_2 group gives '1' as it fails the test.
- Checking the parity for the P_3 group gives '1' as it fails the test.
- Checking the parity for the P_4 group gives '0' as it passes the test.
- The bits resulting from the parity check, written in reverse order, constitute 0110, which is the binary equivalent of '6'. This shows that the bit in error is the sixth from the MSB.
- Therefore, the corrected Hamming code is 00100001001, which is the same as the transmitted code.
- The received data word is 1000001.

Review Questions

1. Distinguish between weighted and unweighted codes. Give two examples each of both types of code.
2. What is an excess-3 BCD code? Which shortcoming of the 8421 BCD code is overcome in the excess-3 BCD code? Illustrate with the help of an example.
3. What is the Gray code? Why is it also known as the binary-reflected Gray code? Briefly outline some of the important applications of the Gray code.
4. Briefly describe salient features of the ASCII and EBCDIC codes in terms of their capability to represent characters and suitability for their use in different platforms.
5. What is the Unicode? Why is it called the most complete character code?

6. What is a parity bit? Define even and odd parity. What is the limitation of the parity code when it comes to detection and correction of bit errors?
7. What is the Hamming distance? What is the role of the Hamming distance in deciding the error detection and correction capability of a code meant for the purpose? How does it influence the information throughput rate?
8. With the help of the generalized form of the Hamming code, explain how the number of parity bits required to transmit a given number of data bits is decided upon.

Problems

1. Write the excess-3 equivalent codes of $(6)_{10}$, $(78)_{10}$ and $(357)_{10}$, all in 16-bit format.

   0011001100111001, 0011001110101011, 0011011010001010

2. Determine the Gray code equivalent of $(10011)_2$ and the binary equivalent of the Gray code number 110011.

   11010, $(100010)_2$

3. A 16-bit data word given by 1001100001110110 is to be transmitted by using a fourfold repetition code. If the data word is broken into four blocks of four bits each, then write the transmitted bit stream.

   1001100110011001100010001000100001110111011101110110011001100110

4. Write (a) the Hamming (7, 4) code for 0000 using even parity and (b) the Hamming (11, 7) code for 1111111 using odd parity.

   (a) 0000000; (b) 00101110111

5. Write the last four of the 16 possible numbers in the two-bit quaternary Gray code with 0, 1, 2 and 3 as its independent digits, beginning with the thirteenth number.

   33, 32, 31, 30

Further Reading

1. Tokheim, R. L. (1994) *Schaum's Outline Series of Digital Principles*, McGraw-Hill Book Companies Inc., USA.
2. Gillam, R. (2002) *Unicode Demystified: A Practical Programmer's Guide to the Encoding Standard*, 1st edition, Addison-Wesley Professional, Boston, MA, USA.
3. MacWilliams, F. J. and Sloane, N. J. A. (2006) *The Theory of Error-Correcting Codes*, North-Holland Mathematical Library, Elsevier Ltd, Oxford, UK.
4. Huffman, W. C. and Pless, V. (2003) *Fundamentals of Error-Correcting Codes*, Cambridge University Press, Cambridge, UK.

3

Digital Arithmetic

Having discussed different methods of numeric and alphanumeric data representation in the first two chapters, the next obvious step is to study the rules of data manipulation. Two types of operation that are performed on binary data include arithmetic and logic operations. Basic arithmetic operations include addition, subtraction, multiplication and division. AND, OR and NOT are the basic logic functions. While the rules of arithmetic operations are covered in the present chapter, those related to logic operations will be discussed in the next chapter.

3.1 Basic Rules of Binary Addition and Subtraction

The basic principles of binary addition and subtraction are similar to what we all know so well in the case of the decimal number system. In the case of addition, adding '0' to a certain digit produces the same digit as the sum, and, when we add '1' to a certain digit or number in the decimal number system, the result is the next higher digit or number, as the case may be. For example, 6 + 1 in decimal equals '7' because '7' immediately follows '6' in the decimal number system. Also, 7 + 1 in octal equals '10' as, in the octal number system, the next adjacent higher number after '7' is '10'. Similarly, 9 + 1 in the hexadecimal number system is 'A'. With this background, we can write the basic rules of binary addition as follows:

1. $0 + 0 = 0$.
2. $0 + 1 = 1$.
3. $1 + 0 = 1$.
4. $1 + 1 = 0$ with a carry of '1' to the next more significant bit.
5. $1 + 1 + 1 = 1$ with a carry of '1' to the next more significant bit.

Table 3.1 summarizes the sum and carry outputs of all possible three-bit combinations. We have taken three-bit combinations as, in all practical situations involving the addition of two larger bit

Table 3.1 Binary addition of three bits.

A	B	Carry-in (C_{in})	Sum	Carry-out (C_o)	A	B	Carry-in (C_{in})	Sum	Carry-out (C_o)
0	0	0	0	0	1	0	0	1	0
0	0	1	1	0	1	0	1	0	1
0	1	0	1	0	1	1	0	0	1
0	1	1	0	1	1	1	1	1	1

numbers, we need to add three bits at a time. Two of the three bits are the bits that are part of the two binary numbers to be added, and the third bit is the carry-in from the next less significant bit column.

The basic principles of binary subtraction include the following:

1. $0 - 0 = 0$.
2. $1 - 0 = 1$.
3. $1 - 1 = 0$.
4. $0 - 1 = 1$ with a borrow of 1 from the next more significant bit.

The above-mentioned rules can also be explained by recalling rules for subtracting decimal numbers. Subtracting '0' from any digit or number leaves the digit or number unchanged. This explains the first two rules. Subtracting '1' from any digit or number in decimal produces the immediately preceding digit or number as the answer. In general, the subtraction operation of larger-bit binary numbers also involves three bits, including the two bits involved in the subtraction, called the minuend (the upper bit) and the subtrahend (the lower bit), and the borrow-in. The subtraction operation produces the difference output and borrow-out, if any. Table 3.2 summarizes the binary subtraction operation. The entries in Table 3.2 can be explained by recalling the basic rules of binary subtraction mentioned above, and that the subtraction operation involving three bits, that is, the minuend (A), the subtrahend (B) and the borrow-in (B_{in}), produces a difference output equal to ($A - B - B_{in}$). It may be mentioned here that, in the case of subtraction of larger-bit binary numbers, the least significant bit column always involves two bits to produce a difference output bit and the borrow-out

Table 3.2 Binary subtraction.

Inputs			Outputs	
Minuend (A)	Subtrahend (B)	Borrow-in (B_{in})	Difference (D)	Borrow-out (B_o)
0	0	0	0	0
0	0	1	1	1
0	1	0	1	1
0	1	1	0	1
1	0	0	1	0
1	0	1	0	0
1	1	0	0	0
1	1	1	1	1

bit. The borrow-out bit produced here becomes the borrow-in bit for the next more significant bit column, and the process continues until we reach the most significant bit column. The addition and subtraction of larger-bit binary numbers is illustrated with the help of examples in sections 3.2 and 3.3 respectively.

3.2 Addition of Larger-Bit Binary Numbers

The addition of larger binary integers, fractions or mixed binary numbers is performed columnwise in just the same way as in the case of decimal numbers. In the case of binary numbers, however, we follow the basic rules of addition of two or three binary digits, as outlined earlier. The process of adding two larger-bit binary numbers can be best illustrated with the help of an example.

Consider two generalized four-bit binary numbers $(A_3 A_2 A_1 A_0)$ and $(B_3 B_2 B_1 B_0)$, with A_0 and B_0 representing the LSB and A_3 and B_3 representing the MSB of the two numbers. The addition of these two numbers is performed as follows. We begin with the LSB position. We add the LSB bits and record the sum S_0 below these bits in the same column and take the carry C_0, if any, to the next column of bits. For instance, if $A_0 = 1$ and $B_0 = 0$, then $S_0 = 1$ and $C_0 = 0$. Next we add the bits A_1 and B_1 and the carry C_0 from the previous addition. The process continues until we reach the MSB bits. The four steps are shown ahead. C_0, C_1, C_2 and C_3 are carrys, if any, produced as a result of adding first, second, third and fourth column bits respectively, starting from LSB and proceeding towards MSB. A similar procedure is followed when the given numbers have both integer as well as fractional parts:

$$
\text{1.}\quad
\begin{array}{cccc}
 & & (C_0) & \\
A_3 & A_2 & A_1 & A_0 \\
B_3 & B_2 & B_1 & B_0 \\
\hline
 & & & S_0 \\
\hline
\end{array}
\qquad
\text{2.}\quad
\begin{array}{ccccc}
 & & (C_1) & (C_0) & \\
 & A_3 & A_2 & A_1 & A_0 \\
 & B_3 & B_2 & B_1 & B_0 \\
\hline
 & & & S_1 & S_0 \\
\hline
\end{array}
$$

$$
\text{3.}\quad
\begin{array}{cccc}
(C_2) & (C_1) & (C_0) & \\
A_3 & A_2 & A_1 & A_0 \\
B_3 & B_2 & B_1 & B_0 \\
\hline
 & S_2 & S_1 & S_0 \\
\hline
\end{array}
\qquad
\text{4.}\quad
\begin{array}{ccccc}
 & (C_2) & (C_1) & (C_0) & \\
 & A_3 & A_2 & A_1 & A_0 \\
 & B_3 & B_2 & B_1 & B_0 \\
\hline
C_3 & S_3 & S_2 & S_1 & S_0 \\
\hline
\end{array}
$$

3.2.1 Addition Using the 2's Complement Method

The 2's complement is the most commonly used code for processing positive and negative binary numbers. It forms the basis of arithmetic circuits in modern computers. When the decimal numbers to be added are expressed in 2's complement form, the addition of these numbers, following the basic laws of binary addition, gives correct results. Final carry obtained, if any, while adding MSBs should be disregarded. To illustrate this, we will consider the following four different cases:

1. Both the numbers are positive.
2. Larger of the two numbers is positive.
3. The larger of the two numbers is negative.
4. Both the numbers are negative.

Case 1

- Consider the decimal numbers +37 and +18.
- The 2's complement of +37 in eight-bit representation = 00100101.
- The 2's complement of +18 in eight-bit representation = 00010010.
- The addition of the two numbers, that is, +37 and +18, is performed as follows

$$\begin{array}{r} 00100101 \\ +\,00010010 \\ \hline 00110111 \end{array}$$

- The decimal equivalent of $(00110111)_2$ is (+55), which is the correct answer.

Case 2

- Consider the two decimal numbers +37 and -18.
- The 2's complement representation of +37 in eight-bit representation = 00100101.
- The 2's complement representation of −18 in eight-bit representation = 11101110.
- The addition of the two numbers, that is, +37 and −18, is performed as follows:

$$\begin{array}{r} 00100101 \\ +\,11101110 \\ \hline 00010011 \end{array}$$

- The final carry has been disregarded.
- The decimal equivalent of $(00010011)_2$ is +19, which is the correct answer.

Case 3

- Consider the two decimal numbers +18 and −37.
- −37 in 2's complement form in eight−bit representation = 11011011.
- +18 in 2's complement form in eight−bit representation = 00010010.
- The addition of the two numbers, that is, −37 and +18, is performed as follows:

$$\begin{array}{r} 11011011 \\ +\,00010010 \\ \hline 11101101 \end{array}$$

- The decimal equivalent of $(11101101)_2$, which is in 2's complement form, is −19, which is the correct answer. 2's complement representation was discussed in detail in Chapter 1 on number systems.

Case 4

- Consider the two decimal numbers −18 and −37.
- −18 in 2's complement form is 11101110.
- −37 in 2's complement form is 11011011.
- The addition of the two numbers, that is, −37 and −18, is performed as follows:

$$\begin{array}{r} 11011011 \\ +\,11101110 \\ \hline 11001001 \end{array}$$

- The final carry in the ninth bit position is disregarded.
- The decimal equivalent of $(11001001)_2$, which is in 2's complement form, is −55, which is the correct answer.

It may also be mentioned here that, in general, 2's complement notation can be used to perform addition when the expected result of addition lies in the range from -2^{n-1} to $+(2^{n-1}-1)$, n being the number of bits used to represent the numbers. As an example, eight-bit 2's complement arithmetic cannot be used to perform addition if the result of addition lies outside the range from −128 to +127. Different steps to be followed to do addition in 2's complement arithmetic are summarized as follows:

1. Represent the two numbers to be added in 2's complement form.
2. Do the addition using basic rules of binary addition.
3. Disregard the final carry, if any.
4. The result of addition is in 2's complement form.

Example 3.1

Perform the following addition operations:

1. $(275.75)_{10} + (37.875)_{10}$.
2. $(AF1.B3)_{16} + (FFF.E)_{16}$.

Solution

1. As a first step, the two given decimal numbers will be converted into their equivalent binary numbers (decimal-to-binary conversion has been covered at length in Chapter 1, and therefore the decimal-to-binary conversion details will not be given here):

$$(275.75)_{10} = (100010011.11)_2 \text{ and } (37.875)_{10} = (100101.111)_2$$

The two binary numbers can be rewritten as $(100010011.110)_2$ and $(000100101.111)_2$ to have the same number of bits in their integer and fractional parts. The addition of two numbers is performed as follows:

$$\begin{array}{r} 100010011.110 \\ 000100101.111 \\ \hline 100111001.101 \end{array}$$

The decimal equivalent of $(100111001.101)_2$ is $(313.625)_{10}$.

2. $(AF1.B3)_{16} = (101011110001.10110011)_2$ and $(FFF.E)_{16} = (111111111111.1110)_2$. $(111111111 11.1110)_2$ can also be written as $(111111111111.11100000)_2$ to have the same number of bits in the integer and fractional parts. The two numbers can now be added as follows:

$$\begin{array}{r} 0101011110001.10110011 \\ 0111111111111.11100000 \\ \hline 1101011110001.10010011 \end{array}$$

The hexadecimal equivalent of $(1101011110001.10010011)_2$ is $(1AF1.93)_{16}$, which is equal to the hex addition of $(AF1.B3)_{16}$ and $(FFF.E)_{16}$.

Example 3.2

Find out whether 16-bit 2's complement arithmetic can be used to add 14 276 and 18 490.

Solution

The addition of decimal numbers 14 276 and 18 490 would yield 32 766. 16-bit 2's complement arithmetic has a range of -2^{15} to $+(2^{15}-1)$, i.e. $-32\ 768$ to $+32\ 767$. The expected result is inside the allowable range. Therefore, 16-bit arithmetic can be used to add the given numbers.

Example 3.3

Add −118 and −32 firstly using eight-bit 2's complement arithmetic and then using 16-bit 2's complement arithmetic. Comment on the results.

Solution

- −118 in eight-bit 2's complement representation = 10001010.
- −32 in eight-bit 2's complement representation = 11100000.
- The addition of the two numbers, after disregarding the final carry in the ninth bit position, is 01101010. Now, the decimal equivalent of $(01101010)_2$, which is in 2's complement form, is +106. The reason for the wrong result is that the expected result, i.e. −150, lies outside the range of eight-bit 2's complement arithmetic. Eight-bit 2's complement arithmetic can be used when the expected result lies in the range from -2^7 to $+(2^7-1)$, i.e. −128 to +127. −118 in 16-bit 2's complement representation = 1111111110001010.
- −32 in 16-bit 2's complement representation = 1111111111100000.
- The addition of the two numbers, after disregarding the final carry in the 17th position, produces 1111111101101010. The decimal equivalent of $(1111111101101010)_2$, which is in 2's complement form, is −150, which is the correct answer. 16-bit 2's complement arithmetic has produced the correct result, as the expected result lies within the range of 16-bit 2's complement notation.

3.3 Subtraction of Larger-Bit Binary Numbers

Subtraction is also done columnwise in the same way as in the case of the decimal number system. In the first step, we subtract the LSBs and subsequently proceed towards the MSB. Wherever the subtrahend (the bit to be subtracted) is larger than the minuend, we borrow from the next adjacent higher bit position having a '1'. As an example, let us go through different steps of subtracting $(1001)_2$ from $(1100)_2$.

In this case, '1' is borrowed from the second MSB position, leaving a '0' in that position. The borrow is first brought to the third MSB position to make it '10'. Out of '10' in this position, '1' is taken to the LSB position to make '10' there, leaving a '1' in the third MSB position. $10-1$ in the LSB column gives '1', $1-0$ in the third MSB column gives '1', $0-0$ in the second MSB column gives '0' and $1-1$ in the MSB also gives '0' to complete subtraction. Subtraction of mixed numbers is also done in the same manner. The above-mentioned steps are summarized as follows:

```
1.  1 1 0 0       2.  1 1 0 0
    1 0 0 1           1 0 0 1
    -------           -------
          1               1 1
    -------           -------

3.  1 1 0 0       4.  1 1 0 0
    1 0 0 1           1 0 0 1
    -------           -------
      0 1 1           0 0 1 1
    -------           -------
```

3.3.1 Subtraction Using 2's Complement Arithmetic

Subtraction is similar to addition. Adding 2's complement of the subtrahend to the minuend and disregarding the carry, if any, achieves subtraction. The process is illustrated by considering six different cases:

1. Both minuend and subtrahend are positive. The subtrahend is the smaller of the two.
2. Both minuend and subtrahend are positive. The subtrahend is the larger of the two.
3. The minuend is positive. The subtrahend is negative and smaller in magnitude.
4. The minuend is positive. The subtrahend is negative and greater in magnitude.
5. Both minuend and subtrahend are negative. The minuend is the smaller of the two.
6. Both minuend and subtrahend are negative. The minuend is the larger of the two.

Case 1

- Let us subtract +14 from +24.
- The 2's complement representation of +24 = 00011000.
- The 2's complement representation of +14 = 00001110.
- Now, the 2's complement of the subtrahend (i.e. +14) is 11110010.
- Therefore, +24 − (+14) is given by

```
  00011000
+ 11110010
  --------
  00001010
```

  with the final carry disregarded.
- The decimal equivalent of $(00001010)_2$ is +10, which is the correct answer.

Case 2

- Let us subtract +24 from +14.
- The 2's complement representation of +14 = 00001110.
- The 2's complement representation of +24 = 00011000.
- The 2's complement of the subtrahend (i.e. +24) = 11101000.
- Therefore, +14 − (+24) is given by

$$\begin{array}{r} 00001110 \\ +\,11101000 \\ \hline 11110110 \\ \hline \end{array}$$

- The decimal equivalent of $(11110110)_2$, which is of course in 2's complement form, is −10 which is the correct answer.

Case 3

- Let us subtract −14 from +24.
- The 2's complement representation of +24 = 00011000 = minuend.
- The 2's complement representation of −14 = 11110010 = subtrahend.
- The 2's complement of the subtrahend (i.e. −14) = 00001110.
- Therefore, +24 − (−14) is performed as follows:

$$\begin{array}{r} 00011000 \\ +\,00001110 \\ \hline 00100110 \\ \hline \end{array}$$

- The decimal equivalent of $(00100110)_2$ is +38, which is the correct answer.

Case 4

- Let us subtract −24 from +14.
- The 2's complement representation of +14 = 00001110 = minuend.
- The 2's complement representation of −24 = 11101000 = subtrahend.
- The 2's complement of the subtrahend (i.e. −24) = 00011000.
- Therefore, +14 − (−24) is performed as follows:

$$\begin{array}{r} 00001110 \\ +\,00011000 \\ \hline 00100110 \\ \hline \end{array}$$

- The decimal equivalent of $(00100110)_2$ is +38, which is the correct answer.

Case 5

- Let us subtract −14 from −24.
- The 2's complement representation of −24 = 11101000 = minuend.

- The 2's complement representation of −14=11110010 = subtrahend.
- The 2's complement of the subtrahend = 00001110.
- Therefore, −24 − (−14) is given as follows:

$$\begin{array}{r} 11101000 \\ +\,00001110 \\ \hline 11110110 \end{array}$$

- The decimal equivalent of $(11110110)_2$, which is in 2's complement form, is −10, which is the correct answer.

Case 6

- Let us subtract −24 from −14.
- The 2's complement representation of −14 = 11110010 = minuend.
- The 2's complement representation of −24=11101000 = subtrahend.
- The 2's complement of the subtrahend = 00011000.
- Therefore, −14 − (−24) is given as follows:

$$\begin{array}{r} 11110010 \\ +\,00011000 \\ \hline 00001010 \end{array}$$

  with the final carry disregarded.
- The decimal equivalent of $(00001010)_2$, which is in 2's complement form, is +10, which is the correct answer.

It may be mentioned that, in 2's complement arithmetic, the answer is also in 2's complement notation, only with the MSB indicating the sign and the remaining bits indicating the magnitude. In 2's complement notation, positive magnitudes are represented in the same way as the straight binary numbers, while the negative magnitudes are represented as the 2's complement of their straight binary counterparts. A '0' in the MSB position indicates a positive sign, while a '1' in the MSB position indicates a negative sign.

The different steps to be followed to do subtraction in 2's complement arithmetic are summarized as follows:

1. Represent the minuend and subtrahend in 2's complement form.
2. Find the 2's complement of the subtrahend.
3. Add the 2's complement of the subtrahend to the minuend.
4. Disregard the final carry, if any.
5. The result is in 2's complement form.
6. 2's complement notation can be used to perform subtraction when the expected result of subtraction lies in the range from -2^{n-1} to $+(2^{n-1}-1)$, n being the number of bits used to represent the numbers.

Example 3.4

Subtract $(1110.011)_2$ from $(11011.11)_2$ using basic rules of binary subtraction and verify the result by showing equivalent decimal subtraction.

Solution

The minuend and subtrahend are first modified to have the same number of bits in the integer and fractional parts. The modified minuend and subtrahend are $(11011.110)_2$ and $(01110.011)_2$ respectively:

$$\begin{array}{r} 11011.110 \\ -\,01110.011 \\ \hline 01101.011 \end{array}$$

The decimal equivalents of $(11011.110)_2$ and $(01110.011)_2$ are 27.75 and 14.375 respectively. Their difference is 13.375, which is the decimal equivalent of $(01101.011)_2$.

Example 3.5

Subtract (a) $(-64)_{10}$ from $(+32)_{10}$ and (b) $(29.A)_{16}$ from $(4F.B)_{16}$. Use 2's complement arithmetic.

Solution:

(a) $(+32)_{10}$in 2's complement notation $= (00100000)_2$.
$(-64)_{10}$ in 2's complement notation $= (11000000)_2$.
The 2's complement of $(-64)_{10} = (01000000)_2$.
$(+32)_{10} - (-64)_{10}$ is determined by adding the 2's complement of $(-64)_{10}$ to $(+32)_{10}$.
Therefore, the addition of $(00100000)_2$ to $(01000000)_2$ should give the result. The operation is shown as follows:

$$\begin{array}{r} 00100000 \\ +\,01000000 \\ \hline 01100000 \end{array}$$

The decimal equivalent of $(01100000)_2$ is +96, which is the correct answer as $+32 - (-64) = +96$.

(b) The minuend $= (4F.B)_{16} = (01001111.1011)_2$.
The minuend in 2's complement notation $= (01001111.1011)_2$.
The subtrahend $= (29.A)_{16} = (00101001.1010)_2$.
The subtrahend in 2's complement notation $= (00101001.1010)_2$.
The 2's complement of the subtrahend $= (11010110.0110)_2$.
$(4F.B)_{16} - (29.A)_{16}$ is given by the addition of the 2's complement of the subtrahend to the minuend.

$$\begin{array}{r} 01001111.1011 \\ +\,11010110.0110 \\ \hline 00100110.0001 \end{array}$$

with the final carry disregarded. The result is also in 2's complement form. Since the result is a positive number, 2's complement notation is the same as it would be in the case of the straight binary code.
The hex equivalent of the resulting binary number $= (26.1)_{16}$, which is the correct answer.

3.4 BCD Addition and Subtraction in Excess-3 Code

Below, we will see how the excess-3 code can be used to perform addition and subtraction operations on BCD numbers.

3.4.1 Addition

The excess-3 code can be very effectively used to perform the addition of BCD numbers. The steps to be followed for excess-3 addition of BCD numbers are as follows:

1. The given BCD numbers are written in excess-3 form by adding '0011' to each of the four-bit groups.
2. The two numbers are then added using the basic laws of binary addition.
3. Add '0011' to all those four-bit groups that produce a carry, and subtract '0011' from all those four-bit groups that do not produce a carry during addition.
4. The result thus obtained is in excess-3 form.

3.4.2 Subtraction

Subtraction of BCD numbers using the excess-3 code is similar to the addition process discussed above. The steps to be followed for excess-3 substraction of BCD numbers are as follows:

1. Express both minuend and subtrahend in excess-3 code.
2. Perform subtraction following the basic laws of binary subtraction.
3. Subtract '0011' from each invalid BCD four-bit group in the answer.
4. Subtract '0011' from each BCD four-bit group in the answer if the subtraction operation of the relevant four-bit groups required a borrow from the next higher adjacent four-bit group.
5. Add '0011' to the remaining four-bit groups, if any, in the result.
6. This gives the result in excess-3 code.

The process of addition and subtraction can be best illustrated with the help of following examples.

Example 3.6

Add $(0011\ 0101\ 0110)_{BCD}$ and $(0101\ 0111\ 1001)_{BCD}$ using the excess-3 addition method and verify the result using equivalent decimal addition.

Solution

The excess-3 equivalents of 0011 0101 0110 and 0101 0111 1001 are 0110 1000 1001 and 1000 1010 1100 respectively. The addition of the two excess-3 numbers is given as follows:

```
 0110 1000 1001
 1000 1010 1100
 --------------
 1111 0011 0101
 --------------
```

After adding 0011 to the groups that produced a carry and subtracting 0011 from the groups that did not produce a carry, we obtain the result of the above addition as 1100 0110 1000. Therefore, 1100

0110 1000 represents the excess-3 code for the true result. The result in BCD code is 1001 0011 0101, which is the BCD equivalent of 935. This is the correct answer as the addition of the given BCD numbers 0011 0101 0110 = $(356)_{10}$ and 0101 0111 1001 = $(579)_{10}$ yields $(935)_{10}$ only.

Example 3.7

Perform $(185)_{10} - (8)_{10}$ using the excess-3 code.

Solution

- $(185)_{10} = (0001\ 1000\ 0101)_{BCD}$.The excess-3 equivalent of $(0001\ 1000\ 0101)_{BCD}$ = 0100 1011 1000.
- $(8)_{10} = (008)_{10} = (0000\ 0000\ 1000)_{BCD}$. The excess-3 equivalent of $(0000\ 0000\ 1000)_{BCD}$ = 0011 0011 1011.
- Subtraction is performed as follows:

```
  0100 1011 1000
- 0011 0011 1011
----------------
  0001 0111 1101
----------------
```

- In the subtraction operation, the least significant column of four-bit groups needed a borrow, while the other two columns did not need any borrow. Also, the least significant column has produced an invalid BCD code group. Subtracting 0011 from the result of this column and adding 0011 to the results of other two columns, we get 0100 1010 1010. This now constitutes the result of subtraction expressed in excess-3 code.
- The result in BCD code is therefore 0001 0111 0111.
- The decimal equivalent of 0001 0111 0111 is 177, which is the correct result.

3.5 Binary Multiplication

The basic rules of binary multiplication are governed by the way an AND gate functions when the two bits to be multiplied are fed as inputs to the gate. Logic gates are discussed in detail in the next chapter. As of now, it would suffice to say that the result of multiplying two bits is the same as the output of the AND gate with the two bits applied as inputs to the gate. The basic rules of multiplication are listed as follows:

1. $0 \times 0 = 0$.
2. $0 \times 1 = 0$.
3. $1 \times 0 = 0$.
4. $1 \times 1 = 1$.

One of the methods for multiplication of larger-bit binary numbers is similar to what we are familiar with in the case of decimal numbers. This is called the 'repeated left-shift and add' algorithm. Microprocessors and microcomputers, however, use what is known as the 'repeated add and right-shift' algorithm to do binary multiplication as it is comparatively much more convenient to implement than the 'repeated left-shift and add' algorithm. The two algorithms are briefly described below. Also, binary multiplication of mixed binary numbers is done by performing multiplication without considering the binary point. Starting from the LSB, the binary point is then placed after n bits, where n is equal to the sum of the number of bits in the fractional parts of the multiplicand and multiplier.

3.5.1 Repeated Left-Shift and Add Algorithm

In the 'repeated left-shift and add' method of binary multiplication, the end-product is the sum of several partial products, with the number of partial products being equal to the number of bits in the multiplier binary number. This is similar to the case of decimal multiplication. Each successive partial product after the first is shifted one digit to the left with respect to the immediately preceding partial product. In the case of binary multiplication too, the first partial product is obtained by multiplying the multiplicand binary number by the LSB of the multiplier binary number. The second partial product is obtained by multiplying the multiplicand binary number by the next adjacent higher bit in the multiplier binary number and so on. We begin with the LSB of the multiplier to obtain the first partial product. If the LSB is a '1', a copy of the multiplicand forms the partial product, and it is an all '0' sequence if the LSB is a '0'. We proceed towards the MSB of the multiplier and obtain various partial products. The second partial product is shifted one bit position to the left relative to the first partial product; the third partial product is shifted one bit position to the left relative to the second partial product and so on. The addition of all partial products gives the final answer. If the multiplicand and multiplier have different signs, the end result has a negative sign, otherwise it is positive. The procedure is further illustrated by showing $(23)_{10} \times (6)_{10}$ multiplication.

$$\begin{array}{lr l}
 & 1\,0\,1\,1\,1 & \\
\textit{Multiplicand}: & \times\,1\,1\,0 & \cdots\cdots\cdots\; (23)_{10} \\
\textit{Multiplier}: & \overline{} & \cdots\cdots\cdots\; (6)_{10} \\
 & 0\,0\,0\,0\,0 & \\
 & 1\,0\,1\,1\,1 & \\
 & 1\,0\,1\,1\,1 & \\
\hline
 & 1\,0\,0\,0\,1\,0\,1\,0 &
\end{array}$$

The decimal equivalent of $(10001010)_2$ is $(138)_{10}$, which is the correct result.

3.5.2 Repeated Add and Right-Shift Algorithm

The multiplication process starts with writing an all '0' bit sequence, with the number of bits equal to the number of bits in the multiplicand. This bit sequence (all '0' sequence) is added to another same-sized bit sequence, which is the same as the multiplicand if the LSB of the multiplier is a '1', and an all '0' sequence if it is a '0'. The result of the first addition is shifted one bit position to the right, and the bit shifted out is recorded. The vacant MSB position is replaced by a '0'. This new sequence is added to another sequence, which is an all '0' sequence if the next adjacent higher bit in the multiplier is a '0', and the same as the multiplicand if it is a '1'. The result of the second addition is also shifted one bit position to the right, and a new sequence is obtained. The process continues until all multiplier bits are exhausted. The result of the last addition together with the recorded bits constitutes the result of multiplication. We will illustrate the procedure by doing $(23)_{10} \times (6)_{10}$ multiplication again, this time by using the 'repeated add and right-shift' algorithm:

- The multiplicand $= (23)_{10} = (10111)_2$ and the multiplier $= (6)_{10} = (110)_2$. The multiplication process is shown in Table 3.3.
- Therefore, $(10111)_2 \times (110)_2 = (10001010)_2$.

Table 3.3 Multiplication using the repeated add and right-shift algorithm.

1 0 1 1 1 1 1 0	Multiplicand Multiplier
0 0 0 0 0 + 0 0 0 0 0	Start
0 0 0 0 0	Result of first addition
0 0 0 0 0 + 1 0 1 1 1	0 (Result of addition shifted one bit to right)
1 0 1 1 1	Result of second addition
0 1 0 1 1 + 1 0 1 1 1	10 (Result of addition shifted one bit to right)
1 0 0 0 1 0	Result of third addition
0 1 0 0 0 1	010 (Result of addition shifted one bit to right)

Example 3.8

Multiply (a) $(100.01)_2 \times (10.1)_2$ *by using the 'repeated add and left-shift' algorithm and (b)* $(2B)_{16} \times (3)_{16}$ *by using the 'add and right-shift' algorithm. Verify the results by showing equivalent decimal multiplication.*

Solution

(a) As a first step, we will multiply $(10001)_2$ by $(101)_2$. The process is shown as follows:

```
      1 0 0 0 1
      × 1 0 1
    -----------
      1 0 0 0 1
    0 0 0 0 0
  1 0 0 0 1
  -------------
  1 0 1 0 1 0 1
```

The multiplication result is then given by placing the binary point three bits after the LSB, which gives $(1010.101)_2$ as the final result. Also, $(100.01)_2 = (4.25)_{10}$ and $(10.1)_2 = (2.5)_{10}$. Moreover, $(4.25)_{10} \times (2.5)_{10} = (10.625)_{10}$ and $(1010.101)_2$ equals $(10.625)_{10}$, which verifies the result.

(b) $(2B)_{16} = 00101011 = 101011$ and $(3)_{16} = 0011 = 11$.
Different steps involved in the multiplication process are shown in Table 3.4.
The result of multiplication is therefore $(10000001)_2$. Also, $(2B)_{16} = (43)_{10}$ and $(3)_{16} = (3)_{10}$. Therefore, $(2B)_{16} \times (3)_{16} = (129)_{10}$. Moreover, $(10000001)_2 = (129)_{10}$, which verifies the result.

3.6 Binary Division

While binary multiplication is the process of repeated addition, binary division is the process of repeated subtraction. Binary division can be performed by using either the 'repeated right-shift and

Table 3.4 Example 3.8.

1 0 1 0 1 1	Multiplicand
1 1	Multiplier
0 0 0 0 0 0	Start
+ 1 0 1 0 1 1	
1 0 1 0 1 1	Result of first addition
0 1 0 1 0 1	1 (Result of addition shifted one bit to right)
+ 1 0 1 0 1 1	
1 0 0 0 0 0 0	Result of second addition
0 1 0 0 0 0 0	01 (Result of addition shifted one bit to right)

subtract' or the 'repeated subtract and left-shift' algorithm. These are briefly described and suitably illustrated in the following sections.

3.6.1 Repeated Right-Shift and Subtract Algorithm

The algorithm is similar to the case of conventional division with decimal numbers. At the outset, starting from MSB, we begin with the number of bits in the dividend equal to the number of bits in the divisor and check whether the divisor is smaller or greater than the selected number of bits in the dividend. If it happens to be greater, we record a '0' in the quotient column. If it is smaller, we subtract the divisor from the dividend bits and record a '1' in the quotient column. If it is greater and we have already recorded a '0', then, as a second step, we include the next adjacent bit in the dividend bits, shift the divisor to the right by one bit position and again make a similar check like the one made in the first step. If it is smaller and we have made the subtraction, then in the second step we append the next MSB of the dividend to the remainder, shift the divisor one bit to the right and again make a similar check. The options are again the same. The process continues until we have exhausted all the bits in the dividend. We will illustrate the algorithm with the help of an example. Let us consider the division of $(100110)_2$ by $(1100)_2$. The sequence of operations needed to carry out the above division is shown in Table 3.5. The quotient = 011 and the remainder = 10.

Table 3.5 Binary division using the repeated right-shift and subtract algorithm.

	Quotient		
First step	0	1 0 0 1 1 0	Dividend
		−1 1 0 0	Divisor
Second step	1	1 0 0 1 1	First five MSBs of dividend
		−1 1 0 0	Divisor shifted to right
		0 1 1 1	First subtraction remainder
Third step	1	0 1 1 1 0	Next MSB appended
		−1 1 0 0	Divisor right shifted
		0 0 1 0	Second subtraction remainder

Table 3.6 Binary division using the repeate subtract and left-shift algorithm.

Quotient	1 0 0 1 −1 1 0 0	1 0
0	1 1 0 1 +1 1 0 0	Borrow exists
	1 0 0 1	Final carry ignored
	1 0 0 1 1 −1 1 0 0	Next MSB appended
1	0 1 1 1	No borrow
	0 1 1 1 0 −1 1 0 0	Next MSB appended
1	0 0 0 1 0	No borrow

3.6.2 Repeated Subtract and Left-Shift Algorithm

The procedure can again be best illustrated with the help of an example. Let us consider solving the above problem using this algorithm. The steps needed to perform the division are as follows. We begin with the first four MSBs of the dividend, four because the divisor is four bits long. In the first step, we subtract the divisor from the dividend. If the subtraction requires borrow in the MSB position, enter a '0' in the quotient column; otherwise, enter a '1'. In the present case there exists a borrow in the MSB position, and so there is a '0' in the quotient column. If there is a borrow, the divisor is added to the result of subtraction. In doing so, the final carry, if any, is ignored. The next MSB is appended to the result of the first subtraction if there is no borrow, or to the result of subtraction, restored by adding the divisor, if there is a borrow. By appending the next MSB, the remaining bits of the dividend are one bit position shifted to the left. It is again compared with the divisor, and the process is repeated. It goes on until we have exhausted all the bits of the dividend. The final remainder can be further processed by successively appending 0s and trying subtraction to get fractional part bits of the quotient. The different steps are summarized in Table 3.6. The quotient = 011 and the remainder = 10.

Example 3.9

Use the 'repeated right-shift and subtract' algorithm to divide $(110101)_2$ by $(1011)_2$. Determine both the integer and the fractional parts of the quotient. The fractional part may be determined up to three bit places.

Solution

The sequence of operations is given in Table 3.7. The operations are self-explanatory.

- The quotient = 100.110.
- Now, $(110101)_2 = (53)_{10}$ and $(1011)_2 = (11)_{10}$.
- $(53)_{10}$ divided by $(11)_{10}$ gives $(4.82)_{10}$.
- $(100.110)_2 = (4.75)_{10}$, which matches with the expected result to a good approximation.

Table 3.7 Example 3.9.

	Quotient		
First step	1	1 1 0 1 0 1 −1 0 1 1	Dividend Divisor
		0 0 1 0	First subtraction
Second step	0	0 0 1 0 0 −1 0 1 1	Next MSB appended Divisor right shifted
Third step	0	0 0 1 0 0 1 −1 0 1 1	Next MSB appended Divisor right shifted
		0 0 1 0 0 1	All bits exhausted
	1	0 0 1 0 0 1 0 −1 0 1 1	'0' appended Divisor right shifted
		0 1 1 1	Second subtraction
Fourth step	1	0 1 1 1 0 −1 0 1 1	'0' appended Divisor right shifted
		0 0 0 1 1	Third subtraction
Fifth step	0	0 0 0 1 1 0 −1 0 1 1	'0' appended Divisor right shifted
		0 0 1 1	Fourth subtraction

Example 3.10

Use the 'repeated subtract and left-shift' algorithm to divide $(100011)_2$ by $(100)_2$ to determine both the integer and fractional parts of the quotient. Verify the result by showing equivalent decimal division. Determine the fractional part to two bit places.

Solution

The sequence of operations is given in Table 3.8. The operations are self-explanatory.

- The quotient $= (1000.11)_2 = (8.75)_{10}$.
- Now, $(100011)_2 = (35)_{10}$ and $(100)_2 = (4)_{10}$.
- $(35)_{10}$ divided by $(4)_{10}$ gives $(8.75)_{10}$ and hence is verified.

Example 3.11

Divide $(AF)_{16}$ by $(09)_{16}$ using the method of 'repeated right shift and subtract', bearing in mind the signs of the given numbers, assuming that we are working in eight-bit 2's complement arithmetic.

Solution

- The dividend $= (AF)_{16}$.
- As it is a negative hexadecimal number, the magnitude of this number is determined by its 2's complement (or more precisely by its 16's complement in hexadecimal number language).

Table 3.8 Example 3.10.

Quotient		
	1 0 0 −1 0 0	0 1 1 Dividend Divisor
1	0 0 0	No borrow
	0 0 0 0 −1 0 0	Next MSB appended
0	1 0 0 +1 0 0	Borrow exists
	0 0 0 0 0 0 1 −1 0 0	Final carry ignored Next MSB appended
0	1 0 1 + 1 0 0	Borrow exists
	0 0 1 0 0 1 1 − 1 0 0	Final carry ignored Next MSB appended
0	1 1 1 +1 0 0	Borrow exists
	0 1 1 0 1 1 0 − 1 0 0	Final carry ignored '0' appended
1	0 1 0	No borrow
	0 1 0 0 −1 0 0	'0' appended
1	0 0 0	No borrow

- The 16's complement of $(AF)_{16} = (51)_{16}$.
- The binary equivalent of $(51)_{16} = 01010001 = 1010001$.
- The divisor $= (09)_{16}$.
- It is a positive number.
- The binary equivalent of $(09)_{16} = 00001001$.
- As the dividend is a negative number and the divisor a positive number, the quotient will be a negative number. The division process using the 'repeated right-shift and subtract' algorithm is given in Table 3.9.
- The quotient $= 1001 = (09)_{16}$.
- As the quotient should be a negative number, its magnitude is given by the 16's complement of $(09)_{16}$, i.e. $(F7)_{16}$.
- Therefore, $(AF)_{16}$ divided by $(09)_{16}$ gives $(F7)_{16}$.

3.7 Floating-Point Arithmetic

Before performing arithmetic operations on floating-point numbers, it is necessary to make a few checks, such as finding the signs of the two mantissas, checking any possible misalignment of exponents, etc.

Table 3.9 Example 3.11

1	1 0 1 0 0 0 1 −1 0 0 1	Divisor less than dividend
	0 0 0 1	
0	0 0 0 1 0 −1 0 0 1	Divisor greater than dividend
0	0 0 0 1 0 0 −1 0 0 1	Divisor still greater
1	0 0 0 1 0 0 1 −1 0 0 1	Divisor less than dividend
	0 0 0 0 0 0 0	

For example, if the exponents of the two numbers are not equal, the addition and subtraction operations necessitate that they be made equal. In that case, the mantissa of the smaller of the two numbers is shifted right, and the exponent is incremented for each shift until the two exponents are equal. Once the binary points are aligned and the exponents made equal, addition and subtraction operations become straightforward. While doing subtraction, of course, a magnitude check is also required to determine the smaller of the two numbers.

3.7.1 Addition and Subtraction

If N_1 and N_2 are two floating-point numbers given by

$$N_1 = m_1 \times 2^e$$

$$N_2 = m_2 \times 2^e$$

then

$$N_1 + N_2 = m_1 \times 2^e + m_2 \times 2^e = (m_1 + m_2) \times 2^e$$

and

$$N_1 - N_2 = m_1 \times 2^e - m_2 \times 2^e = (m_1 - m_2) \times 2^e$$

The subtraction operation assumes that $N_1 > N_2$. Post-normalization of the result may be required after the addition or subtraction operation.

3.7.2 Multiplication and Division

In the case of multiplication of two floating-point numbers, the mantissas of the two numbers are multiplied and their exponents are added. In the case of a division operation, the mantissa of the quotient is given by the division of the two mantissas (i.e. dividend mantissa divided by divisor mantissa) and the exponent of the quotient is given by subtraction of the two exponents (i.e. dividend exponent minus divisor exponent).

If

$$N_1 = m_1 \times 2^{e1} and\ N_2 = m_2 \times 2^{e2}$$

then

$$N_1 \times N_2 = (m_1 \times m_2) \times 2^{(e1+e2)}$$

and

$$N_1/N_2 = (m_1/m_2) \times 2^{(e1-e2)}$$

Again, post-normalization may be required after multiplication or division, as in the case of addition and subtraction operations.

Example 3.12

Add (a) $(39)_{10}$ and $(19)_{10}$ and (b) $(1E)_{16}$ and $(F3)_{16}$ using floating-point numbers. Verify the answers by performing equivalent decimal addition.

Solution

(a) $(39)_{10} = 100111 = 0.100111 \times 2^6$.

$(19)_{10} = 10011 = 0.10011 \times 2^5 = 0.010011 \times 2^6$.

Therefore, $(39)_{10} + (19)_{10} = 0.100111 \times 2^6 + 0.010011 \times 2^6$

$= (0.100111 + 0.010011) \times 2^6 = 0.111010 \times 2^6$

$= 111010 = (58)_{10}$

and hence is verified.

(b) $(1E)_{16} = (00011110)_2 = 0.00011110 \times 2^8$.

$(F3)_{16} = (11110011)_2 = 0.11110011 \times 2^8$.

$(1E)_{16} + (F3)_{16} = (0.00011110 + 0.11110011) \times 2^8 = 100010001$

$= 000100010001$

$= (111)_{16}$.

Also, $(1E)_{16} + (F3)_{16} = (111)_{16}$ and hence is proved.

Example 3.13

Subtract $(17)_8$ from $(21)_8$ using floating-point numbers and verify the answer.

Solution

- $(21)_8 = (010001)_2 = 0.010001 \times 2^6$.
- $(17)_8 = (001111)_2 = 0.001111 \times 2^6$.
- Therefore, $(21)_8 - (17)_8 = (0.010001 - 0.001111) \times 2^6$

  $= 0.000010 \times 2^6 = 000010 = (02)_8$.
- Also, $(21)_8 - (17)_8 = (02)_8$ and hence is verified.

Example 3.14

Multiply $(37)_{10}$ by $(10)_{10}$ using floating-point numbers. Verify by showing equivalent decimal multiplication.

Solution

- The multiplicand $= (37)_{10} = (100101)_2 = 0.100101 \times 2^6$.
- The multiplier $= (10)_{10} = (1010)_2 = 0.1010 \times 2^4$.
- $(37)_{10} \times (10)_{10} = (0.100101 \times 0.1010) \times 2^{10} = 0.0101110010 \times 2^{10} = 101110010$
  $= (370)_{10}$ and hence is verified.

Example 3.15

Perform $(E3B)_{16} \div (1A)_{16}$ using binary floating-point numbers. Verify by showing equivalent decimal division.

Solution

- Dividend $= (\text{E3B})_{16} = (111000111011)_2 = 0.111000111011 \times 2^{12}$.
- Divisor $= (1\text{A})_{16} = (00011010)_2 = (11010)_2 = 0.11010 \times 2^5$.
- Therefore, $(\text{E3B})_{16} \div (1\text{A})_{16} = (0.111000111011 \div 0.11010) \times 2^7$.
- By performing division of the mantissas using either of the two division algorithms described earlier, we obtain the result of division as $(10001100.00011)_2$.
- $(10001100.00011)_2 = (140.093)_{10}$.
- Also, $(\text{E3B})_{16} = (3643)_{10}$ and $(1A)_{16} = (26)_{10}$.
- $(\text{E3B})_{16} \div (1\text{A})_{16} = (3643)_{10} \div (26)_{10} = (140.1)_{10}$, which is the same as the result obtained with binary floating-point arithmetic to a good approximation.

Review Questions

1. Outline the different steps involved in the addition of larger-bit binary numbers for the following two cases:

   (a) The larger of the two numbers is positive and the other number is negative.
   (b) The larger of the two numbers is negative and the other number is positive.

2. Outline the different steps involved in the subtraction of larger-bit binary numbers for the following two cases:

   (a) The minuend is positive. The subtrahend is negative and smaller in magnitude.
   (b) The minuend is positive. The subtrahend is negative and larger in magnitude.

3. What decides whether a particular binary addition or subtraction operation would be possible with 2's complement arithmetic?
4. Why in microprocessors and microcomputers is the 'repeated add and right-shift' algorithm preferred over the 'repeated left-shift and add' algorithm for binary multiplication? Briefly outline the procedure for multiplication in the case of the former.

5. Prove that the largest six-digit hexadecimal number when subtracted from the largest eight-digit octal number yields zero in decimal.

Problems

1. Perform the following operations using 2's complement arithmetic. The numbers are represented using 2's or 10's or 16's complement notation as the case may be. Express the result both in 2's complement binary as well as in decimal.

   (a) $(7F)_{16} + (A1)_{16}$.
   (b) $(110)_{10} + (0111)_{2}$.

   (a) $(00100000)_2, (32)_{10}$; *(b)* $(01110101)_2, (117)_{10}$

2. Evaluate the following to two binary places:

   (a) $(100.0001)_2 \div (10.1)_2$.
   (b) $(111001)_2 \div (1001)_2$.
   (c) $(111.001)_2 \times (1.11)_2$.

   (a) 1.10; (b) 110.01; (c) 1100.01

3. Prove that 16-bit 2's complement arithmetic cannot be used to add +18 150 and +14 618, while it can be used to add −18 150 and −14 618.
4. Add the maximum positive integer to the minimum negative integer, both represented in 16-bit 2's complement binary notation. Express the answer in 2's complement binary.

   1111111111111111

5. The result of adding two BCD numbers represented in excess-3 code is 0111 1011 when the two numbers are added using simple binary addition. If one of the numbers is $(12)_{10}$, find the other.

   $(03)_{10}$

6. Perform the following operations using 2's complement arithmetic:

   (a) $(+43)_{10} - (-53)_{10}$.
   (b) $(1ABC)_{16} + (1DEF)_{16}$.
   (c) $(3E91)_{16} - (1F93)_{16}$.

   (a) 01100000; (b) $(38AB)_{16}$; *(c)* $(1EFE)_{16}$

Further Reading

1. Ercegovac, M. D. and Lang, T. (2003) *Digital Arithmetic*, Morgan Kaufmann Publishers, CA, USA.
2. Tocci, R. J. (2006) *Digital Systems – Principles and Applications*, Prentice-Hall Inc., NJ, USA.
3. Ashmila, E. M., Dlay, S. S. and Hinton, O. R. (2005) 'Adder methodology and design using probabilistic multiple carry estimates'. *IET Computers and Digital Techniques*, **152**(6), pp. 697–703.
4. Lu, M. (2005) *Arithmetic and Logic in Computer Systems*, John Wiley & Sons, Inc., NJ, USA.

4

Logic Gates and Related Devices

Logic gates are electronic circuits that can be used to implement the most elementary logic expressions, also known as Boolean expressions. The logic gate is the most basic building block of combinational logic. There are three basic logic gates, namely the OR gate, the AND gate and the NOT gate. Other logic gates that are derived from these basic gates are the NAND gate, the NOR gate, the EXCLUSIVE-OR gate and the EXCLUSIVE-NOR gate. This chapter deals with logic gates and some related devices such as buffers, drivers, etc., as regards their basic functions. The treatment of the subject matter is mainly with the help of respective truth tables and Boolean expressions. The chapter is adequately illustrated with the help of solved examples. Towards the end, the chapter contains application-relevant information in terms of popular type numbers of logic gates from different logic families and their functional description to help application engineers in choosing the right device for their application. Pin connection diagrams are given on the companion website at http://www.wiley.com/go/maini_digital. Different logic families used to hardware-implement different logic functions in the form of digital integrated circuits are discussed in the following chapter.

4.1 Positive and Negative Logic

The binary variables, as we know, can have either of the two states, i.e. the logic '0' state or the logic '1' state. These logic states in digital systems such as computers, for instance, are represented by two different voltage levels or two different current levels. If the more positive of the two voltage or current levels represents a logic '1' and the less positive of the two levels represents a logic '0', then the logic system is referred to as a *positive logic system*. If the more positive of the two voltage or current levels represents a logic '0' and the less positive of the two levels represents a logic '1', then the logic system is referred to as a *negative logic system*. The following examples further illustrate this concept.

Digital Electronics: Principles, Devices and Applications Anil Kumar Maini

If the two voltage levels are 0 V and +5 V, then in the positive logic system the 0 V represents a logic '0' and the +5 V represents a logic '1'. In the negative logic system, 0 V represents a logic '1' and +5 V represents a logic '0'.

If the two voltage levels are 0 V and −5 V, then in the positive logic system the 0 V represents a logic '1' and the −5 V represents a logic '0'. In the negative logic system, 0 V represents a logic '0' and −5 V represents a logic '1'.

It is interesting to note, as we will discover in the latter part of the chapter, that a positive OR is a negative AND. That is, OR gate hardware in the positive logic system behaves like an AND gate in the negative logic system. The reverse is also true. Similarly, a positive NOR is a negative NAND, and vice versa.

4.2 Truth Table

A truth table lists all possible combinations of input binary variables and the corresponding outputs of a logic system. The logic system output can be found from the logic expression, often referred to as the Boolean expression, that relates the output with the inputs of that very logic system.

When the number of input binary variables is only one, then there are only two possible inputs, i.e. '0' and '1'. If the number of inputs is two, there can be four possible input combinations, i.e. 00, 01, 10 and 11. Figure 4.1(b) shows the truth table of the two-input logic system represented by Fig. 4.1(a). The logic system of Fig. 4.1(a) is such that $Y = 0$ only when both $A = 0$ and $B = 0$. For all other possible input combinations, output $Y = 1$. Similarly, for three input binary variables, the number of possible input combinations becomes eight, i.e. 000, 001, 010, 011, 100, 101, 110 and 111. This statement can be generalized to say that, if a logic circuit has n binary inputs, its truth table will have 2^n possible input combinations, or in other words 2^n rows. Figure 4.2 shows the truth table of a three-input logic circuit, and it has 8 ($= 2^3$) rows. Incidentally, as we will see later in the chapter, this is the truth table of a three-input AND gate. It may be mentioned here that the truth table of a three-input AND gate as given in Fig. 4.2 is drawn following the positive logic system, and also that, in all further discussion throughout the book, we will use a positive logic system unless otherwise specified.

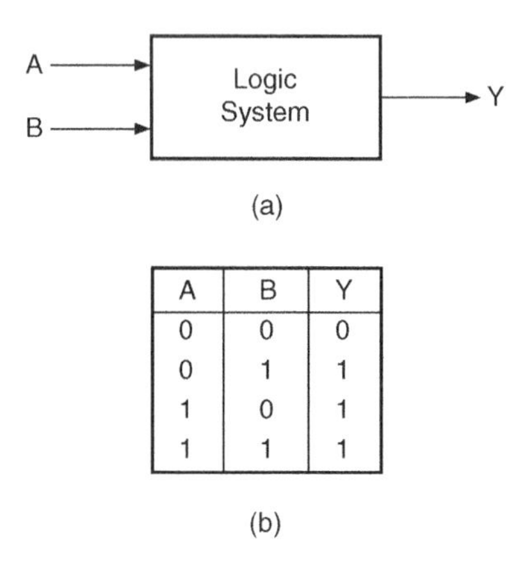

A	B	Y
0	0	0
0	1	1
1	0	1
1	1	1

Figure 4.1 Two-input logic system.

A	B	C	Y
0	0	0	0
0	0	1	0
0	1	0	0
0	1	1	0
1	0	0	0
1	0	1	0
1	1	0	0
1	1	1	1

Figure 4.2 Truth table of a three-input logic system

4.3 Logic Gates

The logic gate is the most basic building block of any digital system, including computers. Each one of the basic logic gates is a piece of hardware or an electronic circuit that can be used to implement some basic logic expression. While laws of Boolean algebra could be used to do manipulation with binary variables and simplify logic expressions, these are actually implemented in a digital system with the help of electronic circuits called logic gates. The three basic logic gates are the OR gate, the AND gate and the NOT gate.

4.3.1 OR Gate

An OR gate performs an ORing operation on two or more than two logic variables. The OR operation on two independent logic variables A and B is written as $Y = A + B$ and reads as Y equals A OR B and not as A plus B. An OR gate is a logic circuit with two or more inputs and one output. The output of an OR gate is LOW only when all of its inputs are LOW. For all other possible input combinations, the output is HIGH. This statement when interpreted for a positive logic system means the following. The output of an OR gate is a logic '0' only when all of its inputs are at logic '0'. For all other possible input combinations, the output is a logic '1'. Figure 4.3 shows the circuit symbol and the truth table of a two-input OR gate. The operation of a two-input OR gate is explained by the logic expression

$$Y = A + B \tag{4.1}$$

As an illustration, if we have four logic variables and we want to know the logical output of $(A + B + C + D)$, then it would be the output of a four-input OR gate with A, B, C and D as its inputs.

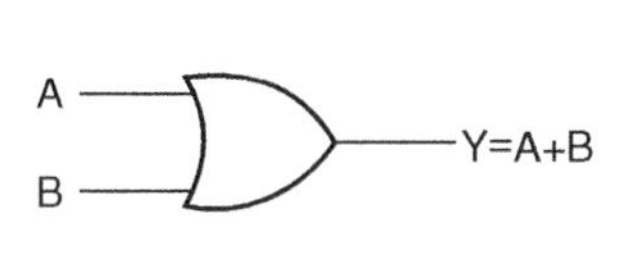

A	B	Y
0	0	0
0	1	1
1	0	1
1	1	1

Figure 4.3 Two-input OR gate.

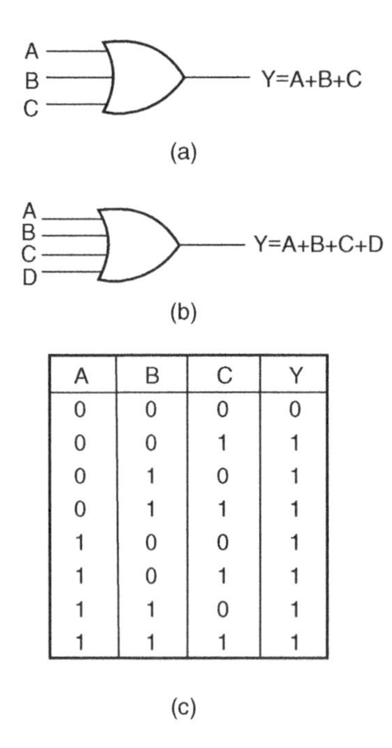

A	B	C	Y
0	0	0	0
0	0	1	1
0	1	0	1
0	1	1	1
1	0	0	1
1	0	1	1
1	1	0	1
1	1	1	1

Figure 4.4 (a) Three-input OR gate, (b) four-input OR gate and (c) the truth table of a three-input OR gate.

Figures 4.4(a) and (b) show the circuit symbol of three-input and four-input OR gates. Figure 4.4(c) shows the truth table of a three-input OR gate. Logic expressions explaining the functioning of three-input and four-input OR gates are $Y = A + B + C$ and $Y = A + B + C + D$.

Example 4.1

How would you hardware-implement a four-input OR gate using two-input OR gates only?

Solution

Figure 4.5(a) shows one possible arrangement of two-input OR gates that simulates a four-input OR gate. A, B, C and D are logic inputs and $Y3$ is the output. Figure 4.5(b) shows another possible arrangement. In the case of Fig. 4.5(a), the output of OR gate 1 is $Y1 = (A + B)$. The second

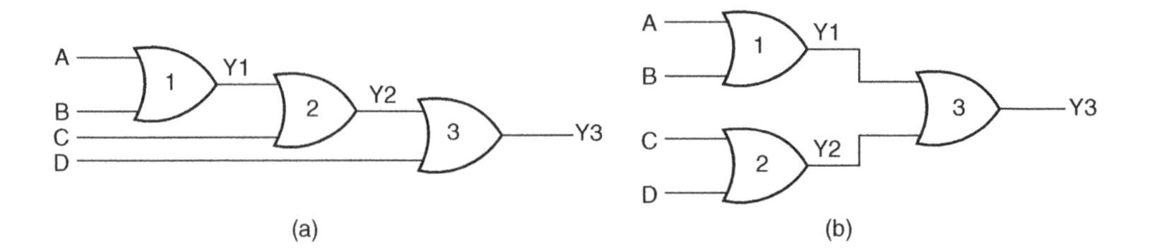

Figure 4.5 Example 4.1.

OR gate produces the output $Y2 = (Y1 + C) = (A + B + C)$. Similarly, the output of OR gate 3 is $Y3 = (Y2 + D) = (A + B + C + D)$. In the case of Fig. 4.5(b), the output of OR gate 1 is $Y1 = (A + B)$. The second OR gate produces the output $Y2 = (C + D)$. Output $Y3$ of the third OR gate is given by $(Y1 + Y2) = (A + B + C + D)$.

Example 4.2

Draw the output waveform for the OR gate and the given pulsed input waveforms of Fig. 4.6(a).

Solution
Figure 4.6(b) shows the output waveform. It can be drawn by following the truth table of the OR gate.

4.3.2 AND Gate

An AND gate is a logic circuit having two or more inputs and one output. The output of an AND gate is HIGH only when all of its inputs are in the HIGH state. In all other cases, the output is LOW. When interpreted for a positive logic system, this means that the output of the AND gate is a logic '1' only when all of its inputs are in logic '1' state. In all other cases, the output is logic '0'. The logic symbol and truth table of a two-input AND gate are shown in Figs 4.7(a) and (b) respectively. Figures 4.8(a) and (b) show the logic symbols of three-input and four-input AND gates respectively. Figure 4.8(c) gives the truth table of a four-input AND gate.

The AND operation on two independent logic variables A and B is written as $Y = A.B$ and reads as Y equals A AND B and not as A multiplied by B. Here, A and B are input logic variables and Y is the output. An AND gate performs an ANDing operation:

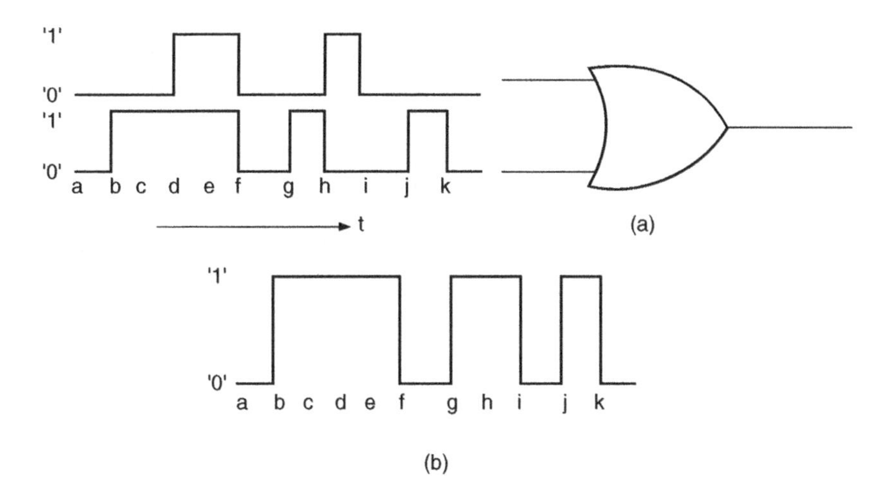

Figure 4.6 Example 4.2.

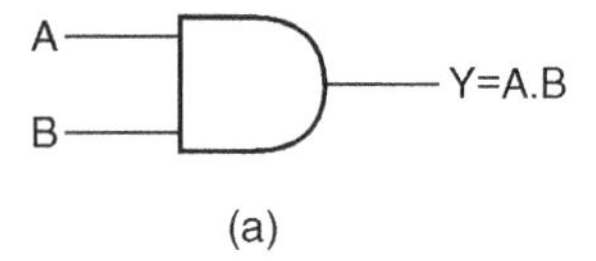

(a)

A	B	Y
0	0	0
0	1	0
1	0	0
1	1	1

(b)

Figure 4.7 Two-input AND gate.

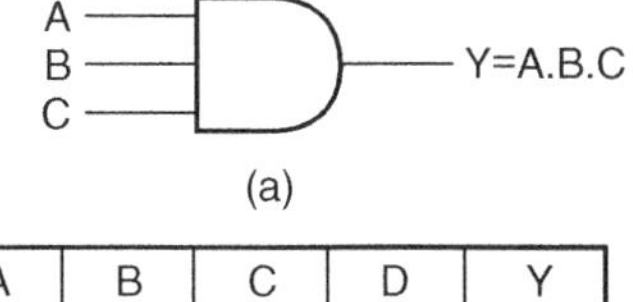

(a)

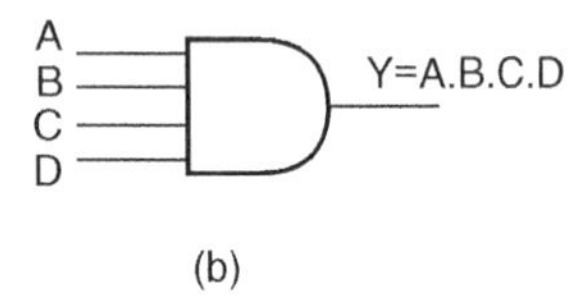

(b)

A	B	C	D	Y
0	0	0	0	0
0	0	0	1	0
0	0	1	0	0
0	0	1	1	0
0	1	0	0	0
0	1	0	1	0
0	1	1	0	0
0	1	1	1	0
1	0	0	0	0
1	0	0	1	0
1	0	1	0	0
1	0	1	1	0
1	1	0	0	0
1	1	0	1	0
1	1	1	0	0
1	1	1	1	1

(c)

Figure 4.8 (a) Three-input AND gate, (b) four-input AND gate and (c) the truth table of a four-input AND gate.

- for a two-input AND gate, $Y = A.B$;
- for a three-input AND gate, $Y = A.B.C$;
- for a four-input AND gate, $Y = A.B.C.D$.

If we interpret the basic definition of OR and AND gates for a negative logic system, we have an interesting observation. We find that an OR gate in a positive logic system is an AND gate in a negative logic system. Also, a positive AND is a negative OR.

Example 4.3

Show the logic arrangement for implementing a four-input AND gate using two-input AND gates only.

Solution

Figure 4.9 shows the hardware implementation of a four-input AND gate using two-input AND gates. The output of AND gate 1 is $Y1 = A.B$. The second AND gate produces an output $Y2$ given by $Y2 = Y1.C = A.B.C$. Similarly, the output of AND gate 3 is $Y = Y2.D = A.B.C.D$ and hence the result.

4.3.3 NOT Gate

A NOT gate is a one-input, one-output logic circuit whose output is always the complement of the input. That is, a LOW input produces a HIGH output, and vice versa. When interpreted for a positive logic system, a logic '0' at the input produces a logic '1' at the output, and vice versa. It is also known as a 'complementing circuit' or an 'inverting circuit'. Figure 4.10 shows the circuit symbol and the truth table.

The NOT operation on a logic variable X is denoted as $\overline{X}$ or X'. That is, if X is the input to a NOT circuit, then its output Y is given by $Y = \overline{X}$ or X' and reads as Y equals NOT X. Thus, if $X = 0$, $Y = 1$ and if $X = 1$, $Y = 0$.

Example 4.4

For the logic circuit arrangements of Figs 4.11(a) and (b), draw the output waveform.

Solution

In the case of the OR gate arrangement of Fig. 4.11(a), the output will be permanently in logic '1' state as the two inputs can never be in logic '0' state together owing to the presence of the inverter. In the case of the AND gate arrangement of Fig. 4.11(b), the output will be permanently in logic '0' state as the two inputs can never be in logic '1' state together owing to the presence of the inverter.

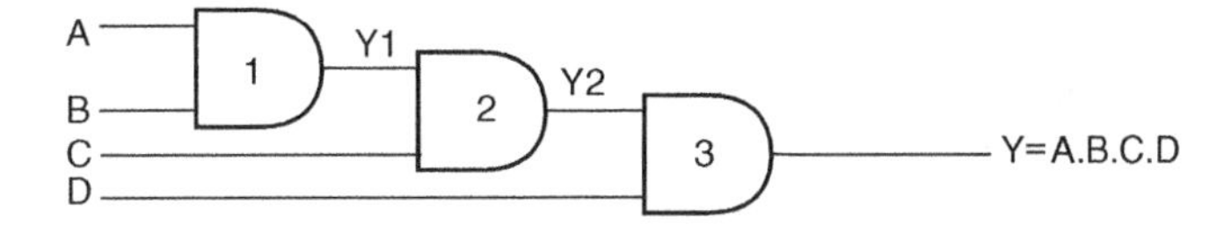

Figure 4.9 Implementation of a four-input AND gate using two-input AND gates.

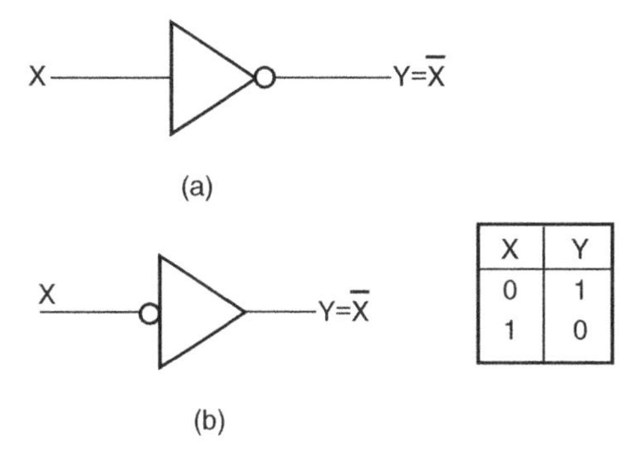

Figure 4.10 (a) Circuit symbol of a NOT circuit and (b) the truth table of a NOT circuit.

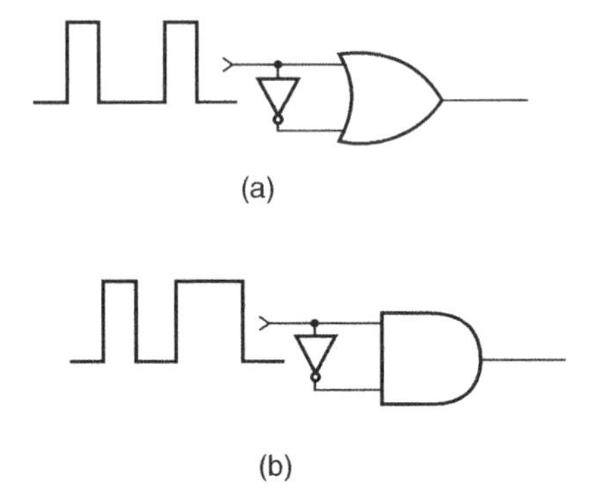

Figure 4.11 Example 4.4.

4.3.4 EXCLUSIVE-OR Gate

The EXCLUSIVE-OR gate, commonly written as EX-OR gate, is a two-input, one-output gate. Figures 4.12(a) and (b) respectively show the logic symbol and truth table of a two-input EX-OR gate. As can be seen from the truth table, the output of an EX-OR gate is a logic '1' when the inputs are unlike and a logic '0' when the inputs are like. Although EX-OR gates are available in integrated circuit form only as two-input gates, unlike other gates which are available in multiple inputs also, multiple-input EX-OR logic functions can be implemented using more than one two-input gates. The truth table of a multiple-input EX-OR function can be expressed as follows. The output of a multiple-input EX-OR logic function is a logic '1' when the number of 1s in the input sequence is odd and a logic '0' when the number of 1s in the input sequence is even, including zero. That is, an all 0s input sequence also produces a logic '0' at the output. Figure 4.12(c) shows the truth table of a four-input EX-OR function. The output of a two-input EX-OR gate is expressed by

$$Y = (A \oplus B) = \overline{A}B + A\overline{B} \tag{4.2}$$

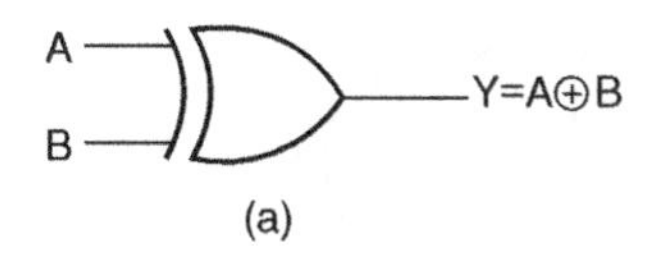

(a)

A	B	Y
0	0	0
0	1	1
1	0	1
1	1	0

(b)

A	B	C	D	Y
0	0	0	0	0
0	0	0	1	1
0	0	1	0	1
0	0	1	1	0
0	1	0	0	1
0	1	0	1	0
0	1	1	0	0
0	1	1	1	1
1	0	0	0	1
1	0	0	1	0
1	0	1	0	0
1	0	1	1	1
1	1	0	0	0
1	1	0	1	1
1	1	1	0	1
1	1	1	1	0

(c)

Figure 4.12 (a) Circuit symbol of a two-input EXCLUSIVE-OR gate, (b) the truth table of a two-input EXCLUSIVE-OR gate and (c) the truth table of a four-input EXCLUSIVE-OR gate

Example 4.5

How do you implement three-input and four-input EX-OR logic functions with the help of two-input EX-OR gates?

Solution

Figures 4.13(a) and (b) show the implementation of a three-input EX-OR logic function and a four-input EX-OR logic function using two-input logic gates:

- For Fig. 4.13(a), the output $Y1$ is given by $A \oplus B$. The final output Y is given by $Y = (Y1 \oplus C) = (A \oplus B) \oplus C = A \oplus B \oplus C$.
- Figure 4.13(b) can be explained on similar lines.

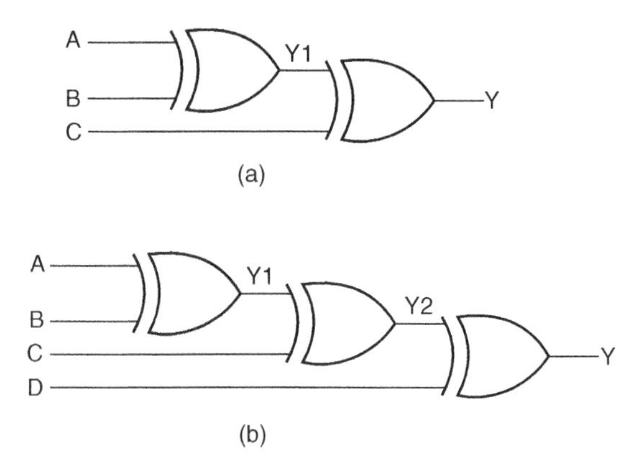

Figure 4.13 (a) Three-input EX-OR gate and (b) a four-input EX-OR gate.

Example 4.6

How can you implement a NOT circuit using a two-input EX-OR gate?

Solution

Refer to the truth table of a two-input EX-OR gate reproduced in Fig. 4.14(a). It is clear from the truth table that, if one of the inputs of the gate is permanently tied to logic '1' level, then the other input and output perform the function of a NOT circuit. Figure 4.14(b) shows the implementation.

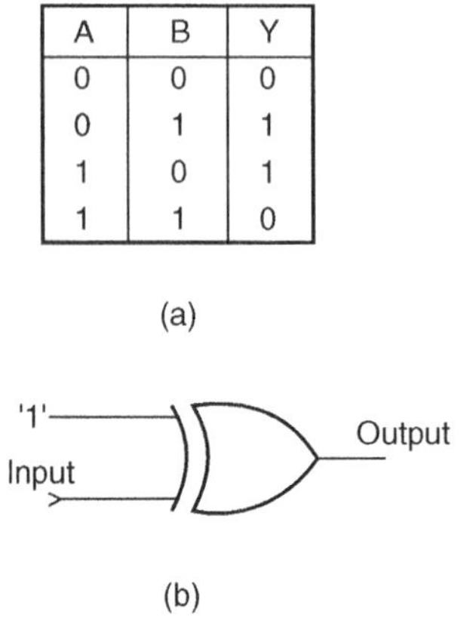

A	B	Y
0	0	0
0	1	1
1	0	1
1	1	0

Figure 4.14 Implementation of a NOT circuit using an EX-OR gate.

4.3.5 NAND Gate

NAND stands for NOT AND. An AND gate followed by a NOT circuit makes it a NAND gate [Fig. 4.15(a)]. Figure 4.15(b) shows the circuit symbol of a two-input NAND gate. The truth table of a NAND gate is obtained from the truth table of an AND gate by complementing the output entries [Fig. 4.15(c)]. The output of a NAND gate is a logic '0' when all its inputs are a logic '1'. For all other input combinations, the output is a logic '1'. NAND gate operation is logically expressed as

$$Y = \overline{A.B} \tag{4.3}$$

In general, the Boolean expression for a NAND gate with more than two inputs can be written as

$$Y = \overline{(A.B.C.D...)} \tag{4.4}$$

4.3.6 NOR Gate

NOR stands for NOT OR. An OR gate followed by a NOT circuit makes it a NOR gate [Fig. 4.16(a)]. The truth table of a NOR gate is obtained from the truth table of an OR gate by complementing the output entries. The output of a NOR gate is a logic '1' when all its inputs are logic '0'. For all other input combinations, the output is a logic '0'. The output of a two-input NOR gate is logically expressed as

$$Y = \overline{(A+B)} \tag{4.5}$$

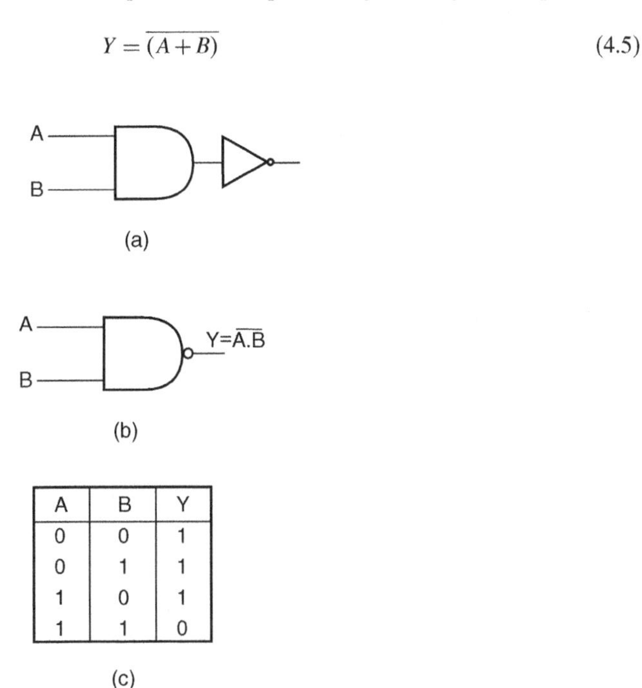

A	B	Y
0	0	1
0	1	1
1	0	1
1	1	0

Figure 4.15 (a) Two-input NAND implementation using an AND gate and a NOT circuit, (b) the circuit symbol of a two-input NAND gate and (c) the truth table of a two-input NAND gate.

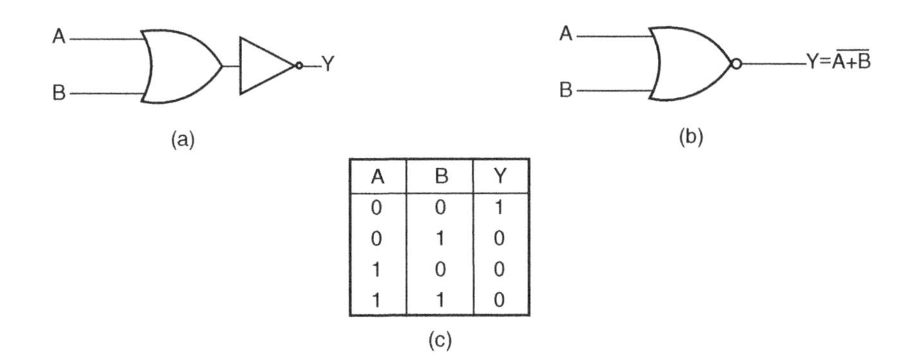

A	B	Y
0	0	1
0	1	0
1	0	0
1	1	0

(c)

Figure 4.16 (a) Two-input NOR implementation using an OR gate and a NOT circuit, (b) the circuit symbol of a two-input NOR gate and (c) the truth table of a two-input NOR gate.

In general, the Boolean expression for a NOR gate with more than two inputs can be written as

$$Y = \overline{(A + B + C + D...)} \tag{4.6}$$

4.3.7 EXCLUSIVE-NOR Gate

EXCLUSIVE-NOR (commonly written as EX-NOR) means NOT of EX-OR, i.e. the logic gate that we get by complementing the output of an EX-OR gate. Figure 4.17 shows its circuit symbol along with its truth table.

The truth table of an EX-NOR gate is obtained from the truth table of an EX-OR gate by complementing the output entries. Logically,

$$Y = (\overline{A \oplus B}) = (A.B + \overline{A}.\overline{B}) \tag{4.7}$$

A
B
$Y=\overline{A \oplus B}$

(a)

A	B	Y
0	0	1
0	1	0
1	0	0
1	1	1

(b)

Figure 4.17 (a) Circuit symbol of a two-input EXCLUSIVE-NOR gate and (b) the truth table of a two-input EXCLUSIVE-NOR gate.

The output of a two-input EX-NOR gate is a logic '1' when the inputs are like and a logic '0' when they are unlike. In general, the output of a multiple-input EX-NOR logic function is a logic '0' when the number of 1s in the input sequence is odd and a logic '1' when the number of 1s in the input sequence is even including zero. That is, an all 0s input sequence also produces a logic '1' at the output.

Example 4.7

Show the logic arrangements for implementing:

(a) a four-input NAND gate using two-input AND gates and NOT gates;
(b) a three-input NAND gate using two-input NAND gates;
(c) a NOT circuit using a two-input NAND gate;
(d) a NOT circuit using a two-input NOR gate;
(e) a NOT circuit using a two-input EX-NOR gate.

Solution

(a) Figure 4.18(a) shows the arrangement. The logic diagram is self-explanatory. The first step is to get a four-input AND gate using two-input AND gates. The output thus obtained is then complemented using a NOT circuit as shown.
(b) Figure 4.18(b) shows the arrangement, which is again self-explanatory. The first step is to get a two-input AND from a two-input NAND. The output of the two-input AND gate and the third input then feed the inputs of another two-input NAND to get the desired output.
(c) Shorting the inputs of the NAND gives a one-input, one-output NOT circuit. This is because when all inputs to a NAND are at logic '0' level the output is a logic '1', and when all inputs to a NAND are at logic '1' level the output is a logic '0'. Figure 4.18(c) shows the implementation.
(d) Again, shorting the inputs of a NOR gate gives a NOT circuit. From the truth table of a NOR gate it is evident that an all 0s input to a NOR gate gives a logic '1' output and an all 1s input gives a logic '0' output. Figure 4.18(d) shows the implementation.
(e) It is evident from the truth table of a two-input EX-NOR gate that, if one of the inputs is permanently tied to a logic '0' level and the other input is treated as the input, then it behaves as a NOT circuit between input and output [Fig. 4.18(e)]. When the input is a logic '0', the two inputs become 00, which produces a logic '1' at the output. When the input is at logic '1' level, a 01 input produces a logic '0' at the output.

Example 4.8

How do you implement a three-input EX-NOR function using only two-input EX-NOR gates?

Solution

Figure 4.19 shows the arrangement. The first two EX-NOR gates implement a two-input EX-OR gate using two-input EX-NOR gates. The second EX-NOR gate here has been wired as a NOT circuit. The output of the second gate and the third input are fed to the two inputs of the third EX-NOR gate.

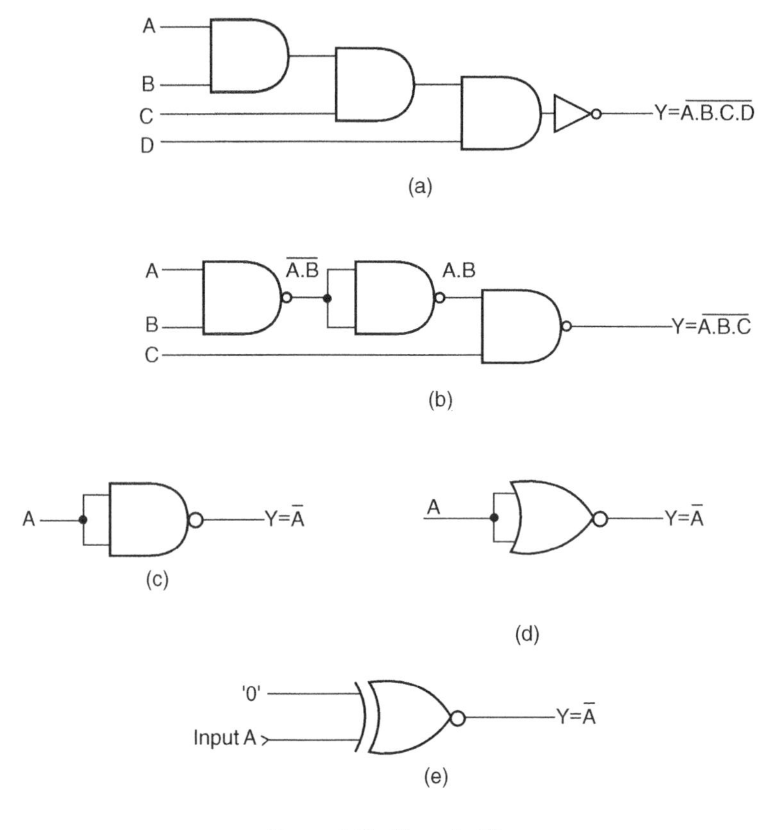

Figure 4.18 Example 4.7.

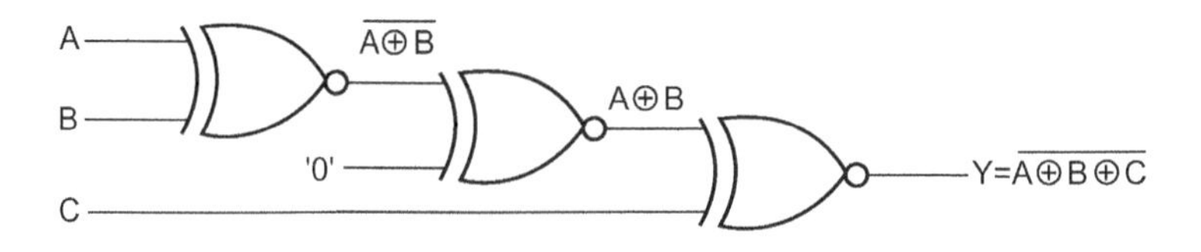

Figure 4.19 Example 4.8.

4.3.8 INHIBIT Gate

There are many situations in digital circuit design where the passage of a logic signal needs to be either enabled or inhibited depending upon certain other control inputs. INHIBIT here means that the gate produces a certain fixed logic level at the output irrespective of changes in the input logic level. As an illustration, if one of the inputs of a four-input NOR gate is permanently tied to logic '1' level, then the output will always be at logic '0' level irrespective of the logic status of other inputs. This gate will behave as a NOR gate only when this control input is at logic '0' level. This is an example of the INHIBIT function. The INHIBIT function is available in integrated circuit form for an AND gate, which is basically an AND gate with one of its inputs negated by an inverter. The negated input acts to inhibit the gate. In other words, the gate will behave like an AND gate only when the negated input is driven to a logic '0'. Figure 4.20 shows the circuit symbol and truth table of a four-input INHIBIT gate.

Example 4.9

Refer to the INHIBIT gate of Fig. 4.21(a). If the waveform of Fig. 4.21(b) is applied to the INHIBIT input, draw the waveform at the output.

Solution

Since all other inputs of the gate have been permanently tied to logic '1' level, a logic '0' at the INHIBIT input would produce a logic '1' at the output and a logic '1' at the INHIBIT input would produce a logic '0' at the output. The output waveform is therefore the inversion of the input waveform and is shown in Fig. 4.22.

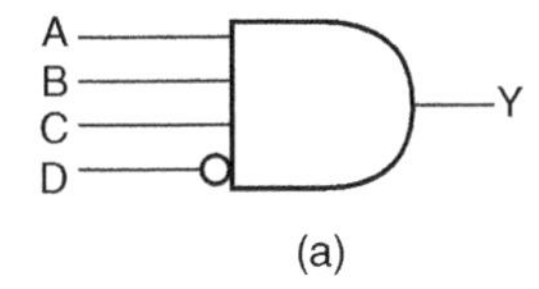

(a)

A	B	C	D	Y
0	0	0	0	0
0	0	0	1	0
0	0	1	0	0
0	0	1	1	0
0	1	0	0	0
0	1	0	1	0
0	1	1	0	0
0	1	1	1	0
1	0	0	0	0
1	0	0	1	0
1	0	1	0	0
1	0	1	1	0
1	1	0	0	0
1	1	0	1	0
1	1	1	0	1
1	1	1	1	0

(b)

Figure 4.20 INHIBIT gate.

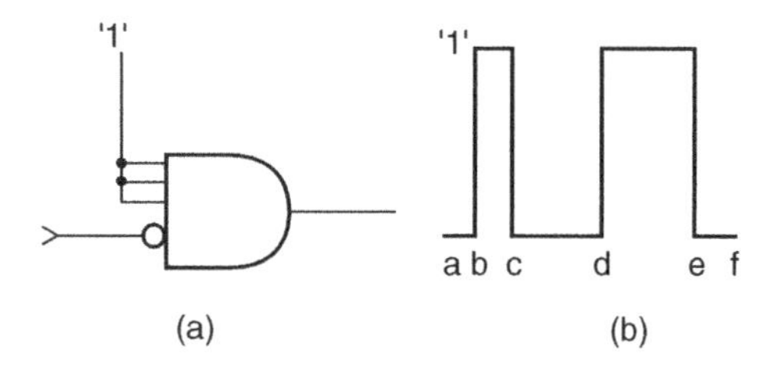

Figure 4.21 Example 4.9.

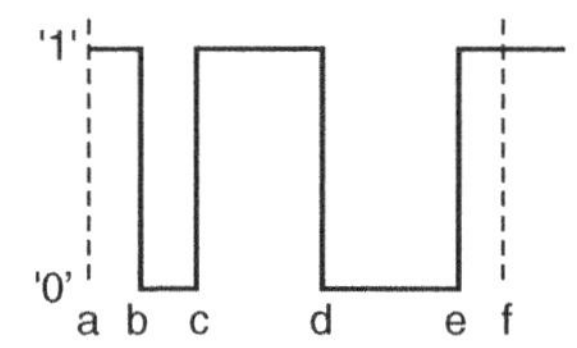

Figure 4.22 Solution to example 4.9.

Example 4.10

Refer to the INHIBIT gate shown in Fig. 4.23(a) and the INHIBIT input waveform shown in Fig. 4.23(b). Sketch the output waveform.

Solution

The output will always be at logic '1' level as two of the inputs of the logic gate, which is a NAND, are permanently tied to logic '0' level. This would have been so even if one of the inputs of the gate were at logic '0' level.

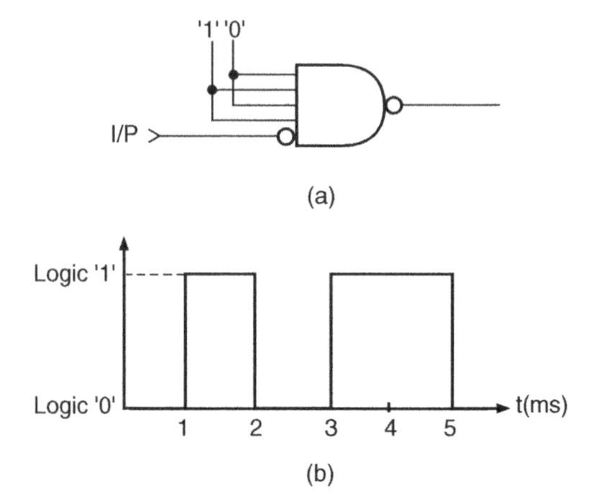

Figure 4.23 Example 4.10.

4.4 Universal Gates

OR, AND and NOT gates are the three basic logic gates as they together can be used to construct the logic circuit for any given Boolean expression. NOR and NAND gates have the property that they individually can be used to hardware-implement a logic circuit corresponding to any given Boolean expression. That is, it is possible to use either only NAND gates or only NOR gates to implement any Boolean expression. This is so because a combination of NAND gates or a combination of NOR gates can be used to perform functions of any of the basic logic gates. It is for this reason that NAND and NOR gates are universal gates.

As an illustration, Fig. 4.24 shows how two-input NAND gates can be used to construct a NOT circuit [Fig. 4.24(a)], a two-input AND gate [Fig. 4.24(b)] and a two-input OR gate [Fig. 4.24(c)]. Figure 4.25 shows the same using NOR gates. Understanding the conversion of NAND to OR and NOR to AND requires the use of DeMorgan's theorem, which is discussed in Chapter 6 on Boolean algebra.

4.5 Gates with Open Collector/Drain Outputs

These are gates where we need to connect an external resistor, called the pull-up resistor, between the output and the DC power supply to make the logic gate perform the intended logic function. Depending on the logic family used to construct the logic gate, they are referred to as gates with open collector output (in the case of the TTL logic family) or open drain output (in the case of the MOS logic family). Logic families are discussed in detail in Chapter 5.

The advantage of using open collector/open drain gates lies in their capability of providing an ANDing operation when outputs of several gates are tied together through a common pull-up resistor,

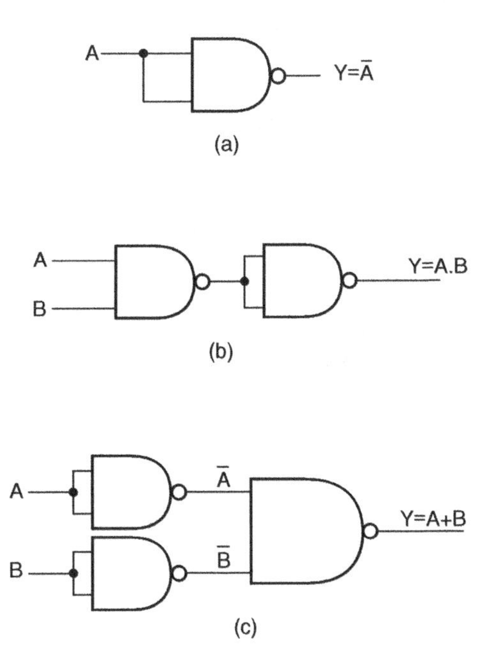

Figure 4.24 Implementation of basic logic gates using only NAND gates.

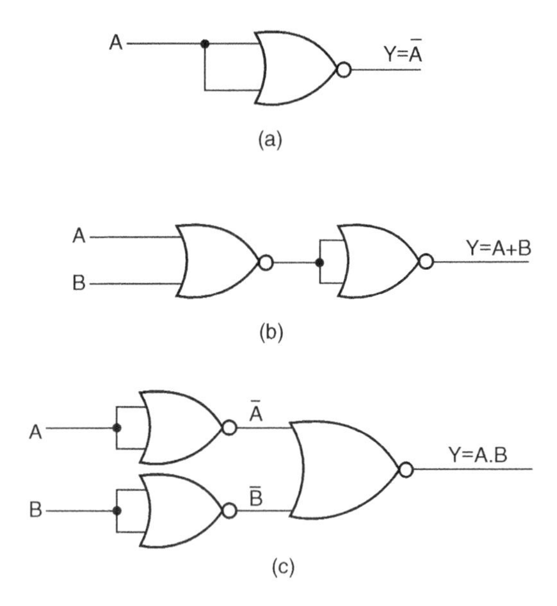

Figure 4.25 Implementation of basic logic gates using only NOR gates.

without having to use an AND gate for the purpose. This connection is also referred to as WIRE-AND connection. Figure 4.26(a) shows such a connection for open collector NAND gates. The output in this case would be

$$Y = \overline{AB}.\overline{CD}.\overline{EF} \tag{4.8}$$

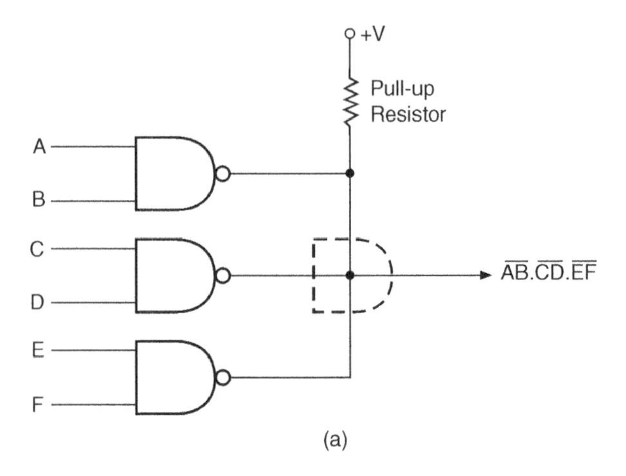

Figure 4.26 WIRE-AND connection with open collector/drain devices.

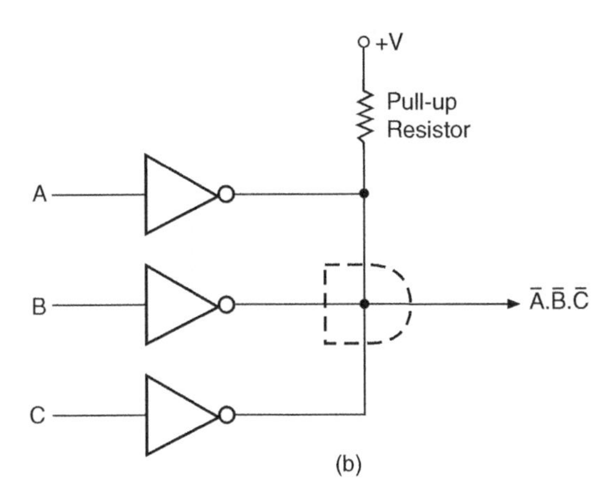

Figure 4.26 (*continued*).

Figure 4.26(b) shows a similar arrangement for NOT gates. The disadvantage is that they are relatively slower and noisier. Open collector/drain devices are therefore not recommended for applications where speed is an important consideration.

4.6 Tristate Logic Gates

Tristate logic gates have three possible output states, i.e. the logic '1' state, the logic '0' state and a high-impedance state. The high-impedance state is controlled by an external ENABLE input. The ENABLE input decides whether the gate is active or in the high-impedance state. When active, it can be '0' or '1' depending upon input conditions. One of the main advantages of these gates is that their inputs and outputs can be connected in parallel to a common bus line. Figure 4.27(a) shows the circuit symbol of a tristate NAND gate with active HIGH ENABLE input, along with its truth table. The one shown in Fig. 4.27(b) has active LOW ENABLE input. When tristate devices are paralleled, only one of them is enabled at a time. Figure 4.28 shows paralleling of tristate inverters having active HIGH ENABLE inputs.

4.7 AND-OR-INVERT Gates

AND-OR and OR-AND gates can be usefully employed to implement sum-of-products and product-of-sums Boolean expressions respectively. Figures 4.29(a) and (b) respectively show the symbols of AND-OR-INVERT and OR-AND-INVERT gates.

Another method for designating the gates shown in Fig. 4.29 is to call them two-wide, two-input AND-OR-INVERT or OR-AND-INVERT gates as the case may be. The gate is two-wide as there are two gates at the input, and two-input as each of the gates has two inputs. Other varieties such as two-wide, four-input AND-OR-INVERT (Fig. 4.30) and four-wide, two-input AND-OR-INVERT (Fig. 4.31) are also available in IC form.

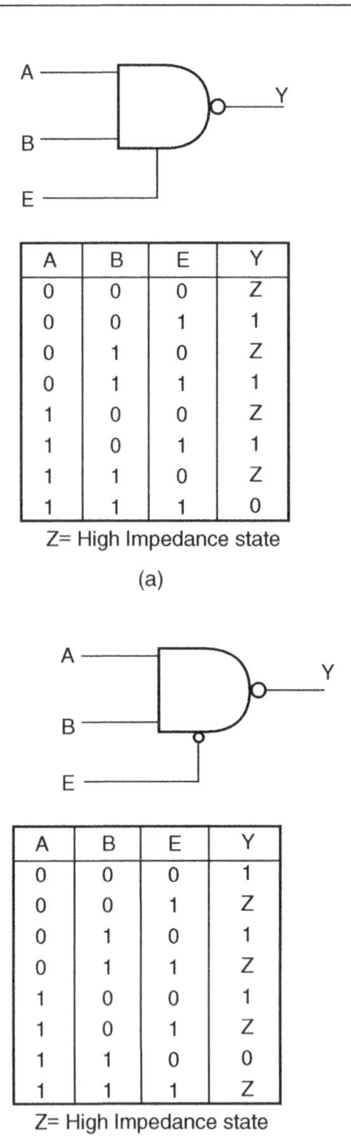

A	B	E	Y
0	0	0	Z
0	0	1	1
0	1	0	Z
0	1	1	1
1	0	0	Z
1	0	1	1
1	1	0	Z
1	1	1	0

Z= High Impedance state

(a)

A	B	E	Y
0	0	0	1
0	0	1	Z
0	1	0	1
0	1	1	Z
1	0	0	1
1	0	1	Z
1	1	0	0
1	1	1	Z

Z= High Impedance state

(b)

Figure 4.27 Tristate devices.

4.8 Schmitt Gates

The logic gates discussed so far have a single-input threshold voltage level. This threshold is the same for both LOW-to-HIGH and HIGH-to-LOW output transitions. This threshold voltage lies somewhere between the highest LOW voltage level and the lowest HIGH voltage level guaranteed by the manufacturer of the device. These logic gates can produce an erratic output when fed with a slow

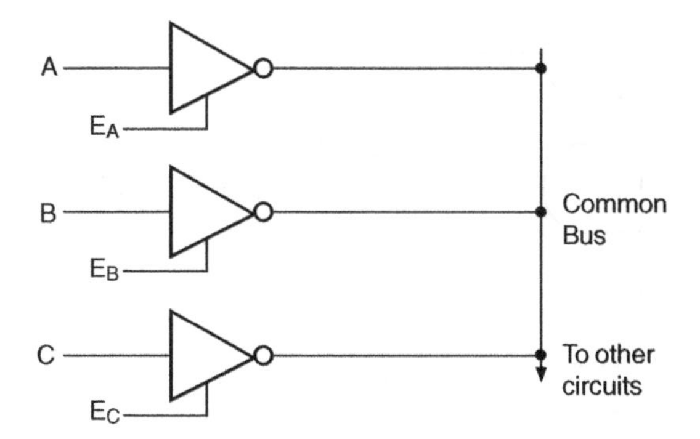

Figure 4.28 Paralleling of tristate inverters.

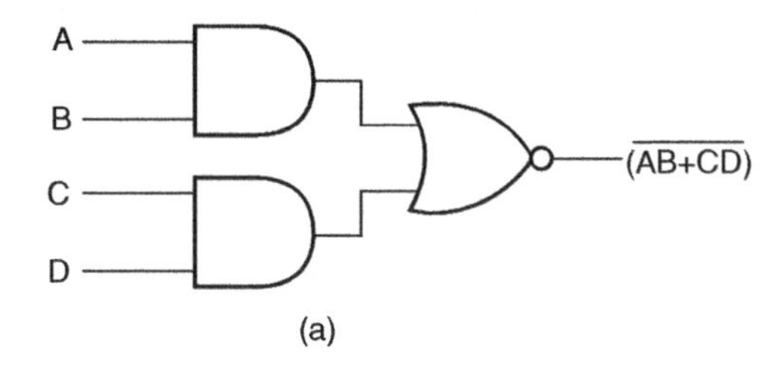

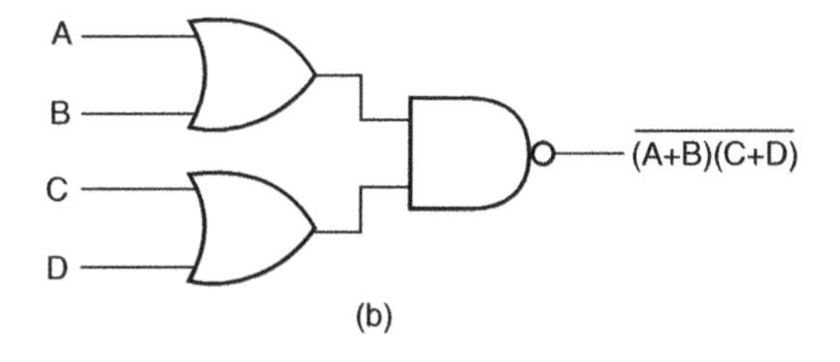

Figure 4.29 AND-OR-INVERT and OR-AND-INVERT gates.

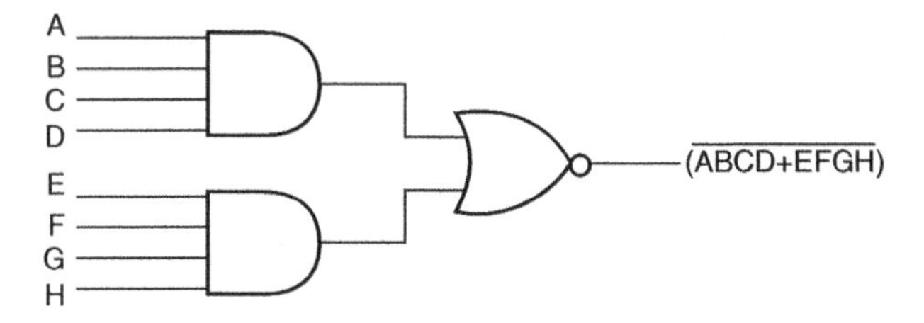

Figure 4.30 Two-wide, four-input AND-OR-INVERT gate.

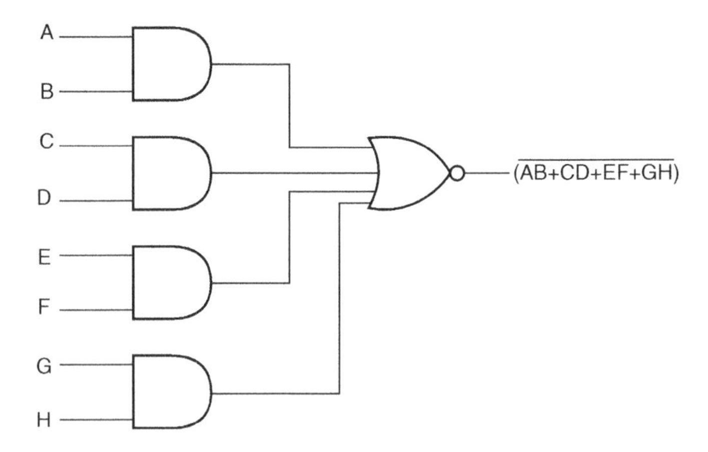

Figure 4.31 Four-wide, two-input AND-OR-INVERT gate.

varying input. Figure 4.32 shows the response of an inverter circuit when fed with a slow varying input both in the case of an ideal signal [Fig. 4.32(a)] and in the case of a practical signal having a small amount of AC noise superimposed on it [Fig. 4.32(b)]. A possible solution to this problem lies in having two different threshold voltage levels, one for LOW-to-HIGH transition and the other for HIGH-to-LOW transition, by introducing some positive feedback in the internal gate circuitry, a phenomenon called hysteresis.

There are some logic gate varieties, mainly in NAND gates and inverters, that are available with built-in hysteresis. These are called Schmitt gates, which interpret varying input voltages according to two threshold voltages, one for LOW-to-HIGH and the other for HIGH-to-LOW output transition. Figures 4.33(a) and (b) respectively show circuit symbols of Schmitt NAND and Schmitt inverter. Schmitt gates are distinguished from conventional gates by the small 'hysteresis' symbol reminiscent of the $B-H$ loop for a ferromagnetic material. Figure 4.33(c) shows typical transfer characteristics for such a device. The difference between the two threshold levels is

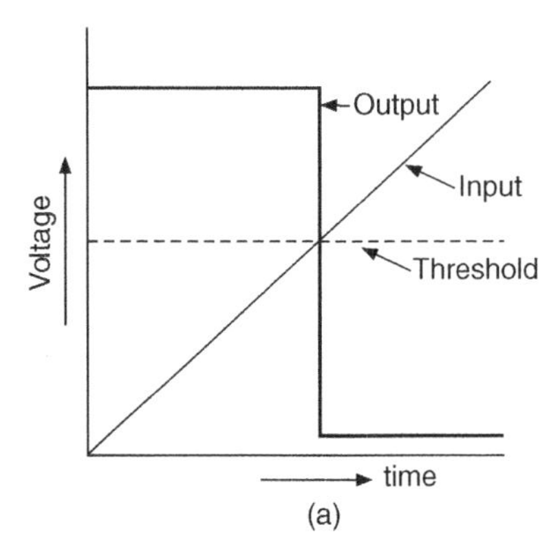

Figure 4.32 Response of conventional inverters to slow varying input.

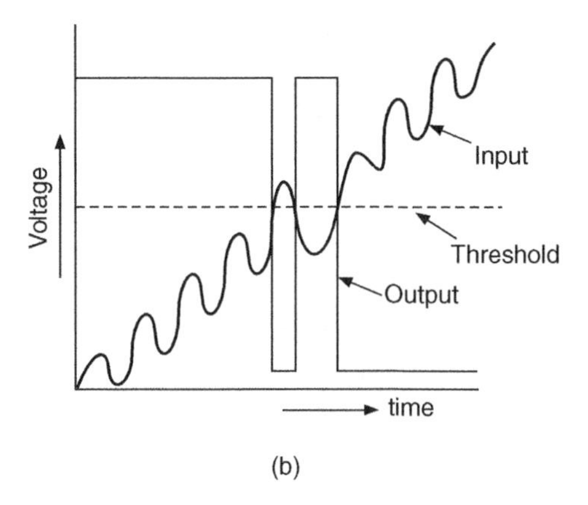

Figure 4.32 (*continued*).

the hysteresis. These characteristics have been reproduced from the data sheet of IC 74LS132, which is a quad two-input Schmitt NAND belonging to the low-power Schottky TTL family. Figure 4.33(d) shows the response of a Schmitt inverter to a slow varying noisy input signal. We will learn more about different logic families in Chapter 5. It may be mentioned here that hysteresis increases noise immunity and is used in applications where noise is expected on input signal lines.

4.9 Special Output Gates

There are many applications where it is desirable to have both noninverted and inverted outputs. Examples include a single-input gate that is both an inverter and a noninverting buffer, or a two-input logic gate that is both an AND and a NAND. Such gates are called complementary output gates and are particularly useful in circuits where PCB space is at a premium. These are also useful in circuits where the addition of an inverter to obtain the inverted output introduces an undesirable time delay between inverted and noninverted outputs. Figure 4.34 shows the circuit symbols of complementary buffer, AND, OR and EX-OR gates.

Example 4.11

Draw the circuit symbols for (a) a two-wide, four-input OR-AND-INVERT gate and (b) a four-wide, two-input OR-AND-INVERT gate.

Solution

(a) Refer to Fig. 4.35(a).
(b) Refer to Fig. 4.35(b).

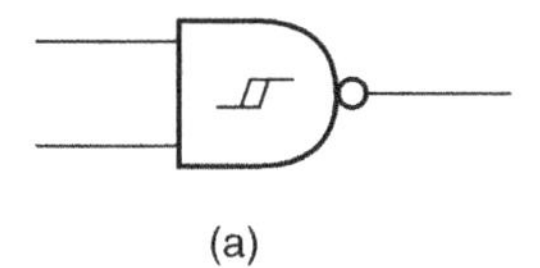

(a)

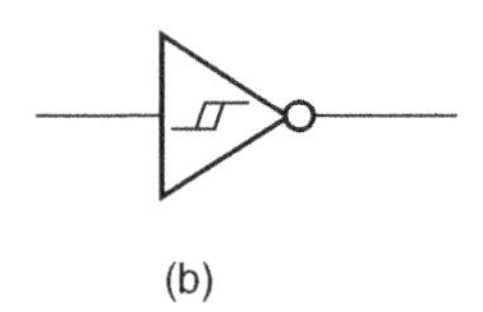

(b)

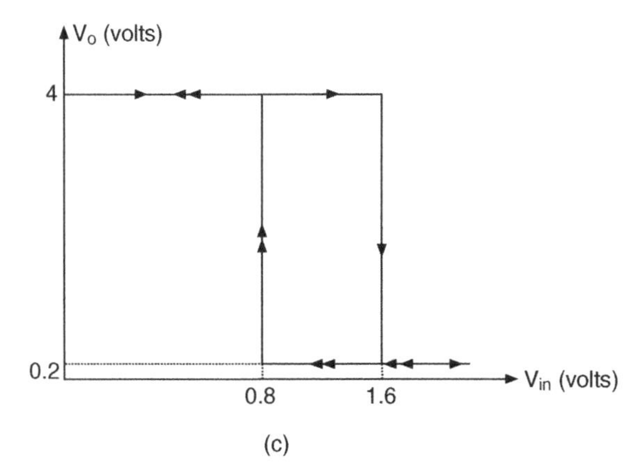

(c)

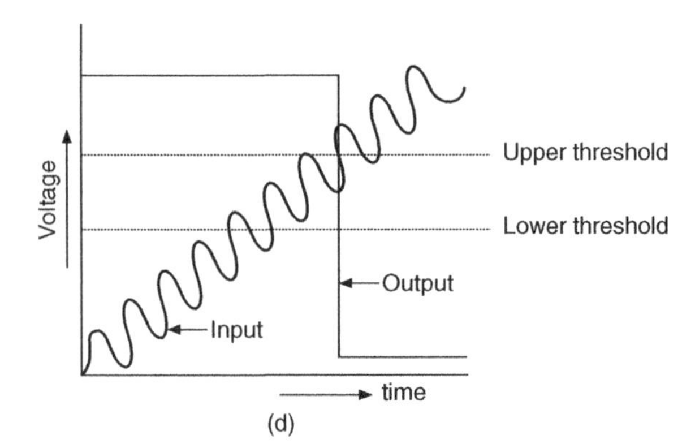

(d)

Figure 4.33 Schmitt gates.

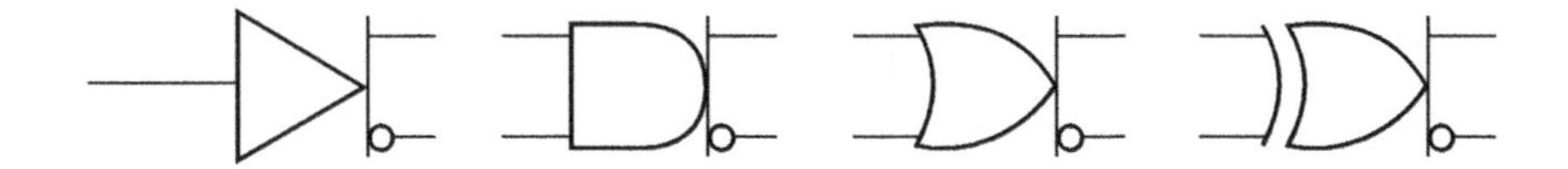

Figure 4.34 Complementary gates.

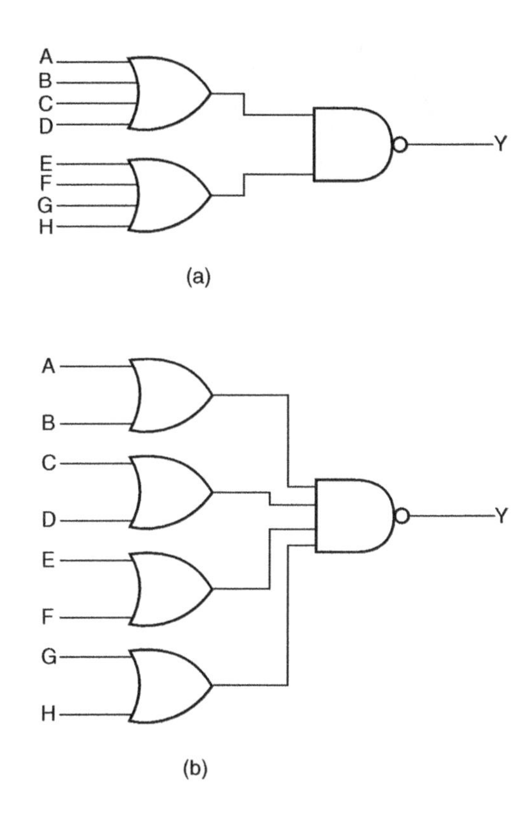

Figure 4.35 Example 4.11.

Example 4.12

Refer to Fig. 4.36(a). If the NAND gate used has the transfer characteristics of Fig. 4.36(b), sketch the expected output waveform.

Solution

The output waveform is shown in Fig. 4.36(c). The output is initially in logic '1' state. It goes from logic '1' to logic '0' state as the input exceeds 2 V. The output goes from logic '0' to logic '1' state as the input drops below 1 V.

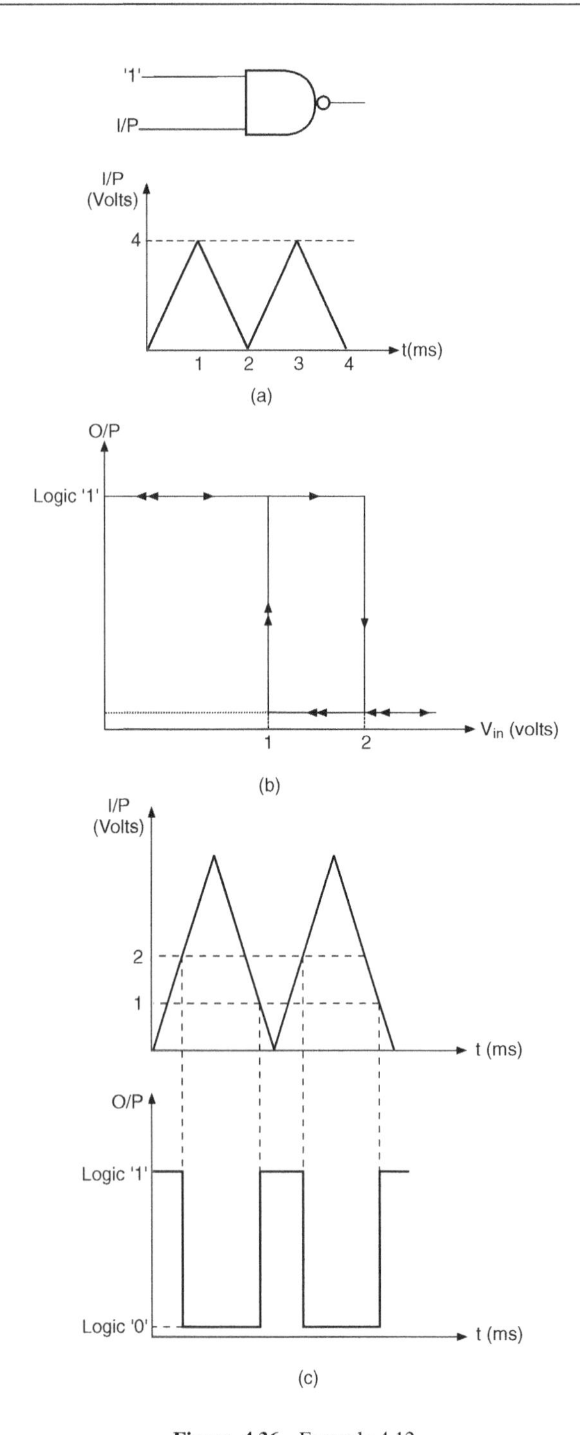

Figure 4.36 Example 4.12.

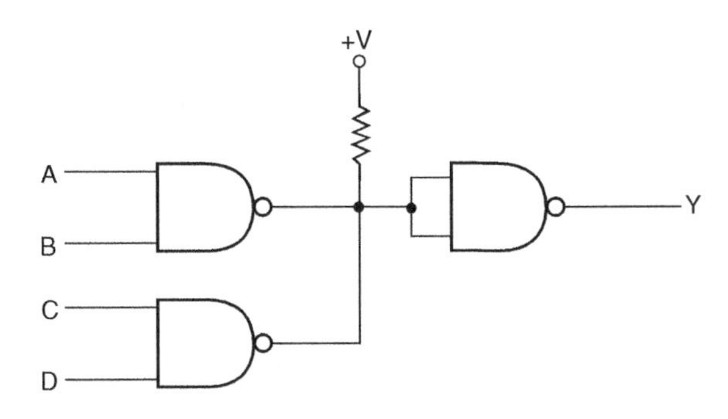

Figure 4.37 Example 4.13.

Example 4.13

Refer to the logic arrangement of Fig. 4.37. Write the logic expression for the output Y.

Solution

The NAND gates used in the circuit are open collector gates. Paralleling of the two NAND gates at the input leads to a WIRE-AND connection. Therefore the logic expression at the point where the two outputs combine is given by the equation

$$(\overline{AB}.\overline{CD}) \tag{4.9}$$

Using DeMorgan's theorem (discussed in Chapter 6 on Boolean algebra),

$$(\overline{AB}.\overline{CD}) = (\overline{AB+CD}) \tag{4.10}$$

The third NAND is wired as an inverter. Therefore, the final output can be written as

$$Y = (AB+CD) \tag{4.11}$$

4.10 Fan-Out of Logic Gates

It is a common occurrence in logic circuits that the output of one logic gate feeds the inputs of several others. It is not practical to drive the inputs of an unlimited number of logic gates from the output of a single logic gate. This is limited by the current-sourcing capability of the output when the output of the logic gate is HIGH and by the current-sinking capability of the output when it is LOW, and also by the requirement of the inputs of the logic gates being fed in the two states.

To illustrate the point further, let us say that the current-sourcing capability of a certain NAND gate is I_{OH} when its output is in the logic HIGH state and that each of the inputs of the logic gate that it is driving requires an input current I_{IH}, as shown in Fig. 4.38(a). In this case, the output of the logic gate will be able to drive a maximum of I_{OH}/I_{IH} inputs when it is in the logic HIGH state. When the output of the driving logic gate is in the logic LOW state, let us say that it has a maximum current-sinking capability I_{OL}, and that each of the inputs of the driven logic gates requires a sinking current I_{IL}, as shown in Fig. 4.38(b). In this case the output of the logic gate will be able to drive a maximum of

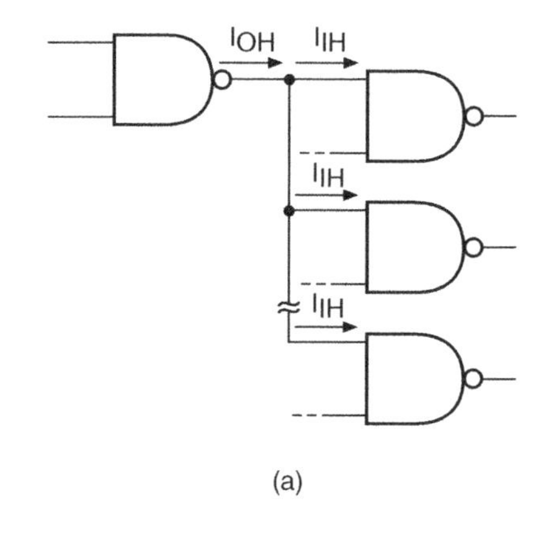

(a)

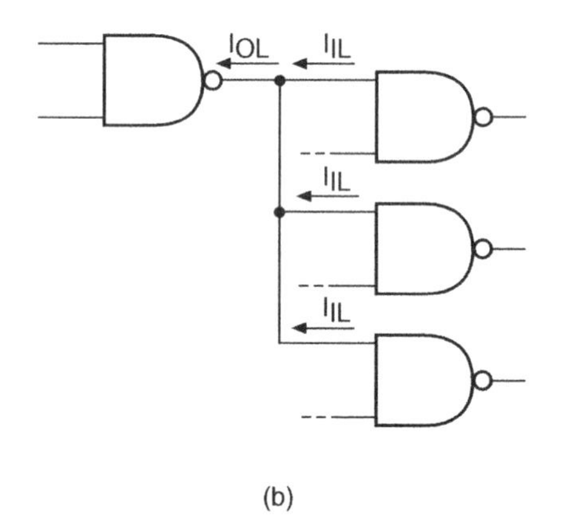

(b)

Figure 4.38 Fan-out of logic gates.

I_{OL}/I_{IL} inputs when it is in the logic LOW state. Thus, the number of logic gate inputs that can be driven from the output of a single logic gate will be I_{OH}/I_{IH} in the logic HIGH state and I_{OL}/I_{IL} in the logic LOW state. The number of logic gate inputs that can be driven from the output of a single logic gate without causing any false output is called fan-out. It is the characteristic of the logic family to which the device belongs. If in a certain case the two values I_{OH}/I_{IH} and I_{OL}/I_{IL} are different, the fan-out is taken as the smaller of the two. Figure 4.39 shows the actual circuit diagram where the output of a single NAND gate belonging to a standard TTL logic family feeds the inputs of multiple NAND gates belonging to the same family when the output of the feeding gate is in the logic HIGH state [Fig. 4.39(a)] and the logic LOW state [Fig. 4.39(b)]. We will learn in Chapter 5 on logic families that the maximum HIGH-state output sourcing current $(I_{OH})_{max}$ and maximum HIGH-state input current $(I_{IH})_{max}$ specifications of standard TTL family devices are 400 μA and 40 μA respectively. Also, the maximum LOW-state output sinking current $(I_{OL})_{max}$ and maximum LOW-state input current $(I_{IL})_{max}$ specifications are 16 mA and 1.6 mA respectively. Considering both the sourcing and sinking capability of standard TTL family devices, we obtain a fan-out figure of 10 both for HIGH and for LOW logic states. If the maximum sourcing and sinking current specifications are exceeded, the output voltage levels in the logic HIGH and LOW states will go out of the specified ranges.

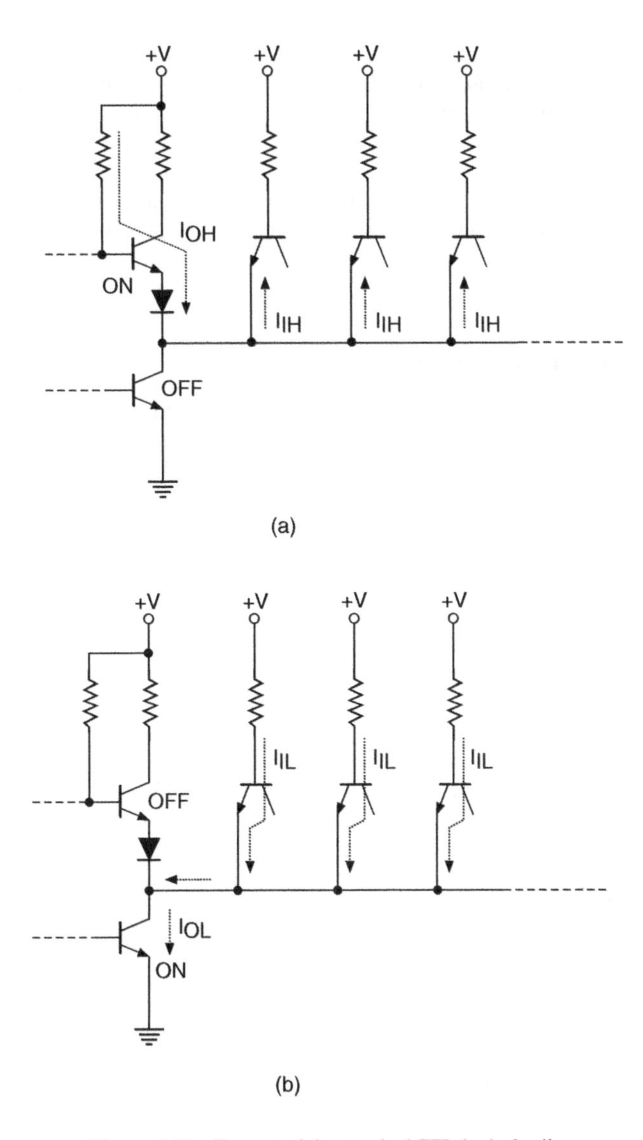

Figure 4.39 Fan-out of the standard TTL logic family.

Example 4.14

A certain logic family has the following input and output current specifications:

1. *The maximum output HIGH-state current = 1 mA.*
2. *The maximum output LOW-state current = 20 mA.*
3. *The maximum input HIGH-state current = 50 μA.*
4. *The maximum input LOW-state current = 2 mA.*

The output of an inverter belonging to this family feeds the clock inputs of various flip-flops belonging to the same family. How many flip-flops can be driven by the output of this inverter providing the clock signal? Incidentally, the data given above are taken from the data sheet of a Schottky TTL family.

Solution

- The HIGH-state fan-out = (1/0.05) = 20 and the LOW-state fan-out = (20/2) = 10.
- Since the lower of the two fan-out values is 10, the inverter output can drive a maximum of 10 flip-flops.

4.11 Buffers and Transceivers

Logic gates, discussed in the previous pages, have a limited load-driving capability. A buffer has a larger load-driving capability than a logic gate. It could be an inverting or noninverting buffer with a single input, a NAND buffer, a NOR buffer, an OR buffer or an AND buffer. 'Driver' is another name for a buffer. A driver is sometimes used to designate a circuit that has even larger drive capability than a buffer. Buffers are usually tristate devices to facilitate their use in bus-oriented systems. Figure 4.40 shows the symbols and functional tables of inverting and noninverting buffers of the tristate type.

A transceiver is a bidirectional buffer with additional direction control and ENABLE inputs. It allows flow of data in both directions, depending upon the logic status of the control inputs. Transceivers, like buffers, are tristate devices to make them compatible with bus-oriented systems. Figures 4.41(a) and (b) respectively show the circuit symbols of inverting and noninverting transceivers. Figure 4.42 shows a typical logic circuit arrangement of a tristate noninverting transceiver with its functional table [Fig. 4.42(b)].

Some of the common applications of inverting and noninverting buffers are as follows. Buffers are used to drive circuits that need more drive current. Noninverting buffers are also used to increase the fan-out of a given logic gate. This means that the buffer can be used to increase the number of logic gate inputs to which the output of a given logic gate can be connected. Yet another application of a noninverting buffer is its use as a delay line. It delays the signal by an amount equal to the propagation delay of the device. More than one device can be connected in cascade to get larger delays.

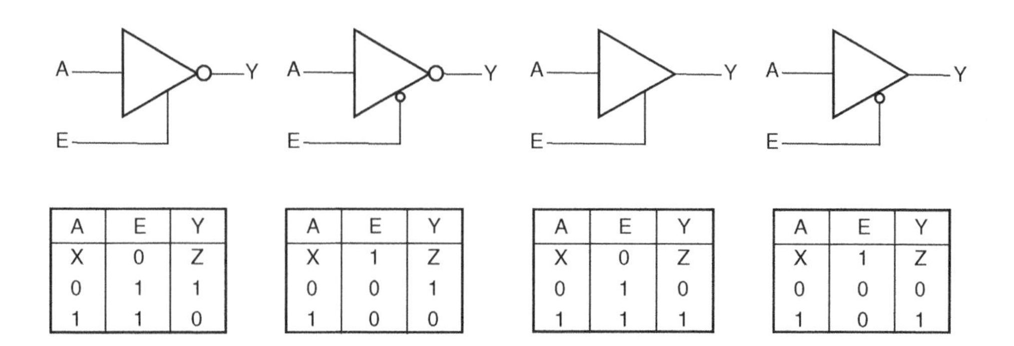

A	E	Y
X	0	Z
0	1	1
1	1	0

A	E	Y
X	1	Z
0	0	1
1	0	0

A	E	Y
X	0	Z
0	1	0
1	1	1

A	E	Y
X	1	Z
0	0	0
1	0	1

Z = High Impedance State

Figure 4.40 (a) Inverting tristate buffers and (b) noninverting tristate buffers.

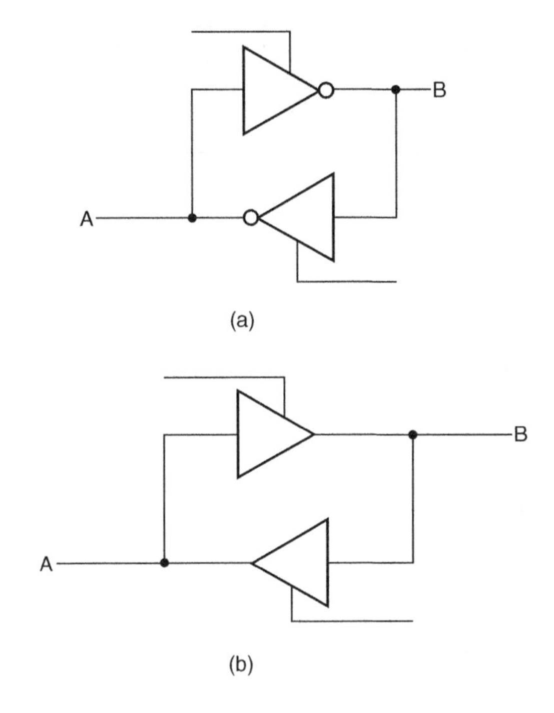

Figure 4.41 (a) Inverting transceivers and (b) noninverting transceivers.

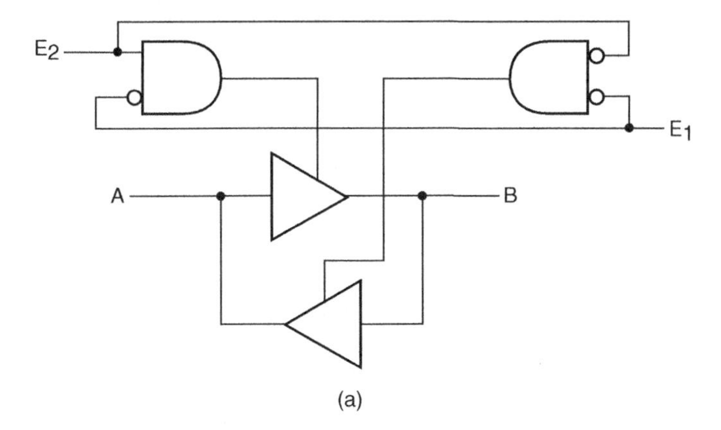

Figure 4.42 Tristate noninverting transceiver.

E_1	E_2	Operation
L	L	Data flow from B to A
L	H	Data flow from A to B
H	X	Isolation

(b)

Figure 4.42 (*continued*).

4.12 IEEE/ANSI Standard Symbols

The symbols used thus far in the chapter for representing different types of gate are the ones that are better known to all of us and have been in use for many years. Each logic gate has a symbol with a distinct shape. However, for more complex logic devices, e.g. sequential logic devices like flip-flops, counters, registers or arithmetic circuits, such as adders, subtractors, etc., these symbols do not carry any useful information. A new set of standard symbols was introduced in 1984 under IEEE/ANSI Standard 91–1984. The logic symbols given under this standard are being increasingly used now and have even started appearing in the literature published by manufacturers of digital integrated circuits. The utility of this new standard will be more evident in the following paragraphs as we go through its salient features and illustrate them with practical examples.

4.12.1 IEEE/ANSI Standards – Salient Features

This standard uses a rectangular symbol for all devices instead of a different symbol shape for each device. For instance, all logic gates (OR, AND, NAND, NOR) will be represented by a rectangular block.

A right triangle is used instead of a bubble to indicate inversion of a logic level. Also, the right triangle is used to indicate whether a given input or output is active LOW. The absence of a triangle indicates an active HIGH input or output. As far as logic gates are concerned, a special notation inside the rectangular block describes the logic relationship between output and inputs. A ‘1’ inside the block indicates that the device has only one input. An AND operation is expressed by ‘&’, and an OR operation is expressed by the symbol ‘$\geq$1’. Figure 4.43 shows the ANSI counterparts of various logic gates. A ‘$\geq$1’ symbol indicates that the output is HIGH when one or more than one input is HIGH. An ‘&’ symbol indicates that the output is HIGH only when all the inputs are HIGH. The two-input EX-OR is represented by the symbol ‘=1’ which implies that the output is HIGH only when one of its inputs is HIGH.

A special dependency notation system is used to indicate how the outputs depend upon the input. This notation contains almost the entire functional information of the logic device in question. This will be more clear as we illustrate this new standard with the help of ANSI symbols for some of the actual devices belonging to the category of flip-flops, counters, etc., in the following chapters. All those control inputs that control the timing of change in output states as per logic status of inputs are designated by the letter ‘C’. These are ENABLE inputs in latches or CLOCK inputs in flip-flops.

Most of the digital ICs contain more than one similar function on one chip such as IC 7400 (quad two-input NAND), IC 7404 (hex inverter), IC 74112 (dual-edge triggered JK flip-flop), IC 7474 (dual D-type latch), IC 7475 (quad D-type latch) and so on. Those inputs to such ICs that are common to

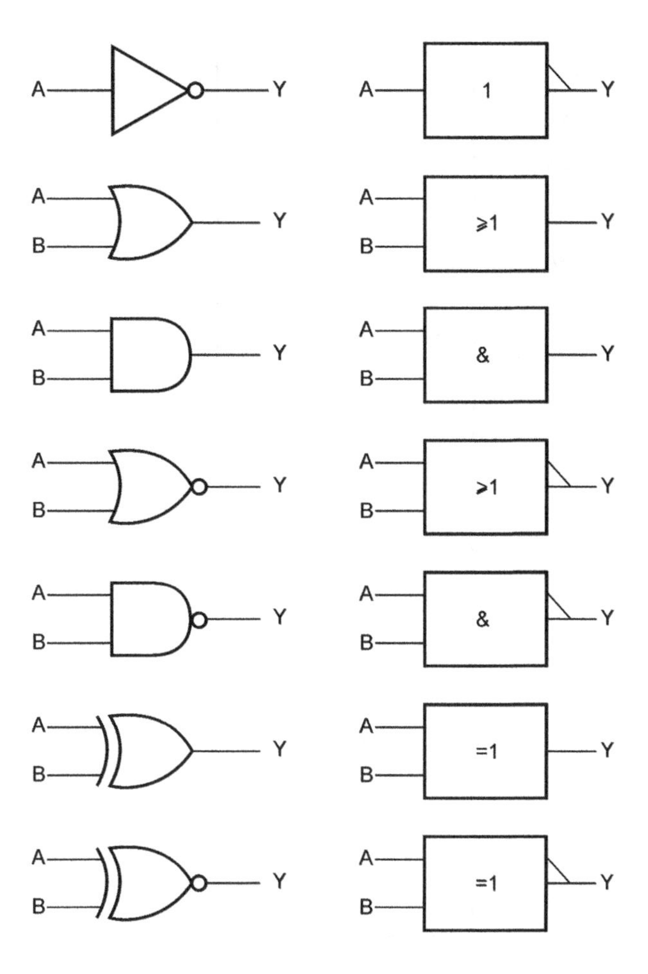

Figure 4.43 IEEE/ANSI symbols.

all the functional blocks or in other words similarly affect various individual but similar functions are represented by a separate notched rectangle on the top of the main rectangle.

4.12.2 ANSI Symbols for Logic Gate ICs

Figure 4.44 shows the ANSI symbol for IC 7400, which is a quad two-input NAND gate. The figure is self-explanatory with the background given in the preceding paragraphs. Any other similar device, i.e. another quad two-input NAND gate belonging to another logic family, would also be represented by the same ANSI symbol. As another illustration, Fig. 4.45 shows the ANSI symbol for IC 7420, which is a dual four-input NAND gate.

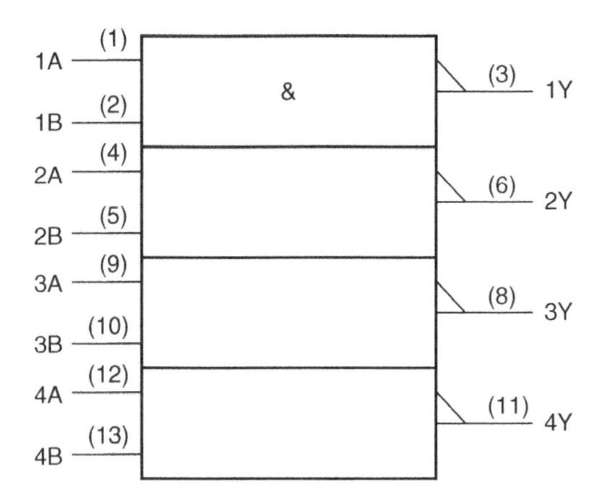

Figure 4.44 ANSI symbol for IC 7400.

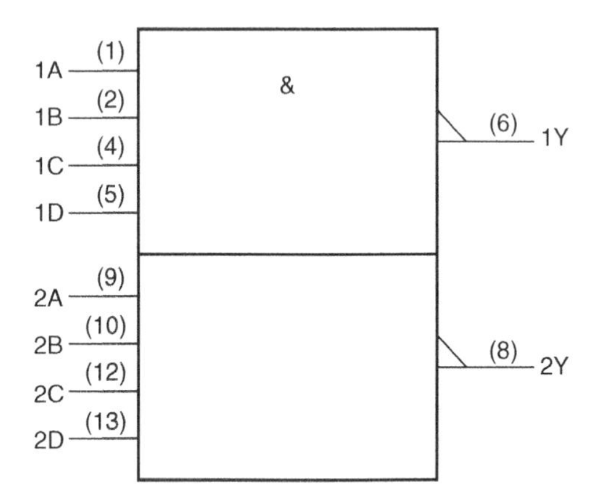

Figure 4.45 ANSI symbol for IC 7420.

Example 4.15

Draw the IEEE/ANSI symbol representation of the logic circuit shown in Fig. 4.46.

Solution

Figure 4.47 shows the circuit using IEEE/ANSI symbols.

4.13 Some Common Applications of Logic Gates

In this section, we will briefly look at some common applications of basic logic gates. The applications discussed here include those where these devices are used to provide a specific function in a larger digital circuit. These also include those where one or more logic gates, along with or without some external components, can be used to build some digital building blocks.

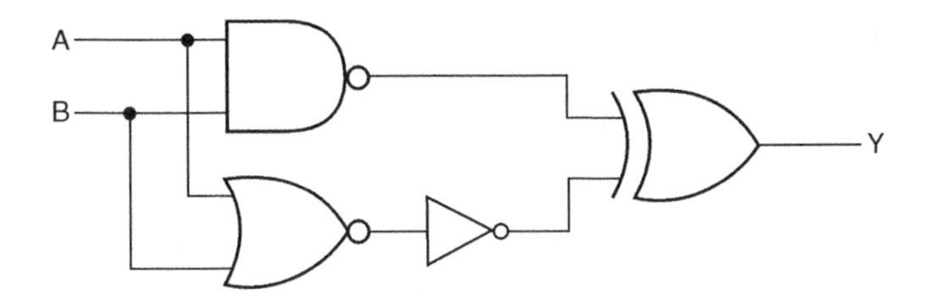

Figure 4.46 Example 4.15.

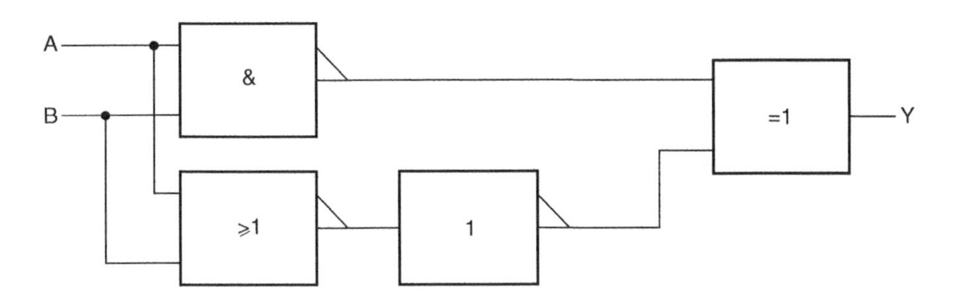

Figure 4.47 Solution to example 4.15.

4.13.1 OR Gate

An OR gate can be used in all those situations where the occurrence of any one or more than one event needs to be detected or acted upon. One such example is an industrial plant where any one or more than one parameter exceeding a preset limiting value should lead to initiation of some kind of protective action. Figure 4.48 shows a typical schematic where the OR gate is used to detect either temperature or pressure exceeding a preset threshold value and produce the necessary command signal for the system.

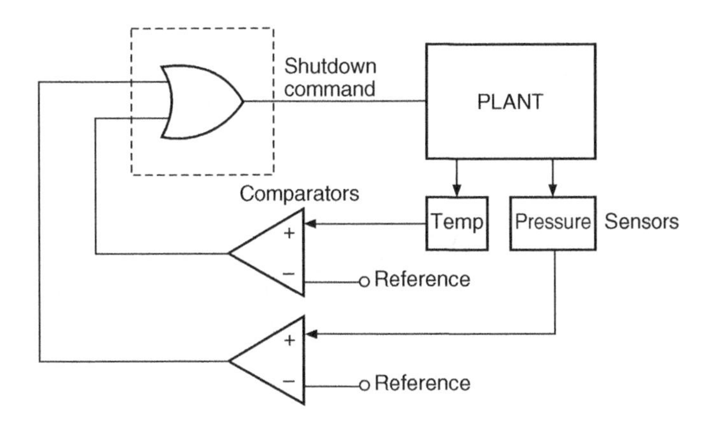

Figure 4.48 Application of an OR gate.

4.13.2 AND Gate

An AND gate is commonly used as an ENABLE or INHIBIT gate to allow or disallow passage of data from one point in the circuit to another. One such application of enabling operation, for instance, is in the measurement of the frequency of a pulsed waveform or the width of a given pulse with the help of a counter. In the case of frequency measurement, a gating pulse of known width is used to enable the passage of the pulse waveform to the counter's clock input. In the case of pulse width measurement, the pulse is used to enable the passage of the clock input to the counter. Figure 4.49 shows the arrangement.

4.13.3 EX-OR/EX-NOR Gate

EX-OR and EX-NOR logic gates are commonly used in parity generation and checking circuits. Figures 4.50(a) and (b) respectively show even and odd parity generator circuits for four-bit data. The circuits are self-explanatory.

The parity check operation can also be performed by similar circuits. Figures 4.51(a) and (b) respectively show simple even and odd parity check circuits for a four-bit data stream. In the circuits shown in Fig. 4.51, a logic '0' at the output signifies correct parity and a logic '1' signifies one-bit error. Parity generator/checker circuits are available in IC form. 74180 in TTL and 40101 in CMOS are nine-bit odd/even parity generator/checker ICs. Parity generation and checking circuits are further discussed in Chapter 7 on arithmetic circuits.

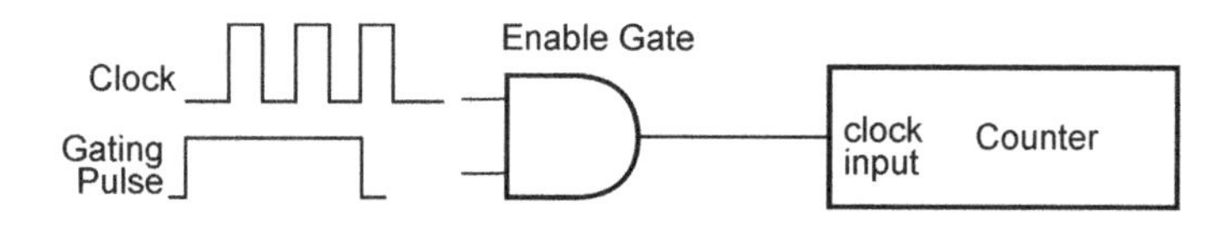

Figure 4.49 Application of an AND gate.

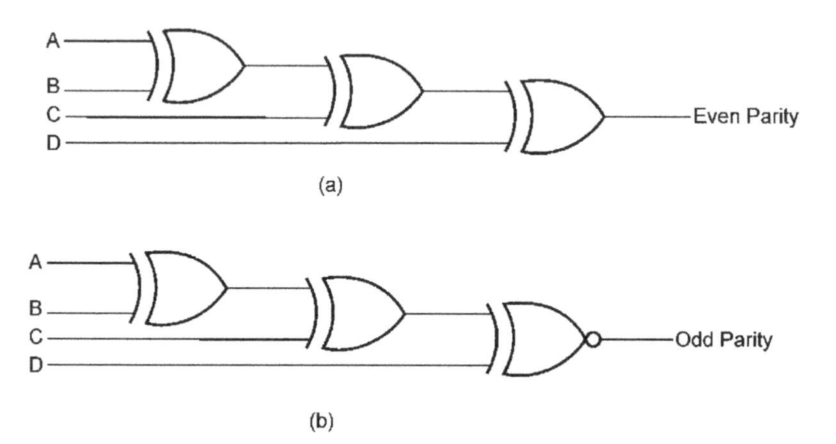

Figure 4.50 Parity generation using EX-OR/EX-NOR gates.

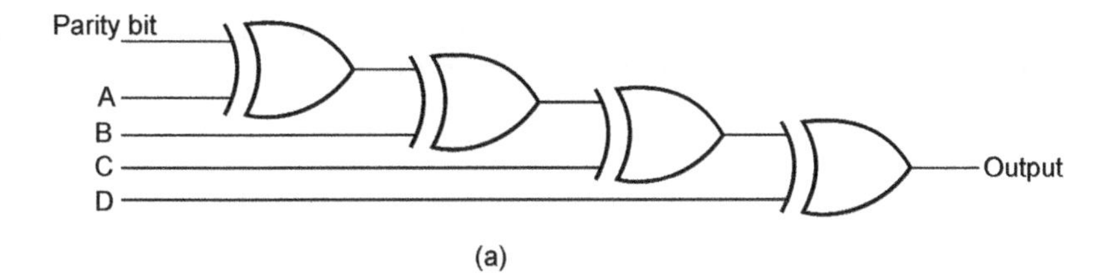

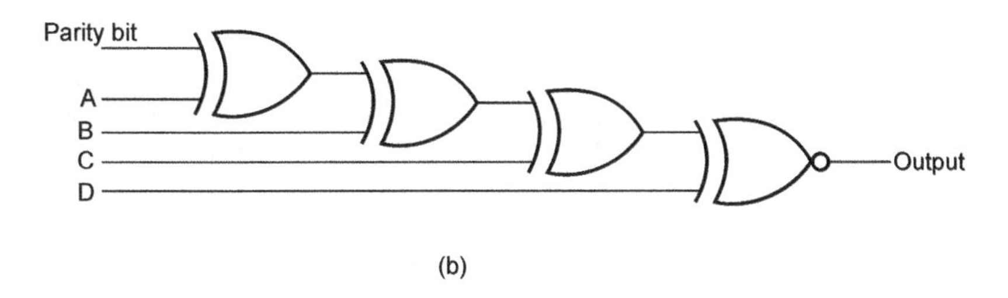

Figure 4.51 Parity check using EX-OR and EX-NOR gates.

4.13.4 Inverter

CMOS inverters are commonly used to build square-wave oscillators for generating clock signals. These clock generators offer good stability, operation over a wide supply voltage range (3–15 V) and frequency range (1 Hz to in excess of 15 MHz), low power consumption and an easy interface to other logic families.

The most fundamental circuit is the ring configuration of any odd number of inverters. Figure 4.52 shows one such circuit using three inverters. Inverting gates such as NAND and NOR gates can also be used instead. This configuration does not make a practical oscillator circuit as its frequency of oscillation is highly susceptible to variation with temperature, supply voltage and external loading. The frequency of oscillation is given by the equation

$$f = 1/(2nt_p) \tag{4.12}$$

where n is the number of inverters and t_p is the propagation delay per gate.

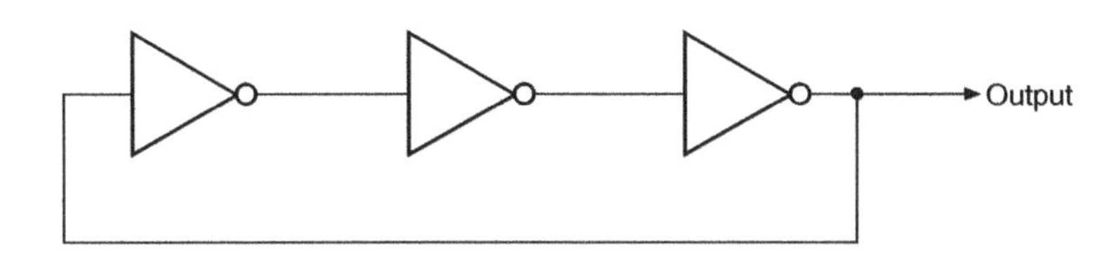

Figure 4.52 Square-wave oscillator using a ring configuration.

Figure 4.53(a) shows a practical oscillator circuit. The frequency of oscillation in this case is given by Equation (4.13) (the duty cycle of the waveform is approximately 50 %):

$$f = 1/2C(0.405R_{eq} + 0.693R_1) \tag{4.13}$$

where $R_{eq} = R_1.R_2/(R_1 + R_2)$.

Figure 4.53(b) shows another circuit using two inverters instead of three inverters. The frequency of oscillation of this circuit is given by the equation

$$f = 1/2.2RC \tag{4.14}$$

The circuits shown in Fig. 4.53 are not as sensitive to supply voltage variations as the one shown in Fig. 4.52. Figure 4.54 shows yet another circuit that is configured around a single Schmitt inverter. The capacitor charges (when the output is HIGH) and discharges (when the output is LOW) between the

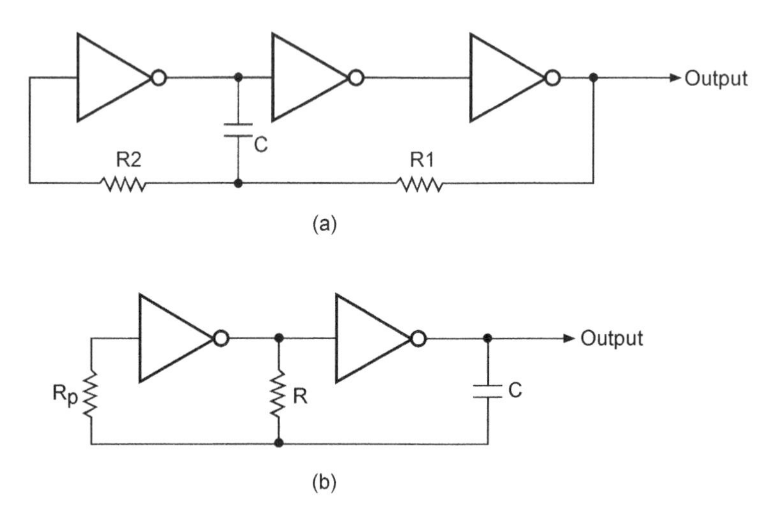

Figure 4.53 Square-wave oscillator with external components.

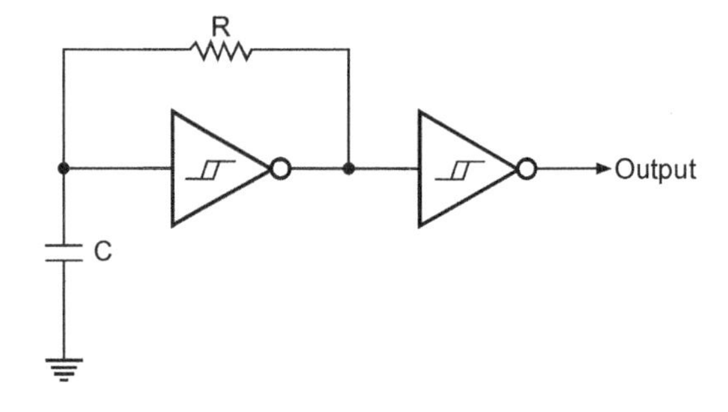

Figure 4.54 Schmitt inverter based oscillator.

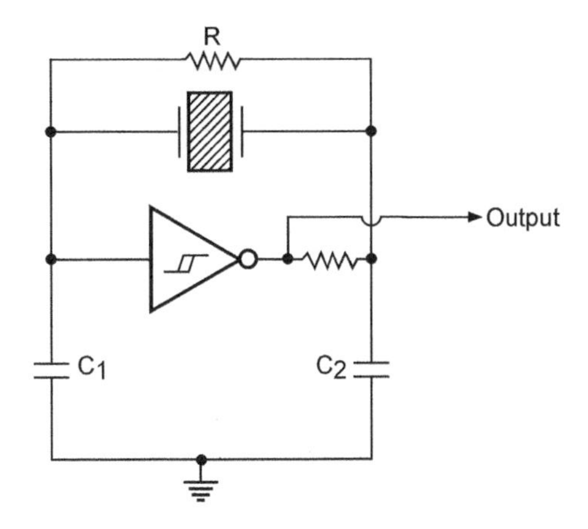

Figure 4.55 Crystal oscillator.

two threshold voltages. The frequency of oscillation, however, is sensitive to supply voltage variations. It is given by the equation

$$f = 1/RC \tag{4.15}$$

Figure 4.55 shows a crystal oscillator configured around a single inverter as the active element. Any odd number of inverters can be used. A larger number of inverters limits the highest attainable frequency of oscillation to a lower value.

4.14 Application-Relevant Information

Table 4.1 lists the commonly used type numbers along with the functional description and the logic family. The pin connection diagrams and the functional tables of the more popular type numbers are given in the companion website.

Table 4.1 Functional index of logic gates.

Type number	Function	Logic family
7400	Quad two-input NAND gate	TTL
7401	Quad two-input NAND gate (open collector)	TTL
7402	Quad two-input NOR gate	TTL
7403	Quad two-input NAND gate (open collector)	TTL
7404	Hex inverter	TTL
7405	Hex inverter (open collector)	TTL
7408	Quad two-input AND gate	TTL
7409	Quad two-input AND gate (open collector)	TTL
7410	Triple three-input NAND gate	TTL

(*Continued overleaf*)

Table 4.1 *(continued).*

Type number	Function	Logic family
7411	Triple three-input AND gate	TTL
7412	Triple three-input NAND gate (open collector)	TTL
7413	Dual four-input Schmitt NAND gate	TTL
7414	Hex Schmitt trigger inverter	TTL
7418	Dual four-input Schmitt NAND gate	TTL
7419	Hex Schmitt trigger inverter	TTL
7420	Dual four-input NAND gate	TTL
7421	Dual four-input AND gate	TTL
7422	Dual four-input NAND gate (open collector)	TTL
7427	Triple three-input NOR gate	TTL
7430	Eight-input NAND gate	TTL
7432	Quad two-input OR gate	TTL
7451	Dual two-wide two-input three-input AND-OR-INVERT gate	TTL
7454	Four-wide two-input AND-OR-INVERT gate	TTL
7455	Two-wide four-input AND-OR-INVERT gate	TTL
7486	Quad two-input EX-OR gate	TTL
74125	Quad tristate noninverting buffer (LOW ENABLE)	TTL
74126	Quad tristate noninverting buffer (HIGH ENABLE)	TTL
74132	Quad two-input Schmitt trigger NAND gate	TTL
74133	13-input NAND gate	TTL
74136	Quad two-input EX-OR gate (open collector)	TTL
74240	Octal tristate inverting bus/line driver	TTL
74241	Octal tristate bus/line driver	TTL
74242	Quad tristate inverting bus transceiver	TTL
74243	Quad tristate noninverting bus transceiver	TTL
74244	Octal tristate noninverting driver	TTL
74245	Octal tristate noninverting bus transceiver	TTL
74266	Quad two-input EXCLUSIVE-NOR gate (open collector)	TTL
74365	Hex tristate noninverting buffer with common ENABLE	TTL
74366	Hex tristate inverting buffer with common ENABLE	TTL
74367	Hex tristate noninverting buffer, four-bit and two-bit	TTL
74368	Hex tristate inverting buffer, four-bit and two-bit	TTL
74386	Quad two-input EX-OR gate	TTL
74465	Octal tristate noninverting buffer Gated ENABLE inverted	TTL
74540	Octal tristate inverting buffer/line driver	TTL
74541	Octal tristate noninverting buffer/line driver	TTL
74640	Octal tristate inverting bus transceiver	TTL
74641	Octal tristate noninverting bus transceiver (open collector)	TTL
74645	Octal tristate noninverting bus transceiver	TTL
4001B	Quad two-input NOR gate	CMOS
4002B	Dual four-input NOR gate	CMOS
4011B	Quad two-input NAND gate	CMOS
4012B	Dual four-input NAND gate	CMOS
4023B	Triple three-input NAND gate	CMOS
4025B	Triple three-input NOR gate	CMOS
4030B	Quad two-input EX-OR gate	CMOS
4049B	Hex inverting buffer	CMOS

Table 4.1 *(continued).*

Type number	Function	Logic family
4050B	Hex noninverting buffer	CMOS
40097B	Tristate hex noninverting buffer	CMOS
40098B	Tristate inverting buffer	CMOS
4069UB	Hex inverter	CMOS
4070B	Quad two-input EX-OR gate	CMOS
4071B	Quad two-input OR gate	CMOS
4081B	Quad two-input AND gate	CMOS
4086B	Four-wide two-input AND-OR-INVERT gate	CMOS
4093B	Quad two-input Schmitt NAND	CMOS
10100	Quad two-input NOR gate with strobe	ECL
10101	Quad two-input OR/NOR gate	ECL
10102	Quad two-input NOR gate	ECL
10103	Quad two-input OR gate	ECL
10104	Quad two-input AND gate	ECL
10113	Quad two-input EX-OR gate	ECL
10114	Triple line receiver	ECL
10115	Quad line Receiver	ECL
10116	Triple Line receiver	ECL
10117	Dual two-wide two- to three-input OR-AND/OR-AND-INVERT gate	ECL
10118	Dual two-wide three-input OR-AND gate	ECL
10123	Triple 4-3-3 input bus driver	
10128	Dual bus driver	ECL
10129	Quad bus driver	ECL
10188	Hex buffer with ENABLE	ECL
10192	Quad bus driver	ECL
10194	Dual simultaneous transceiver	ECL
10195	Hex buffer with invert/noninvert control	ECL

Review Questions

1. How do you distinguish between positive and negative logic systems? Prove that an OR gate in a positive logic system is an AND gate in a negative logic system.
2. Give brief statements that would help one remember the truth table of AND, NAND, OR, NOR, EX-OR and EX-NOR logic gate functions, irrespective of the number of inputs used.
3. Why are NAND and NOR gates called universal gates? Justify your answer with the help of examples.
4. What are Schmitt gates? How does a Schmitt gate overcome the problem of occurrence of an erratic output for slow varying input transitions?
5. What are logic gates with open collector or open drain outputs? What are the major advantages and disadvantages of such devices?
6. Draw the circuit symbol and the associated truth table for the following:

   (a) a tristate noninverting buffer with an active HIGH ENABLE input;
   (b) a tristate inverting buffer with an active LOW ENABLE input;

(c) a three-input NAND with an open collector output;
(d) a four-input INHIBIT gate.

7. Define the fan-out specification of a logic gate. Which parameters would you need to know from the data sheet of a logic gate to determine for yourself the fan-out in case it is not mentioned in the data sheet? Explain the procedure for determining the fan-out specification from those parameters. What are the consequences of exceeding the fan-out specification?
8. What is the main significance of IEEE/ANSI symbols when compared with the conventional ones? Draw the ANSI symbols for four-input OR, two-input AND, two-input EX-OR and two-input NAND gates.

Problems

1. What is the only input combination that:

(a) Will produce a logic '1' at the output of an eight-input AND gate?
(b) Will produce a logic '0' at the output of a four-input NAND gate?
(c) Will produce a logic '1' at the output of an eight-input NOR gate?
(d) Will produce a logic '0' at the output of a four-input OR gate?

(a) 11111111; (b) 1111; (c) 00000000; (d) 0000

2. Draw the truth table of the logic circuit shown in Fig. 4.56.

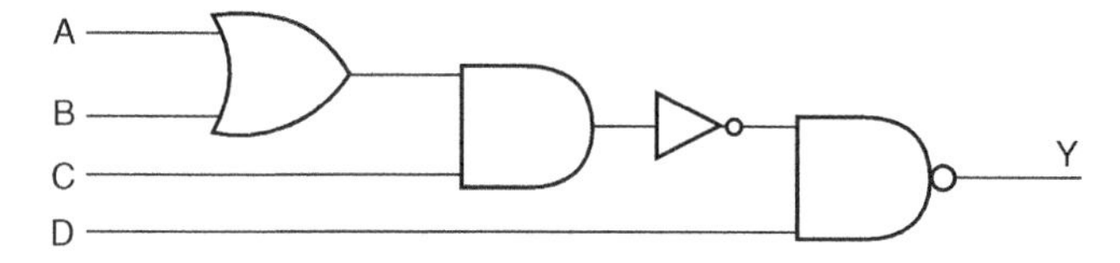

Figure 4.56 Problem 2.

A	B	C	D	Y
0	0	0	0	1
0	0	0	1	0
0	0	1	0	1
0	0	1	1	0
0	1	0	0	1
0	1	0	1	0
0	1	1	0	1
0	1	1	1	1
1	0	0	0	1
1	0	0	1	0
1	0	1	0	1
1	0	1	1	1
1	1	0	0	1
1	1	0	1	0
1	1	1	0	1
1	1	1	1	1

Figure 4.57 Solution of problem 2.

3. Redraw the logic circuit of Fig. 4.56 using IEEE/ANSI symbols.

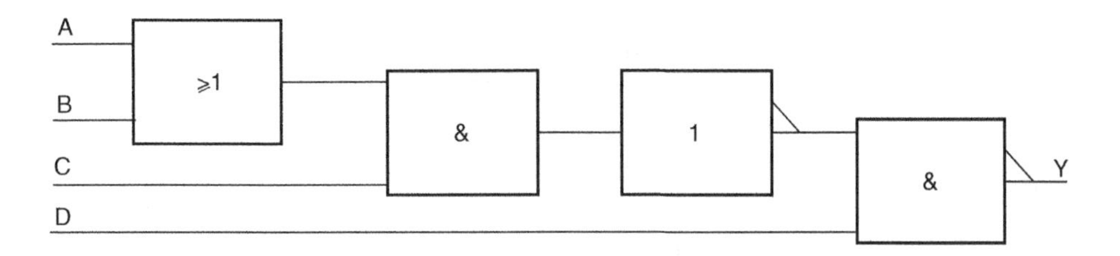

Figure 4.58 Solution to problem 3.

4. Refer to Fig. 4.59(a). The ENABLE waveforms applied at A and B inputs are respectively shown in Figs 4.59(b) and (c). What is the output state of inverter 3 and inverter 4 at (i) $t = 3$ ms and (ii) $t = 5$ ms?

(i) The output of inverter 3 = high Z, while the output of inverter 4 = logic '1'
(ii) The output of inverter 3 = logic '0', while the output of inverter 4 = high Z

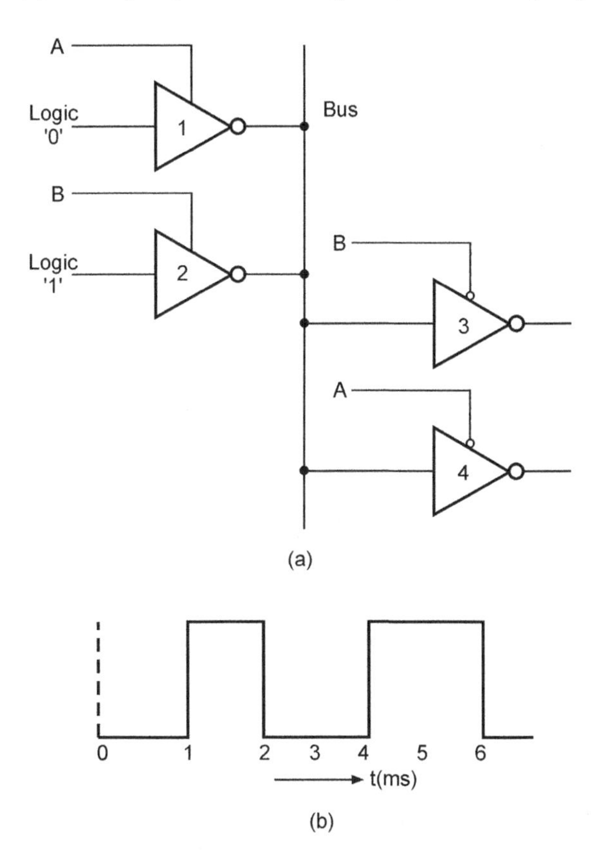

Figure 4.59 Problem 4.

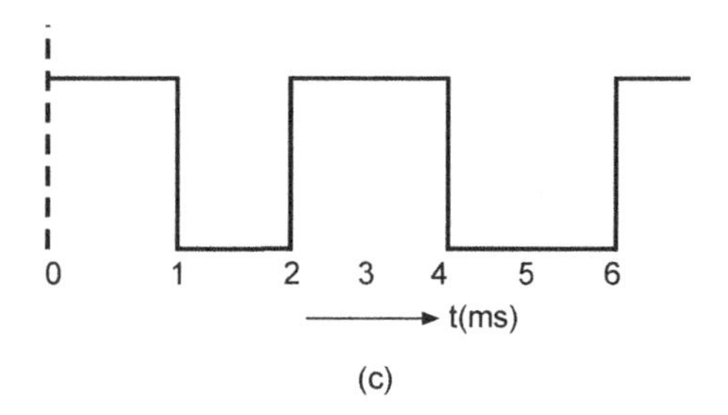

Figure 4.59 (*Continued*)

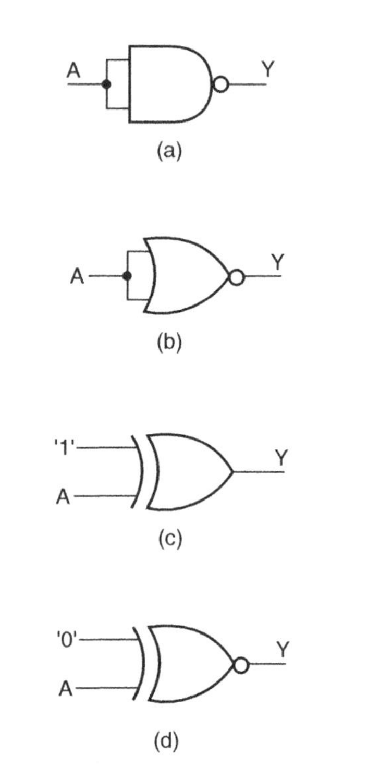

Figure 4.60 Solution to problem 5.

5. Draw logic implementation of an inverter using (i) two-input NAND, (ii) two-input NOR, (iii) two-input EX-OR and (iv) two-input EX-NOR.

(i) Fig. 4.60(a); (ii) Fig. 4.60(b); (iii) Fig. 4.60(c); (iv) Fig. 4.60(d)

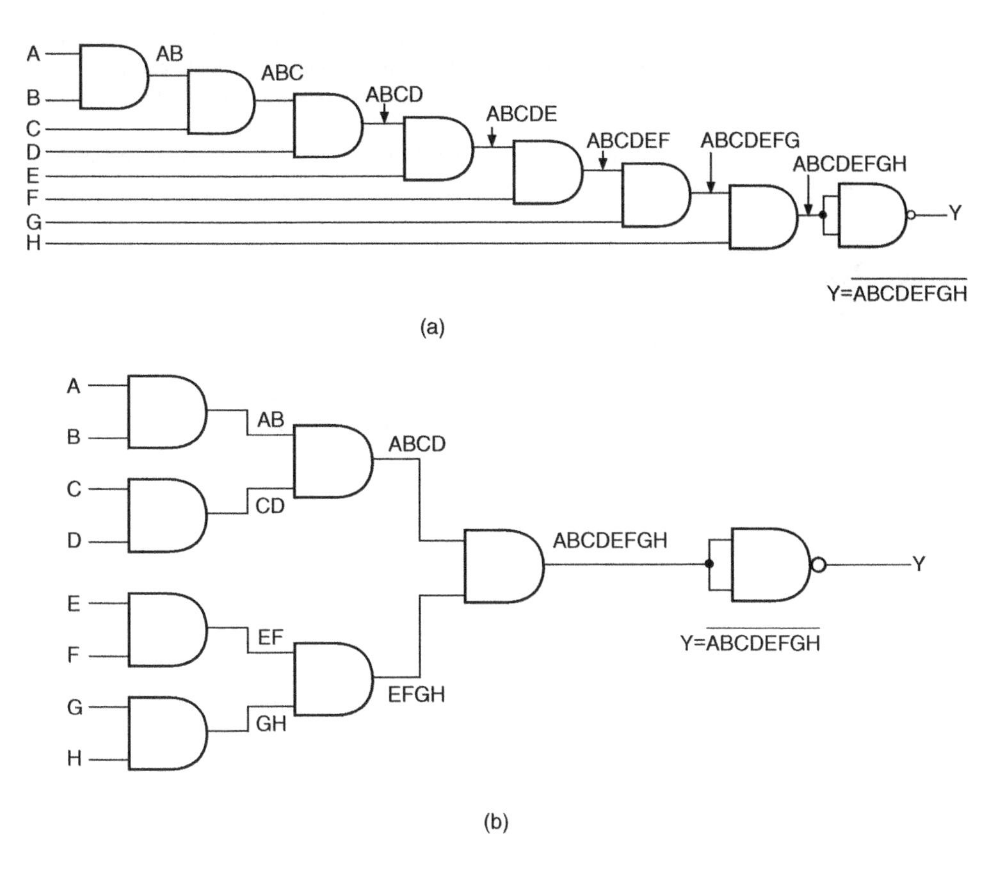

Figure 4.61 Solution to problem 6.

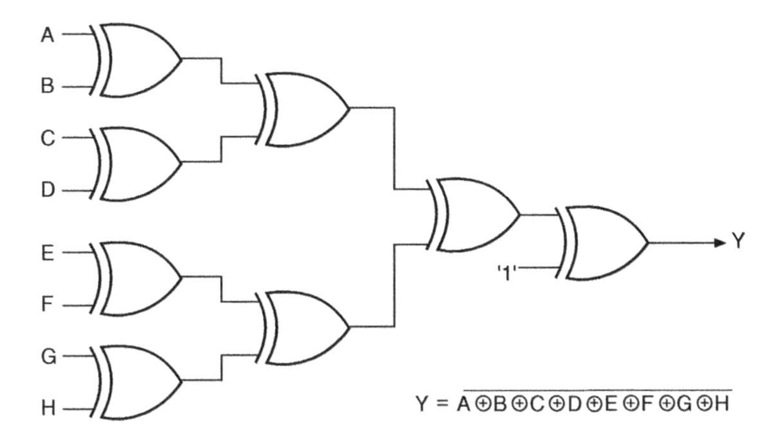

Figure 4.62 Solution to problem 7.

6. It is proposed to construct an eight-input NAND gate using only two-input AND gates and two-input NAND gates. Draw the logic arrangement that uses the minimum number of logic gates.

   The two possible logic circuits are shown in Figs 4.61(a) and (b)

7. Draw the logic diagram to implement an eight-input EX-NOR function using the minimum number of two-input logic gates.

Further Reading

1. Cook, N. P. (2003) *Practical Digital Electronics*, Prentice-Hall, NJ, USA.
2. Fairchild Semiconductor Corporation (October 1974) *CMOS Oscillators*, Application Note 118, South Portland, ME, USA.
3. Holdsworth, B. and Woods, C. (2002) *Digital Logic Design*, Newnes, Oxford, UK.
4. Langholz, G., Mott, J. L. and Kandel, A. (1998) *Foundations of Digital Logic Design*, World Scientific Publ. Co. Inc., Singapore.
5. Chen, W.-K. (2003) *Logic Design*, CRC Press, FL, USA.

5

Logic Families

Digital integrated circuits are produced using several different circuit configurations and production technologies. Each such approach is called a specific logic family. In this chapter, we will discuss different logic families used to hardware-implement different logic functions in the form of digital integrated circuits. The chapter begins with an introduction to logic families and the important parameters that can be used to characterize different families. This is followed by a detailed description of common logic families in terms of salient features, internal circuitry and interface aspects. Logic families discussed in the chapter include transistor transistor logic (TTL), metal oxide semiconductor (MOS) logic, emitter coupled logic (ECL), bipolar-CMOS (Bi-CMOS) logic and integrated injection logic (I^2L).

5.1 Logic Families – Significance and Types

There are a variety of circuit configurations or more appropriately various approaches used to produce different types of digital integrated circuit. Each such fundamental approach is called a *logic family*. The idea is that different logic functions, when fabricated in the form of an IC with the same approach, or in other words belonging to the same logic family, will have identical electrical characteristics. These characteristics include supply voltage range, speed of response, power dissipation, input and output logic levels, current sourcing and sinking capability, fan-out, noise margin, etc. In other words, the set of digital ICs belonging to the same logic family are electrically compatible with each other.

5.1.1 Significance

A digital system in general comprises digital ICs performing different logic functions, and choosing these ICs from the same logic family guarantees that different ICs are compatible with respect to each

Digital Electronics: Principles, Devices and Applications Anil Kumar Maini

other and that the system as a whole performs the intended logic function. In the case where the output of an IC belonging to a certain family feeds the inputs of another IC belonging to a different family, we must use established interface techniques to ensure compatibility. Understanding the features and capabilities of different logic families is very important for a logic designer who is out to make an optimum choice for his new digital design from the available logic family alternatives. A not so well thought out choice can easily underkill or overkill the design with either inadequate or excessive capabilities.

5.1.2 Types of Logic Family

The entire range of digital ICs is fabricated using either bipolar devices or MOS devices or a combination of the two. Different logic families falling in the first category are called bipolar families, and these include diode logic (DL), resistor transistor logic (RTL), diode transistor logic (DTL), transistor transistor logic (TTL), emitter coupled logic (ECL), also known as current mode logic (CML), and integrated injection logic (I^2L). The logic families that use MOS devices as their basis are known as MOS families, and the prominent members belonging to this category are the PMOS family (using P-channel MOSFETs), the NMOS family (using N-channel MOSFETs) and the CMOS family (using both N- and P-channel devices). The Bi-MOS logic family uses both bipolar and MOS devices.

Of all the logic families listed above, the first three, that is, diode logic (DL), resistor transistor logic (RTL) and diode transistor logic (DTL), are of historical importance only. Diode logic used diodes and resistors and in fact was never implemented in integrated circuits. The RTL family used resistors and bipolar transistors, while the DTL family used resistors, diodes and bipolar transistors. Both RTL and DTL suffered from large propagation delay owing to the need for the transistor base charge to leak out if the transistor were to switch from conducting to nonconducting state. Figure 5.1 shows the simplified schematics of a two-input AND gate using DL [Fig. 5.1(a)], a two-input NOR gate using RTL [Fig. 5.1(b)] and a two-input NAND gate using DTL [Fig. 5.1(c)]. The DL, RTL and DTL families, however, were rendered obsolete very shortly after their introduction in the early 1960s owing to the arrival on the scene of transistor transistor logic (TTL).

Logic families that are still in widespread use include TTL, CMOS, ECL, NMOS and Bi-CMOS. The PMOS and I^2L logic families, which were mainly intended for use in custom large-scale integrated (LSI) circuit devices, have also been rendered more or less obsolete, with the NMOS logic family replacing them for LSI and VLSI applications.

5.1.2.1 TTL Subfamilies

The TTL family has a number of subfamilies including standard TTL, low-power TTL, high-power TTL, low-power Schottky TTL, Schottky TTL, advanced low-power Schottky TTL, advanced Schottky TTL and fast TTL. The ICs belonging to the TTL family are designated as 74 or 54 (for standard TTL), 74L or 54L (for low-power TTL), 74H or 54H (for high-power TTL), 74LS or 54LS (for low-power Schottky TTL), 74S or 54S (for Schottky TTL), 74ALS or 54ALS (for advanced low-power Schottky TTL), 74AS or 54AS (for advanced Schottky TTL) and 74F or 54F (for fast TTL). An alphabetic code preceding this indicates the name of the manufacturer (DM for National Semiconductors, SN for Texas Instruments and so on). A two-, three- or four-digit numerical code tells the logic function performed by the IC. It may be mentioned that 74-series devices and 54-series devices are identical except for their operational temperature range. The 54-series devices are MIL-qualified (operational temperature range: −55 °C to +125 °C) versions of the corresponding 74-series ICs (operational temperature range: 0 °C to 70 °C). For example, 7400 and 5400 are both quad two-input NAND gates.

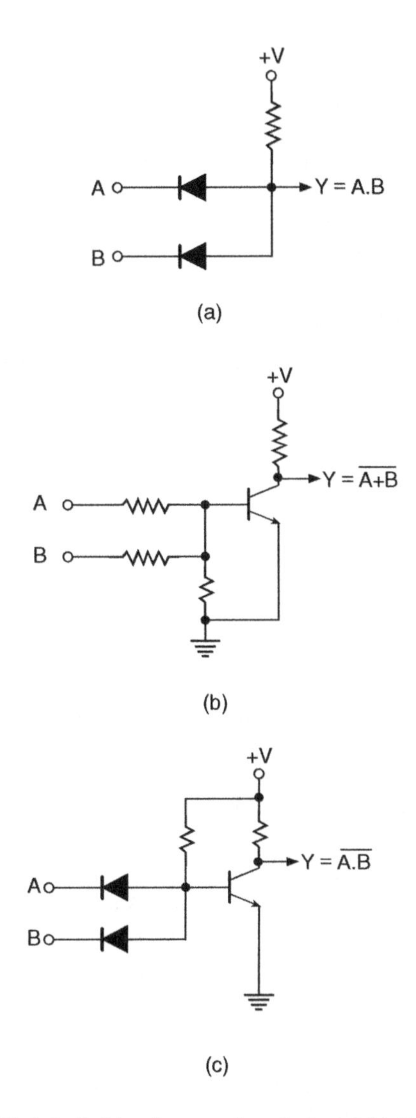

Figure 5.1 (a) Diode logic (b) resistor transistor logic and (c) diode transistor logic.

5.1.2.2 CMOS Subfamilies

The popular CMOS subfamilies include the 4000A, 4000B, 4000UB, 54/74C, 54/74HC, 54/74HCT, 54/74AC and 54/74ACT families. The 4000A CMOS family has been replaced by its high-voltage versions in the 4000B and 4000UB CMOS families, with the former having buffered and the latter having unbuffered outputs. 54/74C, 54/74HC, 54/74HCT, 54/74AC and 54/74ACT are CMOS logic families with pin-compatible 54/74 TTL series logic functions.

5.1.2.3 ECL Subfamilies

The first monolithic emitter coupled logic family was introduced by ON Semiconductor, formerly a division of Motorola, with the MECL-I series of devices in 1962, with the MECL-II series following it up in 1966. Both these logic families have become obsolete. Currently, popular subfamilies of ECL logic include MECL-III (also called the MC 1600 series), the MECL-10K series, the MECL-10H series and the MECL-10E series (ECLinPS and ECLinPSLite). The MECL-10K series further divided into the 10 100-series and 10 200-series devices.

5.2 Characteristic Parameters

In this section, we will briefly describe the parameters used to characterize different logic families. Some of these characteristic parameters, as we will see in the paragraphs to follow, are also used to compare different logic families.

- **HIGH-level input current, I_{IH}.** This is the current flowing into (taken as positive) or out of (taken as negative) an input when a HIGH-level input voltage equal to the minimum HIGH-level output voltage specified for the family is applied. In the case of bipolar logic families such as TTL, the circuit design is such that this current flows into the input pin and is therefore specified as positive. In the case of CMOS logic families, it could be either positive or negative, and only an absolute value is specified in this case.
- **LOW-level input current, I_{IL}.** The LOW-level input current is the maximum current flowing into (taken as positive) or out of (taken as negative) the input of a logic function when the voltage applied at the input equals the maximum LOW-level output voltage specified for the family. In the case of bipolar logic families such as TTL, the circuit design is such that this current flows out of the input pin and is therefore specified as negative. In the case of CMOS logic families, it could be either positive or negative. In this case, only an absolute value is specified.

HIGH-level and LOW-level input current or loading are also sometimes defined in terms of *unit load* (UL). For devices of the TTL family, 1 UL (HIGH) = 40 μA and 1 UL (LOW) = 1.6 mA.

- **HIGH-level output current, I_{OH}.** This is the maximum current flowing out of an output when the input conditions are such that the output is in the logic HIGH state. It is normally shown as a negative number. It tells about the current sourcing capability of the output. The magnitude of I_{OH} determines the number of inputs the logic function can drive when its output is in the logic HIGH state. For example, for the standard TTL family, the minimum guaranteed I_{OH} is −400 μA, which can drive 10 standard TTL inputs with each requiring 40 μA in the HIGH state, as shown in Fig. 5.2(a).
- **LOW-level output current, I_{OL}.** This is the maximum current flowing into the output pin of a logic function when the input conditions are such that the output is in the logic LOW state. It tells about the current sinking capability of the output. The magnitude of I_{OL} determines the number of inputs the logic function can drive when its output is in the logic LOW state. For example, for the standard TTL family, the minimum guaranteed I_{OL} is 16 mA, which can drive 10 standard TTL inputs with each requiring 1.6 mA in the LOW state, as shown in Fig. 5.2(b).
- **HIGH-level off-state (high-impedance state) output current, I_{OZH}.** This is the current flowing into an output of a tristate logic function with the ENABLE input chosen so as to establish a high-impedance state and a logic HIGH voltage level applied at the output. The input conditions are chosen so as to produce logic LOW if the device is enabled.

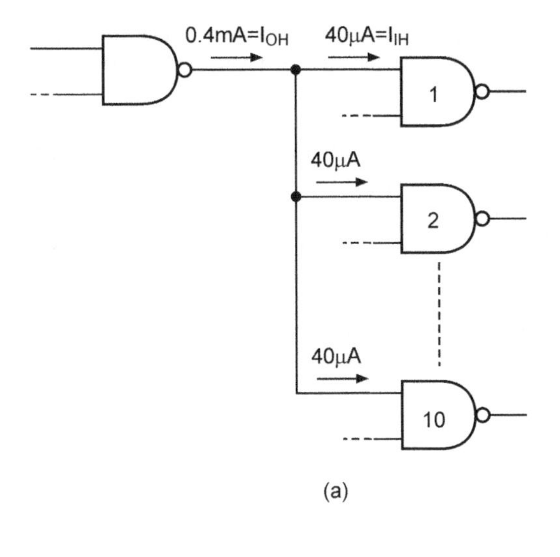

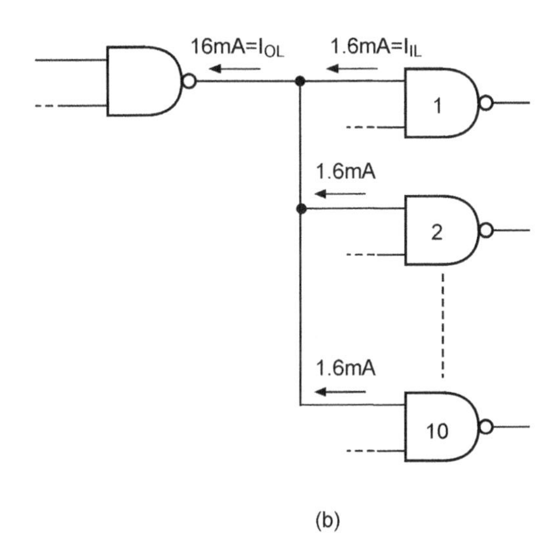

Figure 5.2 Input and output current specifications.

- **LOW-level off-state (high-impedance state) output current, I_{OZL}.** This is the current flowing into an output of a tristate logic function with the ENABLE input chosen so as to establish a high-impedance state and a logic LOW voltage level applied at the output. The input conditions are chosen so as to produce logic HIGH if the device is enabled.
- **HIGH-level input voltage, V_{IH}.** This is the minimum voltage level that needs to be applied at the input to be recognized as a legal HIGH level for the specified family. For the standard TTL family, a 2 V input voltage is a legal HIGH logic state.

- **LOW-level input voltage, V_{IL}.** This is the maximum voltage level applied at the input that is recognized as a legal LOW level for the specified family. For the standard TTL family, an input voltage of 0.8 V is a legal LOW logic state.
- **HIGH-level output voltage, V_{OH}.** This is the minimum voltage on the output pin of a logic function when the input conditions establish logic HIGH at the output for the specified family. In the case of the standard TTL family of devices, the HIGH level output voltage can be as low as 2.4 V and still be treated as a legal HIGH logic state. It may be mentioned here that, for a given logic family, the V_{OH} specification is always greater than the V_{IH} specification to ensure output-to-input compatibility when the output of one device feeds the input of another.
- **LOW-level output voltage, V_{OL}.** This is the maximum voltage on the output pin of a logic function when the input conditions establish logic LOW at the output for the specified family. In the case of the standard TTL family of devices, the LOW-level output voltage can be as high as 0.4 V and still be treated as a legal LOW logic state. It may be mentioned here that, for a given logic family, the V_{OL} specification is always smaller than the V_{IL} specification to ensure output-to-input compatibility when the output of one device feeds the input of another.

The different input/output current and voltage parameters are shown in Fig. 5.3, with HIGH-level current and voltage parameters in Fig. 5.3(a) and LOW-level current and voltage parameters in Fig. 5.3(b). It may be mentioned here that the direction of the LOW-level input and output currents shown in Fig. 5.3(b) is applicable to logic families with current-sinking action such as TTL.

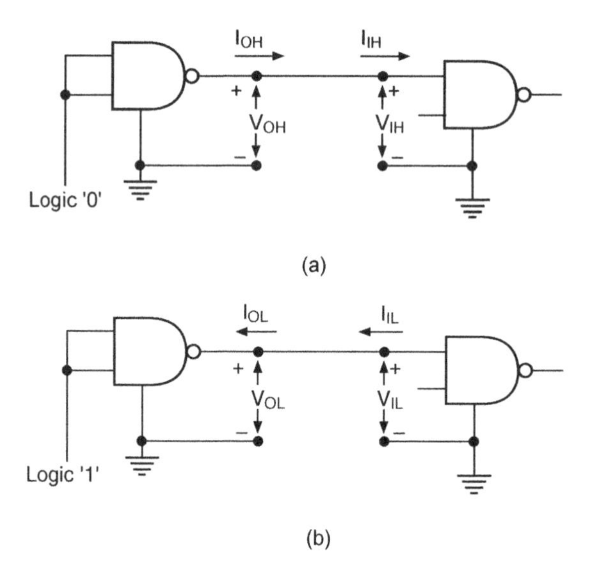

Figure 5.3 (a) HIGH-level current and voltage parameters and (b) LOW-level current and voltage parameters.

- **Supply current, I_{CC}.** The supply current when the output is HIGH, LOW and in the high-impedance state is respectively designated as I_{CCH}, I_{CCL} and I_{CCZ}.
- **Rise time, t_r.** This is the time that elapses between 10 and 90 % of the final signal level when the signal is making a transition from logic LOW to logic HIGH.
- **Fall time, t_f.** This is the time that elapses between 90 and 10 % of the signal level when it is making HIGH to LOW transition.
- **Propagation delay t_p.** The propagation delay is the time delay between the occurrence of change in the logical level at the input and before it is reflected at the output. It is the time delay between the specified voltage points on the input and output waveforms. Propagation delays are separately defined for LOW-to-HIGH and HIGH-to-LOW transitions at the output. In addition, we also define enable and disable time delays that occur during transition between the high-impedance state and defined logic LOW or HIGH states.
- **Propagation delay t_{pLH}.** This is the time delay between specified voltage points on the input and output waveforms with the output changing from LOW to HIGH.
- **Propagation delay t_{pHL}.** This is the time delay between specified voltage points on the input and output waveforms with the output changing from HIGH to LOW. Figure 5.4 shows the two types of propagation delay parameter.
- **Disable time from the HIGH state, t_{pHZ}.** Defined for a tristate device, this is the time delay between specified voltage points on the input and output waveforms with the tristate output changing from the logic HIGH level to the high-impedance state.
- **Disable time from the LOW state, t_{pLZ}.** Defined for a tristate device, this is the time delay between specified voltage points on the input and output waveforms with the tristate output changing from the logic LOW level to the high-impedance state.
- **Enable time from the HIGH state, t_{pZH}.** Defined for a tristate device, this is the time delay between specified voltage points on the input and output waveforms with the tristate output changing from the high-impedance state to the logic HIGH level.

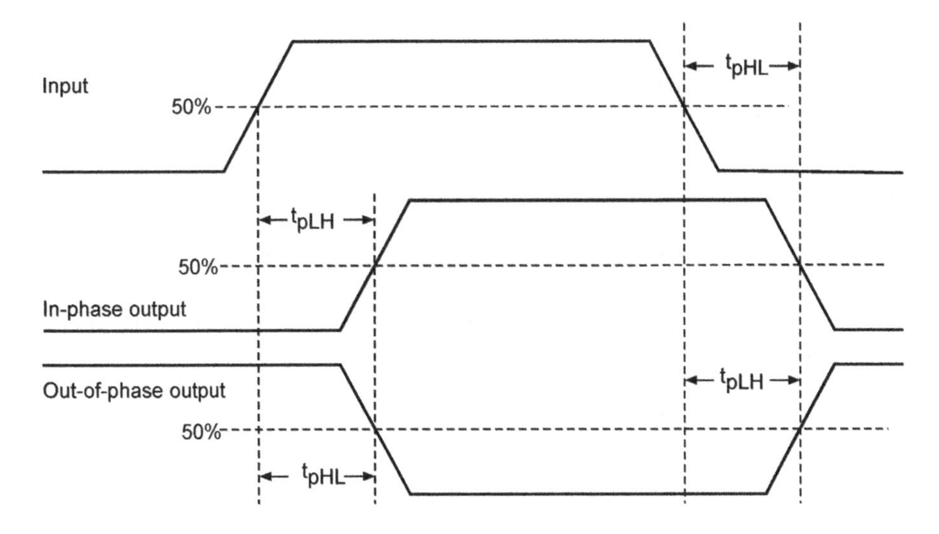

Figure 5.4 Propagation delay parameters.

- **Enable time from the LOW state, t_{pZL}.** Defined for a tristate device, this is the time delay between specified voltage points on the input and output waveforms with the tristate output changing from the high-impedance state to the logic LOW level.
- **Maximum clock frequency, f_{max}.** This is the maximum frequency at which the clock input of a flip-flop can be driven through its required sequence while maintaining stable transitions of logic level at the output in accordance with the input conditions and the product specification. It is also referred to as the maximum toggle rate for a flip-flop or counter device.
- **Power dissipation.** The power dissipation parameter for a logic family is specified in terms of power consumption per gate and is the product of supply voltage V_{CC} and supply current I_{CC}. The supply current is taken as the average of the HIGH-level supply current I_{CCH} and the LOW-level supply current I_{CCL}.
- **Speed–power product.** The speed of a logic circuit can be increased, that is, the propagation delay can be reduced, at the expense of power dissipation. We will recall that, when a bipolar transistor switches between cut-off and saturation, it dissipates the least power but has a large associated switching time delay. On the other hand, when the transistor is operated in the active region, power dissipation goes up while the switching time decreases drastically. It is always desirable to have in a logic family low values for both propagation delay and power dissipation parameters. A useful figure-of-merit used to evaluate different logic families is the speed–power product, expressed in picojoules, which is the product of the propagation delay (measured in nanoseconds) and the power dissipation per gate (measured in milliwatts).
- **Fan-out.** The fan-out is the number of inputs of a logic function that can be driven from a single output without causing any false output. It is a characteristic of the logic family to which the device belongs. It can be computed from I_{OH}/I_{IH} in the logic HIGH state and from I_{OL}/I_{IL} in the logic LOW state. If, in a certain case, the two values I_{OH}/I_{IH} and I_{OL}/I_{IL} are different, the fan-out is taken as the smaller of the two. This description of the fan-out is true for bipolar logic families like TTL and ECL. When determining the fan-out of CMOS logic devices, we should also take into consideration how much input load capacitance can be driven from the output without exceeding the acceptable value of propagation delay.
- **Noise margin.** This is a quantitative measure of noise immunity offered by the logic family. When the output of a logic device feeds the input of another device of the same family, a legal HIGH logic state at the output of the feeding device should be treated as a legal HIGH logic state by the input of the device being fed. Similarly, a legal LOW logic state of the feeding device should be treated as a legal LOW logic state by the device being fed. We have seen in earlier paragraphs while defining important characteristic parameters that legal HIGH and LOW voltage levels for a given logic family are different for outputs and inputs. Figure 5.5 shows the generalized case of legal HIGH and LOW voltage levels for output [Fig. 5.5(a)] and input [Fig. 5.5(b)]. As we can see from the two diagrams, there is a disallowed range of output voltage levels from V_{OL}(max.) to V_{OH}(min.) and an indeterminate range of input voltage levels from V_{IL}(max.) to V_{IH}(min.). Since V_{IL}(max.) is greater than V_{OL}(max.), the LOW output state can therefore tolerate a positive voltage spike equal to V_{IL}(max.) $-$ V_{OL}(max.) and still be a legal LOW input. Similarly, V_{OH}(min.) is greater than V_{IH} (min.), and the HIGH output state can tolerate a negative voltage spike equal to V_{OH}(min.) $-$ V_{IH} (min.) and still be a legal HIGH input. Here, V_{IL}(max.) $-$ V_{OL}(max.) and V_{OH}(min.) $-$ V_{IH} (min.) are respectively known as the LOW-level and HIGH-level noise margin.

Let us illustrate it further with the help of data for the standard TTL family. The minimum legal HIGH output voltage level in the case of the standard TTL is 2.4 V. Also, the minimum legal HIGH input voltage level for this family is 2 V. This implies that, when the output of one device feeds the input of another, there is an available margin of 0.4 V. That is, any negative voltage spikes of amplitude

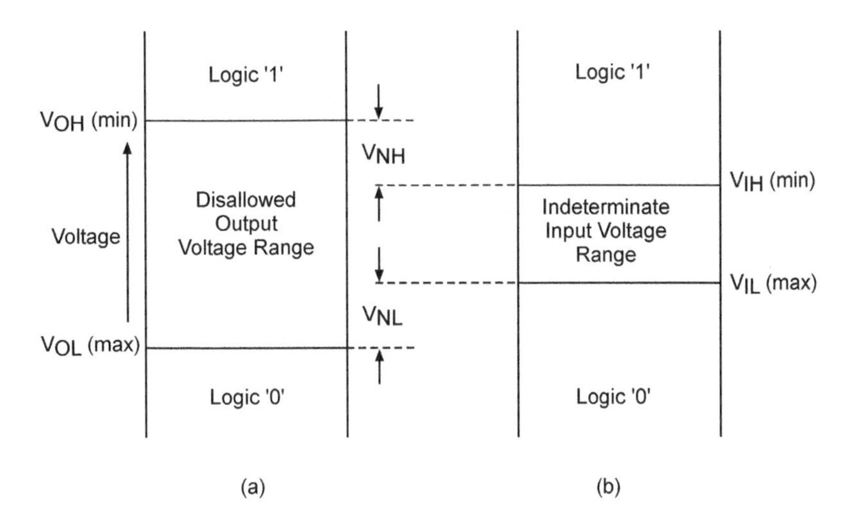

Figure 5.5 Noise margin.

less than or equal to 0.4 V on the signal line do not cause any spurious transitions. Similarly, when the output is in the logic LOW state, the maximum legal LOW output voltage level in the case of the standard TTL is 0.4 V. Also, the maximum legal LOW input voltage level for this family is 0.8 V. This implies that, when the output of one device feeds the input of another, there is again an available margin of 0.4 V. That is, any positive voltage spikes of amplitude less than or equal to 0.4 V on the signal line do not cause any spurious transitions. This leads to the standard TTL family offering a noise margin of 0.4 V. To generalize, the noise margin offered by a logic family, as outlined earlier, can be computed from the HIGH-state noise margin, $V_{NH} = V_{OH}(\text{min.}) - V_{IH}(\text{min.})$, and the LOW-state noise margin, $V_{NL} = V_{IL}(\text{max.}) - V_{OL}(\text{max.})$. If the two values are different, the noise margin is taken as the lower of the two.

Example 5.1

The data sheet of a quad two-input NAND gate specifies the following parameters: $I_{OH}(max.) = 0.4$ mA, $V_{OH}(min.) = 2.7$ V, $V_{IH}(min.) = 2$V, $V_{IL}(max.) = 0.8$ V, $V_{OL}(max.) = 0.4$ V, $I_{OL}(max.) = 8$ mA, $I_{IL}(max.) = 0.4$ mA, I_{IH} (max.) = 20 μA, $I_{CCH}(max.) = 1.6$ mA, $I_{CCL}(max.) = 4.4$ mA, $t_{pLH} = t_{pHL} = 15$ ns and a supply voltage range of 5 V. Determine (a) the average power dissipation of a single NAND gate, (b) the maximum average propagation delay of a single gate, (c) the HIGH-state noise margin and (d) the LOW-state noise margin

Solution

(a) The average supply current $= (I_{CCH} + I_{CCL})/2 = (1.6 + 4.4)/2 = 3$ mA.
    The supply voltage $V_{CC} = 5$ V.
    Therefore, the power dissipation for all four gates in the IC $= 5 \times 3 = 15$ mW.
    The average power dissipation per gate $= 15/4 = 3.75$ mW.
(b) The propagation delay = 15 ns.
(c) The HIGH-state noise margin $= V_{OH}(\text{min.}) - V_{IH}(\text{min.}) = 2.7 - 2 = 0.7$ V.
(d) The LOW-state noise margin $= V_{IL}(\text{max.}) - V_{OL}(\text{max.}) = 0.8 - 0.4 = 0.4$ V.

Example 5.2

Refer to example 5.1. How many NAND gate inputs can be driven from the output of a NAND gate of this type?

Solution

- This figure is given by the worst-case fan-out specification of the device.
- Now, the HIGH-state fan-out = I_{OH}/I_{IH} = 400/20 = 20.
- The LOW-state fan-out = I_{OL}/I_{IL} = 8/0.4 = 20.
- Therefore, the number of inputs that can be driven from a single output = 20.

Example 5.3

Determine the fan-out of IC 74LS04, given the following data: input loading factor (HIGH state) = 0.5 UL, input loading factor (LOW state) = 0.25 UL, output loading factor (HIGH state) = 10 UL, output loading factor (LOW state) = 5 UL, where UL is the unit load.

Solution

- The HIGH-state fan-out can be computed from: fan-out = output loading factor (HIGH)/input loading factor (HIGH) = 10 UL/0.5 UL = 20.
- The LOW-state fan-out can be computed from: fan-out = output loading factor (LOW)/input loading factor (LOW) = 5 UL/0.25 UL = 20.
- Since the fan-out in the two cases turns out to be the same, it follows that the fan-out = 20.

Example 5.4

A certain TTL gate has I_{IH} = 20 μA, I_{IL} = 0.1 mA, I_{OH} = 0.4 mA and I_{OL} = 4 mA. Determine the input and output loading in the HIGH and LOW states in terms of UL.

Solution

- 1 UL (LOW state) = 1.6 mA and 1 UL (HIGH state) = 40 μA.
- The input loading factor (HIGH state) = 20 μA = 20/40 = 0.5 UL.
- The input loading factor (LOW state) = 0.1 mA = 0.1/1.6 = 1/16 UL
- The output loading factor (HIGH state) = 0.4 mA = 0.4/0.04 = 10 UL.
- The output loading factor (LOW state) = 4 mA = 4/1.6 = 2.5 UL.

5.3 Transistor Transistor Logic (TTL)

TTL as outlined above stands for transistor transistor logic. It is a logic family implemented with bipolar process technology that combines or integrates NPN transistors, PN junction diodes and diffused resistors in a single monolithic structure to get the desired logic function. The NAND gate is the basic building block of this logic family. Different subfamilies in this logic family, as outlined earlier, include standard TTL, low-power TTL, high-power TTL, low-power Schottky TTL, Schottky TTL, advanced low-power Schottky TTL, advanced Schottky TTL and fast TTL. In the following paragraphs, we will briefly describe each of these subfamilies in terms of internal structure and characteristic parameters.

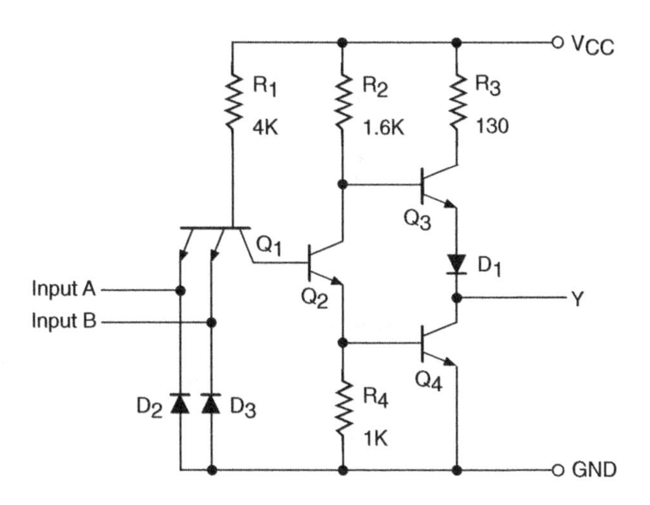

Figure 5.6 Standard TTL NAND gate.

5.3.1 Standard TTL

Figure 5.6 shows the internal schematic of a standard TTL NAND gate. It is one of the four circuits of 5400/7400, which is a quad two-input NAND gate. The circuit operates as follows. Transistor Q_1 is a two-emitter NPN transistor, which is equivalent to two NPN transistors with their base and emitter terminals tied together. The two emitters are the two inputs of the NAND gate. Diodes D_2 and D_3 are used to limit negative input voltages. We will now examine the behaviour of the circuit for various possible logic states at the two inputs.

5.3.1.1 Circuit Operation

When both the inputs are in the logic HIGH state as specified by the TTL family ($V_{IH} = 2$ V minimum), the current flows through the base-collector PN junction diode of transistor Q_1 into the base of transistor Q_2. Transistor Q_2 is turned ON to saturation, with the result that transistor Q_3 is switched OFF and transistor Q_4 is switched ON. This produces a logic LOW at the output, with V_{OL} being 0.4 V maximum when it is sinking a current of 16 mA from external loads represented by inputs of logic functions being driven by the output. The current-sinking action is shown in Fig. 5.7(a). Transistor Q_4 is also referred to as the current-sinking or pull-down transistor, for obvious reasons. Diode D_1 is used to prevent transistor Q_3 from conducting even a small amount of current when the output is LOW. When the output is LOW, Q_4 is in saturation and Q_3 will conduct slightly in the absence of D_1. Also, the input current I_{IH} in the HIGH state is nothing but the reverse-biased junction diode leakage current and is typically 40 μA.

When either of the two inputs or both inputs are in the logic LOW state, the base-emitter region of Q_1 conducts current, driving Q_2 to cut-off in the process. When Q_2 is in the cut-off state, Q_3 is driven to conduction and Q_4 to cut-off. This produces a logic HIGH output with $V_{OH}(\text{min.}) = 2.4$ V guaranteed for minimum supply voltage V_{CC} and a source current of 400 μA. The current-sourcing action is shown in Fig. 5.7(b). Transistor Q_3 is also referred to as the current-sourcing or pull-up transistor. Also, the LOW-level input current I_{IL}, given by $(V_{CC} - V_{BE1})/R_1$, is 1.6 mA (max.) for maximum V_{CC}.

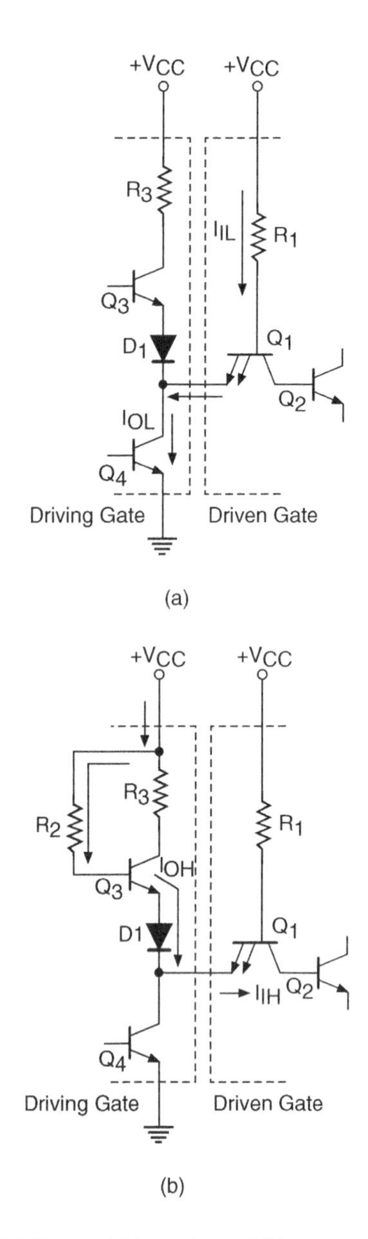

Figure 5.7 (a) Current sinking action and (b) current sourcing action.

5.3.1.2 Totem-Pole Output Stage

Transistors Q_3 and Q_4 constitute what is known as a totem-pole output arrangement. In such an arrangement, either Q_3 or Q_4 conducts at a time depending upon the logic status of the inputs. The totem-pole arrangement at the output has certain distinct advantages. The major advantage of using a totem-pole connection is that it offers low-output impedance in both the HIGH and LOW output states. In the HIGH state, Q_3 acts as an emitter follower and has an output impedance of about 70 Ω. In the LOW state, Q_4 is saturated and the output impedance is approximately 10 Ω. Because of the low output impedance, any stray capacitance at the output can be charged or discharged very rapidly through this low impedance, thus allowing quick transitions at the output from one state to the other. Another advantage is that, when the output is in the logic LOW state, transistor Q_4 would need to conduct a fairly large current if its collector were tied to V_{CC} through R_3 only. A nonconducting Q_3 overcomes this problem. A disadvantage of the totem-pole output configuration results from the switch-off action of Q_4 being slower than the switch-on action of Q_3. On account of this, there will be a small fraction of time, of the order of a few nanoseconds, when both the transistors are conducting, thus drawing heavy current from the supply.

5.3.1.3 Characteristic Features

To sum up, the characteristic parameters and features of the standard TTL family of devices include the following: $V_{IL}=0.8$ V; $V_{IH}=2$ V; $I_{IH}=40$ μA; $I_{IL}=1.6$ mA; $V_{OH}=2.4$ V; $V_{OL}=0.4$ V; $I_{OH}=400$ μA; $I_{OL}=16$ mA; $V_{CC}=4.75$–5.25 V (74-series) and 4.5–5.5 V (54-series); propagation delay (for a load resistance of 400 Ω, a load capacitance of 15 pF and an ambient temperature of 25 °C) = 22 ns (max.) for LOW-to-HIGH transition at the output and 15 ns (max.) for HIGH-to-LOW output transition; worst-case noise margin = 0.4 V; fan-out = 10; I_{CCH} (for all four gates) = 8 mA; I_{CCL} (for all four gates) = 22 mA; operating temperature range = 0–70 °C (74-series) and −55 to +125 °C (54-series); speed–power product = 100 pJ; maximum flip-flop toggle frequency = 35 MHz.

5.3.2 Other Logic Gates in Standard TTL

As outlined earlier, the NAND gate is the fundamental building block of the TTL family. In the following paragraphs we will look at the internal schematics of the other logic gates and find for ourselves their similarity to the schematic of the NAND gate discussed in detail in earlier paragraphs.

5.3.2.1 NOT Gate (or Inverter)

Figure 5.8 shows the internal schematic of a NOT gate (inverter) in the standard TTL family. The schematic shown is that of one of the six inverters in a hex inverter (type 7404/5404). The internal schematic is just the same as that of the NAND gate except that the input transistor is a normal single emitter NPN transistor instead of a multi-emitter one. The circuit is self-explanatory.

5.3.2.2 NOR Gate

Figure 5.9 shows the internal schematic of a NOR gate in the standard TTL family. The schematic shown is that of one of the four NOR gates in a quad two-input NOR gate (type 7402/5402). On the input side there are two separate transistors instead of the multi-emitter transistor of the NAND gate. The inputs are fed to the emitters of the two transistors, the collectors of which again feed the bases of the two transistors with their collector and emitter terminals tied together. The resistance values used are the same as those used in the case of the NAND gate. The output stage is also the same totem-pole output stage. The circuit is self-explanatory. The only input condition for which transistors Q_3 and Q_4

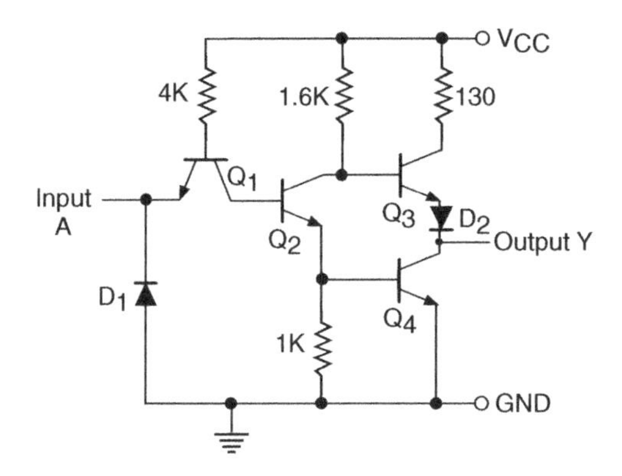

Figure 5.8 Inverter in the standard TTL.

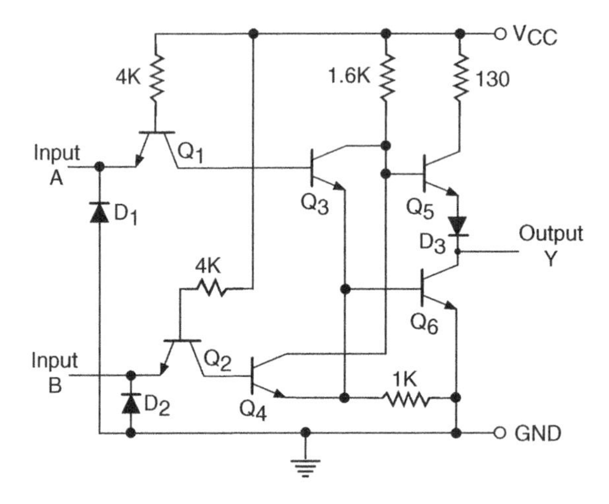

Figure 5.9 NOR gate in the standard TTL.

remain in cut-off, thus driving Q_6 to cut-off and Q_5 to conduction, is the one when both the inputs are in the logic LOW state. The output in such a case is logic HIGH. For all other input conditions, either Q_3 or Q_4 will conduct, driving Q_6 to saturation and Q_5 to cut-off, producing a logic LOW at the output.

5.3.2.3 AND Gate

Figure 5.10 shows the internal schematic of an AND gate in the standard TTL family. The schematic shown is that of one of the four AND gates in a quad two-input AND gate (type 7408/5408). In order to explain how this schematic arrangement behaves as an AND gate, we will begin by investigating the input condition that would lead to a HIGH output. A HIGH output implies Q_6 to be in cut-off and Q_5 to be in conduction. This can happen only when Q_4 is in cut-off. Transistor Q_4 can be in the cut-off

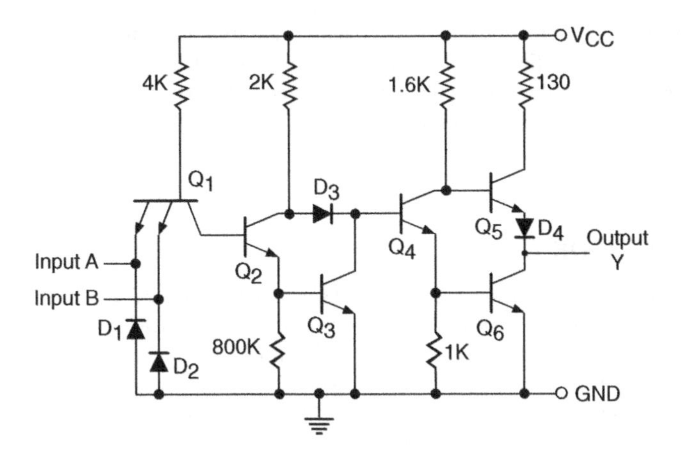

Figure 5.10 AND gate in standard TTL.

state only when both Q_2 and Q_3 are in conduction. This is possible only when both inputs are in the logic HIGH state. Let us now see what happens when either of the two inputs is driven to the LOW state. This drives Q_2 and Q_3 to the cut-off state, which forces Q_4 and subsequently Q_6 to saturation and Q_5 to cut-off.

5.3.2.4 OR Gate

Figure 5.11 shows the internal schematic of an OR gate in the standard TTL family. The schematic shown is that of one of the four OR gates in a quad two-input OR gate (type 7432/5432). We will begin by investigating the input condition that would lead to a LOW output. A LOW output demands a saturated Q_8 and a cut-off Q_7. This in turn requires Q_6 to be in saturation and Q_5, Q_4 and Q_3 to

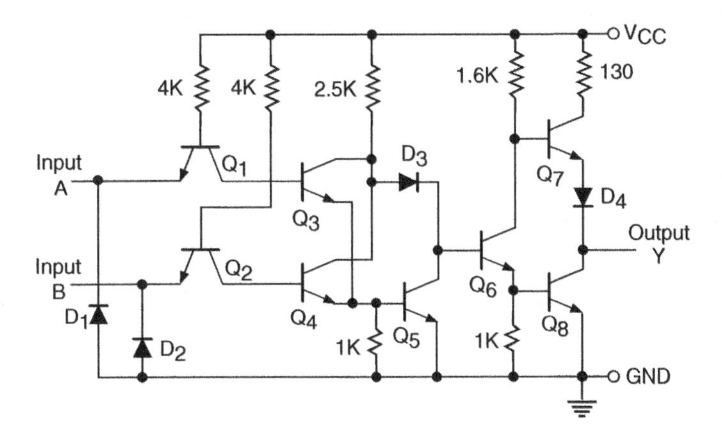

Figure 5.11 OR gate in the standard TTL.

be in cut-off. This is possible only when both Q_1 and Q_2 are in saturation. That is, both inputs are in the logic LOW state. This verifies one of the entries of the truth table of the OR gate. Let us now see what happens when either of the two inputs is driven to the HIGH state. This drives either of the two transistors Q_3 and Q_4 to saturation, which forces Q_5 to saturation and Q_6 to cut-off. This drives Q_7 to conduction and Q_8 to cut-off, producing a logic HIGH output.

5.3.2.5 EXCLUSIVE-OR Gate

Figure 5.12 shows the internal schematic of an EX-OR gate in the standard TTL family. The schematic shown is that of one of the four EX-OR gates in a quad two-input EX-OR gate (type 7486/5486). We will note the similarities between this circuit and that of an OR gate. The only new element is the interconnected pair of transistors Q_7 and Q_8. We will see that, when both the inputs are either HIGH or LOW, both Q_7 and Q_8 remain in cut-off. In the case of inputs being in the logic HIGH state, the base and emitter terminals of both these transistors remain near the ground potential. In the case of inputs being in the LOW state, the base and emitter terminals of both these transistors remain near V_{CC}. The result is conducting Q_9 and Q_{11} and nonconducting Q_{10}, which leads to a LOW output. When either of the inputs is HIGH, either Q_7 or Q_8 conducts. Transistor Q_7 conducts when input B is HIGH, and transistor Q_8 conducts when input A is HIGH. Conducting Q_7 or Q_8 turns off Q_9 and Q_{11} and turns on Q_{10}, producing a HIGH output. This explains how this circuit behaves as an EX-OR gate.

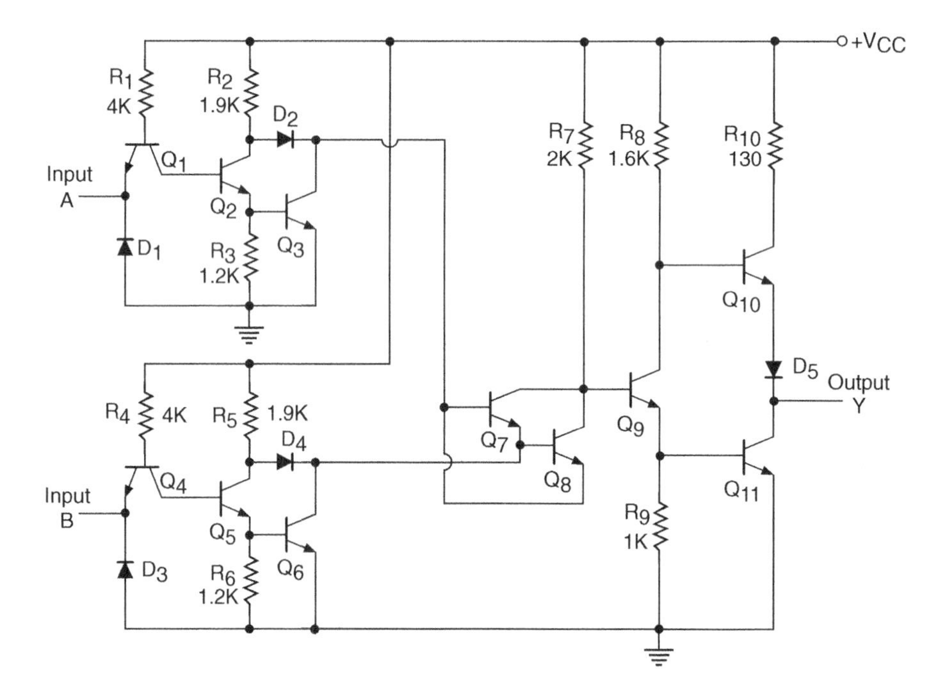

Figure 5.12 EX-OR gate in the standard TTL.

5.3.2.6 AND-OR-INVERT Gate

Figure 5.13 shows the internal schematic of a two-wide, two-input AND-OR-INVERT or AND-NOR gate. The schematic shown is that of one of the two gates in a dual two-wide, two-input AND-OR-INVERT gate (type 7450/5450). The two multi-emitter input transistors Q_1 and Q_2 provide ANDing of their respective inputs. Drive splitters comprising Q_3, Q_4, R_3 and R_4 provide the OR function. The output stage provides inversion. The number of emitters in each of the input transistors determines the number of literals in each of the minterms in the output sum-of-products Boolean expression. How wide the gate is going to be is decided by the number of input transistors, which also equals the number of drive splitter transistors.

5.3.2.7 Open Collector Gate

An open collector gate in TTL is one that is without a totem-pole output stage. The output stage in this case does not have the active pull-up transistor. An external pull-up resistor needs to be connected from the open collector terminal of the pull-down transistor to the V_{CC} terminal. The pull-up resistor is typically 10 kΩ. Figure 5.14 shows the internal schematic of a NAND gate with an open collector output. The schematic shown is that of one of the four gates of a quad two-input NAND (type 74/5401). The advantage of open collector outputs is that the outputs of different gates can be wired together, resulting in ANDing of their outputs. WIRE-AND operation was discussed in Chapter 4 on logic gates.

It may be mentioned here that the outputs of totem-pole TTL devices cannot be tied together. Although a common tied output may end up producing an ANDing of individual outputs, such a connection is impractical. This is illustrated in Fig. 5.15, where outputs of two totem-pole output TTL

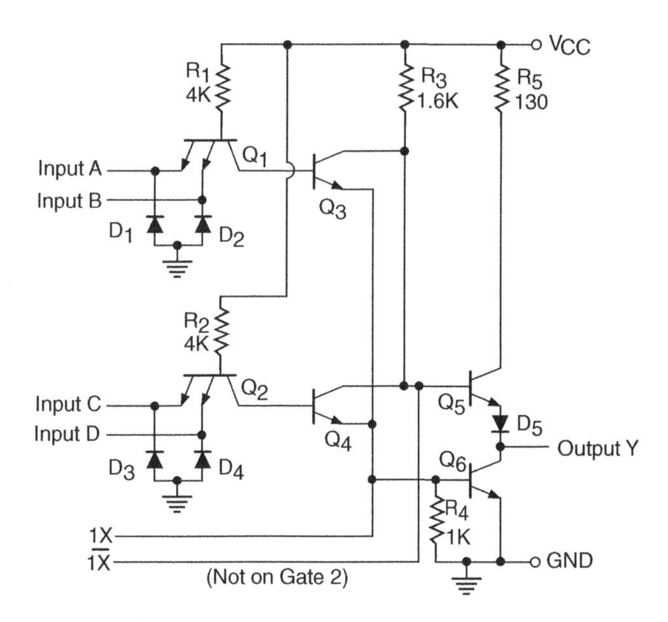

Figure 5.13 Two-input, two-wide AND-OR-INVERT gate.

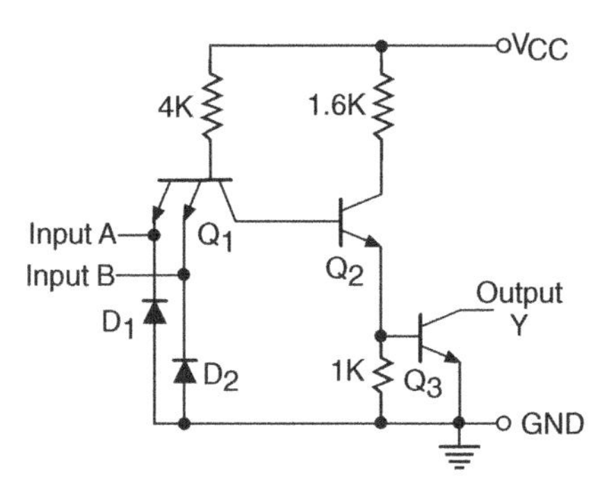

Figure 5.14 NAND gate with an open collector output.

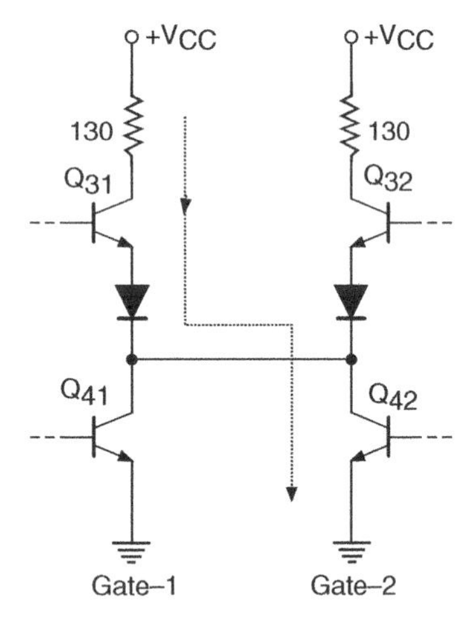

Figure 5.15 Totem-pole output gates tied at the output.

gates have been tied together. Let us assume that the output of one of the gates, say gate-2, is LOW, and the output of the other is HIGH. The result is that a relatively heavier current flows through Q_{31} and Q_{42}. This current, which is of the order of 50–60 mA, exceeds the I_{OL}(max.) rating of Q_{42}. This may eventually lead to both transistors getting damaged. Even if they survive, V_{OL}(max.) of Q_{42} is no longer guaranteed. In view of this, although totem-pole output TTL gates are not tied together, an accidental shorting of outputs is not ruled out. In such a case, both devices are likely to get damaged. In the case of open collector devices, deliberate or nondeliberate, shorting of outputs produces ANDing of outputs with no risk of either damage or compromised performance specifications.

5.3.2.8 Tristate Gate

Tristate gates were discussed in Chapter 4. A tristate gate has three output states, namely the logic LOW state, the logic HIGH state and the high-impedance state. An external enable input decides

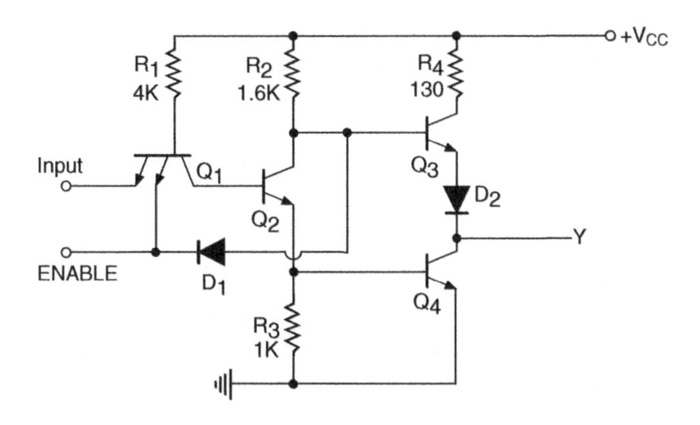

Figure 5.16 Tristate inverter in the TTL.

whether the logic gate works according to its truth table or is in the high-impedance state. Figure 5.16 shows the typical internal schematic of a tristate inverter with an active HIGH enable input. The circuit functions as follows. When the enable input is HIGH, it reverse-biases diode D_1 and also applies a logic HIGH on one of the emitters of the input transistor Q_1. The circuit behaves like an inverter. When the enable input is LOW, diode D_1 becomes forward biased. A LOW enable input forces Q_2 and Q_4 to cut-off. Also, a forward-biased D_1 forces Q_3 to cut-off. With both output transistors in cut-off, the output essentially is an open circuit and thus presents high output impedance.

5.3.3 Low-Power TTL

The low-power TTL is a low-power variant of the standard TTL where lower power dissipation is achieved at the expense of reduced speed of operation. Figure 5.17 shows the internal schematic of a

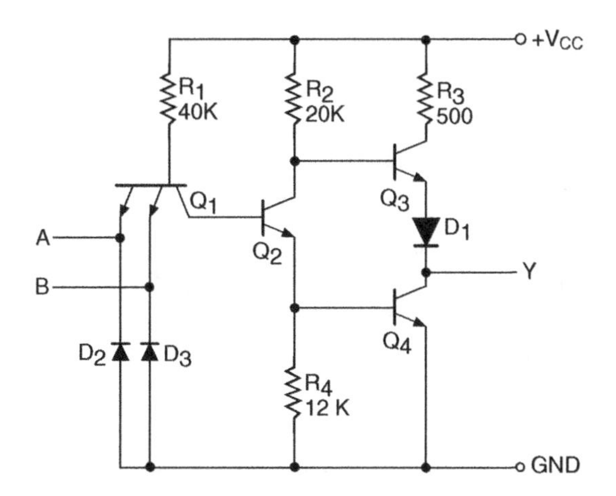

Figure 5.17 NAND gate in the low-power TTL.

low-power TTL NAND gate. The circuit shown is that of one of the four gates inside a quad two-input NAND (type 74L00 or 54L00). The circuit, as we can see, is the same as that of the standard TTL NAND gate except for an increased resistance value of the different resistors used in the circuit. Increased resistance values lead to lower power dissipation.

5.3.3.1 Characteristic Features

Characteristic features of this family are summarized as follows: $V_{IH} = 2$ V; $V_{IL} = 0.7$ V; $I_{IH} = 10$ μA; $I_{IL} = 0.18$ mA; $V_{OH} = 2.4$ V; $V_{OL} = 0.4$ V; $I_{OH} = 200$ μA; $I_{OL} = 3.6$ mA; $V_{CC} = 4.75$–5.25 V (74-series) and 4.5–5.5 V (54-series); propagation delay (for a load resistance of 4000 Ω, a load capacitance of 50 pF, $V_{CC} = 5$ V and an ambient temperature of 25 °C) = 60 ns (max.) for both LOW-to-HIGH and HIGH-to-LOW output transitions; worst-case noise margin = 0.3 V; fan-out = 20; I_{CCH} (for all four gates) = 0.8 mA; I_{CCL} (for all four gates) = 2.04 mA; operating temperature range = 0–70 °C (74-series) and −55 to +125 °C (54-series); speed–power product = 33 pJ; maximum flip-flop toggle frequency = 3 MHz.

5.3.4 High-Power TTL (74H/54H)

The high-power TTL is a high-power, high-speed variant of the standard TTL where improved speed (reduced propagation delay) is achieved at the expense of higher power dissipation. Figure 5.18 shows the internal schematic of a high-power TTL NAND gate. The circuit shown is that of one of the four gates inside a quad two-input NAND (type 74H00 or 54H00). The circuit, as we can see, is nearly the same as that of the standard TTL NAND gate except for the transistor Q_3–diode D_1 combination in the totem-pole output stage having been replaced by a Darlington arrangement comprising Q_3, Q_5 and R_5. The Darlington arrangement does the same job as diode D_1 in the conventional totem-pole arrangement. It ensures that Q_5 does not conduct at all when the output is LOW. The decreased resistance values of different resistors used in the circuit lead to higher power dissipation.

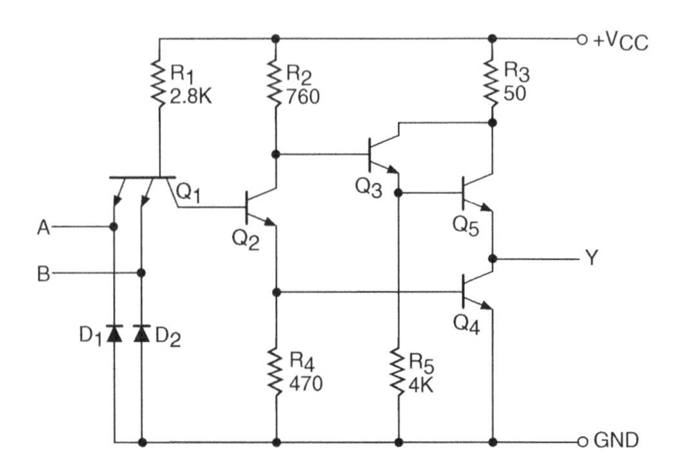

Figure 5.18 NAND gate in the high-power TTL.

5.3.4.1 Characteristic Features

Characteristic features of this family are summarized as follows: $V_{IH} = 2$ V; $V_{IL} = 0.8$ V; $I_{IH} = 50\,\mu$A; $I_{IL} = 2$ mA; $V_{OH} = 2.4$ V; $V_{OL} = 0.4$ V; $I_{OH} = 500\,\mu$A; $I_{OL} = 20$ mA; $V_{CC} = 4.75$–5.25 V (74-series) and 4.5–5.5 V (54-series); propagation delay (for a load resistance of 280 Ω, a load capacitance of 25 pF, $V_{CC} = 5$ V and an ambient temperature of 25 °C) = 10 ns (max.) for both LOW-to-HIGH and HIGH-to-LOW output transitions; worst-case noise margin = 0.4 V; fan-out = 10; I_{CCH} (for all four gates) = 16.8 mA; I_{CCL} (for all four gates) = 40 mA; operating temperature range = 0–70 °C (74-series) and −55 to +125 °C (54-series); speed–power product = 132 pJ; maximum flip-flop frequency = 50 MHz.

5.3.5 Schottky TTL (74S/54S)

The Schottky TTL offers a speed that is about twice that offered by the high-power TTL for the same power consumption. Figure 5.19 shows the internal schematic of a Schottky TTL NAND gate. The circuit shown is that of one of the four gates inside a quad two-input NAND (type 74S00 or 54S00). The circuit, as we can see, is nearly the same as that of the high-power TTL NAND gate. The transistors used in the circuit are all Schottky transistors with the exception of Q_5. A Schottky Q_5 would serve no purpose, with Q_4 being a Schottky transistor. A Schottky transistor is nothing but a conventional bipolar transistor with a Schottky diode connected between its base and collector terminals. The Schottky diode with its metal–semiconductor junction not only is faster but also offers a lower forward voltage drop of 0.4 V as against 0.7 V for a P–N junction diode for the same value of forward current. The presence of a Schottky diode does not allow the transistor to go to deep saturation. The moment the collector voltage of the transistor tends to go below about 0.3 V, the Schottky diode becomes forward biased and bypasses part of the base current through it. The collector voltage is thus not allowed to go to the saturation value of 0.1 V and gets clamped around 0.3 V. While the power consumption of a Schottky TTL gate is almost the same as that of a high-power TTL gate owing to nearly the same values of the resistors used in the circuit, the Schottky TTL offers a higher speed on account of the use of Schottky transistors.

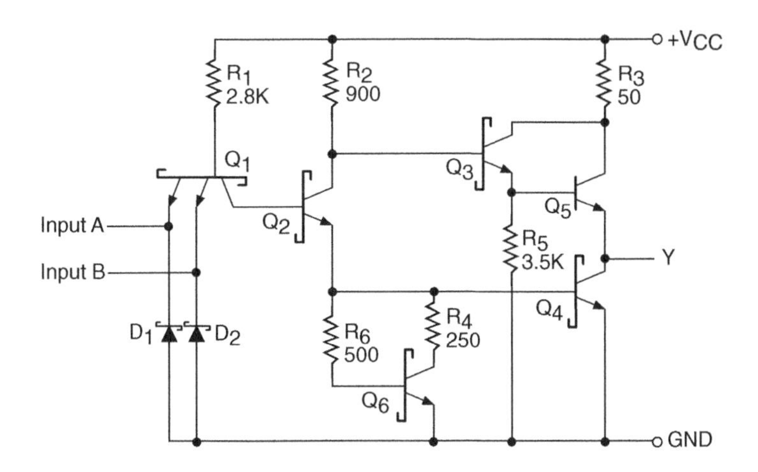

Figure 5.19 NAND gate in the Schottky TTL.

5.3.5.1 Characteristic Features

Characteristic features of this family are summarized as follows: $V_{IH} = 2$ V; $V_{IL} = 0.8$ V; $I_{IH} = 50\,\mu$A; $I_{IL} = 2$ mA; $V_{OH} = 2.7$ V; $V_{OL} = 0.5$ V; $I_{OH} = 1$ mA; $I_{OL} = 20$ mA; $V_{CC} = 4.75$–5.25 V (74-series) and 4.5–5.5 V (54-series); propagation delay (for a load resistance of 280 Ω, a load capacitance of 15 pF, $V_{CC} = 5$ V and an ambient temperature of 25 °C) = 5 ns (max.) for LOW-to-HIGH and 4.5 ns (max.) for HIGH-to-LOW output transitions; worst-case noise margin = 0.3 V; fan-out = 10; I_{CCH} (for all four gates) = 16 mA; I_{CCL} (for all four gates) = 36 mA; operating temperature range = 0–70 °C (74-series) and −55 to +125 °C (54-series); speed–power product = 57 pJ; maximum flip-flop toggle frequency = 125 MHz.

5.3.6 *Low-Power Schottky TTL (74LS/54LS)*

The low-power Schottky TTL is a low power consumption variant of the Schottky TTL. Figure 5.20 shows the internal schematic of a low-power Schottky TTL NAND gate. The circuit shown is that of one of the four gates inside a quad two-input NAND (type 74LS00 or 54LS00). We can notice the significantly increased value of resistors R_1 and R_2 used to achieve lower power consumption. Lower power consumption, of course, occurs at the expense of reduced speed or increased propagation delay. Resistors R_3 and R_5, which primarily affect speed, have not been increased in the same proportion with respect to the corresponding values used in the Schottky TTL as resistors R_1 and R_2. That is why, although the low-power Schottky TTL draws an average maximum supply current of 3 mA (for all four gates) as against 26 mA for the Schottky TTL, the propagation delay is 15 ns in LS-TTL as against 5 ns for S-TTL. Diodes D_3 and D_4 reduce the HIGH-to-LOW propagation delay. While D_3 speeds up the turn-off of Q_4, D_4 sinks current from the load. Another noticeable difference in the internal schematics of the low-power Schottky TTL NAND and Schottky TTL NAND is the replacement of the

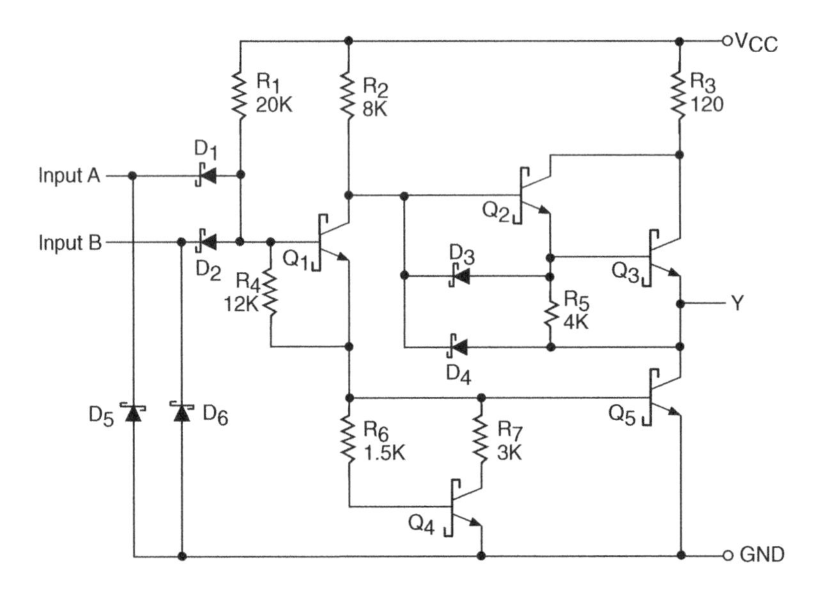

Figure 5.20 NAND gate in the low-power Schottky TTL.

multi-emitter input transistor of the Schottky TTL by diodes D_1 and D_2 and resistor R_1. The junction diodes basically replace the two emitter-base junctions of the multi-emitter input transistor Q_1 of the Schottky TTL NAND (Fig. 5.19). The reason for doing so is that Schottky diodes can be made smaller than the transistor and therefore will have lower parasitic capacitances. Also, since Q_1 of LS-TTL (Fig. 5.20) cannot saturate, it is not necessary to remove its base charge with a bipolar junction transistor.

5.3.6.1 Characteristic Features

Characteristic features of this family are summarized as follows: $V_{IH} = 2\,V$; $V_{IL} = 0.8\,V$; $I_{IH} = 20\,\mu A$; $I_{IL} = 0.4\,mA$; $V_{OH} = 2.7\,V$; $V_{OL} = 0.5\,V$; $I_{OH} = 0.4\,mA$; $I_{OL} = 8\,mA$; $V_{CC} = 4.75–5.25\,V$ (74-series) and 4.5–5.5 V (54-series); propagation delay (for a load resistance of 280 Ω, a load capacitance of 15 pF, $V_{CC} = 5\,V$ and an ambient temperature of 25 °C) = 15 ns (max.) for both LOW-to-HIGH and HIGH-to-LOW output transitions; worst-case noise margin = 0.3 V; fan-out = 20; I_{CCH} (for all four gates) = 1.6 mA; I_{CCL} (for all four gates) = 4.4 mA; operating temperature range = 0–70 °C (74-series) and −55 to +125 °C (54-series); speed–power product = 18 pJ; maximum flip-flop toggle frequency = 45 MHz.

5.3.7 Advanced Low-Power Schottky TTL (74ALS/54ALS)

The basic ideas behind the development of the advanced low-power Schottky TTL (ALS-TTL) and advanced Schottky TTL (AS-TTL) discussed in Section 5.3.8 were further to improve both speed and power consumption performance of the low-power Schottky TTL and Schottky TTL families respectively. In the TTL subfamilies discussed so far, we have seen that different subfamilies achieved improved speed at the expense of increased power consumption, or vice versa. For example, the low-power TTL offered lower power consumption over standard TTL at the cost of reduced speed. The high-power TTL, on the other hand, offered improved speed over the standard TTL at the expense of increased power consumption. ALS-TTL and AS-TTL incorporate certain new circuit design features and fabrication technologies to achieve improvement of both parameters. Both ALS-TTL and AS-TTL offer an improvement in speed–power product respectively over LS-TTL and S-TTL by a factor of 4. Salient features of ALS-TTL and AS-TTL include the following:

1. All saturating transistors are clamped by using Schottky diodes. This virtually eliminates the storage of excessive base charge, thus significantly reducing the turn-off time of the transistors. Elimination of transistor storage time also provides stable switching times over the entire operational temperature range.
2. Inputs and outputs are clamped by Schottky diodes to limit the negative-going excursions.
3. Both ALS-TTL and AS-TTL use ion implantation rather than a diffusion process, which allows the use of small geometries leading to smaller parasitic capacitances and hence reduced switching times.
4. Both ALS-TTL and AS-TTL use oxide isolation rather than junction isolation between transistors. This leads to reduced epitaxial layer–substrate capacitance, which further reduces the switching times.
5. Both ALS-TTL and AS-TTL offer improved input threshold voltage and reduced low-level input current.
6. Both ALS-TTL and AS-TTL feature active turn-off of the LOW-level output transistor, producing a better HIGH-level output voltage and thus a higher HIGH-level noise immunity.

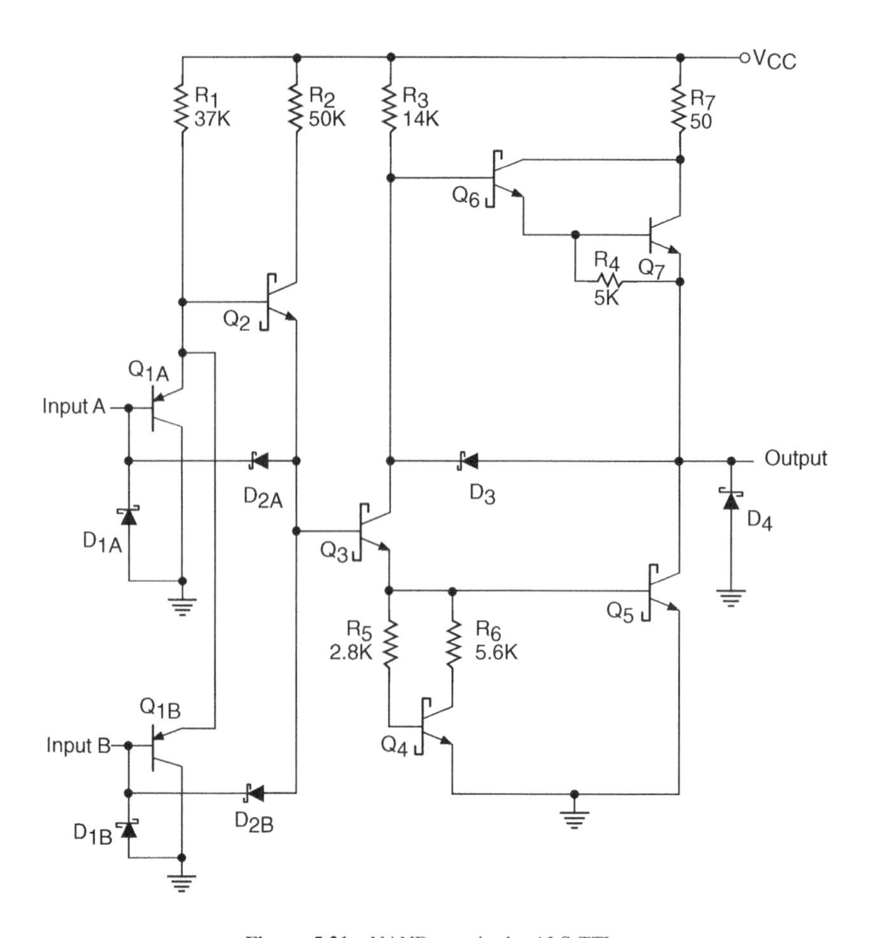

Figure 5.21 NAND gate in the ALS-TTL.

Figure 5.21 shows the internal schematic of an advanced low-power Schottky TTL NAND gate. The circuit shown is that of one of the four gates inside a quad two-input NAND (type 74ALS00 or 54ALS00) The multi-emitter input transistor is replaced by two PNP transistors Q_{1A} and Q_{1B}. Diodes D_{1A} and D_{1B} provide input clamping to negative excursions. Buffering offered by Q_{1A} or Q_{1B} and Q_2 reduces the LOW-level input current by a factor of (1 + h_{FE} of Q_{1A}). HIGH-level output voltage is determined primarily by V_{CC}, transistors Q_6 and Q_7 and resistors R_4 and R_7 and is typically $(V_{CC} - 2)$ V. LOW-level output voltage is determined by the turn-on characteristics of Q_5. Transistor Q_5 gets sufficient base drive through R_3 and a conducting Q_3 whose base terminal in turn is driven by a conducting Q_2 whenever either or both inputs are HIGH. Transistor Q_4 provides active turn-off for Q_5.

5.3.7.1 Characteristic Features

Characteristic features of this family are summarized as follows: $V_{IH} = 2$ V; $V_{IL} = 0.8$ V; $I_{IH} = 20\ \mu$A; $I_{IL} = 0.1$ mA; $V_{OH} = (V_{CC} - 2)$ V; $V_{OL} = 0.5$ V; $I_{OH} = 0.4$ mA; $I_{OL} = 8$ mA (74ALS) and 4 mA (54ALS);

V_{CC} = 4.5–5.5 V; propagation delay (for a load resistance of 500 Ω, a load capacitance of 50 pF, V_{CC} = 4.5–5.5 V and an ambient temperature of minimum to maximum) = 11 ns/16 ns (max.) for LOW-to-HIGH and 8 ns/13 ns for HIGH-to-LOW output transitions (74ALS/54ALS); worst-case noise margin = 0.3 V; fan-out = 20; I_{CCH} (for all four gates) = 0.85 mA; I_{CCL} (for all four gates) = 3 mA; operating temperature range = 0–70 °C (74-series) and −55 to +125 °C (54-series); speed–power product = 4.8 pJ; maximum flip-flop toggle frequency = 70 MHz.

5.3.8 Advanced Schottky TTL (74AS/54AS)

Figure 5.22 shows the internal schematic of an advanced Schottky TTL NAND gate. The circuit shown is that of one of the four gates inside a quad two-input NAND (type 74AS00 or 54AS00). Salient

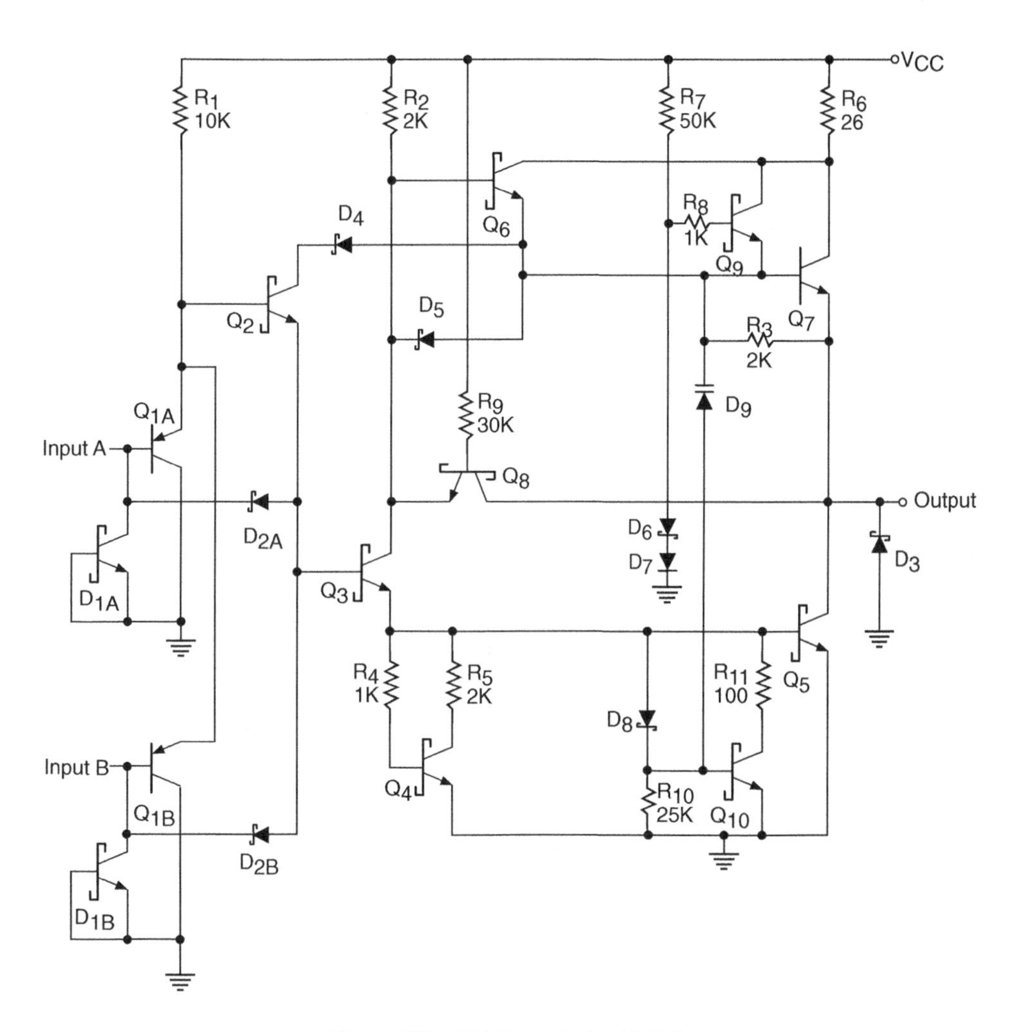

Figure 5.22 NAND gate in the AS-TTL.

features of ALS-TTL and AS-TTL have been discussed at length in the preceding paragraphs. As is obvious from the internal circuit schematic of the AS-TTL NAND gate, it has some additional circuits not found in ALS-TTL devices. These are added to enhance the throughput of AS-TTL family devices. Transistor Q_{10} provides a discharge path for the base-collector capacitance of Q_5. In the absence of Q_{10}, a rising voltage across the output forces current into the base of Q_5 through its base-collector capacitance, thus causing it to turn on. Transistor Q_{10} turns on through D_9, thus keeping transistor Q_5 in the cut-off state.

5.3.8.1 Characteristic Features

Characteristic features of this family are summarized as follows: $V_{IH} = 2$ V; $V_{IL} = 0.8$ V; $I_{IH} = 20\,\mu$A; $I_{IL} = 0.5$ mA; $V_{OH} = (V_{CC} - 2)$ V; $V_{OL} = 0.5$ V; $I_{OH} = 2$ mA; $I_{OL} = 20$ mA; $V_{CC} = 4.5$–5.5 V; propagation delay (for a load resistance of 50 Ω, a load capacitance of 50 pF, $V_{CC} = 4.5$–5.5 V and an ambient temperature of minimum to maximum) = 4.5 ns/5 ns (max.) for LOW-to-HIGH and 4 ns/5 ns (max.) for HIGH-to-LOW output transitions (74AS/54AS); worst-case noise margin = 0.3 V; fan-out = 40; I_{CCH} (for all four gates) = 3.2 mA; I_{CCL} (for all four gates) = 17.4 mA; operating temperature range = 0–70 °C (74-series) and −55 to +125 °C (54-series); speed–power product = 13.6 pJ; maximum flip-flop toggle frequency = 200 MHz.

5.3.9 Fairchild Advanced Schottky TTL (74F/54F)

The Fairchild Advanced Schottky TTL family, commonly known as FAST logic, is similar to the AS-TTL family. Figure 5.23 shows the internal schematic of a Fairchild Advanced Schottky TTL

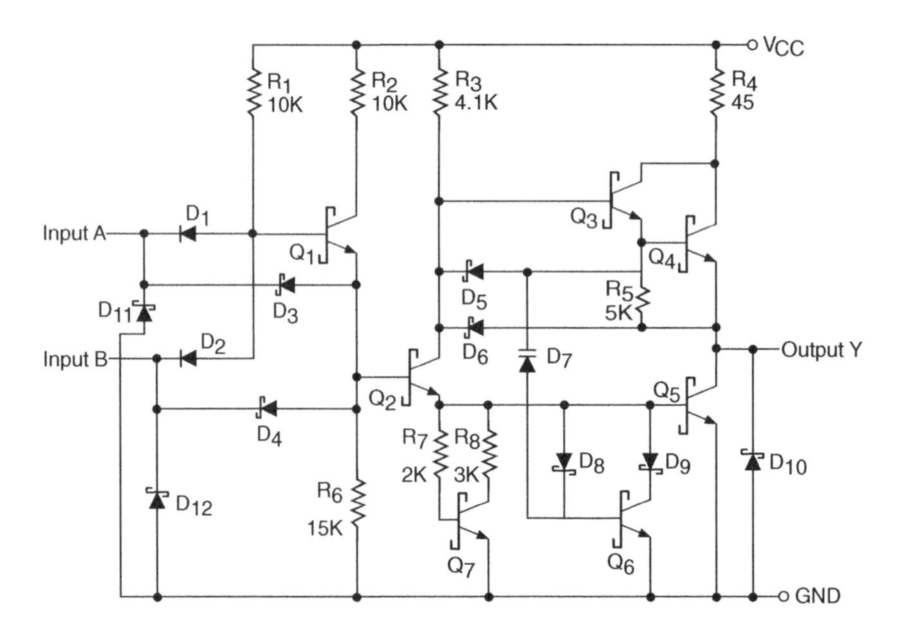

Figure 5.23 NAND gate in the FAST TTL.

NAND gate. The circuit shown is that of one of the four gates inside a quad two-input NAND (type 74F00 or 54F00). The DTL kind of input with emitter follower configuration of Q_1 provides a good base drive to Q_2. The 'Miller killer' configuration comprising varactor diode D_7, transistor Q_6 and associated components speeds up LOW-to-HIGH transition. During LOW-to-HIGH transition, voltage at the emitter terminal of Q_3 begins to rise while Q_5 is still conducting. Varactor diode D_7 conducts, thus supplying base current to Q_6. A conducting Q_6 provides a discharge path for the charge stored in the base-collector capacitance of Q_5, thus expediting its turn-off.

5.3.9.1 Characteristic Features

Characteristic features of this family are summarized as follows: $V_{IH} = 2$ V; $V_{IL} = 0.8$ V; $I_{IH} = 20$ μA; $I_{IL} = 0.6$ mA; $V_{OH} = 2.7$ V; $V_{OL} = 0.5$ V; $I_{OH} = 1$ mA; $I_{OL} = 20$ mA; $V_{CC} = 4.75$–5.25 V (74F) and 4.5–5.5 V (54F); propagation delay (a load resistance of 500 Ω, a load capacitance of 50 pF and full operating voltage and temperature ranges) = 5.3 ns/7 ns (max.) for LOW-to-HIGH and 6 ns/6.5 ns (max.) for HIGH-to-LOW output transitions (74AS/54AS); worst-case noise margin = 0.3 V; fan-out = 40; I_{CCH} (for all four gates) = 2.8 mA; I_{CCL} (for all four gates) = 10.2 mA; operating temperature range = 0–70 °C (74F-series) and −55 to +125 °C (54F-series); speed–power product = 10 pJ; maximum flip-flop toggle frequency = 125 MHz.

5.3.10 Floating and Unused Inputs

The floating input of TTL family devices behaves as if logic HIGH has been applied to the input. Such behaviour is explained from the input circuit of a TTL device. When the input is HIGH, the input emitter-base junction is reverse biased and the current that flows into the input is the reverse-biased diode leakage current. The input diode will be reverse biased even when the input terminal is left unconnected or floating, which implies that a floating input behaves as if there were logic HIGH applied to it.

As an initial thought, we may tend to believe that it should not make any difference if we leave the unused inputs of NAND and AND gates as floating, as logic HIGH like behaviour of the floating input makes no difference to the logical behaviour of the gate, as shown in Figs 5.24(a) and (b). In spite of this, it is strongly recommended that the unused inputs of AND and NAND gates be connected to a logic HIGH input [Fig. 5.24(c)] because floating input behaves as an antenna and may pick up stray noise and interference signals, thus causing the gate to function improperly. 1 kΩ resistance is connected to protect the input from any current spikes caused by any spikes on the power supply line. More than one unused input (up to 50) can share the same 1 kΩ resistance, if needed.

In the case of OR and NOR gates, unused inputs are connected to ground (logic LOW), as shown in Fig. 5.25(c), for obvious reasons. A floating input or an input tied to logic HIGH in this case produces a permanent logic HIGH (for an OR gate) and LOW (for a NOR gate) at the output as shown in Figs 5.25(a) and (b) respectively. An alternative solution is shown in Fig. 5.25(d), where the unused input has been tied to one of the used inputs. This solution works well for all gates, but one has to be conscious of the fact that the fan-out capability of the output driving the tied inputs is not exceeded.

If we recall the internal circuit schematics of AND and NAND gates, we will appreciate that, when more than one input is tied together, the input loading, that is, the current drawn by the tied inputs from the driving gate output, in the HIGH state is n times the loading of one input (Fig. 5.26); n is the number of inputs tied together. When the output is LOW, the input loading is the same as that of a single input. The reason for this is that, in the LOW input state, the current flowing out of the gate is determined by the resistance R_1, as shown in Fig. 5.27. However, the same is not true in the case of

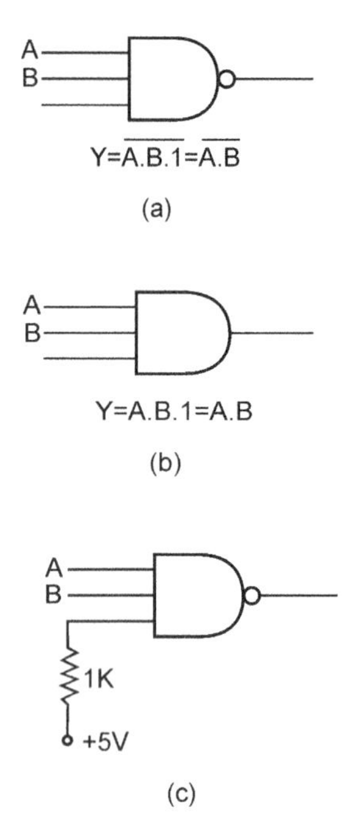

Figure 5.24 Handling unused inputs of AND and NAND gates.

OR and NOR gates, which do not use a multi-emitter input transistor and use separate input transistors instead, as shown in Fig. 5.28. In this case, the input loading is n times the loading of a single input for both HIGH and LOW states.

5.3.11 Current Transients and Power Supply Decoupling

TTL family devices are prone to occurrence of narrow-width current spikes on the power supply line. Current transients are produced when the totem-pole output stage of the device undergoes a transition from a logic LOW to a logic HIGH state. The problem becomes severe when in a digital circuit a large number of gates are likely to switch states at the same time. These current spikes produce voltage spikes due to any stray inductance present on the line. On account of the large rate of change in current in the current spike, even a small value of stray inductance produces voltage spikes large enough adversely to affect the circuit performance.

Figure 5.29 illustrates the phenomenon. When the output changes from LOW to HIGH, there is a small fraction of time when both the transistors are conducting because the pull-up transistor Q_3 has switched on and the pull-down transistor Q_4 has not yet come out of saturation. During this small fraction of time, there is an increase in current drawn from the supply; I_{CCL} experiences a positive spike before it settles down to a usually lower I_{CCH}. The presence of any stray capacitance C across the output owing to any stray wiring capacitance or capacitance loading of the circuit being fed also adds to the problem. The problem of voltage spikes on the power supply line is usually overcome by connecting small-value, low-inductance, high-frequency capacitors between V_{CC} terminal and ground. It is standard practice to use a 0.01 or 0.1 μF ceramic capacitor from V_{CC} to ground. This

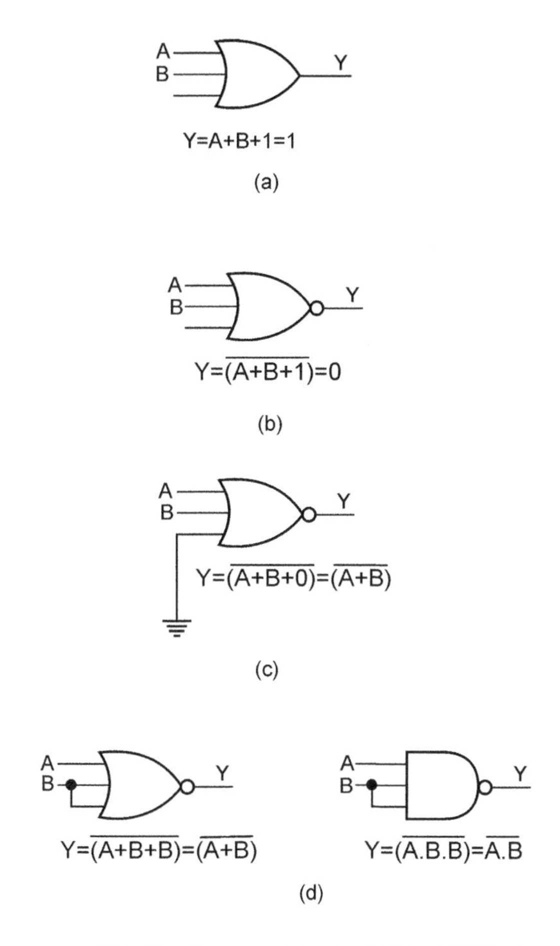

Figure 5.25 Handling unused inputs of OR and NOR gates.

capacitor is also known by the name of power supply decoupling capacitor, and it is recommended to use a separate capacitor for each IC. A decoupling capacitor is connected as close to the V_{CC} terminal as possible, and its leads are kept to a bare minimum to minimize lead inductance. In addition, a single relatively large-value capacitor in the range of 1–22 μF is also connected between V_{CC} and ground on each circuit card to take care of any low-frequency voltage fluctuations in the power supply line.

Example 5.5

Refer to Fig. 5.30. Determine the current being sourced by gate 1 when its output is HIGH and sunk by it when its output is LOW. All gates are from the standard TTL family, given that $I_{IH} = 40\,\mu A$ *and* $I_{IL} = 1.6\,mA$.

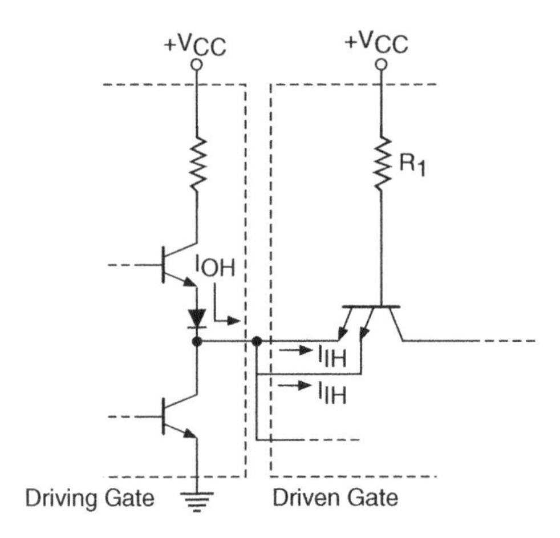

Figure 5.26 Input loading in the case of HIGH tied inputs of NAND and AND gates.

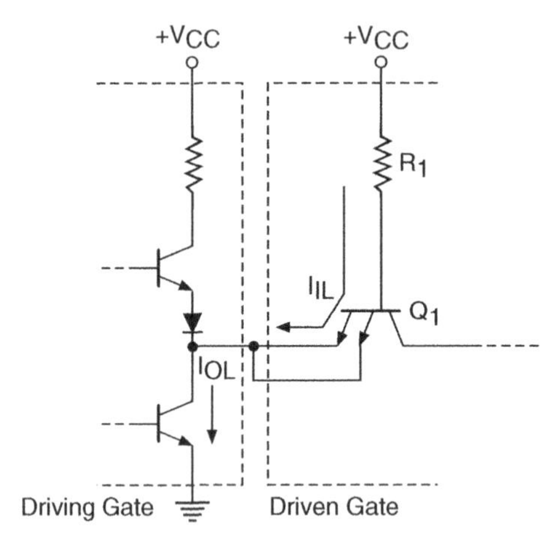

Figure 5.27 Input loading in the case of LOW tied inputs of NAND and AND gates.

Solution

- When the output is HIGH, the inputs of all gates draw current individually.
- Therefore, the input loading factor = equivalent of seven gate inputs = $7 \times 40\,\mu A = 280\,\mu A$.
- The current being sourced by the gate 1 output = $280\,\mu A$.
- When the output is LOW, shorted inputs of AND and NAND gates offer a load equal to that of a single input owing to a multi-emitter transistor at the input of the gate. The inputs of OR and NOR gates draw current individually on account of the use of separate transistors at the input of the gate.
- Therefore, the input loading factor = equivalent of five gate inputs = $5 \times 1.6 = 8$ mA.
- The current being sunk by the gate 1 output = 8 mA.

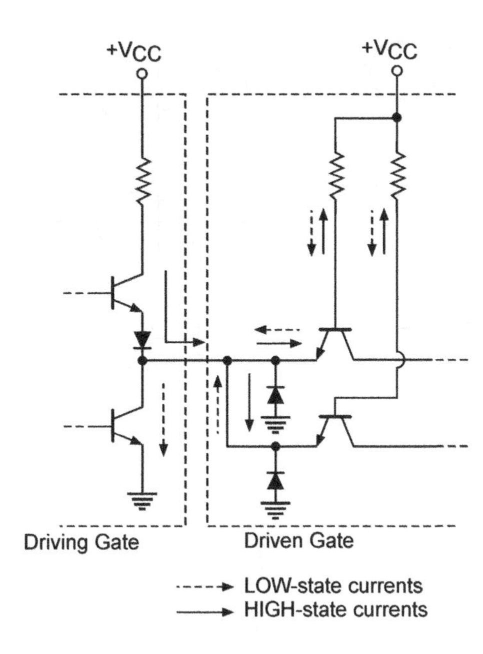

Figure 5.28 Input loading in the case of tied inputs of NOR and OR gates.

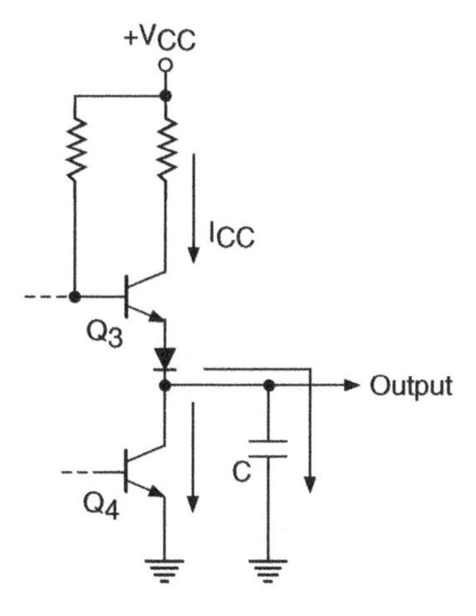

Figure 5.29 Current transients and power supply decoupling.

Example 5.6

Refer to the logic diagram of Fig. 5.31. Gate 1 and gate 4 belong to the standard TTL family, while gate 2 and gate 3 belong to the Schottky TTL family and the low-power Schottky TTL family respectively. Determine whether the fan-out capability of gate 1 is being exceeded. Relevant data for the three logic families are given in Table 5.1.

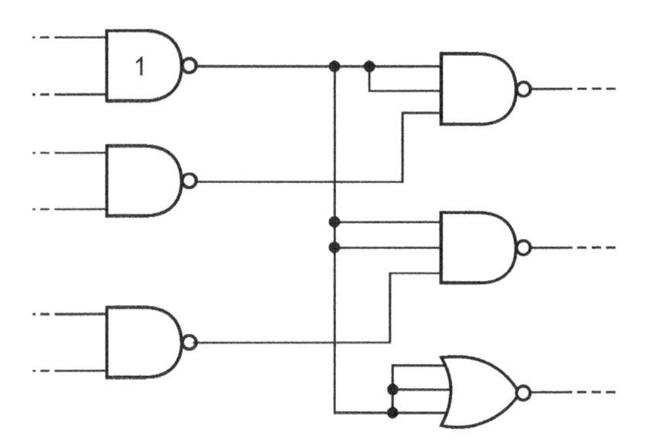

Figure 5.30 Example 5.5.

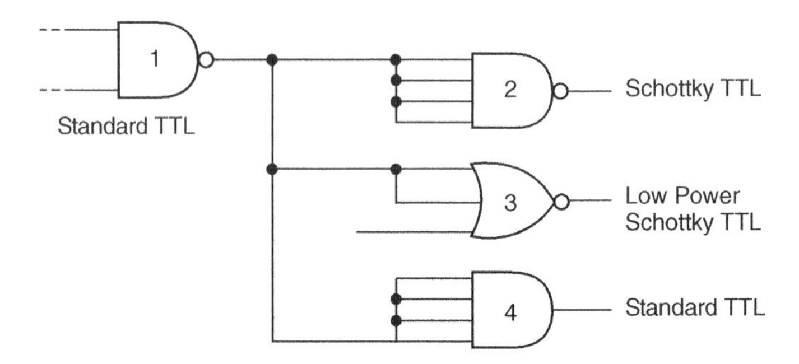

Figure 5.31 Example 5.6.

Table 5.1 Example 5.6

Logic family	$I_{IH}(\mu A)$	$I_{OH}(mA)$	$I_{IL}(mA)$	$I_{OL}(mA)$
Standard TTL	40	0.4	1.6	16
LS-TTL	20	0.4	0.4	8.0
S-TTL	50	1.0	2.0	20

Solution

- In the HIGH-state:

  – the gate 1 output sourcing capability = 400 μA;
  – the gate 2 input requirement = 50 × 4 = 200 μA;
  – the gate 3 input requirement = 20 × 2 = 40 μA;
  – the gate 4 input requirement = 40 × 4 = 160 μA;
  – the total input current requirement = 400 μA;
  – therefore, the fan-out is not exceeded in the HIGH state.

- In the LOW-state,

  - the gate 1 output sinking capability = 16 mA;
  - the gate 2 input sinking requirement = 2 mA;
  - the gate 3 input sinking requirement = 0.4 × 2 = 0.8 mA;
  - the gate 4 input sinking requirement = 1.6 mA;
  - the total input current requirement = 4.4 mA;
  - since the output of gate 1 has a current sinking capability of 16 mA, the fan-out capability is not exceeded in the LOW state either.

5.4 Emitter Coupled Logic (ECL)

The ECL family is the fastest logic family in the group of bipolar logic families. The characteristic features that give this logic family its high speed or short propagation delay are outlined as follows:

1. It is a nonsaturating logic. That is, the transistors in this logic are always operated in the active region of their output characteristics. They are never driven to either cut-off or saturation, which means that logic LOW and HIGH states correspond to different states of conduction of various bipolar transistors.
2. The logic swing, that is, the difference in the voltage levels corresponding to logic LOW and HIGH states, is kept small (typically 0.85 V), with the result that the output capacitance needs to be charged and discharged by a relatively much smaller voltage differential.
3. The circuit currents are relatively high and the output impedance is low, with the result that the output capacitance can be charged and discharged quickly.

5.4.1 Different Subfamilies

Different subfamilies of ECL logic include MECL-I, MECL-II, MECL-III, MECL 10K, MECL 10H and MECL 10E (ECLinPS™and ECLinPS Lite™).

5.4.1.1 MECL-I, MECL-II and MECL-III Series

MECL-I was the first monolithic emitter coupled logic family introduced by ON Semiconductor (formerly a division of Motorola SPS) in 1962. It was subsequently followed up by MECL-II in 1966. Both these logic families have become obsolete and have been replaced by MECL-III (also called the MC1600 series) introduced in 1968. Although, chronologically, MECL-III was introduced before the MECL-10K and MECL-10H families, it features higher speed than both of its successors. With a propagation delay of the order of 1 ns and a flip-flop toggle frequency of 500 MHz, MECL-III is used in high-performance, high-speed systems.

The basic characteristic parameters of MECL-III are as follows: gate propagation delay = 1 ns; output edge speed (indicative of the rise and fall time of output transition) = 1 ns; flip-flop toggle frequency = 500 MHz; power dissipation per gate = 50 mW; speed–power product = 60 pJ; input voltage = 0–V_{EE} (V_{EE} is the negative supply voltage); negative power supply range (for V_{CC} = 0) = −5.1V to −5.3 V; continuous output source current (max.) = 40 mA; surge output source current (max.) = 80 mA; operating temperature range = −30 °C to +85 °C.

5.4.1.2 MECL-10K Series

The MECL-10K family was introduced in 1971 to meet the requirements of more general-purpose high-speed applications. Another important feature of MECL-10K family devices is that they are compatible with MECL-III devices, which facilitates the use of devices of the two families in the same system. The increased propagation delay of 2 ns in the case of MECL-10K comes with the advantage of reduced power dissipation, which is less than half the power dissipation in MECL-III family devices.

The basic characteristic parameters of MECL-10K are as follows: gate propagation delay = 2 ns (10100-series) and 1.5 ns (10200-series); output edge speed = 3.5 ns (10100-series) and 2.5 ns (10200-series); flip-flop toggle frequency = 125 MHz (min.) in the 10100-series and 200 MHz (min.) in the 10200-series; power dissipation per gate = 25 mW; speed–power product = 50 pJ (10100-series) and 37 pJ (10200-series); input voltage = 0–V_{EE} (V_{EE} is the negative supply voltage); negative power supply range (for $V_{CC} = 0$) = −4.68 to −5.72 V; continuous output source current (max.) = 50 mA; surge output source current (max.) = 100 mA; operating temperature range = −30 °C to +85 °C.

5.4.1.3 MECL-10H Series

The MECL-10H family, introduced in 1981, combines the high speed advantage of MECL-III with the lower power dissipation of MECL-10K. That is, it offers the speed of MECL-III with the power economy of MECL-10K. Backed by a propagation delay of 1 ns and a power dissipation of 25 mW per gate, MECL-10H offers one of the best speed–power product specifications in all available ECL subfamilies. Another important aspect of this family is that many of the MECL-10H devices are pin-out/functional replacements of MECL-10K series devices, which allows the users or the designers to enhance the performance of existing systems by increasing speed in critical timing areas.

The basic characteristic parameters of MECL-10H are as follows: gate propagation delay = 1 ns; output edge speed = 1 ns; flip-flop toggle frequency = 250 MHz (min.); power dissipation per gate = 25 mW; speed–power product = 25 pJ; input voltage = 0–V_{EE} (V_{EE} is the negative supply voltage); negative power supply range (for $V_{CC} = 0$) = −4.94 to −5.46 V; continuous output source current (max.) = 50 mA; surge output source current (max.) = 100 mA; operating temperature range = 0 °C to + 75 °C.

5.4.1.4 MECL-10E Series (ECLinPS™ and ECLinPSLite™)

The ECLinPS™ family, introduced in 1987, has a propagation delay of the order of 0.5 ns. ECLinPSLite™ is a recent addition to the ECL family. It offers a propagation delay of the order of 0.2 ns. The ECLPro™ family of devices is a rapidly growing line of high-performance ECL logic, offering a significant speed upgrade compared with the ECLinPSLite™ devices.

5.4.2 Logic Gate Implementation in ECL

OR/NOR is the fundamental logic gate of the ECL family. Figure 5.32 shows a typical internal schematic of an OR/NOR gate in the 10K-series MECL family. The circuit in essence comprises a differential amplifier input circuit with one side of the differential pair having multiple transistors depending upon the number of inputs to the gate, a voltage- and temperature-compensated bias network and emitter follower outputs. The internal schematic of the 10H-series gate is similar, except that the bias network is replaced with a voltage regulator circuit and the source resistor R_{EE} of the differential amplifier is replaced with a constant current source. Typical values of power supply voltages are

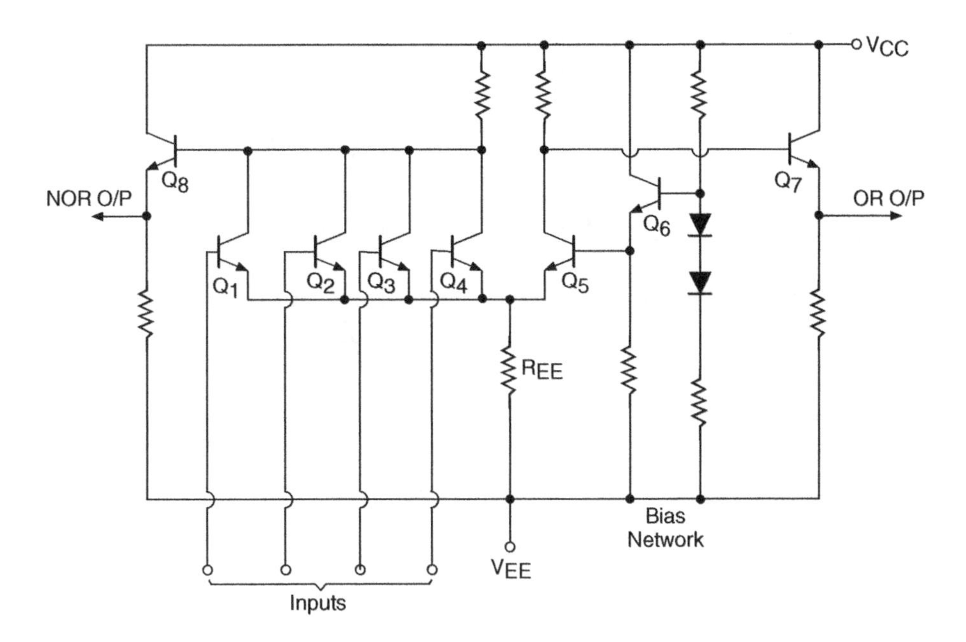

Figure 5.32 OR/NOR in ECL.

$V_{CC} = 0$ and V_{EE}=–5.2 V. The nominal logic levels are logic LOW = logic '0' = −1.75 V and logic HIGH = logic '1' = −0.9 V, assuming a positive logic system. The circuit functions as follows.

The bias network configured around transistor Q_6 produces a voltage of typically −1.29 V at its emitter terminal. This leads to a voltage of −2.09 V at the junction of all emitter terminals of various transistors in the differential amplifier, assuming 0.8 V to be the required forward-biased P–N junction voltage. Now, let us assume that all inputs are in a logic '0' state, that is, the voltage at the base terminals of various input transistors is −1.75 V. This means that the transistors Q_1, Q_2, Q_3 and Q_4 will remain in cut-off as their base-emitter junctions are not forward biased by the required voltage. This leads us to say that transistor Q_7 is conducting, producing a logic '0' output, and transistor Q_8 is in cut-off, producing a logic '1' output.

In the next step, let us see what happens if any one or all of the inputs are driven to logic '1' status, that is, a nominal voltage of −0.9 V is applied to the inputs. The base-emitter voltage differential of transistors Q_1–Q_4 exceeds the required forward-biasing threshold, with the result that these transistors start conducting. This leads to a rise in voltage at the common-emitter terminal, which now becomes approximately −1.7 V as the common-emitter terminal is now 0.8 V more negative than the base-terminal voltage. With rise in the common-emitter terminal voltage, the base-emitter differential voltage of Q_5 becomes 0.31 V, driving Q_5 to cut-off. The Q_7 and Q_8 emitter terminals respectively go to logic '1' and logic '0'.

This explains how this basic schematic functions as an OR/NOR gate. We will note that the differential action of the switching transistors (where one section is ON while the other is OFF) leads to simultaneous availability of complementary signals at the output. Figure 5.33 shows the circuit symbol and switching characteristics of this basic ECL gate. It may be mentioned here that positive ECL (called PECL) devices operating at +5 V and ground are also available. When used in PECL mode, ECL devices must have their input/output DC parameters adjusted for proper operation. PECL DC parameters can be computed by adding ECL levels to the new V_{CC}.

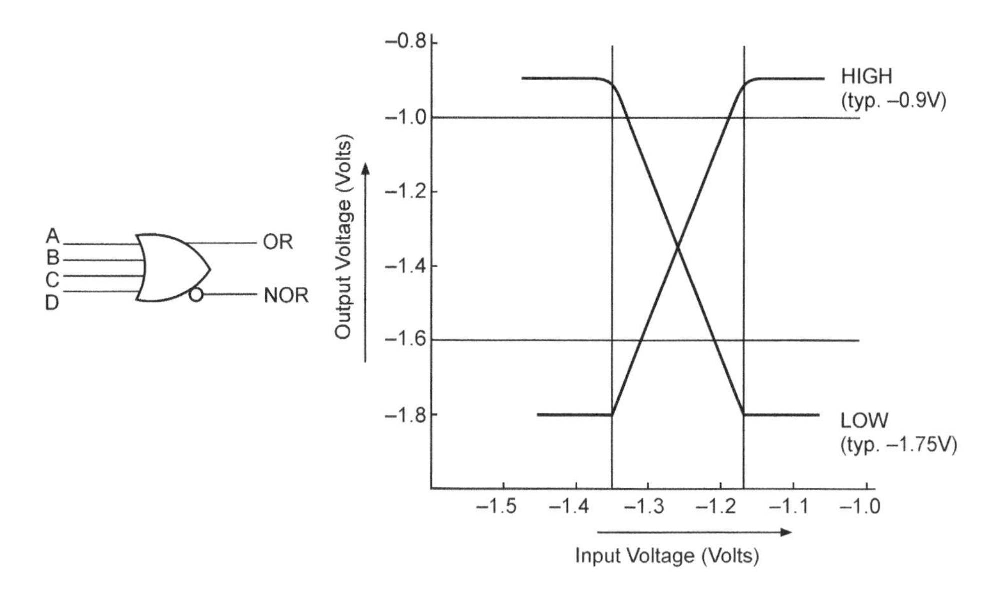

Figure 5.33 ECL input/output characteristics.

We will also note that voltage changes in ECL are small, largely governed by V_{BE} of the various conducting transistors. In fact, the magnitude of the currents flowing through various conducting transistors is of greater relevance to the operation of the ECL circuits. It is for this reason that emitter coupled logic is also sometimes called *current mode logic* (CML).

5.4.3 Salient Features of ECL

There are many features possessed by MECL family devices other than their high speed characteristics that make them attractive for many high-performance applications. The major ones are as follows:

1. ECL family devices produce the true and complementary output of the intended function simultaneously at the outputs without the use of any external inverters. This in turn reduces package count, reduces power requirements and also minimizes problems arising out of time delays that would be caused by external inverters.
2. The ECL gate structure inherently has high input impedance and low output impedance, which is very conducive to achieving large fan-out and drive capability.
3. ECL devices with open emitter outputs allow them to have transmission line drive capability. The outputs match any line impedance. Also, the absence of any pull-down resistors saves power.
4. ECL devices produce a near-constant current drain on the power supply, which simplifies power supply design.
5. On account of the differential amplifier design, ECL devices offer a wide performance flexibility, which allows ECL circuits to be used both as linear and as digital circuits.
6. Termination of unused inputs is easy. Resistors of approximately 50 kΩ allow unused inputs to remain unconnected.

5.5 CMOS Logic Family

The CMOS (Complementary Metal Oxide Semiconductor) logic family uses both N-type and P-type MOSFETs (enhancement MOSFETs, to be more precise) to realize different logic functions. The two types of MOSFET are designed to have matching characteristics. That is, they exhibit identical characteristics in switch-OFF and switch-ON conditions. The main advantage of the CMOS logic family over bipolar logic families discussed so far lies in its extremely low power dissipation, which is near-zero in static conditions. In fact, CMOS devices draw power only when they are switching. This allows integration of a much larger number of CMOS gates on a chip than would have been possible with bipolar or NMOS (to be discussed later) technology. CMOS technology today is the dominant semiconductor technology used for making microprocessors, memory devices and application-specific integrated circuits (ASICs). The CMOS logic family, like TTL, has a large number of subfamilies. The prominent members of CMOS logic were listed in an earlier part of the chapter. The basic difference between different CMOS logic subfamilies such as 4000A, 4000B, 4000UB, 74C, 74HC, 74HCT, 74AC and 74ACT is in the fabrication process used and not in the design of the circuits employed to implement the intended logic function. We will firstly look at the circuit implementation of various logic functions in CMOS and then follow this up with a brief description of different subfamilies of CMOS logic.

5.5.1 Circuit Implementation of Logic Functions

In the following paragraphs, we will briefly describe the internal schematics of basic logic functions when implemented in CMOS logic. These include inverter, NAND, NOR, AND, OR, EX-OR, EX-NOR and AND-OR-INVERT functions.

5.5.1.1 CMOS Inverter

The inverter is the most fundamental building block of CMOS logic. It consists of a pair of N-channel and P-channel MOSFETs connected in cascade configuration as shown in Fig. 5.34. The circuit

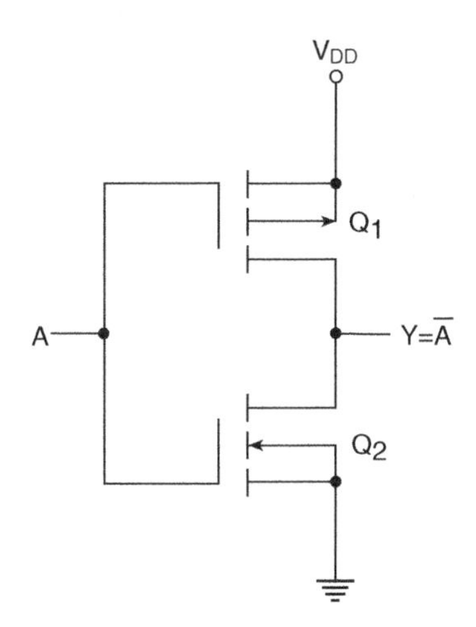

Figure 5.34 CMOS inverter.

functions as follows. When the input is in the HIGH state (logic '1'), P-channel MOSFET Q_1 is in the cut-off state while the N-channel MOSFET Q_2 is conducting. The conducting MOSFET provides a path from ground to output and the output is LOW (logic '0'). When the input is in the LOW state (logic '0'), Q_1 is in conduction while Q_2 is in cut-off. The conducting P-channel device provides a path for V_{DD} to appear at the output, so that the output is in HIGH or logic '1' state. A floating input could lead to conduction of both MOSFETs and a short-circuit condition. It should therefore be avoided. It is also evident from Fig. 5.34 that there is no conduction path between V_{DD} and ground in either of the input conditions, that is, when input is in logic '1' and '0' states. That is why there is practically zero power dissipation in static conditions. There is only dynamic power dissipation, which occurs during switching operations as the MOSFET gate capacitance is charged and discharged. The power dissipated is directly proportional to the switching frequency.

5.5.1.2 NAND Gate

Figure 5.35 shows the basic circuit implementation of a two-input NAND. As shown in the figure, two P-channel MOSFETs (Q_1 and Q_2) are connected in parallel between V_{DD} and the output terminal, and two N-channel MOSFETs (Q_3 and Q_4) are connected in series between ground and output terminal. The circuit operates as follows. For the output to be in a logic '0' state, it is essential that both the series-connected N-channel devices conduct and both the parallel-connected P-channel devices remain in the cut-off state. This is possible only when both the inputs are in a logic '1' state. This verifies one of the entries of the NAND gate truth table. When both the inputs are in a logic '0' state, both the N-channel devices are nonconducting and both the P-channel devices are conducting, which produces a logic '1' at the output. This verifies another entry of the NAND truth table. For the remaining two input combinations, either of the two N-channel devices will be nonconducting and either of the two parallel-connected P-channel devices will be conducting. We have either Q_3 OFF and Q_2 ON or Q_4 OFF and Q_1 ON. The output in both cases is a logic '1', which verifies the remaining entries of the truth table.

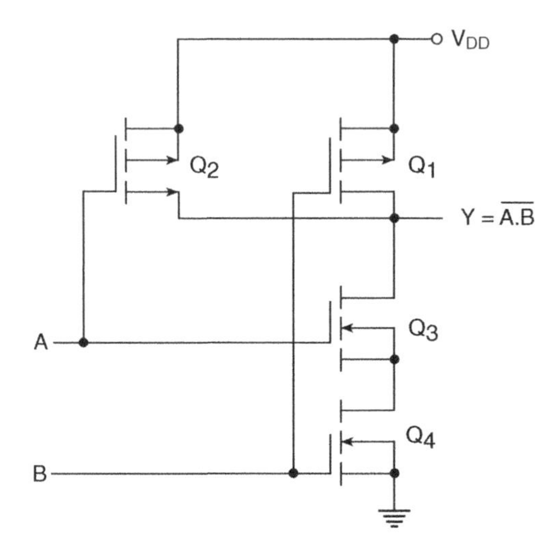

Figure 5.35 CMOS NAND.

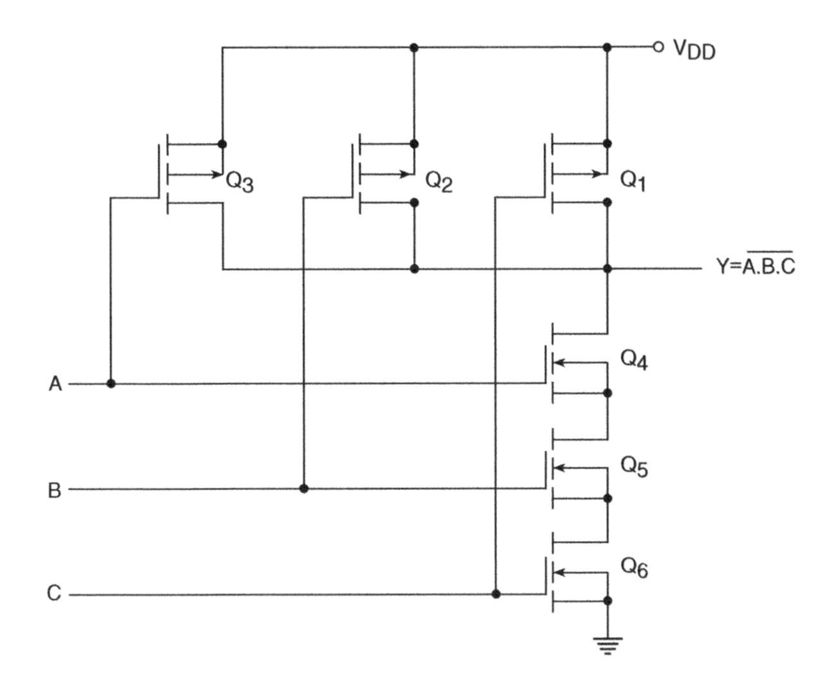

Figure 5.36 Three-input NAND in CMOS.

From the circuit schematic of Fig. 5.35 we can visualize that under no possible input combination of logic states is there a direct conduction path between V_{DD} and ground. This further confirms that there is near-zero power dissipation in CMOS gates under static conditions. Figure 5.36 shows how the circuit of Fig. 5.35 can be extended to build a three-input NAND gate. Operation of this circuit can be explained on similar lines. It may be mentioned here that series connection of MOSFETs adds to the propagation delay, which is greater in the case of P-channel devices than it is in the case of N-channel devices. As a result, the concept of extending the number of inputs as shown in Fig. 5.36 is usually limited to four inputs in the case of NAND and to three inputs in the case of NOR. The number is one less in the case of NOR because it uses series-connected P-channel devices. NAND and NOR gates with larger inputs are realized as a combination of simpler gates.

5.5.1.3 NOR Gate

Figure 5.37 shows the basic circuit implementation of a two-input NOR. As shown in the figure, two P-channel MOSFETs (Q_1 and Q_2) are connected in series between V_{DD} and the output terminal, and two N-channel MOSFETs (Q_3 and Q_4) are connected in parallel between ground and output terminal. The circuit operates as follows. For the output to be in a logic '1' state, it is essential that both the series-connected P-channel devices conduct and both the parallel-connected N-channel devices remain in the cut-off state. This is possible only when both the inputs are in a logic '0' state. This verifies one of the entries of the NOR gate truth table. When both the inputs are in a logic '1' state, both the N-channel devices are conducting and both the P-channel devices are nonconducting, which produces a logic '0' at the output. This verifies another entry of the NOR truth table. For the remaining two

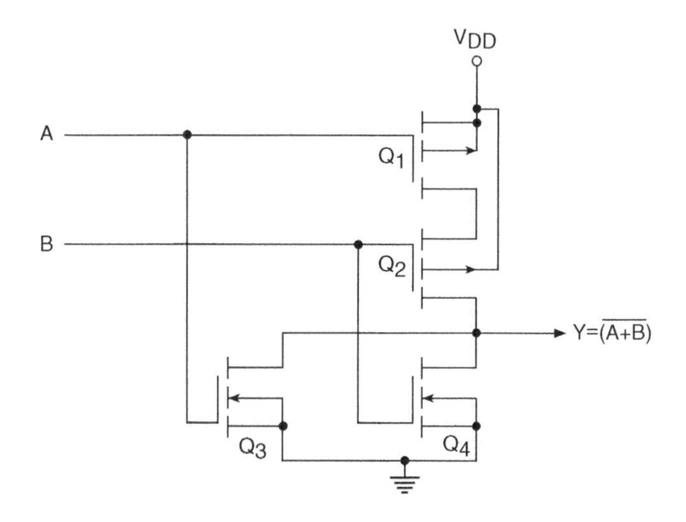

Figure 5.37 Two-input NOR in CMOS.

input combinations, either of the two parallel N-channel devices will be conducting and either of the two series-connected P-channel devices will be nonconducting. We have either Q_1 OFF and Q_3 ON or Q_2 OFF and Q_4 ON. The output in both cases is logic '0', which verifies the remaining entries of the truth table.

Figure 5.38 shows how the circuit of Fig. 5.37 can be extended to build a three-input NOR gate. The operation of this circuit can be explained on similar lines. As already explained, NOR gates with more than three inputs are usually realized as a combination of simpler gates.

5.5.1.4 AND Gate

An AND gate is nothing but a NAND gate followed by an inverter. Figure 5.39 shows the internal schematic of a two-input AND in CMOS. A buffered AND gate is fabricated by using a NOR gate schematic with inverters at both of its inputs and its output feeding two series-connected inverters.

5.5.1.5 OR Gate

An OR gate is nothing but a NOR gate followed by an inverter. Figure 5.40 shows the internal schematic of a two-input OR in CMOS. A buffered OR gate is fabricated by using a NAND gate schematic with inverters at both of its inputs and its output feeding two series-connected inverters.

5.5.1.6 EXCLUSIVE-OR Gate

An EXCLUSIVE-OR gate is implemented using the logic diagram of Fig. 5.41(a). As is evident from the figure, the output of this logic arrangement can be expressed by

$$\overline{[(\overline{A+B}) + A.B} = (\overline{\overline{A}.\overline{B} + A.B})] = \text{EX}-\text{OR function} \tag{5.1}$$

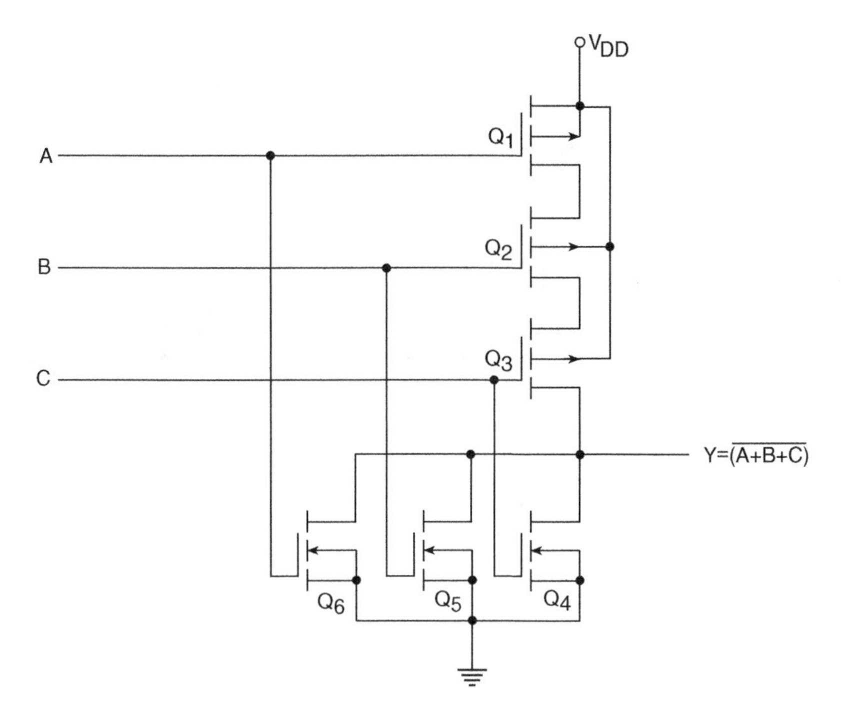

Figure 5.38 Three-input NOR.

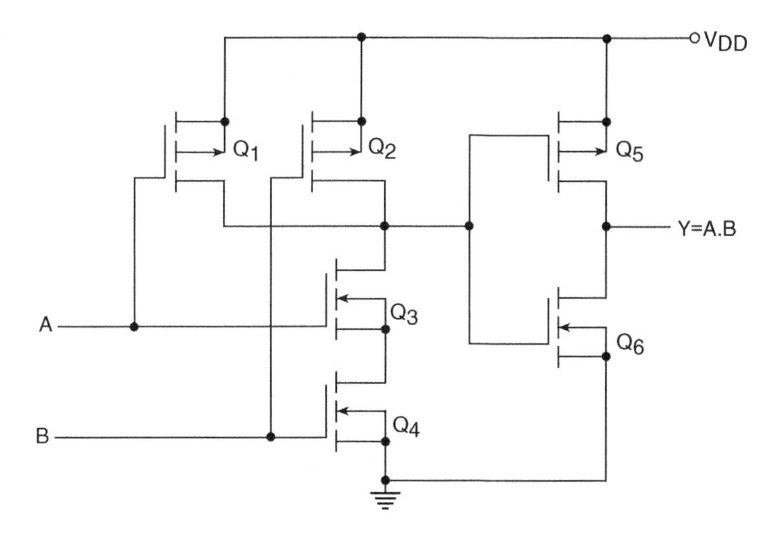

Figure 5.39 Two-input AND in CMOS.

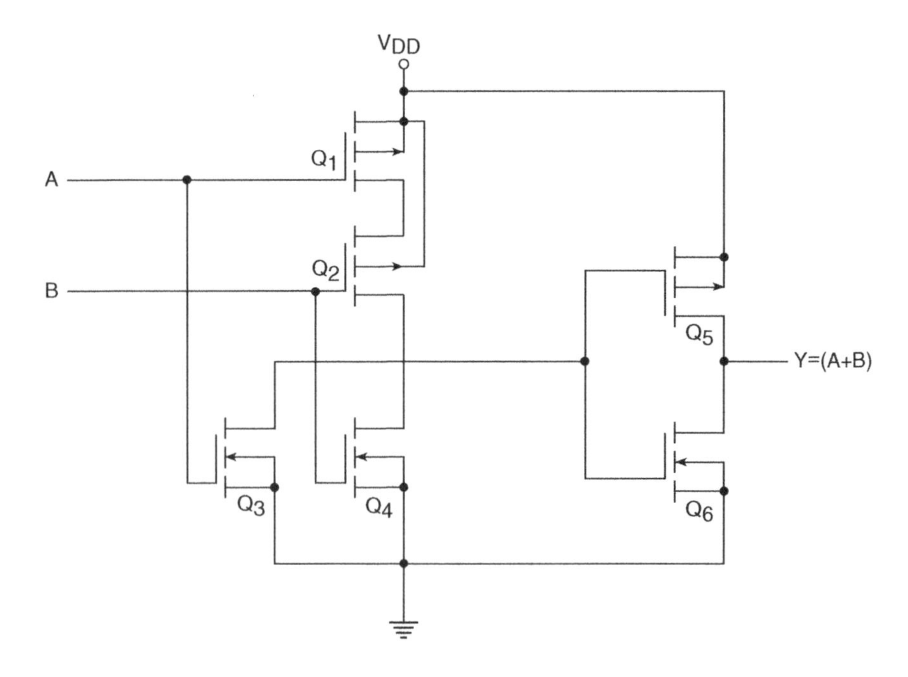

Figure 5.40 Two-input OR in CMOS.

Figure 5.41(b) shows the internal schematic of a two-input EX-OR gate. MOSFETs Q_1–Q_4 constitute the NOR gate. MOSFETS Q_5 and Q_6 simulate ANDing of A and B, and MOSFET Q_7 provides ORing of the NOR output with ANDed output. Since MOSFETs Q_8–Q_{10} make up the complement of the arrangement of MOSFETs Q_5–Q_7, the final output is inverted. Thus, the schematic of Fig. 5.41(b) implements the logic arrangement of Fig. 5.41(a) and hence a two-input EX-OR gate.

5.5.1.7 EXCLUSIVE-NOR Gate

An EXCLUSIVE-NOR gate is implemented using the logic diagram of Fig. 5.42(a). As is evident from the figure, the output of this logic arrangement can be expressed by

$$\overline{[(\overline{A.B}).(A+B)]} = \overline{[(\overline{A}+\overline{B}).(A+B)]} = \text{EX}-\text{NOR function} \tag{5.2}$$

Figure 5.42(b) shows the internal schematic of a two-input EX-NOR gate. MOSFETs Q_1–Q_4 constitute the NAND gate. MOSFETS Q_5 and Q_6 simulate ORing of A and B, and MOSFET Q_7 provides ANDing of the NAND output with ORed output. Since MOSFETs Q_8–Q_{10} make up the complement of the arrangement of MOSFETs Q_5–Q_7, the final output is inverted. Thus, the schematic of Fig. 5.42(b) implements the logic arrangement of Fig. 5.42(a) and hence a two-input EX-NOR gate.

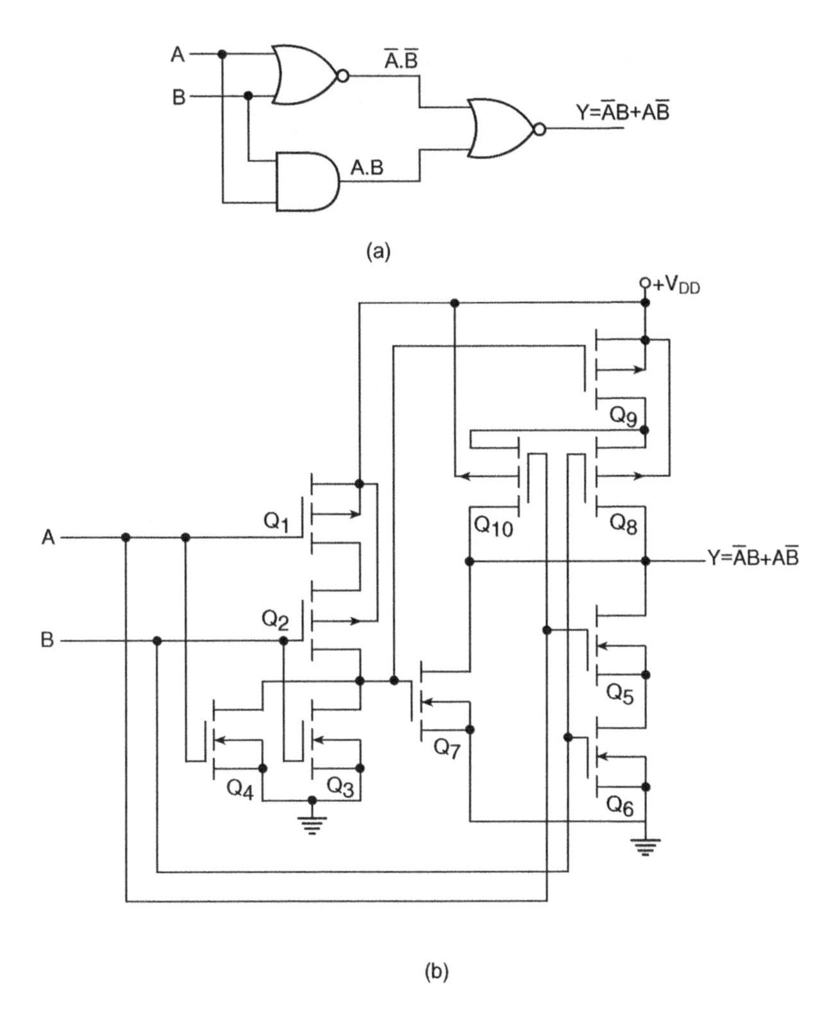

Figure 5.41 Two-input EX-OR in CMOS.

5.5.1.8 AND-OR-INVERT and OR-AND-INVERT Gates

Figure 5.43 shows the internal schematic of a typical two-wide, two-input AND-OR-INVERT gate. The output of this gate can be logically expressed by the Boolean equation

$$Y = (\overline{A.B + C.D}) \tag{5.3}$$

From the above expression, we can say that the output should be in a logic '0' state for the following input conditions:

1. When either $A.B$ = logic '1' or $C.D$ = logic '1'
2. When both $A.B$ and $C.D$ equal logic '1'.

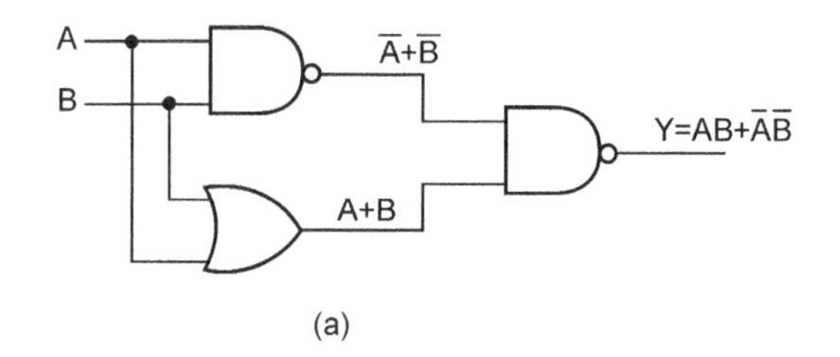

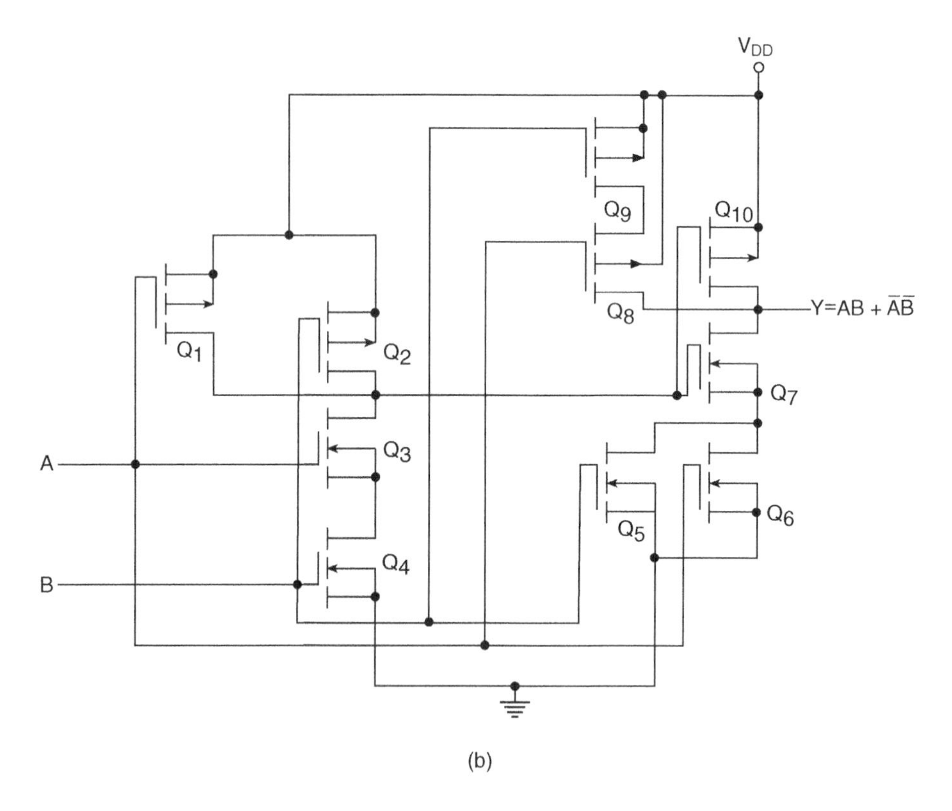

Figure 5.42 Two-input EX-NOR in CMOS.

For both these conditions there is a conduction path available from ground to output, which verifies that the circuit satisfies the logic expression. Also, according to the logic expression for the AND-OR-INVERT gate, the output should be in a logic '1' state when both $A.B$ and $C.D$ equal logic '0'. This implies that:

1. Either A or B or both are in a logic '0' state.
2. Either C or D or both are in a logic '0' state.

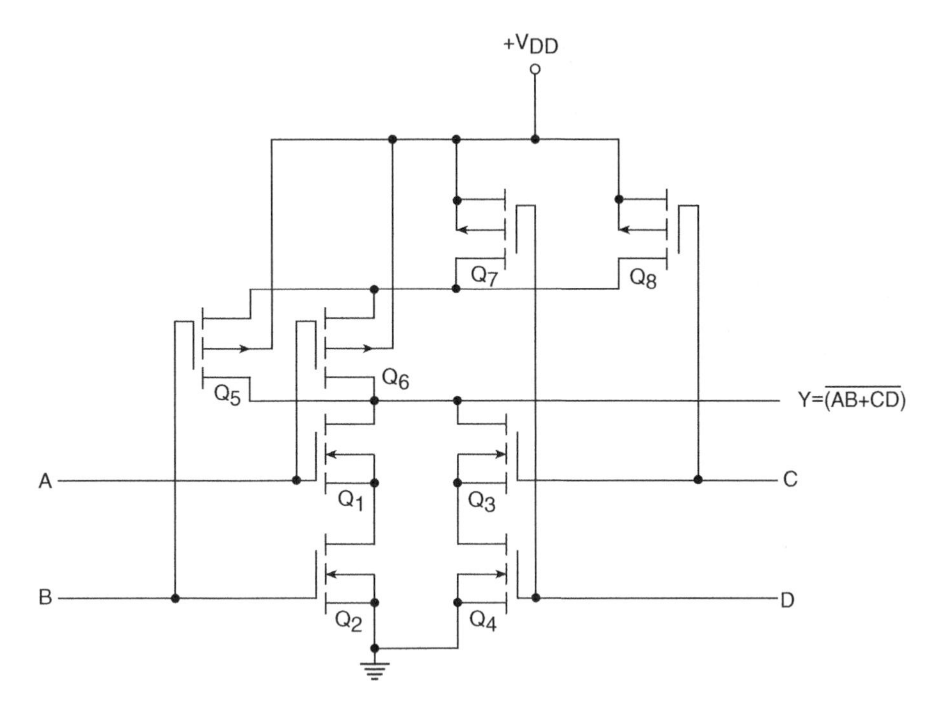

Figure 5.43 Two-wide, two-input AND-OR-INVERT gate in CMOS.

If these conditions are applied to the circuit of Fig. 5.43, we find that the ground will remain disconnected from the output and also that there is always a path from V_{DD} to output. This leads to a logic '1' at the output. Thus, we have proved that the given circuit implements the intended logic expression for the AND-OR-INVERT gate.

The OR-AND-INVERT gate can also be implemented in the same way. Figure 5.44 shows a typical internal schematic of a two-wide, two-input OR-AND-INVERT gate. The output of this gate can be expressed by the Boolean equation

$$Y = \overline{(A+B).(C+D)} \tag{5.4}$$

It is very simple to draw the internal schematic of an AND-OR-INVERT or OR-AND-INVERT gate. The circuit has two parts, that is, the N-channel MOSFET part of the circuit and the P-channel part of the circuit. Let us see, for instance, how Boolean equation (5.4) relates to the circuit of Fig. 5.44. The fact that we need (A OR B) AND (C OR D) explains why the N-channel MOSFETs representing A and B inputs are in parallel and also why the N-channel MOSFETs representing C and D are also in parallel. The two parallel arrangements are then connected in series to achieve an ANDing operation. The complementary P-channel MOSFET section achieves inversion. Note that the P-channel section is the complement of the N-channel section with N-channel MOSFETs replaced by P-channel MOSFETs and parallel connection replaced by series connection, and vice versa. The operation of an AND-OR-INVERT gate can be explained on similar lines to the case of an OR-AND-INVERT gate. Expansion of both AND-OR-INVERT and OR-AND-INVERT gates should be obvious, ensuring that we do not have more than three devices in series.

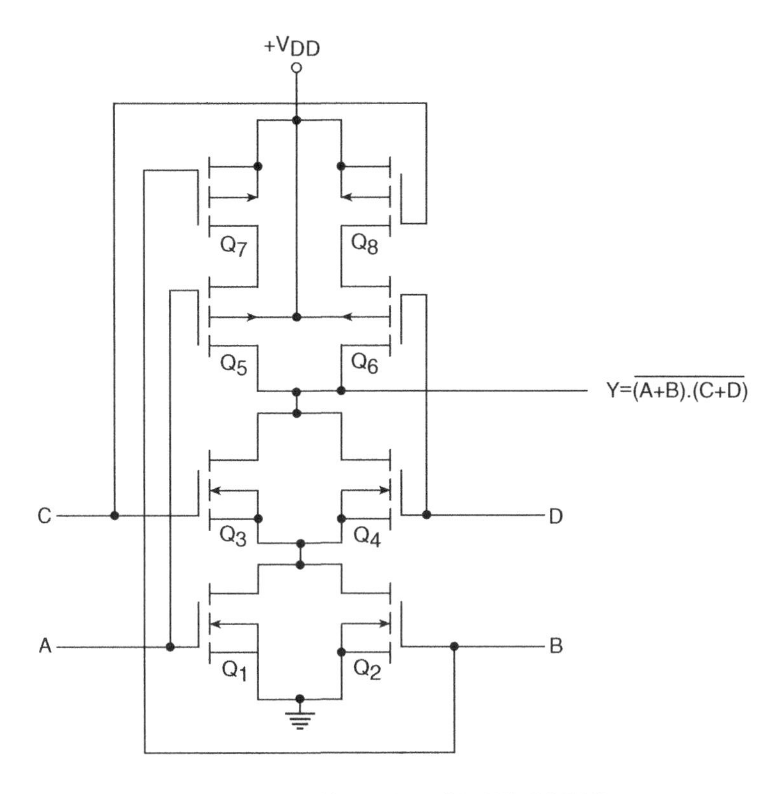

Figure 5.44 Two-wide, two-input OR-AND-INVERT gate.

5.5.1.9 Transmission Gate

The transmission gate, also called the *bilateral switch*, is exclusive to CMOS logic and does not have a counterpart in the TTL and ECL families. It is essentially a single-pole, single-throw (SPST) switch. The opening and closing operations can be controlled by externally applied logic levels. Figure 5.45(a) shows the circuit symbol. If a logic '0' at the control input corresponds to an open switch, then a logic '1' corresponds to a closed switch, and vice versa. The internal schematic of a transmission gate is nothing but a parallel connection of an N-channel MOSFET and a P-channel MOSFET with the control input applied to the gates, as shown in Fig. 5.45(b). Control inputs to the gate terminals of two MOSFETs are the complement of each other. This is ensured by an inbuilt inverter.

When the control input is HIGH (logic '1'), both devices are conducting and the switch is closed. When the control input is LOW (logic '0'), both devices are open and therefore the switch is open. It may be mentioned here that there is no discrimination between input and output terminals. Either of the two can be treated as the input terminal for the purpose of applying input. This is made possible by the symmetry of the two MOSFETs.

It may also be mentioned here that the ON-resistance of a conducting MOSFET depends upon drain and source voltages. In the case of an N-channel MOSFET, if the source voltage is close to V_{DD}, there is an increase in ON-resistance, leading to an increased voltage drop across the switch. A similar phenomenon is observed when the source voltage of a P-channel MOSFET is close to

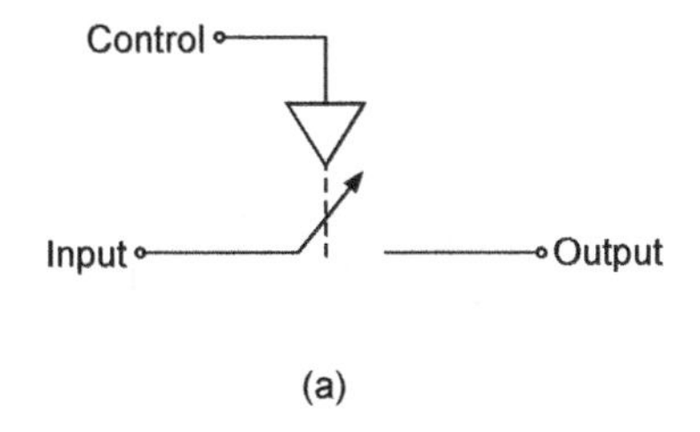

(a)

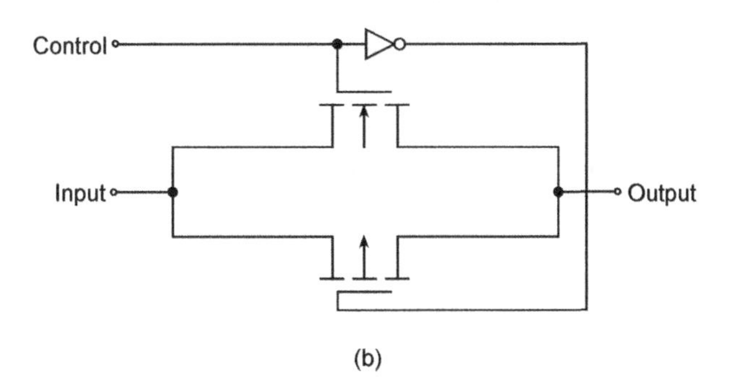

(b)

Figure 5.45 Transmission gate.

ground. Such behaviour causes no problem in static CMOS logic gates, where source terminals of all N-channel MOSFETs are connected to ground and source terminals of all P-channel MOSFETs are connected to V_{DD}. This would cause a problem if a single N-channel or P-channel device were used as a switch. Such a problem is overcome with the use of parallel connection of N-channel and P-channel devices. Transmission gate devices are available in 4000-series as well as 74HC series of CMOS logic.

5.5.1.10 CMOS with Open Drain Outputs

The outputs of conventional CMOS gates should never be shorted together, as illustrated by the case of two inverters shorted at the output terminals (Fig. 5.46). If the input conditions are such that the output of one inverter is HIGH and that of the other is LOW, the output circuit is then like a voltage divider network with two identical resistors equal to the ON-resistance of a conducting MOSFET. The output is then approximately equal to $V_{DD}/2$, which lies in the indeterminate range and is therefore unacceptable. Also, an arrangement like this draws excessive current and could lead to device damage.

This problem does not exist in CMOS gates with open drain outputs. Such a device is the counterpart to gates with open collector outputs in the TTL family. The output stage of a CMOS gate with an open drain output is a single N-channel MOSFET with an open drain terminal, and there is no P-channel MOSFET. The open drain terminal needs to be connected to V_{DD} through an external pull-up resistor. Figure 5.47 shows the internal schematic of a CMOS inverter with an open drain output. The pull-up resistor shown in the circuit is external to the device.

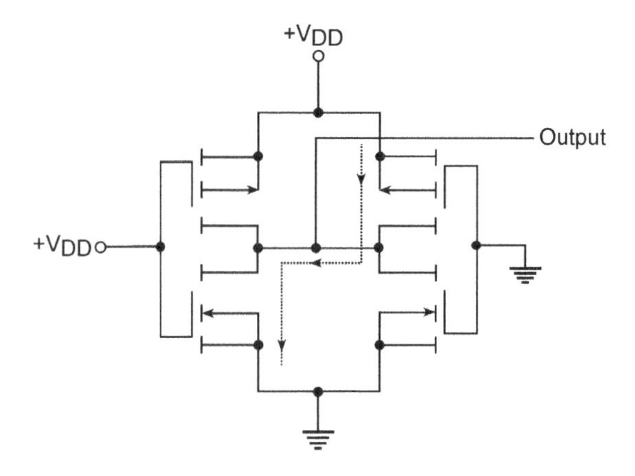

Figure 5.46 CMOS inverters with shorted outputs.

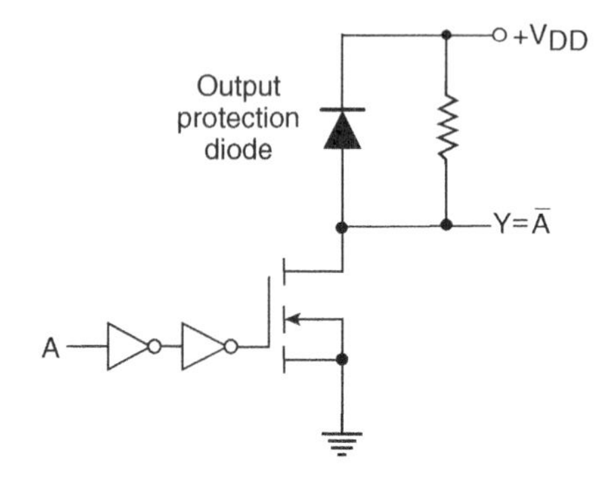

Figure 5.47 CMOS inverter with an open drain output.

5.5.1.11 CMOS with Tristate Outputs

Like tristate TTL, CMOS devices are also available with tristate outputs. The operation of tristate CMOS devices is similar to that of tristate TTL. That is, when the device is enabled it performs its intended logic function, and when it is disabled its output goes to a high-impedance state. In the high-impedance state, both N-channel and P-channel MOSFETs are driven to an OFF-state. Figure 5.48 shows the internal schematic of a tristate buffer with active LOW ENABLE input. The circuit shown is that of one of the buffers in CMOS hex buffer type CD4503B. The outputs of tristate CMOS devices can be connected together in a bus arrangement, like tristate TTL devices with the same condition that only one device is enabled at a time.

5.5.1.12 Floating or Unused Inputs

Unused inputs of CMOS devices should never be left floating or unconnected. A floating input is highly susceptible to picking up noise and accumulating static charge. This can often lead to simultaneous

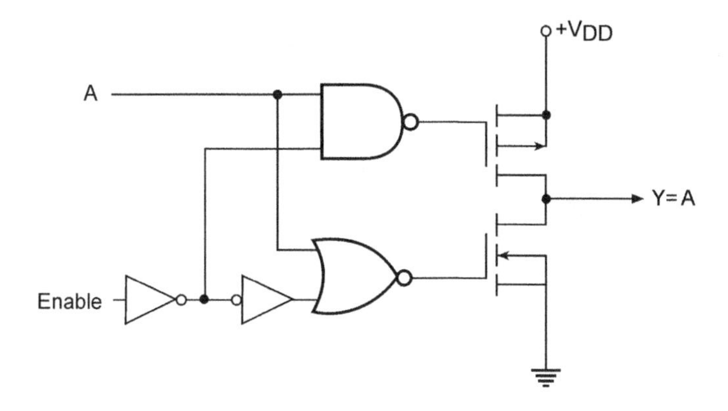

Figure 5.48 Tristate buffer in CMOS.

conduction of P-channel and N-channel devices on the chip, which causes increased power dissipation and overheating. Unused inputs of CMOS gates should either be connected to ground or V_{DD} or shorted to another input. The same is applicable to the inputs of all those gates that are not in use. For example, we may be using only two of the four gates available on an IC having four gates. The inputs of the remaining two gates should be tied to either ground or V_{DD}.

5.5.1.13 Input Protection

Owing to the high input impedance of CMOS devices, they are highly susceptible to static charge build-up. As a result of this, voltage developed across the input terminals could become sufficiently high to cause dielectric breakdown of the gate oxide layer. In order to protect the devices from this static charge build-up and its damaging consequences, the inputs of CMOS devices are protected by using a suitable resistor–diode network, as shown in Fig. 5.49(a). The protection circuit shown is typically used in metal-gate MOSFETs such as those used in 4000-series CMOS devices. Diode D_2 limits the positive voltage surges to $V_{DD} + 0.7$ V, while diode D_3 clamps the negative voltage surges to -0.7 V. Resistor R_1 limits the static discharge current amplitude and thus prevents any damagingly large voltage from being directly applied to the input terminals. Diode D_1 does not contribute to input protection. It is a distributed P–N junction present owing to the diffusion process used for fabrication of resistor R_1. The protection diodes remain reverse biased for the normal input voltage range of 0 to V_{DD}, and therefore do not affect normal operation.

Figure 5.49(b) shows a typical input protection circuit used for silicon-gate MOSFETs used in 74C, 74HC, etc., series CMOS devices. A distributed P–N junction is absent owing to R_1 being a polysilicon resistor. Diodes D_1 and D_2 do the same job as diodes D_2 and D_3 in the case of metal-gate devices. Diode D_2 is usually fabricated in the form of a bipolar transistor with its collector and base terminals shorted.

5.5.1.14 Latch-up Condition

This is an undesired condition that can occur in CMOS devices owing to the existence of parasitic bipolar transistors (NPN and PNP) embedded in the substrate. While N-channel MOSFETs lead to the

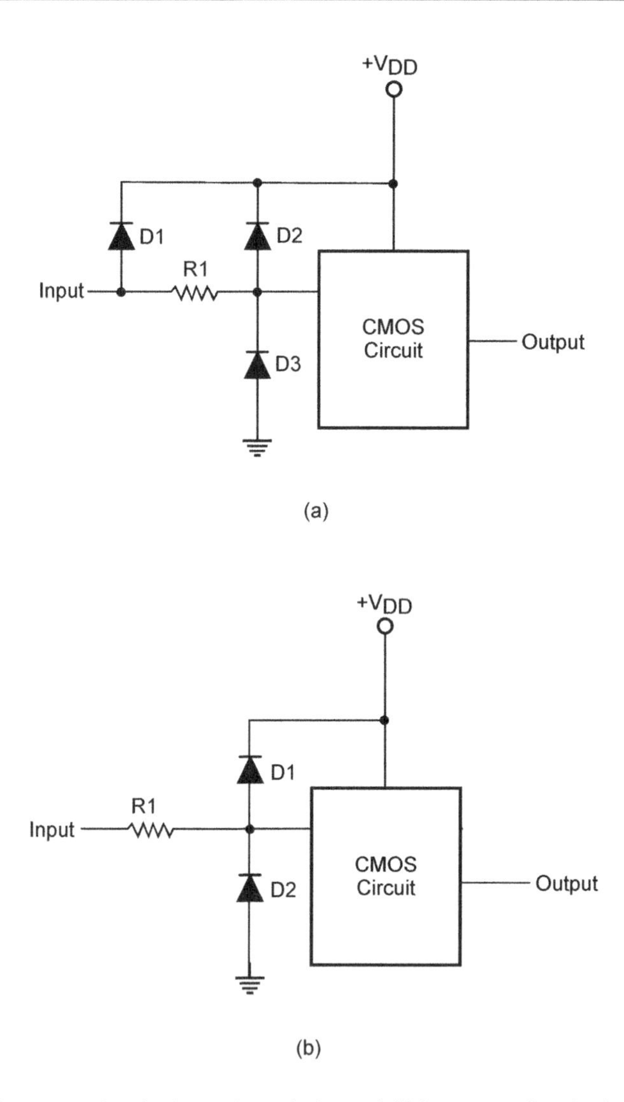

Figure 5.49 (a) Input protection circuit-metal-gate devices and (b) input protection circuit-silicon-gate devices.

presence of NPN transistors, P-channel MOSFETs are responsible for the existence of PNP transistors. If we look into the arrangement of different semiconductor regions in the most basic CMOS building block, that is, the inverter, we will find that these parasitic NPN and PNP transistors find themselves interconnected in a back-to-back arrangement, with the collector of one transistor connected to the base of the other, and vice versa. Two such pairs of transistors connected in series exist between V_{DD} and ground in the case of an inverter, as shown in Fig. 5.50. If for some reason these parasitic elements are triggered into conduction, on account of inherent positive feedback they get into a latch-up condition and remain in conduction permanently. This can lead to the flow of large current and subsequently to destruction of the device. A latch-up condition can be triggered by high voltage spikes and ringing

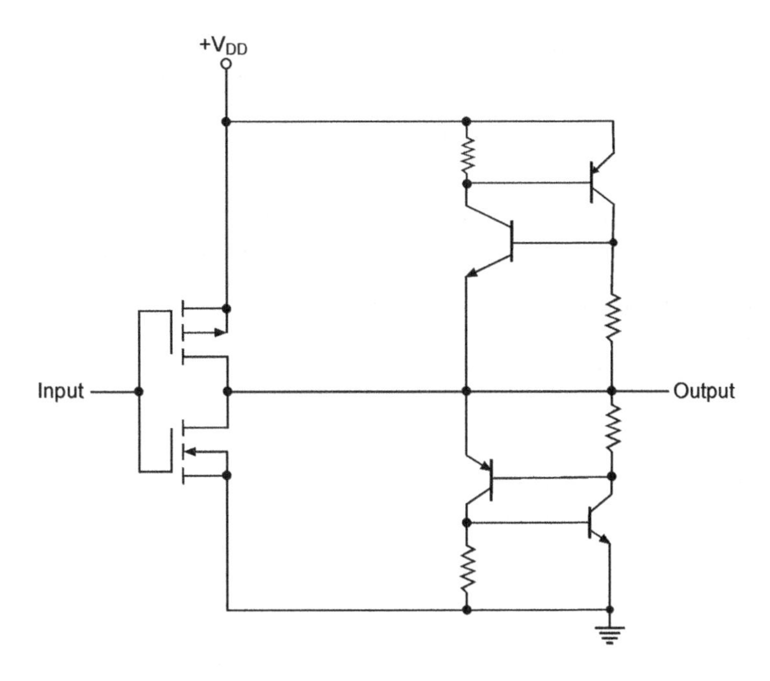

Figure 5.50 CMOS inverter with parasitic elements.

present at the device inputs and outputs. The device can also be prone to latch-up if its maximum ratings are exceeded. Modern CMOS devices use improved fabrication techniques so as to minimize factors that can cause this undesired effect. The use of external clamping diodes at inputs and outputs, proper termination of unused inputs and regulated power supply with a current-limiting feature also helps in minimizing the chances of occurrence of the latch-up condition and in minimizing its effects if it occurs.

5.5.2 CMOS Subfamilies

In the following paragraphs, we will briefly describe various subfamilies of CMOS logic, including subfamilies of the 4000 series and those of TTL pin-compatible 74C series.

5.5.2.1 4000-series

The 4000A-series CMOS ICs, introduced by RCA, were the first to arrive on the scene from the CMOS logic family. The 4000A CMOS subfamily is obsolete now and has been replaced by 4000B and 4000UB subfamilies. We will therefore not discuss it in detail. The 4000B series is a high-voltage version of the 4000A series, and also all the outputs in this series are buffered. The 4000UB series is also a high-voltage version of the 4000A series, but here the outputs are not buffered. A buffered CMOS device is one that has constant output impedance irrespective of the logic status of the inputs. If we recall the internal schematics of the basic CMOS logic gates described in the previous pages, we will see that, with the exception of the inverter, the output impedance of other gates depends upon the logic status of the inputs. This variation in output impedance occurs owing to the varying combination of MOSFETs that conduct for a given input combination. All buffered devices are designated by the suffix 'B' and referred to as the 4000B series. The 4000-series devices that meet 4000B series specifications except for the V_{IL} and V_{IH} specifications and that the outputs are not buffered are called unbuffered devices and are said to belong to the 4000UB series.

Figures 5.51 and 5.52 show a comparison between the internal schematics of a buffered two-input NOR (Fig. 5.51) and an unbuffered two-input NOR (Fig. 5.52). A buffered gate has been implemented by using inverters at the inputs to a two-input NAND whose output feeds another inverter. This is the typical arrangement followed by various manufacturers, as the inverters at the input enhance noise immunity. Another possible arrangement would be a two-input NOR whose output feeds two series-connected inverters.

Variation in the output impedance of unbuffered gates is larger for gates with a larger number of inputs. For example, unbuffered gates have an output impedance of 200–400 Ω in the case of two-input gates, 133–400 Ω for three-input gates and 100–400 Ω for gates with four inputs. Buffered gates have an output impedance of 400 Ω. Since they have the same maximum output impedance, their minimum I_{OL} and I_{OH} specifications are the same.

Characteristic features of 4000B and 4000UB CMOS devices are as follows: V_{IH} (buffered devices) = 3.5 V (for V_{DD} = 5 V), 7.0 V (for V_{DD} = 10 V) and 11.0 V (for V_{DD} = 15 V); V_{IH} (unbuffered devices) = 4.0 V (for V_{DD} = 5 V), 8.0 V (for V_{DD} = 10 V) and 12.5 V (for V_{DD} = 15 V); I_{IH} = 1.0 μA; I_{IL} = 1.0 μA; I_{OH} = 0.2 mA (for V_{DD} = 5 V), 0.5 mA (for V_{DD} = 10 V) and 1.4 mA (for V_{DD} = 15 V); I_{OL} = 0.52 mA (for V_{DD} = 5 V), 1.3 mA (for V_{DD} = 10 V) and 3.6 mA (for V_{DD} = 15 V); V_{IL} (buffered devices) = 1.5 V (for V_{DD} = 5 V), 3.0 V (for V_{DD} = 10 V) and 4.0 V (for V_{DD} = 15 V); V_{IL} (unbuffered devices) = 1.0 V (for V_{DD} = 5 V), 2.0 V (for V_{DD} = 10 V) and 2.5 V (for V_{DD} = 15 V); V_{OH} = 4.95 V

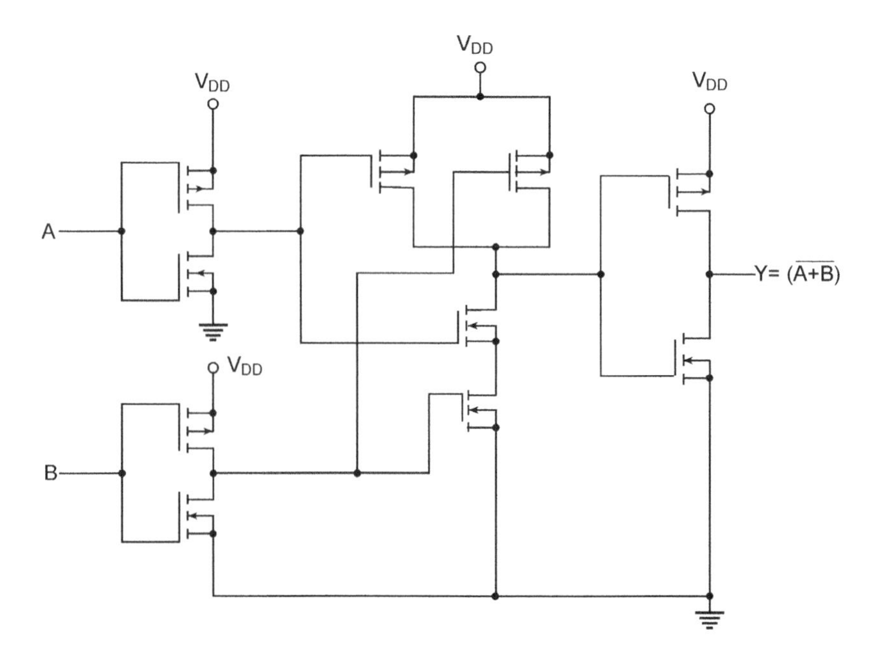

Figure 5.51 Buffered two-input NOR.

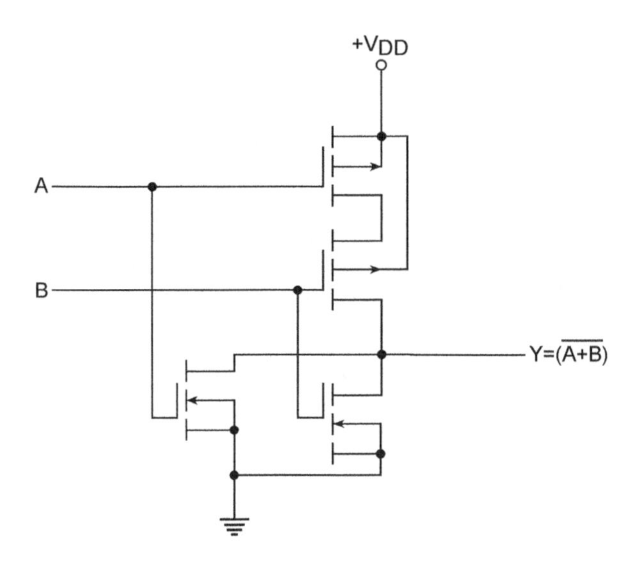

Figure 5.52 Unbuffered two-input NOR.

(for $V_{DD} = 5$ V), 9.95 V (for $V_{DD} = 10$ V) and 14.95 V (for $V_{DD} = 15$ V); $V_{OL} = 0.05$ V; $V_{DD} = 3$–15 V; propagation delay (buffered devices) = 150 ns (for $V_{DD} = 5$ V), 65 ns (for $V_{DD} = 10$ V) and 50 ns (for $V_{DD} = 15$ V); propagation delay (unbuffered devices) = 60 ns (for $V_{DD} = 5$ V), 30 ns (for $V_{DD} = 10$ V) and 25 ns (for $V_{DD} = 15$ V); noise margin (buffered devices) = 1.0 V (for $V_{DD} = 5$ V), 2.0 V (for $V_{DD} = 10$ V) and 2.5 V (for $V_{DD} = 15$ V); noise margin (unbuffered devices) = 0.5 V (for $V_{DD} = 5$ V), 1.0 V (for $V_{DD} = 10$ V) and 1.5 V (for $V_{DD} = 15$ V); output transition time (for $V_{DD} = 5$ V and $C_L = 50$ pF) = 100 ns (buffered devices) and 50–100 ns (for unbuffered devices); power dissipation per gate (for $f = 100$ kHz) = 0.1 mW; speed–power product (for $f = 100$ kHz) = 5 pJ; maximum flip-flop toggle rate = 12 MHz.

5.5.2.2 74C Series

The 74C CMOS subfamily offers pin-to-pin replacement of the 74-series TTL logic functions. For instance, if 7400 is a quad two-input NAND in standard TTL, then 74C00 is a quad two-input NAND with the same pin connections in CMOS. The characteristic parameters of the 74C series CMOS are more or less the same as those of 4000-series devices.

5.5.2.3 74HC/HCT Series

The 74HC/HCT series is the high-speed CMOS version of the 74C series logic functions. This is achieved using silicon-gate CMOS technology rather than the metal-gate CMOS technology used in earlier 4000-series CMOS subfamilies. The 74HCT series is only a process variation of the 74HC series. The 74HC/HCT series devices have an order of magnitude higher switching speed and also a much higher output drive capability than the 74C series devices. This series also offers pin-to-pin replacement of 74-series TTL logic functions. In addition, the 74HCT series devices have TTL-compatible inputs.

5.5.2.4 74AC/ACT Series

The 74AC series is presently the fastest CMOS logic family. This logic family has the best combination of high speed, low power consumption and high output drive capability. Again, 74ACT is only a process variation of 74AC. In addition, 74ACT series devices have TTL-compatible inputs.

The characteristic parameters of the 74C/74HC/74HCT/74AC/74ACT series CMOS are summarized as follows (for V_{DD}= 5 V): V_{IH} (min.) = 3.5 V (74C), 3.5 V (74HC and 74AC) and 2.0 V (74HCT and 74ACT); V_{OH} (min.) = 4.5 V (74C) and 4.9 V (74HC, 74HCT, 74AC and 74ACT); V_{IL}(max.) = 1.5 V (74C), 1.0 V (74HC), 0.8 V (74HCT), 1.5 V (74AC) and 0.8 V (74ACT); V_{OL} (max.) = 0.5 V (74C) and 0.1 V (74HC, 74HCT, 74AC and 74ACT); I_{IH}(max.) = 1 μA; I_{IL} (max.) = 1 μA; I_{OH} (max.) = 0.4 mA (74C), 4.0 mA (74HC and 74HCT) and 24 mA (74AC and 74ACT); I_{OL} (max.) = 0.4 mA (74C), 4.0 mA (74HC and 74HCT) and 24 mA (74AC and 74ACT); V_{NH} = 1.4 V (74C, 74HC and 74AC) and 2.9 V (74HCT and 74ACT); V_{NL} = 1.4 V (74C), 0.9 V (74HC), 0.7 V (74HCT and 74ACT) and 1.4 V (74AC); propagation delay = 50 ns (74C), 8 ns (74HC and 74HCT) and 4.7 ns (74AC and 74ACT); power dissipation per gate (for f = 100 kHz) = 0.1 mW (74C), 0.17 mW (74HC and 74HCT) and 0.08 mW (74AC and 74ACT); speed–power product (for f = 100 kHz) = 5 pJ (74C), 1.4 pJ (74HC and 74HCT) and 0.37 pJ (74AC and 74ACT); maximum flip-flop toggle rate = 12 MHz (74C), 40 MHz (74HC and 74HCT) and 100 MHz (74AC and 74ACT).

Example 5.7

Draw the internal schematic of: (a) a two-wide, four-input AND-OR-INVERT logic function in CMOS and (b) a two-wide, four-input OR-AND-INVERT logic function in CMOS.

Solution

(a) Let us assume that A, B, C, D, E, F, G and H are the logic variables. The output Y of this logic function can then be expressed by the equation

$$Y = \overline{A.B.C.D + E.F.G.H})\qquad(5.5)$$

Following the principles explained earlier in the text, the internal schematic is shown in Fig. 5.53(a). Series connection of N-channel MOSFETs on the left simulates ANDing of A, B, C and D, whereas series connection of N-channel MOSFETs on the right simulates ANDing of E, F, G and H. Parallel connection of two branches produces ORing of the ANDed outputs. Since the P-channel MOSFET arrangement is the complement of the N-channel MOSFET arrangement, the final output is what is given by Equation (5.5).

(b) The output Y of this logic function can be expressed by the equation

$$Y = \overline{(A + B + C + D).(E + F + G + H)}\qquad(5.6)$$

Figure 5.53(b) shows the internal schematic, which can be explained on similar lines.

Example 5.8

Determine the logic function performed by the CMOS digital circuit of Fig. 5.54.

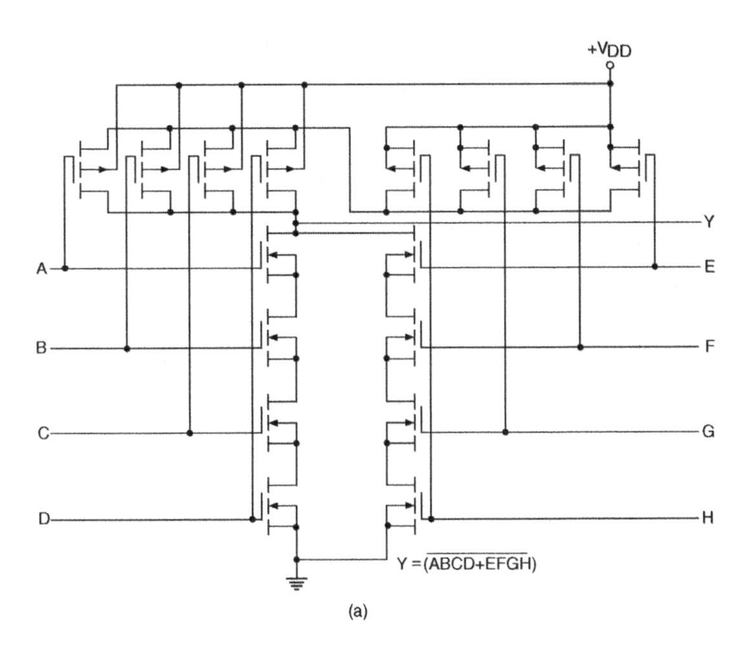

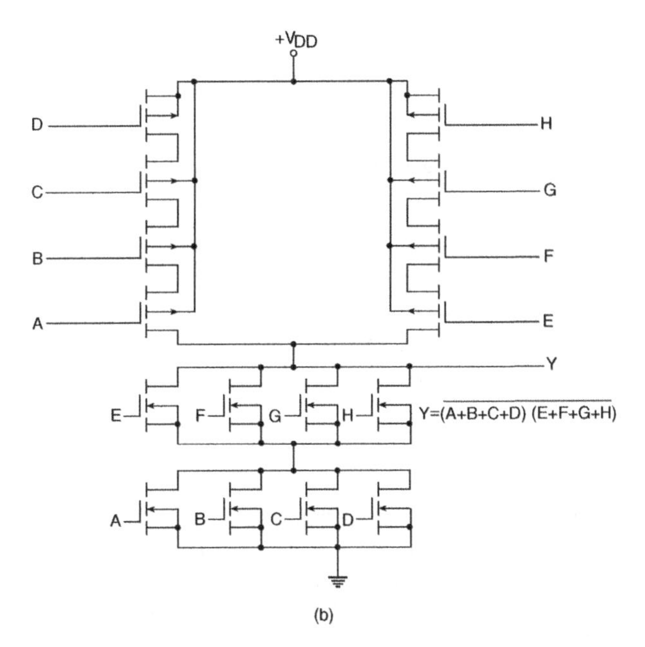

Figure 5.53 Example 5.7.

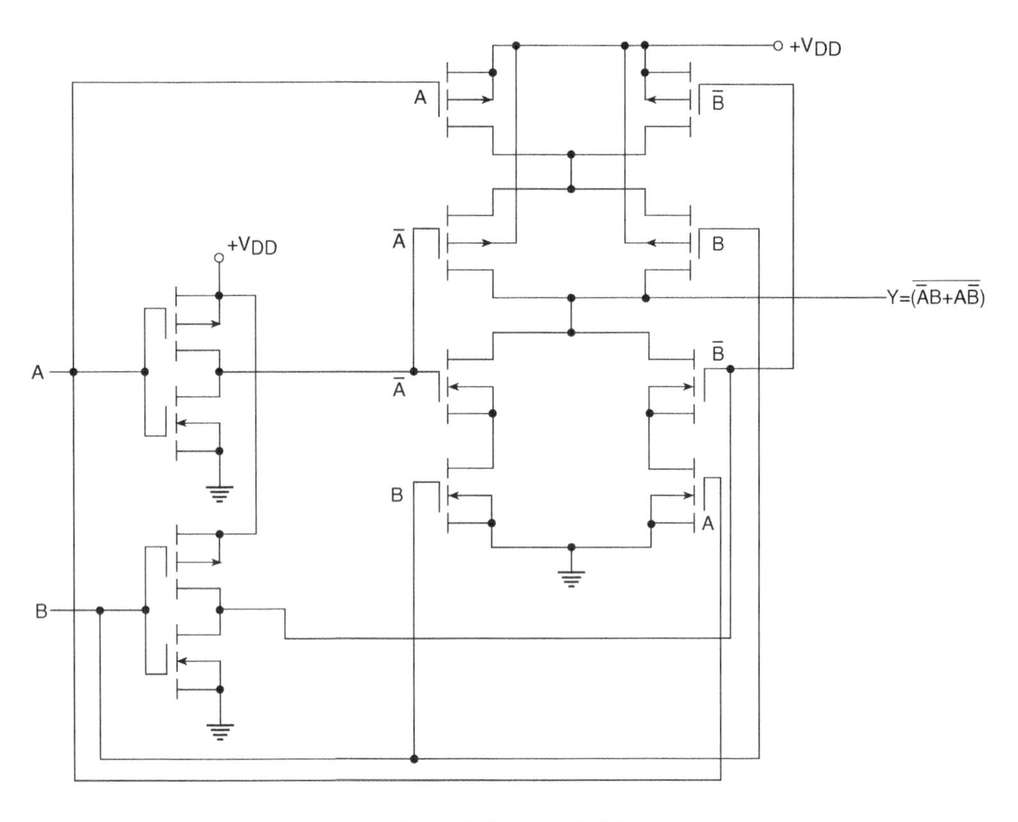

Figure 5.54 Example 5.8.

Solution

The given circuit can be divided into two stages. The first stage comprises two inverters that produce $\overline{A}$ and $\overline{B}$. The second stage is a two-wide, two-input AND-OR-INVERT circuit. Inputs to the first AND are $\overline{A}$ and B, and inputs to the second AND are A and $\overline{B}$. The final output is therefore given by $Y=\overline{(A.\overline{B}+\overline{A}.B)}$, which is an EX-NOR function.

5.6 BiCMOS Logic

The BiCMOS logic family integrates bipolar and CMOS devices on a single chip with the objective of deriving the advantages individually present in bipolar and CMOS logic families. While bipolar logic families such as TTL and ECL have the advantages of faster switching speed and larger output drive current capability, CMOS logic scores over bipolar counterparts when it comes to lower power dissipation, higher noise margin and larger packing density. BiCMOS logic attempts to get the best of both worlds. Two major categories of BiCMOS logic devices have emerged over the years since its introduction in 1985. In one type of device, moderate-speed bipolar circuits are combined with high-performance CMOS circuits. Here, CMOS circuitry continues to provide low power dissipation and larger packing density. Selective use of bipolar circuits gives improved performance. In the other category, the bipolar component is optimized to produce high-performance circuitry. In the following paragraphs, we will briefly describe the basic BiCMOS inverter and NAND circuits.

5.6.1 BiCMOS Inverter

Figure 5.55 shows the internal schematic of a basic BiCMOS inverter. When the input is LOW, N-channel MOSFETs Q_2 and Q_3 are OFF. P-channel MOSFET Q_1 and N-channel MOSFET Q_4 are ON. This leads transistors Q_5 and Q_6 to be in the ON and OFF states respectively. Transistor Q_6 is OFF because it does not get the required forward-biased base-emitter voltage owing to a conducting Q_4. Conducting Q_5 drives the output to a HIGH state, sourcing a large drive current to the load. The HIGH-state output voltage is given by the equation

$$V_{OH} = V_{DD} - V_{BE}(Q_5) \tag{5.7}$$

When the input is driven to a HIGH state, Q_2 and Q_3 turn ON. Initially, Q_4 is also ON and the output discharges through Q_3 and Q_4. When Q_4 turns OFF owing to its gate-source voltage falling below the required threshold voltage, the output continues to discharge until the output voltage equals the forward-biased base-emitter voltage drop of Q_6 in the active region. The LOW-state output voltage is given by the equation

$$V_{OL} = V_{BE}(Q_6 \text{ in active mode}) = 0.7V \tag{5.8}$$

5.6.2 BiCMOS NAND

Figure 5.56 shows the internal schematic of a two-input NAND in BiCMOS logic. The operation of this circuit can be explained on similar lines to the case of an inverter. Note that MOSFETs Q_1–Q_4

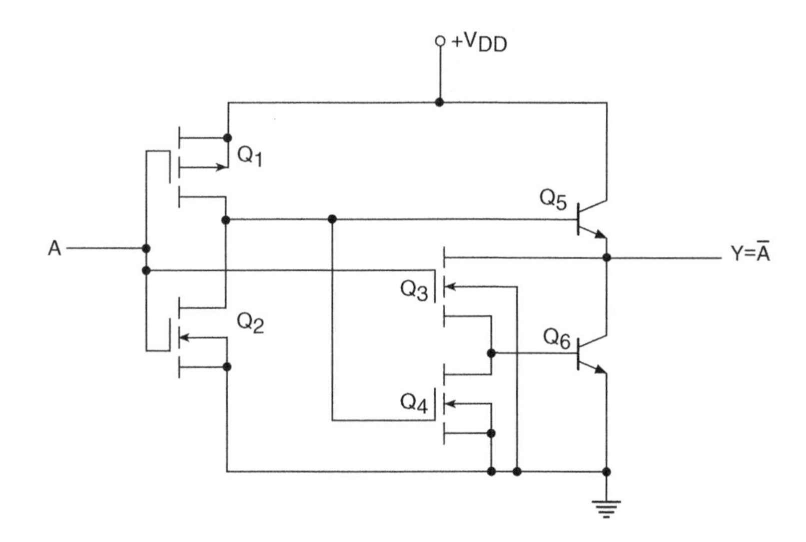

Figure 5.55 BiCMOS inverter.

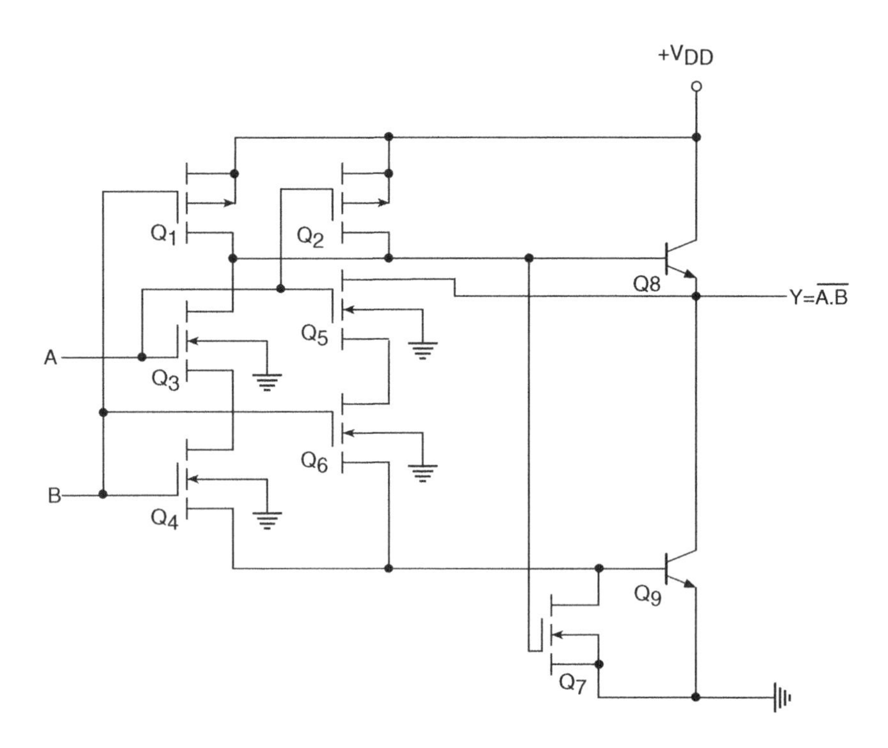

Figure 5.56 BiCMOS two-input NAND.

constitute a two-input NAND in CMOS. Also note the similarity of this circuit to the one shown in Fig. 5.55. The CMOS inverter stage of Fig. 5.55 is replaced by CMOS NAND in Fig. 5.56. N-channel MOSFET Q_3 in Fig. 5.55 is replaced by a series connection of N-channel MOSFETs Q_5 and Q_6 to accommodate the two inputs. The HIGH-state and LOW-state output voltage levels of this circuit are given by the equations

$$V_{OH} = (V_{DD} - 0.7) \tag{5.9}$$

$$V_{OL} = 0.7 \tag{5.10}$$

5.7 NMOS and PMOS Logic

Logic families discussed so far are the ones that are commonly used for implementing discrete logic functions such as logic gates, flip-flops, counters, multiplexers, demultiplexers, etc., in relatively less complex digital ICs belonging to the small-scale integration (SSI) and medium-scale integration (MSI) level of inner circuit complexities. The TTL, the CMOS and the ECL logic families are not suitable for implementing digital ICs that have a large-scale integration (LSI) level of inner circuit complexity and above. The competitors for LSI-class digital ICs are the PMOS, the NMOS and the integrated injection logic (I^2L). The first two are briefly discussed in this section, and the third is discussed in Section 5.8.

5.7.1 PMOS Logic

The PMOS logic family uses P-channel MOSFETS. Figure 5.57(a) shows an inverter circuit using PMOS logic. MOSFET Q_1 acts as an active load for the MOSFET switch Q_2. For the circuit shown, GND and $-V_{DD}$ respectively represent a logic '1' and a logic '0' for a positive logic system. When the input is grounded (i.e. logic '1'), Q_2 remains in cut-off and $-V_{DD}$ appears at the output through

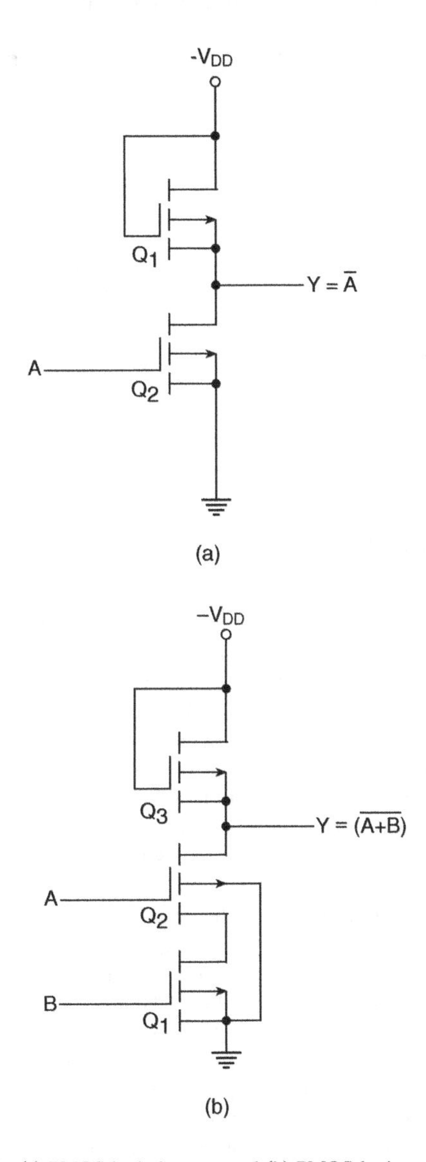

Figure 5.57 (a) PMOS logic inverter and (b) PMOS logic two-input NOR.

the conducting Q_1. When the input is at $-V_{DD}$ or near $-V_{DD}$, Q_2 conducts and the output goes to near-zero potential (i.e. logic '1').

Figure 5.57(b) shows a PMOS logic based two-input NOR gate. In the logic arrangement of Fig. 5.57(b), the output goes to logic '1' state (i.e. ground potential) only when both Q_1 and Q_2 are conducting. This is possible only when both the inputs are in logic '0' state. For all other possible input combinations, the output is in logic '0' state, because, with either Q_1 or Q_2 nonconducting, the output is nearly $-V_{DD}$ through the conducting Q_3. The circuit of Fig. 5.57(b) thus behaves like a two-input NOR gate in positive logic. It may be mentioned here that the MOSFET being used as load [Q_1 in Fig. 5.57(a) and Q_3 in Fig. 5.57(b)] is designed so as to have an ON-resistance that is much greater than the total ON-resistance of the MOSFETs being used as switches [Q_2 in Fig. 5.57(a) and Q_1 and Q_2 in Fig. 5.57(b)].

5.7.2 NMOS Logic

The NMOS logic family uses N-channel MOSFETS. N-channel MOS devices require a smaller chip area per transistor compared with P-channel devices, with the result that NMOS logic offers a higher density. Also, owing to the greater mobility of the charge carriers in N-channel devices, the NMOS logic family offers higher speed too. It is for this reason that most of the MOS memory devices and microprocessors employ NMOS logic or some variation of it such as VMOS, DMOS and HMOS. VMOS, DMOS and HMOS are only structural variations of NMOS, aimed at further reducing the propagation delay. Figures 5.58(a), (b) and (c) respectively show an inverter, a two-input NOR and a two-input NAND using NMOS logic. The logic circuits are self-explanatory.

5.8 Integrated Injection Logic (I^2L) Family

Integrated injection logic (I^2L), also known as current injection logic, is well suited to implementing LSI and VLSI digital functions and is a close competitor to the NMOS logic family. Figure 5.59 shows the basic I^2L family building block, which is a multicollector bipolar transistor with a current source driving its base. Transistors Q_3 and Q_4 constitute current sources. The magnitude of current depends upon externally connected R and applied $+V$. This current is also known as the injection current, which gives it its name of injection logic. If input A is HIGH, the injection current through Q_3 flows through the base-emitter junction of Q_1. Transistor Q_1 saturates and its collector drops to a low voltage, typically 50–100 mV. When A is LOW, the injection current is swept away from the base-emitter junction of Q_1. Transistor Q_1 becomes open and the injection current through Q_4 saturates Q_2, with the result that the Q_1 collector potential equals the base-emitter saturation voltage of Q_2, typically 0.7 V.

The speed of I^2L family devices is a function of the injection current I and improves with increase in current, as a higher current allows a faster charging of capacitive loads present at bases of transistors. The programmable injection current feature is made use of in the I^2L family of digital ICs to choose the desired speed depending upon intended application. The logic '0' level is V_{CE}(sat.) of the driving transistor (Q_1 in the present case), and the logic '1' level is V_{BE}(sat.) of the driven transistor (Q_2 in the present case). Typically, the logic '0' and logic '1' levels are 0.1 and 0.7 V respectively. The speed–power product of the I^2L family is typically under 1 pJ.

Multiple collectors of different transistors can be connected together to form wired logic. Figure 5.60 shows one such arrangement, depicting the generation of OR and NOR outputs of two logic variables A and B.

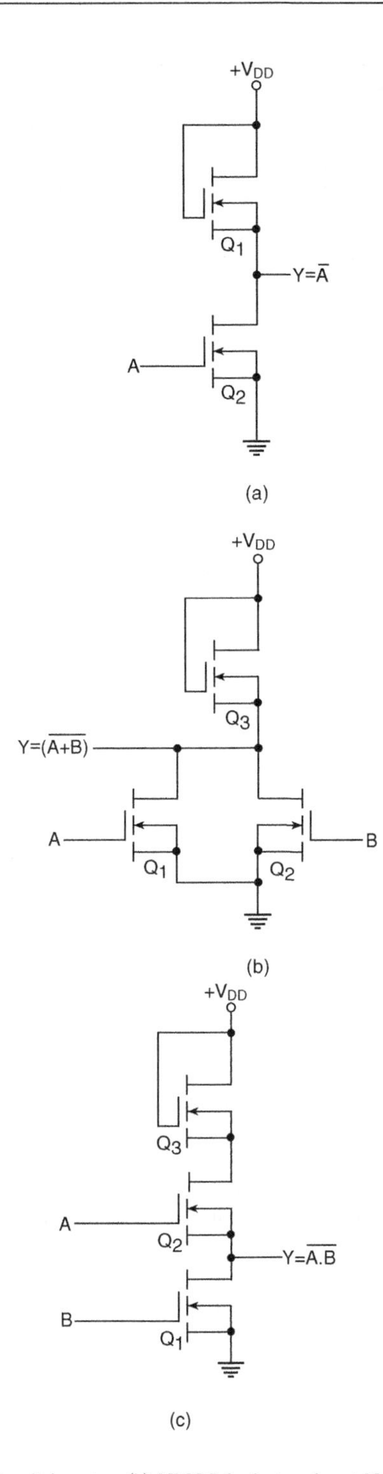

Figure 5.58 (a) NMOS logic circuit inverter, (b) NMOS logic two-input NOR and (c) NMOS logic two-input NAND.

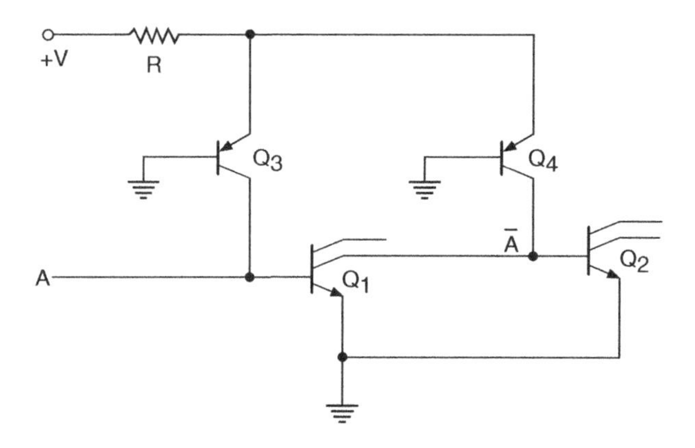

Figure 5.59 Integrated injection logic (I^2L).

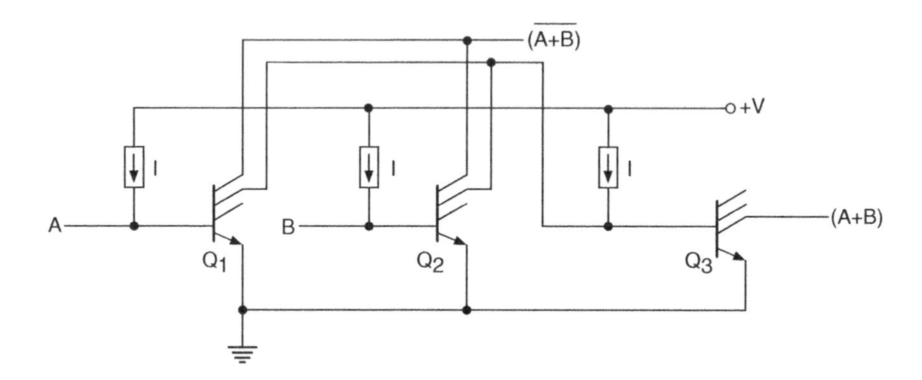

Figure 5.60 Wired logic in I^2L.

5.9 Comparison of Different Logic Families

Table 5.2 gives a comparison of various performance characteristics of important logic families for quick reference. The data given in the case of CMOS families are for $V_{DD} = 5$ V. In the case of ECL families, the data are for $V_{EE} = -5.2$ V. The values of various parameters given in the table should be used only for rough comparison. It is recommended that designers refer to the relevant data books for detailed information on these parameters along with the conditions under which those values are valid.

5.10 Guidelines to Using TTL Devices

The following guidelines should be adhered to while using TTL family devices:

1. Replacing a TTL IC of one TTL subfamily with another belonging to another subfamily (the type numbers remaining the same) should not be done blindly. The designer should ensure that

Table 5.2 Comparison of various performance characteristics of important logic families.

Logic family		Supply voltage (V)	Typical propagation delay (ns)	Worst-case noise margin (V)	Speed–power product (pJ)	Maximum flip-flop toggle frequency (MHz)
TTL	Standard	4.5 to 5.5	17	0.4	100	35
	L	4.5 to 5.5	60	0.3	33	3
	H	4.5 to 5.5	10	0.4	132	50
	S	4.5 to 5.5	5	0.3	57	125
	LS	4.5 to 5.5	15	0.3	18	45
	ALS	4.5 to 5.5	10	0.3	4.8	70
	AS	4.5 to 5.5	4.5	0.3	13.6	200
	F	4.5 to 5.5	6	0.3	10	125
CMOS	4000	3 to 15	150	1.0	5	12
	74C	3 to 13	50	1.4	5	12
	74HC	2 to –6	8	0.9	1.4	40
	74HCT	4.5 to 5.5	8	1.4	1.4	40
	74AC	2 to 6	4.7	0.7	0.37	100
	74ACT	4.5 to 5.5	4.7	0.72.9	0.37	100
ECL	MECL III	–5.1 to –5.3	1	0.2	60	500
	MECL 10K	–4.68 to –5.72	2.5	0.2	50	200
	MECL 10H	–4.94 to –5.46	1	0.15	25	250
	ECLINPS™	–4.2 to –5.5	0.5	0.15	10	1000
	ECLINPS LITE™	–4.2 to –5.5	0.2	0.15	10	2800

the replacement device is compatible with the existing circuit with respect to parameters such as output drive capability, input loading, speed and so on. As an illustration, let us assume that we are using 74S00 (quad two-input NAND), the output of which drives 20 different NAND inputs implemented using 74S00, as shown in Fig. 5.61. This circuit works well as the Schottky TTL family has a fan-out of 20 with an output HIGH drive capability of 1 mA and an input HIGH current requirement of 50 μA. If we try replacing the 74S00 driver with a 74LS00 driver, the circuit fails to work as 74LS00 NAND has an output HIGH drive capability of 0.4 mA only. It cannot feed 20 NAND input loads implemented using 74S00. By doing so, we will be exceeding the HIGH-state fan-out capability of the device. Also, 74LS00 has an output current-sinking specification of 8 mA, whereas the input current-sinking requirement of 74S00 is 2 mA. This implies that 74LS00 could reliably feed only four inputs of 74S00 in the LOW state. By feeding as many as 20 inputs, we will be exceeding the LOW-state fan-out capability of 74LS00 by a large margin.

2. None of the inputs and outputs of TTL ICs should be driven by more than 0.5 V below ground reference.
3. Proper grounding techniques should be used while designing the PCB layout. If the grounding is improper, the ground loop currents give rise to voltage drops, with the result that different ICs will not be at the same reference. This effectively reduces the noise immunity.

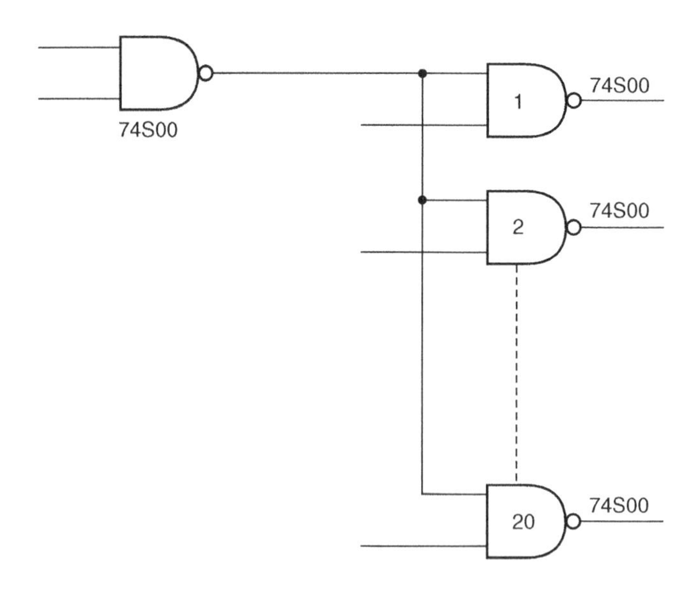

Figure 5.61 Output of one TTL subfamily driving another.

4. The power supply rail must always be properly decoupled with appropriate capacitors so that there is no drop in V_{CC} rail as the inputs and outputs make logic transitions. Usually, two capacitors are used at the V_{CC} point of each IC. A 0.1 μF ceramic disc should be used to take care of high-frequency noise, while typically a 10–20 μF electrolytic is good enough to eliminate any low-frequency variations resulting from variations in I_{CC} current drawn from V_{CC}, depending upon logic states of inputs and outputs. To be effective, the decoupling capacitors should be wired as close as feasible to the V_{CC} pin of the IC.
5. The unused inputs should not be left floating. All unused inputs should be tied to logic HIGH in the case of AND and NAND gates, and to ground in the case of OR and NOR gates. An alternative is to connect the unused input to one of the used inputs.
6. While using open collector devices, resistive pull-up should be used. The value of pull-up resistance should be determined from the following equations:

$$R_X = [V_{CC}(\text{max.}) - V_{OL}]/[I_{OL} - N_2(\text{LOW}) \times 1.6] \tag{5.11}$$

$$R_X(\text{max.}) = [V_{CC}(\text{min.}) - V_{OH}]/[N_1 \times I_{OH} + N_2(\text{HIGH}) \times 40] \tag{5.12}$$

where R_X is the external pull-up resistor; R_X(max.) is the maximum value of the external pull-up resistor; N_1 is the number of WIRED-OR outputs; N_2 is the number of unit input loads being driven; I_{OH} is the output HIGH leakage current (in mA); I_{OL} is the LOW-level output current of the driving element (in mA); V_{OL} is the LOW-level output voltage; and V_{OH} is the HIGH-level output voltage. One TTL unit load in the HIGH state = 40 mA, and one TTL unit load in the LOW-state = 1.6 mA.

5.11 Guidelines to Handling and Using CMOS Devices

The following guidelines should be adhered to while using CMOS family devices:

1. Proper handling of CMOS ICs before they are used and also after they have been mounted on the PC boards is very important as these ICs are highly prone to damage by electrostatic discharge. Although all CMOS ICs have inbuilt protection networks to guard them against electrostatic discharge, precautions should be taken to avoid such an eventuality. While handling unmounted chips, potential differences should be avoided. It is good practice to cover the chips with a conductive foil. Once the chips have been mounted on the PC board, it is good practice again to put conductive clips or conductive tape on the PC board terminals. Remember that PC board is nothing but an extension of the leads of the ICs mounted on it unless it is integrated with the overall system and proper voltages are present.
2. All unused inputs must always be connected to either V_{SS} or V_{DD} depending upon the logic involved. A floating input can result in a faulty logic operation. In the case of high-current device types such as buffers, it can also lead to the maximum power dissipation of the chip being exceeded, thus causing device damage. A resistor (typically 220 kΩ to 1 MΩ) should preferably be connected between input and the V_{SS} or V_{DD} if there is a possibility of device terminals becoming temporarily unconnected or open.
3. The recommended operating supply voltage ranges are 3–12 V for A-series (3–15 V being the maximum rating) and 3–15 V for B-series and UB-series (3–18 V being the maximum). For CMOS IC application circuits that are operated in a linear mode over a portion of the voltage range, such as RC or crystal oscillators, a minimum V_{DD} of 4 V is recommended.
4. Input signals should be maintained within the power supply voltage range $V_{SS} < V_i < V_{DD}$ (-0.5 V $< V_i < V_{DD} + 0.5$ V being the absolute maximum). If the input signal exceeds the recommended input signal range, the input current should be limited to ±100 mA.
5. CMOS ICs like active pull-up TTL ICs cannot be connected in WIRE-OR configuration. Paralleling of inputs and outputs of gates is also recommended for ICs in the same package only.
6. The majority of CMOS clocked devices have maximum rise and fall time ratings of normally 5–15 μs. The device may not function properly with larger rise and fall times. The restriction, however, does not apply to those CMOS ICs that have inbuilt Schmitt trigger shaping in the clock circuit.

5.12 Interfacing with Different Logic Families

CMOS and TTL are the two most widely used logic families. Although ICs belonging to the same logic family have no special interface requirements, that is, the output of one can directly feed the input of the other, the same is not true if we have to interconnect digital ICs belonging to different logic families. Incompatibility of ICs belonging to different families mainly arises from different voltage levels and current requirements associated with LOW and HIGH logic states at the inputs and outputs. In this section, we will discuss simple interface techniques that can be used for CMOS-to-TTL and TTL-to-CMOS interconnections. Interface guidelines for CMOS–ECL, ECL–CMOS, TTL–ECL and ECL–TTL are also given.

5.12.1 CMOS-to-TTL Interface

The first possible type of CMOS-to-TTL interface is the one where both ICs are operated from a common supply. We have read in earlier sections that the TTL family has a recommended supply voltage of 5 V, whereas the CMOS family devices can operate over a wide supply voltage range of 3–18 V. In the present case, both ICs would operate from 5 V. As far as the voltage levels in the two logic states are concerned, the two have become compatible. The CMOS output has a V_{OH}(min.) of 4.95 V (for $V_{CC} = 5$ V) and a V_{OL}(max.) of 0.05 V, which is compatible with V_{IH}(min.) and V_{IL}(max.) requirements of approximately 2 and 0.8 V respectively for TTL family devices. In fact, in a CMOS-to-TTL interface, with the two devices operating on the same V_{CC}, voltage level compatibility is always there. It is the current level compatibility that needs attention. That is, in the LOW state, the output current-sinking capability of the CMOS IC in question must at least equal the input current-sinking requirement of the TTL IC being driven. Similarly, in the HIGH state, the HIGH output current drive capability of the CMOS IC must equal or exceed the HIGH-level input current requirement of TTL IC. For a proper interface, both the above conditions must be met. As a rule of thumb, a CMOS IC belonging to the 4000B family (the most widely used CMOS family) can feed one LS TTL or two low-power TTL unit loads. When a CMOS IC needs to drive a standard TTL or a Schottky TTL device, a CMOS buffer (4049B or 4050B) is used. 4049B and 4050B are hex buffers of inverting and noninverting types respectively, with each buffer capable of driving two standard TTL loads. Figure 5.62(a) shows a CMOS-to-TTL interface with both devices operating from 5 V supply and the CMOS IC driving a low-power TTL or a low-power Schottky TTL device. Figure 5.62(b) shows a CMOS-to-TTL interface where the TTL device in use is either a standard TTL or a Schottky TTL. The CMOS-to-TTL interface when the two are operating on different power supply voltages can be achieved in several ways. One such scheme is shown in Fig. 5.62(c). In this case, there is both a voltage level as well as a current level compatibility problem.

5.12.2 TTL-to-CMOS Interface

In the TTL-to-CMOS interface, current compatibility is always there. The voltage level compatibility in the two states is a problem. V_{OH} (min.) of TTL devices is too low as regards the V_{IH} (min.) requirement of CMOS devices. When the two devices are operating on the same power supply voltage, that is, 5 V, a pull-up resistor of 10 kΩ achieves compatibility [Fig. 5.63(a)]. The pull-up resistor causes the TTL output to rise to about 5 V when HIGH. When the two are operating on different power supplies, one of the simplest interface techniques is to use a transistor (as a switch) in-between the two, as shown in Fig. 5.63(b). Another technique is to use an open collector type TTL buffer [Fig. 5.63(c)].

5.12.3 TTL-to-ECL and ECL-to-TTL Interfaces

TTL-to-ECL and ECL-to-TTL interface connections are not as straightforward as TTL-to-CMOS and CMOS-to-TTL connections owing to widely different power supply requirements for the two types and also because ECL devices have differential inputs and differential outputs. Nevertheless, special chips are available that can take care of all these aspects. These are known as level translators. MC10124 is one such quad TTL-to-ECL level translator. That is, there are four independent single-input and complementary-output translators inside the chip. Figure 5.64(a) shows a TTL-to-ECL interface using MC10124.

MC10125 is a level translator for ECL-to-TTL interfaces; it has differential inputs and a single-ended output. Figure 5.64(b) shows a typical interface schematic using MC10125. Note that in the interface schematics of Figs 5.64(a) and (b), only one of the available four translators has been used.

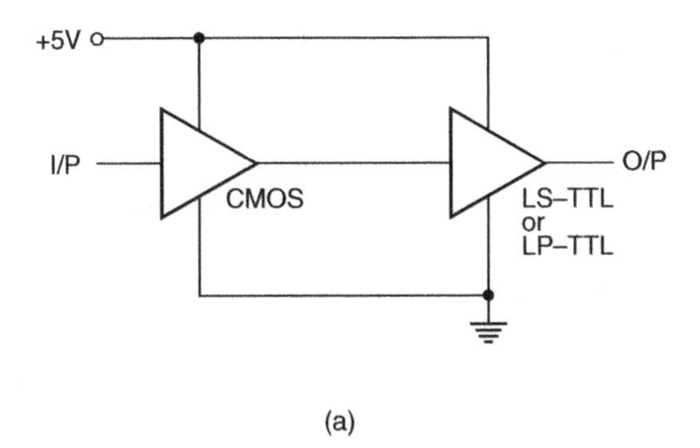

(a)

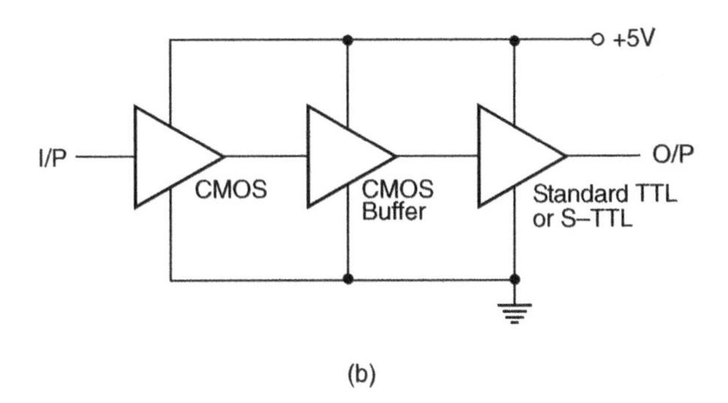

(b)

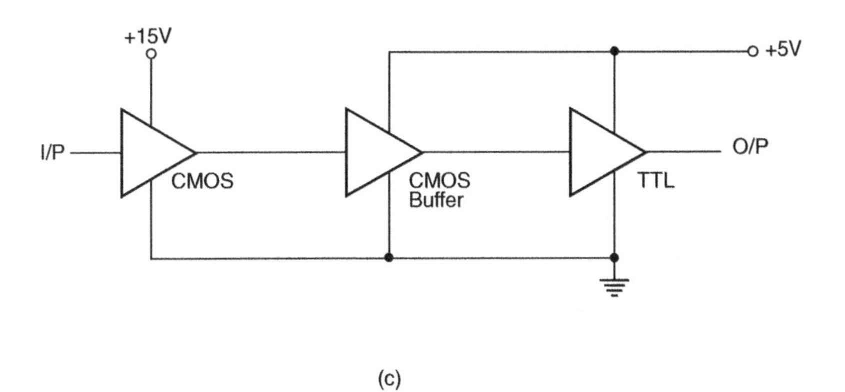

(c)

Figure 5.62 CMOS-to-TTL interface.

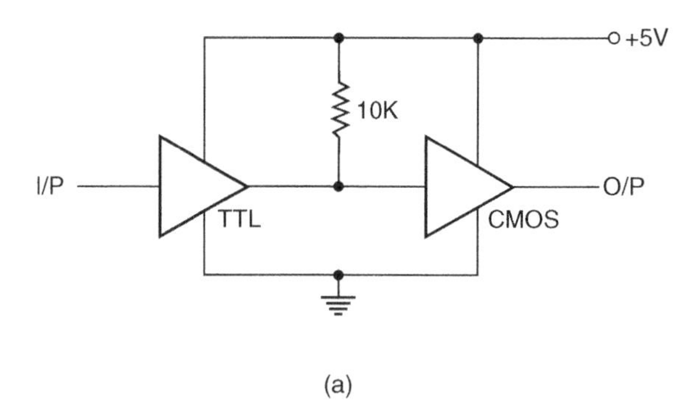

(a)

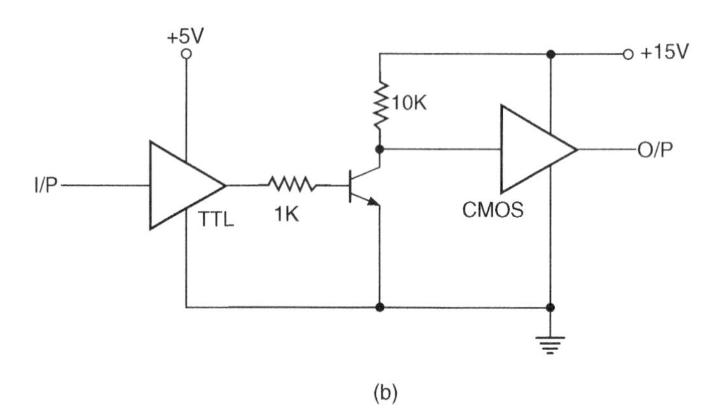

(b)

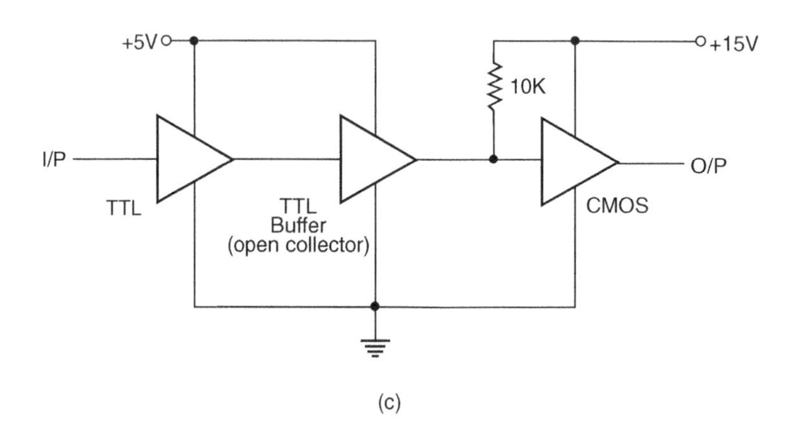

(c)

Figure 5.63 TTL-to-CMOS interface.

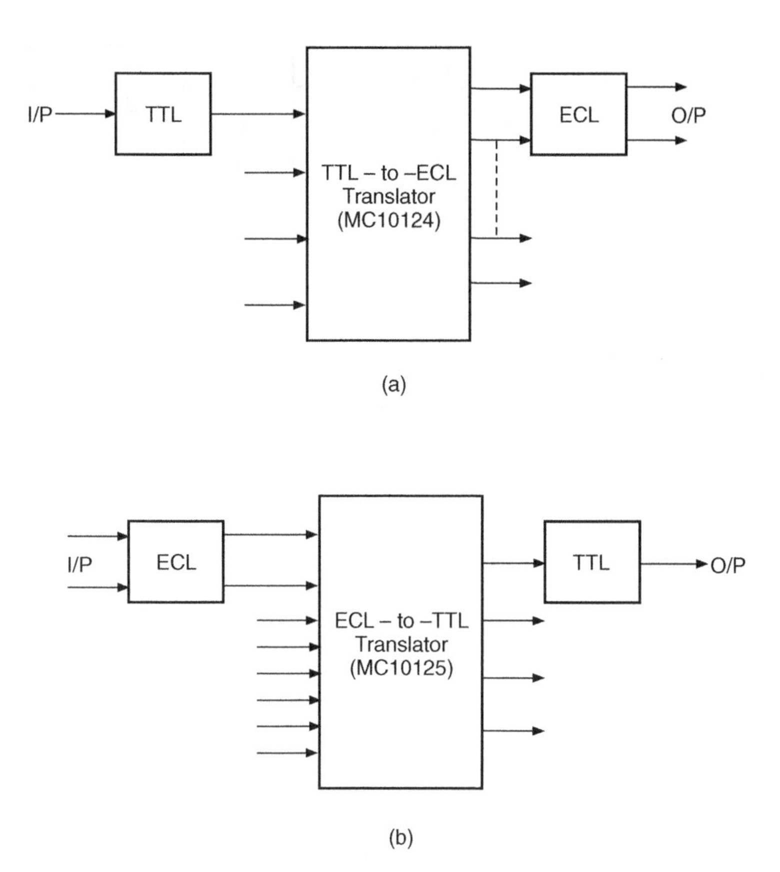

Figure 5.64 TTL-to-ECL and ECL-to-TTL interfaces.

5.12.4 CMOS-to-ECL and ECL-to-CMOS Interfaces

CMOS-to-ECL and ECL-to-CMOS interfaces are similar to the TTL-to-ECL and ECL-to-TTL interfaces described. Again, dedicated level translators are available. MC10352, for instance, is a quad CMOS-to-ECL level translator chip. A CMOS-to-ECL interface is also possible by having firstly a CMOS-to-TTL interface followed by a TTL-to-ECL interface using MC10124 or a similar chip. Figure 5.65(a) shows the arrangement. Similarly, an ECL-to-CMOS interface is possible by having an ECL-to-TTL interface using MC10125 or a similar chip followed by a TTL-to-CMOS interface. Figure 5.65(b) shows a typical interface schematic.

5.13 Classification of Digital ICs

We are all familiar with terms like SSI, MSI, LSI, VLSI and ULSI being used with reference to digital integrated circuits. These terms refer to groups in which digital ICs are divided on the basis of the complexity of the circuitry integrated on the chip. It is common practice to consider the complexity of

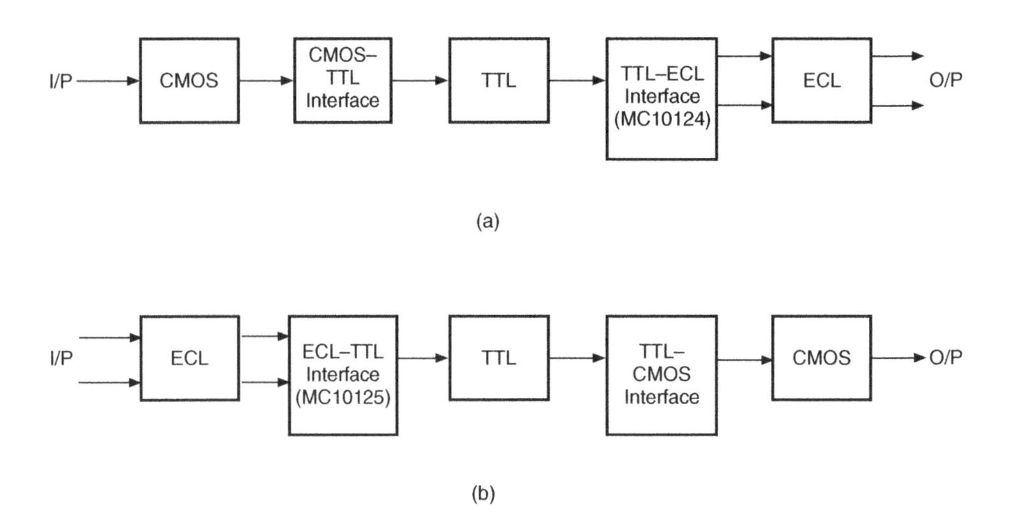

Figure 5.65 CMOS-to-ECL and ECL-to-CMOS interfaces.

a logic gate as a reference for defining the complexities of the other digital IC functions. A broadly accepted definition of different groups of ICs mentioned above is as follows.

A small-scale integration (SSI) chip is one that contains circuitry equivalent in complexity to less than or equal to 10 logic gates. This category of digital ICs includes basic logic gates and flip-flops. A medium-scale integration (MSI) chip is one that contains circuitry equivalent in complexity to 10–100 gates. This category of digital ICs includes multiplexers, demultiplexers, counters, registers, small memories, arithmetic circuits and others. A large-scale integration (LSI) chip is one that contains circuitry equivalent in complexity to 100–10 000 gates. A very-large-scale integration (VLSI) chip contains circuitry equivalent in complexity to 10 000–100 000 gates. Large-sized memories and microprocessors come in the category of LSI and VLSI chips. An ultralarge-scale integration (ULSI) chip contains circuitry equivalent in complexity to more than 100 000 gates. Very large memories, larger microprocessors and larger single-chip computers come into this category.

5.14 Application-Relevant Information

Table 5.3 lists the commonly used type numbers of level translator ICs, along with the functional description. The pin connection diagrams and functional tables for TTL-to-ECL level translator IC type MC10124 and ECL-to-TTL level translator IC type MC10125 are given in the companion website.

Table 5.3 Functional index of level translators

Type number	Function
10124	Quad TTL-to-ECL translator
10125	Quad ECL-to-TTL translator
10177	Triple ECL-to-CMOS translator
10352	Quad CMOS-to-ECL translator

Review Questions

1. What do you understand by the term logic family? What is the significance of the logic family with reference to digital integrated circuits (ICs)?
2. Briefly describe propagation delay, power dissipation, speed–power product, fan-out and noise margin parameters, with particular reference to their significance as regards the suitability of the logic family for a given application.
3. Compare the standard TTL, low-power Schottky TTL and Schottky TTL on the basis of speed, power dissipation and fan-out capability.
4. What is the totem-pole output stage? What are its advantages?
5. What are the basic differences between buffered and unbuffered CMOS devices? How is a buffered NAND usually implemented in 4000B-series CMOS logic?
6. With the help of relevant circuit schematics, briefly describe the operation of CMOS NAND and NOR gates.
7. Compare standard TTL and 4000B CMOS families on the basis of speed and power dissipation parameters.
8. Why is ECL called nonsaturating logic? What is the main advantage accruing from this? With the help of a relevant circuit schematic, briefly describe the operation of ECL OR/NOR logic.
9. What is the main criterion for the suitability of a logic family for use in fabricating LSI and VLSI logic functions? Name any two popular candidates and compare their features.
10. Why is it not recommended to leave unused logic inputs floating? What should we do to such inputs in the case of TTL and CMOS logic gates?
11. What special precautions should we observe in handling and using CMOS ICs?
12. With the help of suitable schematics, briefly describe how you would achieve TTL-to-CMOS and CMOS-to-TTL interfaces?
13. What is Bi-CMOS logic? What are its advantages?
14. What in a logic family decides the fan-out, speed of operation, noise immunity and power dissipation?

Problems

1. The data sheet of a quad two-input AND gate (type 74S08) specifies the propagation delay and power supply parameters as $V_{CC} = 5.0$ V (typical), I_{CCH} (for all four gates) = 18 mA, I_{CCL} (for all four gates) = 32 mA, t_{pLH}= 4.5 ns and t_{pHL} = 5.0 ns. Determine the speed–power product specification.

   148.4 pJ

2. How many inputs of a low-power Schottky TTL NAND can be reliably driven from a single output of a Schottky TTL NAND, given the following relevant specifications for the devices of two TTL subfamilies:
   Schottky TTL: I_{OH} = 1.0 mA; I_{IH}= 0.05 mA; I_{OL} = 20.0 mA; I_{IL} = 2.0 mA
   Low-power Schottky TTL: I_{OH} = 0.4 mA; I_{IH}= 0.02 mA; I_{OL} = 8.0 mA; I_{IL} = 0.4 mA

   50

3. Refer to the logic diagram in Fig. 5.66. Determine the current being sourced by the NAND gate when its output is HIGH and also the current sunk by it when its output is LOW, given that I_{IH} (AND gate) = 0.02 mA, I_{IL} (AND gate) = 0.4 mA, I_{IH} (OR gate) = 0.04 mA, I_{IL} (OR gate) = 1.6 mA, I_{OH}(NAND gate) = 1.0 mA, I_{OL}(NAND gate) = 20.0 mA.

   HIGH-state current = 0.08 mA; LOW-state current = 2.0 mA

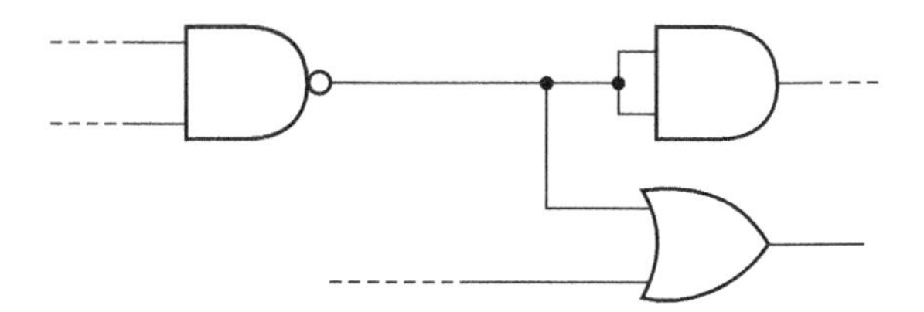

Figure 5.66 Problem 3.

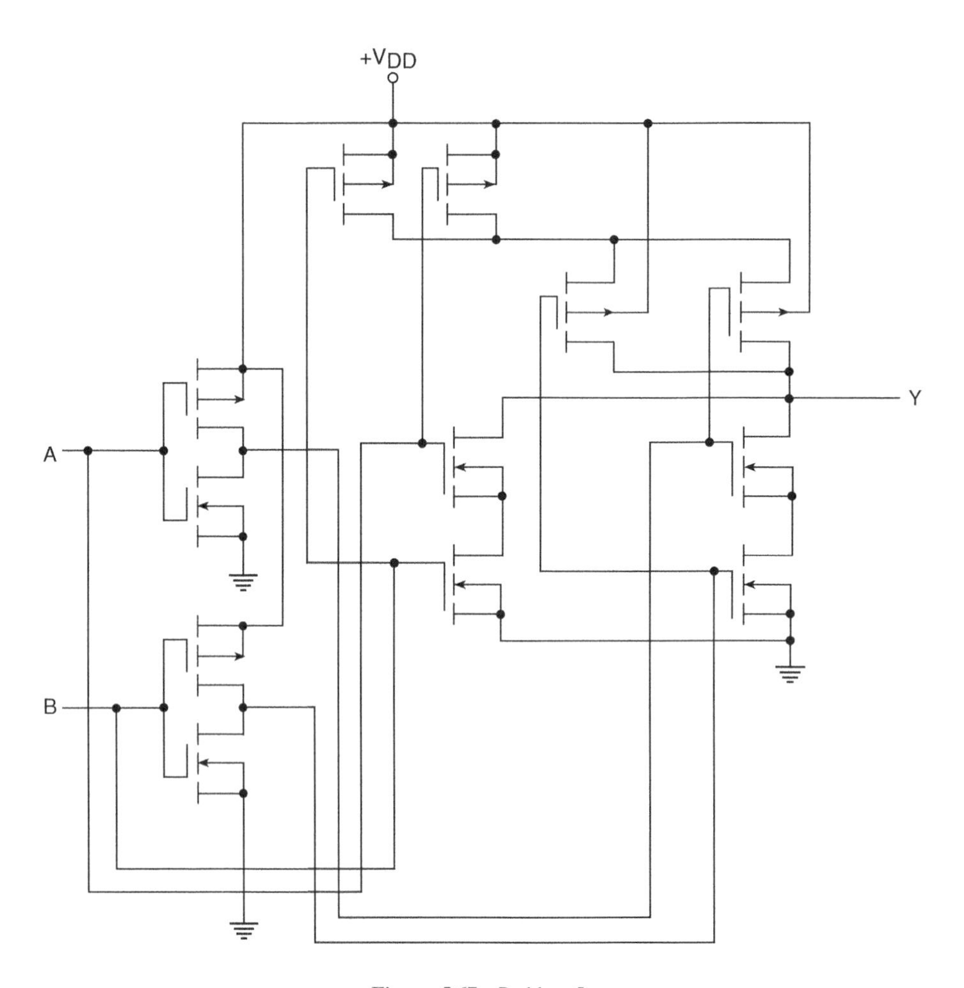

Figure 5.67 Problem 5.

4. Write the logic expression for the CMOS circuit of Fig. 5.67.

$$Y = (A.B + \overline{A}.\overline{B})$$

5. Refer to the data given for 4000B-series CMOS, 74LS-TTL and 74HCT CMOS logic. Determine:

(a) the number of 74LS-TTL inputs that can be reliably driven from a single 4000B output;
(b) the number of 74LS-TTL inputs that can be reliably driven from a single 74HCT output.
4000B: $I_{OH} = 0.4\,mA$; $I_{IH} = 1.0\,\mu A$; $I_{OL} = 0.4\,\mu A$; $I_{IL} = 1.0\,\mu A$
74HCT: $I_{OH} = 4.0\,mA$; $I_{IH} = 1.0\,\mu A$; $I_{OL} = 4.0\,\mu A$; $I_{IL} = 1.0\,\mu A$
74LS-TTL: $I_{OH} = 0.4\,mA$; $I_{IH} = 20.0\,\mu A$; $I_{OL} = 8.0\,\mu A$; $I_{IL} = 0.4\,mA$

(a) 1; (b) 10

Further Reading

1. Tocci, R. J. (2006) *Digital Systems – Principles and Applications*, Prentice-Hall Inc., NJ, USA.
2. Demassa, T. A. and Ciccone, Z. (1995) *Digital Integrated Circuits,* John Wiley & Sons, Inc., New York, USA.
3. Fairchild Semiconductor (August 1973) *74C Family Characteristics*, Application Note 90, South Portland, ME, USA.
4. Wakeman, L. (April 1998) *DC Electrical Characteristics of MM74HC High-speed CMOS Logic*, Application Note 313, Fairchild Semiconductor, South Portland, ME, USA.
5. Funk, R. E. (October 2002) *Understanding Buffered and Unbuffered CD4XXXB-series Device Characteristics*, Application Report SCHA004, Texas Instruments, USA.
6. Buchanan, J. E. and Buchanan, B. D. (1995) *Signal and Power Integrity in Digital Systems: TTL, CMOS, and BiCMOS,* McGraw-Hill Companies, NJ, USA.
7. Lancaster, D. E. (1997) *CMOS Cookbook*, Butterworth-Heinemann, USA.
8. Elmasry, M. I. (1994) *BiCMOS Integrated Circuit Design,* IEEE Press, USA.

6

Boolean Algebra and Simplification Techniques

Boolean algebra is mathematics of logic. It is one of the most basic tools available to the logic designer and thus can be effectively used for simplification of complex logic expressions. Other useful and widely used techniques based on Boolean theorems include the use of Karnaugh maps in what is known as the mapping method of logic simplification and the tabular method given by Quine–McCluskey. In this chapter, we will have a closer look at the different postulates and theorems of Boolean algebra and their applications in minimizing Boolean expressions. We will also discuss at length the mapping and tabular methods of minimizing fairly complex and large logic expressions.

6.1 Introduction to Boolean Algebra

Boolean algebra, quite interestingly, is simpler than ordinary algebra. It is also composed of a set of symbols and a set of rules to manipulate these symbols. However, this is the only similarity between the two. The differences are many. These include the following:

1. In ordinary algebra, the letter symbols can take on any number of values including infinity. In Boolean algebra, they can take on either of two values, that is, 0 and 1.
2. The values assigned to a variable have a numerical significance in ordinary algebra, whereas in its Boolean counterpart they have a logical significance.
3. While '.' and '+' are respectively the signs of multiplication and addition in ordinary algebra, in Boolean algebra '.' means an AND operation and '+' means an OR operation. For instance, $A+B$ in ordinary algebra is read as A plus B, while the same in Boolean algebra is read as A OR B. Basic logic operations such as AND, OR and NOT have already been discussed at length in Chapter 4.

Digital Electronics: Principles, Devices and Applications Anil Kumar Maini

4. More specifically, Boolean algebra captures the essential properties of both logic operations such as AND, OR and NOT and set operations such as intersection, union and complement. As an illustration, the logical assertion that both a statement and its negation cannot be true has a counterpart in set theory, which says that the intersection of a subset and its complement is a null (or empty) set.
5. Boolean algebra may also be defined to be a set A supplied with two binary operations of logical AND ($\wedge$), logical OR ($\vee$), a unary operation of logical NOT ($\neg$) and two elements, namely logical FALSE (0) and logical TRUE (1). This set is such that, for all elements of this set, the postulates or axioms relating to the associative, commutative, distributive, absorption and complementation properties of these elements hold good. These postulates are described in the following pages.

6.1.1 Variables, Literals and Terms in Boolean Expressions

Variables are the different symbols in a Boolean expression. They may take on the value '0' or '1'. For instance, in expression (6.1), A, B and C are the three variables. In expression (6.2), P, Q, R and S are the variables:

$$\overline{A}+A.B+A.\overline{C}+\overline{A}.B.C \tag{6.1}$$

$$(\overline{P}+Q).(R+\overline{S}).(P+\overline{Q}+R) \tag{6.2}$$

The complement of a variable is not considered as a separate variable. Each occurrence of a variable or its complement is called a *literal*. In expressions (6.1) and (6.2) there are eight and seven literals respectively. A term is the expression formed by literals and operations at one level. Expression (6.1) has five terms including four AND terms and the OR term that combines the first-level AND terms.

6.1.2 Equivalent and Complement of Boolean Expressions

Two given Boolean expressions are said to be *equivalent* if one of them equals '1' only when the other equals '1' and also one equals '0' only when the other equals '0'. They are said to be the *complement* of each other if one expression equals '1' only when the other equals '0', and vice versa. The complement of a given Boolean expression is obtained by complementing each literal, changing all '.' to '+' and all '+' to '.', all 0s to 1s and all 1s to 0s. The examples below give some Boolean expressions and their complements:

Given Boolean expression

$$\overline{A}.B+A.\overline{B} \tag{6.3}$$

Corresponding complement

$$(A+\overline{B}).(\overline{A}+B) \tag{6.4}$$

Given Boolean expression

$$(A+B).(\overline{A}+\overline{B}) \tag{6.5}$$

Corresponding complement

$$\overline{A}.\overline{B}+A.B \tag{6.6}$$

When ORed with its complement the Boolean expression yields a '1', and when ANDed with its complement it yields a '0'. The '.' sign is usually omitted in writing Boolean expressions and is implied merely by writing the literals in juxtaposition. For instance, $A.B$ would normally be written as AB.

6.1.3 Dual of a Boolean Expression

The dual of a Boolean expression is obtained by replacing all '.' operations with '+' operations, all '+' operations with '.' operations, all 0s with 1s and all 1s with 0s and leaving all literals unchanged. The examples below give some Boolean expressions and the corresponding dual expressions:

Given Boolean expression

$$\overline{A}.B+A.\overline{B} \tag{6.7}$$

Corresponding dual

$$(\overline{A}+B).(A+\overline{B}) \tag{6.8}$$

Given Boolean expression

$$(A+B).(\overline{A}+\overline{B}) \tag{6.9}$$

Corresponding dual

$$A.B+\overline{A}.\overline{B} \tag{6.10}$$

Duals of Boolean expressions are mainly of interest in the study of Boolean postulates and theorems. Otherwise, there is no general relationship between the values of dual expressions. That is, both of them may equal '1' or '0'. One may even equal '1' while the other equals '0'. The fact that the dual of a given logic equation is also a valid logic equation leads to many more useful laws of Boolean algebra. The principle of duality has been put to ample use during the discussion on postulates and theorems of Boolean algebra. The postulates and theorems, to be discussed in the paragraphs to follow, have been presented in pairs, with one being the dual of the other.

Example 6.1

Find (a) the dual of $A.\overline{B}+B.\overline{C}+C.\overline{D}$ *and (b) the complement of* $[(A.\overline{B}+\overline{C}).D+\overline{E}].F$.

Solution

(a) The dual of $A.\overline{B}+B.\overline{C}+C.\overline{D}$ is given by $(A+\overline{B}).(B+\overline{C}).(C+\overline{D})$.
(b) The complement of $[(A.\overline{B}+\overline{C}).D+\overline{E}].F$ is given by $[(\overline{A}+B).C+\overline{D}].E+\overline{F}$.

Example 6.2

Simplify $(A.B+C.D).[(\overline{A}+\overline{B}).(\overline{C}+\overline{D})]$.

Solution

- Let $(A.B+C.D)=X$.
- Then the given expression reduces to $X.\overline{X}$.
- Therefore, $(A.B+C.D).[(\overline{A}+\overline{B}).(\overline{C}+\overline{D})]=0$.

6.2 Postulates of Boolean Algebra

The following are the important postulates of Boolean algebra:

1. $1.1=1, 0+0=0$.
2. $1.0=0.1=0, 0+1=1+0=1$.
3. $0.0=0, 1+1=1$.
4. $\overline{1}=0$ and $\overline{0}=1$.

Many theorems of Boolean algebra are based on these postulates, which can be used to simplify Boolean expressions. These theorems are discussed in the next section.

6.3 Theorems of Boolean Algebra

The theorems of Boolean algebra can be used to simplify many a complex Boolean expression and also to transform the given expression into a more useful and meaningful equivalent expression. The theorems are presented as pairs, with the two theorems in a given pair being the dual of each other. These theorems can be very easily verified by the method of 'perfect induction'. According to this method, the validity of the expression is tested for all possible combinations of values of the variables involved. Also, since the validity of the theorem is based on its being true for all possible combinations of values of variables, there is no reason why a variable cannot be replaced with its complement, or vice versa, without disturbing the validity. Another important point is that, if a given expression is valid, its dual will also be valid. Therefore, in all the discussion to follow in this section, only one of the theorems in a given pair will be illustrated with a proof. Proof of the other being its dual is implied.

6.3.1 Theorem 1 (Operations with '0' and '1')

$$\text{(a) } 0.X=0 \quad \text{and} \quad \text{(b) } 1+X=1 \tag{6.11}$$

where X is not necessarily a single variable – it could be a term or even a large expression.

Theorem 1(a) can be proved by substituting all possible values of X, that is, 0 and 1, into the given expression and checking whether the LHS equals the RHS:

- For $X=0$, LHS $=0.X=0.0=0=$ RHS.
- For $X=1$, LHS $=0.1=0=$ RHS.

Thus, $0.X=0$ irrespective of the value of X, and hence the proof.

Theorem 1(b) can be proved in a similar manner. In general, according to theorem 1, 0.(Boolean expression) = 0 and 1 + (Boolean expression) = 1. For example, $0.(A.B+B.C+C.D)=0$ and $1+(A.B+B.C+C.D)=1$, where A, B and C are Boolean variables.

6.3.2 Theorem 2 (Operations with '0' and '1')

$$\text{(a) } 1.X = X \quad \text{and} \quad \text{(b) } 0+X = X \tag{6.12}$$

where X could be a variable, a term or even a large expression. According to this theorem, ANDing a Boolean expression to '1' or ORing '0' to it makes no difference to the expression:

- For $X = 0$, LHS $= 1.0 = 0 =$ RHS.
- For $X = 1$, LHS $= 1.1 = 1 =$ RHS.

Also, 1.(Boolean expression) = Boolean expression and 0 + (Boolean expression) = Boolean expression. For example,

$$1.\ (A+B.C+C.D) = 0+(A+B.C+C.D) = A+B.C+C.D.$$

6.3.3 Theorem 3 (Idempotent or Identity Laws)

$$\text{(a) } X.X.X.\ldots.X = X \quad \text{and} \quad \text{(b)} X+X+X+\cdots+X = X \tag{6.13}$$

Theorems 3(a) and (b) are known by the name of *idempotent laws*, also known as *identity laws*. Theorem 3(a) is a direct outcome of an AND gate operation, whereas theorem 3(b) represents an OR gate operation when all the inputs of the gate have been tied together. The scope of idempotent laws can be expanded further by considering X to be a term or an expression. For example, let us apply idempotent laws to simplify the following Boolean expression:

$$\begin{aligned}(A.\overline{B}.\overline{B}+C.C).(A.\overline{B}.\overline{B}+A.\overline{B}+C.C) &= (A.\overline{B}+C).(A.\overline{B}+A.\overline{B}+C) \\ &= (A.\overline{B}+C).(A.\overline{B}+C) = A.\overline{B}+C\end{aligned}$$

6.3.4 Theorem 4 (Complementation Law)

$$\text{(a) } X.\overline{X} = 0 \quad \text{and} \quad \text{(b) } X+\overline{X} = 1 \tag{6.14}$$

According to this theorem, in general, any Boolean expression when ANDed to its complement yields a '0' and when ORed to its complement yields a '1', irrespective of the complexity of the expression:

- For $X = 0$, $\overline{X} = 1$. Therefore, $X.\overline{X} = 0.1 = 0$.
- For $X = 1$, $\overline{X} = 0$. Therefore, $X.\overline{X} = 1.0 = 0$.

Hence, theorem 4(a) is proved. Since theorem 4(b) is the dual of theorem 4(a), its proof is implied.

The example below further illustrates the application of complementation laws:

$$(A+B.C)(\overline{A+B.C}) = 0 \quad \text{and} \quad (A+B.C)+(\overline{A+B.C}) = 1$$

Example 6.3

Simplify the following:

$$[1+L.M+L.\overline{M}+\overline{L}.M].[(L+\overline{M}).(\overline{L}.M)+\overline{L}.\overline{M}.(L+M)].$$

Solution

- We know that (1 + Boolean expression) = 1.
- Also, $(\overline{L}.M)$ is the complement of $(L+\overline{M})$ and $(\overline{L}.\overline{M})$ is the complement of $(L+M)$.
- Therefore, the given expression reduces to 1.(0 + 0) = 1.0 = 0.

6.3.5 Theorem 5 (Commutative Laws)

$$\text{(a) } X+Y=Y+X \quad \text{and} \quad \text{(b) } X.Y=Y.X \tag{6.15}$$

Theorem 5(a) implies that the order in which variables are added or ORed is immaterial. That is, the result of *A* OR *B* is the same as that of *B* OR *A*. Theorem 5(b) implies that the order in which variables are ANDed is also immaterial. The result of *A* AND *B* is same as that of *B* AND *A*.

6.3.6 Theorem 6 (Associative Laws)

$$\text{(a) } X+(Y+Z)=Y+(Z+X)=Z+(X+Y)$$

and

$$\text{(b) } X.(Y.Z)=Y.(Z.X)=Z.(X.Y) \tag{6.16}$$

Theorem 6(a) says that, when three variables are being ORed, it is immaterial whether we do this by ORing the result of the first and second variables with the third variable or by ORing the first variable with the result of ORing of the second and third variables or even by ORing the second variable with the result of ORing of the first and third variables. According to theorem 6(b), when three variables are being ANDed, it is immaterial whether you do this by ANDing the result of ANDing of the first and second variables with the third variable or by ANDing the result of ANDing of the second and third variables with the first variable or even by ANDing the result of ANDing of the third and first variables with the second variable.

For example,

$$\overline{A}.B+(C.\overline{D}+\overline{E}.\overline{F})=C.\overline{D}+(\overline{A}.B+\overline{E}.\overline{F})=\overline{E}.\overline{F}+(\overline{A}.B+C.\overline{D})$$

Also

$$\overline{A}.B.(C.\overline{D}.\overline{E}.\overline{F})=C.\overline{D}.(\overline{A}.B.\overline{E}.\overline{F})=\overline{E}.\overline{F}.(\overline{A}.B.C.\overline{D})$$

Theorems 6(a) and (b) are further illustrated by the logic diagrams in Figs 6.1(a) and (b).

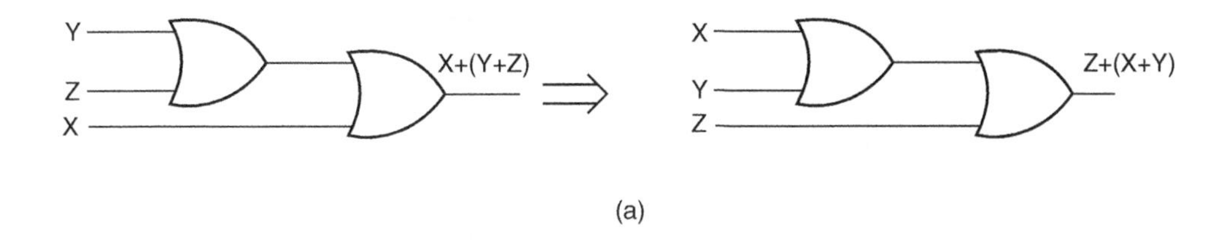

(a)

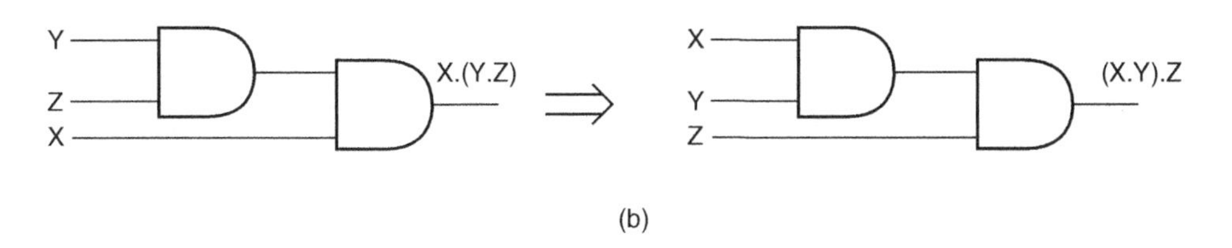

(b)

Figure 6.1 Associative laws.

6.3.7 Theorem 7 (Distributive Laws)

$$(a)\ X.(Y+Z)=X.Y+X.Z \quad \text{and} \quad (b)\ X+Y.Z=(X+Y).(X+Z) \qquad (6.17)$$

Theorem 7(b) is the dual of theorem 7(a). The distribution law implies that a Boolean expression can always be expanded term by term. Also, in the case of the expression being the sum of two or more than two terms having a common variable, the common variable can be taken as common as in the case of ordinary algebra. Table 6.1 gives the proof of theorem 7(a) using the method of perfect induction. Theorem 7(b) is the dual of theorem 7(a) and therefore its proof is implied. Theorems 7(a) and (b) are further illustrated by the logic diagrams in Figs 6.2(a) and (b). As an illustration, theorem 7(a) can be used to simplify $\overline{A}.\overline{B}+\overline{A}.B+A.\overline{B}+A.B$ as follows:

$$\overline{A}.\overline{B}+\overline{A}.B+A.\overline{B}+A.B=\overline{A}.(\overline{B}+B)+A.(\overline{B}+B)=\overline{A}.1+A.1=\overline{A}+A=1$$

Table 6.1 Proof of distributive law.

X	Y	Z	Y+Z	XY	XZ	X(Y+Z)	XY+XZ
0	0	0	0	0	0	0	0
0	0	1	1	0	0	0	0
0	1	0	1	0	0	0	0
0	1	1	1	0	0	0	0
1	0	0	0	0	0	0	0
1	0	1	1	0	1	1	1
1	1	0	1	1	0	1	1
1	1	1	1	1	1	1	1

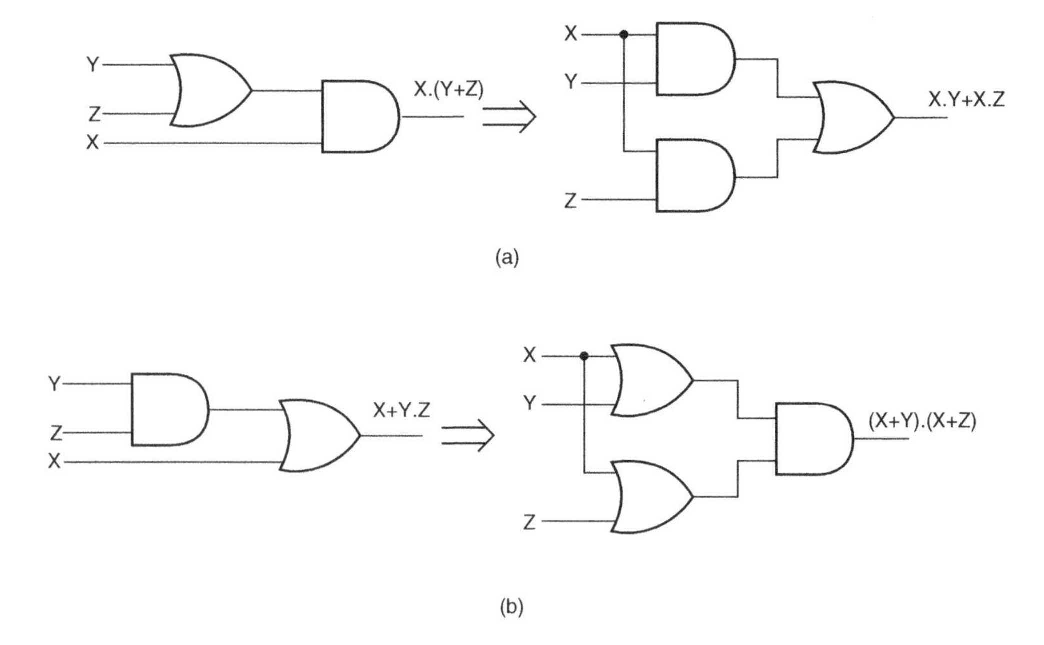

Figure 6.2 Distributive laws.

Theorem 7(b) can be used to simplify $(\overline{A}+\overline{B}).(\overline{A}+B).(A+\overline{B}).(A+B)$ as follows:

$$(\overline{A}+\overline{B}).(\overline{A}+B).(A+\overline{B}).(A+B) = (\overline{A}+\overline{B}.B).(A+\overline{B}.B) = (\overline{A}+0).(A+0) = \overline{A}.A = 0$$

6.3.8 Theorem 8

$$\text{(a) } X.Y + X.\overline{Y} = X \quad \text{and} \quad \text{(b) } (X+Y).(X+\overline{Y}) = X$$

This is a special case of theorem 7 as

$$X.Y + X.\overline{Y} = X.(Y+\overline{Y}) = X.1 = X \quad \text{and} \quad (X+Y).(X+\overline{Y}) = X + Y.\overline{Y} = X + 0 = X$$

This theorem, however, has another very interesting interpretation. Referring to theorem 8(a), there are two two-variable terms in the LHS expression. One of the variables, Y, is present in all possible combinations in this expression, while the other variable, X, is a common factor. The expression then reduces to this common factor. This interpretation can be usefully employed to simplify many a complex Boolean expression.

As an illustration, let us consider the following Boolean expression:

$$A.\overline{B}.\overline{C}.\overline{D} + A.\overline{B}.\overline{C}.D + A.\overline{B}.C.\overline{D} + A.\overline{B}.C.D + A.B.\overline{C}.\overline{D} + A.B.\overline{C}.D + A.B.C.\overline{D} + A.B.C.D$$

In the above expression, variables B, C and D are present in all eight possible combinations, and variable A is the common factor in all eight product terms. With the application of theorem 8(a), this expression reduces to A. Similarly, with the application of theorem 8(b), $(A+\overline{B}+\overline{C}).(A+\overline{B}+C).(A+B+\overline{C}).(A+B+C)$ also reduces to A as the variables B and C are present in all four possible combinations in sum terms and variable A is the common factor in all the terms.

6.3.9 Theorem 9

$$\text{(a) } (X+\overline{Y}).Y = X.Y \quad \text{and} \quad \text{(b) } X.\overline{Y}+Y = X+Y \tag{6.18}$$

$$(X+\overline{Y}).Y = X.Y+\overline{Y}.Y = X.Y$$

Theorem 9(b) is the dual of theorem 9(a) and hence stands proved.

6.3.10 Theorem 10 (Absorption Law or Redundancy Law)

$$\text{(a) } X+X.Y = X \quad \text{and} \quad \text{(b) } X.(X+Y) = X \tag{6.19}$$

The proof of absorption law is straightforward:

$$X+X.Y = X.(1+Y) = X.1 = X$$

Theorem 10(b) is the dual of theorem 10(a) and hence stands proved.

The crux of this simplification theorem is that, if a smaller term appears in a larger term, then the larger term is redundant. The following examples further illustrate the underlying concept:

$$A+A.\overline{B}+A.\overline{B}.\overline{C}+A.\overline{B}.C+\overline{C}.B.A = A$$

and

$$(\overline{A}+B+\overline{C}).(\overline{A}+B).(C+B+\overline{A}) = \overline{A}+B$$

6.3.11 Theorem 11

$$\text{(a) } Z.X+Z.\overline{X}.Y = Z.X+Z.Y$$

and

$$\text{(b) } (Z+X).(Z+\overline{X}+Y) = (Z+X).(Z+Y) \tag{6.20}$$

Table 6.2 gives the proof of theorem 11(a) using the method of perfect induction. Theorem 11(b) is the dual of theorem 11(a) and hence stands proved. A useful interpretation of this theorem is that, when

Table 6.2 Proof of theorem 11(a).

X	Y	Z	ZX	ZY	$Z\overline{X}$	$Z\overline{X}Y$	$ZX+Z\overline{X}Y$	ZX+ZY
0	0	0	0	0	0	0	0	0
0	0	1	0	0	1	0	0	0
0	1	0	0	0	0	0	0	0
0	1	1	0	1	1	1	1	1
1	0	0	0	0	0	0	0	0
1	0	1	1	0	0	0	1	1
1	1	0	0	0	0	0	0	0
1	1	1	1	1	0	0	1	1

a smaller term appears in a larger term except for one of the variables appearing as a complement in the larger term, the complemented variable is redundant.

As an example, $(A+\overline{B}).(\overline{A}+\overline{B}+C).(\overline{A}+\overline{B}+D)$ can be simplified as follows:

$$(A+\overline{B}).(\overline{A}+\overline{B}+C).(\overline{A}+\overline{B}+D)$$
$$=(A+\overline{B}).(\overline{B}+C).(\overline{A}+\overline{B}+D)=(A+\overline{B}).(\overline{B}+C).(\overline{B}+D)$$

6.3.12 Theorem 12 (Consensus Theorem)

$$\text{(a) } X.Y+\overline{X}.Z+Y.Z=X.Y+\overline{X}.Z$$

and

$$\text{(b) } (X+Y).(\overline{X}+Z).(Y+Z)=(X+Y).(\overline{X}+Z) \quad (6.21)$$

Table 6.3 shows the proof of theorem 12(a) using the method of perfect induction. Theorem 12(b) is the dual of theorem 12(a) and hence stands proved.

A useful interpretation of theorem 12 is as follows. If in a given Boolean expression we can identify two terms with one having a variable and the other having its complement, then the term that is formed by the product of the remaining variables in the two terms in the case of a sum-of-products expression

Table 6.3 Proof of theorem 12(a).

X	Y	Z	XY	$\overline{X}Z$	YZ	$XY+\overline{X}Z+YZ$	$XY+\overline{X}Z$
0	0	0	0	0	0	0	0
0	0	1	0	1	0	1	1
0	1	0	0	0	0	0	0
0	1	1	0	1	1	1	1
1	0	0	0	0	0	0	0
1	0	1	0	0	0	0	0
1	1	0	1	0	0	1	1
1	1	1	1	0	1	1	1

or by the sum of the remaining variables in the case of a product-of-sums expression will be redundant. The following example further illustrates the point:

$$A.B.C+\overline{A}.C.D+\overline{B}.C.D+B.C.D+A.C.D=A.B.C+\overline{A}.C.D+\overline{B}.C.D$$

If we consider the first two terms of the Boolean expression, $B.C.D$ becomes redundant. If we consider the first and third terms of the given Boolean expression, $A.C.D$ becomes redundant.

Example 6.4

Prove that $A.B.C.D+A.B.\overline{C}.\overline{D}+A.B.C.\overline{D}+A.B.\overline{C}.D+A.B.C.D.E+A.B.\overline{C}.\overline{D}.\overline{E}+A.B.\overline{C}.D.E$ *can be simplified to* $A.B$.

Solution

$$A.B.C.D+A.B.\overline{C}.\overline{D}+A.B.C.\overline{D}+A.B.\overline{C}.D+A.B.C.D.E+A.B.\overline{C}.\overline{D}.\overline{E}+A.B.\overline{C}.D.E$$

$$=A.B.C.D+A.B.\overline{C}.\overline{D}+A.B.C.\overline{D}+A.B.\overline{C}.D$$

$$=A.B.(C.D+\overline{C}.\overline{D}+C.\overline{D}+\overline{C}.D)=A.B$$

- $A.B.C.D$ appears in $A.B.C.D.E$, $A.B.\overline{C}.\overline{D}$ appears in $A.B.\overline{C}.\overline{D}.\overline{E}$ and $A.B.\overline{C}.D$ appears in $A.B.\overline{C}.D.E$.
- As a result, all three five-variable terms are redundant.
- Also, variables C and D appear in all possible combinations and are therefore redundant.

6.3.13 Theorem 13 (DeMorgan's Theorem)

$$\text{(a) } [\overline{X_1+X_2+X_3+\ldots+X_n}]=\overline{X_1}.\overline{X_2}.\overline{X_3}.\ldots.\overline{X_n} \qquad (6.22)$$

$$\text{(b) } [\overline{X_1.X_2.X_3.\ldots.X_n}]=[\overline{X_1}+\overline{X_2}+\overline{X_3}+\ldots+\overline{X_n}] \qquad (6.23)$$

According to the first theorem the complement of a sum equals the product of complements, while according to the second theorem the complement of a product equals the sum of complements. Figures 6.3(a) and (b) show logic diagram representations of De Morgan's theorems. While the first theorem can be interpreted to say that a multi-input NOR gate can be implemented as a multi-input bubbled AND gate, the second theorem, which is the dual of the first, can be interpreted to say that a multi-input NAND gate can be implemented as a multi-input bubbled OR gate.

DeMorgan's theorem can be proved as follows. Let us assume that all variables are in a logic '0' state. In that case

$$\text{LHS}=[\overline{X_1+X_2+X_3+\cdots+X_n}]=[\overline{0+0+0+\cdots+0}]=\overline{0}=1$$

$$\text{RHS}=\overline{X_1}.\overline{X_2}.\overline{X_3}.\ldots.\overline{X_n}=\overline{0}.\overline{0}.\overline{0}.\ldots.\overline{0}=1.1.1.\ldots.1=1$$

Therefore, LHS = RHS.

Now, let us assume that any one of the n variables, say X_1, is in a logic HIGH state:

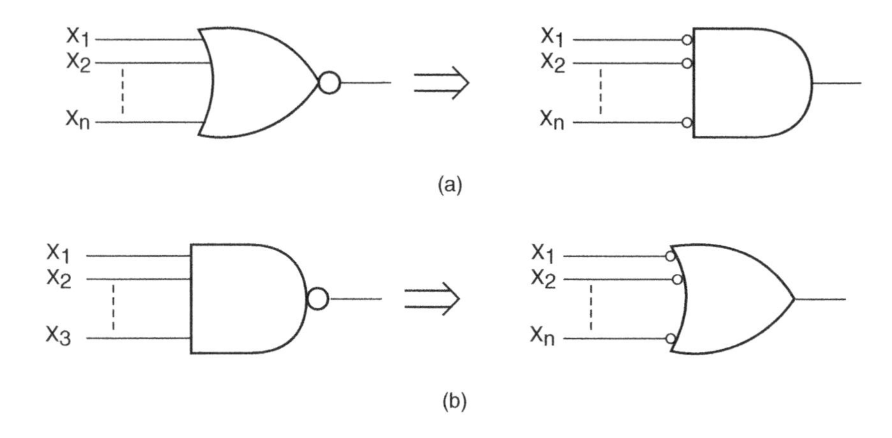

Figure 6.3 DeMorgan's theorem.

$$\text{LHS} = [\overline{X_1 + X_2 + X_3 + \cdots X_n}] = [\overline{1+0+0+\cdots+0}] = \overline{1} = 0$$

$$\text{RHS} = \overline{X_1}.\overline{X_2}.\overline{X_3}.\ \ldots\ .\overline{X_n} = \overline{1}.\overline{0}.\overline{0}.\ \ldots\ .\overline{0} = 0.1.1.\ \ldots\ .1 = 0$$

Therefore, again LHS = RHS.

The same holds good when more than one or all variables are in the logic '1' state. Therefore, theorem 13(a) stands proved. Since theorem 13(b) is the dual of theorem 13(a), the same also stands proved. Theorem 13(b), though, can be proved on similar lines.

6.3.14 Theorem 14 (Transposition Theorem)

$$\text{(a) } X.Y + \overline{X}.Z = (X+Z).(\overline{X}+Y)$$

and

$$\text{(b) } (X+Y).(\overline{X}+Z) = X.Z + \overline{X}.Y \tag{6.24}$$

This theorem can be applied to any sum-of-products or product-of-sums expression having two terms, provided that a given variable in one term has its complement in the other. Table 6.4 gives the proof of theorem 14(a) using the method of perfect induction. Theorem 14(b) is the dual of theorem 14(a) and hence stands proved.

As an example,

$$\overline{A}.B + A.\overline{B} = (A+B).(\overline{A}+\overline{B}) \quad \text{and} \quad A.B + \overline{A}.\overline{B} = (A+\overline{B}).(\overline{A}+B)$$

Incidentally, the first expression is the representation of a two-input EX-OR gate, while the second expression gives two forms of representation of a two-input EX-NOR gate.

Table 6.4 Proof of theorem 13(a).

X	Y	Z	XY	$\overline{X}Z$	X+Z	$\overline{X}+Y$	$XY+\overline{X}Z$	(X+Z)($\overline{X}+Y$)
0	0	0	0	0	0	1	0	0
0	0	1	0	1	1	1	1	1
0	1	0	0	0	0	1	0	0
0	1	1	0	1	1	1	1	1
1	0	0	0	0	1	0	0	0
1	0	1	0	0	1	0	0	0
1	1	0	1	0	1	1	1	1
1	1	1	1	0	1	1	1	1

6.3.15 Theorem 15

$$\text{(a) } X.f(X,\overline{X},Y,Z,\ldots)=X.f(1,0,Y,Z,\ldots) \quad (6.25)$$

$$\text{(b) } X+f(X,\overline{X},Y,Z,\ldots)=X+f(0,1,Y,Z,\ldots) \quad (6.26)$$

According to theorem 15(a), if a variable X is multiplied by an expression containing X and $\overline{X}$ in addition to other variables, then all Xs and $\overline{X}$s can be replaced with 1s and 0s respectively. This would be valid as $X.X=X$ and $X.1=X$. Also, $X.\overline{X}=0$ and $X.0=0$. According to theorem 15(b), if a variable X is added to an expression containing terms having X and $\overline{X}$ in addition to other variables, then all Xs can be replaced with 0s and all $\overline{X}$s can be replaced with ls. This is again permissible as $X+X$ as well as $X+0$ equals X. Also, $X+\overline{X}$ and $\overline{X}+1$ both equal 1.

This pair of theorems is very useful in eliminating redundancy in a given expression. An important corollary of this pair of theorems is that, if the multiplying variable is $\overline{X}$ in theorem 15(a), then all Xs will be replaced by 0s and all $\overline{X}$s will be replaced by ls. Similarly, if the variable being added in theorem 15(b) is $\overline{X}$, then Xs and $\overline{X}$s in the expression are replaced by 1s and 0s respectively. In that case the two theorems can be written as follows:

$$\text{(a) } \overline{X}.f(X,\overline{X},Y,Z,\ldots)=\overline{X}.f(0,1,Y,Z,\ldots) \quad (6.27)$$

$$\text{(b) } \overline{X}+f(X,\overline{X},Y,Z,\ldots)=\overline{X}+f(1,0,Y,Z,\ldots) \quad (6.28)$$

The theorems are further illustrated with the help of the following examples:

1. $A.[\overline{A}.B+A.\overline{C}+(\overline{A}+D).(A+\overline{E})]=A.[0.B+1.\overline{C}+(0+D).(1+\overline{E})]=A.(\overline{C}+D)$.
2. $\overline{A}+[\overline{A}.B+A.\overline{C}+(\overline{A}+B).(A+\overline{E})]=\overline{A}+[0.B+1.\overline{C}+(0+B).(1+\overline{E})]=\overline{A}+\overline{C}+B$.

6.3.16 Theorem 16

$$\text{(a) } f(X,\overline{X},Y,\ldots,Z)=X.f(1,0,Y,\ldots,Z)+\overline{X}.f(0,1,Y,\ldots,Z) \quad (6.29)$$

$$\text{(b) } f(X,\overline{X},Y,\ldots,Z)=[X+f(0,1,Y,\ldots,Z)][\overline{X}+f(1,0,Y,\ldots,Z)] \quad (6.30)$$

The proof of theorem 16(a) is straightforward and is given as follows:

$$f(X,\overline{X},Y,\ldots,Z) = X.f(X,\overline{X},Y,\ldots,Z)+\overline{X}.f(X,\overline{X},Y,\ldots,Z)$$
$$= X.f(1,0,Y,\ldots,Z)+\overline{X}.f(0,1,Y,\ldots,Z)[(\text{Theorem 15(a)}]$$

Also

$$f(X,\overline{X},Y,\ldots,Z) = [X+f(X,\overline{X},Y,\ldots,Z)][\overline{X}+.f(X,\overline{X},Y,\ldots,Z)]$$
$$= [X+f(0,1,Y,\ldots,Z)][\overline{X}+f(1,0,Y,\ldots,Z)][\text{Theorem 15(b)}]$$

6.3.17 Theorem 17 (Involution Law)

$$\overline{\overline{X}} = X \tag{6.31}$$

Involution law says that the complement of the complement of an expression leaves the expression unchanged. Also, the dual of the dual of an expression is the original expression. This theorem forms the basis of finding the equivalent product-of-sums expression for a given sum-of-products expression, and vice versa.

Example 6.5

Prove the following:

1. $L.(M+\overline{N})+\overline{L}.\overline{P}.Q = (L+\overline{P}.Q).(\overline{L}+M+\overline{N})$.
2. $[A.\overline{B}+\overline{C}+\overline{D}].[D+(E+\overline{F}).G] = D.(A.\overline{B}+\overline{C})+\overline{D}.G.(E+\overline{F})$.

Solution

1. Let us assume that $L = X$, $(M+\overline{N}) = Y$ and $\overline{P}.Q = Z$.
   The LHS of the given Boolean equation then reduces to $X.Y+\overline{X}.Z$.
   Applying the transposition theorem,

$$X.Y+\overline{X}.Z = (X+Z).(\overline{X}+Y) = (L+\overline{P}.Q)(\overline{L}+M+\overline{N}) = \text{RHS}$$

2. Let us assume $\overline{D} = X$, $A.\overline{B}+\overline{C} = Y$ and $(E+\overline{F}).G = Z$.
   The LHS of given the Boolean equation then reduces to $(X+Y).(\overline{X}+Z)$.
   Applying the transposition theorem,

$$(X+Y).(\overline{X}+Z) = X.Z+\overline{X}.Y = \overline{D}.G.(E+\overline{F})+D.(A.\overline{B}+\overline{C}) = \text{RHS}$$

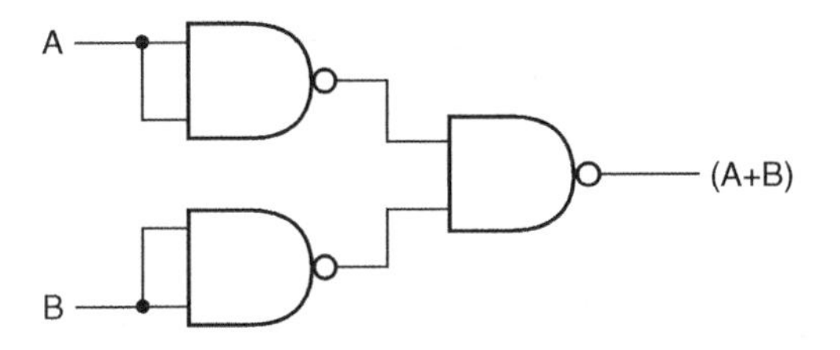

Figure 6.4 Example 6.6.

Example 6.6

Starting with the Boolean expression for a two-input OR gate, apply Boolean laws and theorems to modify it in such a way as to facilitate the implementation of a two-input OR gate by using two-input NAND gates only.

Solution

- A two-input OR gate is represented by the Boolean equation $Y = (A + B)$, where A and B are the input logic variables and Y is the output.
- Now, $(A + B) = (\overline{\overline{A}} + \overline{\overline{B}})$ Involution law

  $= (\overline{\overline{A}.\overline{B}})$ DeMorgan's theorem

  $= [\overline{(\overline{A.A}).(\overline{B.B})}]$ Idempotent law
- Figure 6.4 shows the NAND gate implementation of a two-input OR gate.

Example 6.7

Apply suitable Boolean laws and theorems to modify the expression for a two-input EX-OR gate in such a way as to implement a two-input EX-OR gate by using the minimum number of two-input NAND gates only.

Solution

- A two-input EX-OR gate is represented by the Boolean expression $Y = \overline{A}.B + A.\overline{B}$.
- Now, $\overline{A}.B + A.\overline{B} = \overline{\overline{\overline{A}.B}} + \overline{\overline{A.\overline{B}}}$ Involution law

  $= \overline{\overline{\overline{A}.B}.\overline{A.\overline{B}}}$ DeMorgan's law

  $= \overline{[\overline{B.(\overline{A}+\overline{B})}].[\overline{A.(\overline{A}+\overline{B})}]}$

  $$= \overline{(\overline{B.\overline{A.B}}).(\overline{A.\overline{A.B}})} \qquad (6.32)$$
- Equation (6.32) is in a form that can be implemented with NAND gates only.
- Figure 6.5 shows the logic diagram.

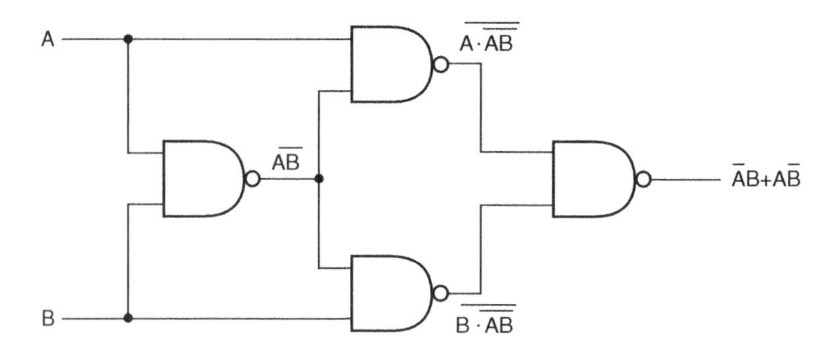

Figure 6.5 Example 6.7.

6.4 Simplification Techniques

In this section, we will discuss techniques other than the application of laws and theorems of Boolean algebra discussed in the preceding paragraphs of this chapter for simplifying or more precisely minimizing a given complex Boolean expression. The primary objective of all simplification procedures is to obtain an expression that has the minimum number of terms. Obtaining an expression with the minimum number of literals is usually the secondary objective. If there is more than one possible solution with the same number of terms, the one having the minimum number of literals is the choice. The techniques to be discussed include:

(a) the Quine–McCluskey tabular method;
(b) the Karnaugh map method.

Before we move on to discuss these techniques in detail, it would be relevant briefly to describe sum-of-products and product-of-sums Boolean expressions. The given Boolean expression will be in either of the two forms, and the objective will be to find a minimized expression in the same or the other form.

6.4.1 *Sum-of-Products Boolean Expressions*

A sum-of-products expression contains the sum of different terms, with each term being either a single literal or a product of more than one literal. It can be obtained from the truth table directly by considering those input combinations that produce a logic '1' at the output. Each such input combination produces a term. Different terms are given by the product of the corresponding literals. The sum of all terms gives the expression. For example, the truth table in Table 6.5 can be represented by the Boolean expression

$$Y=\overline{A}.\overline{B}.\overline{C}+\overline{A}.B.C+A.B.\overline{C}+A.\overline{B}.C \tag{6.33}$$

Considering the first term, the output is '1' when $A=0$, $B=0$ and $C=0$. This is possible only when $\overline{A}$, $\overline{B}$ and $\overline{C}$ are ANDed. Also, for the second term, the output is '1' only when B, C and $\overline{A}$ are ANDed. Other terms can be explained similarly. A sum-of-products expression is also known as a *minterm expression*.

Table 6.5 truth table of boolean expression of equation 6.33.

A	B	C	Y
0	0	0	1
0	0	1	0
0	1	0	0
0	1	1	1
1	0	0	0
1	0	1	1
1	1	0	1
1	1	1	0

6.4.2 Product-of-Sums Expressions

A product-of-sums expression contains the product of different terms, with each term being either a single literal or a sum of more than one literal. It can be obtained from the truth table by considering those input combinations that produce a logic '0' at the output. Each such input combination gives a term, and the product of all such terms gives the expression. Different terms are obtained by taking the sum of the corresponding literals. Here, '0' and '1' respectively mean the uncomplemented and complemented variables, unlike sum-of-products expressions where '0' and '1' respectively mean complemented and uncomplemented variables.

To illustrate this further, consider once again the truth table in Table 6.5. Since each term in the case of the product-of-sums expression is going to be the sum of literals, this implies that it is going to be implemented using an OR operation. Now, an OR gate produces a logic '0' only when all its inputs are in the logic '0' state, which means that the first term corresponding to the second row of the truth table will be $A+B+\overline{C}$. The product-of-sums Boolean expression for this truth table is given by $(A+B+\overline{C}).(A+\overline{B}+C).(\overline{A}+B+C).(\overline{A}+\overline{B}+\overline{C})$.

Transforming the given product-of-sums expression into an equivalent sum-of-products expression is a straightforward process. Multiplying out the given expression and carrying out the obvious simplification provides the equivalent sum-of-products expression:

$$
\begin{aligned}
&(A+B+\overline{C}).(A+\overline{B}+C).(\overline{A}+B+C).(\overline{A}+\overline{B}+\overline{C})\\
&=(A.A+A.\overline{B}+A.C+B.A+B.\overline{B}+B.C+\overline{C}.A+\overline{C}.\overline{B}+\overline{C}.C).(\overline{A}.\overline{A}+\overline{A}.\overline{B}+\overline{A}.\overline{C}+B.\overline{A}+B.\overline{B}\\
&\quad+B.\overline{C}+C.\overline{A}+C.\overline{B}+C.\overline{C}\\
&=(A+B.C+\overline{B}.\overline{C}).(\overline{A}+B.\overline{C}+C.\overline{B})=A.B.\overline{C}+A.\overline{B}.C+\overline{A}.B.C+\overline{A}.\overline{B}.\overline{C}
\end{aligned}
$$

A given sum-of-products expression can be transformed into an equivalent product-of-sums expression by (a) taking the dual of the given expression, (b) multiplying out different terms to get the sum-of-products form, (c) removing redundancy and (d) taking a dual to get the equivalent product-of-sums expression. As an illustration, let us find the equivalent product-of-sums expression of the sum-of-products expression

$$A.B+\overline{A}.\overline{B}$$

The dual of the given expression $=(A+B).(\overline{A}+\overline{B})$:

$$(A+B).(\overline{A}+\overline{B})=A.\overline{A}+A.\overline{B}+B.\overline{A}+B.\overline{B}=0+A.\overline{B}+B.\overline{A}+0=A.\overline{B}+\overline{A}.B$$

The dual of $(A.\overline{B}+\overline{A}.B)=(A+\overline{B}).(\overline{A}+B)$. Therefore

$$A.B+\overline{A}.\overline{B}=(A+\overline{B}).(\overline{A}+B)$$

6.4.3 Expanded Forms of Boolean Expressions

Expanded sum-of-products and product-of-sums forms of Boolean expressions are useful not only in analysing these expressions but also in the application of minimization techniques such as the Quine–McCluskey tabular method and the Karnaugh mapping method for simplifying given Boolean expressions. The expanded form, sum-of-products or product-of-sums, is obtained by including all possible combinations of missing variables.

As an illustration, consider the following sum-of-products expression:

$$A.\overline{B}+B.\overline{C}+A.B.\overline{C}+\overline{A}.C$$

It is a three-variable expression. Expanded versions of different minterms can be written as follows:

- $A.\overline{B}=A.\overline{B}.(C+\overline{C})=A.\overline{B}.C+A.\overline{B}.\overline{C}$.
- $B.\overline{C}=B.\overline{C}.(A+\overline{A})=B.\overline{C}.A+B.\overline{C}.\overline{A}$.
- $A.B.\overline{C}$ is a complete term and has no missing variable.
- $\overline{A}.C=\overline{A}.C.(B+\overline{B})=\overline{A}.C.B+\overline{A}.C.\overline{B}$.

The expanded sum-of-products expression is therefore given by

$$A.\overline{B}.C+A.\overline{B}.\overline{C}+A.B.\overline{C}+\overline{A}.B.\overline{C}+A.B.\overline{C}+\overline{A}.B.C+\overline{A}.\overline{B}.C=A.\overline{B}.C+A.\overline{B}.\overline{C}$$
$$+A.B.\overline{C}+\overline{A}.B.\overline{C}+\overline{A}.B.C+\overline{A}.\overline{B}.C$$

As another illustration, consider the product-of-sums expression

$$(\overline{A}+B).(\overline{A}+B+\overline{C}+\overline{D})$$

It is four-variable expression with A, B, C and D being the four variables. $\overline{A}+B$ in this case expands to $(\overline{A}+B+C+D).(\overline{A}+B+C+\overline{D}).(\overline{A}+B+\overline{C}+D).(\overline{A}+B+\overline{C}+\overline{D})$.

The expanded product-of-sums expression is therefore given by

$$(\overline{A}+B+C+D).(\overline{A}+B+C+\overline{D}).(\overline{A}+B+\overline{C}+D).(\overline{A}+B+\overline{C}+\overline{D}).(\overline{A}+B+\overline{C}+\overline{D})$$
$$=(\overline{A}+B+C+D).(\overline{A}+B+C+\overline{D}).(\overline{A}+B+\overline{C}+D).(\overline{A}+B+\overline{C}+\overline{D})$$

6.4.4 Canonical Form of Boolean Expressions

An expanded form of Boolean expression, where each term contains all Boolean variables in their true or complemented form, is also known as the *canonical form* of the expression.

As an illustration, $f(A.B,C)=\overline{A}.\overline{B}.\overline{C}+\overline{A}.\overline{B}.C+A.B.C$ is a Boolean function of three variables expressed in canonical form. This function after simplification reduces to $\overline{A}.\overline{B}+A.B.C$ and loses its canonical form.

6.4.5 Σ and Π Nomenclature

Σ and Π notations are respectively used to represent sum-of-products and product-of-sums Boolean expressions. We will illustrate these notations with the help of examples. Let us consider the following Boolean function:

$$f(A, B, C, D) = A.\overline{B}.\overline{C} + A.B.C.D + \overline{A}.B.\overline{C}.D + \overline{A}.\overline{B}.\overline{C}.D$$

We will represent this function using Σ notation. The first step is to write the expanded sum-of-products given by

$$\begin{aligned} f(A, B, C, D) &= A.\overline{B}.\overline{C}.(D + \overline{D}) + A.B.C.D + \overline{A}.B.\overline{C}.D + \overline{A}.\overline{B}.\overline{C}.D \\ &= A.\overline{B}.\overline{C}.D + A.\overline{B}.\overline{C}.\overline{D} + A.B.C.D + \overline{A}.B.\overline{C}.D + \overline{A}.\overline{B}.\overline{C}.D \end{aligned}$$

Different terms are then arranged in ascending order of the binary numbers represented by various terms, with true variables representing a '1' and a complemented variable representing a '0'. The expression becomes

$$f(A, B, C, D) = \overline{A}.\overline{B}.\overline{C}.D + \overline{A}.B.\overline{C}.D + A.\overline{B}.\overline{C}.\overline{D} + A.\overline{B}.\overline{C}.D + A.B.C.D$$

The different terms represent 0001, 0101, 1000, 1001 and 1111. The decimal equivalent of these terms enclosed in the Σ then gives the Σ notation for the given Boolean function. That is, $f(A, B, C, D) = \sum 1, 5, 8, 9, 15$.

The complement of $f(A, B, C, D)$, that is, $f'(A, B, C, D)$, can be directly determined from Σ notation by including the left-out entries from the list of all possible numbers for a four-variable function. That is,

$$f'(A, B, C, D) = \sum 0, 2, 3, 4, 6, 7, 10, 11, 12, 13, 14$$

Let us now take the case of a product-of-sums Boolean function and its representation in Π nomenclature. Let us consider the Boolean function

$$f(A, B, C, D) = (B + \overline{C} + \overline{D}).(\overline{A} + \overline{B} + C + D).(A + \overline{B} + \overline{C} + \overline{D})$$

The expanded product-of-sums form is given by

$$(A + B + \overline{C} + \overline{D}).(\overline{A} + B + \overline{C} + \overline{D}).(\overline{A} + \overline{B} + C + D).(A + \overline{B} + \overline{C} + \overline{D})$$

The binary numbers represented by the different sum terms are 0011, 1011, 1100 and 0111 (true and complemented variables here represent 0 and 1 respectively). When arranged in ascending order, these numbers are 0011, 0111, 1011 and 1100. Therefore,

$$f(A, B, C, D) = \prod 3, 7, 11, 12 \quad \text{and} \quad f'(A, B, C, D) = \prod 0, 1, 2, 4, 5, 6, 8, 9, 10, 13, 14, 15$$

An interesting corollary of what we have discussed above is that, if a given Boolean function $f(A,B,C)$ is given by $f(A, B, C) = \sum 0, 1, 4, 7$, then

$$f(A, B, C) = \prod 2, 3, 5, 6 \quad \text{and} \quad f'(A, B, C) = \sum 2, 3, 5, 6 = \prod 0, 1, 4, 7$$

Optional combinations can also be incorporated into Σ and Π nomenclature using suitable identifiers; ϕ or d are used as identifiers. For example, if $f(A, B, C) = \overline{A}.\overline{B}.\overline{C} + A.\overline{B}.\overline{C} + A.\overline{B}.C$ and $\overline{A}.B.C$, $A.B.C$ are optional combinations, then

$$f(A,B,C) = \sum 0,4,5 + \sum_{\phi} 3,7 = \sum 0,4,5 + \sum_{d} 3,7$$

$$f(A,B,C) = \prod 1,2,6 + \prod_{\phi} 3,7 = \prod 1,2,6 + \prod_{d} 3,7$$

Example 6.8

For a Boolean function $f(A, B) = \sum 0, 2$, *prove that* $f(A, B) = \prod 1, 3$ *and* $f'(A, B) = \sum 1, 3 = \prod 0, 2$.

Solution

- $f(A, B) = \sum 0, 2 = \overline{A}.\overline{B} + A.\overline{B} = \overline{B}.(A + \overline{A}) = \overline{B}$.
- Now, $\prod 1, 3 = (A + \overline{B}).(\overline{A} + \overline{B}) = A.\overline{A} + A.\overline{B} + \overline{B}.\overline{A} + \overline{B}.\overline{B} = A.\overline{B} + \overline{A}.\overline{B} + \overline{B} = \overline{B}$.
- Now, $\sum 1, 3 = \overline{A}.B + A.B = B.(\overline{A} + A) = B$.
  and $\prod 0, 2 = (A + B).(\overline{A} + B) = A.\overline{A} + A.B + B.\overline{A} + B.B = A.B + \overline{A}.B + B = B$.
- Therefore, $\sum 1, 3 = \prod 0, 2$.
- Also, $f(A, B) = \overline{B}$.
- Therefore, $f'(A, B) = B$ or $f'(A, B) = \sum 1, 3 = \prod 0, 2$.

6.5 Quine–McCluskey Tabular Method

The Quine–McCluskey tabular method of simplification is based on the complementation theorem, which says that

$$X.Y + X.\overline{Y} = X \tag{6.34}$$

where X represents either a variable or a term or an expression and Y is a variable. This theorem implies that, if a Boolean expression contains two terms that differ only in one variable, then they can be combined together and replaced with a term that is smaller by one literal. The same procedure is applied for the other pairs of terms wherever such a reduction is possible. All these terms reduced by one literal are further examined to see if they can be reduced further. The process continues until the terms become irreducible. The irreducible terms are called *prime implicants*. An optimum set of prime implicants that can account for all the original terms then constitutes the minimized expression. The technique can be applied equally well for minimizing sum-of-products and product-of-sums expressions and is particularly useful for Boolean functions having more than six variables as it can be mechanized and run on a computer. On the other hand, the Karnaugh mapping method, to be discussed later, is a graphical method and becomes very cumbersome when the number of variables exceeds six.

The step-by-step procedure for application of the tabular method for minimizing Boolean expressions, both sum-of-products and product-of-sums, is outlined as follows:

1. The Boolean expression to be simplified is expanded if it is not in expanded form.
2. Different terms in the expression are divided into groups depending upon the number of 1s they have. True and complemented variables in a sum-of-products expression mean '1' and '0' respectively.

The reverse is true in the case of a product-of-sums expression. The groups are then arranged, beginning with the group having the least number of 1s in its included terms. Terms within the same group are arranged in ascending order of the decimal numbers represented by these terms.

As an illustration, consider the expression

$$A.B.C+\overline{A}.B.C+A.\overline{B}.\overline{C}+A.\overline{B}.C+\overline{A}.\overline{B}.\overline{C}$$

The grouping of different terms and the arrangement of different terms within the group are shown below:

Term		Binary	Group
$\overline{A}.\overline{B}.\overline{C}$		000	First group
$A.\overline{B}.\overline{C}$	$\longrightarrow$	100	Second group
$\overline{A}.B.C$		011	Third group
$A.\overline{B}.C$		101	
ABC		111	Fourth group

As another illustration, consider a product-of-sums expression given by

$$(\overline{A}+\overline{B}+\overline{C}+\overline{D}).(\overline{A}+\overline{B}+\overline{C}+D).(\overline{A}+B+\overline{C}+D).(A+B+\overline{C}+\overline{D}).(A+B+C+D).$$
$$(A+\overline{B}+\overline{C}+\overline{D}.(A+\overline{B}+C+\overline{D})$$

The formation of groups and the arrangement of terms within different groups for the product-of-sums expression are as follows:

Term		Binary
$A.B.C.D$		0000
$A.B.\overline{C}.\overline{D}$		0011
$A.\overline{B}.C.\overline{D}$		0101
$\overline{A}.B.\overline{C}.D$	$\longrightarrow$	1010
$A.\overline{B}.\overline{C}.\overline{D}$		0111
$\overline{A}.\overline{B}.\overline{C}.D$		1110
$\overline{A}.\overline{B}.\overline{C}.\overline{D}$		1111

It may be mentioned here that the Boolean expressions that we have considered above did not contain any optional terms. If there are any, they are also considered while forming groups. This completes the first table.

3. The terms of the first group are successively matched with those in the next adjacent higher-order group to look for any possible matching and consequent reduction. The terms are considered matched when all literals except for one match. The pairs of matched terms are replaced with a

single term where the position of the unmatched literals is replaced with a dash (—). These new terms formed as a result of the matching process find a place in the second table. The terms in the first table that do not find a match are called the prime implicants and are marked with an asterisk (∗). The matched terms are ticked (✓).
4. Terms in the second group are compared with those in the third group to look for a possible match. Again, terms in the second group that do not find a match become the prime implicants.
5. The process continues until we reach the last group. This completes the first round of matching. The terms resulting from the matching in the first round are recorded in the second table.
6. The next step is to perform matching operations in the second table. While comparing the terms for a match, it is important that a dash (—) is also treated like any other literal, that is, the dash signs also need to match. The process continues on to the third table, the fourth table and so on until the terms become irreducible any further.
7. An optimum selection of prime implicants to account for all the original terms constitutes the terms for the minimized expression. Although optional (also called 'don't care') terms are considered for matching, they do not have to be accounted for once prime implicants have been identified.

Let us consider an example. Consider the following sum-of-products expression:

$$\overline{A}.B.C+\overline{A}.\overline{B}.D+A.\overline{C}.D+B.\overline{C}.\overline{D}+\overline{A}.B.\overline{C}.D \tag{6.35}$$

In the first step, we write the expanded version of the given expression. It can be written as follows:

$$\overline{A}.B.C.D+\overline{A}.B.C.\overline{D}+\overline{A}.\overline{B}.C.D+\overline{A}.\overline{B}.\overline{C}.D+A.B.\overline{C}.D+A.\overline{B}.\overline{C}.D+A.B.\overline{C}.\overline{D}$$
$$+\overline{A}.B.\overline{C}.\overline{D}+\overline{A}.B.\overline{C}.D$$

The formation of groups, the placement of terms in different groups and the first-round matching are shown as follows:

A	B	C	D
0	0	0	1
0	0	1	1
0	1	0	0
0	1	0	1
0	1	1	0
0	1	1	1
1	0	0	1
1	1	0	0
1	1	0	1

A	B	C	D	
0	0	0	1	✓
0	1	0	0	✓
0	0	1	1	✓
0	1	0	1	✓
0	1	1	0	✓
1	0	0	1	✓
1	1	0	0	✓
0	1	1	1	✓
1	1	0	1	✓

A	B	C	D	
0	0	–	1	✓
0	–	0	1	✓
–	0	0	1	✓
0	1	0	–	✓
0	1	–	0	✓
–	1	0	0	✓
0	–	1	1	✓
0	1	–	1	✓
–	1	0	1	✓
0	1	1	–	✓
1	–	0	1	✓
1	1	0	–	✓

The second round of matching begins with the table shown on the previous page. Each term in the first group is compared with every term in the second group. For instance, the first term in the first group 00–1 matches with the second term in the second group 01–1 to yield 0––1, which is recorded in the table shown below. The process continues until all terms have been compared for a possible match. Since this new table has only one group, the terms contained therein are all prime implicants. In the present example, the terms in the first and second tables have all found a match. But that is not always the case.

A	B	C	D	
0	–	–	1	*
–	–	0	1	*
0	1	–	–	*
–	1	0	–	*

The next table is what is known as the prime implicant table. The prime implicant table contains all the original terms in different columns and all the prime implicants recorded in different rows as shown below:

0001	0011	0100	0101	0110	0111	1001	1100	1101		
✓	✓		✓		✓				0––1	$P \rightarrow \overline{A}.D$
✓			✓			✓		✓	––01	$Q \rightarrow \overline{C}.D$
		✓	✓	✓	✓				01––	$R \rightarrow \overline{A}.B$
		✓	✓				✓	✓	–10–	$S \rightarrow B.\overline{C}$

Each prime implicant is identified by a letter. Each prime implicant is then examined one by one and the terms it can account for are ticked as shown. The next step is to write a product-of-sums expression using the prime implicants to account for all the terms. In the present illustration, it is given as follows.

$$(P+Q).(P).(R+S).(P+Q+R+S).(R).(P+R).(Q).(S).(Q+S)$$

Obvious simplification reduces this expression to $PQRS$ which can be interpreted to mean that all prime implicants, that is, P, Q, R and S, are needed to account for all the original terms.

Therefore, the minimized expression $= \overline{A}.D+\overline{C}.D+\overline{A}.B+B.\overline{C}$.

What has been described above is the formal method of determining the optimum set of prime implicants. In most of the cases where the prime implicant table is not too complex, the exercise can be done even intuitively. The exercise begins with identification of those terms that can be accounted for by only a single prime implicant. In the present example, 0011, 0110, 1001 and 1100 are such terms. As a result, P, Q, R and S become the essential prime implicants. The next step is to find out if any terms have not been covered by the essential prime implicants. In the present case, all terms have been covered by essential prime implicants. In fact, all prime implicants are essential prime implicants in the present example.

As another illustration, let us consider a product-of-sums expression given by

$$(\overline{A}+\overline{B}+\overline{C}+\overline{D}).(\overline{A}+\overline{B}+\overline{C}+D).(\overline{A}+\overline{B}+C+\overline{D}).(A+\overline{B}+\overline{C}+\overline{D}).(A+\overline{B}+C+\overline{D})$$

The procedure is similar to that described for the case of simplification of sum-of-products expressions. The resulting tables leading to identification of prime implicants are as follows:

A	B	C	D
0	1	0	1
0	1	1	1
1	1	0	1
1	1	1	0
1	1	1	1

A	B	C	D	
0	1	0	1	✓
0	1	1	1	✓
1	1	0	1	✓
1	1	1	0	✓
1	1	1	1	✓

A	B	C	D	
0	1	–	1	✓
–	1	0	1	✓
–	1	1	1	✓
1	1	–	1	✓
1	1	1	–	*

A	B	C	D	
–	1	–	1	*

The prime implicant table is constructed after all prime implicants have been identified to look for the optimum set of prime implicants needed to account for all the original terms. The prime implicant table shows that both the prime implicants are the essential ones:

0101	0111	1101	1110	1111	Prime implicants
			✓	✓	111–
✓	✓	✓		✓	–1–1

The minimized expression = $(\overline{A}+\overline{B}+\overline{C}).(\overline{B}+\overline{D})$.

6.5.1 *Tabular Method for Multi-Output Functions*

When it comes to a multi-output logic network, a network that has more than one output, sharing of some logic blocks between different functions is highly probable. For an optimum logic implementation of the multi-output function, different functions cannot be and should not be minimized in isolation because a possible common term that could have been shared may not turn out to be a prime implicant if the functions are worked out individually. The method of applying the tabular approach to multi-output functions is to get a minimized set of expressions that would lead to an optimum overall system. The method is illustrated by the following example.

Consider a logic system with two outputs that is described by the following Boolean expressions:

$$Y_1 = \overline{A}.B.D + \overline{A}.C.D + \overline{A}.\overline{C}.\overline{D} \tag{6.36}$$

$$Y_2 = \overline{A}.B.C + A.C.D + A.\overline{B}.C.\overline{D} + \overline{A}.\overline{B}.C.\overline{D} \tag{6.37}$$

The expanded forms of the two functions are as follows:

$$Y_1 = \overline{A}.B.C.D + \overline{A}.B.\overline{C}.D + \overline{A}.B.C.D + \overline{A}.\overline{B}.C.D + \overline{A}.B.\overline{C}.\overline{D} + \overline{A}.\overline{B}.\overline{C}.\overline{D}$$

$$Y_1 = \overline{A}.B.C.D + \overline{A}.B.\overline{C}.D + \overline{A}.\overline{B}.C.D + \overline{A}.B.\overline{C}.\overline{D} + \overline{A}.\overline{B}.\overline{C}.\overline{D}$$

$$Y_2 = \overline{A}.B.C.D + \overline{A}.B.C.\overline{D} + A.B.C.D + A.\overline{B}.C.D + A.\overline{B}.C.\overline{D} + \overline{A}.\overline{B}.C.\overline{D}$$

The rows representing different terms are arranged in the usual manner, with all the terms contained in the two functions finding a place without repetition, as shown in the table below:

ABCD	1	2	
0000	✓		✓
0010		✓	✓
0100	✓		✓
0011	✓		✓
0101	✓		✓
0110		✓	✓
1010		✓	✓
0111	✓	✓	✓
1011		✓	✓
1111		✓	✓

Each term is checked under the column or columns depending upon the functions in which it is contained. For instance, if a certain term is contained in the logic expressions for both output 1 and output 2, it will be checked in both output columns. The matching process begins in the same way as described earlier for the case of single-output functions, with some modifications outlined as follows:

1. Only those terms can be combined that have at least one check mark in the output column in common. For instance, 0000 cannot combine with 0010 but can combine with 0100.
2. In the resulting row, only the common outputs are checked. For instance, when 0101 is matched with 0111, then, in the resulting term 01–1, only output 1 will be checked.
3. A combining term can be checked off only if the resulting term accounts for all the outputs in which the term is contained.

The table below shows the results of the first round of matching:

ABCD	1	2	
0–00	✓		*
0–10		✓	*
–010		✓	*
010–	✓		*
0–11	✓		*
01–1	✓		*
011–		✓	*
101–		✓	*
–111		✓	*
1–11		✓	*

No further matching is possible. The prime implicant table is shown below:

Output 1					Output 2						
0000	0011	0100	0101	0111	0010	0110	0111	1010	1011	1111	ABCD
✓		✓									0–00
					✓	✓					0–10
					✓			✓			–010
		✓	✓								010–
	✓			✓							0–11
			✓	✓							01–1
						✓	✓				011–
								✓	✓		101–
							✓			✓	–111
									✓	✓	1–11

For each prime implicant, check marks are placed only in columns that pertain to the outputs checked off for this prime implicant. For instance, 0-00 has only output 1 checked off. Therefore, the relevant terms under output 1 will be checked off. The completed table is treated as a whole while marking the required prime implicants to be considered for writing the minimized expressions. The minimized expressions are as follows:

$$Y_1 = \overline{A}.\overline{C}.\overline{D} + \overline{A}.C.D + \overline{A}.B.\overline{C} \quad \text{and} \quad Y_2 = B.C.D + A.\overline{B}.C + \overline{A}.C.\overline{D}$$

Example 6.9

Using the Quine–McCluskey tabular method, find the minimum sum of products for $f(A, B, C, D) = \sum(1, 2, 3, 9, 12, 13, 14) + \sum_{\phi}(0, 7, 10, 15)$.

Solution

The different steps to finding the solution to the given problem are tabulated below. As we can see, eight prime implicants have been identified. These prime implicants along with the inputs constitute the prime implicant table. Remember that optional inputs are not considered while constructing the prime implicant table:

A	*B*	*C*	*D*	
0	0	0	0	✓
0	0	0	1	✓
0	0	1	0	✓
0	0	1	1	✓
1	0	0	1	✓
1	0	1	0	✓
1	1	0	0	✓
0	1	1	1	✓
1	1	0	1	✓
1	1	1	0	✓
1	1	1	1	✓

A	*B*	*C*	*D*	
0	0	0	–	✓
0	0	–	0	✓
0	0	–	1	✓
–	0	0	1	*
0	0	1	–	✓
–	0	1	0	*
0	–	1	1	*
1	–	0	1	*
1	–	1	0	*
1	1	0	–	✓
1	1	–	0	✓
–	1	1	1	*
1	1	–	1	✓
1	1	1	–	✓

A	*B*	*C*	*D*	
0	0	–	–	*
1	1	–	–	*

The product-of-sums expression that tells about the combination of prime implicants required to account for all the terms is given by the expression

$$(L+S).(M+S).(N+S).(L+P).(T).(P+T).(Q+T) \quad (6.38)$$

After obvious simplification, this reduces to the expression

$$\begin{aligned} &T.(L+S).(M+S).(N+S).(L+P)\\ &= T.(LM+LS+MS+S).(LN+PN+LS+PS)\\ &= T.(LM+S).(LN+PN+LS+PS)\\ &= T.(LMN+LMPN+LMS+LMPS+LNS+PNS+LS+PS)\\ &= T.(LMN+LMPN+LS+PS)\\ &= TLMN+TLMPN+TLS+TPS \end{aligned} \quad (6.39)$$

0001	0010	0011	1001	1100	1101	1110	Prime implicants	
✓			✓				–001	L
	✓						–010	M
		✓					0–11	N
			✓		✓		1–01	P
						✓	1–10	Q
							–111	R
✓	✓	✓					00––	S
				✓	✓	✓	11––	T

The sum-of-products Boolean expression (6.39) states that all the input combinations can be accounted for by the prime implicants (T, L, M, N) or (T, L, M, P, N) or (T, L, S) or (T, P, S). The most optimum expression would result from either *TLS* or *TPS*. Therefore, the minimized Boolean function is given by

$$f(A,B,C,D) = A.B+\overline{B}.\overline{C}.D+\overline{A}.\overline{B} \quad (6.40)$$

or by

$$f(A,B,C,D) = A.B+\overline{A}.\overline{B}+A.\overline{C}.D \quad (6.41)$$

Example 6.10

A logic system has three inputs A, B and C and two outputs Y_1 and Y_2. The output functions Y_1 and Y_2 are expressed by $Y_1 = \overline{A}.B.C+B.\overline{C}+\overline{A}.\overline{C}+A.\overline{B}.C+A.B.C$ and $Y_2 = \overline{A}.B+A.\overline{C}+A.B.C$. Determine the minimized output logic functions using the Quine–McCluskey tabular method.

Solution

The expanded forms of Y_1 and Y_2 are written as follows:

$$\begin{aligned} Y_1 &= \overline{A}.B.C+A.B.\overline{C}+\overline{A}.B.\overline{C}+\overline{A}.B.\overline{C}+\overline{A}.\overline{B}.\overline{C}+A.\overline{B}.C+A.B.C\\ &= \overline{A}.B.C+A.B.\overline{C}+\overline{A}.B.\overline{C}+\overline{A}.\overline{B}.\overline{C}+A.\overline{B}.C+A.B.C\\ Y_2 &= \overline{A}.B.C+\overline{A}.B.\overline{C}+A.B.C+A.B.\overline{C}+A.\overline{B}.\overline{C} \end{aligned}$$

The different steps leading to construction of the prime implicant table are given in tabular form below:

A	B	C	1	2	
0	0	0	✓		✓
0	1	0	✓	✓	✓
1	0	0		✓	✓
0	1	1	✓	✓	✓
1	0	1	✓		✓
1	1	0	✓	✓	✓
1	1	1	✓	✓	✓

A	B	C	1	2	
0	–	0	✓		*
0	1	–	✓	✓	✓
–	1	0	✓	✓	✓
1	–	0		✓	*
–	1	1	✓	✓	✓
1	–	1	✓		*
1	1	–	✓	✓	✓

A	B	C	1	2	
–	1	–	✓	✓	*

Y_1						Y_2					ABC
000	**010**	**011**	**101**	**110**	**111**	**010**	**011**	**100**	**110**	**111**	
✓	✓										**0–0**
								✓	✓		**1–0**
			✓		✓						**1–1**
	✓	✓		✓	✓	✓	✓		✓	✓	**–1–**

From the prime implicant table, the minimized output Boolean functions can be written as follows:

$$Y_1 = B + \overline{A}.\overline{C} + A.C \tag{6.42}$$

$$Y_2 = B + A.\overline{C} \tag{6.43}$$

6.6 Karnaugh Map Method

A Karnaugh map is a graphical representation of the logic system. It can be drawn directly from either minterm (sum-of-products) or maxterm (product-of-sums) Boolean expressions. Drawing a Karnaugh map from the truth table involves an additional step of writing the minterm or maxterm expression depending upon whether it is desired to have a minimized sum-of-products or a minimized product-of-sums expression.

6.6.1 Construction of a Karnaugh Map

An n-variable Karnaugh map has 2^n squares, and each possible input is allotted a square. In the case of a minterm Karnaugh map, '1' is placed in all those squares for which the output is '1', and '0' is placed in all those squares for which the output is '0'. 0s are omitted for simplicity. An 'X' is placed in squares corresponding to 'don't care' conditions. In the case of a maxterm Karnaugh map, a '1' is placed in all those squares for which the output is '0', and a '0' is placed for input entries corresponding to a '1' output. Again, 0s are omitted for simplicity, and an 'X' is placed in squares corresponding to 'don't care' conditions.

The choice of terms identifying different rows and columns of a Karnaugh map is not unique for a given number of variables. The only condition to be satisfied is that the designation of adjacent rows and adjacent columns should be the same except for one of the literals being complemented. Also, the extreme rows and extreme columns are considered adjacent. Some of the possible designation styles for two-, three- and four-variable minterm Karnaugh maps are given in Figs 6.6, 6.7 and 6.8 respectively.

The style of row identification need not be the same as that of column identification as long as it meets the basic requirement with respect to adjacent terms. It is, however, accepted practice to adopt a uniform style of row and column identification. Also, the style shown in Figs 6.6(a), 6.7(a) and 6.8(a) is more commonly used. Some more styles are shown in Fig. 6.9. A similar discussion applies for maxterm Karnaugh maps.

Having drawn the Karnaugh map, the next step is to form groups of 1s as per the following guidelines:

1. Each square containing a '1' must be considered at least once, although it can be considered as often as desired.
2. The objective should be to account for all the marked squares in the minimum number of groups.
3. The number of squares in a group must always be a power of 2, i.e. groups can have 1, 2, 4, 8, 16, ... squares.
4. Each group should be as large as possible, which means that a square should not be accounted for by itself if it can be accounted for by a group of two squares; a group of two squares should not be made if the involved squares can be included in a group of four squares and so on.
5. 'Don't care' entries can be used in accounting for all of 1-squares to make optimum groups. They are marked 'X' in the corresponding squares. It is, however, not necessary to account for all 'don't care' entries. Only such entries that can be used to advantage should be used.

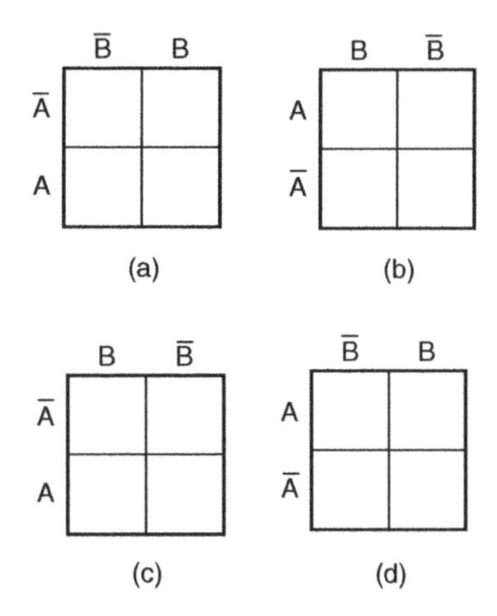

Figure 6.6 Two-variable Karnaugh map.

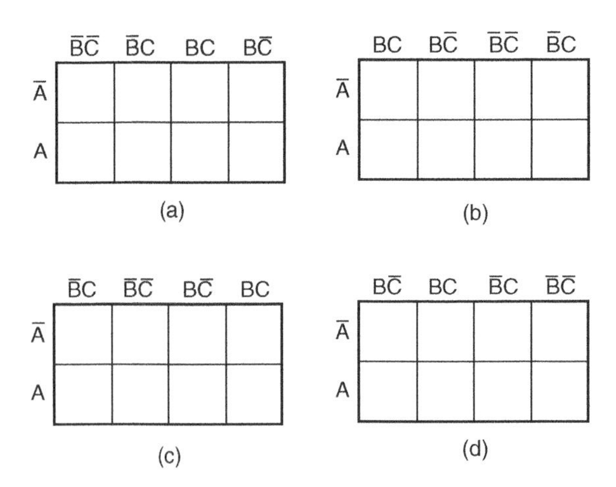

Figure 6.7 Three-variable Karnaugh map.

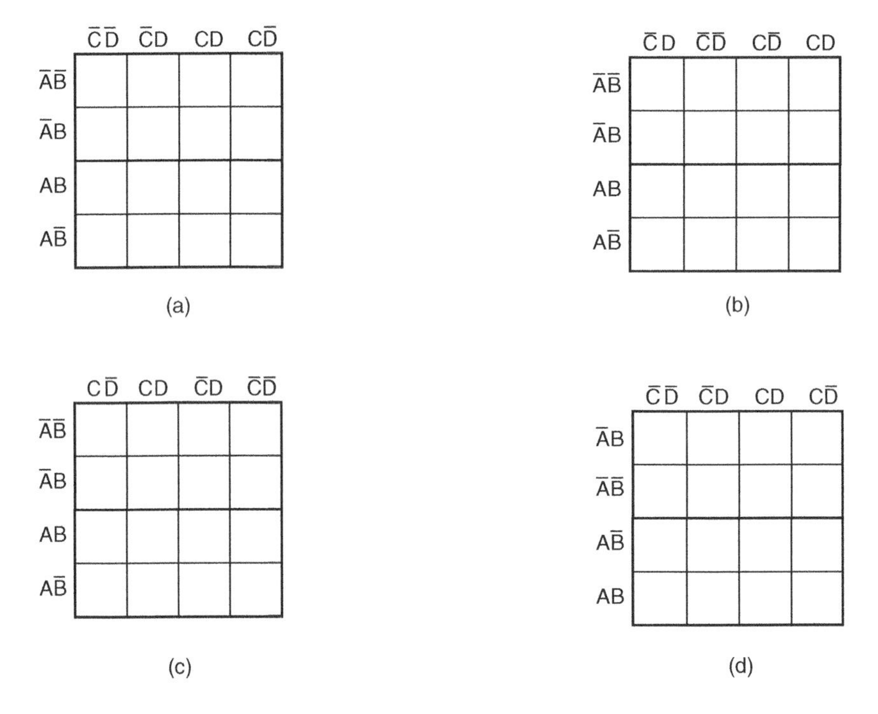

Figure 6.8 Four-variable Karnaugh map.

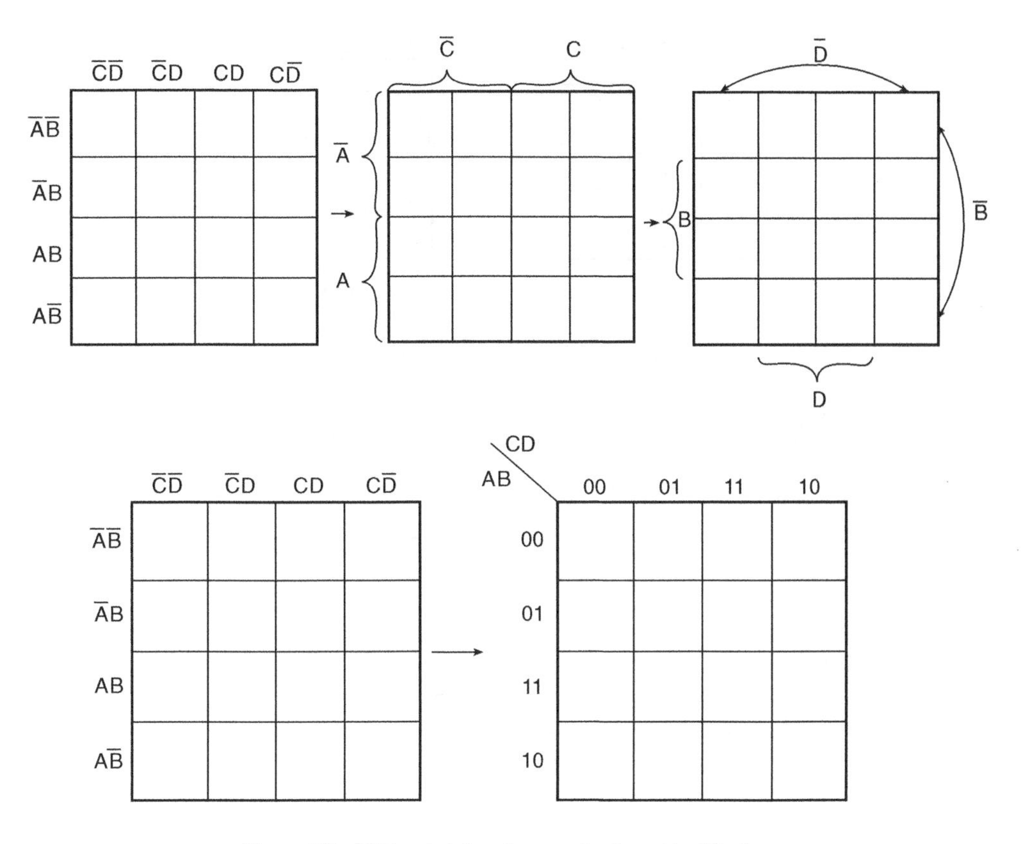

Figure 6.9 Different styles of row and column identification.

Having accounted for groups with all 1s, the minimum 'sum-of-products' or 'product-of-sums' expressions can be written directly from the Karnaugh map.

Figure 6.10 shows the truth table, minterm Karnaugh map and maxterm Karnaugh map of the Boolean function of a two-input OR gate. The minterm and maxterm Boolean expressions for the two-input OR gate are as follows:

$$Y = A + B \text{ (maxterm or product-of-sums)} \tag{6.44}$$

$$Y = \overline{A}.B + A.\overline{B} + A.B \text{ (minterm or sum-of-products)} \tag{6.45}$$

Figure 6.11 shows the truth table, minterm Karnaugh map and maxterm Karnaugh map of the three-variable Boolean function

$$Y = \overline{A}.\overline{B}.\overline{C} + \overline{A}.B.\overline{C} + A.\overline{B}.\overline{C} + A.B.\overline{C} \tag{6.46}$$

$$Y = (\overline{A} + \overline{B} + \overline{C}).(\overline{A} + B + \overline{C}).(A + \overline{B} + \overline{C}).(A + B + \overline{C}) \tag{6.47}$$

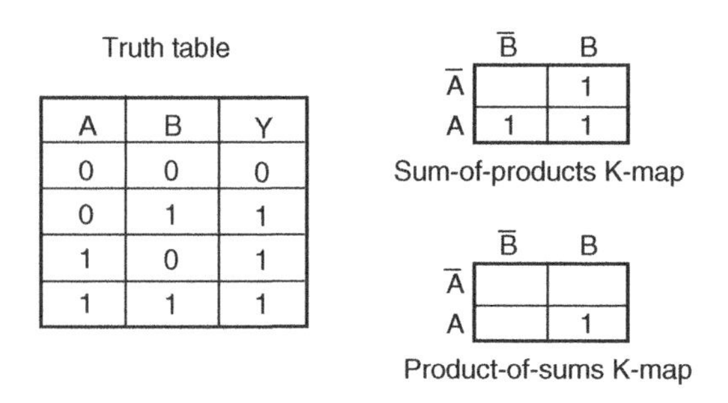

A	B	Y
0	0	0
0	1	1
1	0	1
1	1	1

Figure 6.10 Two-variable Karnaugh maps.

A	B	C	Y
0	0	0	1
0	0	1	0
0	1	0	1
0	1	1	0
1	0	0	1
1	0	1	0
1	1	0	1
1	1	1	0

	$\overline{B}\overline{C}$	$\overline{B}C$	BC	$B\overline{C}$
$\overline{A}$	1			1
A	1			1

Sum-of-products K-map

	$\overline{B}+\overline{C}$	$\overline{B}+C$	$B+C$	$B+\overline{C}$
$\overline{A}$	1			1
A	1			1

Product-of-sums K-map

Figure 6.11 Three-variable Karnaugh maps.

Figure 6.12 shows the truth table, minterm Karnaugh map and maxterm Karnaugh map of the four-variable Boolean function

$$Y=\overline{A}.\overline{B}.\overline{C}.\overline{D}+\overline{A}.\overline{B}.\overline{C}.D+\overline{A}.B.\overline{C}.\overline{D}+\overline{A}.B.\overline{C}.D+A.\overline{B}.\overline{C}.\overline{D}+A.\overline{B}.\overline{C}.D+A.B.\overline{C}.\overline{D}+A.B.\overline{C}.D \quad (6.48)$$

$$\begin{aligned}Y=&(A+B+\overline{C}+D).(A+B+\overline{C}+\overline{D}).(A+\overline{B}+\overline{C}+D).(A+\overline{B}+\overline{C}+\overline{D})\\&.(\overline{A}+B+\overline{C}+D).(\overline{A}+B+\overline{C}+\overline{D}).(\overline{A}+\overline{B}+\overline{C}+D).(\overline{A}+\overline{B}+\overline{C}+\overline{D})\end{aligned} \quad (6.49)$$

To illustrate the process of forming groups and then writing the corresponding minimized Boolean expression, Figs 6.13(a) and (b) respectively show minterm and maxterm Karnaugh maps for the Boolean functions expressed by equations (6.50) and (6.51). The minimized expressions as deduced from Karnaugh maps in the two cases are given by Equation (6.52) in the case of the minterm Karnaugh map and Equation (6.53) in the case of the maxterm Karnaugh map:

$$Y=\overline{A}.\overline{B}.\overline{C}.\overline{D}+\overline{A}.\overline{B}.C.\overline{D}+\overline{A}.B.\overline{C}.D+\overline{A}.B.C.D+A.\overline{B}.\overline{C}.\overline{D}+A.\overline{B}.C.\overline{D}+A.B.\overline{C}.D+A.B.C.D \quad (6.50)$$

$$\begin{aligned}Y=&(A+B+C+\overline{D}).(A+B+\overline{C}+\overline{D}).(A+\overline{B}+C+D).(A+\overline{B}+C+\overline{D}).(A+\overline{B}+\overline{C}+\overline{D})\\&.(A+\overline{B}+\overline{C}+D).(\overline{A}+\overline{B}+C+\overline{D}).(\overline{A}+\overline{B}+\overline{C}+\overline{D}).(\overline{A}+B+C+\overline{D}).(\overline{A}+B+\overline{C}+\overline{D})\end{aligned} \quad (6.51)$$

$$Y=\overline{B}.\overline{D}+B.D \quad (6.52)$$

$$Y=\overline{D}.(A+\overline{B}) \quad (6.53)$$

Truth table

A	B	C	D	Y
0	0	0	0	1
0	0	0	1	1
0	0	1	0	0
0	0	1	1	0
0	1	0	0	1
0	1	0	1	1
0	1	1	0	0
0	1	1	1	0
1	0	0	0	1
1	0	0	1	1
1	0	1	0	0
1	0	1	1	0
1	1	0	0	1
1	1	0	1	1
1	1	1	0	0
1	1	1	1	0

	$\bar{C}\bar{D}$	$\bar{C}D$	CD	$C\bar{D}$
$\bar{A}\bar{B}$	1	1		
$\bar{A}B$	1	1		
AB	1	1		
$A\bar{B}$	1	1		

Sum-of-products K-map

	$\bar{C}+\bar{D}$	$\bar{C}+D$	$C+D$	$C+\bar{D}$
$\bar{A}+\bar{B}$	1	1		
$\bar{A}+B$	1	1		
$A+B$	1	1		
$A+\bar{B}$	1	1		

Product-of-sums K-map

Figure 6.12 Four-variable Karnaugh maps.

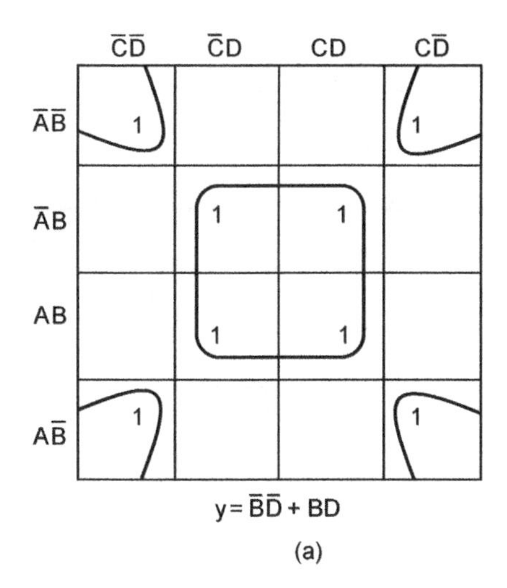

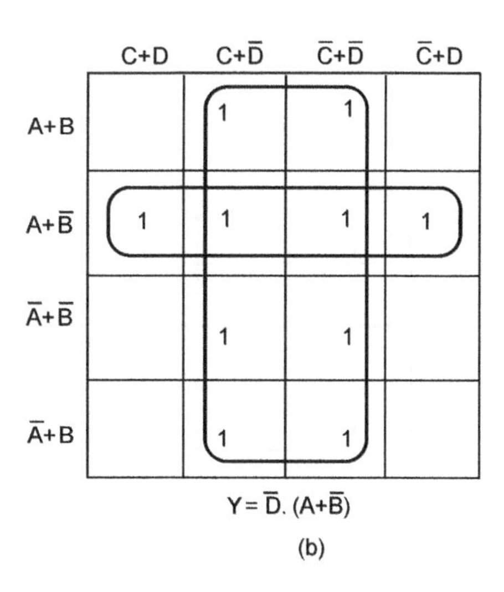

Figure 6.13 Group formation in minterm and maxterm Karnaugh maps.

6.6.2 Karnaugh Map for Boolean Expressions with a Larger Number of Variables

The construction of Karnaugh maps for a larger number of variables is a complex and cumbersome exercise, although manageable up to six variables. Five- and six-variable representative Karnaugh maps are shown in Figs 6.14(a) and (b) respectively. One important point to remember while forming groups in Karnaugh maps involving more than four variables is that terms equidistant from the central horizontal and central vertical lines are considered adjacent. These lines are shown thicker in Figs 6.14(a) and (b). Squares marked 'X' in Figs 6.14(a) and (b) are adjacent and therefore can be grouped.

Boolean expressions with more than four variables can also be represented by more than one four-variable map. Five-, six-, seven- and eight-variable Boolean expressions can be represented by two, four, eight and 16 four-variable maps respectively. In general, an n-variable Boolean expression can be represented by 2^{n-4} four-variable maps. In such multiple maps, groups are made as before, except that, in addition to adjacencies discussed earlier, corresponding squares in two adjacent maps are also considered adjacent and can therefore be grouped. We will illustrate the process of formation of groups in multiple Karnaugh maps with a larger number of variables with the help of examples. Consider the five-variable Boolean function given by the equation

$$Y = A.\overline{B}.\overline{C}.D.E + A.\overline{B}.C.D.E + \overline{A}.B.C.\overline{D}.E + A.B.\overline{C}.D.E + \overline{A}.\overline{B}.C.D.\overline{E} + \overline{A}.B.C.D.\overline{E} + A.B.C.D.\overline{E} + A.\overline{B}.C.D.\overline{E} + \overline{A}.B.C.\overline{D}.\overline{E} \quad (6.54)$$

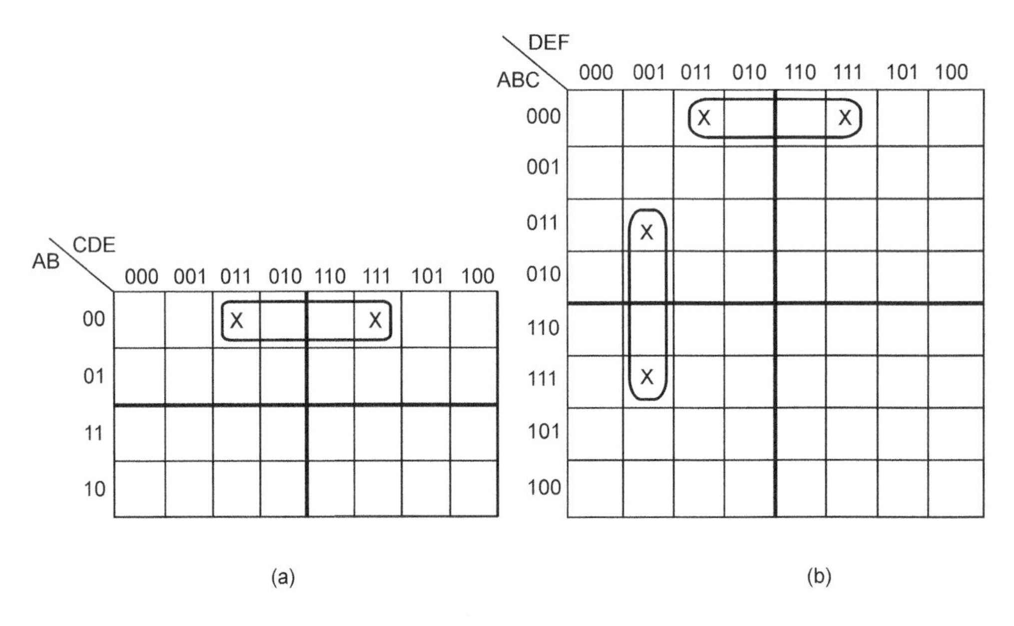

Figure 6.14 Five-variable and six-variable Karnaugh maps.

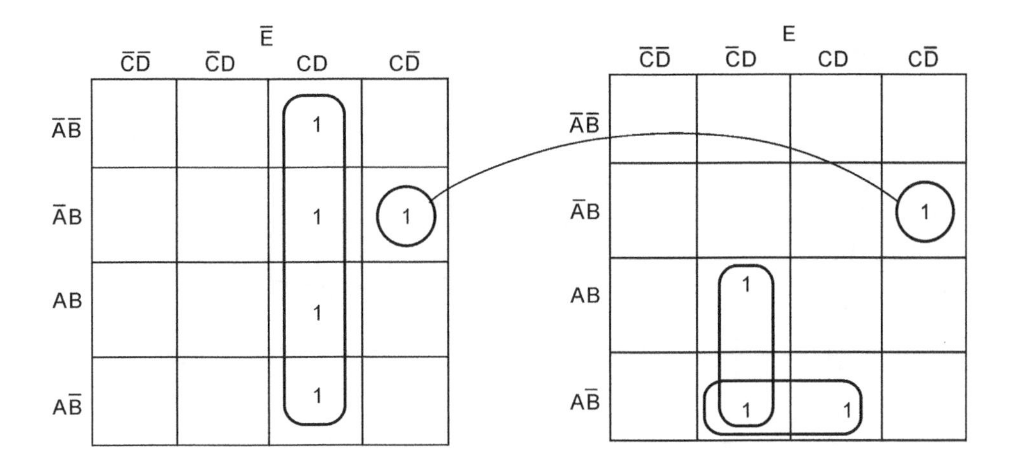

Figure 6.15 Multiple Karnaugh map for a five-variable Boolean function.

The multiple Karnaugh map for this five-variable expression is shown in Fig. 6.15. The construction of the Karnaugh map and the formation of groups are self-explanatory.

The minimized expression is given by the equation

$$Y = C.D.\overline{E} + \overline{A}.B.C.\overline{D} + A.\overline{C}.D.E + A.\overline{B}.D.E \tag{6.55}$$

As another illustration, consider a six-variable Boolean function given by the equation

$$\begin{aligned}Y =& \overline{A}.B.C.\overline{D}.\overline{E}.\overline{F} + A.B.\overline{C}.D.\overline{E}.F + \overline{A}.\overline{B}.\overline{C}.\overline{D}.\overline{E}.\overline{F} + A.B.C.D.E.F + A.\overline{B}.C.D.E.\overline{F}\\ &+ \overline{A}.\overline{B}.\overline{C}.\overline{D}.\overline{E}.F + \overline{A}.B.C.\overline{D}.E.\overline{F}\end{aligned} \tag{6.56}$$

Figure 6.16 gives the Karnaugh map for this six-variable Boolean function, comprising four four-variable Karnaugh maps. The figure also shows the formation of groups. The minimized expression is given by the equation

$$Y = \overline{A}.\overline{B}.\overline{C}.\overline{D}.\overline{E} + \overline{A}.B.C.\overline{D}.\overline{F} + A.\overline{B}.C.D.E.\overline{F} + A.B.\overline{C}.D.\overline{E}.F + A.B.C.D.E.F \tag{6.57}$$

Example 6.11

Minimize the Boolean function

$$f(A, B, C) = \sum 0, 1, 3, 5 + \sum_{\phi} 2, 7$$

using the mapping method in both minimized sum-of-products and product-of-sums forms.

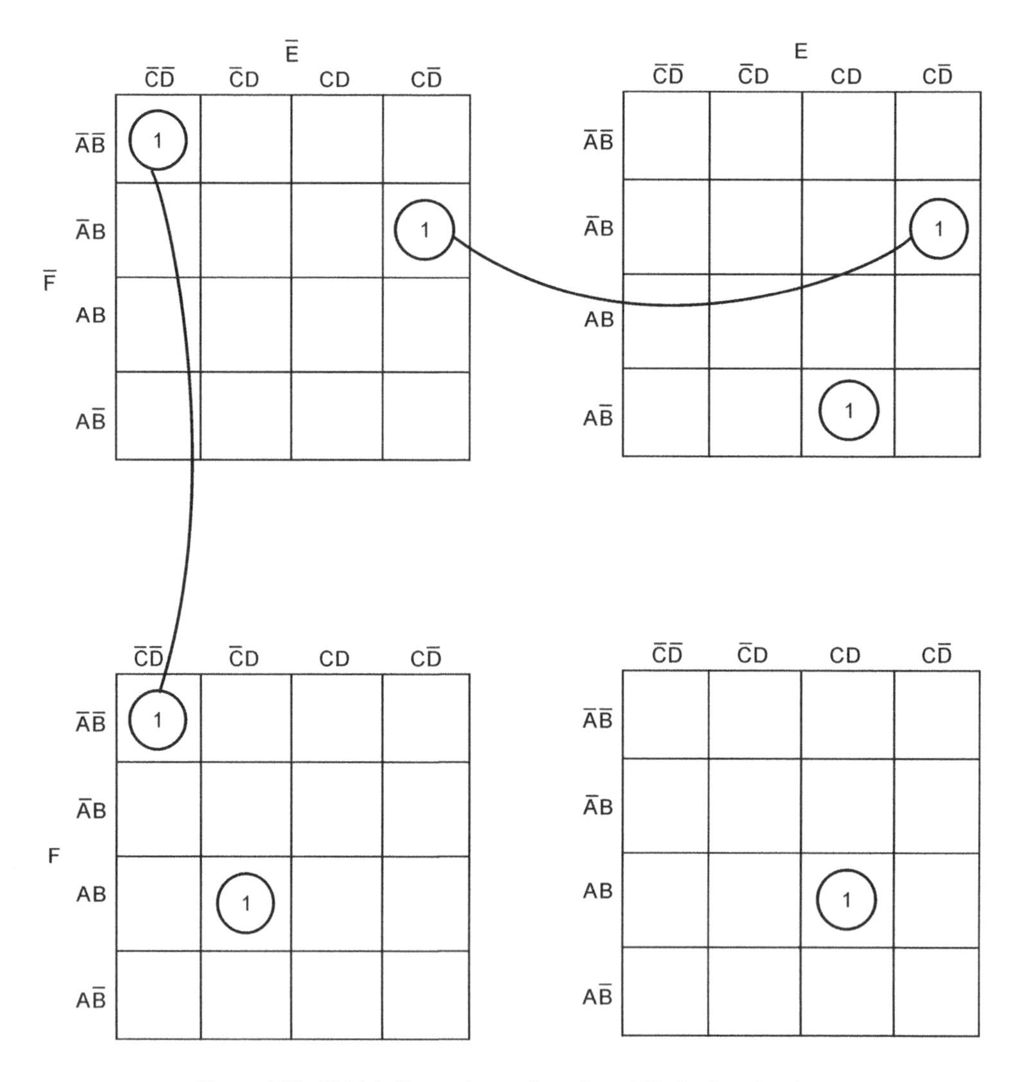

Figure 6.16 Multiple Karnaugh map for a six-variable Boolean function.

Solution

- $f(A, B, C) = \sum 0, 1, 3, 5 + \sum_{\phi} 2, 7 = \prod 4, 6 + \prod_{\phi} 2, 7.$
- From given Boolean functions in Σ and Π notation, we can write sum-of-products and product-of-sums Boolean expressions as follows:

$$f(A, B, C) = \overline{A}.\overline{B}.\overline{C} + \overline{A}.\overline{B}.C + \overline{A}.B.C + A.\overline{B}.C \tag{6.58}$$

$$f(A, B, C) = (\overline{A} + B + C).(\overline{A} + \overline{B} + C) \tag{6.59}$$

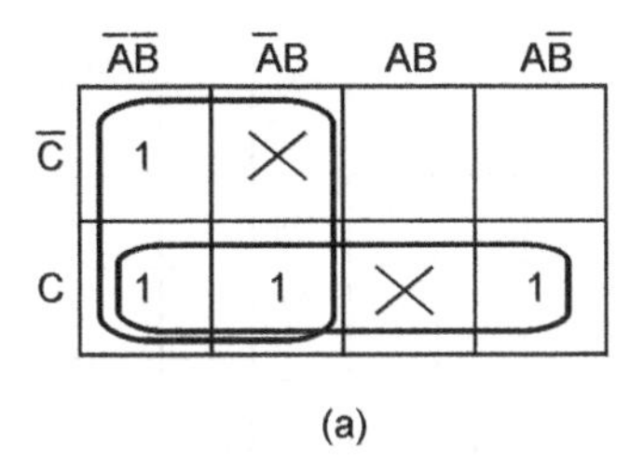

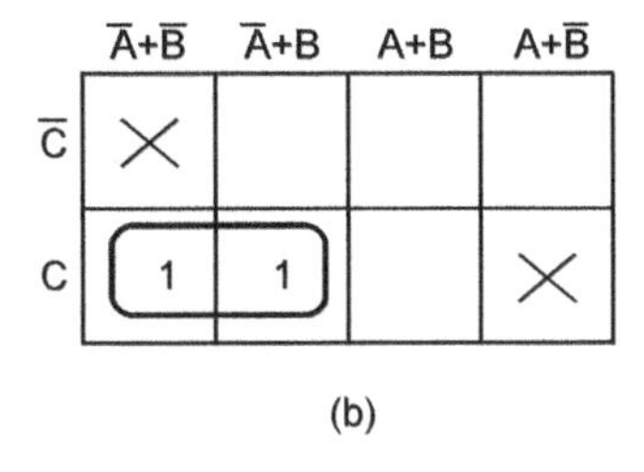

Figure 6.17 Example 6.11.

- The 'don't care' input combinations for the sum-of-products Boolean expression are $\overline{A}.B.\overline{C}$, $A.B.C$.
- The 'don't care' input combinations for the product-of-sums expression are $(A+\overline{B}+C).(\overline{A}+\overline{B}+\overline{C})$.
- The Karnaugh maps for the two cases are shown in Figs 6.17(a) and (b).
- The minimized sum-of-products and product-of-sums Boolean functions are respectively given by the equations

$$f(A,B,C)=C+\overline{A} \tag{6.60}$$

$$f(A,B,C)=\overline{A}+C \tag{6.61}$$

6.6.3 Karnaugh Maps for Multi-Output Functions

Karnaugh maps can be used for finding minimized Boolean expressions for multi-output functions. To begin with, a Karnaugh map is drawn for each function following the guidelines described in the earlier pages. In the second step, two-function Karnaugh maps are drawn. In the third step, three-function Karnaugh maps are drawn. The process continues until we have a single all-function Karnaugh map. As an illustration, for a logic system having four outputs, the first step would give four Karnaugh maps for individual functions. The second step would give six two-function Karnaugh maps (1–2, 1–3, 1–4, 2–3, 2–4 and 3–4). The third step would yield four three–function Karnaugh maps (1–2–3, 1–2–4, 1–3–4 and 2–3–4) and lastly we have one four-function Karnaugh map. A multifunction Karnaugh map is basically an intersection of the Karnaugh maps of the functions involved. That is, a '1' appears in a square of a multifunction map only if a '1' appears in the corresponding squares of the maps of all the relevant functions. To illustrate further, a two-function map involving functions 1 and 2 would be an intersection of maps for functions 1 and 2. In the two-function map, squares will have a '1' only when the corresponding squares in functions 1 and 2 also have a '1'. Figure 6.18 illustrates the formation of a three-function Karnaugh map from three given individual functions.

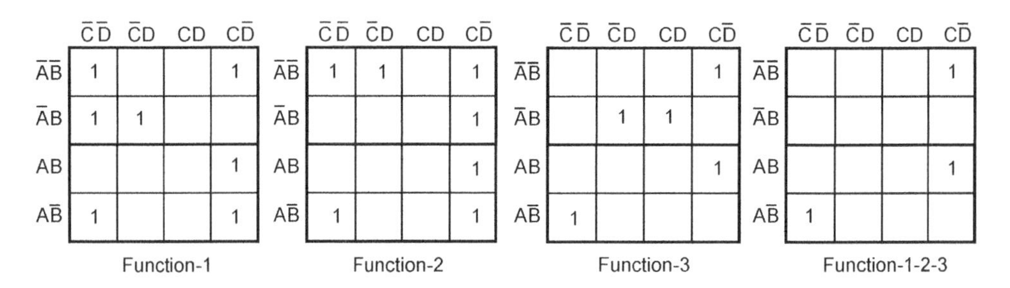

Figure 6.18 Three-function Karnaugh map.

The formation of groups begins with the largest multifunction map, which is nothing but the intersection of maps of all individual functions. Then we move to the Karnaugh maps one step down the order. The process continues until we reach the maps corresponding to individual functions. The groups in all the Karnaugh maps other than the largest map are formed subject to the condition that, once a group is identified in a certain function, then the same cannot be identified in any map of a subset of that function. For example, a group identified in a four-function map cannot be identified in a three-, two- or one-function map. With the formation of groups, prime implicants are identified. These prime implicants can be compiled in the form of a table along with input combinations of different output functions in the same way as for the tabular method to write minimized expressions. If the expressions corresponding to different output functions are not very complex, then the minimized expressions can even be written directly from the set of maps.

Example 6.12

Using Karnaugh maps, write the minimized Boolean expressions for the output functions of a two-output logic system whose outputs Y_1 and Y_2 are given by the following Boolean functions:

$$Y_1 = \overline{A}.B.\overline{C} + A.\overline{B}.\overline{C} + A.B.C + \overline{A}.\overline{B}.\overline{C} \tag{6.62}$$

$$Y_2 = \overline{A}.\overline{B}.C + A.B.\overline{C} + \overline{A}.\overline{B}.\overline{C} + A.\overline{B}.C + A.B.C \tag{6.63}$$

Solution

The individual Karnaugh maps and the two-function map are shown in Fig. 6.19 along with the formation of groups. The prime implicant table along with the input combinations for the two output functions is given below:

Y_1				Y_2					Prime implicants		
000	010	100	111	000	001	101	110	111			
✓				✓					0	0	0
			✓					✓	1	1	1
✓	✓								0	–	0
✓		✓							–	0	0
							✓	✓	1	1	–
					✓	✓			–	0	1

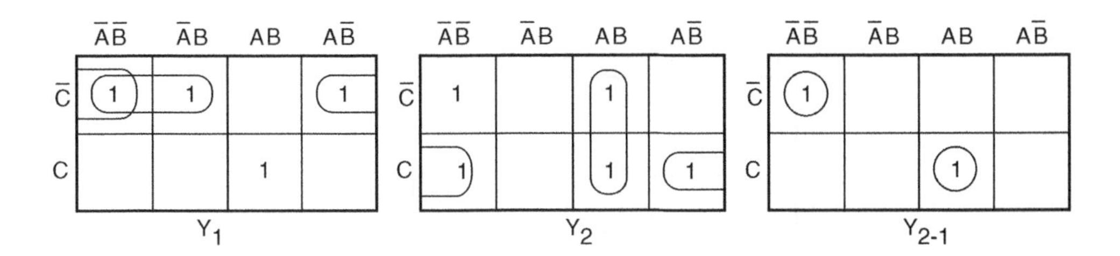

Figure 6.19 Example 6.12.

The minimized expressions for Y_1 and Y_2 are as follows:

$$Y_1 = \overline{B}.\overline{C} + \overline{A}.\overline{C} + A.B.C \tag{6.64}$$

$$Y_2 = A.B + \overline{A}.\overline{B}.\overline{C} + \overline{B}.C \tag{6.65}$$

Example 6.13

Write the simplified Boolean expression given by the Karnaugh map shown in Fig. 6.20.

Solution

- The Karnaugh map is shown in Fig. 6.21.
- Consider the group of four 1s at the top left of the map. It yields a term $\overline{A}.\overline{C}$.
- Consider the group of four 1s, two on the extreme left and two on the extreme right. This group yields a term $\overline{A}.\overline{D}$.
- The third group of two 1s is in the third row of the map. The third row corresponds to the intersection of A and B, as is clear from the map. Therefore, this group yields a term ABC.
- The simplified Boolean expression is given by $\overline{A}.\overline{C} + \overline{A}.\overline{D} + A.B.C$.

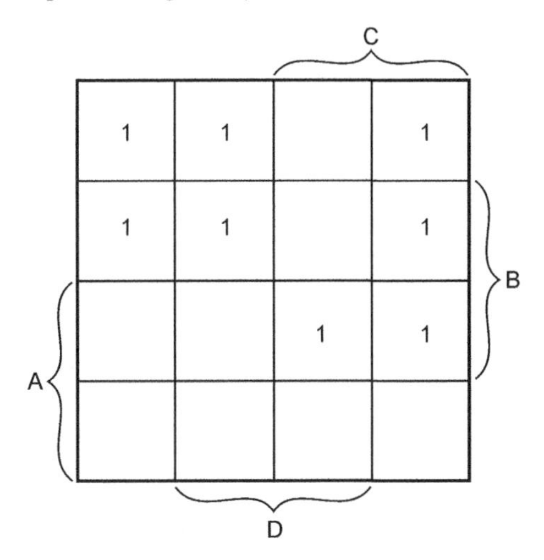

Figure 6.20 Example 6.13.

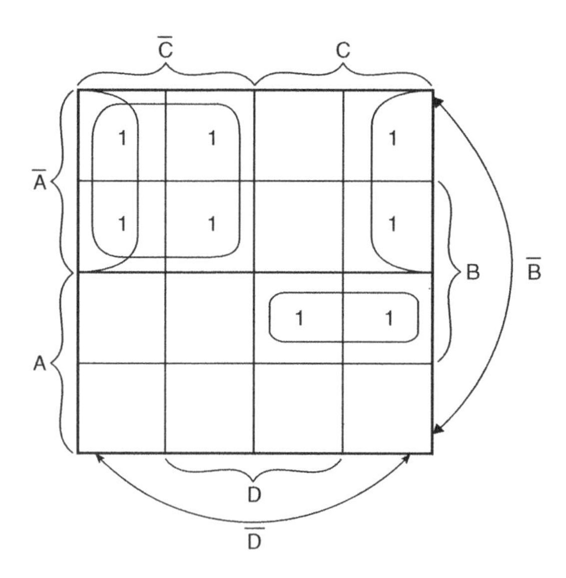

Figure 6.21 Solution to example 6.13.

Example 6.14

Minimizing a given Boolean expression using the Quine–McCluskey tabular method yields the following prime implicants: −0−0, −1−1, 1−10 and 0−00. Draw the corresponding Karnaugh map.

Solution

- As is clear from the prime implicants, the expression has four variables. If the variables are assumed to be A, B, C and D, then the given prime implicants correspond to the following terms:

1. $-0-0 \rightarrow \overline{B}.\overline{D}$.
2. $-1-1 \rightarrow B.D$.
3. $1-10 \rightarrow A.C.\overline{D}$.
4. $0-00 \rightarrow \overline{A}.\overline{C}.\overline{D}$.

- The Karnaugh map can now be drawn as shown in Fig. 6.22.

Example 6.15

$\overline{A}.B + C.D$ is a simplified Boolean expression of the expression $A.B.C.D + \overline{A}.\overline{B}.C.D + \overline{A}.B$. Determine if there are any 'don't care' entries.

Solution

The expanded version of the given expression is given by the equation

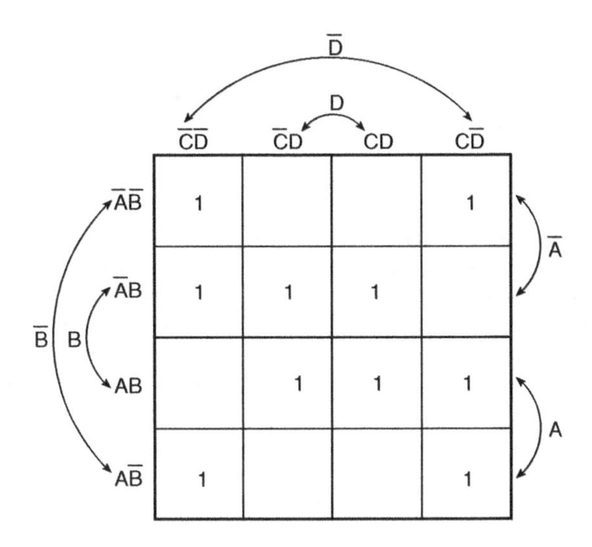

Figure 6.22 Solution to example 6.14.

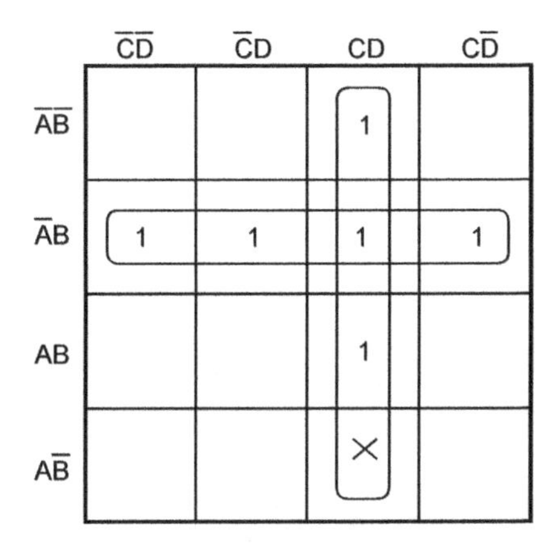

Figure 6.23 Example 6.15.

$$A.B.C.D+\overline{A}.\overline{B}.C.D+\overline{A}.B.(\overline{C}.\overline{D}+\overline{C}.D+C.D+C.\overline{D}) \qquad (6.66)$$

$$=A.B.C.D+\overline{A}.\overline{B}.C.D+\overline{A}.B.\overline{C}.\overline{D}+\overline{A}.B.\overline{C}.D+\overline{A}.B.C.D+\overline{A}.B.C.\overline{D} \qquad (6.63)$$

The Karnaugh map for this Boolean expression is shown in Fig. 6.23. Now, if it is to be a simplified version of the expression $\overline{A}.B+C.D$, then the lowermost square in the *CD* column should not be empty. This implies that there is a 'don't care' entry. This has been reflected in the map by putting X in the relevant square. With the groups formed along with the 'don't care' entry, the simplified expression becomes the one stated in the problem.

Review Questions

1. Read the following statements carefully. For each one of these, identify the law associated with it. Define the law and illustrate the same with one or two examples.

   (a) While a NAND gate is equivalent to a bubbled OR gate, a NOR gate is equivalent to a bubbled AND gate.
   (b) When all the inputs of an AND gate or an OR gate are tied together to get a single-input, single-output gate, both AND and OR gates with all their inputs tied together produce an output that is the same as the input.
   (c) When a variable is ORed with its complement the result is a logic '1', and when it is ANDed with its complement the result is a logic '0', irrespective of the logic status of the variable.
   (d) When two variables are ANDed and the result of the AND operation is ORed with one of the variables, the result is that variable. Also, when two variables are ORed and the result of the OR operation is ANDed with one of the variables, the result is that variable.

2. Write both sum-of-products and product-of-sums Boolean expressions for (a) a two-input AND gate, (b) a two-input NAND-gate, (c) a two-input EX-OR gate and (d) a two-input NOR gate from their respective truth tables.
3. What do you understand by canonical and expanded forms of Boolean expressions? Illustrate with examples.
4. With the help of an example, prove that in an n-variable Karnaugh map, a group formed with 2^{n-m} 1s will yield a term having m literals, where m = 1, 2, 3, ..., n.
5. With the help of an example, prove that the dual of the complement of a Boolean expression is the same as the complement of the dual of the same.

Problems

1. Simplify the following Boolean expressions:

   (a) $A.B.C+A.B.\overline{C}+A.\overline{B}.C+A.\overline{B}.\overline{C}+\overline{A}.B.C+\overline{A}.B.\overline{C}+\overline{A}.\overline{B}.\overline{C}+\overline{A}.\overline{B}.C$;
   (b) $(\overline{A}+B+\overline{C}).(\overline{A}+B+C).(C+D).(C+D+E)$.

   (*a*) 1; (*b*) $(\overline{A}+B).(C+D)$

2. (a) Find the dual of $A.B.C.\overline{D}+A.\overline{B}.\overline{C}.D+\overline{A}.\overline{B}.\overline{C}.\overline{D}$.
   (b) Find the complement of $A+[(B+\overline{C}).D+\overline{E}].F$.

   (*a*) $(A+B+C+\overline{D}).(A+\overline{B}+\overline{C}+D).(\overline{A}+\overline{B}+\overline{C}+\overline{D})$; (*b*)$\overline{A}.[(\overline{B}.C+\overline{D}).E+\overline{F}]$

3. The dual of the complement of a certain Boolean expression is given by $A.B.C+\overline{D}.E+B.\overline{C}.E$. Find the expression.

   $\overline{A}.\overline{B}.\overline{C}+D.\overline{E}+\overline{B}.C.\overline{E}$

4. Consider the Boolean expression given by

$$\overline{B}.\overline{C}.\overline{D}.\overline{E}+B.\overline{C}.\overline{D}.E+\overline{A}.B.C.E+A.B.C.D.E+A.\overline{B}.C.\overline{D}.\overline{E}+\overline{A}.B.\overline{C}.D.E+\overline{A}.\overline{B}.D.\overline{E}$$
$$+\overline{A}.\overline{B}.C.\overline{D}.\overline{E}+A.\overline{B}.\overline{C}.D.\overline{E}$$

The simplified version of this Boolean expression is given by $B.E+\overline{B}.D.\overline{E}+\overline{B}.\overline{D}.\overline{E}$. Determine if there are any 'don't care' entries. If yes, find them.

Yes, $A.B.\overline{C}.D.E, A.B.C.\overline{D}.E, A.\overline{B}.C.D.\overline{E}$

5. Write minterm and maxterm Boolean functions expressed by $f(A, B, C) = \Pi\ 0, 3, 7$

minterm: $\overline{A}.\overline{B}.C+\overline{A}.B.\overline{C}+A.\overline{B}.\overline{C}+A.\overline{B}.C+A.B.\overline{C}$

maxterm: $(A+B+C).(A+\overline{B}+\overline{C}).(\overline{A}+\overline{B}+\overline{C})$

6. Write a simplified maxterm Boolean expression for Π 0, 4, 5, 6, 7, 10, 14 using the Karnaugh mapping method.

$(A+\overline{B}).(A+B+C+D).(\overline{A}+\overline{C}+D)$

7. Simplify the following Boolean functions using the Quine–McCluskey tabulation method:

   (a) $f(A, B, C, D, E, F, G) = \Sigma$ (20, 21, 28, 29, 52, 53, 60, 61);
   (b) $f(A, B, C, D, E, F) = \Sigma$ (6, 9, 13,18,19, 25, 26, 27, 29, 41, 45, 57, 61).

(a) $\overline{A}.C.E.\overline{F}$*; (b)* $C.\overline{E}.F+\overline{A}.B.\overline{D}.E+\overline{A}.\overline{B}.\overline{C}.D.E.\overline{F}$

8. (a) Simplify the Boolean function $f(X, Y, Z) = Y.Z+\overline{X}.\overline{Z}$ for the 'don't care' condition expressed as $X.\overline{Y}+X.Y.\overline{Z}+\overline{X}.\overline{Y}.Z$.
   (b) Simplify the Boolean function given by $f(A, B, C) = (A+B+C).(\overline{A}+B+\overline{C}).(A+\overline{B}+C)$ for the don't care condition expressed as $(\overline{A}+\overline{B}).(\overline{A}+B+C)$.

(a) 1; (b) $\overline{A}.C$

Further Reading

1. Holdsworth, B. and Woods, C. (2002) *Digital Logic Design*, Newnes, Oxford, UK.
2. Chen, W.-K. (2003) *Logic Design*, CRC Press, FL, USA.
3. Floyd, T. L. (2005) *Digital Fundamentals*, Prentice-Hall Inc., USA.
4. Tokheim, R. L. (1994) *Schaum's Outline Series of Digital Principles*, McGraw-Hill Companies Inc., USA.

7

Arithmetic Circuits

Beginning with this chapter, and in the two chapters following, we will take a comprehensive look at various building blocks used to design more complex combinational circuits. A combinational logic circuit is one where the output or outputs depend upon the present state of combination of the logic inputs. The logic gates discussed in Chapter 4 constitute the most fundamental building block of a combinational circuit. More complex combinational circuits such as adders and subtractors, multiplexers and demultiplexers, magnitude comparators, etc., can be implemented using a combination of logic gates. However, the aforesaid combinational logic functions and many more, including more complex ones, are available in monolithic IC form. A still more complex combinational circuit may be implemented using a combination of these functions available in IC form. In this chapter, we will cover devices used to perform arithmetic and other related operations. These include adders, subtractors, magnitude comparators and look-ahead carry generators. Particular emphasis is placed upon the functioning and design of these combinational circuits. The text has been adequately illustrated with the help of a large number of solved problems, the majority of which are design oriented.

7.1 Combinational Circuits

A *combinational circuit* is one where the output at any time depends only on the present combination of inputs at that point of time with total disregard to the past state of the inputs. The logic gate is the most basic building block of combinational logic. The logical function performed by a combinational circuit is fully defined by a set of Boolean expressions. The other category of logic circuits, called *sequential logic circuits*, comprises both logic gates and memory elements such as flip-flops. Owing to the presence of memory elements, the output in a sequential circuit depends upon not only the present but also the past state of inputs. Basic building blocks of sequential logic circuits are described in detail in Chapters 10 and 11.

Figure 7.1 shows the block schematic representation of a generalized combinational circuit having n input variables and m output variables or simply outputs. Since the number of input variables is

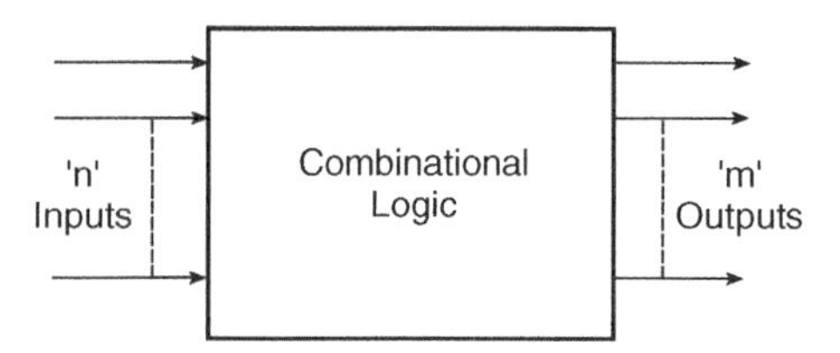

Figure 7.1 Generalized combinational circuit.

n, there are 2^n possible combinations of bits at the input. Each output can be expressed in terms of input variables by a Boolean expression, with the result that the generalized system of Fig. 7.1 can be expressed by m Boolean expressions. As an illustration, Boolean expressions describing the function of a four-input OR/NOR gate are given as

$$Y_1 \text{ (OR output)} = A+B+C+D \quad \text{and} \quad Y_2 \text{ (NOR output)} = \overline{A+B+C+D}$$

Also, each of the input variables may be available as only the normal input on the input line designated for the purpose. In that case, the complemented input, if desired, can be generated by using an inverter, as shown in Fig. 7.2(a), which illustrates the case of a four-input, two-output combinational function. Also, each of the input variables may appear in two wires, one representing the normal literal and the other representing the complemented one, as shown in Fig. 7.2(b).

In combinational circuits, input variables come from an external source and output variables feed an external destination. Both source and destination in the majority of cases are storage registers, and these

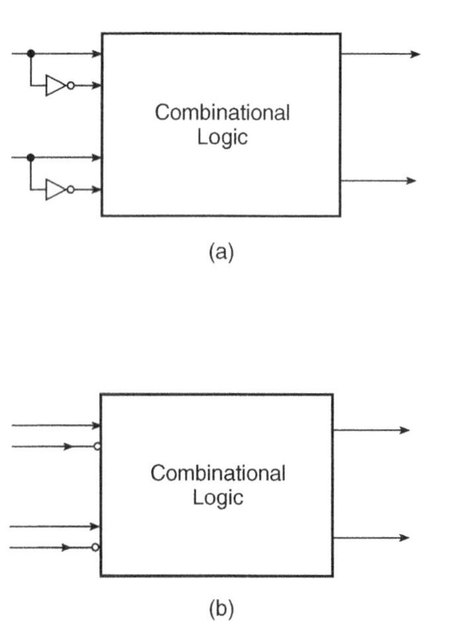

Figure 7.2 Combinational circuit with normal and complemented inputs.

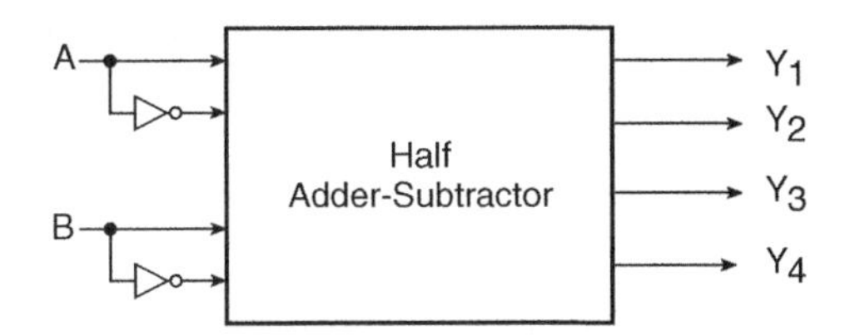

Figure 7.3 Two-input, four-output combinational circuit.

storage devices provide both normal as well as complemented outputs of the stored binary variable. As an illustration, Fig. 7.3 shows a simple two-input (A, B), four-output (Y_1, Y_2, Y_3, Y_4) combinational logic circuit described by the following Boolean expressions

$$Y_1 = A.\overline{B} + \overline{A}.B \tag{7.1}$$

$$Y_2 = A.\overline{B} + \overline{A}.B \tag{7.2}$$

$$Y_3 = A.B \tag{7.3}$$

$$Y_4 = \overline{A}.B \tag{7.4}$$

The implementation of these Boolean expressions needs both normal as well as complemented inputs. Incidentally, the combinational circuit shown is that of a half-adder–subtractor, with A and B representing the two bits to be added or subtracted and Y_1, Y_2, Y_3, Y_4 representing SUM, DIFFERENCE, CARRY and BORROW outputs respectively. Adder and subtractor circuits are discussed in Sections 7.3, 7.4 and 7.5.

7.2 Implementing Combinational Logic

The different steps involved in the design of a combinational logic circuit are as follows:

1. Statement of the problem.
2. Identification of input and output variables.
3. Expressing the relationship between the input and output variables.
4. Construction of a truth table to meet input–output requirements.
5. Writing Boolean expressions for various output variables in terms of input variables.
6. Minimization of Boolean expressions.
7. Implementation of minimized Boolean expressions.

These different steps are self-explanatory. One or two points, however, are worth mentioning here. There are various simplification techniques available for minimizing Boolean expressions, which have been discussed in the previous chapter. These include the use of theorems and identities, Karnaugh mapping, the Quinne–McCluskey tabulation method and so on. Also, there are various possible minimized forms of Boolean expressions. The following guidelines should be followed while choosing the preferred form for hardware implementation:

1. The implementation should have the minimum number of gates, with the gates used having the minimum number of inputs.
2. There should be a minimum number of interconnections, and the propagation time should be the shortest.
3. Limitation on the driving capability of the gates should not be ignored.

It is difficult to generalize as to what constitutes an acceptable simplified Boolean expression. The importance of each of the above-mentioned aspects is governed by the nature of application.

7.3 Arithmetic Circuits – Basic Building Blocks

In this section, we will discuss those combinational logic building blocks that can be used to perform addition and subtraction operations on binary numbers. Addition and subtraction are the two most commonly used arithmetic operations, as the other two, namely multiplication and division, are respectively the processes of repeated addition and repeated subtraction, as was outlined in Chapter 2 dealing with binary arithmetic. We will begin with the basic building blocks that form the basis of all hardware used to perform the aforesaid arithmetic operations on binary numbers. These include half-adder, full adder, half-subtractor, full subtractor and controlled inverter.

7.3.1 Half-Adder

A *half-adder* is an arithmetic circuit block that can be used to add two bits. Such a circuit thus has two inputs that represent the two bits to be added and two outputs, with one producing the SUM output and the other producing the CARRY. Figure 7.4 shows the truth table of a half-adder, showing all possible input combinations and the corresponding outputs.

The Boolean expressions for the SUM and CARRY outputs are given by the equations

$$\text{SUM } S = A.\overline{B} + \overline{A}.B \tag{7.5}$$

$$\text{CARRY } C = A.B \tag{7.6}$$

An examination of the two expressions tells that there is no scope for further simplification. While the first one representing the SUM output is that of an EX-OR gate, the second one representing the

A	B	S	C
0	0	0	0
0	1	1	0
1	0	1	0
1	1	0	1

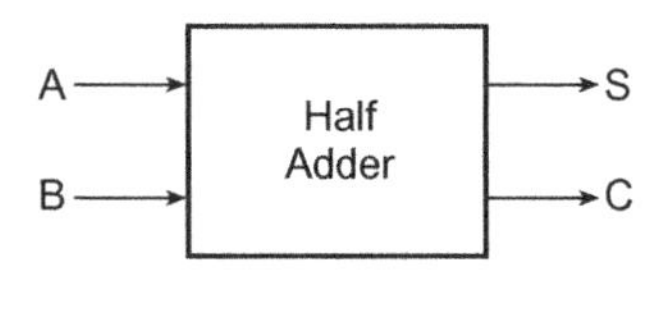

Figure 7.4 Truth table of a half-adder.

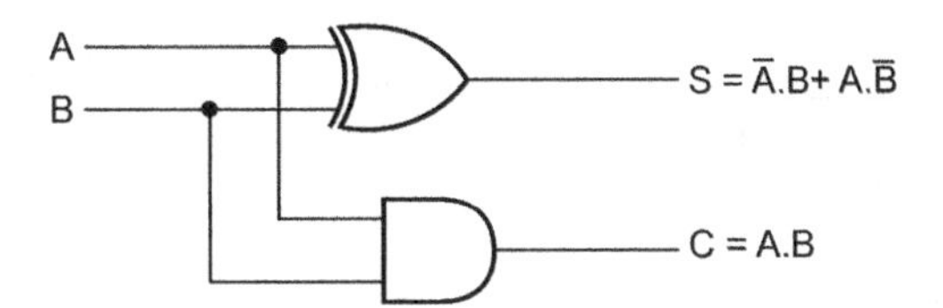

Figure 7.5 Logic implementation of a half-adder.

CARRY output is that of an AND gate. However, these two expressions can certainly be represented in different forms using various laws and theorems of Boolean algebra to illustrate the flexibility that the designer has in hardware-implementing as simple a combinational function as that of a half-adder. We have studied in Chapter 6 on Boolean algebra how various logic gates can be implemented in the form of either only NAND gates or NOR gates. Although the simplest way to hardware-implement a half-adder would be to use a two-input EX-OR gate for the SUM output and a two-input AND gate for the CARRY output, as shown in Fig. 7.5, it could also be implemented by using an appropriate arrangement of either NAND or NOR gates. Figure 7.6 shows the implementation of a half-adder with NAND gates only.

A close look at the logic diagram of Fig. 7.6 reveals that one part of the circuit implements a two-input EX-OR gate with two-input NAND gates. EX-OR implementation using NAND was discussed in the previous chapter. The AND gate required to generate CARRY output is implemented by complementing an already available NAND output of the input variables.

7.3.2 Full Adder

A *full adder* circuit is an arithmetic circuit block that can be used to add three bits to produce a SUM and a CARRY output. Such a building block becomes a necessity when it comes to adding binary numbers with a large number of bits. The full adder circuit overcomes the limitation of the half-adder, which can be used to add two bits only. Let us recall the procedure for adding larger binary numbers. We begin with the addition of LSBs of the two numbers. We record the sum under the LSB column and take the carry, if any, forward to the next higher column bits. As a result, when we add the next adjacent higher column bits, we would be required to add three bits if there were a carry from the previous addition. We have a similar situation for the other higher column bits

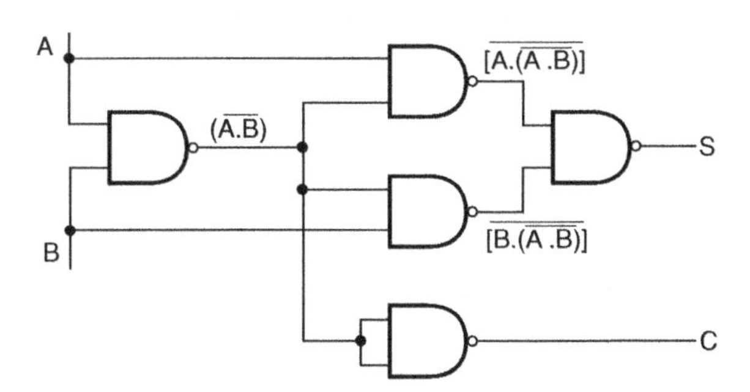

Figure 7.6 Half-adder implementation using NAND gates.

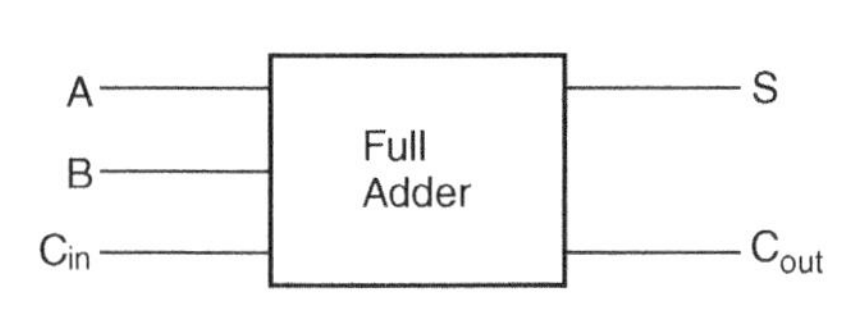

A	B	C_{in}	SUM (S)	C_{out}
0	0	0	0	0
0	0	1	1	0
0	1	0	1	0
0	1	1	0	1
1	0	0	1	0
1	0	1	0	1
1	1	0	0	1
1	1	1	1	1

Figure 7.7 Truth table of a full adder.

also until we reach the MSB. A full adder is therefore essential for the hardware implementation of an adder circuit capable of adding larger binary numbers. A half-adder can be used for addition of LSBs only.

Figure 7.7 shows the truth table of a full adder circuit showing all possible input combinations and corresponding outputs. In order to arrive at the logic circuit for hardware implementation of a full adder, we will firstly write the Boolean expressions for the two output variables, that is, the SUM and CARRY outputs, in terms of input variables. These expressions are then simplified by using any of the simplification techniques described in the previous chapter. The Boolean expressions for the two output variables are given in Equation (7.7) for the SUM output (S) and in Equation (6.6) for the CARRY output (C_{out}):

$$S=\overline{A}.\overline{B}.C_{in}+\overline{A}.B.\overline{C}_{in}+A.\overline{B}.\overline{C}_{in}+A.B.C_{in} \tag{7.7}$$

$$C_{out}=\overline{A}.B.C_{in}+A.\overline{B}.C_{in}+A.B.\overline{C}_{in}+A.B.C_{in} \tag{7.8}$$

The next step is to simplify the two expressions. We will do so with the help of the Karnaugh mapping technique. Karnaugh maps for the two expressions are given in Fig. 7.8(a) for the SUM output and Fig. 7.8(b) for the CARRY output. As is clear from the two maps, the expression for the SUM (S) output cannot be simplified any further, whereas the simplified Boolean expression for C_{out} is given by the equation

$$C_{out}=B.C_{in}+A.B+A.C_{in} \tag{7.9}$$

Figure 7.9 shows the logic circuit diagram of the full adder. A full adder can also be seen to comprise two half-adders and an OR gate. The expressions for SUM and CARRY outputs can be rewritten as follows:

$$S=\overline{C}_{in}.(\overline{A}.B+A.\overline{B})+C_{in}.(A.B+\overline{A}.\overline{B})$$

$$S=\overline{C}_{in}.(\overline{A}.B+A.\overline{B})+C_{in}.(\overline{\overline{A}.B+A.\overline{B}}) \tag{7.10}$$

Similarly, the expression for CARRY output can be rewritten as follows:

$$\begin{aligned}C_{out}&=B.C_{in}.(A+\overline{A})+A.B+A.C_{in}.(B+\overline{B})\\&=A.B+A.B.C_{in}+\overline{A}.B.C_{in}+A.B.C_{in}+A.\overline{B}.C_{in}=A.B+A.B.C_{in}+\overline{A}.B.C_{in}+A.\overline{B}.C_{in}\\&=A.B.(1+C_{in})+C_{in}.(\overline{A}.B+A.\overline{B})\end{aligned}$$

$AB \backslash C_{in}$	$\overline{C_{in}}$	C_{in}
$\overline{A}\,\overline{B}$		1
$\overline{A}\,B$	1	
$A\,B$		1
$A\,\overline{B}$	1	

(a)

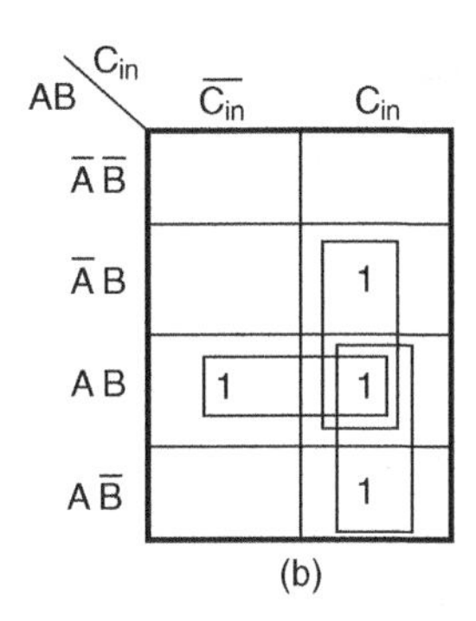

Figure 7.8 Karnaugh maps for the sum and carry-out of a full adder.

$$C_{out} = A.B + C_{in}.(\overline{A}.B + A.\overline{B}) \tag{7.11}$$

Boolean expression (7.10) can be implemented with a two-input EX-OR gate provided that one of the inputs is C_{in} and the other input is the output of another two-input EX-OR gate with A and B as its inputs. Similarly, Boolean expression (7.11) can be implemented by ORing two minterms. One of them is the AND output of A and B. The other is also the output of an AND gate whose inputs are C_{in} and the output of an EX-OR operation on A and B. The whole idea of writing the Boolean expressions in this modified form was to demonstrate the use of a half-adder circuit in building a full adder. Figure 7.10(a) shows logic implementation of Equations (7.10) and (7.11). Figure 7.10(b) is nothing but Fig. 7.10(a) redrawn with the portion of the circuit representing a half-adder replaced with a block.

The full adder of the type described above forms the basic building block of binary adders. However, a single full adder circuit can be used to add one-bit binary numbers only. A cascade arrangement of these adders can be used to construct adders capable of adding binary numbers with a larger number of bits. For example, a four-bit binary adder would require four full adders of the type shown in Fig. 7.10 to be connected in cascade. Figure 7.11 shows such an arrangement. $(A_3A_2A_1A_0)$ and $(B_3B_2B_1B_0)$ are the two binary numbers to be added, with A_0 and B_0 representing LSBs and A_3 and B_3 representing MSBs of the two numbers.

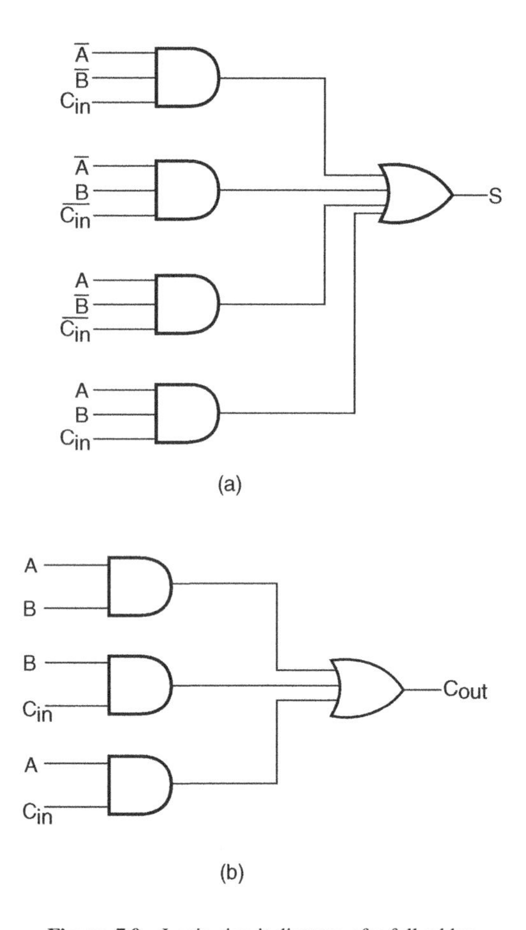

Figure 7.9 Logic circuit diagram of a full adder.

7.3.3 Half-Subtractor

We have seen in Chapter 3 on digital arithmetic how subtraction of two given binary numbers can be carried out by adding 2's complement of the subtrahend to the minuend. This allows us to do a subtraction operation with adder circuits. We will study the use of adder circuits for subtraction operations in the following pages. Before we do that, we will briefly look at the counterparts of half-adder and full adder circuits in the half-subtractor and full subtractor for direct implementation of subtraction operations using logic gates.

A *half-subtractor* is a combinational circuit that can be used to subtract one binary digit from another to produce a DIFFERENCE output and a BORROW output. The BORROW output here specifies whether a '1' has been borrowed to perform the subtraction. The truth table of a half-subtractor, as shown in Fig. 7.12, explains this further. The Boolean expressions for the two outputs are given by the equations

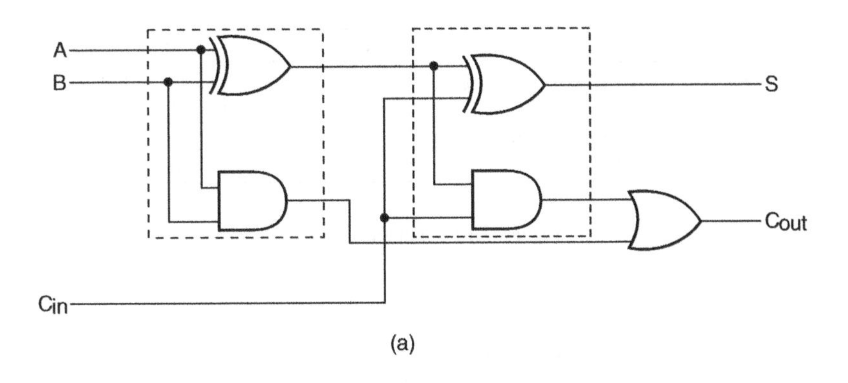

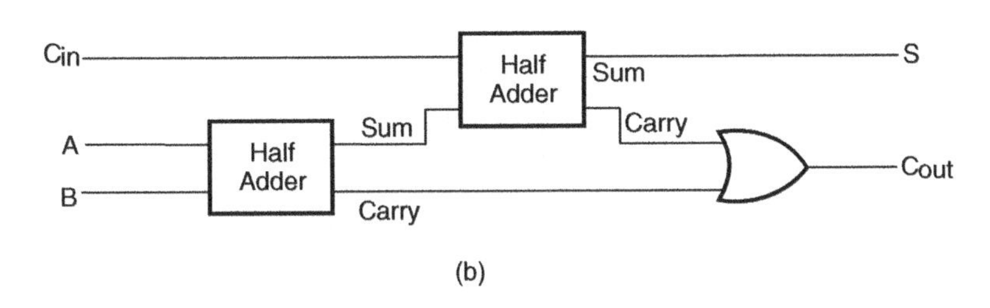

Figure 7.10 Logic implementation of a full adder with half-adders.

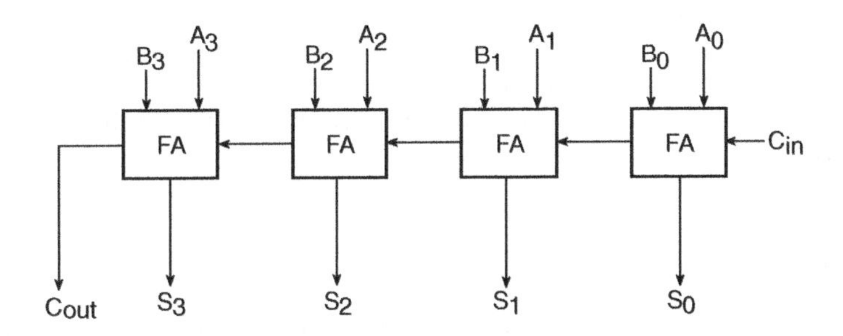

Figure 7.11 Four-bit binary adder.

$$D = \overline{A}.B + A.\overline{B} \tag{7.12}$$

$$B_o = \overline{A}.B \tag{7.13}$$

It is obvious that there is no further scope for any simplification of the Boolean expressions given by Equations (7.12) and (7.13). While the expression for the DIFFERENCE (D) output is that of

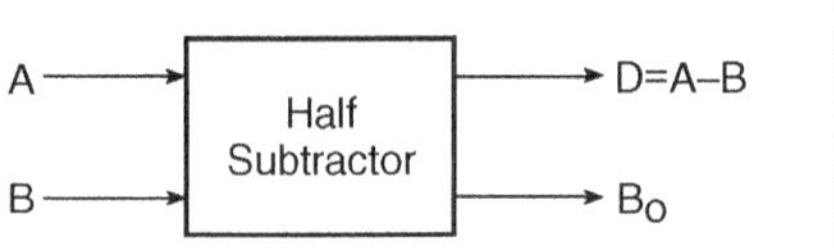

A	B	D	B_o
0	0	0	0
0	1	1	1
1	0	1	0
1	1	0	0

Figure 7.12 Half-subtractor.

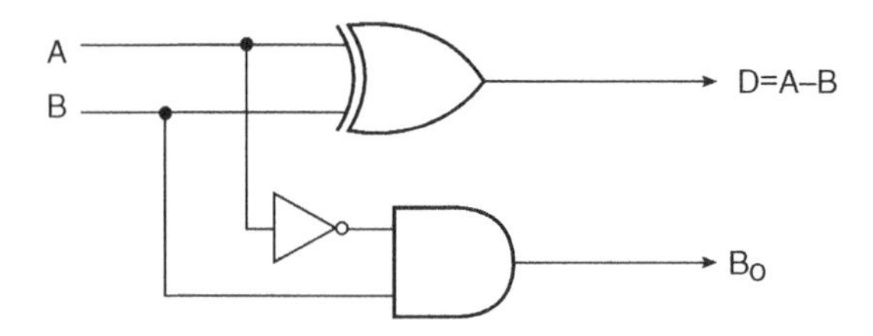

Figure 7.13 Logic diagram of a half-subtractor.

an EX-OR gate, the expression for the BORROW output (B_o) is that of an AND gate with input A complemented before it is fed to the gate. Figure 7.13 shows the logic implementation of a half-subtractor. Comparing a half-subtractor with a half-adder, we find that the expressions for the SUM and DIFFERENCE outputs are just the same. The expression for BORROW in the case of the half-subtractor is also similar to what we have for CARRY in the case of the half-adder. If the input A, that is, the minuend, is complemented, an AND gate can be used to implement the BORROW output. Note the similarities between the logic diagrams of Fig. 7.5 (half-adder) and Fig. 7.13 (half-subtractor).

7.3.4 Full Subtractor

A *full subtractor* performs subtraction operation on two bits, a minuend and a subtrahend, and also takes into consideration whether a '1' has already been borrowed by the previous adjacent lower minuend bit or not. As a result, there are three bits to be handled at the input of a full subtractor, namely the two bits to be subtracted and a borrow bit designated as B_{in}. There are two outputs, namely the DIFFERENCE output D and the BORROW output B_o. The BORROW output bit tells whether the minuend bit needs to borrow a '1' from the next possible higher minuend bit. Figure 7.14 shows the truth table of a full subtractor.

The Boolean expressions for the two output variables are given by the equations

$$D = \overline{A}.\overline{B}.B_{in} + \overline{A}.B.\overline{B}_{in} + A.\overline{B}.\overline{B}_{in} + A.B.B_{in} \tag{7.14}$$

$$B_o = \overline{A}.\overline{B}.B_{in} + \overline{A}.B.\overline{B}_{in} + \overline{A}.B.B_{in} + A.B.B_{in} \tag{7.15}$$

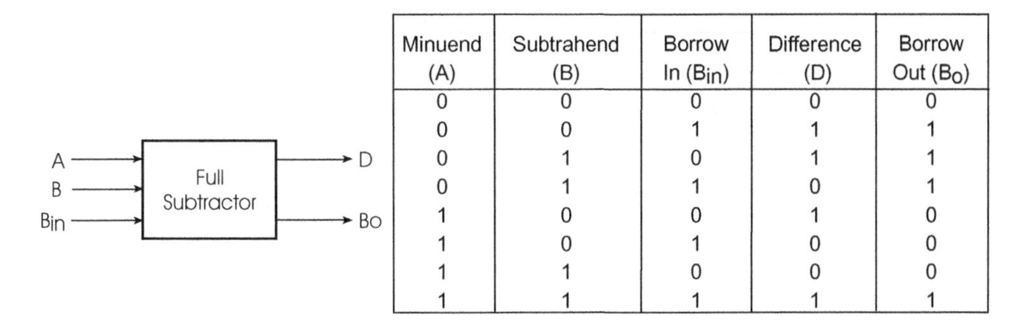

Minuend (A)	Subtrahend (B)	Borrow In (B_{in})	Difference (D)	Borrow Out (B_o)
0	0	0	0	0
0	0	1	1	1
0	1	0	1	1
0	1	1	0	1
1	0	0	1	0
1	0	1	0	0
1	1	0	0	0
1	1	1	1	1

Figure 7.14 Truth table of a full subtractor.

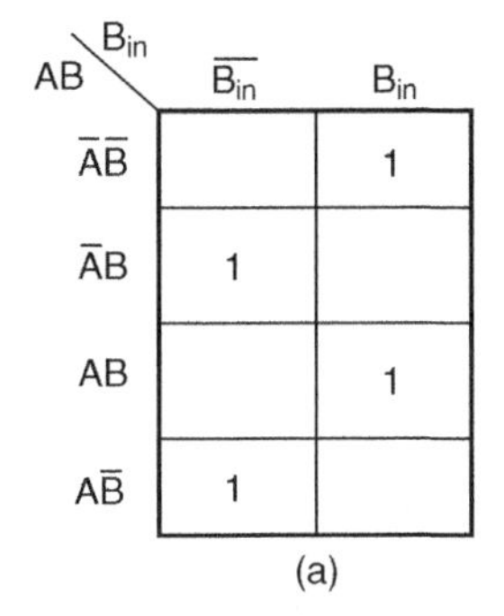

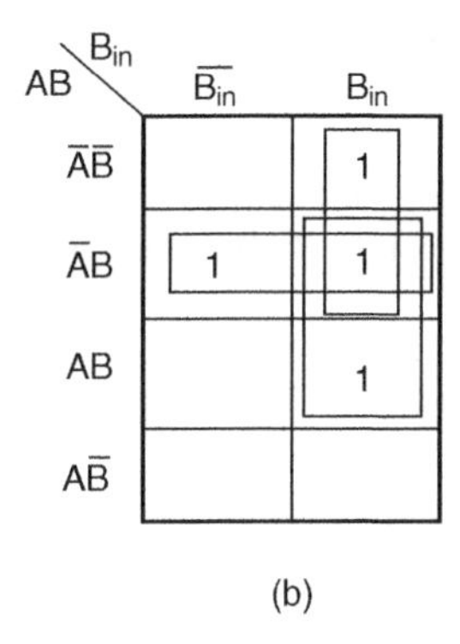

Figure 7.15 Karnaugh maps for difference and borrow outputs.

The Karnaugh maps for the two expressions are given in Fig. 7.15(a) for DIFFERENCE output D and in Fig. 7.15(b) for BORROW output B_o. As is clear from the two Karnaugh maps, no simplification is possible for the difference output D. The simplified expression for B_o is given by the equation

$$B_o = \overline{A}.B + \overline{A}.B_{in} + B.B_{in} \tag{7.16}$$

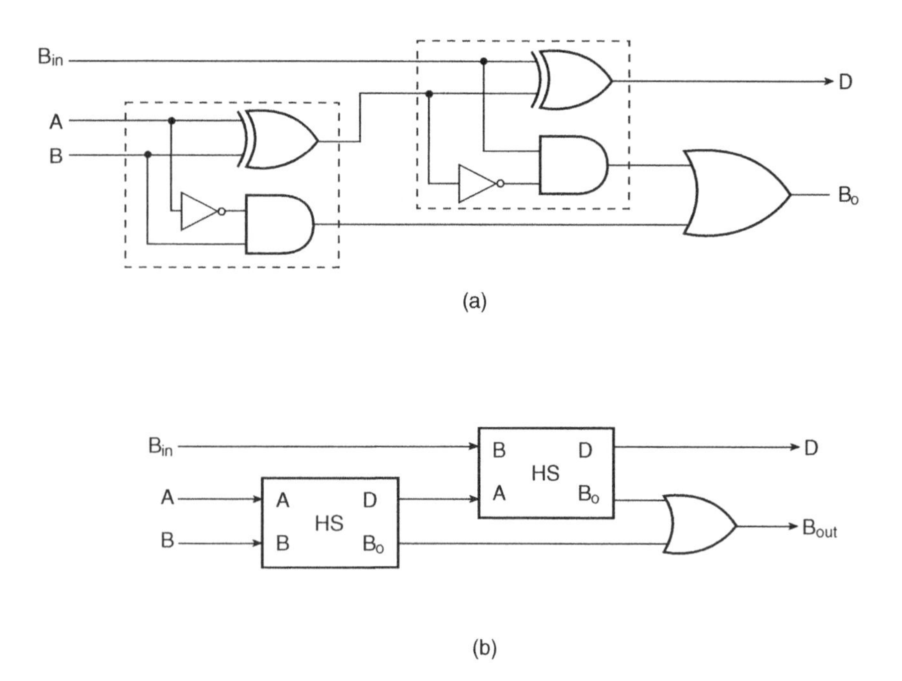

Figure 7.16 Logic implementation of a full subtractor with half-subtractors.

If we compare these expressions with those derived earlier in the case of a full adder, we find that the expression for DIFFERENCE output D is the same as that for the SUM output. Also, the expression for BORROW output B_o is similar to the expression for CARRY-OUT C_o. In the case of a half-subtractor, the A input is complemented. By a similar analysis it can be shown that a full subtractor can be implemented with half-subtractors in the same way as a full adder was constructed using half-adders. Relevant logic diagrams are shown in Figs 7.16(a) and (b) corresponding to Figs 7.10(a) and (b) respectively for a full adder.

Again, more than one full subtractor can be connected in cascade to perform subtraction on two larger binary numbers. As an illustration, Fig. 7.17 shows a four-bit subtractor.

7.3.5 Controlled Inverter

A *controlled inverter* is needed when an adder is to be used as a subtractor. As outlined earlier, subtraction is nothing but addition of the 2's complement of the subtrahend to the minuend. Thus, the first step towards practical implementation of a subtractor is to determine the 2's complement of the subtrahend. And for this, one needs firstly to find 1's complement. A controlled inverter is used to find 1's complement. A one-bit controlled inverter is nothing but a two-input EX-OR gate with one of its inputs treated as a control input, as shown in Fig. 7.18(a). When the control input is LOW, the input bit is passed as such to the output. (Recall the truth table of an EX-OR gate.) When the control input is HIGH, the input bit gets complemented at the output. Figure 7.18(b) shows an eight-bit controlled inverter of this type. When the control input is LOW, the output (Y_7 Y_6 Y_5 Y_4 Y_3 Y_2 Y_1 Y_0) is the same as the input (A_7 A_6 A_5 A_4 A_3 A_2 A_1 A_0). When the control input is HIGH, the output is 1's complement

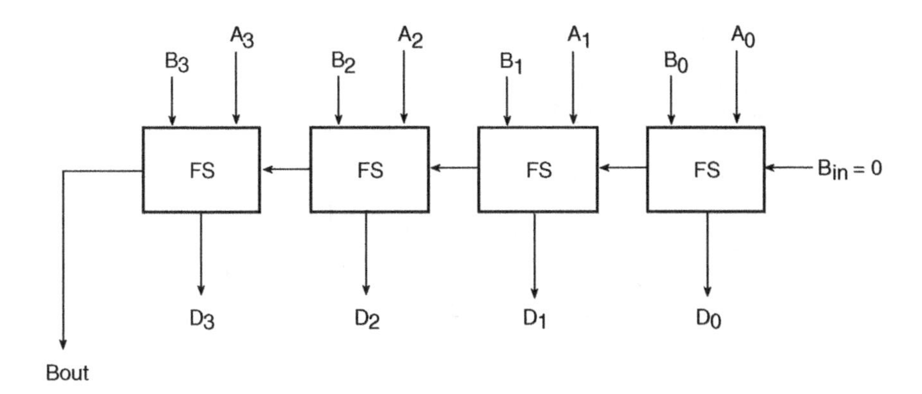

Figure 7.17 Four-bit subtractor.

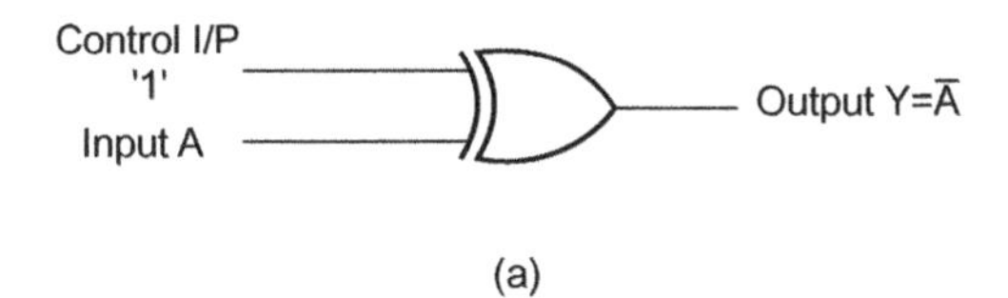

(a)

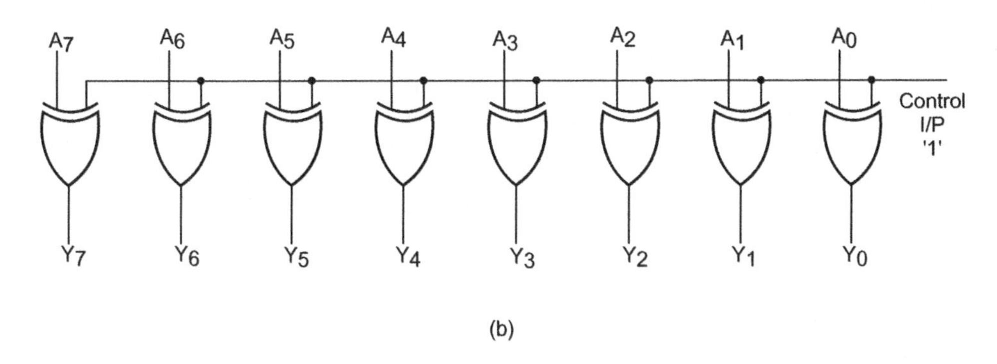

(b)

Figure 7.18 (a) One-bit controlled inverter and (b) eight-bit controlled inverter.

of the input. As an example, 11010010 at the input would produce 00101101 at the output when the control input is in a logic '1' state.

7.4 Adder–Subtractor

Subtraction of two binary numbers can be accomplished by adding 2's complement of the subtrahend to the minuend and disregarding the final carry, if any. If the MSB bit in the result of addition is

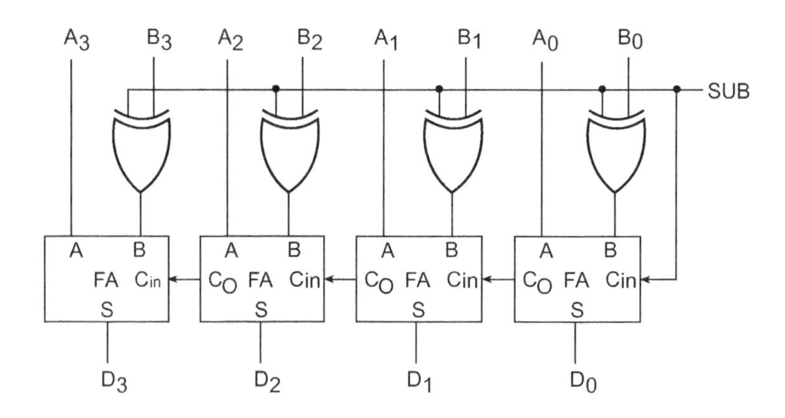

Figure 7.19 Four-bit adder-subtractor.

a '0', then the result of addition is the correct answer. If the MSB bit is a '1', this implies that the answer has a negative sign. The true magnitude in this case is given by 2's complement of the result of addition.

Full adders can be used to perform subtraction provided we have the necessary additional hardware to generate 2's complement of the subtrahend and disregard the final carry or overflow. Figure 7.19 shows one such hardware arrangement. Let us see how it can be used to perform subtraction of two four-bit binary numbers. A close look at the diagram would reveal that it is the hardware arrangement for a four-bit binary adder, with the exception that the bits of one of the binary numbers are fed through controlled inverters. The control input here is referred to as the SUB input. When the SUB input is in logic '0' state, the four bits of the binary number ($B_3\,B_2\,B_1\,B_0$) are passed on as such to the B inputs of the corresponding full adders. The outputs of the full adders in this case give the result of addition of the two numbers. When the SUB input is in logic '1' state, four bits of one of the numbers, ($B_3\,B_2\,B_1\,B_0$) in the present case, get complemented. If the same '1' is also fed to the CARRY-IN of the LSB full adder, what we finally achieve is the addition of 2's complement and not 1's complement. Thus, in the adder arrangement of Fig. 7.19, we are basically adding 2's complement of ($B_3\,B_2\,B_1\,B_0$) to ($A_3\,A_2\,A_1\,A_0$). The outputs of the full adders in this case give the result of subtraction of the two numbers. The arrangement shown achieves $A-B$. The final carry (the CARRY-OUT of the MSB full adder) is ignored if it is not displayed.

For implementing an eight-bit adder–subtractor, we will require eight full adders and eight two-input EX-OR gates. Four-bit full adders and quad two-input EX-OR gates are individually available in integrated circuit form. A commonly used four-bit adder in the TTL family is the type number 7483. Also, type number 7486 is a quad two-input EX-OR gate in the TTL family. Figure 7.20 shows a four-bit binary adder–subtractor circuit implemented with 7483 and 7486. Two each of 7483 and 7486 can be used to construct an eight-bit adder–subtractor circuit.

7.5 BCD Adder

A *BCD adder* is used to perform the addition of BCD numbers. A BCD digit can have any of the ten possible four-bit binary representations, that is, 0000, 0001, . . . , 1001, the equivalent of decimal numbers 0, 1, . . . , 9. When we set out to add two BCD digits and we assume that there is an input carry too, the highest binary number that we can get is the equivalent of decimal number 19 ($9+9+1$).

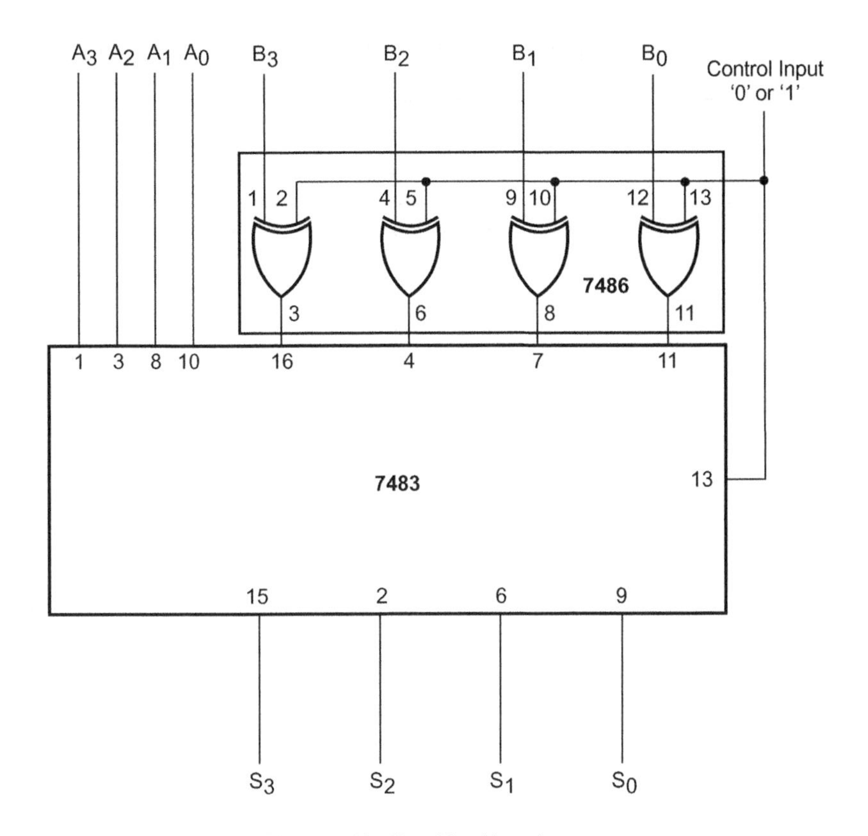

Figure 7.20 Four-bit adder-subtractor.

This binary number is going to be $(10011)_2$. On the other hand, if we do BCD addition, we would expect the answer to be $(0001\ 1001)_{BCD}$. And if we restrict the output bits to the minimum required, the answer in BCD would be $(1\ 1001)_{BCD}$. Table 7.1 lists the possible results in binary and the expected results in BCD when we use a four-bit binary adder to perform the addition of two BCD digits. It is clear from the table that, as long as the sum of the two BCD digits remains equal to or less than 9, the four-bit adder produces the correct BCD output.

The binary sum and the BCD sum in this case are the same. It is only when the sum is greater than 9 that the two results are different. It can also be seen from the table that, for a decimal sum greater than 9 (or the equivalent binary sum greater than 1001), if we add 0110 to the binary sum, we can get the correct BCD sum and the desired carry output too. The Boolean expression that can apply the necessary correction is written as

$$C = K + Z_3.Z_2 + Z_3.Z_1 \tag{7.17}$$

Equation (7.17) implies the following. A correction needs to be applied whenever $K = 1$. This takes care of the last four entries. Also, a correction needs to be applied whenever both Z_3 and Z_2 are '1'. This takes care of the next four entries from the bottom, corresponding to a decimal sum equal to

Table 7.1 Results in binary and the expected results in BCD using a four-bit binary adder to perform the addition of two BCD digits.

Decimal sum	Binary sum					BCD sum				
	K	Z_3	Z_2	Z_1	Z_0	C	S_3	S_2	S_1	S_0
0	0	0	0	0	0	0	0	0	0	0
1	0	0	0	0	1	0	0	0	0	1
2	0	0	0	1	0	0	0	0	1	0
3	0	0	0	1	1	0	0	0	1	1
4	0	0	1	0	0	0	0	1	0	0
5	0	0	1	0	1	0	0	1	0	1
6	0	0	1	1	0	0	0	1	1	0
7	0	0	1	1	1	0	0	1	1	1
8	0	1	0	0	0	0	1	0	0	0
9	0	1	0	0	1	0	1	0	0	1
10	0	1	0	1	0	1	0	0	0	0
11	0	1	0	1	1	1	0	0	0	1
12	0	1	1	0	0	1	0	0	1	0
13	0	1	1	0	1	1	0	0	1	1
14	0	1	1	1	0	1	0	1	0	0
15	0	1	1	1	1	1	0	1	0	1
16	1	0	0	0	0	1	0	1	1	0
17	1	0	0	0	1	1	0	1	1	1
18	1	0	0	1	0	1	1	0	0	0
19	1	0	0	1	1	1	1	0	0	1

12, 13, 14 and 15. For the remaining two entries corresponding to a decimal sum equal to 10 and 11, a correction is applied for both Z_3 and Z_1, being '1'. While hardware-implementing, 0110 can be added to the binary sum output with the help of a second four-bit binary adder. The correction logic as dictated by the Boolean expression (7.17) should ensure that (0110) gets added only when the above expression is satisfied. Otherwise, the sum output of the first binary adder should be passed on as such to the final output, which can be accomplished by adding (0000) in the second adder. Figure 7.21 shows the logic arrangement of a BCD adder capable of adding two BCD digits with the help of two four-bit binary adders and some additional combinational logic.

The BCD adder described in the preceding paragraphs can be used to add two single-digit BCD numbers only. However, a cascade arrangement of single-digit BCD adder hardware can be used to perform the addition of multiple-digit BCD numbers. For example, an n-digit BCD adder would require n such stages in cascade. As an illustration, Fig. 7.22 shows the block diagram of a circuit for the addition of two three-digit BCD numbers. The first BCD adder, labelled LSD (Least Significant Digit), handles the least significant BCD digits. It produces the sum output ($S_3 S_2 S_1 S_0$), which is the BCD code for the least significant digit of the sum. It also produces an output carry that is fed as an input carry to the next higher adjacent BCD adder. This BCD adder produces the sum output ($S_7 S_6 S_5 S_4$), which is the BCD code for the second digit of the sum, and a carry output. This output carry serves as an input carry for the BCD adder representing the most significant digits. The sum outputs ($S_{11} S_{10} S_9 S_8$) represent the BCD code for the MSD of the sum.

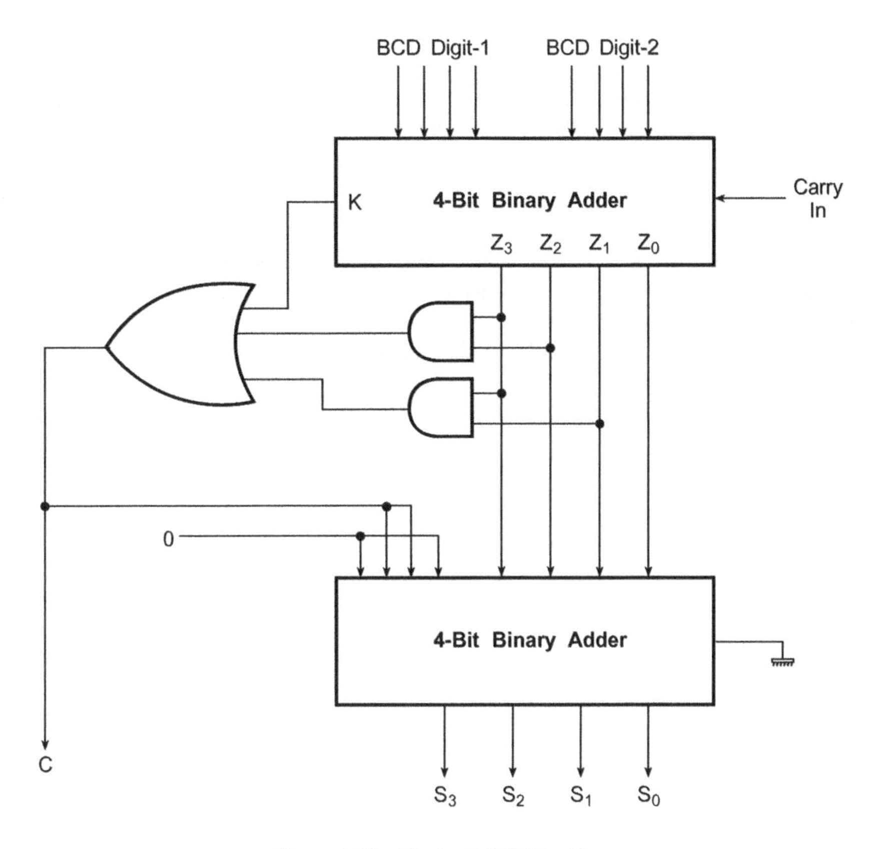

Figure 7.21 Single-digit BCD adder.

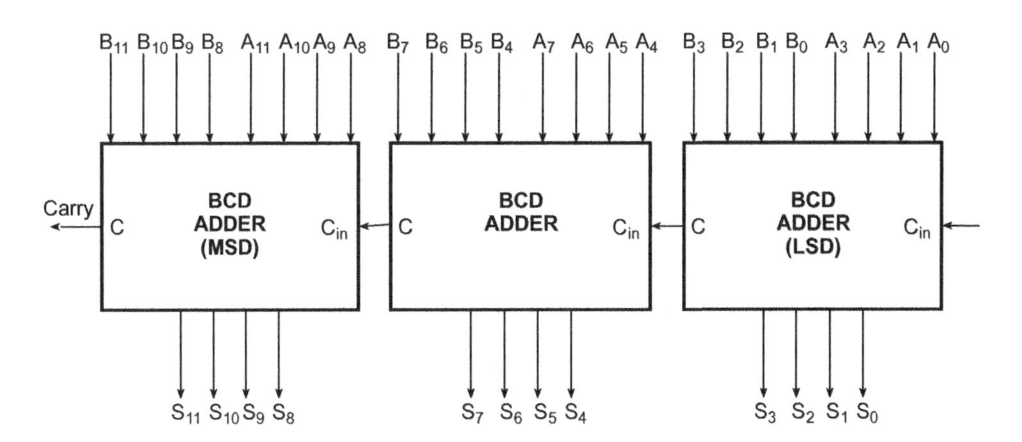

Figure 7.22 Three-digit BCD adder.

Example 7.1

For the half-adder circuit of Fig. 7.23(a), the inputs applied at A and B are as shown in Fig. 7.23(b). Plot the corresponding SUM and CARRY outputs on the same scale.

Solution

The SUM and CARRY waveforms can be plotted from our knowledge of the truth table of the half-adder. All that we need to remember to solve this problem is that 0 + 0 yields a '0' as the SUM output and a '0' as the CARRY. 0 + 1 or 1 + 0 yield '1' as the SUM output and '0' as the CARRY. 1 + 1 produces a '0' as the SUM output and a '1' as the CARRY. The output waveforms are as shown in Fig. 7.24.

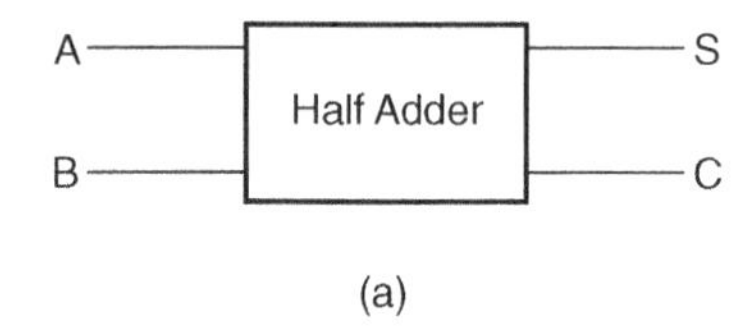

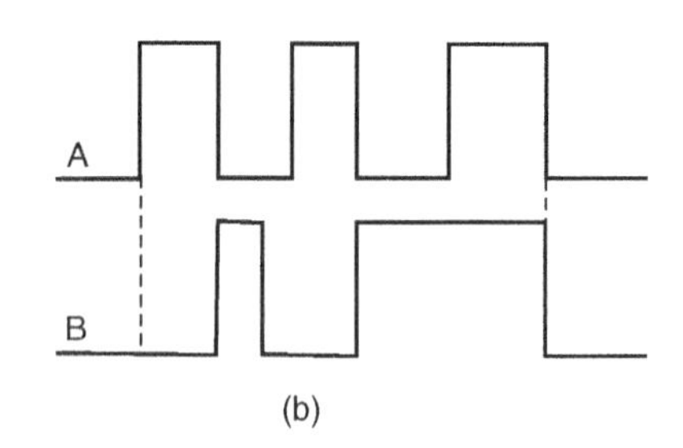

Figure 7.23 Example 7.1.

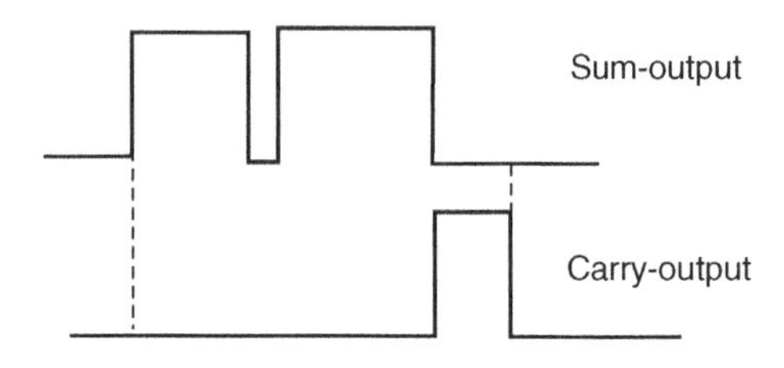

Figure 7.24 Solution to example 7.1.

Example 7.2

Given the relevant Boolean expressions for half-adder and half-subtractor circuits, design a half-adder–subtractor circuit that can be used to perform either addition or subtraction on two one-bit numbers. The desired arithmetic operation should be selectable from a control input.

Solution

Boolean expressions for the half-adder and half-subtractor are given as follows:

Half-adder

$$\text{SUM output} = \overline{A}B + A\overline{B} \quad \text{and} \quad \text{CARRY output} = AB$$

Half-subtractor

$$\text{DIFFERENCE output} = \overline{A}B + A\overline{B} \quad \text{and} \quad \text{BORROW output} = \overline{A}B$$

If we use a controlled inverter for complementing A in the case of the half-subtractor circuit, then the same hardware can also be used to add two one-bit numbers. Figure 7.25 shows the logic circuit diagram. When the control input is '0', input variable A is passed uncomplemented to the input of the NAND gate. In this case, the AND gate generates the CARRY output of the addition operation. The EX-OR gate generates the SUM output. On the other hand, when the control input is '1', the AND gate generates the BORROW output and the EX-OR gate generates the DIFFERENCE output. Thus, '0' at the control input makes it a half-adder, while '1' at the control input makes it a half-subtractor.

Example 7.3

Refer to Fig. 7.26. Write the simplified Boolean expressions for DIFFERENCE and BORROW outputs.

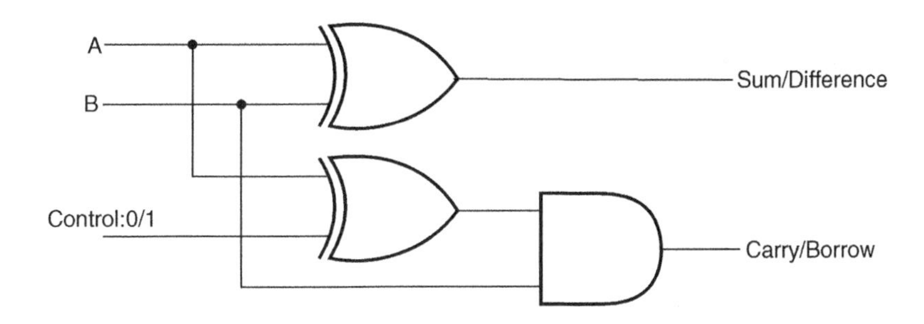

Figure 7.25 Solution to example 7.2.

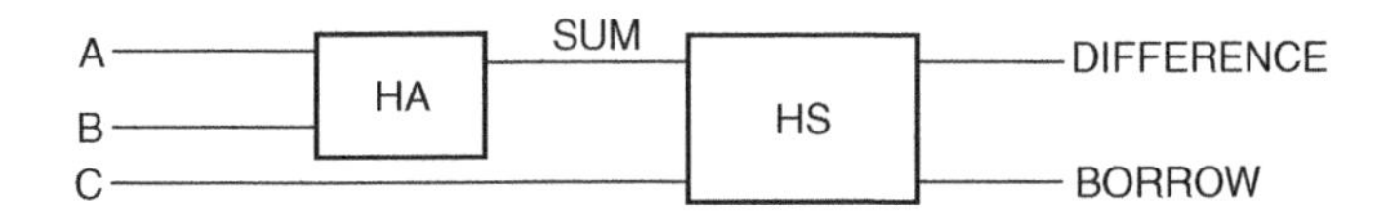

Figure 7.26 Example 7.3.

Solution

Let us assume that the two inputs to the half-subtractor circuit are X and Y, with X equal to the SUM output of the half-adder and Y equal to C. DIFFERENCE and BORROW outputs can then be expressed as follows:

$$\text{DIFFERENCE output} = X \oplus Y = \overline{X}.Y + X.\overline{Y} \quad \text{and} \quad \text{BORROW output} = \overline{X}.Y$$

Also, $X = \overline{A}.B + A.\overline{B}$ and $Y = C$.

Substituting the values of X and Y, we obtain

$$\text{DIFFERENCE output} = \overline{(\overline{A}.B + A.\overline{B})}.C + (\overline{A}.B + A.\overline{B}).\overline{C} = (A.B + \overline{A}.\overline{B}).C + (\overline{A}.B + A.\overline{B}).\overline{C}$$
$$= A.B.C + \overline{A}.\overline{B}.C + \overline{A}.B.\overline{C} + A.\overline{B}.\overline{C}$$

$$\text{BORROW output} = \overline{X}.Y = \overline{(\overline{A}.B + A.\overline{B})}.C = (A.B + \overline{A}.\overline{B}).C = A.B.C + \overline{A}.\overline{B}.C$$

Example 7.4

Design an eight-bit adder–subtractor circuit using four-bit binary adders, type number 7483, and quad two-input EX-OR gates, type number 7486. Assume that pin connection diagrams of these ICs are available to you.

Solution

IC 7483 is a four-bit binary adder, which means that it can add two four-bit binary numbers. In order to add two eight-bit numbers, we need to use two 7483s in cascade. That is, the CARRY-OUT (pin 14) of the 7483 handling less significant four bits is fed to the CARRY-IN (pin 13) of the 7483 handling more significant four bits. Also, if $(A_0 \ldots A_7)$ and $(B_0 \ldots B_7)$ are the two numbers to be operated upon, and if the objective is to compute $A - B$, bits B_0, B_1, B_2, B_3, B_4, B_5, B_6 and B_7 are complemented using EX-OR gates. One of the inputs of all EX-OR gates is tied together to form the control input. When the control input is in logic '1' state, bits B_0 to B_7 get complemented. Also, feeding this logic '1' to the CARRY-IN of lower 7483 ensures that we get 2's complement of bits $(B_0 \ldots B_7)$. Therefore, when the control input is in logic '1' state, the two's complement of $(B_0 \ldots B_7)$ is added to $(A_0 \ldots A_7)$. The output is therefore $A - B$. A logic '0' at the control input allows $(B_0 \ldots B_7)$ to pass through EX-OR gates uncomplemented, and the output in that case is $A + B$. Figure 7.27 shows the circuit diagram.

Example 7.5

The logic diagram of Fig. 7.28 performs the function of a very common arithmetic building block. Identify the logic function.

Solution

Writing Boolean expressions for X and Y,

$$X = \overline{(\overline{\overline{A}.B}).(\overline{A.\overline{B}})} = (\overline{\overline{\overline{A}.B}} + \overline{\overline{A.\overline{B}}}) = \overline{A}.B + A.\overline{B} \quad \text{and} \quad Y = \overline{(\overline{A} + \overline{B})} = A.B$$

Boolean expressions for X and Y are those of a half-adder. X and Y respectively represent SUM and CARRY outputs.

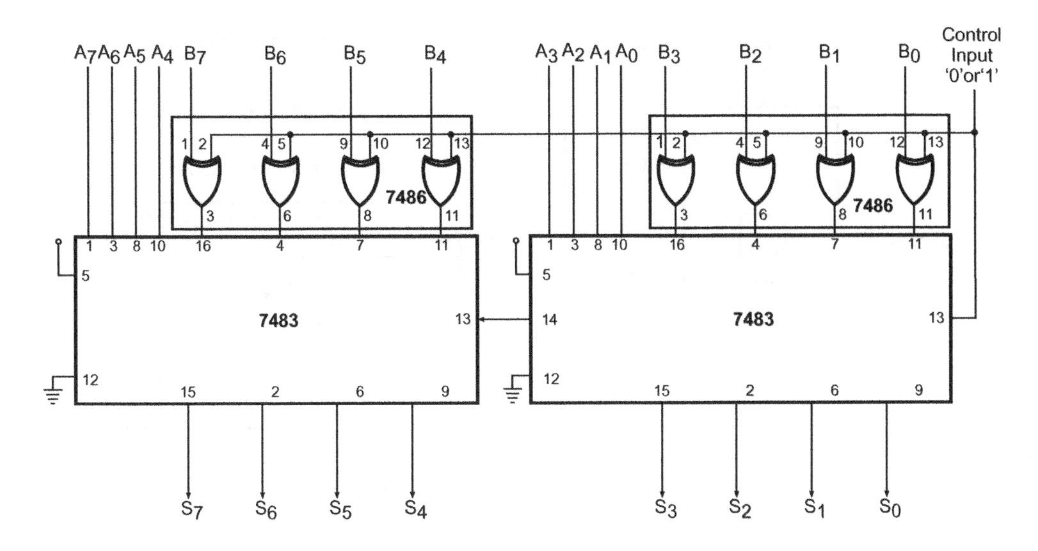

Figure 7.27 Solution to example 7.4.

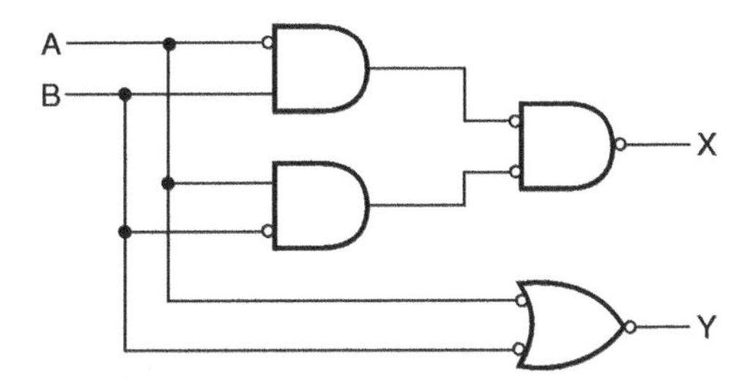

Figure 7.28 Example 7.5.

Example 7.6

Design a BCD adder circuit capable of adding BCD equivalents of two-digit decimal numbers. Indicate the IC type numbers used if the design has to be TTL logic family compatible.

Solution

The desired BCD adder is a cascaded arrangement of two stages of the type of BCD adder discussed in the previous pages. Figure 7.29 shows the logic diagram, and it follows the generalized cascaded arrangement discussed earlier and shown in Fig. 7.22 for a three-digit BCD adder. The BCD adder of Fig. 7.21 can be used to add four-bit BCD equivalents of two single-digit decimal numbers. A cascaded arrangement of two such stages, where the output C of Fig. 7.21 (CARRY-OUT) is fed to the CARRY-IN of the second stage, is shown in Fig. 7.29. In terms of IC type numbers, IC 7483 can be used for four-bit binary adders as shown in the diagram, IC 7408 can be used for implementing

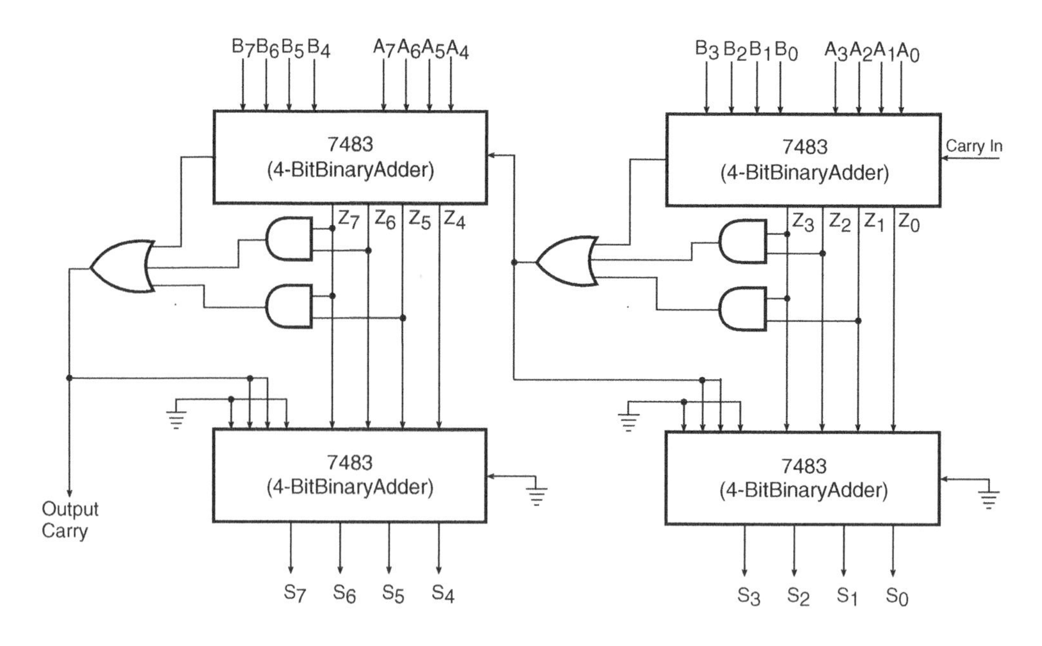

Figure 7.29 Example 7.6.

the required four two-input AND gates (IC 7408 is a quad two-input AND) and IC 7432 can be used to implement the required two three-input OR gates. IC 7432 is a quad two-input OR. Two two-input OR gates can be connected in cascade to get a three-input OR gate.

7.6 Carry Propagation–Look-Ahead Carry Generator

The four-bit binary adder described in the previous pages can be used to add two four-bit binary numbers. Multiple numbers of such adders are used to perform addition operations on larger-bit binary numbers. Each of the adders is composed of four full adders (FAs) connected in cascade. The block schematic arrangement of a four-bit adder is reproduced in Fig. 7.30(a) for reference and further discussion. This type of adder is also called a parallel binary adder because all the bits of the augend and addend are present and are fed to the full adder blocks simultaneously. Theoretically, the addition operation in various full adders takes place simultaneously. What is of importance and interest to users, more so when they are using a large number of such adders in their overall computation system, is whether the result of addition and carry-out are available to them at the same time. In other words, we need to see if this addition operation is truly parallel in nature. We will soon see that it is not. It is in fact limited by what is known as *carry propagation time*. Refer to Figs 7.30(a) and (b). Figure 7.30(b) shows the logic diagram of a full adder. Here, C_i and C_{i+1} are the input and output CARRY; P_i and G_i are two new binary variables called CARRY PROPAGATE and CARRY GENERATE and will be addressed a little later.

For $i=1$, the diagram in Fig. 7.30(b) is that of the LSB full adder of Fig. 7.30(a). We can see here that C_2, which is the output CARRY of FA (1) and the input CARRY for FA (2), will appear at the output after a minimum of two gate delays plus delay due to the half adder after application of A_i, B_i and C_i inputs.

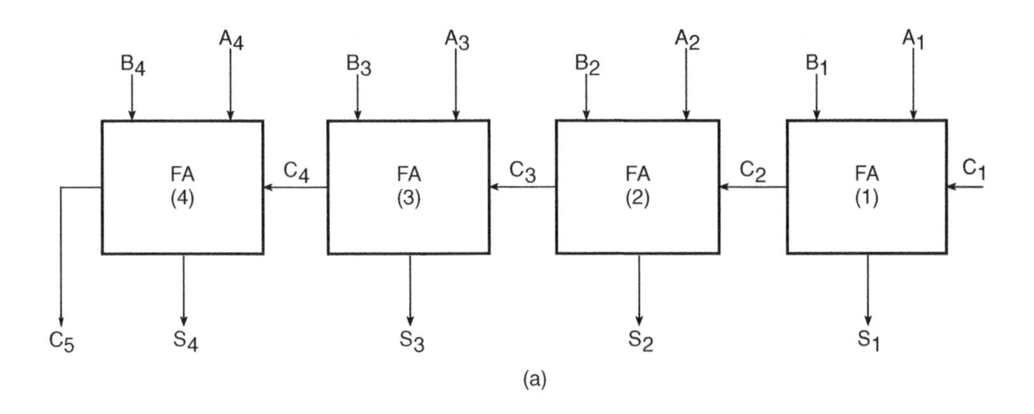

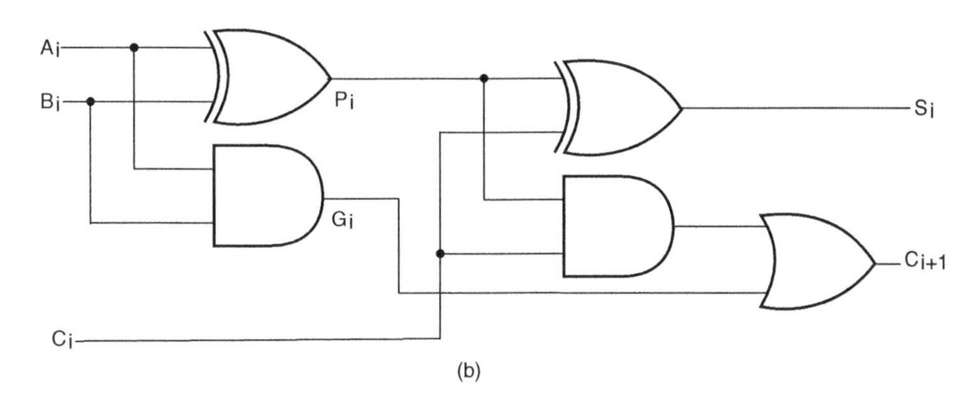

Figure 7.30 Four-bit binary adder.

The steady state of C_2 will be delayed by two gate delays after the appearance of C_1. Similarly, C_3 and C_4 steady state will be four and six gate delays respectively after C_1. And final carry C_5 will appear after eight gate delays.

Extending it a little further, let us assume that we are having a cascade arrangement of two four-bit adders to be able to handle eight-bit numbers. Now, C_5 will form the input CARRY to the second four-bit adder. The final output CARRY C_9 will now appear after 16 gate delays. This carry propagation delay limits the speed with which two numbers are added. The outputs of any such adder arrangement will be correct only if signals are given enough time to propagate through gates connected between input and output. Since subtraction is also an addition process and operations like multiplication and division are also processes involving successive addition and subtraction, the time taken by an addition process is very critical.

One of the possible methods for reducing carry propagation delay time is to use faster logic gates. But then there is a limit below which the gate delay cannot be reduced. There are other hardware-related techniques, the most widely used of which is the concept of look-ahead carry. This concept attempts to look ahead and generate the carry for a certain given addition operation that would otherwise have resulted from some previous operation. In order to explain the concept, let us define two new binary variables: P_i called CARRY PROPAGATE and G_i called CARRY GENERATE. Binary variable G_i is so called as it generates a carry whenever A_i and B_i are '1'. Binary variable P_i is called CARRY PROPAGATE as it is instrumental in propagation of C_i to C_{i+1}. CARRY, SUM, CARRY GENERATE and CARRY PROPAGATE parameters are given by the following expressions:

$$P_i = A_i \oplus B_i \tag{7.18}$$

$$G_i = A_i.B_i \tag{7.19}$$

$$S_i = P_i \oplus C_i \tag{7.20}$$

$$C_{i+1} = P_i.C_i + G_i \tag{7.21}$$

In the next step, we write Boolean expressions for the CARRY output of each full adder stage in the four-bit binary adder. We obtain the following expressions:

$$C_2 = G_1 + P_1.C_1 \tag{7.22}$$

$$C_3 = G_2 + P_2.C_2 = G_2 + P_2.(G_1 + P_1.C_1) = G_2 + P_2.G_1 + P_1.P_2.C_1 \tag{7.23}$$

$$C_4 = G_3 + P_3.C_3 = G_3 + P_3.(G_2 + P_2.G_1 + P_1.P_2.C_1)$$

$$C_4 = G_3 + P_3.G_2 + P_3.P_2.G_1 + P_1.P_2.P_3.C_1 \tag{7.24}$$

From the expressions for C_2, C_3 and C_4 it is clear that C_4 need not wait for C_3 and C_2 to propagate. Similarly, C_3 does not wait for C_2 to propagate. Hardware implementation of these expressions gives us a kind of look-ahead carry generator. A look-ahead carry generator that implements the above expressions using AND-OR logic is shown in Fig. 7.31.

Figure 7.32 shows the four-bit adder with the look-ahead carry concept incorporated. The block labelled *look-ahead carry generator* is similar to that shown in Fig. 7.31. The logic gates shown to the left of the block represent the input half-adder portion of various full adders constituting the four-bit adder. The EX-OR gates shown on the right are a portion of the output half-adders of various full adders.

All sum outputs in this case will be available at the output after a delay of two levels of logic gates. 74182 is a typical look-ahead carry generator IC of the TTL logic family. This IC can be used to generate relevant carry inputs for four four-bit binary adders connected in cascade to perform operation on two 16-bit numbers. Of course, the four-bit adders should be of the type so as to produce CARRY GENERATE and CARRY PROPAGATE outputs. Figure 7.33 shows the arrangement. In the figure shown, C_n is the CARRY input, G_0, G_1, G_2 and G_3 are CARRY GENERATE inputs for 74182 and P_0, P_1, P_2 and P_3 are CARRY PROPAGATE inputs for 74182. C_{n+x}, C_{n+y} and C_{n+z} are the CARRY outputs generated by 74182 for the four-bit adders. The G and P outputs of 74182 need to be cascaded. Figure 7.34 shows the arrangement needed for adding two 64-bit numbers.

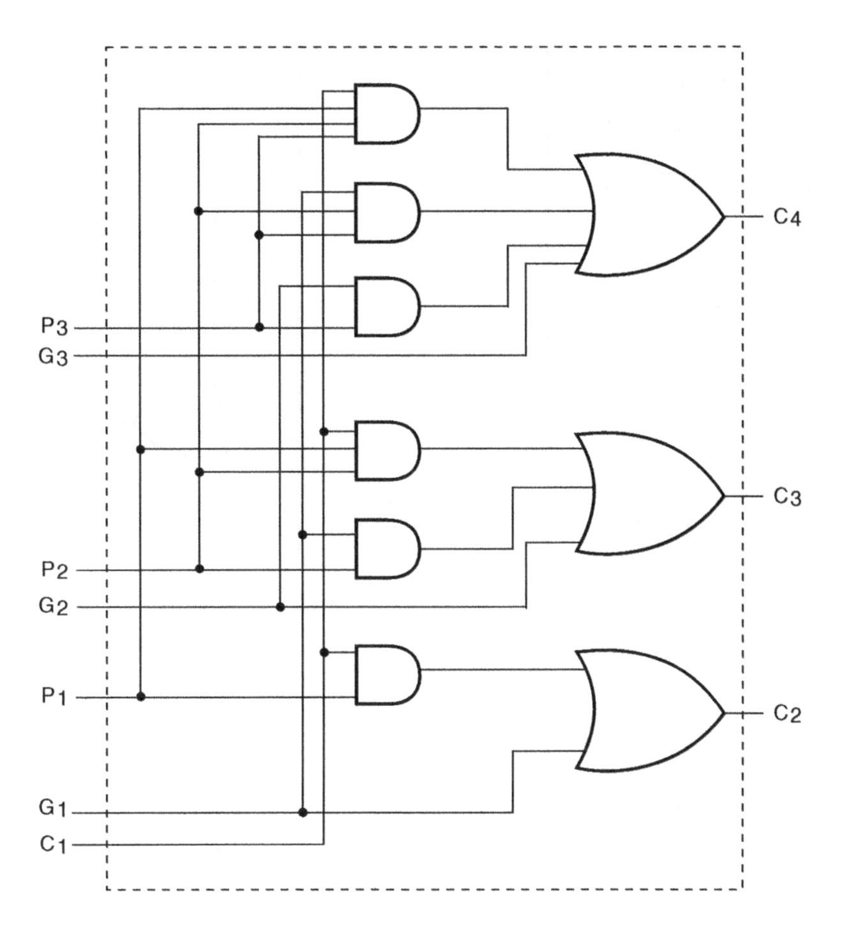

Figure 7.31 Look-ahead carry generator.

Example 7.7

If the CARRY GENERATE G_i and CARRY PROPAGATE P_i are redefined as $P_i = (A_i + B_i)$ and $G_i = A_i B_i$, show that the CARRY output C_{i+1} and the SUM output S_i of a full adder can be expressed by the following Boolean functions:

$$C_{i+1} = (\overline{\overline{C_i}.\overline{G_i} + \overline{P_i}}) = G_i + P_i.C_i \quad \text{and} \quad S_i = (P_i.\overline{G_i}) \oplus C_i$$

Solution

$$C_{i+1} = (\overline{\overline{C_i}.\overline{G_i} + \overline{P_i}}) = [\overline{\overline{C_i}.(\overline{A_{i.}B_i}) + (\overline{A_i + B_i})}]$$

$$= [\overline{\overline{C_i}.(\overline{A_i.B_i})}.(A_i + B_i)]$$

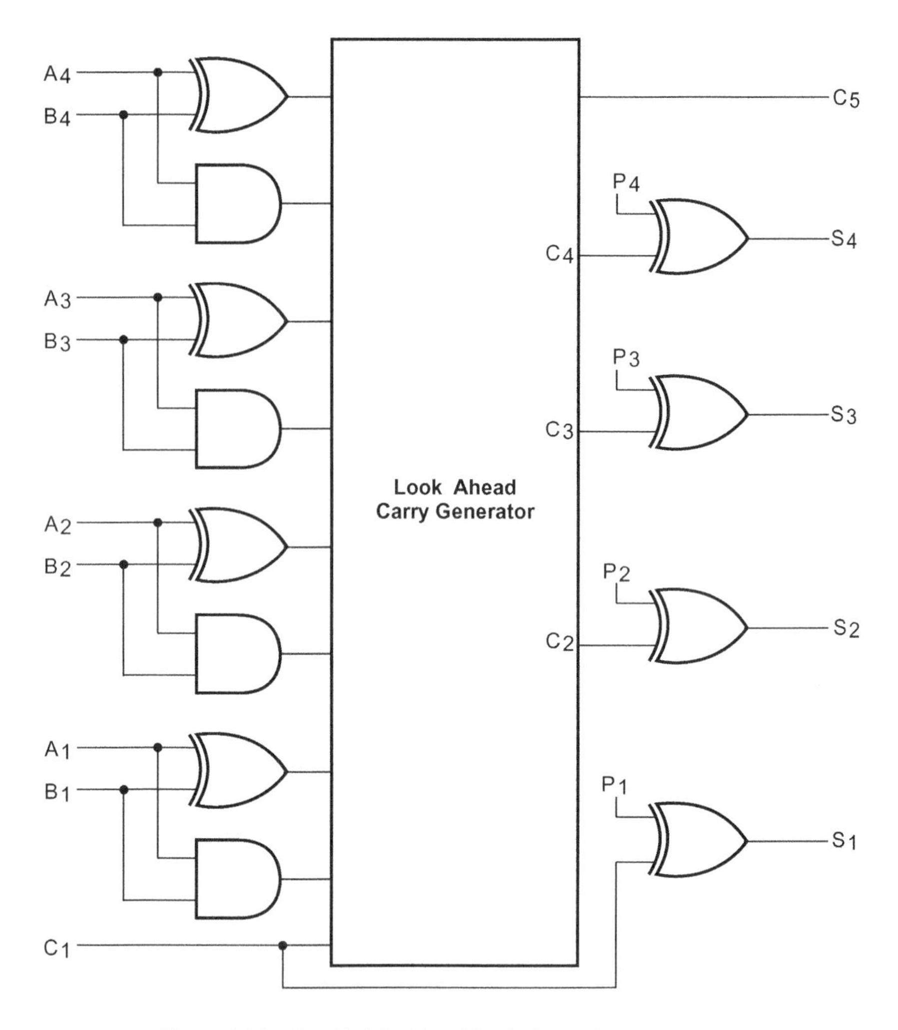

Figure 7.32 Four-bit full adder with a look-ahead carry generator.

$$= (C_i + A_i.B_i).(A_i + B_i) = C_i.(A_i + B_i) + A_i.B_i.(A_i + B_i)$$
$$= C_i.(A_i + B_i) + A_i.B_i = P_i.C_i + G_i$$

$$S_i = (A_i \oplus B_i) \oplus C_i = (\overline{A_i}.B_i + A_i.\overline{B_i}) \oplus C_i$$

Also

$$(P_i.\overline{G_i}) \oplus C_i = [(A_i + B_i).(\overline{A_i.B_i})] \oplus C_i$$
$$= [(A_i + B_i).(\overline{A_i} + \overline{B_i})] \oplus C_i = (\overline{A_i}.B_i + A_i.\overline{B_i}) \oplus C_i$$

Therefore, $S_i = (P_i.\overline{G_i}) \oplus C_i$.

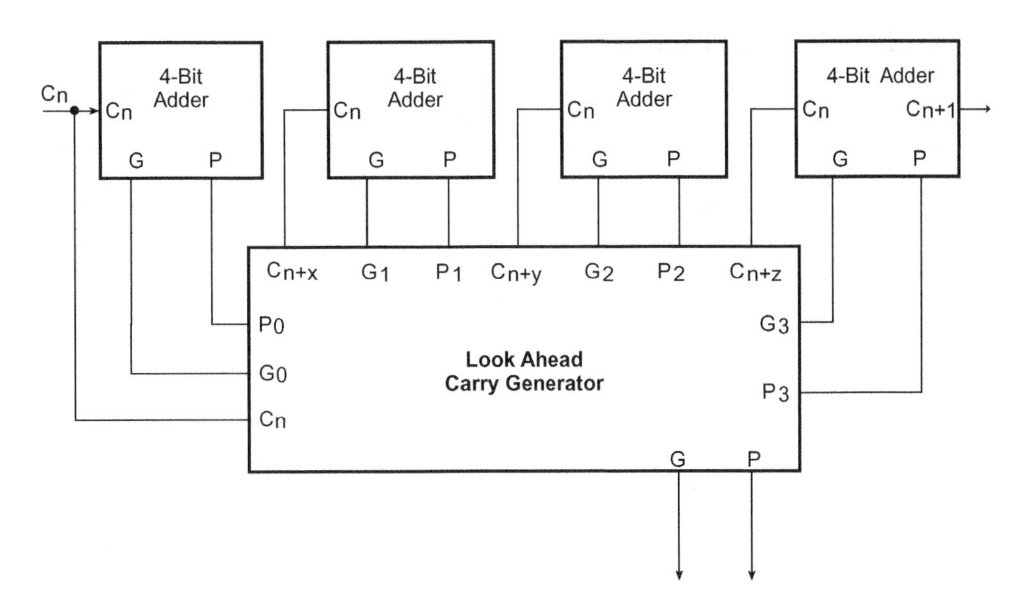

Figure 7.33 IC 74182 interfaced with four four-bit adders.

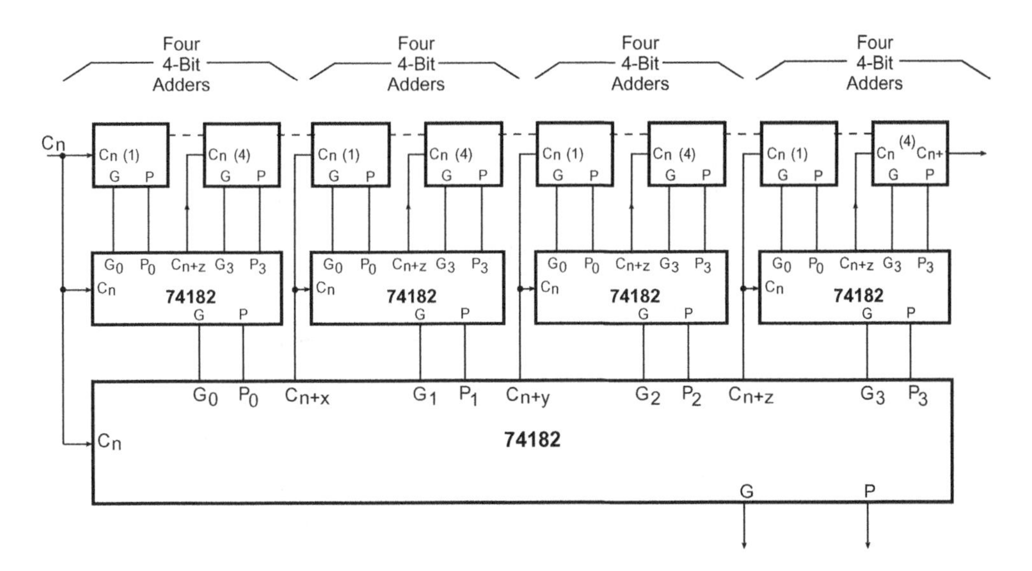

Figure 7.34 Look-ahead carry generation for adding 64-bit numbers.

7.7 Arithmetic Logic Unit (ALU)

The *arithmetic logic unit* (ALU) is a digital building block capable of performing both arithmetic as well as logic operations. Arithmetic logic units that can perform a variety of arithmetic operations such as addition, subtraction, etc., and logic functions such as ANDing, ORing, EX-ORing, etc., on two four-bit numbers are usually available in IC form. The function to be performed is selectable from *function select* pins. Some of the popular type numbers of ALU include 74181, 74381, 74382, 74582 (all from the TTL logic family) and 40181 (from the CMOS logic family). Functional details of these ICs are given in the latter part of the chapter under the heading of *Application-Relevant Information*. More than one such IC can always be connected in cascade to perform arithmetic and logic operations on larger bit numbers.

7.8 Multipliers

Multiplication of binary numbers is usually implemented in microprocessors and microcomputers by using *repeated addition and shift* operations. Since the binary adders are designed to add only two binary numbers at a time, instead of adding all the partial products at the end, they are added two at a time and their sum is accumulated in a register called the *accumulator register*. Also, when the multiplier bit is '0', that very partial product is ignored, as an all '0' line does not affect the final result. The basic hardware arrangement of such a binary multiplier would comprise shift registers for the multiplicand and multiplier bits, an accumulator register for storing partial products, a binary parallel adder and a clock pulse generator to time various operations.

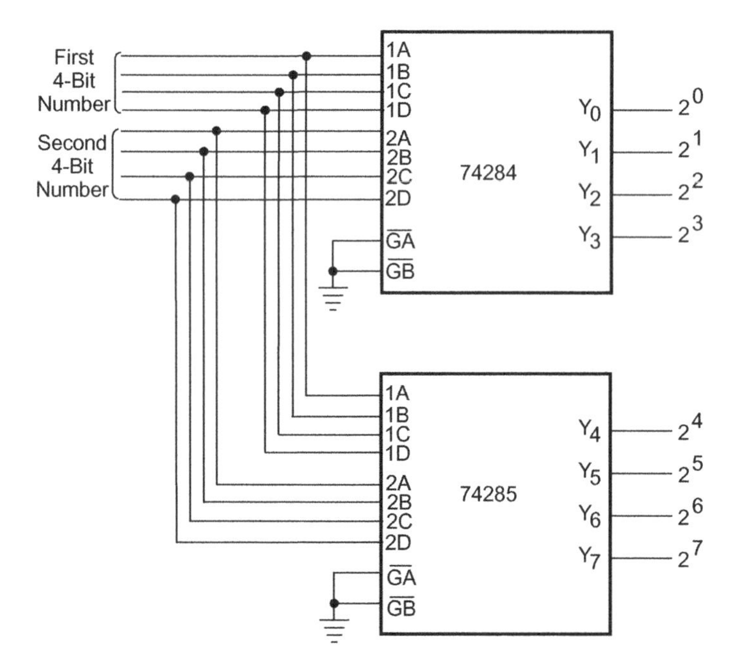

Figure 7.35 4 × 4 bit multiplier.

Binary multipliers are also available in IC form. Some of the popular type numbers in the TTL family include 74261 which is a 2 × 4 bit multiplier (a four-bit multiplicand designated as B_0,B_1,B_2, B_3 and B_4, and a two-bit multiplier designated as M_0, M_1 and M_2).

The MSBs B_4 and M_2 are used to represent signs. 74284 and 74285 are 4 × 4 bit multipliers. They can be used together to perform high-speed multiplication of two four-bit numbers. Figure 7.35 shows the arrangement. The result of multiplication is often required to be stored in a register. The size of this register (accumulator) depends upon the number of bits in the result, which at the most can be equal to the sum of the number of bits in the multiplier and multiplicand. Some multiplier ICs have an in-built register.

Many microprocessors do not have in their ALU the hardware that can perform multiplication or other complex arithmetic operations such as division, determining the square root, trigonometric functions, etc. These operations in these microprocessors are executed through software. For example, a multiplication operation may be accomplished by using a software program that does multiplication through repeated execution of addition and shift instructions. Other complex operations mentioned above can also be executed with similar programs. Although the use of software reduces the hardware needed in the microprocessor, the computation time in general is higher in the case of software-executed operations when compared with the use of hardware to perform those operations.

7.9 Magnitude Comparator

A *magnitude comparator* is a combinational circuit that compares two given numbers and determines whether one is equal to, less than or greater than the other. The output is in the form of three binary variables representing the conditions $A = B$, $A > B$ and $A < B$, if A and B are the two numbers being compared. Depending upon the relative magnitude of the two numbers, the relevant output changes state. If the two numbers, let us say, are four-bit binary numbers and are designated as $(A_3\,A_2\,A_1\,A_0)$ and $(B_3\,B_2\,B_1\,B_0)$, the two numbers will be equal if all pairs of significant digits are equal, that is, $A_3 = B_3$, $A_2 = B_2$, $A_1 = B_1$ and $A_0 = B_0$. In order to determine whether A is greater than or less than B, we inspect the relative magnitude of pairs of significant digits, starting from the most significant position. The comparison is done by successively comparing the next adjacent lower pair of digits if the digits of the pair under examination are equal. The comparison continues until a pair of unequal digits is reached. In the pair of unequal digits, if $A_i = 1$ and $B_i = 0$, then $A > B$, and if $A_i = 0$, $B_i = 1$ then $A < B$. If X, Y and Z are three variables respectively representing the $A = B$, $A > B$ and $A < B$ conditions, then the Boolean expression representing these conditions are given by the equations

$$X = x_3.x_2.x_1.x_0 \quad \text{where } x_i = A_i.B_i + \overline{A_i}.\overline{B_i} \tag{7.25}$$

$$Y = A_3.\overline{B_3} + x_3.A_2.\overline{B_2} + x_3.x_2.A_1.\overline{B_1} + x_3.x_2.x_1.A_0.\overline{B_0} \tag{7.26}$$

$$Z = \overline{A_3}.B_3 + x_3.\overline{A_2}.B_2 + x_3.x_2.\overline{A_1}.B_1 + x_3.x_2.x_1.\overline{A_0}.B_0 \tag{7.27}$$

Let us examine equation (7.25). x_3 will be '1' only when both A_3 and B_3 are equal. Similarly, conditions for x_2, x_1 and x_0 to be '1' respectively are equal A_2 and B_2, equal A_1 and B_1 and equal A_0 and B_0. ANDing of x_3, x_2, x_1 and x_0 ensures that X will be '1' when x_3, x_2, x_1 and x_0 are in the logic '1' state. Thus, $X = 1$ means that $A = B$. On similar lines, it can be visualized that equations (7.26) and

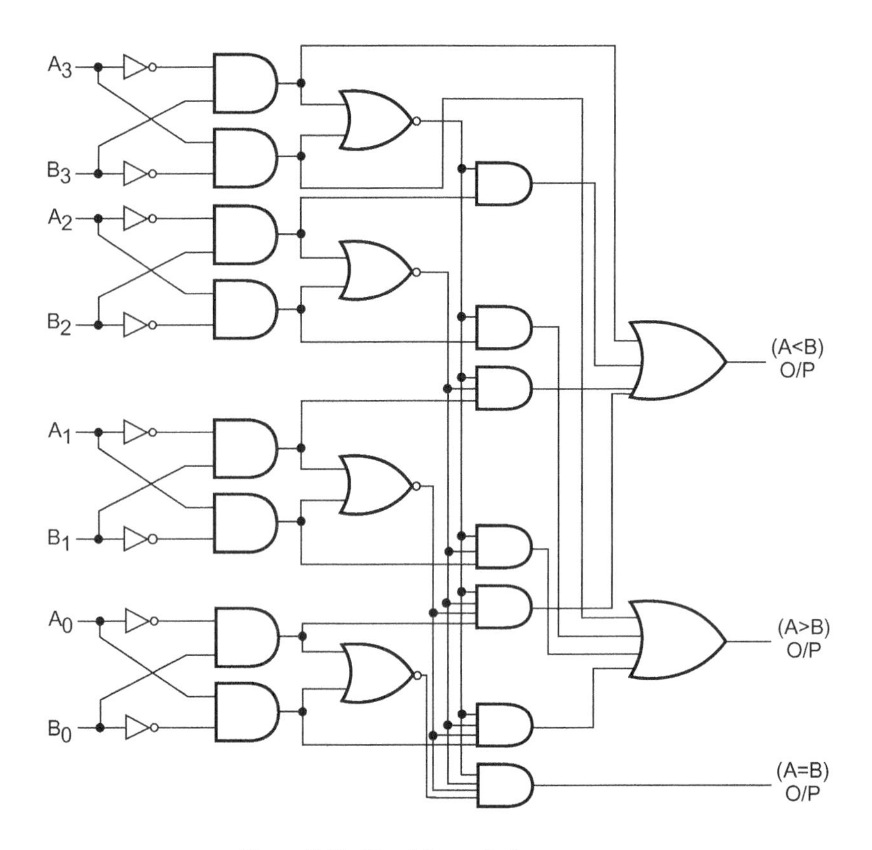

Figure 7.36 Four-bit magnitude comparator.

(7.27) respectively represent $A > B$ and $A < B$ conditions. Figure 7.36 shows the logic diagram of a four-bit magnitude comparator.

Magnitude comparators are available in IC form. For example, 7485 is a four-bit magnitude comparator of the TTL logic family. IC 4585 is a similar device in the CMOS family. 7485 and 4585 have the same pin connection diagram and functional table. The logic circuit inside these devices determines whether one four-bit number, binary or BCD, is *less than*, *equal to* or *greater than* a second four-bit number. It can perform comparison of straight binary and straight BCD (8-4-2-1) codes. These devices can be cascaded together to perform operations on larger bit numbers without the help of any external gates. This is facilitated by three additional inputs called cascading or expansion inputs available on the IC. These cascading inputs are also designated as $A = B$, $A > B$ and $A < B$ inputs. Cascading of individual magnitude comparators of the type 7485 or 4585 is discussed in the following paragraphs. IC 74AS885 is another common magnitude comparator. The device is an eight-bit magnitude comparator belonging to the advanced Schottky TTL family. It can perform high-speed arithmetic or logic comparisons on two eight-bit binary or 2's complement numbers and produces two fully decoded decisions at the output about one number being either greater than or less than the other. More than one of these devices can also be connected in a cascade arrangement to perform comparison of numbers of longer lengths.

7.9.1 Cascading Magnitude Comparators

As outlined earlier, magnitude comparators available in IC form are designed in such a way that they can be connected in a cascade arrangement to perform comparison operations on numbers of longer lengths. In cascade arrangement, the $A = B$, $A > B$ and $A < B$ outputs of a stage handling less significant bits are connected to corresponding inputs of the next adjacent stage handling more significant bits. Also, the stage handling least significant bits must have a HIGH level at the $A = B$ input. The other two cascading inputs ($A > B$ and $A < B$) may be connected to a LOW level. We will illustrate the concept by showing the arrangement of an eight-bit magnitude comparator using two four-bit magnitude comparators of the type 7485 or 4585. Figure 7.37 shows the cascaded arrangement of the two comparators. We can see the three comparison outputs of the comparator handling less significant four bits of the two numbers being connected to the corresponding cascading inputs of the comparator handling more significant four bits of the two numbers. Also, cascading inputs of the less significant comparator have been connected to a HIGH or LOW level as per the guidelines mentioned in the previous paragraph.

Operation of this circuit can be explained by considering the functional table of IC 7485 or IC 4585 as shown in Table 7.2. The two numbers being compared here are ($A_7 \ldots A_0$) and ($B_7 \ldots B_0$). The less significant comparator handles (A_3, A_2, A_1, A_0) and (B_3, B_2, B_1, B_0), and the more significant comparator handles (A_7, A_6, A_5, A_4) and (B_7, B_6, B_5, B_4). Let us take the example of the two numbers being such that $A_7 > B_7$. From the first-row entry of the function table it is clear that, irrespective of the status of other bits of the more significant comparator, and also regardless of the status of its cascading inputs, the final output produces a HIGH at the $A > B$ output and a LOW at the $A < B$ and $A = B$ outputs. Since the status of cascading inputs of the more significant comparator depends upon the status of comparison bits of the less significant comparator, the cascade arrangement produces the correct output for $A_7 > B_7$ regardless of the status of all other comparison bits. On similar lines, the circuit produces a valid output for any given status of comparison bits.

Example 7.8

Design a two-bit magnitude comparator. Also, write relevant Boolean expressions.

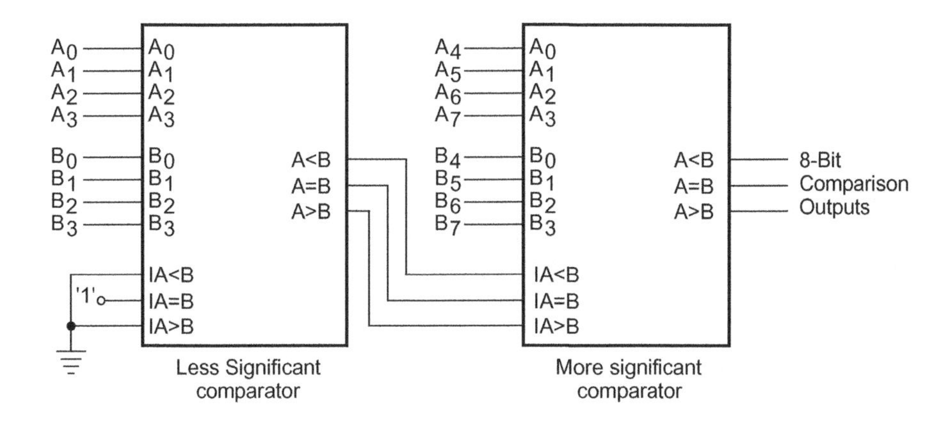

Figure 7.37 Cascading of individual magnitude comparators.

Table 7.2 Functional table of IC 7485 or IC 4585.

Comparison inputs				Cascading inputs			Outputs		
A_3,B_3	A_2,B_2	A_1,B_1	A_0,B_0	A>B	A<B	A=B	A>B	A<B	A=B
$A_3 > B_3$	X	X	X	X	X	X	HIGH	LOW	LOW
$A_3 < B_3$	X	X	X	X	X	X	LOW	HIGH	LOW
$A_3 = B_3$	$A_2 > B_2$	X	X	X	X	X	HIGH	LOW	LOW
$A_3 = B_3$	$A_2 < B_2$	X	X	X	X	X	LOW	HIGH	LOW
$A_3 = B_3$	$A_2 = B_2$	$A_1 > B_1$	X	X	X	X	HIGH	LOW	LOW
$A_3 = B_3$	$A_2 = B_2$	$A_1 < B_1$	X	X	X	X	LOW	HIGH	LOW
$A_3 = B_3$	$A_2 = B_2$	$A_1 = B_1$	$A_0 > B_0$	X	X	X	HIGH	LOW	LOW
$A_3 = B_3$	$A_2 = B_2$	$A_1 = B_1$	$A_0 < B_0$	X	X	X	LOW	HIGH	LOW
$A_3 = B_3$	$A_2 = B_2$	$A_1 = B_1$	$A_0 = B_0$	HIGH	LOW	LOW	HIGH	LOW	LOW
$A_3 = B_3$	$A_2 = B_2$	$A_1 = B_1$	$A_0 = B_0$	LOW	HIGH	LOW	LOW	HIGH	LOW
$A_3 = B_3$	$A_2 = B_2$	$A_1 = B_1$	$A_0 = B_0$	LOW	LOW	HIGH	LOW	LOW	HIGH
$A_3 = B_3$	$A_2 = B_2$	$A_1 = B_1$	$A_0 = B_0$	X	X	HIGH	LOW	LOW	HIGH
$A_3 = B_3$	$A_2 = B_2$	$A_1 = B_1$	$A_0 = B_0$	HIGH	HIGH	LOW	LOW	LOW	LOW
$A_3 = B_3$	$A_2 = B_2$	$A_1 = B_1$	$A_0 = B_0$	LOW	LOW	LOW	HIGH	HIGH	LOW

Solution

Let $A\,(A_1A_0)$ and $B\,(B_1B_0)$ be the two numbers. If X, Y and Z represent the conditions $A = B$, $A > B$ and $A < B$ respectively (that is, $X = 1$, $Y = 0$ and $Z = 0$ for A =B; X= 0, $Y = 1$ and $Z = 0$ for $A > B$; and $X = 0$, $Y = 0$ and $Z = 1$ for $A < B$), then expressions for X, Y and Z can be written as follows:

$$X = x_1.x_0 \text{ where } x_1 = A_1.B_1 + \overline{A_1}.\overline{B_1} \quad \text{and} \quad x_0 = A_0.B_0 + \overline{A_0}.\overline{B_0}$$

$$Y = A_1.\overline{B_1} + x_1.A_0.\overline{B_0}$$

$$Z = \overline{A_1}.B_1 + x_1.\overline{A_0}.B_0$$

Figure 7.38 shows the logic diagram of the two-bit comparator.

Example 7.9

Hardware-implement a three-bit magnitude comparator having one output that goes HIGH when the two three-bit numbers are equal. Use only NAND gates.

Solution

The equivalence condition of the two three-bit numbers is given by the equation $X = x_2.x_1.x_0$, where $x_2 = A_2.B_2 + \overline{A_2}.\overline{B_2}$, $x_1 = A_1.B_1 + \overline{A_1}.\overline{B_1}$ and $x_0 = A_0.B_0 + \overline{A_0}.\overline{B_0}$.

Figure 7.39 shows the logic diagram. x_2, x_1 and x_0 are respectively given by EX-NOR operation of (A_2, B_2), (A_1, B_1) and (A_0, B_0). These are then ANDed to get X.

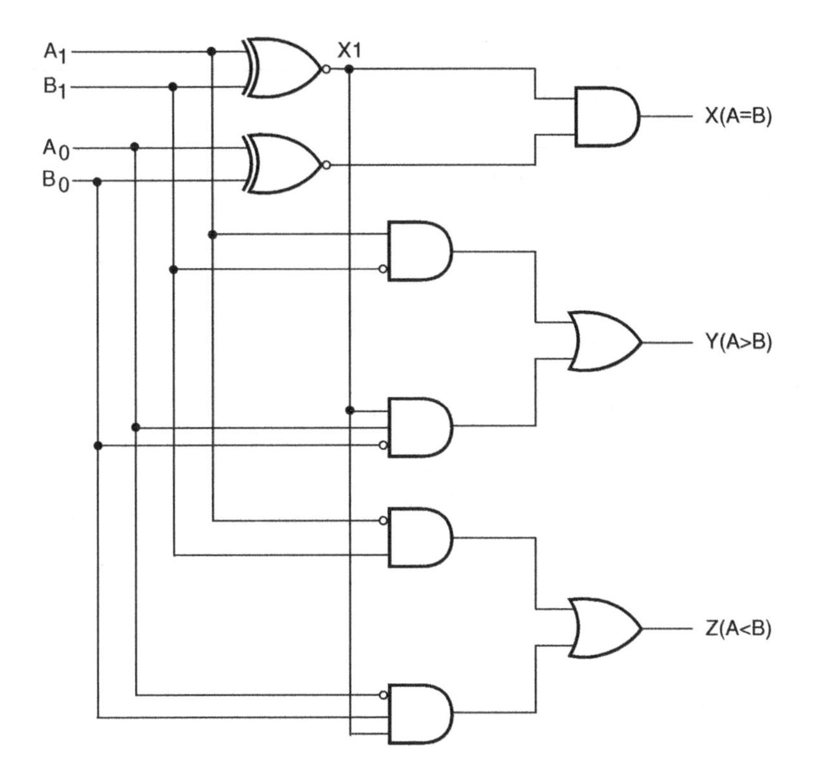

Figure 7.38 Solution to example 7.8.

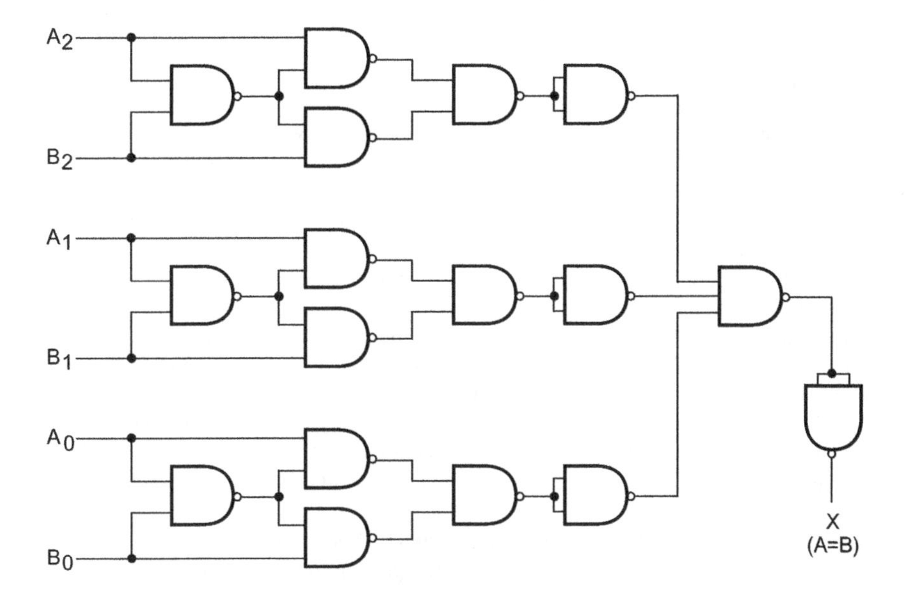

Figure 7.39 Solution to example 7.9.

Table 7.3 Commonly used IC type numbers used for arithmetic operations.

IC type number	Function	Logic family
7483	Four-bit full adder	TTL
7485	Four-bit magnitude comparator	TTL
74181	Four-Bit ALU and function generator	TTL
74182	Look-ahead carry generator	TTL
74183	Dual carry save full adder	TTL
74283	Four-bit full binary adder	TTL
74885	Eight-bit magnitude comparator	TTL
4008	Four-bit binary full adder	CMOS
4527	BCD rate multiplier	CMOS
4585	Four-bit magnitude comparator	CMOS
40181	Four-bit arithmetic logic unit	CMOS
40182	Look-ahead carry generator	CMOS
10179	Look-ahead carry block	ECL
10180	Dual high-speed two-bit adder/subtractor	ECL
10181	Four-bit arithmetic logic unit/function generator	ECL
10182	Four-bit arithmetic logic unit/function generator	ECL
10183	4 × 2 multiplier	ECL

7.10 Application-Relevant Information

Table 7.3 lists commonly used IC type numbers used for arithmetic operations. Application-relevant information such as pin connection diagrams, truth tables, etc., in respect of the more popular of these type numbers is given on the companion website.

Review Questions

1. How do you characterize or define a combinational circuit? How does it differ from a sequential circuit? Give two examples each of combinational and sequential logic devices.
2. Beginning with the statement of the problem, outline different steps involved in the design of a suitable combinational logic circuit to implement the hardware required to solve the given problem.
3. Write down Boolean expressions representing the SUM and CARRY outputs in terms of three input binary variables to be added. Design a suitable combinational circuit to hardware-implement the design using NAND gates only.
4. Draw the truth table of a full subtractor circuit. Write a minterm Boolean expression for DIFFERENCE and BORROW outputs in terms of minuend variable, subtrahend variable and BORROW-IN. Minimize the expressions and implement them in hardware.
5. Draw the logic diagram of a three-digit BCD adder and briefly describe its functional principle.
6. Briefly describe the concept of look-ahead carry generation with respect to its use in adder circuits. What is its significance while implementing hardware for addition of binary numbers of longer lengths?
7. With the help of a block schematic of the logic circuit, briefly describe how individual four-bit magnitude comparators can be used in a cascade arrangement to perform magnitude comparison of binary numbers of longer lengths.

Problems

1. A, B, B_{in}, D and B_{out} are respectively the minuend, the subtrahend, the BORROW-IN, the DIFFERENCE output and the BORROW-OUT in the case of a full subtractor. Determine the bit status of D and B_{out} for the following values of A, B and B_{in}:

   (a) $A = 0$, $B = 1$, $B_{in} = 1$
   (b) $A = 1$, $B = 1$, $B_{in} = 0$
   (c) $A = 1$, $B = 1$, $B_{in} = 1$
   (d) $A = 0$, $B = 0$, $B_{in} = 1$

   (a) $D = 0$, $B_{out} = 1$; (b) $D = 0$, $B_{out} = 0$; (c) $D = 1$, $B_{out} = 1$; (d) $D = 1$, $B_{out} = 1$

2. Determine the number of half and full adder circuit blocks required to construct a 64-bit binary parallel adder. Also, determine the number and type of additional logic gates needed to transform this 64-bit adder into a 64-bit adder–subtractor.

   For a 64-bit adder: HA=1, FA=63
   For a 64-bit adder–subtractor: HA = 1, FA = 63, EX-OR gates = 64

3. If the minuend, subtrahend and BORROW-IN bits are respectively applied to the Augend, Addend and the CARRY-IN inputs of a full adder, prove that the SUM output of the full adder will produce the correct DIFFERENCE output.
4. Prove that the logic diagram of Fig. 7.40 performs the function of a half-subtractor provided that Y represents the DIFFERENCE output and X represents the BORROW output.
5. Determine the number of 7483s (four-bit binary adders) and 7486s (quad two-input EX-OR gates) required to design a 16-bit adder–subtractor circuit.

   Number of 7483 = 4; number of 7486 = 4

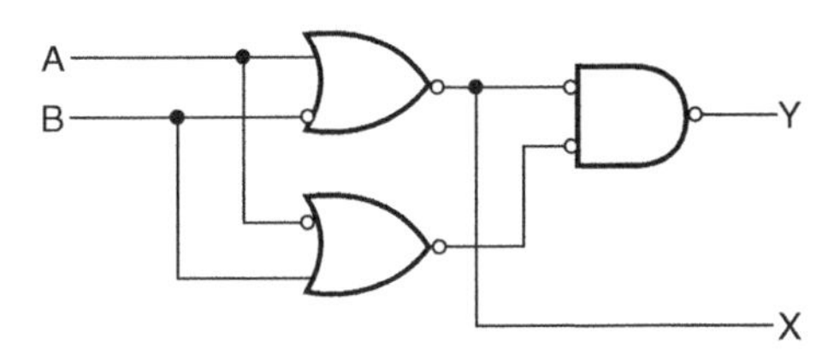

Figure 7.40 Problem 4.

6. The objective is to design a BCD adder circuit using four-bit binary adders and additional combinational logic. If the decimal numbers to be added can be anywhere in the range from 0 to 9999, determine the number of four-bit binary adder circuit blocks of type IC 7483 required to do the job.

   Number of four-bit adders = 8

Further Reading

1. Koren, I. (2001) *Computer Arithmetic Algorithms*, A. K. Peters Ltd, Natick, MA, USA.
2. Ercegovac, M. D. and Lang, T. (2003) *Digital Arithmetic*, Morgan Kaufmann Publishers, CA, USA.
3. Rafiquzzaman, M. (2005) *Fundamentals of Digital Logic and Microcomputer Design*, Wiley-Interscience, New York, USA.
4. Morris Mano, M. and Kime, C. R. (2003) *Logic and Computer Design Fundamentals*, Prentice-Hall, USA.
5. Tokheim, R. L. (1994) *Schaum's Outline Series of Digital Principles*, McGraw-Hill Companies Inc., USA.
6. Tocci, R. J. (2006) *Digital Systems – Principles and Applications*, Prentice-Hall Inc., NJ, USA.
7. Malvino, A. P. and Leach, D. P. (1994) *Digital Principles and Applications*, McGraw-Hill Book Company, USA.

8

Multiplexers and Demultiplexers

In the previous chapter, we described at length those combinational logic circuits that can be used to perform arithmetic and related operations. This chapter takes a comprehensive look at yet another class of building blocks used to design more complex combinational circuits, and covers building blocks such as multiplexers and demultiplexers and other derived devices such as encoders and decoders. Particular emphasis is given to the operational basics and use of these devices to design more complex combinational circuits. Application-relevant information in terms of the list of commonly used integrated circuits available in this category, along with their functional description is given towards the end of the chapter. The text has been adequately illustrated with the help of a large number of solved examples.

8.1 Multiplexer

A *multiplexer* or *MUX*, also called a *data selector*, is a combinational circuit with more than one input line, one output line and more than one selection line. There are some multiplexer ICs that provide complementary outputs. Also, multiplexers in IC form almost invariably have an ENABLE or STROBE input, which needs to be active for the multiplexer to be able to perform its intended function. A multiplexer selects binary information present on any one of the input lines, depending upon the logic status of the selection inputs, and routes it to the output line. If there are n selection lines, then the number of maximum possible input lines is 2^n and the multiplexer is referred to as a 2^n-to-1 multiplexer or $2^n \times 1$ multiplexer. Figures 8.1(a) and (b) respectively show the circuit representation and truth table of a basic 4-to-1 multiplexer.

To familiarize readers with the practical multiplexer devices available in IC form, Figs 8.2 and 8.3 respectively show the circuit representation and function table of 8-to-1 and 16-to-1 multiplexers. The 8-to-1 multiplexer of Fig. 8.2 is IC type number 74151 of the TTL family. It has an active LOW ENABLE input and provides complementary outputs. Figure 8.3 refers to IC type number 74150 of the TTL family. It is a 16-to-1 multiplexer with active LOW ENABLE input and active LOW output.

Digital Electronics: Principles, Devices and Applications Anil Kumar Maini

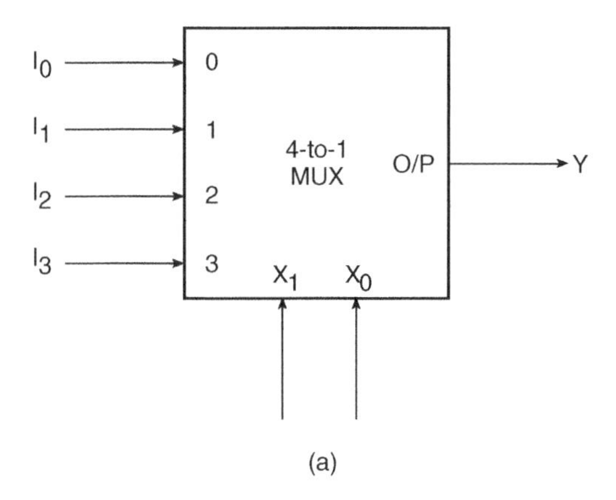

(a)

X_1	X_0	Y
0	0	I_0
0	1	I_1
1	0	I_2
1	1	I_3

(b)

Figure 8.1 (a) 4-to-1 multiplexer circuit representation and (b) 4-to-1 multiplexer truth table.

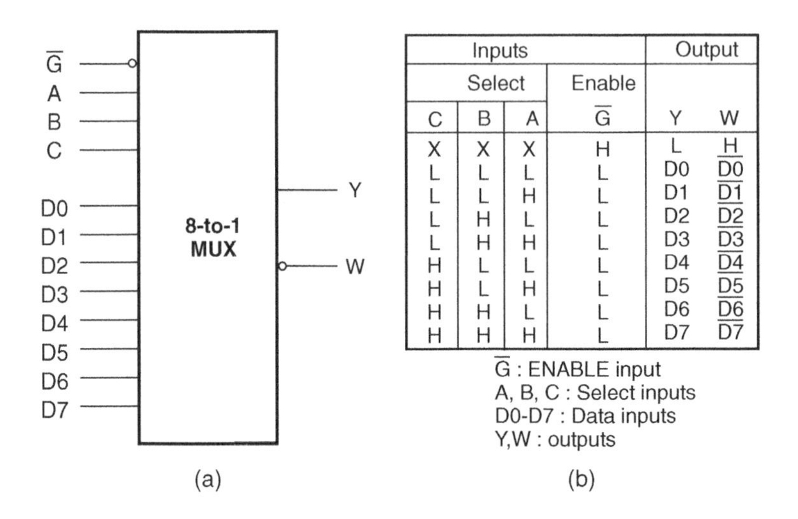

Inputs				Output	
Select			Enable		
C	B	A	$\overline{G}$	Y	W
X	X	X	H	L	H
L	L	L	L	D0	$\overline{D0}$
L	L	H	L	D1	$\overline{D1}$
L	H	L	L	D2	$\overline{D2}$
L	H	H	L	D3	$\overline{D3}$
H	L	L	L	D4	$\overline{D4}$
H	L	H	L	D5	$\overline{D5}$
H	H	L	L	D6	$\overline{D6}$
H	H	H	L	D7	$\overline{D7}$

$\overline{G}$: ENABLE input
A, B, C : Select inputs
D0-D7 : Data inputs
Y,W : outputs

(b)

Figure 8.2 (a) 8-to-1 multiplexer circuit representation and (b) 8-to-1 multiplexer truth table.

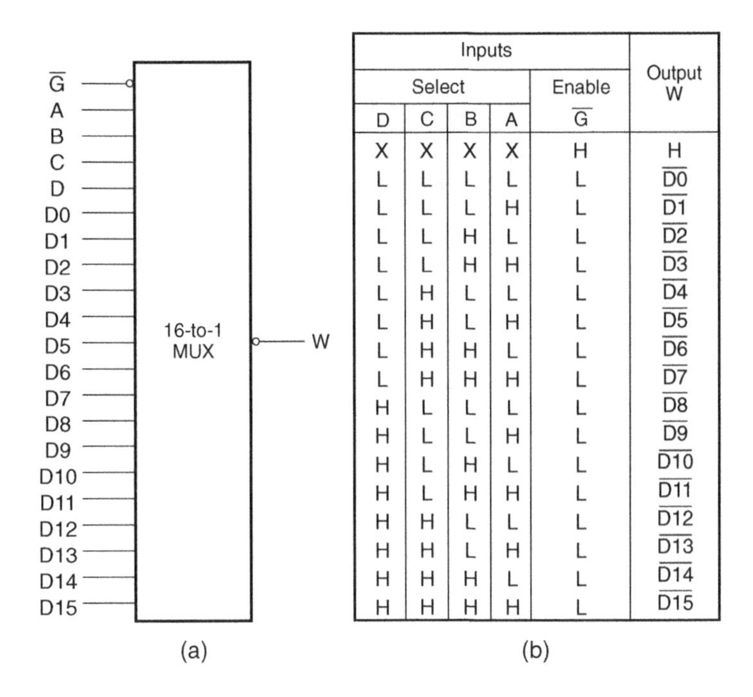

Inputs					Output W
Select				Enable	
D	C	B	A	$\overline{G}$	
X	X	X	X	H	H
L	L	L	L	L	$\overline{D0}$
L	L	L	H	L	$\overline{D1}$
L	L	H	L	L	$\overline{D2}$
L	L	H	H	L	$\overline{D3}$
L	H	L	L	L	$\overline{D4}$
L	H	L	H	L	$\overline{D5}$
L	H	H	L	L	$\overline{D6}$
L	H	H	H	L	$\overline{D7}$
H	L	L	L	L	$\overline{D8}$
H	L	L	H	L	$\overline{D9}$
H	L	H	L	L	$\overline{D10}$
H	L	H	H	L	$\overline{D11}$
H	H	L	L	L	$\overline{D12}$
H	H	L	H	L	$\overline{D13}$
H	H	H	L	L	$\overline{D14}$
H	H	H	H	L	$\overline{D15}$

Figure 8.3 (a) 16-to-1 multiplexer circuit representation and (b) 16-to-1 multiplexer truth table.

8.1.1 Inside the Multiplexer

We will briefly describe the type of combinational logic circuit found inside a multiplexer by considering the 2-to-1 multiplexer in Fig. 8.4(a), the functional table of which is shown in Fig. 8.4(b). Figure 8.4(c) shows the possible logic diagram of this multiplexer. The circuit functions as follows:

- For $S = 0$, the Boolean expression for the output becomes $Y = I_0$.
- For $S = 1$, the Boolean expression for the output becomes $Y = I_1$.

Thus, inputs I_0 and I_1 are respectively switched to the output for $S = 0$ and $S = 1$. Extending the concept further, Fig. 8.5 shows the logic diagram of a 4-to-1 multiplexer. The input combinations 00, 01, 10 and 11 on the select lines respectively switch I_0, I_1, I_2 and I_3 to the output.The operation of the circuit is governed by the Boolean function (8.1). Similarly, an 8-to-1 multiplexer can be represented by the Boolean function (8.2):

$$Y = I_0.\overline{S_1}.\overline{S_0} + I_1.\overline{S_1}.S_0 + I_2.S_1.\overline{S_0} + I_3.S_1.S_0 \tag{8.1}$$

$$Y = I_0.\overline{S_2}.\overline{S_1}.\overline{S_0} + I_1.\overline{S_2}.\overline{S_1}.S_0 + I_2.\overline{S_2}.S_1.\overline{S_0} + I_3.\overline{S_2}.S_1.S_0 + I_4.S_2.\overline{S_1}.\overline{S_0} + I_5.S_2.\overline{S_1}.S_0 + I_6.S_2.S_1.\overline{S_0} + I_7.S_2.S_1.S_0 \tag{8.2}$$

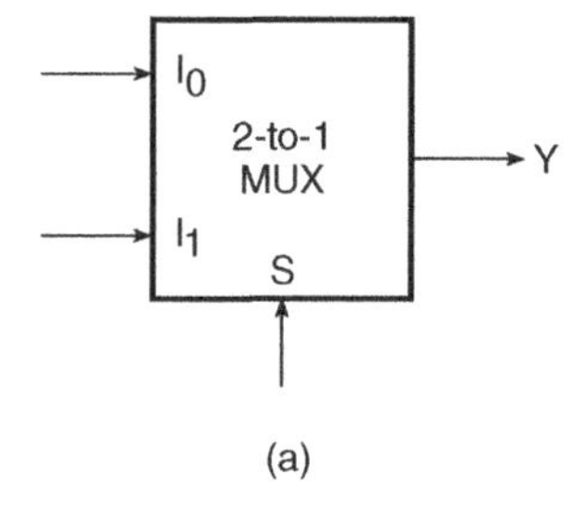

(a)

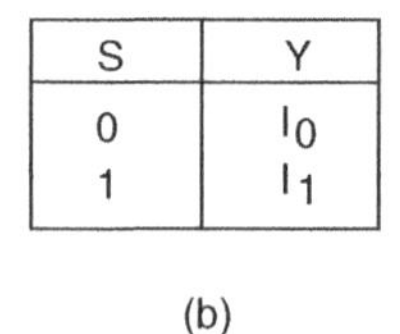

S	Y
0	I_0
1	I_1

(b)

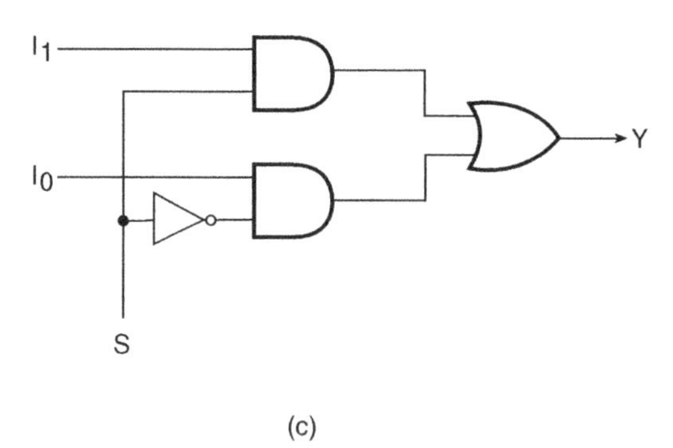

(c)

Figure 8.4 (a) 2-to-1 multiplexer circuit representation, (b) 2-to-1 multiplexer truth table and (c) 2-to-1 multiplexer logic diagram.

As outlined earlier, multiplexers usually have an ENABLE input that can be used to control the multiplexing function. When this input is enabled, that is, when it is in logic '1' or logic '0' state, depending upon whether the ENABLE input is active HIGH or active LOW respectively, the output is enabled. The multiplexer functions normally. When the ENABLE input is inactive, the output is disabled and permanently goes to either logic '0' or logic '1' state, depending upon whether the output is uncomplemented or complemented. Figure 8.6 shows how the 2-to-1 multiplexer of Fig. 8.4 can be modified to include an ENABLE input. The functional table of this modified multiplexer is also shown in Fig. 8.6. The ENABLE input here is active when HIGH. Some IC packages have more than one multiplexer. In that case, the ENABLE input and selection inputs are common to all multiplexers within the same IC package. Figure 8.7 shows a 4-to-1 multiplexer with an active LOW ENABLE input.

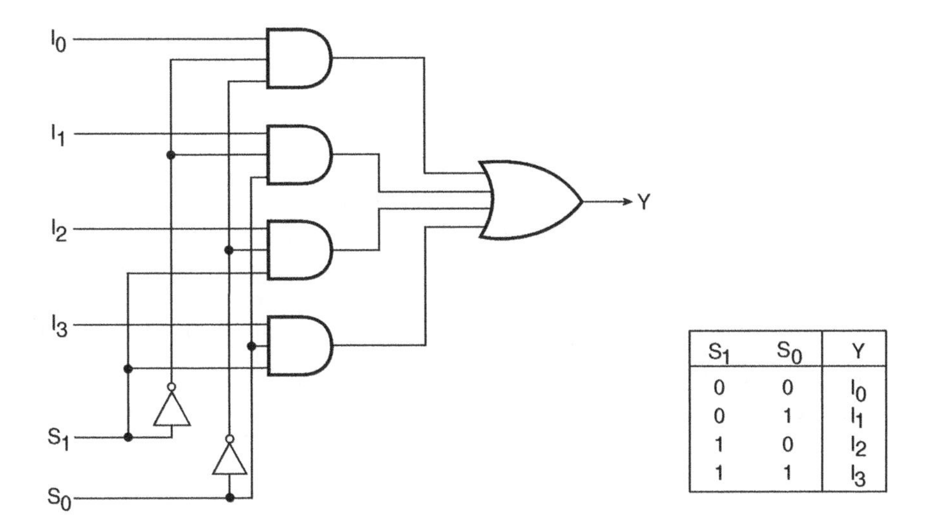

S_1	S_0	Y
0	0	I_0
0	1	I_1
1	0	I_2
1	1	I_3

Figure 8.5 Logic diagram of a 4-to-1 multiplexer.

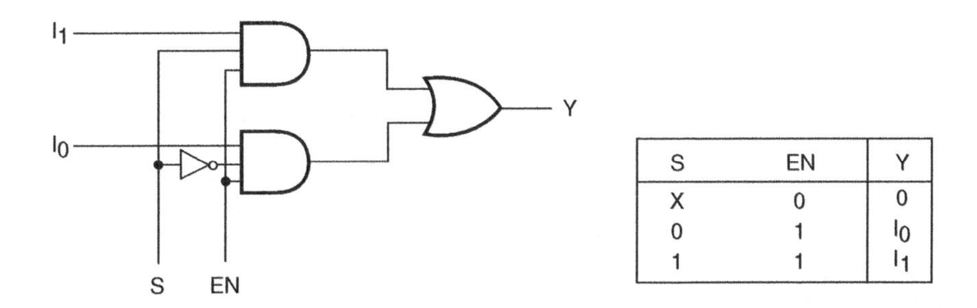

S	EN	Y
X	0	0
0	1	I_0
1	1	I_1

Figure 8.6 2-to-1 multiplexer with an ENABLE input.

8.1.2 Implementing Boolean Functions with Multiplexers

One of the most common applications of a multiplexer is its use for implementation of combinational logic Boolean functions. The simplest technique for doing so is to employ a 2^n-to-1 MUX to implement an n-variable Boolean function. The input lines corresponding to each of the minterms present in the Boolean function are made equal to logic '1' state. The remaining minterms that are absent in the Boolean function are disabled by making their corresponding input lines equal to logic '0'. As an example, Fig. 8.8(a) shows the use of an 8-to-1 MUX for implementing the Boolean function given by the equation

$$f(A, B, C) = \sum 2, 4, 7 \quad (8.3)$$

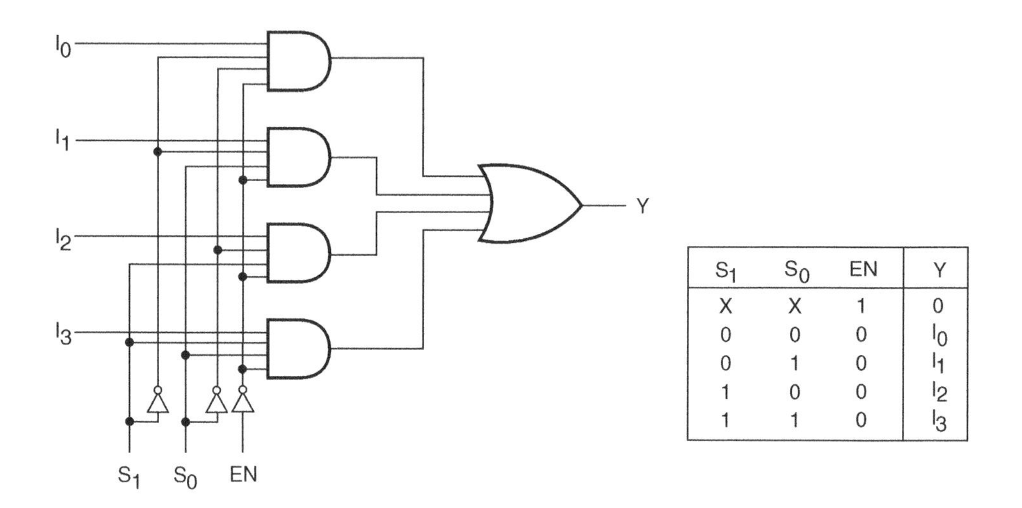

S_1	S_0	EN	Y
X	X	1	0
0	0	0	I_0
0	1	0	I_1
1	0	0	I_2
1	1	0	I_3

Figure 8.7 4-to-1 multiplexer with an ENABLE input.

In terms of variables A, B and C, equation (8.3) can be written as follows:

$$f(A, B, C) = \overline{A}.B.\overline{C} + A.\overline{B}.\overline{C} + A.B.C \tag{8.4}$$

As shown in Fig. 8.8, the input lines corresponding to the three minterms present in the given Boolean function are tied to logic '1'. The remaining five possible minterms absent in the Boolean function are tied to logic '0'.

However, there is a better technique available for doing the same. In this, a 2^n-to-1 MUX can be used to implement a Boolean function with $n + 1$ variables. The procedure is as follows. Out of n + 1 variables, n are connected to the n selection lines of the 2^n-to-1 multiplexer. The left-over variable is used with the input lines. Various input lines are tied to one of the following: '0', '1', the left-over variable and the complement of the left-over variable. Which line is given what logic status can be easily determined with the help of a simple procedure. The complete procedure is illustrated for the Boolean function given by equation (8.3).

It is a three-variable Boolean function. Conventionally, we will need to use an 8-to-1 multiplexer to implement this function. We will now see how this can be implemented with a 4-to-1 multiplexer. The chosen multiplexer has two selection lines. The first step here is to determine the truth table of the given Boolean function, which is shown in Table 8.1.

In the next step, two of the three variables are connected to the two selection lines, with the higher-order variable connected to the higher-order selection line. For instance, in the present case, variables B and C are the chosen variables for the selection lines and are respectively connected to selection lines S_1 and S_0. In the third step, a table of the type shown in Table 8.2 is constructed. Under the inputs to the multiplexer, minterms are listed in two rows, as shown. The first row lists those terms where remaining variable A is complemented, and second row lists those terms where A is uncomplemented. This is easily done with the help of the truth table.

The required minterms are identified or marked in some manner in this table. In the given table, these entries have been highlighted. Each column is inspected individually. If neither minterm of a certain column is highlighted, a '0' is written below that. If both are highlighted, a '1' is

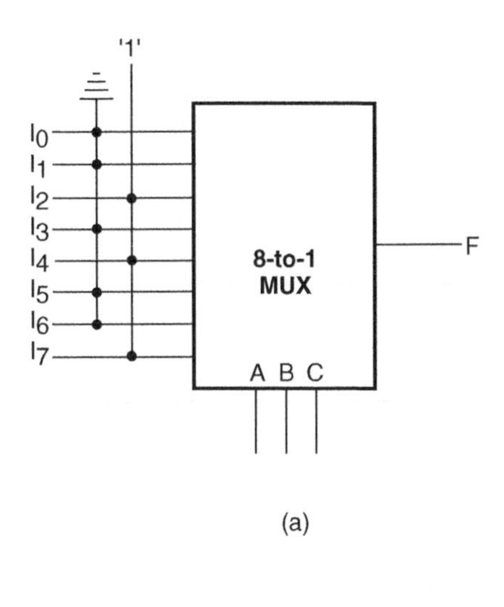

(a)

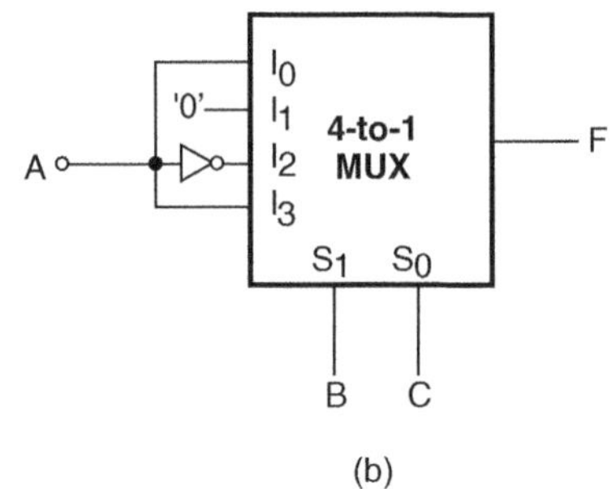

(b)

Figure 8.8 Hardware implementation of the Boolean function given by equation (8.3).

Table 8.1 Truth table.

Minterm	A	B	C	$f(A,B,C)$
0	0	0	0	0
1	0	0	1	0
2	0	1	0	1
3	0	1	1	0
4	1	0	0	1
5	1	0	1	0
6	1	1	0	0
7	1	1	1	1

written. If only one is highlighted, the corresponding variable (complemented or uncomplemented) is written. The input lines are then given appropriate logic status. In the present case, I_0, I_1, I_2 and I_3 would be connected to A, 0, $\overline{A}$ and A respectively. Figure 8.8(b) shows the logic implementation.

Table 8.2 Implementation table for multiplexers.

	I_0	I_1	I_2	I_3
$\overline{A}$	0	1	**2**	3
A	**4**	5	6	**7**
	A	0	$\overline{A}$	A

Table 8.3 Implementation table for multiplexers.

	I_0	I_1	I_2	I_3
$\overline{C}$	0	**2**	**4**	6
C	1	3	5	**7**
	0	$\overline{C}$	$\overline{C}$	C

It is not necessary to choose only the leftmost variable in the sequence to be used as input to the multiplexer. Any of the variables can be used provided the implementation table is constructed accordingly. In the problem illustrated above, A was chosen as the variable for the input lines, and accordingly the first row of the implementation table contained those entries where 'A' was complemented and the second row contained those entries where A was uncomplemented. If we consider C as the left-out variable, the implementation table will be as shown in Table 8.3.

Figure 8.9 shows the hardware implementation. For the case of B being the left-out variable, the implementation table is shown in Table 8.4 and the hardware implementation is shown in Fig. 8.10.

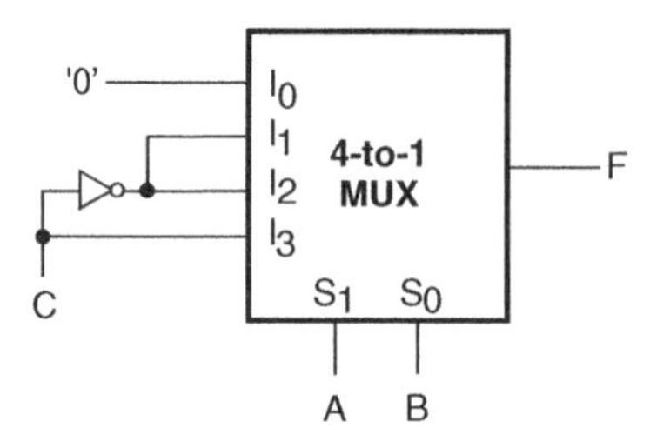

Figure 8.9 Hardware implementation using a 4-to-1 multiplexer.

Table 8.4 Implementation table for multiplexers.

	I_0	I_1	I_2	I_3
$\overline{B}$	0	1	**4**	5
B	**2**	3	6	**7**
	B	0	$\overline{B}$	B

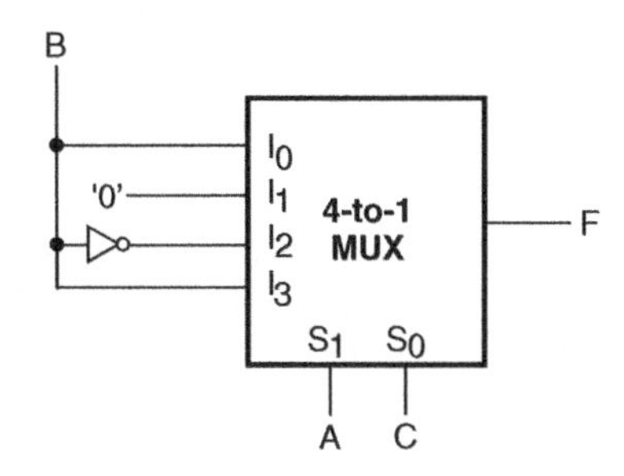

Figure 8.10 Hardware implementation using a 4-to-1 multiplexer.

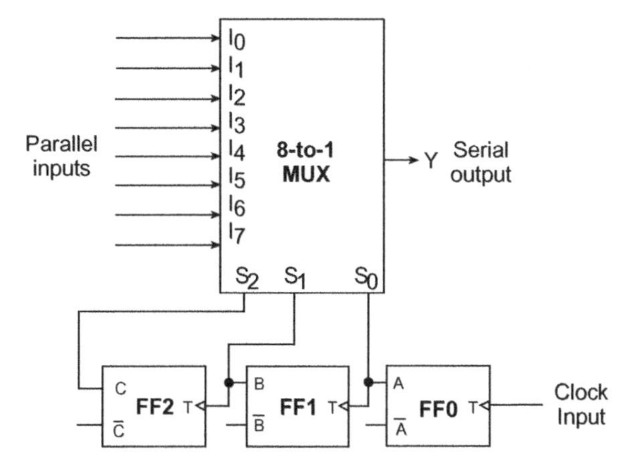

Figure 8.11 Multiplexer for parallel-to-serial conversion.

8.1.3 Multiplexers for Parallel-to-Serial Data Conversion

Although data are processed in parallel in many digital systems to achieve faster processing speeds, when it comes to transmitting these data relatively large distances, this is done serially. The parallel arrangement in this case is highly undesirable as it would require a large number of transmission lines. Multiplexers can possibly be used for parallel-to-serial conversion. Figure 8.11 shows one such arrangement where an 8-to-1 multiplexer is used to convert eight-bit parallel binary data to serial form. A three-bit counter controls the selection inputs. As the counter goes through 000 to 111, the multiplexer output goes through I_0 to I_7. The conversion process takes a total of eight clock cycles. In the figure shown, the three-bit counter has been constructed with the help of three toggle flip-flops. A variety of counter circuits of various types and complexities are, however, available in IC form. Flip-flops and counters are discussed in detail in Chapters 10 and 11 respectively.

Example 8.1

Implement the product-of-sums Boolean function expressed by $\Pi 1,2,5$ by a suitable multiplexer.

Solution

- Let the Boolean function be $f(A, B, C) = \prod 1, 2, 5$.
- The equivalent sum-of-products expression can be written as $f(A, B, C) = \sum 0, 3, 4, 6, 7$.

The truth table for the given Boolean function is given in Table 8.5. The given function can be implemented with a 4-to-1 multiplexer with two selection lines. Variables A and B are chosen for the selection lines. The implementation table as drawn with the help of the truth table is given in Table 8.6. Figure 8.12 shows the hardware implementation.

Table 8.5 Truth table.

C	B	A	$f(A,B,C)$
0	0	0	1
0	0	1	0
0	1	0	0
0	1	1	1
1	0	0	1
1	0	1	0
1	1	0	1
1	1	1	1

Table 8.6 Implementation table.

	I_0	I_1	I_2	I_3
$\overline{C}$	**0**	1	2	**3**
C	**4**	5	**6**	**7**
	1	0	C	1

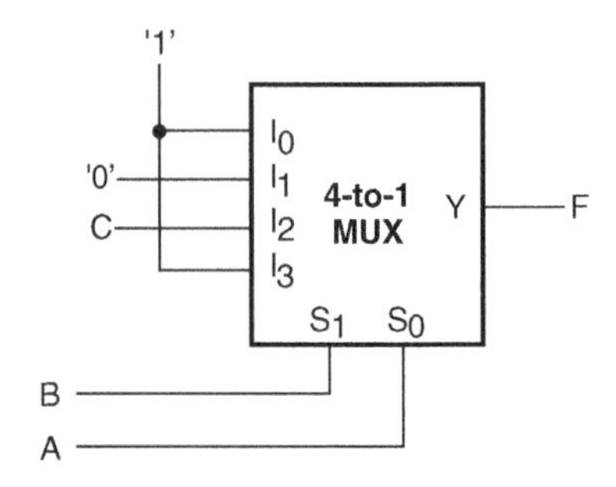

Figure 8.12 Example 8.1.

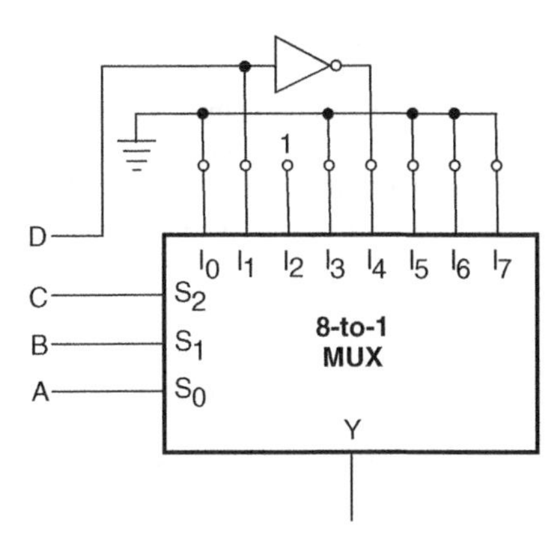

Figure 8.13 Example 8.2.

Example 8.2

Figure 8.13 shows the use of an 8-to-1 multiplexer to implement a certain four-variable Boolean function. From the given logic circuit arrangement, derive the Boolean expression implemented by the given circuit.

Solution

This problem can be solved by simply working backwards in the procedure outlined earlier for designing the multiplexer-based logic circuit for a given Boolean function. Here, the hardware implementation is known and the objective is to determine the corresponding Boolean expression.

From the given logic circuit, we can draw the implementation table as given in Table 8.7. The entries in the first row (0, 1, 2, 3, 4, 5, 6, 7) and the second row (8, 9, 10, 11, 12, 13, 14, 15) are so because the selection variable chosen for application to the inputs is the MSB variable D. Entries in the first row include all those minterms that contain $\overline{D}$, and entries in the second row include all those minterms that contain D. After writing the entries in the first two rows, the entries in the third row can be filled in by examining the logic status of different input lines in the given logic circuit diagram. Having completed the third row, relevant entries in the first and second rows are highlighted. The Boolean expression can now be written as follows:

$$Y = \sum 2, 4, 9, 10 = \overline{D}.\overline{C}.B.\overline{A} + \overline{D}.C.\overline{B}.\overline{A} + D.\overline{C}.\overline{B}.A + D.\overline{C}.B.\overline{A}$$

$$= \overline{C}.B.\overline{A}.(\overline{D} + D) + \overline{D}.C.\overline{B}.\overline{A} + D.\overline{C}.\overline{B}.A$$

$$= \overline{C}.B.\overline{A} + \overline{D}.C.\overline{B}.\overline{A} + D.\overline{C}.\overline{B}.A$$

Table 8.7 Implementation table.

	I_0	I_1	I_2	I_3	I_4	I_5	I_6	I_7
$\overline{D}$	0	1	**2**	3	**4**	5	6	7
D	8	**9**	**10**	11	12	13	14	15
	0	D	1	0	$\overline{D}$	0	0	0

8.1.4 Cascading Multiplexer Circuits

There can possibly be a situation where the desired number of input channels is not available in IC multiplexers. A multiple number of devices of a given size can be used to construct multiplexers that can handle a larger number of input channels. For instance, 8-to-1 multiplexers can be used to construct 16-to-1 or 32-to-1 or even larger multiplexer circuits. The basic steps to be followed to carry out the design are as follows:

1. If 2^n is the number of input lines in the available multiplexer and 2^N is the number of input lines in the desired multiplexer, then the number of individual multiplexers required to construct the desired multiplexer circuit would be 2^{N-n}.
2. From the knowledge of the number of selection inputs of the available multiplexer and that of the desired multiplexer, connect the less significant bits of the selection inputs of the desired multiplexer to the selection inputs of the available multiplexer.
3. The left-over bits of the selection inputs of the desired multiplexer circuit are used to enable or disable the individual multiplexers so that their outputs when ORed produce the final output. The procedure is illustrated in solved example 8.3.

Example 8.3

Design a 16-to-1 multiplexer using two 8-to-1 multiplexers having an active LOW ENABLE input.

Solution

A 16-to-1 multiplexer can be constructed from two 8-to-1 multiplexers having an ENABLE input. The ENABLE input is taken as the fourth selection variable occupying the MSB position. Figure 8.14 shows the complete logic circuit diagram. IC 74151 can be used to implement an 8-to-1 multiplexer.

The circuit functions as follows. When S_3 is in logic '0' state, the upper multiplexer is enabled and the lower multiplexer is disabled. If we recall the truth table of a four-variable Boolean function, S_3 would be '0' for the first eight entries and '1' for the remaining eight entries. Therefore, when $S_3 = 0$ the final output will be any of the inputs from D_0 to D_7, depending upon the logic status of S_2, S_1 and S_0. Similarly, when $S_3 = 1$ the final output will be any of the inputs from D_8 to D_{15}, again depending upon the logic status of S_2, S_1 and S_0. The circuit therefore implements the truth table of a 16-to-1 multiplexer.

8.2 Encoders

An *encoder* is a multiplexer without its single output line. It is a combinational logic function that has 2^n (or fewer) input lines and n output lines, which correspond to n selection lines in a multiplexer. The n output lines generate the binary code for the possible 2^n input lines. Let us take the case of an octal-to-binary encoder. Such an encoder would have eight input lines, each representing an octal digit, and three output lines representing the three-bit binary equivalent. The truth table of such an encoder is given in Table 8.8. In the truth table, D_0 to D_7 represent octal digits 0 to 7. A, B and C represent the binary digits.

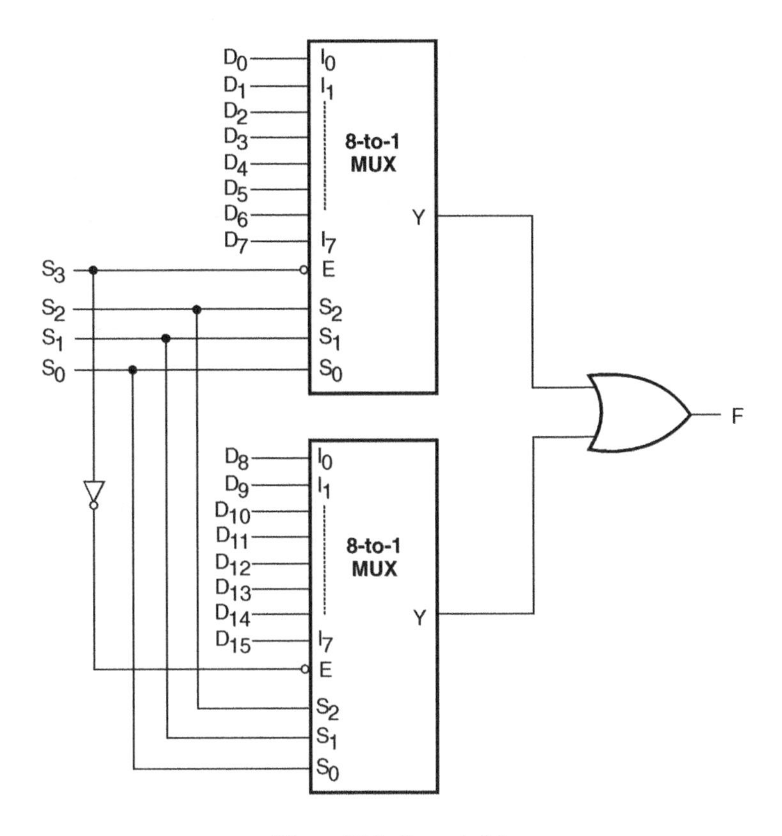

Figure 8.14 Example 8.3.

The eight input lines would have $2^8 = 256$ possible combinations. However, in the case of an octal-to-binary encoder, only eight of these 256 combinations would have any meaning. The remaining combinations of input variables are 'don't care' input combinations. Also, only one of the input lines at a time is in logic '1' state. Figure 8.15 shows the hardware implementation of the octal-to-binary encoder described by the truth table in Table 8.8. This circuit has the shortcoming that it produces an all 0s output sequence when all input lines are in logic '0' state. This can be overcome by having an additional line to indicate an all 0s input sequence.

8.2.1 Priority Encoder

A *priority encoder* is a practical form of an encoder. The encoders available in IC form are all priority encoders. In this type of encoder, a priority is assigned to each input so that, when more than one input is simultaneously active, the input with the highest priority is encoded. We will illustrate the concept of priority encoding with the help of an example. Let us assume that the octal-to-binary encoder described in the previous paragraph has an input priority for higher-order digits. Let us also assume that input lines D_2, D_4 and D_7 are all simultaneously in logic '1' state. In that case, only D_7 will be encoded and the output will be 111. The truth table of such a priority

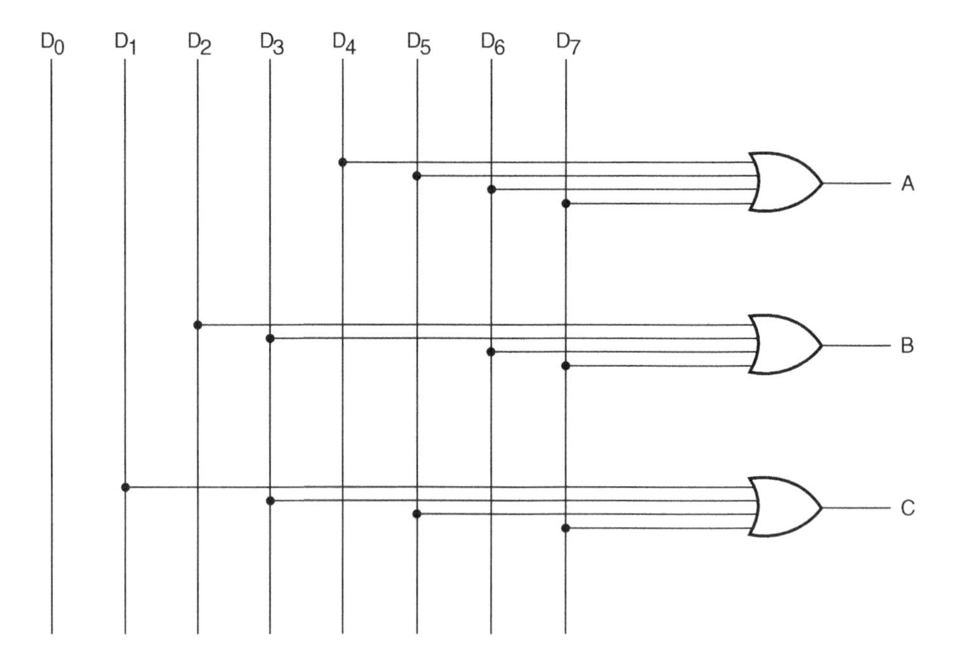

Figure 8.15 Octal-to-binary encoder.

Table 8.8 Truth table of an encoder.

D_0	D_1	D_2	D_3	D_4	D_5	D_6	D_7	A	B	C
1	0	0	0	0	0	0	0	0	0	0
0	1	0	0	0	0	0	0	0	0	1
0	0	1	0	0	0	0	0	0	1	0
0	0	0	1	0	0	0	0	0	1	1
0	0	0	0	1	0	0	0	1	0	0
0	0	0	0	0	1	0	0	1	0	1
0	0	0	0	0	0	1	0	1	1	0
0	0	0	0	0	0	0	1	1	1	1

encoder will then be modified to what is shown in Table 8.9. Looking at the last row of the table, it implies that, if $D_7 = 1$, then, irrespective of the logic status of other inputs, the output is 111 as D_7 will only be encoded. As another example, Fig. 8.16 shows the logic symbol and truth table of a 10-line decimal to four-line BCD encoder providing priority encoding for higher-order digits, with digit 9 having the highest priority. In the functional table shown, the input line with highest priority having a LOW on it is encoded irrespective of the logic status of the other input lines.

Table 8.9 Priority encoder.

D_0	D_1	D_2	D_3	D_4	D_5	D_6	D_7	A	B	C
1	0	0	0	0	0	0	0	0	0	0
X	1	0	0	0	0	0	0	0	0	1
X	X	1	0	0	0	0	0	0	1	0
X	X	X	1	0	0	0	0	0	1	1
X	X	X	X	1	0	0	0	1	0	0
X	X	X	X	X	1	0	0	1	0	1
X	X	X	X	X	X	1	0	1	1	0
X	X	X	X	X	X	X	1	1	1	1

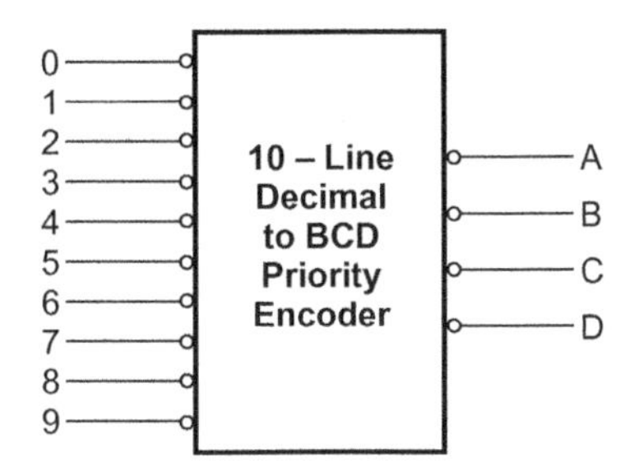

Inputs										Outputs			
0	1	2	3	4	5	6	7	8	9	D	C	B	A
X	X	X	X	X	X	X	X	X	0	0	1	1	0
X	X	X	X	X	X	X	X	0	1	0	1	1	1
X	X	X	X	X	X	X	0	1	1	1	0	0	0
X	X	X	X	X	X	0	1	1	1	1	0	0	1
X	X	X	X	X	0	1	1	1	1	1	0	1	0
X	X	X	X	0	1	1	1	1	1	1	0	1	1
X	X	X	0	1	1	1	1	1	1	1	1	0	0
X	X	0	1	1	1	1	1	1	1	1	1	0	1
X	0	1	1	1	1	1	1	1	1	1	1	1	0
0	1	1	1	1	1	1	1	1	1	1	1	1	1

Figure 8.16 10-line decimal to four-line BCD priority encoder.

Some of the encoders available in IC form provide additional inputs and outputs to allow expansion. IC 74148, which is an eight-line to three -line priority encoder, is an example. ENABLE-IN (EI) and ENABLE-OUT (EO) terminals on this IC allow expansion. For instance, two 74148s can be cascaded to build a 16-line to four-line priority encoder.

Example 8.4

We have an eight-line to three-line priority encoder circuit with D_0, D_1, D_2, D_3, D_4, D_5, D_6 and D_7 as the data input lines. the output bits are A (MSB), B and C (LSB). Higher-order data bits have been assigned a higher priority, with D_7 having the highest priority. If the data inputs and outputs are active when LOW, determine the logic status of output bits for the following logic status of data inputs:

(a) All inputs are in logic '0' state.
(b) D_1 to D_4 are in logic '1' state and D_5 to D_7 are in logic '0' state.
(c) D_7 is in logic '0' state. The logic status of the other inputs is not known.

Solution

(a) Since all inputs are in logic '0' state, it implies that all inputs are active. Since D_7 has the highest priority and all inputs and outputs are active when LOW, the output bits are $A = 0$, $B = 0$ and $C = 0$.
(b) Inputs D_5 to D_7 are the ones that are active. among these, D_7 has the highest priority. Therefore, the output bits are $A = 0$, $B = 0$ and $C = 0$.
(c) D_7 is active. Since D_7 has the highest priority, it will be encoded irrespective of the logic status of other inputs. Therefore, the output bits are $A = 0$, $B = 0$ and $C = 0$.

Example 8.5

Design a four-line to two-line priority encoder with active HIGH inputs and outputs, with priority assigned to the higher-order data input line.

Solution

The truth table for such a priority encoder is given in Table 8.10, with D_0, D_1, D_2 and D_3 as data inputs and X and Y as outputs.

The Boolean expressions for the two output lines X and $Y$are given by the equations

$$X = D_2.\overline{D}_3 + D_3 = D_2 + D_3 \tag{8.5}$$

$$Y = D_1.\overline{D}_2.\overline{D}_3 + D_3 = D_1.\overline{D}_2 + D_3 \tag{8.6}$$

Figure 8.17 shows the logic diagram that implements the Boolean functions given in equations (8.5) and (8.6).

Table 8.10 Example 8.5.

D_0	D_1	D_2	D_3	X	Y
1	0	0	0	0	0
X	1	0	0	0	1
X	X	1	0	1	0
X	X	X	1	1	1

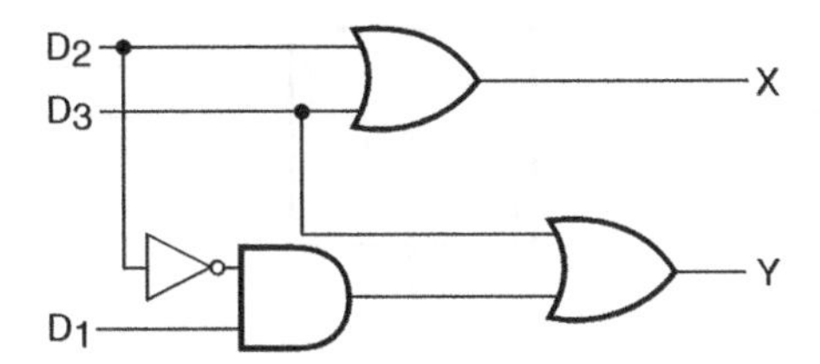

Figure 8.17 Example 8.5.

8.3 Demultiplexers and Decoders

A *demultiplexer* is a combinational logic circuit with an input line, 2^n output lines and n select lines. It routes the information present on the input line to any of the output lines. The output line that gets the information present on the input line is decided by the bit status of the selection lines. A *decoder* is a special case of a demultiplexer without the input line. Figure 8.18(a) shows the circuit representation of a 1-to-4 demultiplexer. Figure 8.18(b) shows the truth table of the demultiplexer when the input line is held HIGH.

A decoder, as mentioned earlier, is a combinational circuit that decodes the information on n input lines to a maximum of 2^n unique output lines. Figure 8.19 shows the circuit representation of 2-to-4, 3-to-8 and 4-to-16 line decoders. If there are some unused or 'don't care' combinations in the n-bit code, then there will be fewer than 2^n output lines. As an illustration, if there are three input lines, it

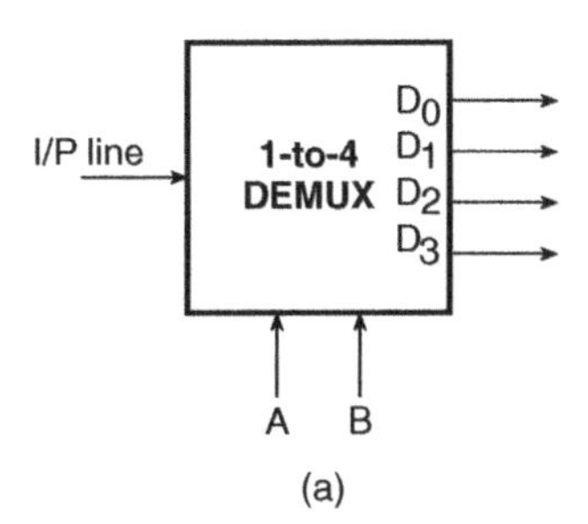

(a)

I/P	Select		O/P			
	A	B	D_0	D_1	D_2	D_3
1	0	0	1	0	0	0
1	0	1	0	1	0	0
1	1	0	0	0	1	0
1	1	1	0	0	0	1

(b)

Figure 8.18 1-to-4 demultiplexer.

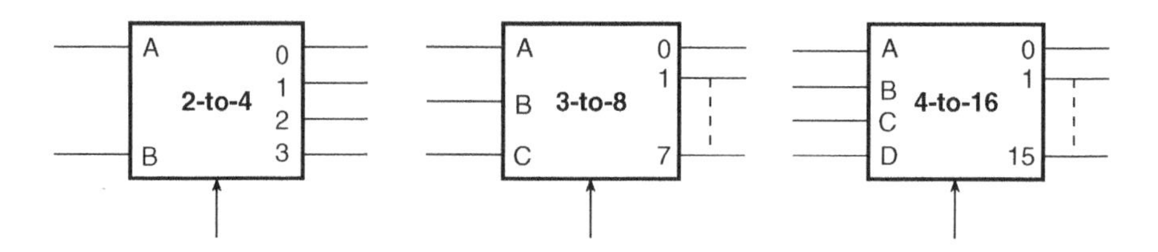

Figure 8.19 Circuit representation of 2-to-4, 3-to-8 and 4-to-16 line decoders.

can have a maximum of eight unique output lines. If, in the three-bit input code, the only used three-bit combinations are 000, 001, 010, 100, 110 and 111 (011 and 101 being either unused or don't care combinations), then this decoder will have only six output lines. In general, if n and m are respectively the numbers of input and output lines, then $m \leq 2^n$.

A decoder can generate a maximum of 2^n possible minterms with an n-bit binary code. In order to illustrate further the operation of a decoder, consider the logic circuit diagram in Fig. 8.20. This logic circuit, as we will see, implements a 3-to-8 line decoder function. This decoder has three inputs designated as A, B and C and eight outputs designated as D_0, D_1, D_2, D_3, D_4, D_5, D_6 and D_7. From the truth table given along with the logic diagram it is clear that, for any given input combination, only one of the eight outputs is in logic '1' state. Thus, each output produces a certain minterm that corresponds to the binary number currently present at the input. In the present case, D_0, D_1, D_2, D_3, D_4, D_5, D_6 and D_7 respectively represent the following minterms:

$$D_0 \rightarrow \overline{A}.\overline{B}.\overline{C},\ D_1 \rightarrow \overline{A}.\overline{B}.C,\ D_2 \rightarrow \overline{A}.B.\overline{C},\ D_3 \rightarrow \overline{A}.B.C$$

$$D_4 \rightarrow \overline{A}.\overline{B}.\overline{C},\ D_5 \rightarrow A.\overline{B}.C,\ D_6 \rightarrow A.B.\overline{C},\ D_7 \rightarrow A.B.C$$

8.3.1 Implementing Boolean Functions with Decoders

A decoder can be conveniently used to implement a given Boolean function. The decoder generates the required minterms and an external OR gate is used to produce the sum of minterms. Figure 8.21 shows the logic diagram where a 3-to-8 line decoder is used to generate the Boolean function given by the equation

$$Y = A.\overline{B}.\overline{C} + \overline{A}.B.\overline{C} + A.B.C + \overline{A}.\overline{B}.\overline{C} \tag{8.7}$$

In general, an n-to-2^n decoder and m external OR gates can be used to implement any combinational circuit with n inputs and m outputs. We can appreciate that a Boolean function with a large number of minterms, if implemented with a decoder and an external OR gate, would require an OR gate with an equally large number of inputs. Let us consider the case of implementing a four-variable Boolean function with 12 minterms using a 4-to-16 line decoder and an external OR gate. The OR gate here needs to be a 12-input gate. In all such cases, where the number of minterms in a given Boolean function with n variables is greater than $2^n/2$ (or 2^{n-1}), the complement Boolean function will have fewer minterms. In that case it would be more advantageous to do NORing of minterms of the complement Boolean function using a NOR gate rather than doing ORing of the given function using an OR gate. The output will be nothing but the given Boolean function.

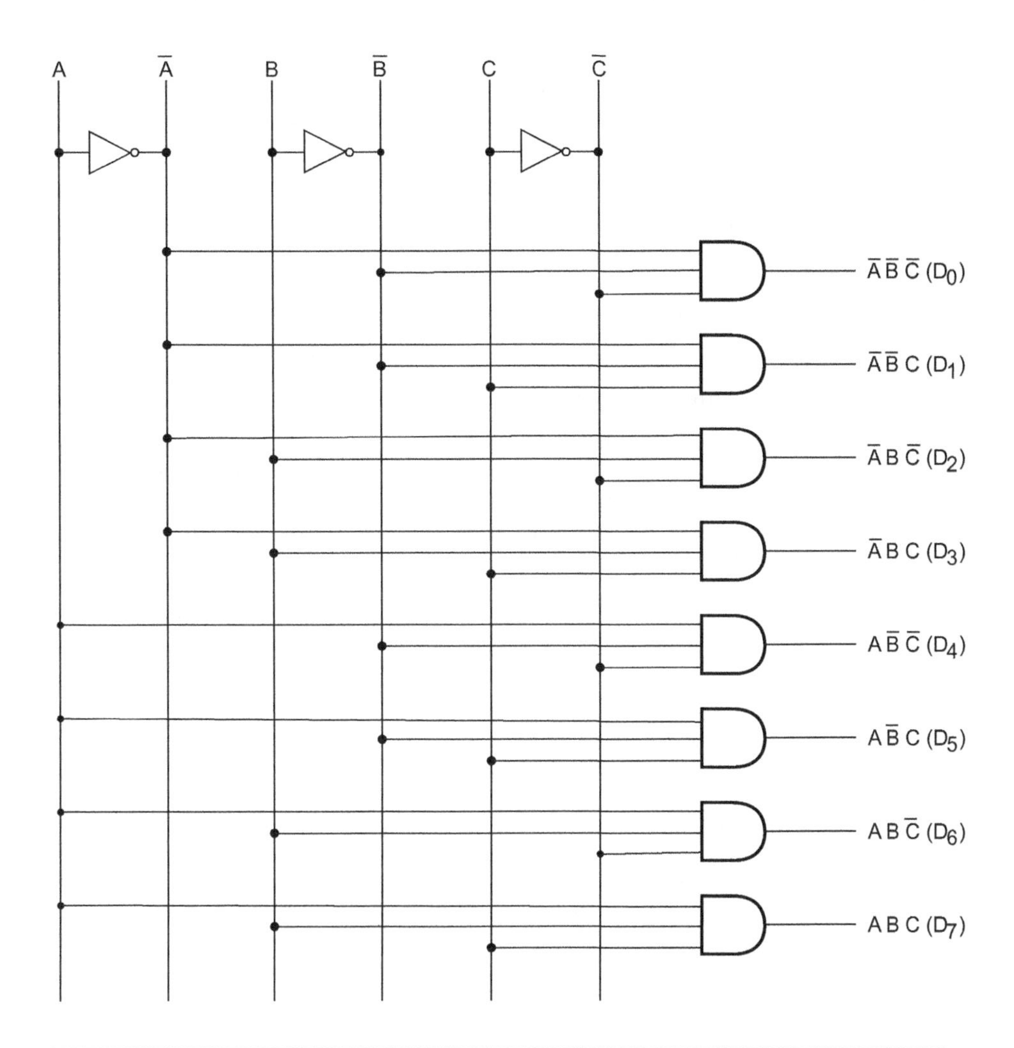

INPUTS			OUTPUTS							
A	B	C	D_0	D_1	D_2	D_3	D_4	D_5	D_6	D_7
0	0	0	1	0	0	0	0	0	0	0
0	0	1	0	1	0	0	0	0	0	0
0	1	0	0	0	1	0	0	0	0	0
0	1	1	0	0	0	1	0	0	0	0
1	0	0	0	0	0	0	1	0	0	0
1	0	1	0	0	0	0	0	1	0	0
1	1	0	0	0	0	0	0	0	1	0
1	1	1	0	0	0	0	0	0	0	1

Figure 8.20 Logic diagram of a 3-to-8 line decoder.

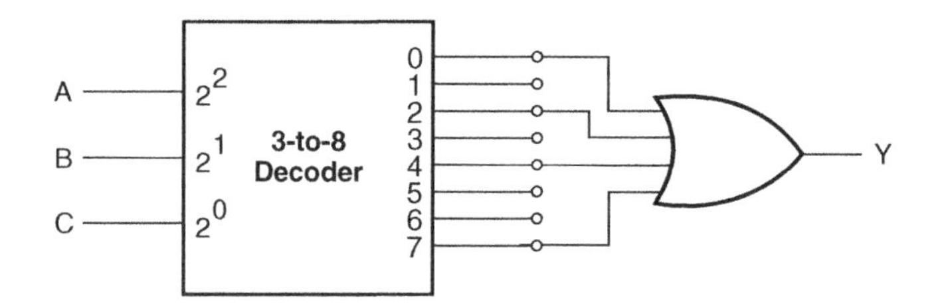

Figure 8.21 Implementing Boolean functions with decoders.

8.3.2 Cascading Decoder Circuits

There can possibly be a situation where the desired number of input and output lines is not available in IC decoders. More than one of these devices of a given size may be used to construct a decoder that can handle a larger number of input and output lines. For instance, 3-to-8 line decoders can be used to construct 4-to-16 or 5-to-32 or even larger decoder circuits. The basic steps to be followed to carry out the design are as follows:

1. If n is the number of input lines in the available decoder and N is the number of input lines in the desired decoder, then the number of individual decoders required to construct the desired decoder circuit would be 2^{N-n}.
2. Connect the less significant bits of the input lines of the desired decoder to the input lines of the available decoder.
3. The left-over bits of the input lines of the desired decoder circuit are used to enable or disable the individual decoders.
4. The output lines of the individual decoders together constitute the output lines, with the outputs of the less significant decoder constituting the less significant output lines and those of the higher–order decoders constituting the more significant output lines. The concept is further illustrated in solved example 8.8, which gives the design of a 4-to-16 decoder using 3-to-8 decoders.

Example 8.6

Implement a full adder circuit using a 3-to-8 line decoder.

Solution

A decoder with an OR gate at the output can be used to implement the given Boolean function. The decoder should at least have as many input lines as the number of variables in the Boolean function to be implemented. The truth table of the full adder is given in Table 8.11, and Fig. 8.22 shows the hardware implementation.

From the truth table, Boolean functions for SUM and CARRY outputs are given by the following equations:

$$\text{Sum output } S = \Sigma\ 1, 2, 4, 7 \tag{8.8}$$

$$\text{Carry output } C_o = \Sigma\ 3, 5, 6, 7 \tag{8.9}$$

Table 8.11 Example 8.6.

A	B	C	S	C_o
0	0	0	0	0
0	0	1	1	0
0	1	0	1	0
0	1	1	0	1
1	0	0	1	0
1	0	1	0	1
1	1	0	0	1
1	1	1	1	1

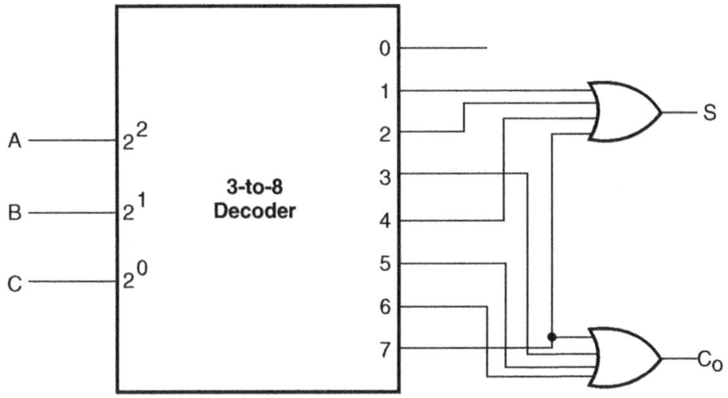

Figure 8.22 Example 8.6.

Example 8.7

A combinational circuit is defined by $F = \Sigma$ 0, 2, 5, 6, 7. Hardware implement the Boolean function F with a suitable decoder and an external OR/NOR gate having the minimum number of inputs.

Solution

The given Boolean function has five three-variable minterms. This implies that the function can be implemented with a 3-to-8 line decoder and a five-input OR gate. Also, $\overline{F}$ will have only three three-variable minterms, which means that F could also be implemented by considering minterms corresponding to the complement function and using a three-input NOR gate at the output. The second option uses a NOR gate with fewer inputs and therefore is used instead. $F = \Sigma$ 0, 2, 5, 6, 7. Therefore, $\overline{F} = \Sigma$ 1, 3, 4.

Figure 8.23 shows the hardware implementation of Boolean function F.

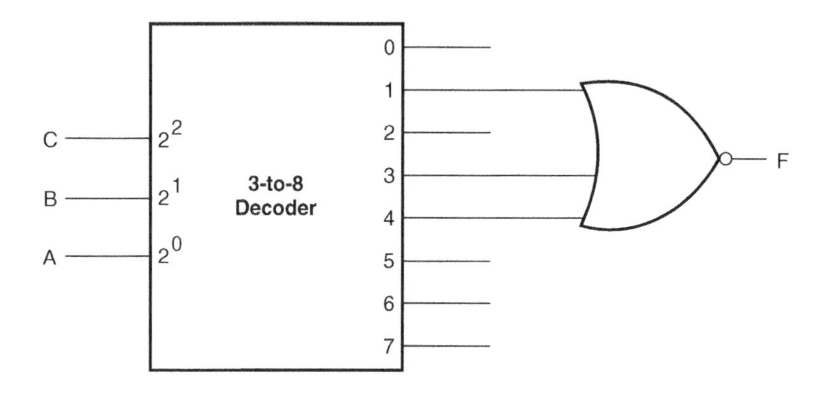

Figure 8.23 Example 8.7.

Example 8.8

Construct a 4-to-16 line decoder with two 3-to-8 line decoders having active LOW ENABLE inputs.

Solution

Let us assume that A (LSB), B, C and D (MSB) are the input variables for the 4-to-16 line decoder. Following the steps outlined earlier, A (LSB), B and C (MSB) will then be the input variables for the two 3-to-8 line decoders. If we recall the 16 possible input combinations from 0000 to 1111 in the case of a 4-to-16 line decoder, we find that the first eight combinations have $D = 0$, with CBA going through 000 to 111. The higher-order eight combinations all have $D = 1$, with CBA going through 000 to 111. If we use the D-bit as the ENABLE input for the less significant 3-to-8 line decoder and the $\overline{D}$-bit as the ENABLE input for the more significant 3-to-8 line decoder, the less significant 3-to-8 line decoder will be enabled for the less significant eight of the 16 input combinations, and the more significant 3-to-8 line decoder will be enabled for the more significant of the 16 input combinations. Figure 8.24 shows the hardware implementation. One of the output lines D_0 to D_{15} is activated as the input bit sequence $DCBA$ goes through 0000 to 1111.

Example 8.9

Figure 8.25 shows the logic symbol of IC 74154, which is a 4-to-16 line decoder/demultiplexer. The logic symbol is in ANSI/IEEE format. Determine the logic status of all 16 output lines for the following conditions:

(a) $D = HIGH$, $C = HIGH$, $B = LOW$, $A = HIGH$, $\overline{G_1} = LOW$ and $\overline{G_2} = LOW$.
(b) $D = HIGH$, $C = HIGH$, $B = LOW$, $A = HIGH$, $\overline{G_1} = HIGH$ and $\overline{G_2} = HIGH$.
(c) $D = HIGH$, $C = HIGH$, $B = LOW$, $A = HIGH$, $\overline{G_1} = HIGH$ and $\overline{G_2} = HIGH$.

Solution

It is clear from the given logic symbol that the device has active HIGH inputs, active LOW outputs and two active LOW ENABLE inputs. Also, both ENABLE inputs need to be active for the decoder to function owing to the indicated ANDing of the two ENABLE inputs.

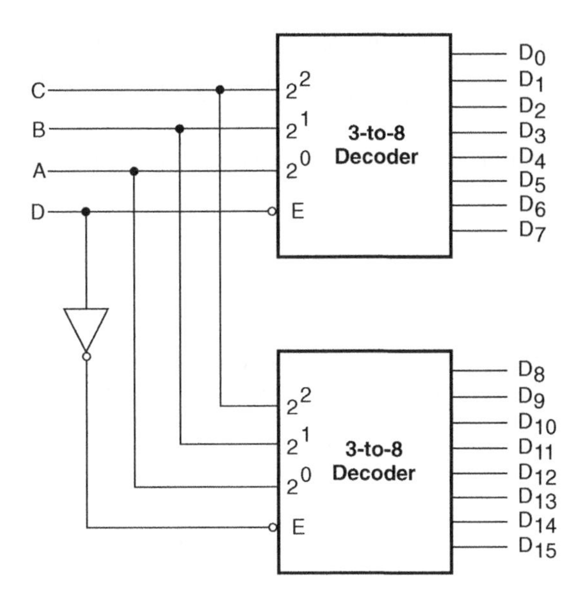

Figure 8.24 Example 8.8.

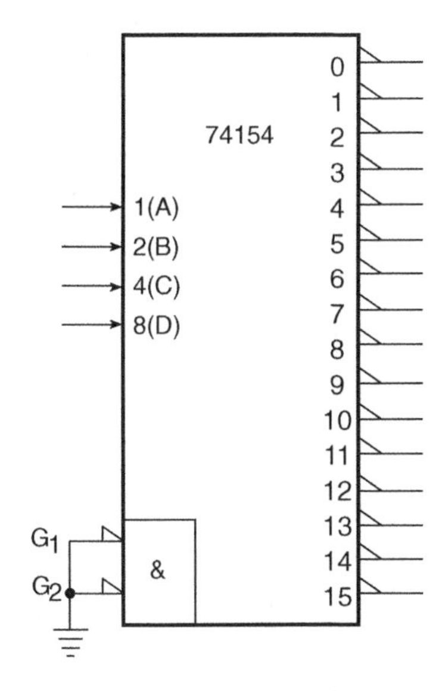

Figure 8.25 Example 8.9.

(a) Since both ENABLE inputs are active, the decoder outputs will therefore be active depending upon the logic status of the input lines. For the given logic status of the input lines, decoder output line 13 will be active and therefore LOW. All other output lines will be inactive and therefore in the logic HIGH state.
(b) Since neither ENABLE input is active, all decoder outputs will be inactive and in the logic HIGH state.
(c) The same as (b).

Example 8.10

The decoder of example 8.9 is to be used as a 1-of-16 demultiplexer. A logically compatible pulsed waveform is to be switched between output line 9 and line 15 when the logic status of an external control input is LOW and HIGH respectively. Draw the logic diagram indicating the logic status of ENABLE inputs and DCBA inputs and the point of application of the pulsed waveform.

Solution

Figure 8.26 shows the logic diagram. When the external control input is in the logic LOW state, $D =$ HIGH, $C =$ LOW, $B =$ LOW and $A =$ HIGH. This means that output line 9 is activated. When the external control input is in the logic HIGH state, $D =$ HIGH, $C =$ HIGH, $B =$ HIGH and $A =$ HIGH. This means that output line 15 is activated. In the logic diagram shown in Fig. 8.26, the two ENABLE inputs are tied together and the pulsed waveform is applied to a common point. This means that either both ENABLE inputs are active (when the input waveform is in the logic LOW state) or inactive (when the input waveform is in the logic HIGH state). Thus, when the input waveform is in the logic LOW state, output line 9 will be in the logic LOW state and all other output lines will be in the logic HIGH state provided the external control input is also in the logic LOW state. If the external

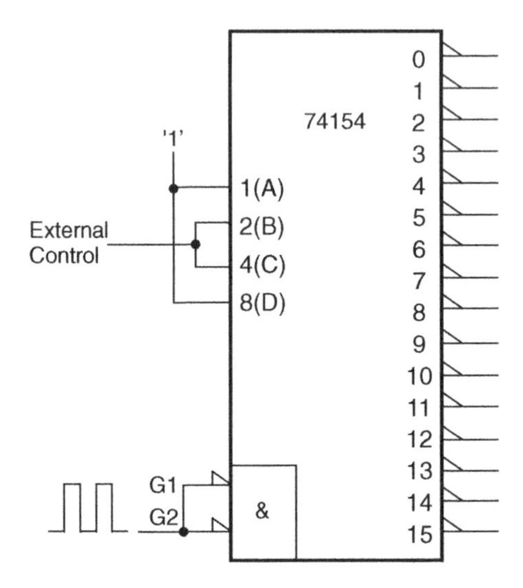

Figure 8.26 Example 8.10.

control input is in the logic HIGH state, logic LOW in the input waveform appears at output line 15. In essence, the logic status of the input waveform is reproduced at either line 9 or line 15, depending on whether the external control signal is LOW or HIGH.

8.4 Application-Relevant Information

Table 8.12 lists commonly used IC type numbers used as multiplexers, encoders, demultiplexers and decoders. Application-relevant information such as the pin connection diagram, truth table, etc., in respect of the more popular of these type numbers is given in the companion website.

Table 8.12 Commonly used IC type numbers used as multiplexers, encoders, demultiplexers and decoders.

IC Type number	Function	Logic family
7442	1-of-10 decoder	TTL
74138	1-of-8 decoder/demultiplexer	TTL
74139	Dual 1-of-4 decoder/demultiplexer	TTL
74145	1-of-10 decoder/driver (open collector)	TTL
74147	10-line to four-line priority encoder	TTL
74148	Eight-line to three-line priority encoder	TTL
74150	16-input multiplexer	TTL
74151	Eight-input multiplexer	TTL
74152	Eight-input multiplexer	TTL
74153	Dual four-input multiplexer	TTL
74154	4-of-16 decoder/demultiplexer	TTL
74155	Dual 1-of-4 decoder/demultiplexer	TTL
74156	Dual 1-of-4 decoder/demultiplexer (open collector)	TTL
74157	Quad two-input noninverting multiplexer	TTL
74158	Quad two-input inverting multiplexer	TTL
74247	BCD to seven-segment decoder/driver (open collector)	TTL
74248	BCD to seven-segment decoder/driver with Pull-ups	TTL
74251	Eight-input three-state multiplexer	TTL
74253	Dual four-input three-state multiplexer	TTL
74256	Dual four-bit addressable latch	TTL
74257	Quad two-input non-inverting three-state multiplexer	TTL
74258	Quad two-input inverting three-state multiplexer	TTL
74259	Eight-bit addressable latch	TTL
74298	Dual two-input multiplexer with output latches	TTL
74348	Eight-line to three-line priority encoder (three-state)	TTL
74353	Dual four-input multiplexer	TTL
74398	Quad two-input multiplexer with output register	TTL
74399	Quad two-input multiplexer with output register	TTL
4019	Quad two-input multiplexer	CMOS
4028	1-of-10 decoder	CMOS
40147	10-line to four-line BCD priority encoder	CMOS
4511	BCD to seven-segment latch/decoder/driver	CMOS
4512	Eight-input three-state multiplexer	CMOS
4514	1-of-16 decoder/demultiplexer with input latch	CMOS

(*continued overleaf*)

Table 8.12 *(continued).*

IC Type number	Function	Logic family
4515	1-of-16 decoder/demultiplexer with input latch	CMOS
4532	Eight-line to three-line priority encoder	CMOS
4539	Dual four-input multiplexer	CMOS
4543	BCD to seven-segment latch/decoder/driver for LCD displays	CMOS
4555	Dual 1-of-4 decoder/demultiplexers	CMOS
4556	Dual 1-of-4 decoder/demultiplexers	CMOS
4723	Dual four-bit addressable latch	CMOS
4724	Eight-bit addressable latch	CMOS
10132	Dual two-input multiplexer with latch and common reset	ECL
10134	Dual multiplexer with latch	ECL
10158	Quad two-input multiplexer (non-inverting)	ECL
10159	Quad two-input multiplexer (inverting)	ECL
10161	3-to-8 line decoder (LOW)	ECL
10162	3-to-8 line decoder (HIGH)	ECL
10164	Eight-line multiplexer	ECL
10165	Eight-input priority encoder	ECL
10171	Dual 2-to-4 line decoder (LOW)	ECL
10172	Dual 2-to-4 line decoder (HIGH)	ECL
10173	Quad two-input multiplexer/latch	ECL
10174	Dual 4-to-1 multiplexer	ECL

Review Questions

1. What is a multiplexer circuit? Briefly describe one or two applications of a multiplexer?
2. Is it possible to enhance the capability of an available multiplexer in terms of the number of input lines it can handle by using more than one device? If yes, briefly describe the procedure to do so, with the help of an example.
3. What is an encoder? How does a priority encoder differ from a conventional encoder? With the help of a truth table, briefly describe the functioning of a 10-line to four-line priority encoder with active LOW inputs and outputs and priority assigned to the higher-order inputs.
4. What is a demultiplexer and how does it differ from a decoder? Can a decoder be used as a demultiplexer? If yes, from where do we get the required input line?
5. Briefly describe how we can use a decoder optimally to implement a given Boolean function? Illustrate your answer with the help of an example.
6. Draw truth tables for the following:

   (a) an 8-to-1 multiplexer with active LOW inputs and an active LOW ENABLE input;
   (b) a four-line to 16-line decoder with active HIGH inputs and active LOW outputs and an active LOW ENABLE input;
   (c) an eight-line to three-line priority encoder with active LOW inputs and outputs and an active LOW ENABLE input.

Problems

1. Implement the three-variable Boolean function $F(A, B, C) = \overline{A}.C + A.\overline{B}.C + A.B.\overline{C}$ using (i) an 8-to-1 multiplexer and (ii) a 4-to-1 multiplexer.

   (i) Fig. 8.27(a); (ii) Fig. 8.27(b)

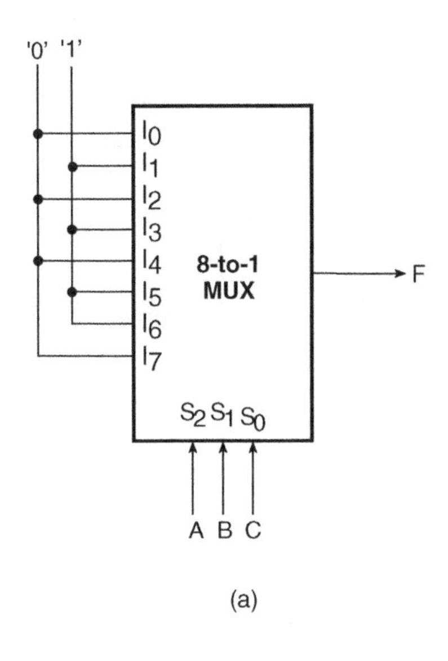

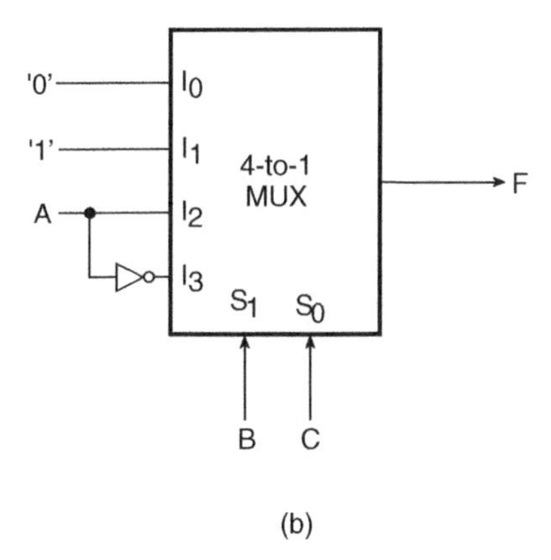

Figure 8.27 Problem 1.

2. Design a 32-to-1 multiplexer using 8-to-1 multiplexers having an active LOW ENABLE input and a 2-to-4 decoder.

   Fig. 8.28

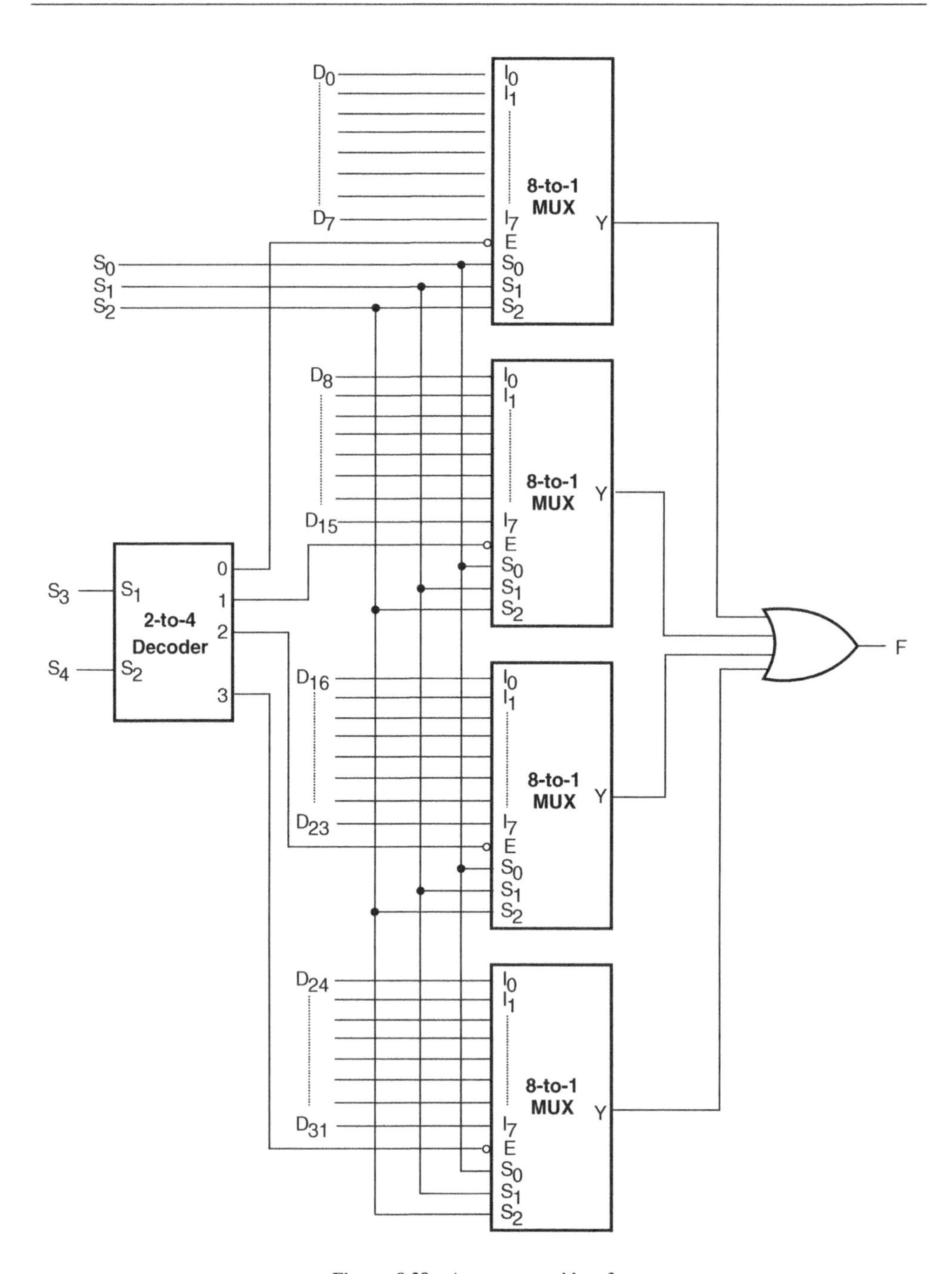

Figure 8.28 Answer to problem 2.

3. Determine the function performed by the combinational circuit of Fig. 8.29.

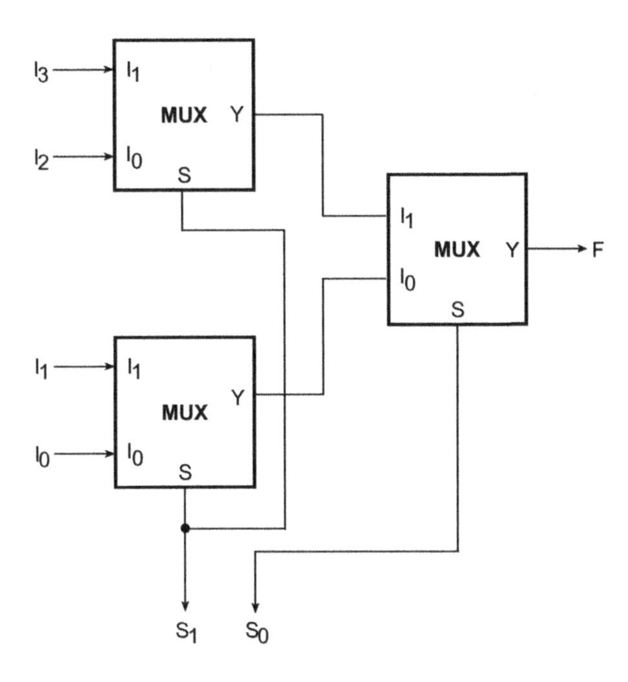

Figure 8.29 Problem 3.

4-to-1 multiplexer

4. Implement a full subtractor combinational circuit using a 3-to-8 decoder and external NOR gates.
Fig. 8.30

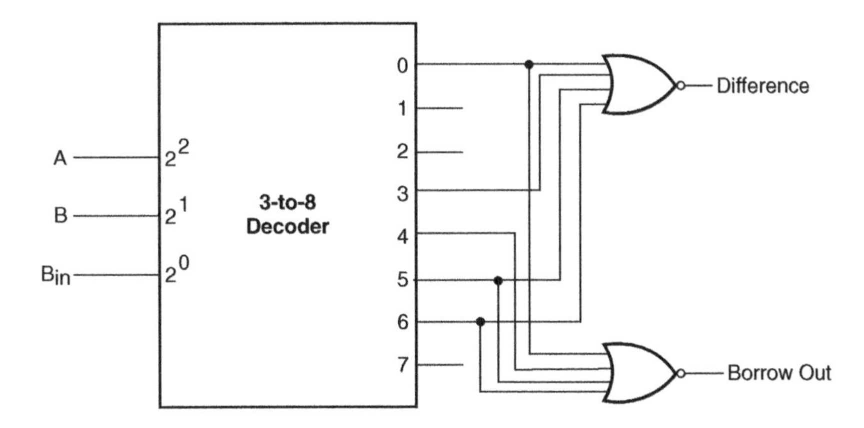

Figure 8.30 Answer to problem 4.

Further Reading

1. Floyd, T. L. (2005) *Digital Fundamentals*, Prentice-Hall Inc., USA.
2. Tokheim, R. L. (1994) *Schaum's Outline Series of Digital Principles*, McGraw-Hill Companies Inc., USA.
3. Tocci, R. J. (2006) *Digital Systems – Principles and Applications*, Prentice-Hall Inc., NJ, USA.
4. Cook, N. P. (2003) *Practical Digital Electronics*, Prentice-Hall, NJ, USA.
5. Rafiquzzaman, M. (2005) *Fundamentals of Digital Logic and Microcomputer Design*, Wiley-Interscience, New York, USA.
6. Morris Mano, M. and Kime, C. R. (2003) *Logic and Computer Design Fundamentals*, Prentice-Hall Inc., USA.

9

Programmable Logic Devices

Logic devices constitute one of the three important classes of devices used to build digital electronics systems, memory devices and microprocessors being the other two. Memory devices such as ROM and RAM are used to store information such as the software instructions of a program or the contents of a database, and microprocessors execute software instructions to perform a variety of functions, from running a word-processing program to carrying out far more complex tasks. Logic devices implement almost every other function that the system must perform, including device-to-device interfacing, data timing, control and display operations and so on. So far, we have discussed those logic devices that perform fixed logic functions decided upon at the manufacturing stage. Logic gates, multiplexers, demultiplexers, arithmetic circuits, etc., are some examples. Sequential logic devices such as flip-flops, counters, registers, etc., to be discussed in the following chapters, also belong to this category of logic devices. In the present chapter, we will discuss a new category of logic devices called *programmable logic devices* (PLDs). The function to be performed by a programmable logic device is undefined at the time of its manufacture. These devices are programmed by the user to perform a range of functions depending upon the logic capacity and other features offered by the device. We will begin with a comparison of fixed and programmable logic, and then follow this up with a detailed description of different types of PLDs in terms of operational fundamentals, salient features, architecture and typical applications. A brief introduction to the devices offered by some of the major manufacturers of PLDs and PLD programming languages is given towards the end of the chapter.

9.1 Fixed Logic Versus Programmable Logic

As outlined in the introduction, there are two broad categories of logic devices, namely fixed logic devices and programmable logic devices. Whereas a fixed logic device such as a logic gate or a multiplexer or a flip-flop performs a given logic function that is known at the time of device manufacture, a programmable logic device can be configured by the user to perform a large variety of

Digital Electronics: Principles, Devices and Applications Anil Kumar Maini

logic functions. In terms of the internal schematic arrangement of the two types of device, the circuits or building blocks and their interconnections in a fixed logic device are permanent and cannot be altered after the device is manufactured.

A *programmable logic device* offers to the user a wide range of logic capacity in terms of digital building blocks, which can be configured by the user to perform the intended function or set of functions. This configuration can be modified or altered any number of times by the user by reprogramming the device. Figure 9.1 shows a simple logic circuit comprising four three-input AND gates and a four-input OR gate. This circuit produces an output that is the sum output of a full adder. Here, A and B are the two bits to be added, and C is the carry-in bit. It is a fixed logic device as the circuit is unalterable from outside owing to fixed interconnections between the various building blocks.

Figure 9.2 shows the logic diagram of a simple programmable device. The device has an array of four six-input AND gates at the input and a four-input OR gate at the output. Each AND gate can handle three variables and thus can produce a product term of three variables. The three variables (A, B and C in this case) or their complements can be programmed to appear at the inputs of any of the four AND gates through fusible links called antifuses. This means that each AND gate can produce the desired three-variable product term. It may be mentioned here that an antifuse performs a function that is opposite to that performed by a conventional electrical fuse. A fuse has a low initial resistance and permanently breaks an electrically conducting path when current through it exceeds a certain limiting value. In the case of an antifuse, the initial resistance is very high and it is designed to create a low-resistance electrically conducting path when voltage across it exceeds a certain level. As a result, this circuit can be programmed to generate any three-variable sum-of-products Boolean function having four minterms by activating the desired fusible links. For example, the circuit could be programmed to produce the sum output resulting from the addition of three bits (the sum output in the case of a full adder) or to produce difference outputs resulting from subtraction of two bits with a borrow-in (the difference output in the case of a full subtractor).

We can visualize that the logic circuit of Fig. 9.2 has a programmable AND array at the input and a fixed OR gate at the output. Incidentally, this is the architecture of programmable logic devices called programmable array logic (PAL). Practical PAL devices have a much larger number of programmable AND gates and fixed OR gates to have enhanced logic capacity and performance capability. PAL devices are discussed in detail in the latter part of the chapter.

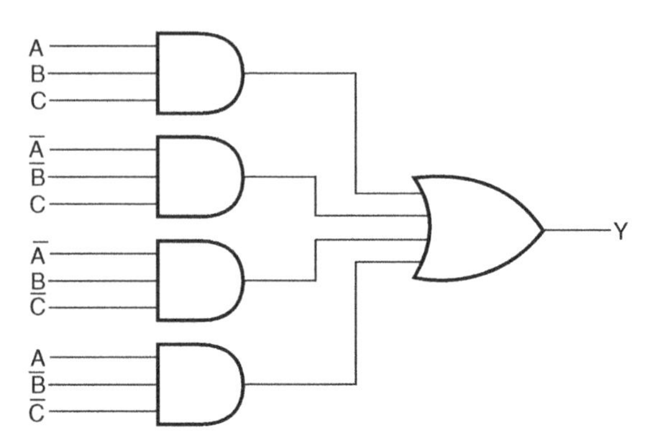

Figure 9.1 Fixed logic circuit.

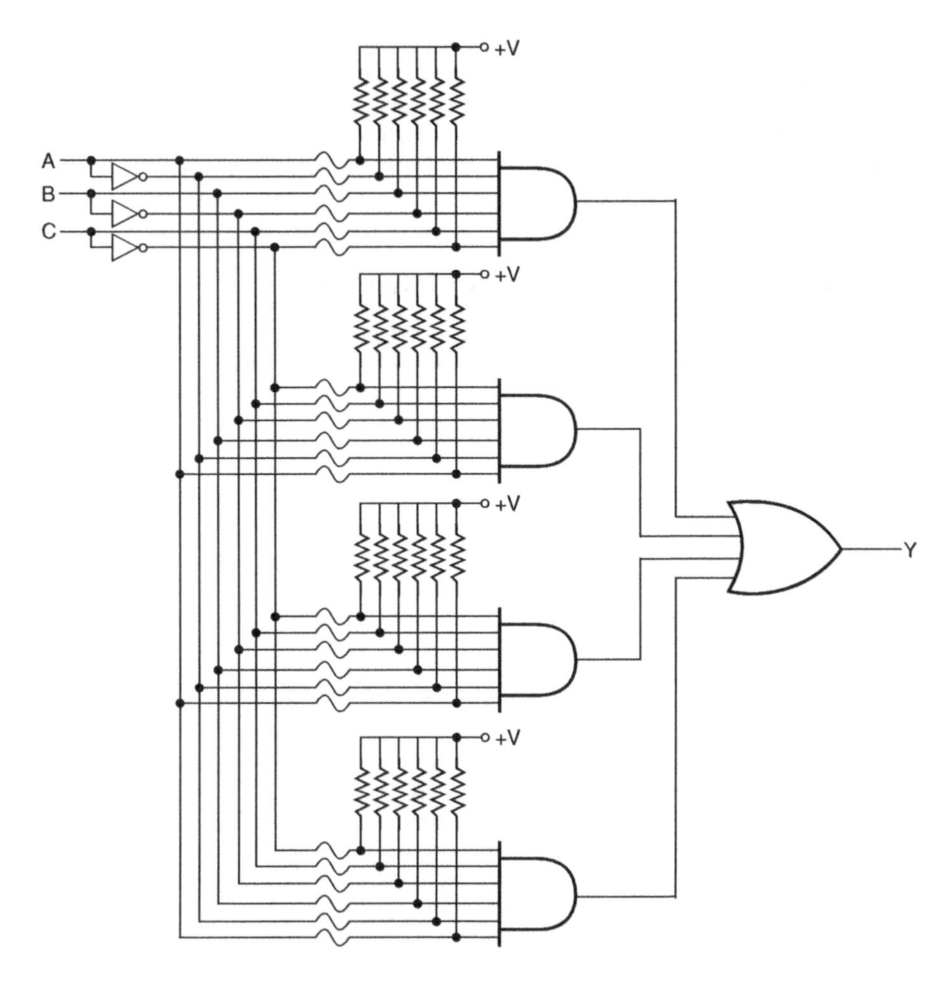

Figure 9.2 Simple programmable logic circuit.

9.1.1 Advantages and Disadvantages

1. If we want to build a fixed logic device to perform a certain specific function, the time required from design to the final stage when the manufactured device is actually available for use could easily be several months to a year or so. PLD-based design requires much less time from design cycle to production run.
2. In the case of fixed logic devices, the process of design validation followed by incorporation of changes, if any, involves substantial nonrecurring engineering (NRE) costs, which leads to an enhanced cost of the initial prototype device. In the case of PLDs, inexpensive software tools can be used for quick validation of designs. The programmable feature of these devices allows quick incorporation of changes and also a quick testing of the device in an actual application environment. In this case, the device used for prototyping is the same as the one that would qualify for use in the end equipment.

3. In the case of programmable logic devices, users can change the circuit as often as they want to until the design operates to their satisfaction. PLDs offer to the users much more flexibility during the design cycle. Design iterations are nothing but changes to the programming file.
4. Fixed logic devices have an edge for large-volume applications as they can be mass produced more economically. They are also the preferred choice in applications requiring the highest performance level.

9.2 Programmable Logic Devices – An Overview

There are many types of programmable logic device, distinguishable from one another in terms of architecture, logic capacity, programmability and certain other specific features. In this section, we will briefly discuss commonly used PLDs and their salient features. A detailed description of each of them will follow in subsequent sections.

9.2.1 Programmable ROMs

PROM (Programmable Read Only Memory) and EPROM (Erasable Programmable Read Only Memory) can be considered to be predecessors to PLDs. The architecture of a programmable ROM allows the user to hardware-implement an arbitrary combinational function of a given number of inputs. When used as a memory device, n inputs of the ROM (called address lines in this case) and m outputs (called data lines) can be used to store $2^n m$-bit words. When used as a PLD, it can be used to implement m different combinational functions, with each function being a chosen function of n variables. Any conceivable n-variable Boolean function can be made to appear at any of the m output lines. A generalized ROM device with n inputs and m outputs has 2^n hard-wired AND gates at the input and m programmable OR gates at the output. Each AND gate has n inputs, and each OR gate has 2^n inputs. Thus, each OR gate can be used to generate any conceivable Boolean function of n variables, and this generalized ROM can be used to produce m arbitrary n-variable Boolean functions. The AND array produces all possible minterms of a given number of input variables, and the programmable OR array allows only the desired minterms to appear at their inputs. Figure 9.3 shows the internal architecture of a PROM having four input lines, a hard-wired array of 16 AND gates and a programmable array of four OR gates. A cross ($\times$) indicates an intact (or unprogrammed) fusible link or interconnection, and a dot (•) indicates a hard-wired interconnection. PROMs, EPROMs and EEPROMs (Electrically Erasable Programmable Read Only Memory) can be programmed using standard PROM programmers. One of the major disadvantages of PROMs is their inefficient use of logic capacity. It is not economical to use PROMs for all those applications where only a few minterms are needed. Other disadvantages include relatively higher power consumption and an inability to provide safe covers for asynchronous logic transitions. They are usually much slower than the dedicated logic circuits. Also, they cannot be used to implement sequential logic owing to the absence of flip-flops.

9.2.2 Programmable Logic Array

A *programmable logic array* (PLA) device has a programmable AND array at the input and a programmable OR array at the output, which makes it one of the most versatile PLDs. Its architecture differs from that of a PROM in the following respects. It has a programmable AND array rather than a hard-wired AND array. The number of AND gates in an m-input PROM is always equal to 2^m. In the case of a PLA, the number of AND gates in the programmable AND array for m input variables

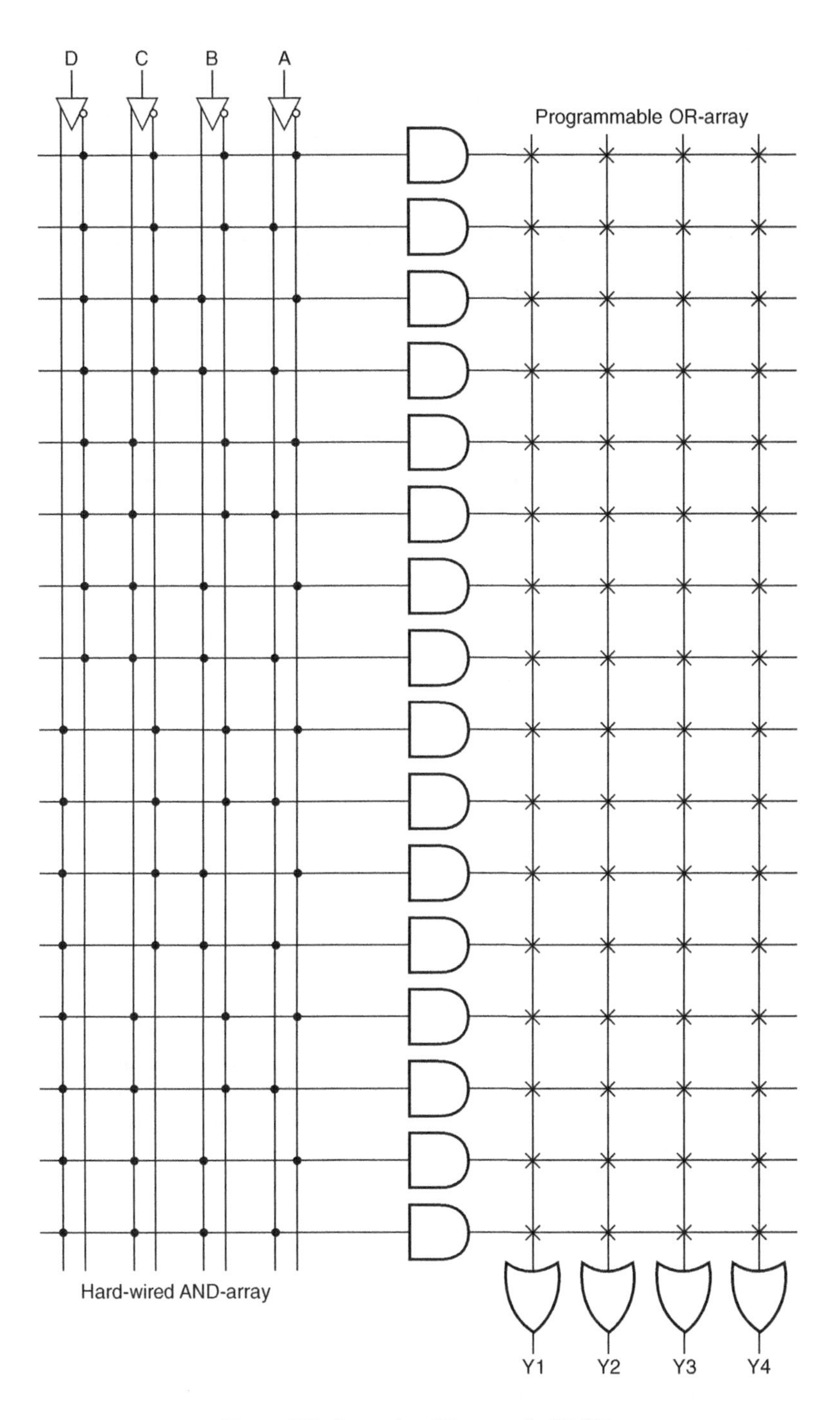

Figure 9.3 Internal architecture of a PROM.

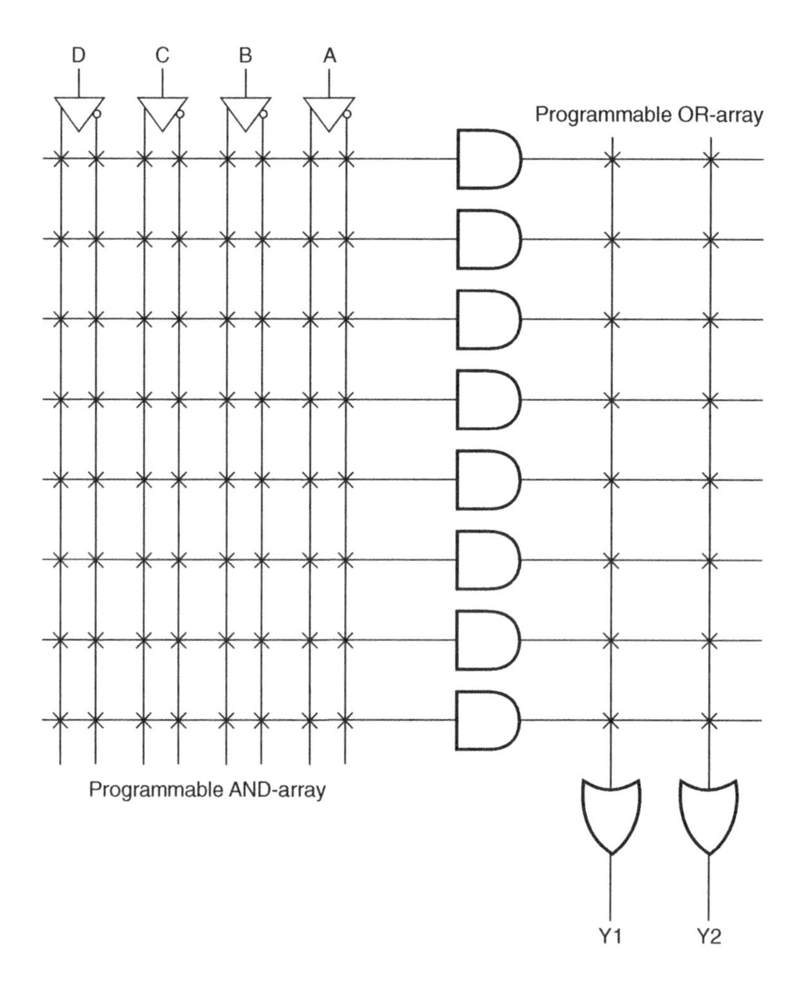

Figure 9.4 Internal architecture of a PLA device.

is usually much less than 2^m, and the number of inputs of each of the OR gates equals the number of AND gates. Each OR gate can generate an arbitrary Boolean function with a maximum of minterms equal to the number of AND gates. Figure 9.4 shows the internal architecture of a PLA device with four input lines, a programmable array of eight AND gates at the input and a programmable array of two OR gates at the output. A PLA device makes more efficient use of logic capacity than a PROM. However, it has its own disadvantages resulting from two sets of programmable fuses, which makes it relatively more difficult to manufacture, program and test.

9.2.3 Programmable Array Logic

Programmable array logic (PAL) architecture has a programmable AND array at the input and a fixed OR array at the output. The programmable AND array of a PAL device is similar to that of a PLA device. That is, the number of programmable AND gates is usually smaller than the number required to generate all possible minterms of the given number of input variables. The OR array is fixed and the AND outputs are equally divided between available OR gates. For instance, a practical PAL device may have eight input variables, 64 programmable AND gates and four fixed OR gates, with each OR gate having 16 inputs. That is, each OR gate is fed from 16 of the 64 AND outputs. Figure 9.5 shows the internal architecture of a PAL device that has four input lines, an array of eight AND gates at the input and two OR gates at the output, to introduce readers to the arrangement of various building blocks inside a PAL device and allow them a comparison between different programmable logic devices.

9.2.4 Generic Array Logic

A *generic array logic* (GAL) device is similar to a PAL device and was invented by Lattice Semiconductor. It differs from a PAL device in that the programmable AND array of

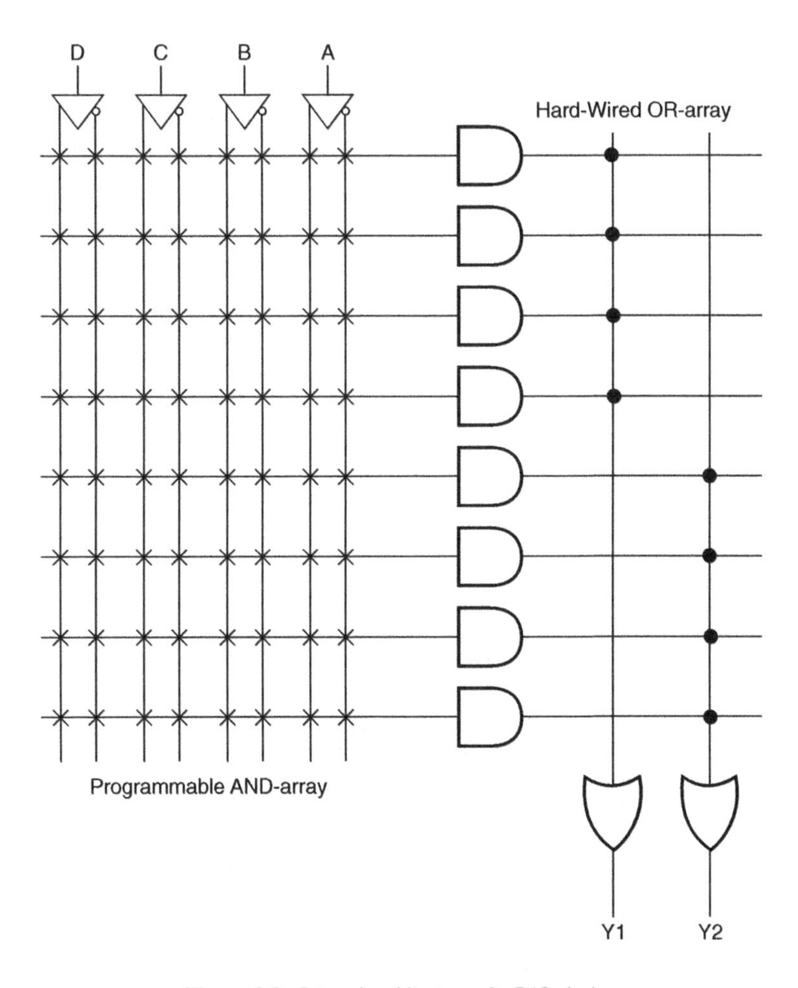

Figure 9.5 Internal architecture of a PAL device.

a GAL device can be erased and reprogrammed. Also, it has reprogrammable output logic. This feature makes it particularly attractive at the device prototyping stage, as any bugs in the logic can be corrected by reprogramming. A similar device called PEEL (Programmable Electrically Erasable Logic) was introduced by the International CMOS Technology (ICT) Corporation.

9.2.5 Complex Programmable Logic Device

Programmable logic devices such as PLAs, PALs, GALs and other PAL-like devices are often grouped into a single category called *simple programmable logic devices* (SPLDs) to distinguish them from the ones that are far more complex. A *complex programmable logic device* (CPLD), as the name suggests, is a much more complex device than any of the programmable logic devices discussed so far. A CPLD may contain circuitry equivalent to that of several PAL devices linked to each other by programmable interconnections. Figure 9.6 shows the internal structure of a typical CPLD. Each of the four logic blocks is equivalent to a PLD such as a PAL device. The number of logic blocks in a CPLD could be more or less than four. Each of the logic blocks has programmable interconnections. A switch matrix is used for logic block to logic block interconnections. Also, the switch matrix in a CPLD may or may not be fully connected. That is, some of the possible connections between logic block outputs and inputs may not be supported by a given CPLD. While the complexity of a typical PAL device may be of the order of a few hundred logic gates, a CPLD may have a complexity equivalent to tens of thousands of logic gates. When compared with FPGAs, CPLDs offer predictable timing characteristics owing to their less flexible internal architecture and are thus ideal for critical control applications and other applications where a high performance level is required. Also, because of their relatively much lower power consumption and lower cost, CPLDs are an ideal solution for battery-operated portable applications such as mobile phones, digital assistants and so on. A CPLD can be programmed either by using a PAL programmer or by feeding it with a serial data stream from a PC after soldering it on the PC board. A circuit on the CPLD decodes the data stream and configures it to perform the intended logic function.

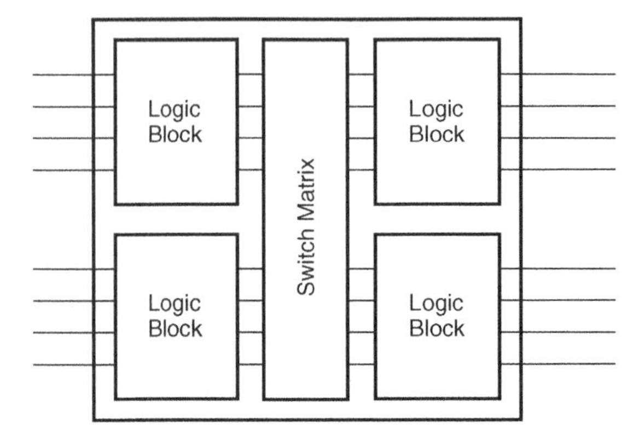

Figure 9.6 CPLD architecture.

9.2.6 Field-Programmable Gate Array

A *field-programmable gate array* (FPGA) uses an array of logic blocks, which can be configured by the user. The term 'field-programmable' here signifies that the device is programmable outside the factory where it is manufactured. The internal architecture of an FPGA device has three main parts, namely the array of logic blocks, the programmable interconnects and the I/O blocks. Figure 9.7 shows the architecture of a typical FPGA. Each of the I/O blocks provides an individually selectable input, output or bidirectional access to one of the general-purpose I/O pins on the FPGA package. The logic blocks in an FPGA are no more complex than a couple of logic gates or a look-up table feeding a flip-flop. The programmable interconnects connect logic blocks to logic blocks and also I/O blocks to logic blocks.

FPGAs offer a much higher logic density and much larger performance features compared with CPLDs. Some of the contemporary FPGA devices offer a logic complexity equivalent to that of eight million system gates. Also, these devices offer features such as built-in hard-wired processors,

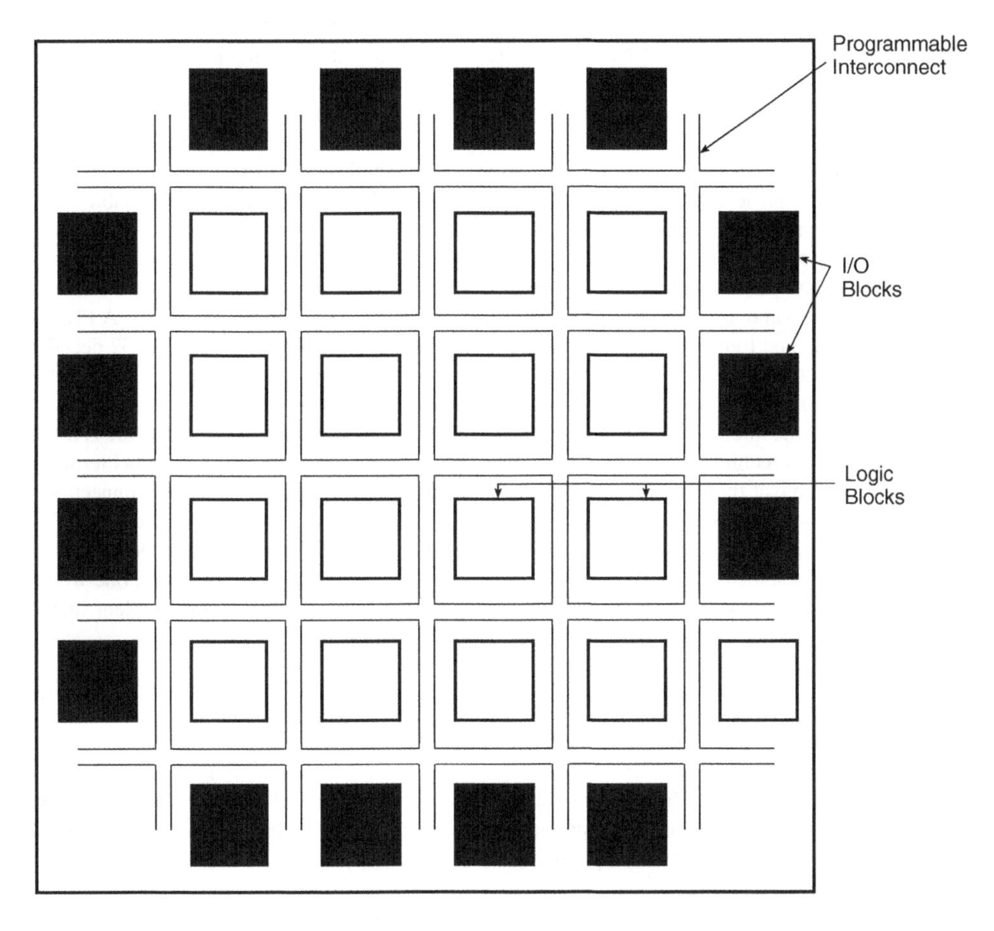

Figure 9.7 FPGA architecture.

large memory, clock management systems and support for many of the contemporary device-to-device signalling technologies. FPGAs find extensive use in a variety of applications, which include data processing and storage, digital signal processing, instrumentation and telecommunications.

FPGAs are also programmed like CPLDs after they are soldered onto the PC board. In the case of FPGAs, the programmed configuration is usually volatile and therefore needs to be reloaded whenever power is applied or a different functionality is required.

9.3 Programmable ROMs

A *read only memory* (ROM) is essentially a memory device that can be used to store a certain fixed set of binary information. As outlined earlier, these devices have certain inherent links that can be made or broken depending upon the type of fusible link to store any user-specified binary information in the device. While, in the case of a conventional fusible link, relevant interconnections are broken to program the device, in the case of an antifuse the relevant interconnections are made to do the same job. This is illustrated in Fig. 9.8. Figure 9.8(a) shows the internal logic diagram of a 4×2 PROM. The figure shows an unprogrammed PROM. Figures 9.8(b) and (c) respectively show the use of a fuse and an antifuse to produce output-1 $= AB$. Note that in the case of a fuse an unprogrammed interconnection is a 'make' connection, whereas in the case of an antifuse it is a 'break' connection.

Once a given pattern is formed, it remains as such even if power is turned off and on. In the case of PROMs, the user can erase the data already stored on the ROM chip and load it with fresh data. Memory-related issues of ROMs are discussed in detail in Chapter 15 on microcomputer fundamentals. In the present section, we will discuss the use of a PROM as a programmable logic device for implementation of combinational logic functions, which is one of the most widely exploited applications of PROMs. A PROM in general has n input lines and m output lines and is designated as a $2^n \times m$ PROM. Looking at the internal architecture of a PROM device, it is a combinational circuit with the AND gates wired as a decoder and having OR gates equal to the number of outputs. A PROM with five input lines and four output lines, for instance, would have the equivalent of a 5×32 decoder at the input that would generate 32 possible minterms or product terms. Each of these four OR gates would be a 32-input gate fed from 32 outputs of the decoder through fusible links.

Figure 9.9 shows the internal architecture of a 32×4 PROM. We can see that the input side is hard-wired to produce all possible 32 product terms corresponding to five variables. All 32 product terms or minterms are available at the inputs of each of the OR gates through programmable interconnections. This allows the users to have four different five-variable Boolean functions of their choice. Very complex combinational functions can be generated with PROMs by suitably making or breaking these links.

To sum up, for implementing an n-input or n-variable, $m-$output combinational circuit, one would need a $2^n \times m$ PROM. As an illustration, let us see how a PROM can be used to implement the following Boolean function with two outputs given by the equations

$$F_1(A, B, C) = \Sigma 0, 2 \tag{9.1}$$

$$F_2(A, B, C) = \Sigma 1, 4, 7 \tag{9.2}$$

Implementation of this Boolean function would require an 8×2 PROM. The internal logic diagram of the PROM in this case, after it is programmed, would be as shown in Fig. 9.10. Note that, in the programmed PROM of Fig. 9.10, an unprogrammed interconnection indicated by a cross ($\times$) is a 'make' connection.

It may be mentioned here that in practice a PROM would not be used to implement as simple a Boolean function as that illustrated above. The purpose here is to indicate to readers how a PROM

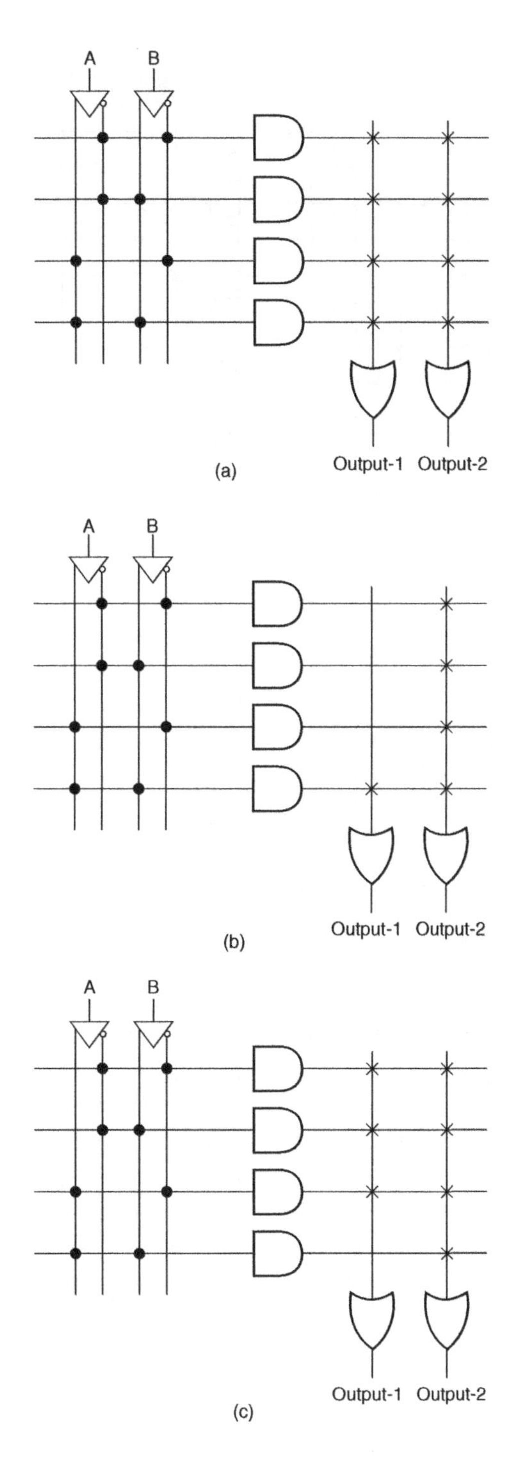

Figure 9.8 Use of fuse and antifuse.

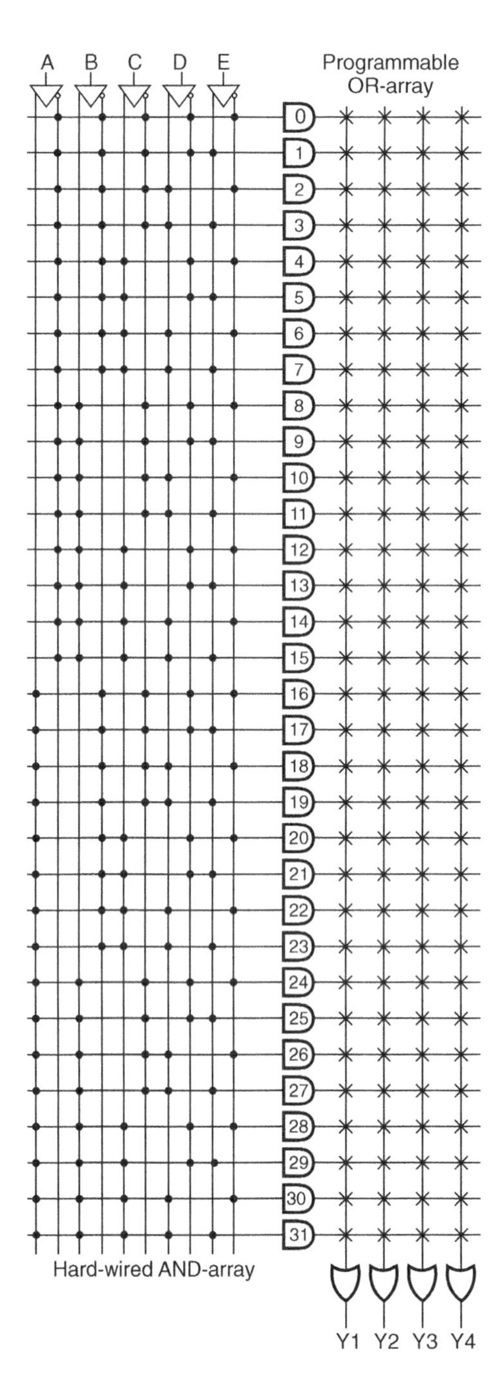

Figure 9.9 Internal architecture of a 32×4 PROM.

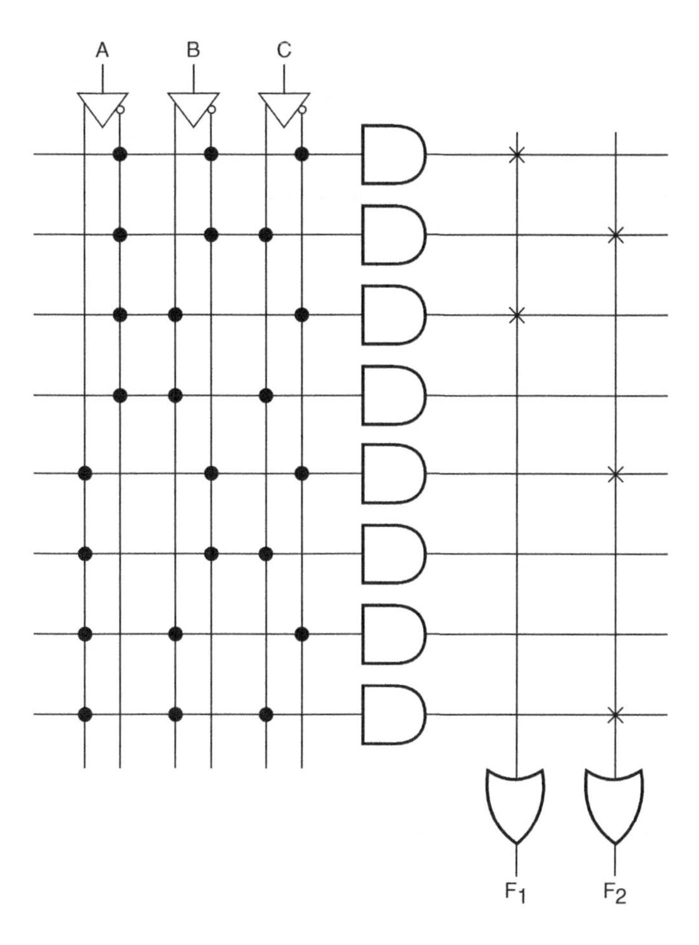

Figure 9.10 8 × 2 PROM internal logic diagram to implement given Boolean function.

implements a Boolean function. In actual practice, PROMs would be used only in the case of very complex Boolean functions.

Another noteworthy point is that, when it comes to implementing Boolean functions with PROMs, it is not economical to use PROM for those Boolean functions that have a large number of 'don't care' conditions. In the case of a PROM, each 'don't care' condition would have either all 0s or all 1s. In other words, the space on the chip is not optimally utilized. Other programmable logic devices such as a PLA or PAL are more suitable in such situations.

Example 9.1

Determine the size of the PROM required for implementing the following logic circuits:

(a) a binary multiplier that multiplies two four-bit numbers;
(b) a dual 8-to-1 multiplexer with common selection inputs;
(c) a single-digit BCD adder/subtractor with a control input for selection of operation.

Solution

(a) The number of inputs required here would be eight. The result of multiplication would be in eight bits. Therefore, the size of the PROM $= 2^8 \times 8 = 256 \times 8$.

(b) The number of inputs $= 8+8+3 = 19$ (the number of selection inputs $= 3$). The number of outputs $= 2$. Therefore, the size of the PROM $= 2^{19} \times 2 = 512\text{K} \times 2$.

(c) The number of inputs $= 4$ (augend bits) $+ 4$ (addend bits) $+ 1$ (carry-in) $+ 1$ (control input) $= 10$. The number of outputs $= 4$ (sum or subtraction output bits) $+ 1$ (carry or borrow bit) $= 5$. The size of the PROM $= 2^{10} \times 5 = 1024 \times 5 = 1\text{K} \times 5$.

9.4 Programmable Logic Array

A *programmable logic array* (PLA) enables logic functions expressed in sum-of-products form to be implemented directly. It is similar in concept to a PROM. However, unlike a PROM, the PLA does not provide full decoding of the input variables and does not generate all possible minterms. While a PROM has a fixed AND gate array at the input and a programmable OR gate array at the output, a PLA device has a programmable AND gate array at the input and a programmable OR gate array at the output. In a PLA device, each of the product terms of the given Boolean function is generated by an AND gate which can be programmed to form the AND of any subset of inputs or their complements. The product terms so produced can be summed up in an array of programmable OR gates. Thus, we have a programmable OR gate array at the output. The input and output gates are constructed in the form of arrays with input lines orthogonal to product lines and product lines orthogonal to output lines.

Figure 9.11 shows the internal architecture of a PLA device with four input lines, eight product lines and four output lines. That is, the programmable AND gate array has eight AND gates. Each of the AND gates here has eight inputs, corresponding to four input variables and their complements. The input to each of the AND gates can be programmed to be any of the possible 16 combinations of four input variables and their complements. Four OR gates at the output can generate four different Boolean functions, each having a maximum of eight minterms out of 16 minterms possible with four variables. The logic diagram depicts the unprogrammed state of the device. The internal architecture shown in Fig. 9.11 can also be represented by the schematic form of Fig. 9.12. PLAs usually have inverters at the output of OR gates to enable them to implement a given Boolean function in either AND-OR or AND-OR-INVERT form.

Figure 9.13 shows a generalized block schematic representation of a PLA device having n inputs, m outputs and k product terms, with n, m and k respectively representing the number of input variables, the number of OR gates and the number of AND gates. The number of inputs to each OR gate and each AND gate are k and $2n$ respectively.

A PLA is specified in terms of the number of inputs, the number of product terms and the number of outputs. As is clear from the description given in the preceding paragraph, the PLA would have a total of $2\text{K}n + \text{K}m$ programmable interconnections. A ROM with the same number of input and output lines would have $2^n \times \text{m}$ programmable interconnections.

A PLA could be either mask programmable or field programmable. In the case of a mask-programmable PLA, the customer submits a program table to the manufacturer to produce a custom-made PLA having the desired internal paths between inputs and outputs. A *field-programmable logic array* (FPLA) is programmed by the users themselves by means of a hardware programmer unit available commercially.

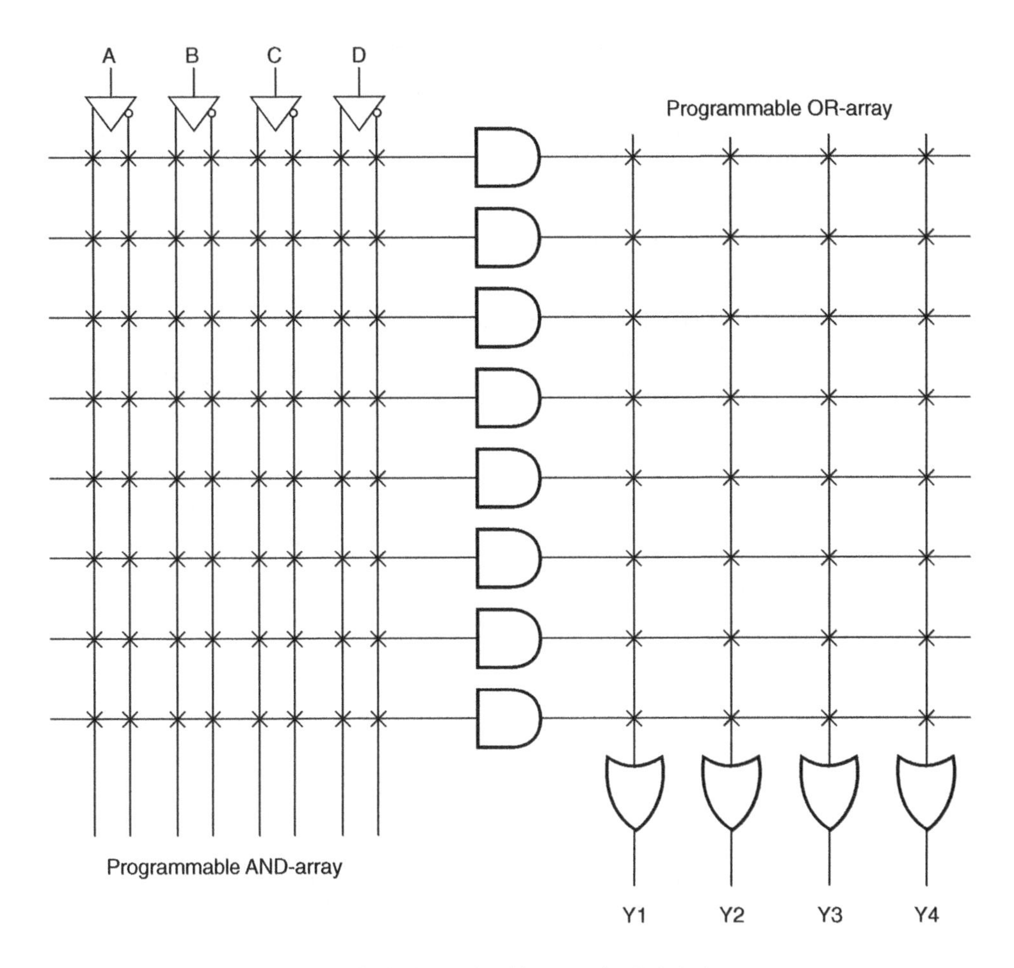

Figure 9.11 Internal architecture of a PLA device.

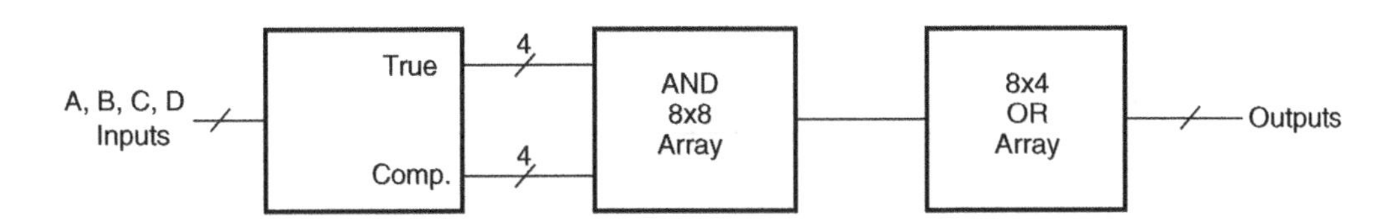

Figure 9.12 Alternative representation of PLA architecture.

While implementing a given Boolean function with a PLA, it is important that each expression is simplified to a minimum number of product terms which would minimize the number of AND gates required for the purpose. Since all input variables are available to different AND gates, simplification of Boolean functions to reduce the number of literals in various product terms is not important. In fact,

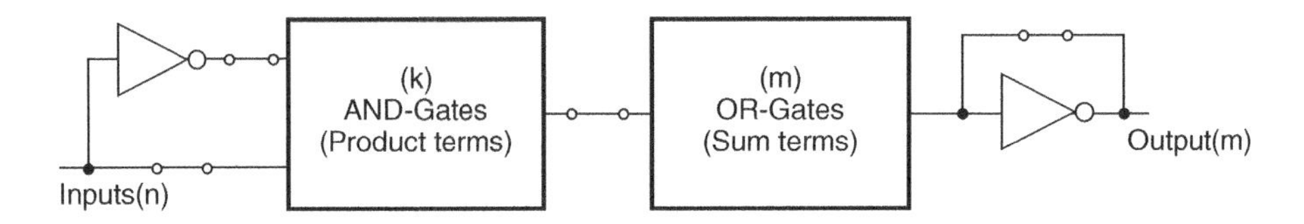

Figure 9.13 Generalized representation of PLA architecture.

each of the Boolean functions and their complements should be simplified. What is desirable is to have fewer product terms and product terms that are common to other functions. We would recall that PLAs offer the flexibility of implementing Boolean functions in both AND-OR and AND-OR-INVERT forms.

Example 9.2

Show the logic arrangement of both a PROM and a PLA required to implement a binary full adder.

Solution

The truth table of a full adder is given in Table 9.1. The Boolean expressions for sum S and carry-out C_o can be written as follows:

$$S = \Sigma 1, 2, 4, 7 \tag{9.3}$$

$$C_o = \Sigma 3, 5, 6, 7 \tag{9.4}$$

Figure 9.14 shows the implementation with an 8 × 2 PROM.

If we simplify the Boolean expressions for the sum and carry outputs, we will find that the expression for the sum output cannot be simplified any further, and also that the expression for carry-out can be simplified to three product terms with fewer literals. If we examine even the existing expressions, we find that we would need seven AND gates in the PLA implementation. And if we use the simplified expressions, even then we would require the same number of AND gates. Therefore, the simplification here would not help as far as its implementation with a PLA is concerned. Figure 9.15 shows the implementation of a full adder with a PLA device.

Table 9.1 Truth table for example 9.2.

A	B	Carry-in (C_i)	Sum (S)	Carry-out (C_o)
0	0	0	0	0
0	0	1	1	0
0	1	0	1	0
0	1	1	0	1
1	0	0	1	0
1	0	1	0	1
1	1	0	0	1
1	1	1	1	1

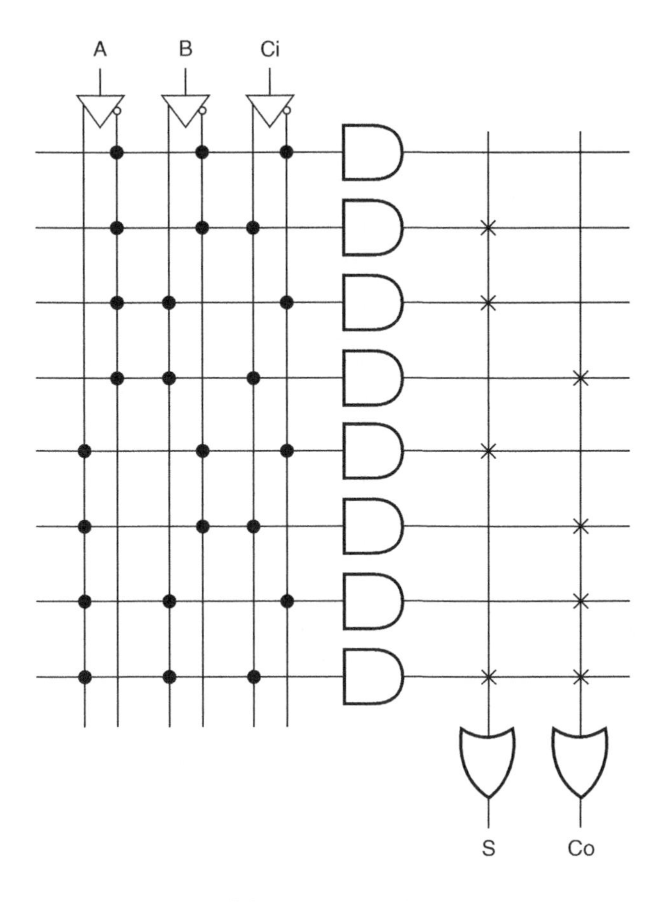

Figure 9.14 Solution to problem 9.2 using a PROM.

Example 9.3

We have two two-bit binary numbers A_1A_0 and B_1B_0. Design a PLA device to implement a magnitude comparator to produce outputs for A_1A_0 being 'equal to', 'not equal to', 'less than' and 'greater than' B_1B_0.

Solution

Table 9.2 shows the function table with inputs and desired outputs. The Boolean expressions for the desired outputs are given in the following equations:

$$\text{Output 1(equal to)} = \overline{A_1}.\overline{A_0}.\overline{B_1}.\overline{B_0} + \overline{A_1}.A_0.\overline{B_1}.B_0 + A_1.A_0.B_1.B_0 + A_1.\overline{A_0}.B_1.\overline{B_0} \quad (9.5)$$

Output 2 (not equal to)

$$= \overline{A_1}.\overline{A_0}.\overline{B_1}.B_0 + \overline{A_1}.\overline{A_0}.B_1.\overline{B_0} + \overline{A_1}.\overline{A_0}.B_1.B_0 + \overline{A_1}.A_0.\overline{B_1}.\overline{B_0} + \overline{A_1}.A_0.B_1.\overline{B_0} + \overline{A_1}.A_0.B_1.B_0 + A_1.\overline{A_0}.\overline{B_1}.\overline{B_0}$$
$$+ A_1.\overline{A_0}.\overline{B_1}.B_0 + A_1.\overline{A_0}.B_1.B_0 + A_1.A_0.\overline{B_1}.\overline{B_0} + A_1.A_0.\overline{B_1}.B_0 + A_1.A_0.B_1.\overline{B_0} \quad (9.6)$$

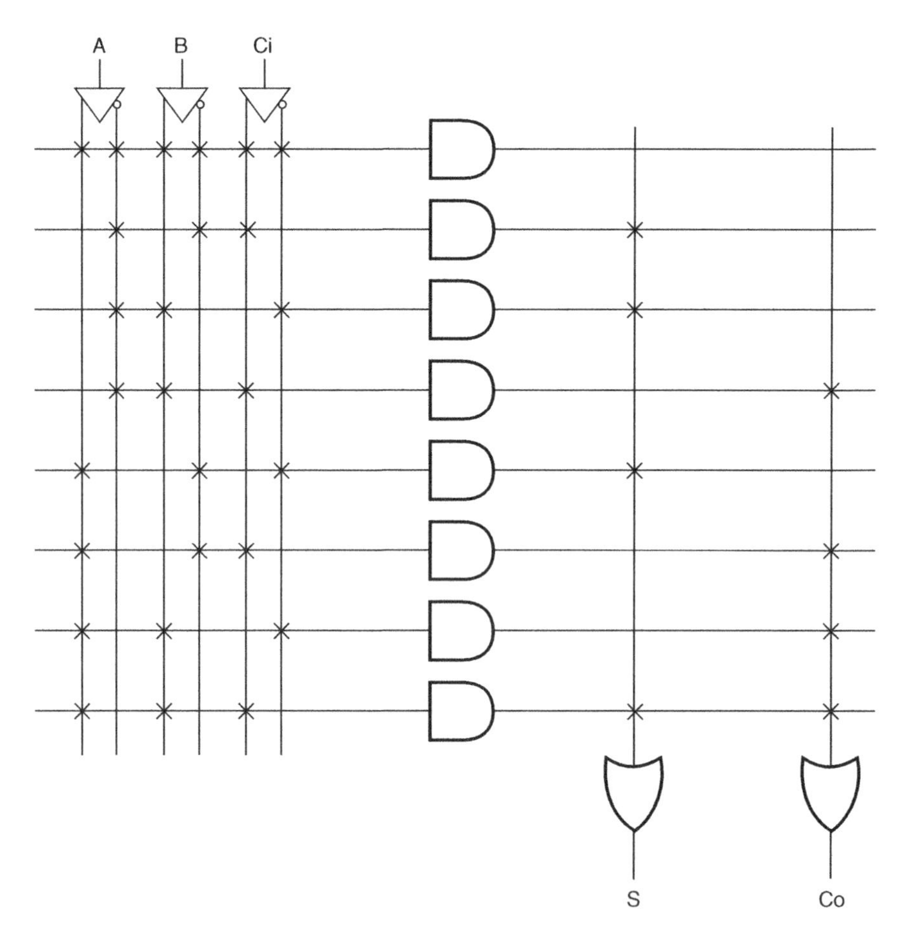

Figure 9.15 Solution to problem 9.2 using a PLA.

Output 3 (less than)

$$= \overline{A_1}.\overline{A_0}.\overline{B_1}.B_0 + \overline{A_1}.\overline{A_0}.B_1.\overline{B_0} + \overline{A_1}.\overline{A_0}.B_1.B_0 + \overline{A_1}.A_0.B_1.\overline{B_0} + \overline{A_1}.A_0.B_1.B_0 + A_1.\overline{A_0}.B_1.B_0 \quad (9.7)$$

Output 4 (greater than)

$$= \overline{A_1}.A_0.\overline{B_1}.\overline{B_0} + A_1.\overline{A_0}.\overline{B_1}.\overline{B_0} + A_1.\overline{A_0}.\overline{B_1}.B_0 + A_1.A_0.\overline{B_1}.\overline{B_0} + A_1.A_0.\overline{B_1}.B_0 + A_1.A_0.B_1.\overline{B_0} \quad (9.8)$$

Figures 9.16(a) to (d) show the Karnaugh maps for the four outputs. The minimized Boolean expressions can be written from the Karnaugh maps as follows:

$$\text{Output 1(equal to)} = \overline{A_1}.\overline{A_0}.\overline{B_1}.\overline{B_0} + \overline{A_1}.A_0.\overline{B_1}.B_0 + A_1.A_0.B_1.B_0 + A_1.\overline{A_0}.B_1.\overline{B_0} \quad (9.9)$$

$$\text{Output 2(not equal to)} = \overline{A_1}.B_1 + A_1.\overline{B_1} + \overline{A_0}.B_0 + A_0.\overline{B_0} \quad (9.10)$$

Table 9.2 Function table for example 9.3.

A_1	A_0	B_1	B_0	Output 1	Output 2	Output 3	Output 4
0	0	0	0	1	0	0	0
0	0	0	1	0	1	1	0
0	0	1	0	0	1	1	0
0	0	1	1	0	1	1	0
0	1	0	0	0	1	0	1
0	1	0	1	1	0	0	0
0	1	1	0	0	1	1	0
0	1	1	1	0	1	1	0
1	0	0	0	0	1	0	1
1	0	0	1	0	1	0	1
1	0	1	0	1	0	0	0
1	0	1	1	0	1	1	0
1	1	0	0	0	1	0	1
1	1	0	1	0	1	0	1
1	1	1	0	0	1	0	1
1	1	1	1	1	0	0	0

$$\text{Output 3(less than)} = \overline{A_1}.B_1 + \overline{A_1}.\overline{A_0}.B_0 + \overline{A_0}.B_1.B_0 \tag{9.11}$$

$$\text{Output 4(Greater than)} = A_1.\overline{B_1} + A_1.A_0.\overline{B_0} + A_0.\overline{B_1}.\overline{B_0} \tag{9.12}$$

Examination of minimized Boolean expressions (9.9) to (9.12) reveals that there are 12 different product terms to be accounted for. Therefore, a PLA device with 12 AND gates will meet the requirement. Also, since there are four outputs, we need to have four OR gates at the output. Figure 9.17 shows the programmed PLA device. Note that, in the programmed PLA device, an unprogrammed interconnection indicated by a cross ($\times$) is a 'make' connection.

9.5 Programmable Array Logic

The *programmable array logic* (PAL) device is a variant of the PLA device. As outlined in Section 9.2, it has a programmable AND gate array at the input and a fixed OR gate array at the output. The idea to have a fixed OR gate array at the output and make the device less complex originated from the fact that there were many applications where the product-term sharing capability of the PLA was not fully utilized and thus wasted. The PAL device is a trademark of Advanced Micro Devices Inc. PAL devices are however less flexible than PLA devices. The flexibility of a PAL device can be enhanced by having different output logic configurations including the availability of both OR (also called active HIGH) and NOR (also called active LOW) outputs and bidirectional pins that can act both as inputs and outputs, having clocked flip-flops at the outputs to provide what is called registered outputs. These features allow the device to be used in a wider range of applications than would be possible with a device with fixed input and output allocations. The mask-programmed version of PAL is known as the HAL (Hard Array Logic) device. A HAL device is pin-to-pin compatible with its PAL counterpart.

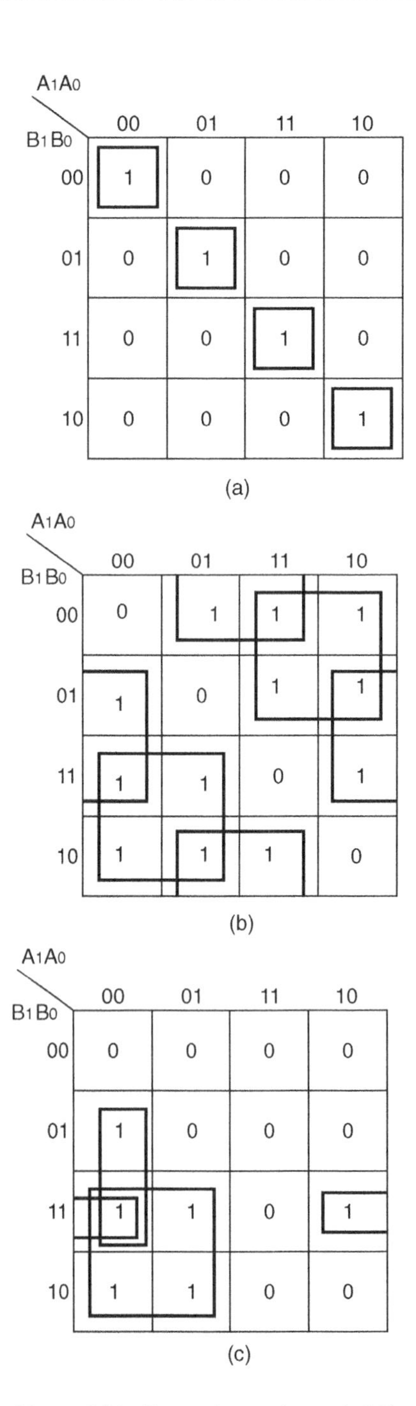

Figure 9.16 Karnaugh maps (example 9.3).

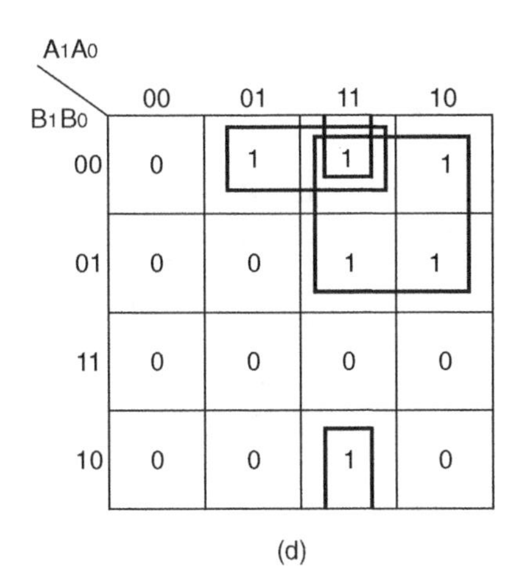

Figure 9.16 (*continued*).

9.5.1 PAL Architecture

Figure 9.18 shows the block schematic representation of the generalized architecture of a PAL device. As we can see from the arrangement shown, the device has a programmable AND gate array that is fed with various input variables and their complements. Programmable input connections allow any of the input variables or their complements to appear at the inputs of any of the AND gates in the array. Each of the AND gates generates a minterm of a user-defined combination of input variables and their complements. As an illustration, Fig. 9.19 gives an example of the generation of minterms.

Outputs from the programmable AND array feed an array of hard-wired OR gates. Here, the output of each of the AND gates does not feed the input of each of the OR gates. Each OR gate is fed from a subset of AND gates in the array. This implies that the sum-of-product Boolean functions generated by each of the OR gates at the output will have only a restricted number of minterms depending upon the number of AND gates from which it is being fed. Outputs from the PAL device, as is clear from the generalized form of representation shown in Fig. 9.18, are available both as OR outputs as well as complemented (or NOR) outputs.

Practical PAL devices offer various output logic arrangements. One of them, of course, is the availability of both OR and NOR outputs as mentioned in the previous paragraph. Another feature available with many PAL devices is that of registered outputs. In the case of registered outputs, the OR gate output drives the D-input of a *D*-type flip-flop, which is loaded with the data on either the LOW-to-HIGH or the HIGH-to-LOW edge of a clock signal. Yet another feature is the availability of bidirectional pins, which can be used both as outputs and inputs. This facility allows the user to feed a product term back to the programmable AND array. It helps particularly in those multi-output function logic circuits that share some common minterms. Some of the common output logic arrangements available with PAL devices are shown in Fig. 9.20.

Some PAL devices offer an EX-OR gate following the OR gate at each output. One of the inputs to the EX-OR gate is programmable, which allows the user to configure it as either an inverter or a noninverting buffer or as a two-input EX-OR gate. This feature is particularly useful while implementing parity and arithmetic operations.

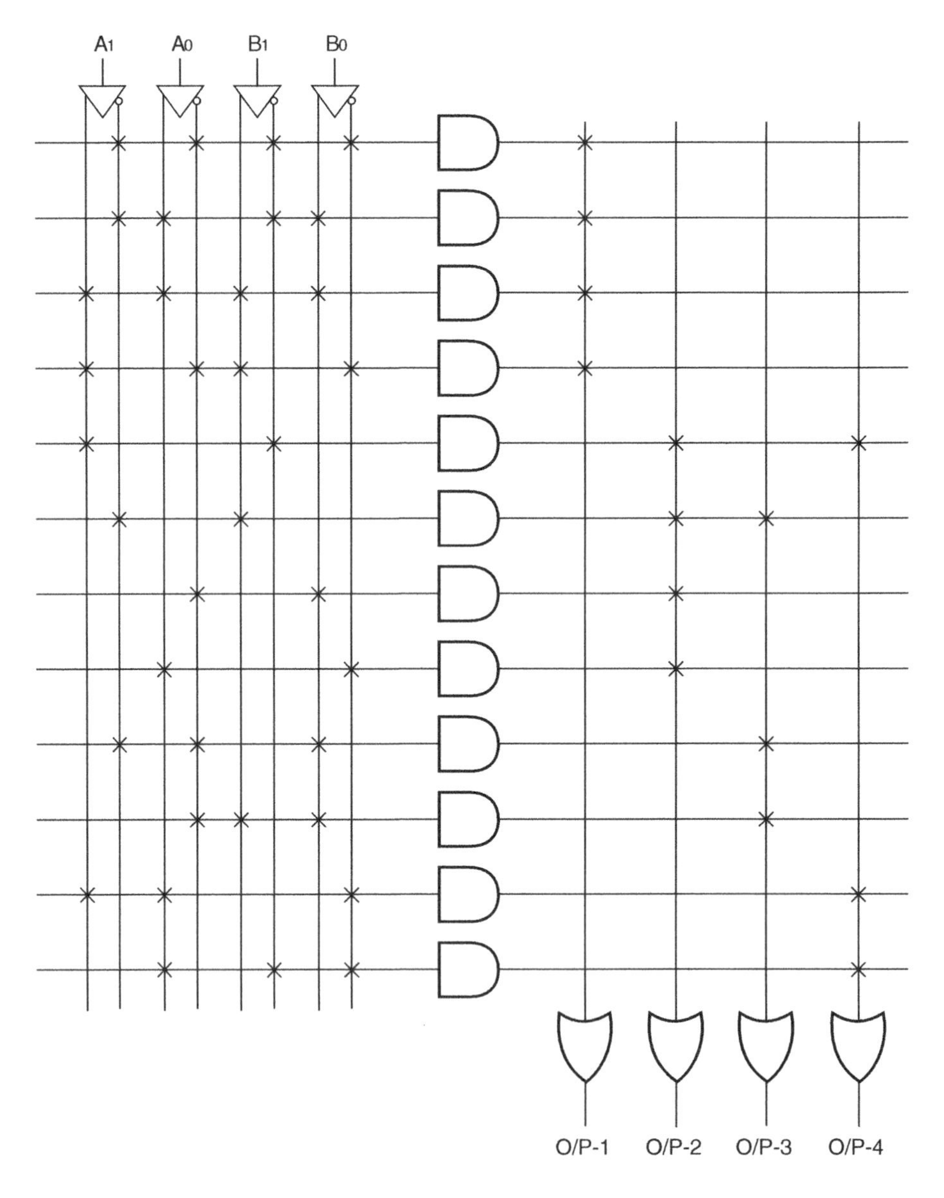

Figure 9.17 Programmed PLA device (example 9.3).

9.5.2 PAL Numbering System

The standard PAL numbering system uses an alphanumeric designation comprising a two-digit number indicating the number of inputs followed by a letter that tells about the architecture/type of logic output. Table 9.3 gives an interpretation of different letter designations in use. Another number following the

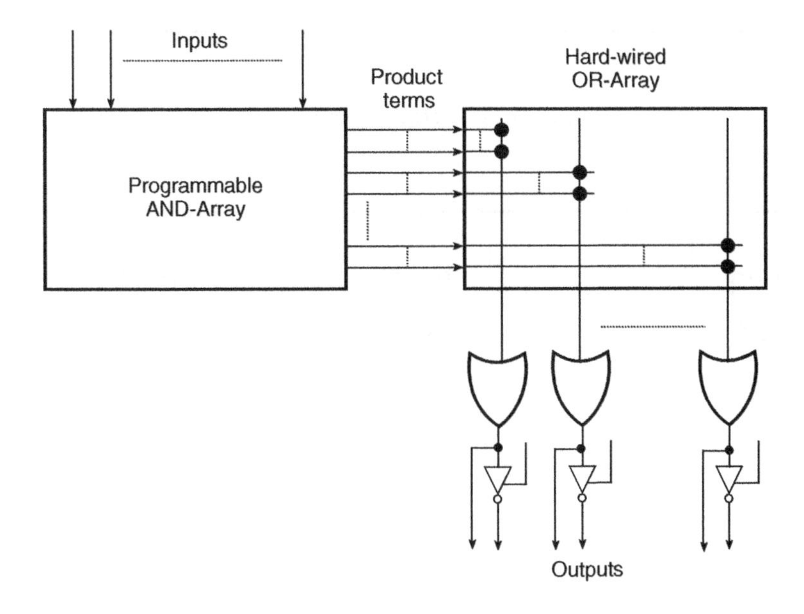

Figure 9.18 Generalized PAL device.

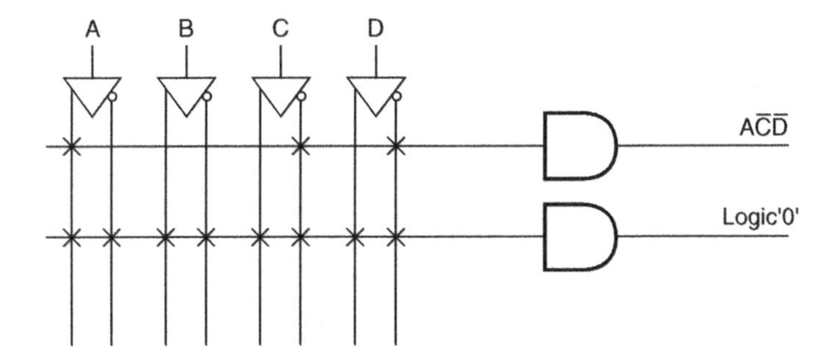

Figure 9.19 Programmability of inputs in a PAL device.

letter indicates the number of outputs. In the case of PAL devices offering a combination of different types of logic output, the rightmost number indicates the number of the output type implied by the letter used in the designation. For example, a PAL device designated PAL-16L8 will have 16 inputs and eight active LOW outputs. Another PAL device designated PAL-16R4 has 16 inputs and four registered outputs. Also, the number of inputs as given by the number designation includes dedicated inputs, user-programmable inputs accessible from combinational I/O pins and any feedback inputs

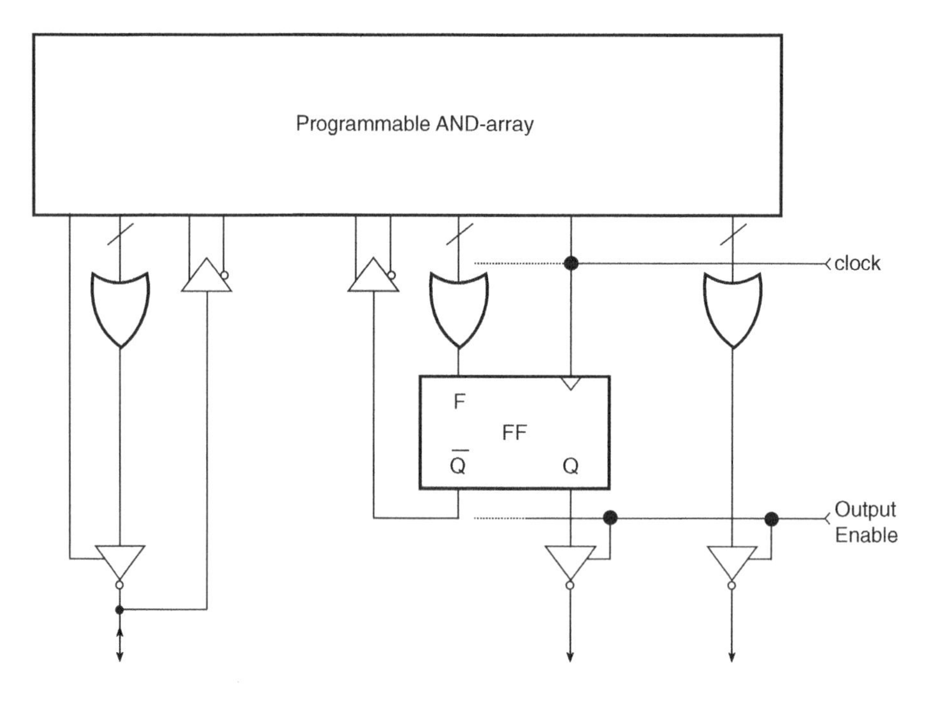

Figure 9.20 Output logic arrangements in a PAL device.

Table 9.3 PAL numbering system.

Architecture – Combinational devices		Architecture – Registered devices	
Code Letter	Description	Code letter	Description
H	Active HIGH outputs	R	Registered outputs
L	Active LOW outputs	X	EXCLUSIVE-OR gates
P	Programmable output polarity	RP	Registered polarity Programmable
C	Complementary outputs	RS	Registered-term steering
XP	EXCLUSIVE-OR gate-Programmable	V	Versatile varied Product terms
S	Product term steering	RX	Registered EX-OR
		MA	Macrocell

from combinational and registered outputs. For example, PAL-16L8 has 10 dedicated inputs and six inputs accessible from I/O pins.

In addition to the numbering system described above, an alphanumeric designation on the extreme left may be used to indicate the technology used. 'C' stands for CMOS, '10H' for 10KH ECL and '100' for 100K ECL. TTL is represented by a blank. A letter on the extreme right may be used to indicate the power level, with 'L' and 'Q' respectively indicating low and quarter power levels and a blank representing full power.

Example 9.4

Table 9.4 shows the function table of a converter. Starting with the Boolean expressions for the four outputs (P, Q, R, S), minimize them using Karnaugh maps and then hardware-implement this converter with a suitable PLD with PAL architecture.

Solution

From the given function table, we can write the Boolean expressions for the four outputs as follows:

$$P=\overline{A}.B.\overline{C}.D+\overline{A}.B.C.\overline{D}+\overline{A}.B.C.D+A.\overline{B}.\overline{C}.\overline{D}+A.\overline{B}.\overline{C}.D \tag{9.13}$$

$$Q=\overline{A}.B.\overline{C}.\overline{D}+\overline{A}.B.\overline{C}.D \tag{9.14}$$

$$R=\overline{A}.\overline{B}.C.\overline{D}+\overline{A}.\overline{B}.C.D+\overline{A}.B.\overline{C}.\overline{D}+\overline{A}.B.\overline{C}.D+\overline{A}.B.C.\overline{D}+\overline{A}.B.C.D \tag{9.15}$$

$$S=\overline{A}.\overline{B}.\overline{C}.D+\overline{A}.\overline{B}.C.\overline{D}+\overline{A}.B.C.D+A.\overline{B}.\overline{C}.\overline{D} \tag{9.16}$$

Karnaugh maps for the four outputs P,Q,R and S are respectively shown in Figs 9.21(a) to (d). The minimized Boolean expressions are given by the equations

Table 9.4 Function table in example 9.4.

A	B	C	D	P	Q	R	S
0	0	0	0	0	0	0	0
0	0	0	1	0	0	0	1
0	0	1	0	0	0	1	1
0	0	1	1	0	0	1	0
0	1	0	0	0	1	1	0
0	1	0	1	1	1	1	0
0	1	1	0	1	0	1	0
0	1	1	1	1	0	1	1
1	0	0	0	1	0	0	1
1	0	0	1	1	0	0	0
1	0	1	0	X	X	X	X
1	0	1	1	X	X	X	X
1	1	0	0	X	X	X	X
1	1	0	1	X	X	X	X
1	1	1	0	X	X	X	X
1	1	1	1	X	X	X	X

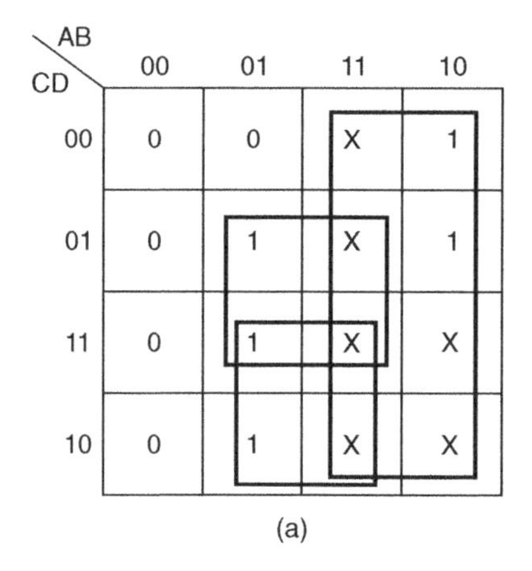

(a)

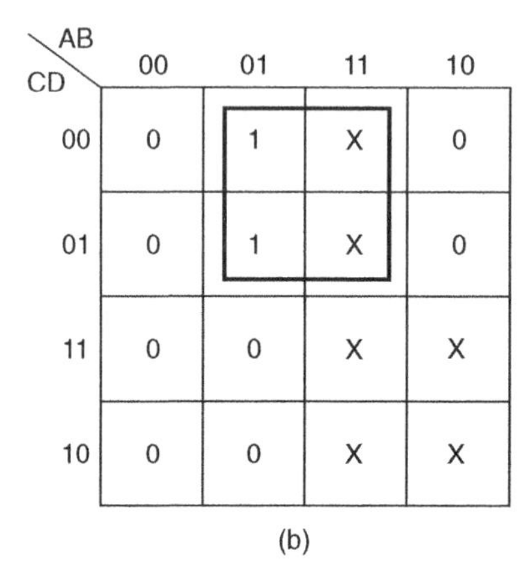

(b)

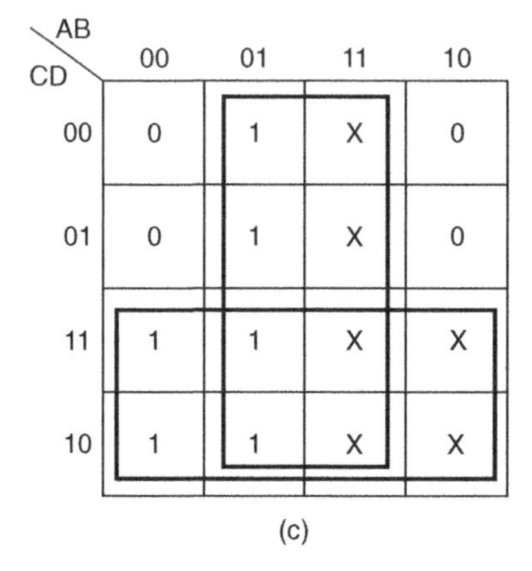

(c)

Figure 9.21 Karnaugh maps (example 9.4).

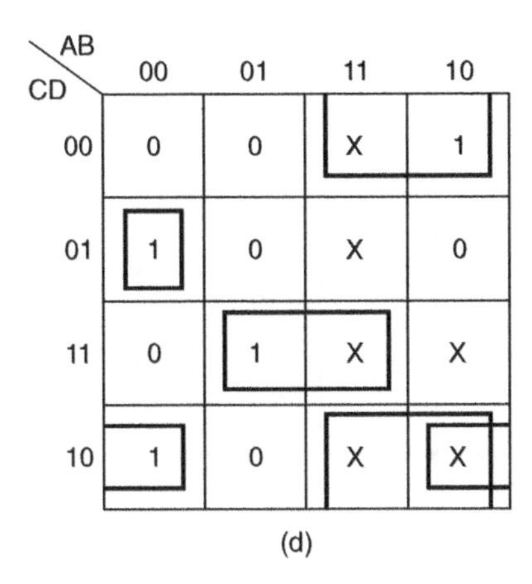

Figure 9.21 (*continued*).

$$P = B.D + B.C + A \tag{9.17}$$

$$Q = B.\overline{C} \tag{9.18}$$

$$R = B + C \tag{9.19}$$

$$S = \overline{A}.\overline{B}.\overline{C}.D + B.C.D + A.\overline{D} + \overline{B}.C.\overline{D} \tag{9.20}$$

The next step is to choose a suitable PAL device. Since there are four output functions, we will need a PAL device with at least four OR gates at the output. Since each of the OR gates is to be hard wired to only a subset of programmable AND arrays, and also because one of the output functions has four product terms, we will need an AND array of 16 AND gates. Since there are four input variables, we need each AND gate in the array to have eight inputs to cater for four variables and their complements. To sum up, we choose a PAL device that has eight inputs, 16 AND gates in the programmable AND array and four OR gates at the output. Each OR gate has four inputs.

Figure 9.22 shows the architecture of the programmed PAL device. We can see that the *P* output has only three product terms. The fourth input to the relevant OR gate needs to be applied a logic '0' input. This is achieved by feeding the inputs of the corresponding AND gate with all four variables and their complements. Logic 0s, wherever required, are implemented in the same manner. Note that, in the programmed PAL device of Fig. 9.22, an unprogrammed interconnection indicated by a cross (×) is a 'make' connection.

9.6 Generic Array Logic

Generic array logic (GAL) is characterized by a reprogrammable AND array, a fixed OR array and a reprogrammable output logic. It is similar to a PAL device, with the difference that the AND

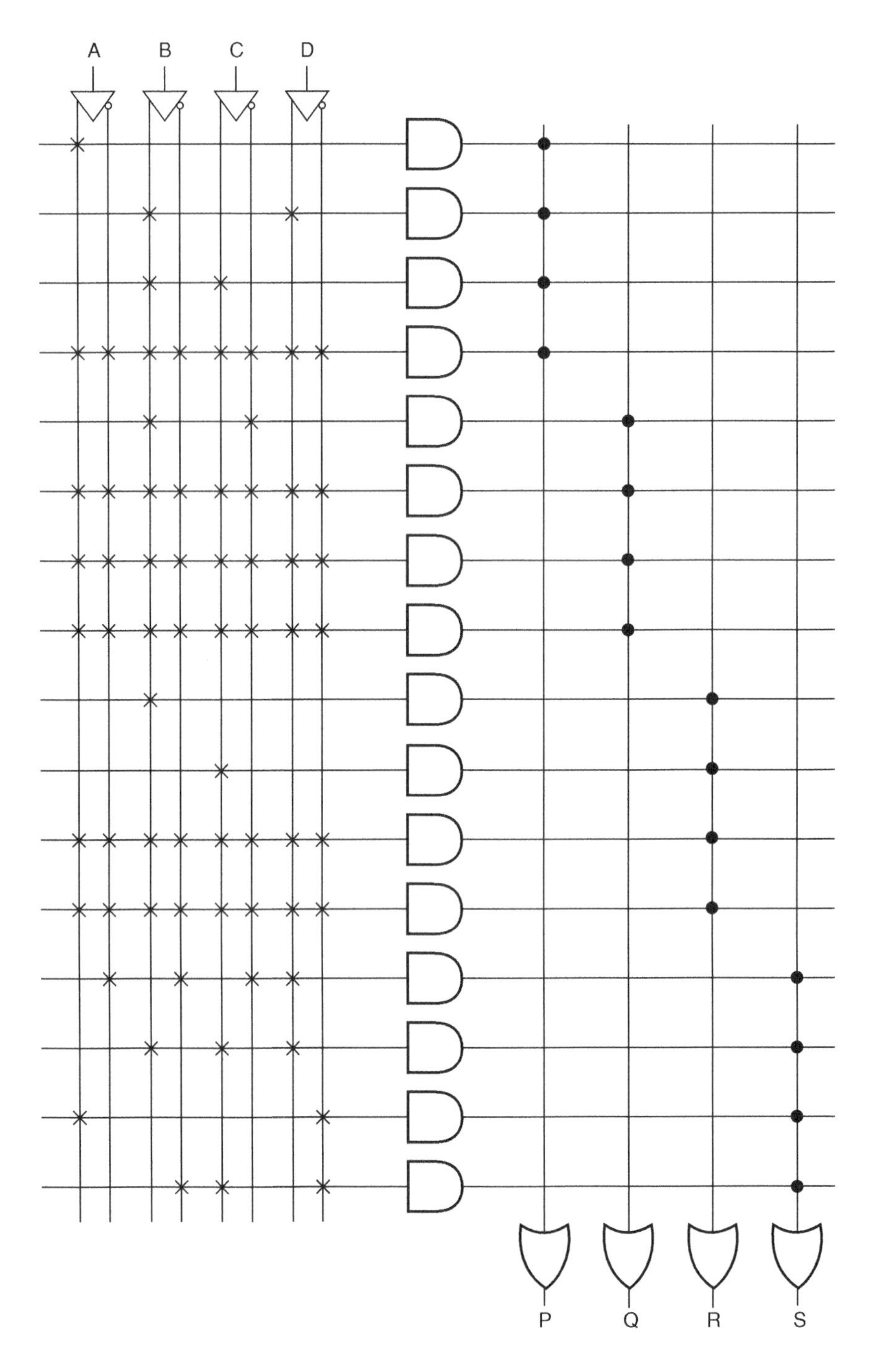

Figure 9.22 Programmed PAL (example 9.4).

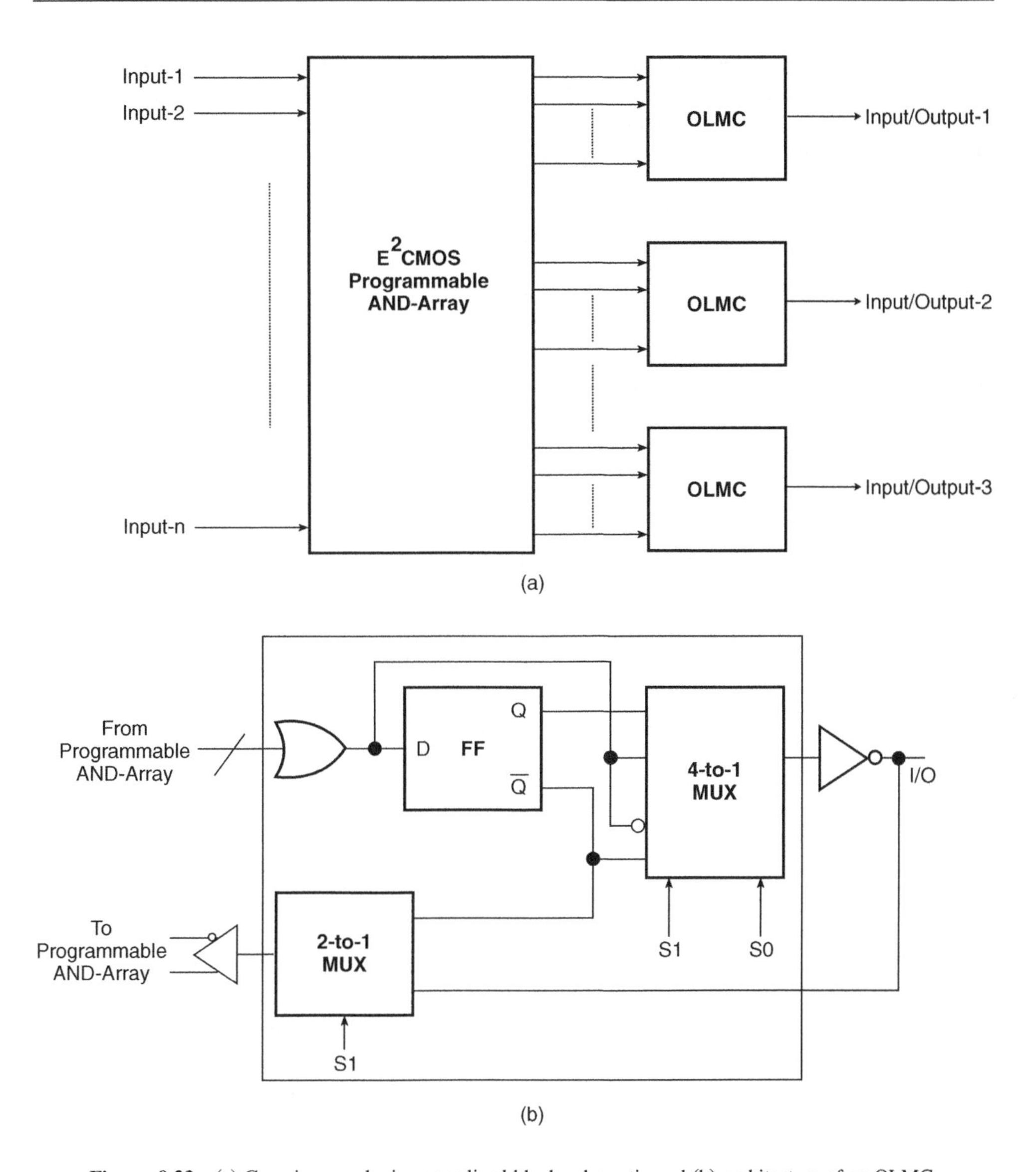

Figure 9.23 (a) Generic array logic generalized block schematic and (b) architecture of an OLMC.

array is not just programmable as is the case in a PAL device but is reprogrammable. That is, it can be reprogrammed any number of times. This has been made possible by the use of electrically erasable PROM cells for storing the programming pattern. The other difference is in the use of reprogrammable output logic, which provides more flexibility to the designer. GAL devices employ output logic macrocells (OLMCs) at the output, which allows the designer to configure the outputs either as combinational outputs or registered outputs.

Figures 9.23(a) and (b) respectively show the block schematic representation of a GAL device and the architecture of a typical OLMC used with GAL devices. The OLMC of the type shown in Fig. 9.23(b) can be configured to produce four different outputs depending upon the selection inputs. These include the following:

1. $S_1S_0 = 00$: registered mode with active LOW output.
2. $S_1S_0 = 01$: registered mode with active HIGH output.
3. $S_1S_0 = 10$: combinational mode with active LOW output.
4. $S_1S_0 = 11$: combinational mode with active HIGH output.

We can see that two of the four inputs to the 4-to-1 multiplexer are combinational outputs, and the other two are the registered outputs. Also, of the two combinational outputs, one is an active HIGH output while the other is an active LOW output. The same is the case with registered outputs. Of the four inputs to the multiplexer, the one appearing at the output depends upon selection inputs. The 2-to-1 multiplexer ensures that the final output is also available as feedback to the programmable AND array.

9.7 Complex Programmable Logic Devices

If we examine the internal architecture of simple programmable logic devices (SPLDs) such as PLAs and PALs, we find that it is not practical to increase their complexity beyond a certain level. This is because the size of the programmable plane (such as the programmable AND plane in a PLA or PAL device) increases too rapidly with increase in the number of inputs to make it a practically viable device. One way to increase the logic capacity of simple programmable logic devices is to integrate multiple SPLDs on a single chip with a programmable interconnect between them. These devices have the same basic internal structure that we see in the case of SPLDs and are grouped together in the category of complex programmable logic devices (CPLDs). Typically, CPLDs may offer a logic capacity equivalent to that of about 50 SPLDs. Programmable logic devices with much higher logic capacities would require a different approach rather than simple extension of the concept of SPLDs.

9.7.1 Internal Architecture

As outlined in the previous paragraph, a CPLD is nothing but the integration of multiple PLDs, a programmable interconnect matrix and an I/O control block on a single chip. Each of the identical PLDs is referred to as a *logic block* or *function block*. Figure 9.24 shows the architecture of a typical CPLD. As is evident from the block schematic arrangement, the programmable interconnect matrix is capable of connecting the input or output of any of the logic blocks to any other logic block. Also, input and output pins connect directly to both the interconnect matrix as well as logic blocks.

Logic blocks may further comprise smaller logic units called macrocells, where each of the macrocells is a subset of a PLD-like logic block. Figure 9.25 shows the structure of a logic block along with its interconnections with the programmable interconnect matrix and I/O block. The horizontal grey-coloured bars inside the logic block constitute an array of macrocells. Typically, each macrocell comprises a set of product terms generated by a subset of the programmable AND array and feeding a configurable output logic. The output logic typically comprises an OR gate, an EX-OR gate and a flip-flop. The flip-flop in the case of most contemporary CPLDs is configurable as a D-type, J-K, T, or R-S flip-flop or can even be transparent. Also, the OR gate can be fed with any or all of the product terms generated within the macrocell. Most contemporary CPLDs also offer an architecture where the

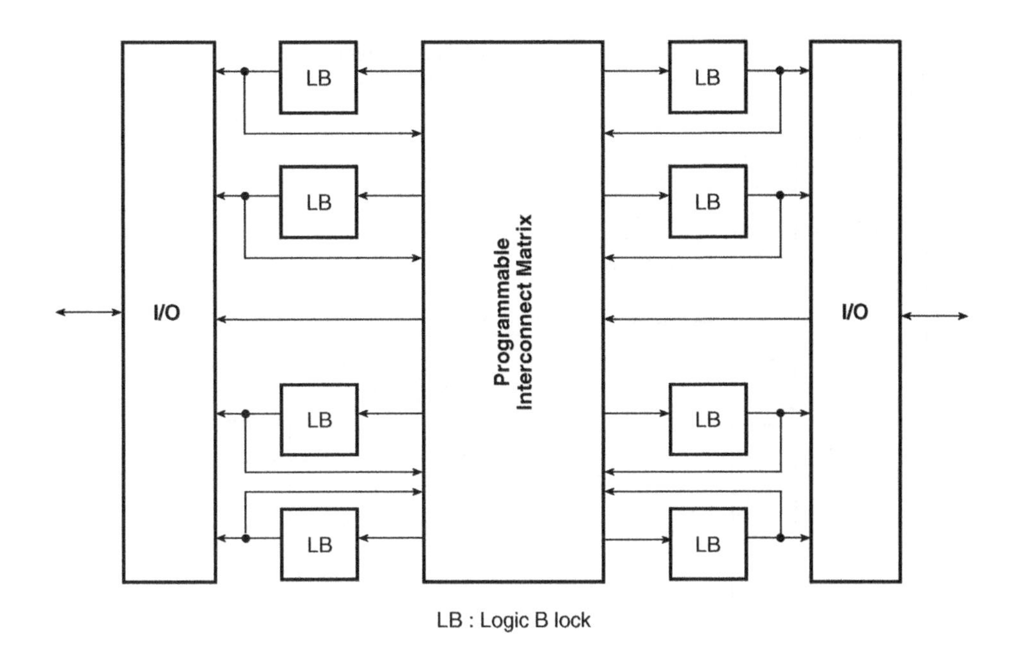

Figure 9.24 CPLD architecture.

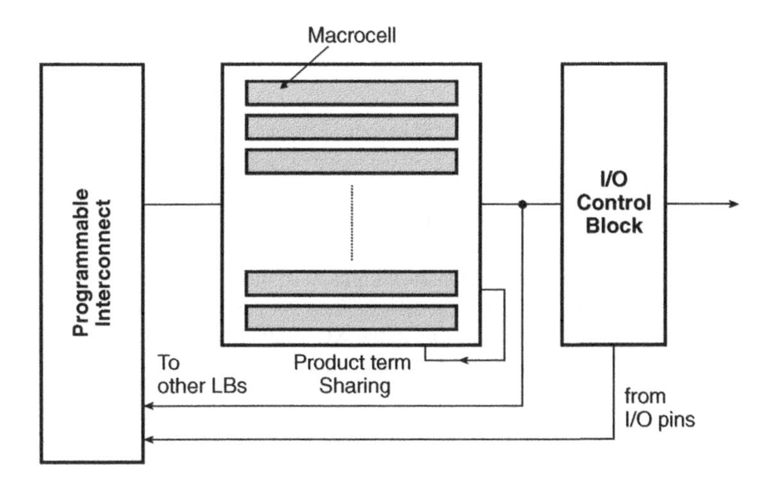

Figure 9.25 Logic block structure.

OR gate can also be fed with some additional product terms generated within other macrocells of the same logic block. For example, a logic block in the case of the MAX-7000 series of CPLDs from Altera offers this product-term flexibility, where the OR gate of each macrocell can have up to 15

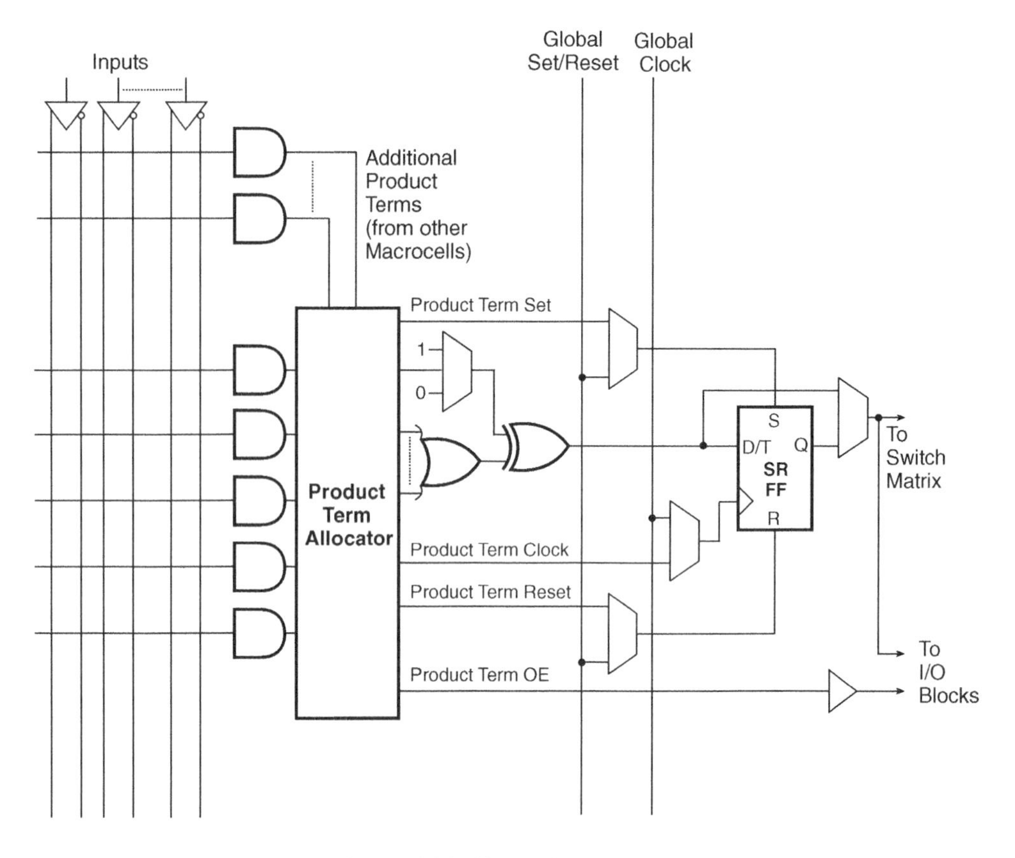

Figure 9.26 Macrocell architecture.

additional product terms from other macrocells in the same logic block, apart from a maximum of five product terms from within the same macrocell.

Figure 9.26 shows the logic diagram of a macrocell typical of macrocells in the logic blocks of most contemporary CPLDs. The diagram is self-explanatory. There may be minor variations in devices from different manufacturers. For example, macrocells in the XC-7000 series CPLDs from Xilinx have two OR gates fed from a two-bit arithmetic logic unit (ALU) and its output feeds a configurable flip-flop.

9.7.2 Applications

Owing to their less flexible internal architecture leading to predictable timing performance, high speed and a range of logic capacities, CPLDs find extensive use in a wide assortment of applications. These include the implementation of random glue logic in prototyping small gate arrays, implementing critical control designs such as graphics controllers, cache control, UARTs, LAN controllers and many more.

CPLDs are fast replacing SPLDs in complex designs. Complex designs using a large number of SPLDs can be replaced with a CPLD-based design with a much smaller number of devices. This is particularly attractive in portable applications such as mobile phones, digital assistants and so on.

CPLD architecture particularly suits those designs that exploit wide AND/OR gates and do not require a large number of flip-flops.

The reprogramming feature of CPLDs makes the incorporation of design changes very easy. With the availability of CPLDs having an in-circuit programming feature, it is even possible to reconfigure the hardware without power down. Changing protocol in a communication circuit could be one such example. One of the most significant advantages of CPLD architecture comes from its simple SPLD-like structure, which allows the design to partition naturally into SPLD-like blocks. This leads to a much more predictable timing or speed performance than would be possible if the design were split into many pieces and mapped into different areas of the chip.

9.8 Field-Programmable Gate Arrays

As outlined earlier, it is not practical to increase the logic capacity with a CPLD architecture beyond a certain point. The highest-capacity general-purpose logic chips available today are the traditional gate arrays, which comprise an array of prefabricated transistors. The chip can be customized during fabrication as per the user's logic design by specifying the metal interconnect pattern. These chips are also referred to as *mask-programmable gate arrays* (MPGAs). These, however, are not field-programmable devices. A field-programmable gate array (FPGA) chip is the user-programmable equivalent of an MPGA chip.

9.8.1 Internal Architecture

An FPGA consists of an array of uncommitted configurable logic blocks, programmable interconnects and I/O blocks. The basic architecture of an FPGA was shown earlier in Fig. 9.7 when presenting an overview of programmable logic devices. As outlined earlier, the basic difference between a CPLD and an FPGA lies in their internal architecture. CPLD architecture is dominated by a relatively smaller number of programmable sum-of-products logic arrays feeding a small number of clocked flip-flops, which makes the architecture less flexible but with more predictable timing characteristics. On the other hand, FPGA architecture is dominated by programmable interconnects, and the configurable logic blocks are relatively simpler. Logic blocks within an FPGA can be as small as the macrocells in a PLD, called fine-grained architecture, or larger and more complex, called coarse-grained architecture. However, they are never as large as the entire PLD like the logic blocks of a CPLD. This feature makes these devices far more flexible in terms of the range of designs that can be implemented with these devices.

Contemporary FPGAs have an on-chip presence of higher-level embedded functions and embedded memories. Some of them even come with an on-chip microprocessor and related peripherals to constitute what is called a complete 'system on a programmable chip'. Virtex-II Pro and Virtex-4 FPGA devices from Xilinx are examples. These devices have one or more PowerPC processors embedded within the FPGA logic fabric.

Figure 9.27 shows a typical logic block of an FPGA. It consists of a four-input look-up table (LUT) whose output feeds a clocked flip-flop. The output can either be a registered output or an unregistered LUT output. Selection of the output takes place in the multiplexer. An LUT is nothing but a small one-bit wide memory array with its address lines representing the inputs to the logic block and a one-bit output acting as the LUT output. An LUT with n inputs can realize any logic function of n inputs by programming the truth table of the desired logic function directly into the memory.

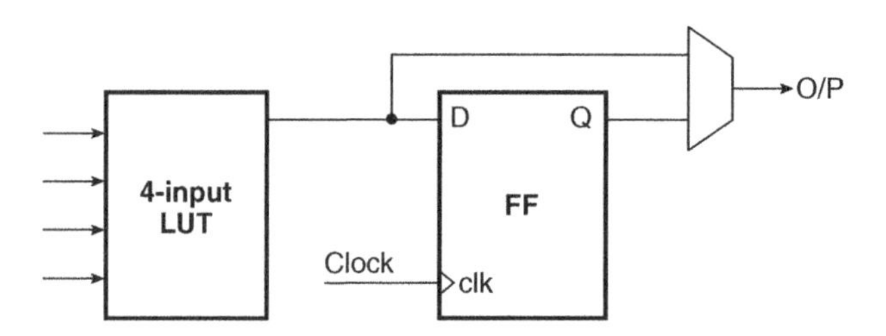

Figure 9.27 Logic block of a typical FPGA.

Logic blocks can have more than one LUT and flip-flops also to give them the capability of realizing more complex logic functions. Figure 9.28 shows the architecture of one such logic block. The architecture shown in Fig. 9.28 is that of a logic block of the XC4000 series of FPGAs from Xilinx. This logic block has two four-input LUTs fed with logic block inputs and a third LUT that can be used in conjunction with the two LUTs to offer a wide range of functions. These include two separate logic functions of four inputs each, a single logic function of up to nine inputs and many more. The logic block contains two flip-flops.

Figure 9.29 shows another similar LUT-based architecture that uses multiple LUTs and flip-flops. The architecture shown in Fig. 9.29 is that of a logic block called a programmable function unit

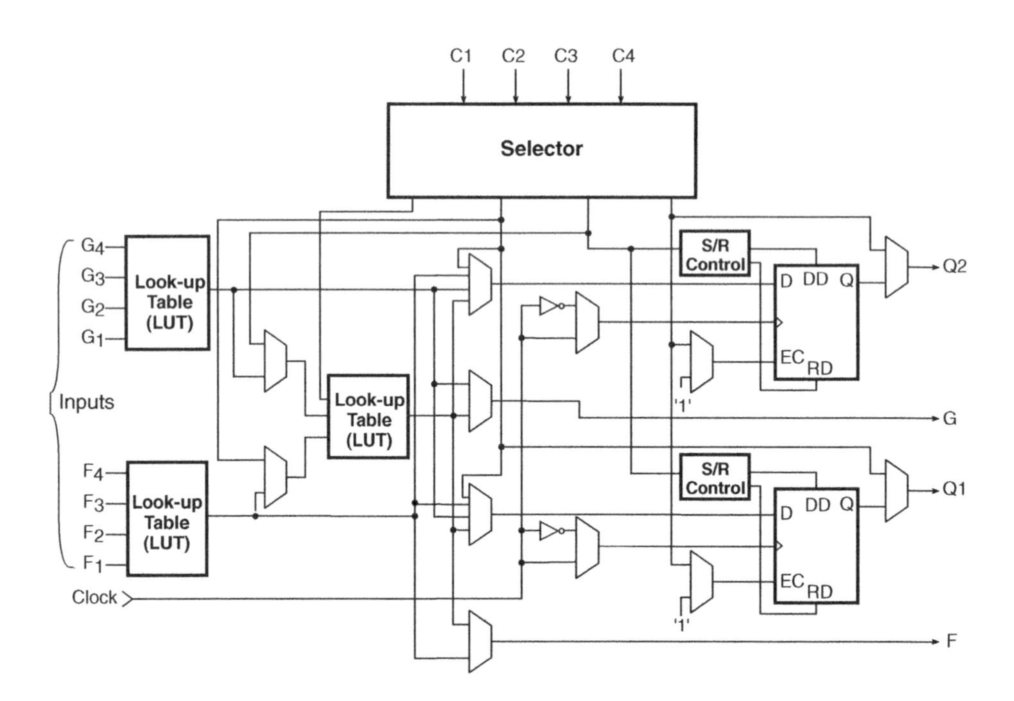

Figure 9.28 Logic block architecture of the XC4000 FPGA from Xilinx.

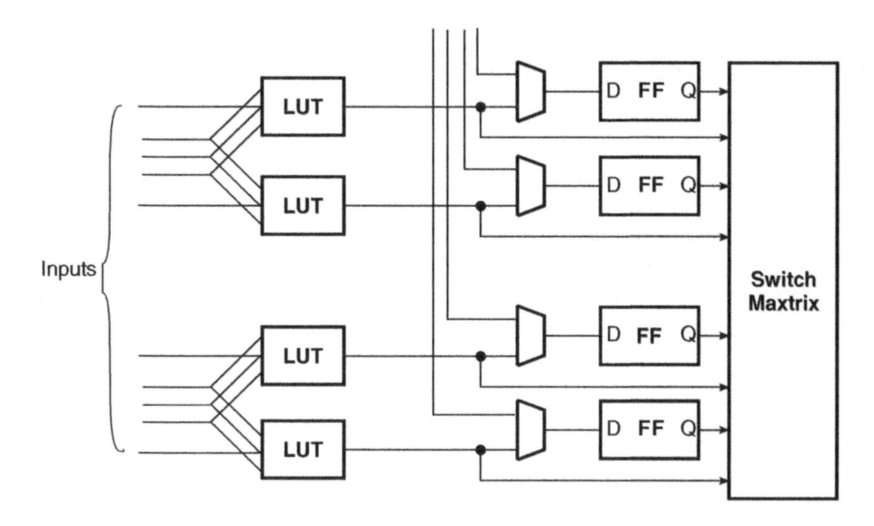

Figure 9.29 Logic block architecture of an AT&T FPGA.

(PFU) by the manufacturer of AT&T FPGA devices. This logic block can be configured either as four four-input LUTs or two five-input LUTs or one six-input LUT.

9.8.2 Applications

In the early days of their arrival on the scene, FPGAs began as competitors to CPLDs for applications such as glue logic for PCBs. With increase in their logic capacity and capability, the availability of a large embedded memory, higher-level embedded functions such as adders and multipliers, the emergence of hybrid technologies combining the logic blocks and interconnects of traditional FPGAs with embedded microprocessors and the facility of full or partial in-system reconfiguration have immensely widened the scope of applications of FPGAs. FPGAs today offer a complete system solution on a single chip, although very complex systems might be implemented with more than one FPGA device.

Some of the major application areas of FPGA devices include digital signal processing, data storage and processing, software-defined radio, ASIC prototyping, speech recognition, computer vision, cryptography, medical imaging, defence systems, bioinformatics, computer hardware emulation and reconfigurable computing. Reconfigurable computing, also called customized computing, involves the use of programmable parts to execute software rather than compiling the software to be run on a regular CPU. This has been made possible by in-system reconfiguration, which allows the internal design to be altered on-the-fly.

9.9 Programmable Interconnect Technologies

The programmable features of every PLD, be it simple programmable logic devices (SPLDs) such as PLAs, PALs and GALs or complex programmable logic devices (CPLDs) or even field-programmable gate arrays (FPGAs), come from their programmable interconnect structure. Interconnect technologies that have evolved over the years for programming PLDs include fuses, EPROM or EEPROM floating-gate transistors, static RAM and antifuses.

Each one of these is briefly described in the following paragraphs.

9.9.1 Fuse

A fuse is an electrical device that has a low initial resistance and is designed permanently to break an electrically conducting path when current through it exceeds a specified limit. It uses bipolar technology and is nonvolatile and one-time programmable. It was the first user-programmable switch developed for use in PLAs. They were earlier used in smaller PLDs and are now being rapidly replaced by newer technologies.

9.9.2 Floating-Gate Transistor Switch

This interconnect technology is based on the principle of placing a floating-gate transistor between two wires in such a way as to facilitate a WIRE-AND function. This concept is used in EPROM and EEPROM devices, and that is why the floating-gate transistor is sometimes referred to as an EPROM or EEPROM transistor. Figure 9.30 shows the use of floating-gate transistor interconnects in the AND plane of a CPLD or SPLD. All those inputs that are required to be part of a particular product term are activated to drive the product wire to a logic '0' level through the EPROM transistor. For inputs that are not part of the product term, relevant transistors are switched off.

This technology is commonly used in SPLDs and CPLDs. A floating-gate transistor based switch matrix, however, requires a large number of interconnects and therefore transistors. For example, a CPLD with 128 macrocells with four inputs and one output each would require as many as 65 536 interconnects for 100 % routability. A large number of interconnects also adds to the propagation delay.

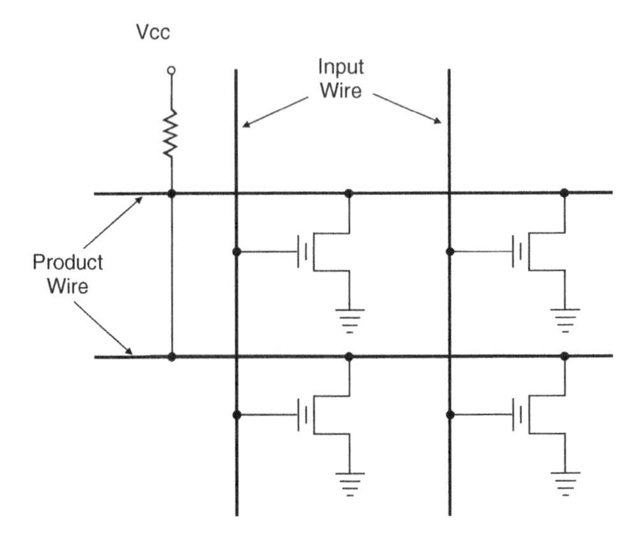

Figure 9.30 Floating-gate transistor interconnect.

The use of multiplexers can reduce this number significantly and can also address the problem of increased propagation delay. An MUX-based interconnect matrix is being used in CPLDs. CPLD type XPLA3 from Xilinx is an example.

9.9.3 Static RAM-Controlled Programmable Switches

Static RAM (SRAM) is basically a semiconductor memory, and the word 'static' implies that it is a nonvolatile memory. That is, the memory retains its contents as long as power is on. A SRAM with m address lines and n data lines is referred to as a $2^m \times n$ memory and is capable of storing 2^m n-bit words. Figure 9.31 shows the basic SRAM cell comprising six MOSFET switches, with four of them connected as cross-coupled inverters. A basic SRAM cell can store one bit of information. The reading operation is carried out by precharging both the bit lines (BL and $\overline{BL}$) to logic '1' and then asserting the WL line. The writing operation is done by giving the desired logic status to the BL line and its complement to the $\overline{BL}$ line and then asserting the WL line.

Figure 9.32 shows the use of SRAM-controlled switches. SRAMs are used to control not only the gate nodes but also the select inputs of multiplexers that drive the logic block inputs. The figure illustrates the routing scheme for feeding the output of one logic block to the input of another via SRAM-controlled pass transistor switches and a SRAM-controlled multiplexer. It may be mentioned here that a SRAM-controlled programmable interconnect matrix does not necessarily use both pass transistors and multiplexers. Whether it uses pass transistors or multiplexers or both is product specific.

9.9.4 Antifuse

An antifuse is an electrical device with a high initial resistance and is designed permanently to create an electrically conducting path typically when voltage across it exceeds a certain level. Antifuses

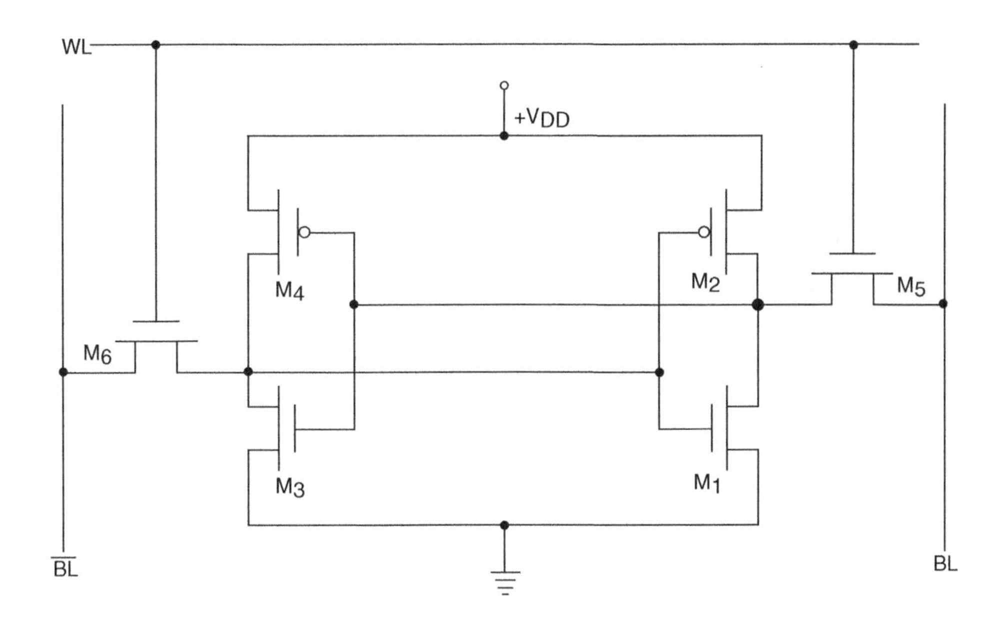

Figure 9.31 SRAM cell.

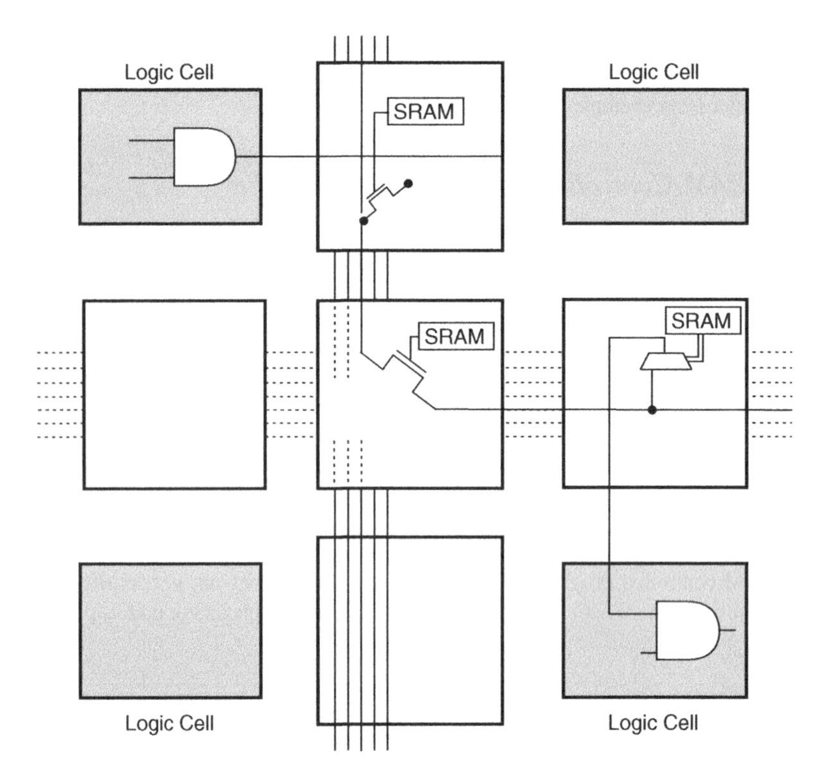

Figure 9.32 SRAM-controlled interconnect.

use CMOS technology, which is one of the main reasons for their wide use in PLDs, FPGAs in particular. A typical antifuse consists of an insulating layer sandwiched between two conducting layers. In the unprogrammed state, the insulating layer isolates the top and bottom conducting layers. When programmed, the insulating layer is transformed into a low-resistance link. Typically, metal is used for conductors and amorphous silicon for the insulator. The application of high voltage across amorphous silicon permanently transforms it into a polycrystalline silicon–metal alloy having a low resistance. There are other antifuse structures too, such as that used in the Actel antifuse. This antifuse, known as PLICE, uses polysilicon and n+ diffusion as conductors and ONO as insulator. Figure 9.33(a) shows the construction. This type of antifuse is usually triggered by a small current of the order of a few milliamperes. The high current density produced in the thin insulating layer produces heat, thus melting the insulating layer and creating an irreversible resistive silicon link.

Antifuses are widely used as programmable interconnects in PLDs [Fig. 9.33(b)]. Antifuse PLDs are one-time programmable, in contrast to SRAM-controlled interconnect-based PLDs, which are reprogrammable. It may be mentioned here that the reprogrammable feature helps the designers fix logic bugs or add new functions. Antifuse PLDs have advantages of nonvolatility and usually higher speeds. Antifuses may also be used in PROMs. In that case, each bit contains both a fuse and an antifuse. The device is programmed by triggering one of the two.

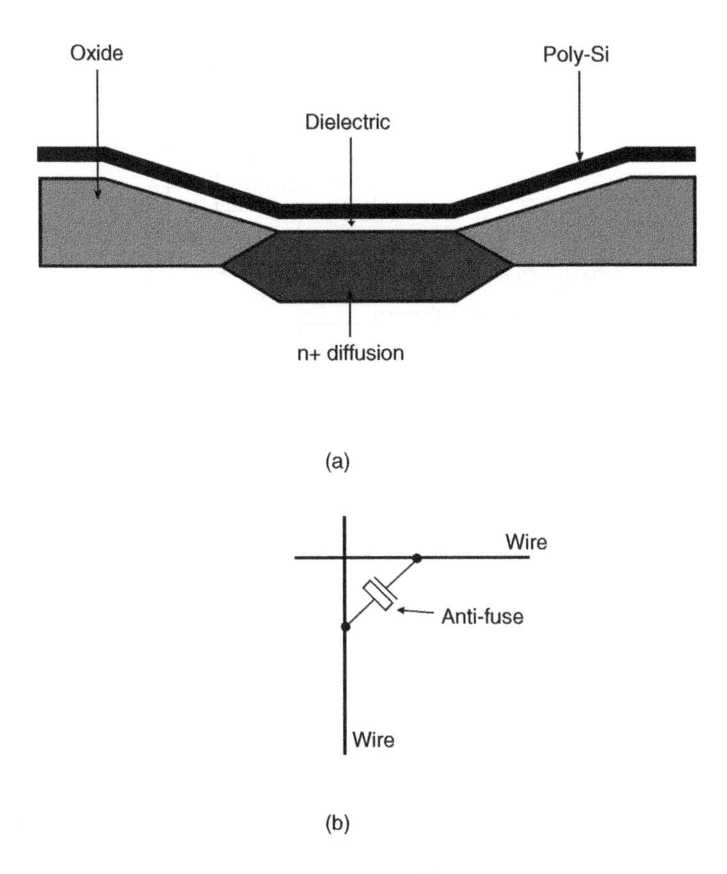

Figure 9.33 (a) Actel's antifuse and (b) the antifuse as a programmable interconnect.

9.10 Design and Development of Programmable Logic Hardware

In this section, we will briefly discuss the various steps involved in the design and development of programmable logic hardware. Figure 9.34 shows a block diagram representation of the sequence of steps involved, in the order in which they are executed.

The process begins with a description of behavioural aspects and the architecture of the intended hardware. This is done by writing a source code in a high-level *hardware description language* (HDL) such as VHDL or Verilog. This step is known as *design entry*. Although schematic capture is also an option for design entry, it has been replaced with language-based tools owing to the designs becoming more and more complex, and also owing to advances in language-based tools.

The most important difference between a hardware and software design is as follows. While software developers tend to think sequentially, hardware designers must think and program in parallel. All input signals are processed in parallel as they travel through a series of macrocells and associated interconnects towards their destination. As a result, statements of HDL create structures, which are executed at the same time. It may be mentioned here that the transfer of information from macrocell to macrocell is synchronized to another signal such as a clock.

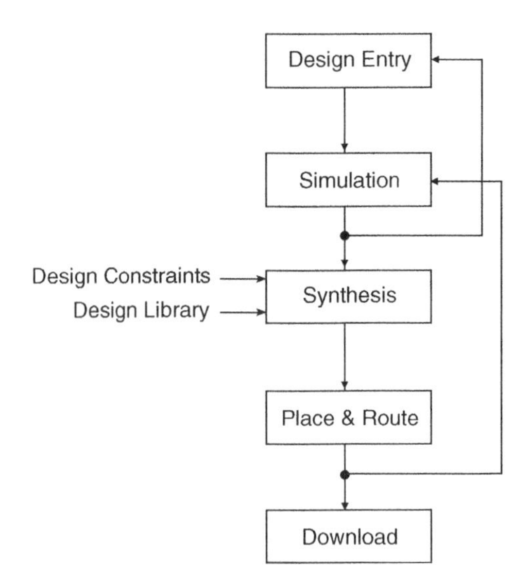

Figure 9.34 Programmable logic design and development process.

The design entry step is either followed by or interspersed with periodic functional *simulation*. The simulator executes the design for a given set of inputs and confirms that the logic is functionally correct.

Hardware compilation comes next. It involves two steps. The first step is *synthesis*, and the result of that is a hardware representation called a netlist. The netlist is device independent and its contents do not depend on the parameters of the PLD to be programmed. It is usually stored in a standard format called the electronic design interchange format (EDIF). The second step, called *place and route*, involves mapping of the logical structure described in the netlist onto actual logic blocks, interconnects and inputs/outputs. The place and route process produces a bit stream, which is nothing but the binary data that must be loaded into CPLD/FPGA to make the chip execute the intended hardware design. It may be mentioned here that each device family has its own proprietary bit stream format.

9.11 Programming Languages

During the PLD development cycle, from design entry to the generation of a bit stream that can be loaded onto the chip using some kind of electronic programming system, two types of software program are needed to perform two different functions.

The first is a *hardware description language* (HDL), which is needed at the design entry stage. HDL is a software programming language that is used to model or describe the intended operation of a piece of hardware. In the present case, this is the function that the PLD chip is intended to perform after it is programmed. It may be worth mentioning here that modern computer languages, including both hardware description languages and high-level programming languages, almost invariably contain declarative and executable statements, and the hardware description languages are particularly rich in the former. If we compare the results of a high-level programming language such as C++ and an HDL, it will be an executable program in the case of the former and declarative in the case of the latter. Hardware description languages that have evolved over the years include ABEL-HDL,

VHDL (VHSIC HDL), Verilog and JHDL (Java HDL). VHSIC stands for Very High-Speed Integrated Circuit.

The second type of software program is a computer program, called a logic compiler, that is used to transform a source code written in HDL into a bit stream. Logic compilers are available from manufacturers or third-party vendors. In the paragraphs to follow, we will briefly describe each of the hardware description languages mentioned above.

9.11.1 ABEL-Hardware Description Language

ABEL-HDL from DATA I/O was intended for relatively simpler PLD circuit designs that could be implemented on SPLDs. ABEL allows the designers to describe the digital circuit designs expressed in the form of truth tables, Boolean functions, state diagrams or any combination of these. It also allows the designer to optimize the design through design validation without specifying a device. In other words, ABEL-HDL facilitates writing hardware-independent programs, and it is only after the design verification and optimization have taken place that the PLD device is chosen. The source code written in the ABEL environment is in standard format to have interface compatibility with other tools.

9.11.2 VHDL-VHSIC Hardware Description Language

VHDL is the most widely used hardware description language used for the purpose of describing complex digital circuit designs that would be implemented on CPLDs and FPGAs. VHDL was originally developed to document the behaviour of ASICs used by various manufacturers in their equipment. It was subsequently followed by the development of logic simulation and synthesis tools that could read VHDL files and output a definition of the physical implementation of the circuit. With modern synthesis tools capable of extracting various digital building blocks such as counters, RAMs, arithmetic blocks, etc., and implementing them as specified by the user, the same VHDL code could be synthesized differently for optimum performance.

VHDL is a strongly typed language. One of the key features of VHDL is that it allows the behaviour of the intended hardware to be described and then verified before the design is translated into actual hardware with the help of synthesis tools. Another feature of VHDL that makes it attractive for digital system design is that it allows description of a concurrent system.

9.11.3 Verilog

Verilog, like VHDL, supports design, design validation and subsequent implementation of analogue, digital and mixed signal circuits at various levels of abstraction. Verilog-based design consists of a hierarchy of modules whose behaviour is defined by concurrent and sequential statements. Sequential statements are placed inside a 'begin/end' block and sequential statements contained inside the block are executed sequentially. All concurrent statements and all 'begin/end' blocks in the design are executed in parallel. A subset of statements in Verilog is synthesizable. Therefore, if in a given design the different modules use only synthesizable statements, the design can be translated into a netlist, which can further be translated into a bit stream.

Verilog has some similarities and dissimilarities with C-language. It has a similar preprocessor, similar major control keywords like 'if', 'while', etc., and also a similar formatting mechanism in the printing routines and language operators. Dissimilarities include the use of 'begin/end' instead of curly braces to define a block of code, and also that Verilog does not have structures, pointers and recursive subroutines. Also, the definition of constants in Verilog requires bit width along with their base.

9.11.4 Java HDL

Java HDL (JHDL) was developed in the Configurable Computing Laboratory of Brigham Young University (BYU). It is a low-level hardware description language that primarily uses an object-oriented approach to build circuits. It was developed primarily for the design of FPGA-based hardware, and developers have paid particular attention to supporting the Xilinx series of FPGA chips.

9.12 Application Information on PLDs

In this section, we will look at salient features of some of the commonly used programmable logic devices including SPLDs such as PALs/GALs, CPLDs and FPGAs covering a wide spectrum of devices from leading international manufacturers. Other application-relevant information such as internal architecture, pin connection diagram, etc., is also given for some of the more popular type numbers.

9.12.1 SPLDs

Some of the famous companies that offer SPLDs include Advanced Micro Devices (AMD), Altera, Philips-Signetics, Cypress, Lattice Semiconductor Corporation and ICT. A large range of SPLD products are available from these companies. All of these SPLDs share some common features in terms of the nature of the programmable logic planes, configurable output logic, etc. However, each of these logic devices does offer some unique features that make it particularly attractive for some applications. Some of the widely exploited SPLDs include the 16XX series (16L8, 16R8,16R6 and 16R4) and 22V10 from AMD and EP610 from Altera. These devices are also widely second-sourced by many companies. The Plus 16XX series from Philips is 100 % pin and functional compatible with the 16XX series. 16R8 in the 16XX series and 22V10 PAL devices are industry standards and are widely second-sourced. We will discuss 16XX and 22V10 in a little more detail in the following paragraphs.

The 16XX family of PAL devices employs the familiar sum-of-products implementation comprising a programmable AND array and a fixed OR array. The family offers four PAL-type devices including 16L8, 16R8, 16R6 and 16R4.

Each of the devices in the 16XX family is characterized by a certain number of combinational and registered outputs available to the designer. The devices have three-state output buffers on each output pin, which can be programmed for individual control of all outputs. Other features include the availability of programmable bidirectional pins and output registers. These devices are capable of replacing an equivalent of four or more SSI/MSI integrated circuits. The I/O configuration of the four devices in the 16XX family is summarized in Table 9.5. Figures 9.35(a) to (d) give the basic architecture/pin connections of 16L8, 16R8, 16R6 and 16R4 respectively.

As outlined earlier, many companies offer 22V10 PAL devices. These are available in both bipolar and CMOS technologies. One such contemporary device is GAL 22V10 from Lattice Semiconductor Corporation. As inherent in the type number, the device offers a maximum of 22 inputs and 10 outputs. The outputs are versatile. That is, each one of them can be configured by the user to be either a combinational or registered output. Also, the outputs can be configured to be either active HIGH or active LOW.

Table 9.5 Input/output configuration of the 16XX family.

Device number	Dedicated inputs	Combinational outputs	Registered outputs
16L8	10	8 (6 I/O)	0
16R8	8	0	8
16R6	8	2 I/O	6
16R4	8	4 I/O	4

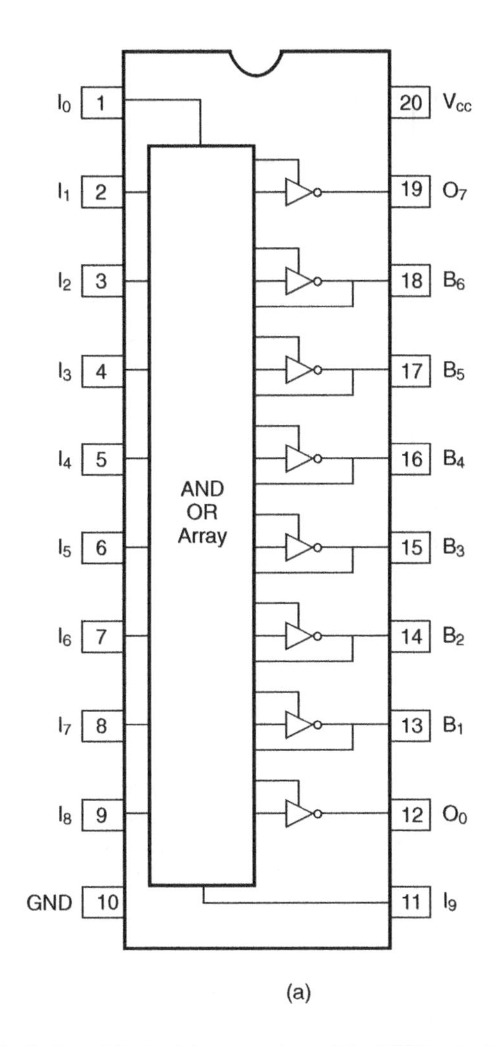

Figure 9.35 Basic architecture/pin connections of the 16XX-series PAL devices.

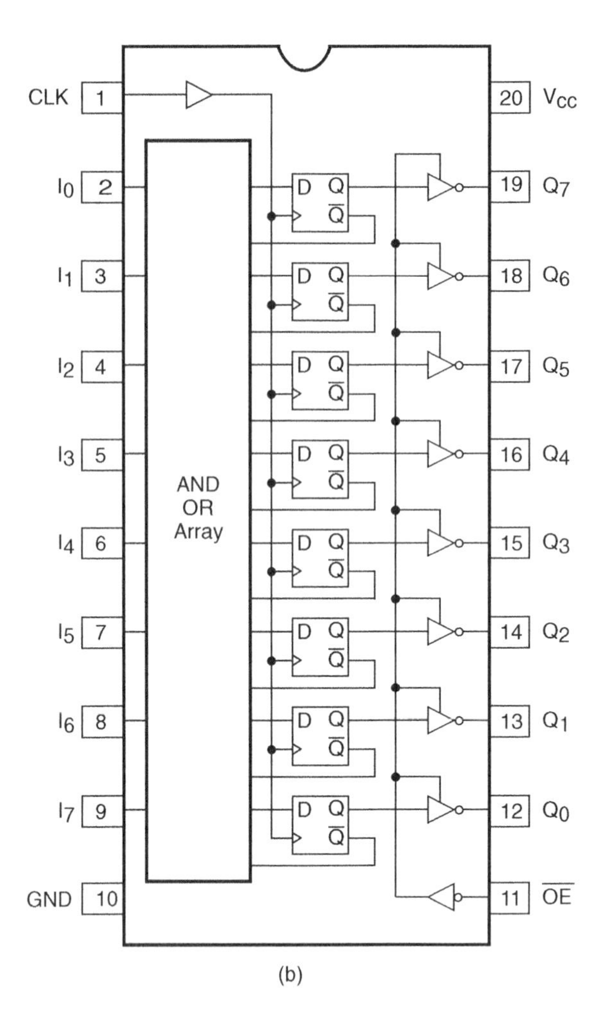

(b)

Figure 9.35 (*continued*).

GAL 22V10 uses E^2CMOS (electrically erasable CMOS) technology which allows the device to be reprogrammable through the use of an electrically erasable (E^2) floating-gate technology and consume much less power compared with bipolar 22V10 devices owing to the use of advanced CMOS technology. The device specifies 100 erase/write cycles, a 50–75 % saving in power consumption compared with bipolar equivalents and a maximum propagation delay of 4 ns. Each of the output logic macrocells offers two primary functional modes, which include combinational I/O and registered modes. The type of mode (whether combinational I/O or registered) and the output polarity (whether active HIGH or active LOW) are decided by the selection inputs S_0 and S_1, which are normally controlled by the logic compiler. For S_1S_0 equal to 00, 01, 10 and 11, outputs are active LOW registered, active LOW combinational, active HIGH registered and active HIGH combinational respectively.

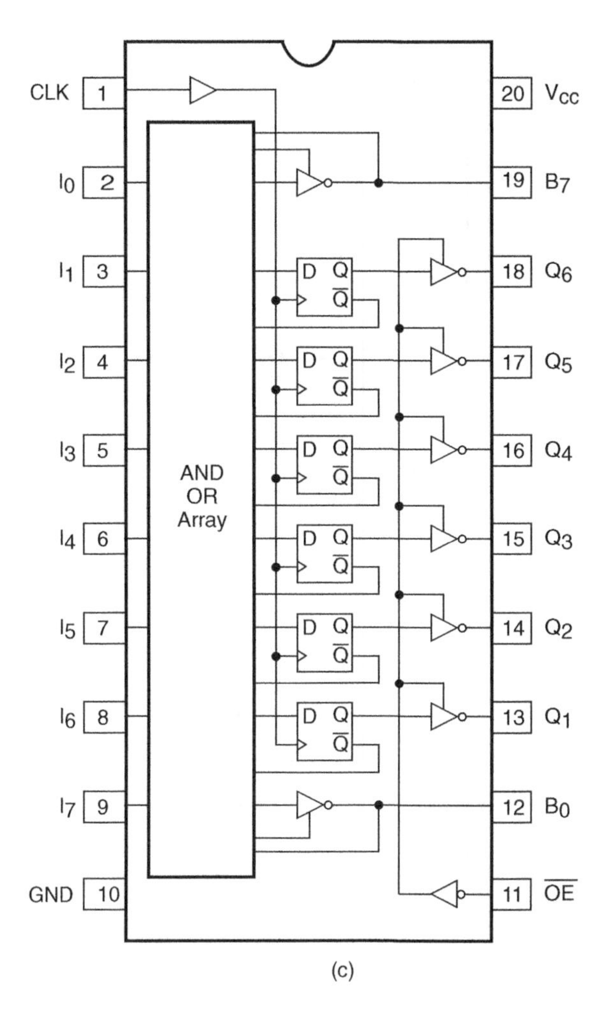

Figure 9.35 (*continued*).

Figure 9.36 shows the basic architecture and pin connection diagram of GAL 22V10. The internal architecture of the output logic macrocell (OLMC) shown as a block in Fig. 9.36 is given in Fig. 9.37.

9.12.2 CPLDs

Major CPLD manufacturers include Altera, Lattice Semiconductor Corporation, Advanced Micro Devices, ICT, Cypress and Xilinx. A large variety of CPLD devices are available from these companies. In the following paragraphs, some of the popular type numbers of CPLDs offered by some of these companies are examined in terms of their characteristic features.

We will begin with CPLDs from Altera. Altera offers three families of CPLDs. These include MAX-5000, MAX-7000 and MAX-9000. MAX-5000 uses an older technology and is used in applications

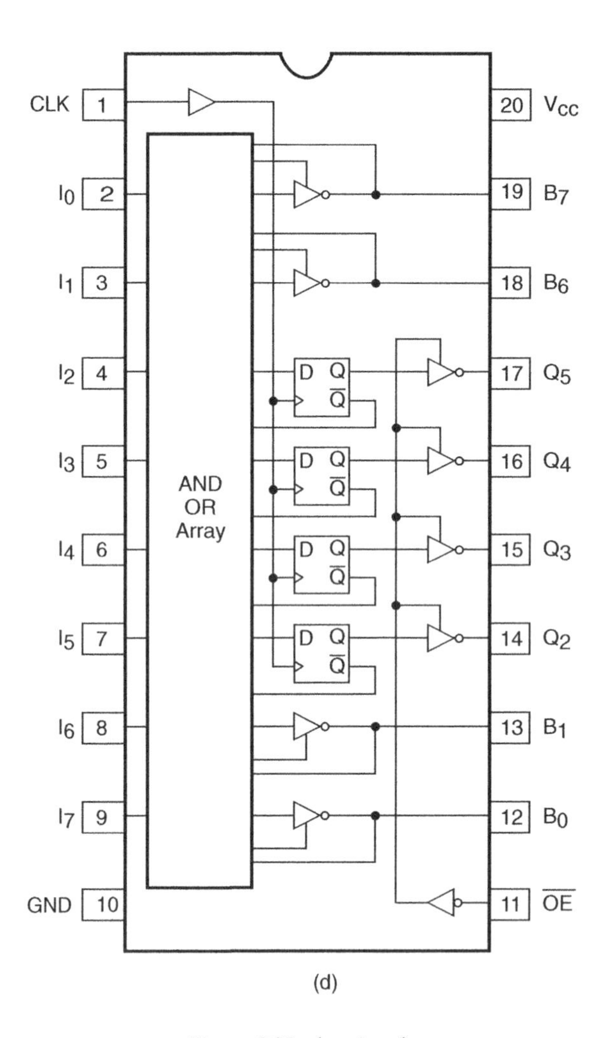

(d)

Figure 9.35 (*continued*).

where the designer is looking for cost-effective solutions. The MAX-7000 series of CPLDs are the most widely used ones. MAX-9000 is similar to MAX-7000 except for its higher logic capacity. MAX-7000 series devices use advanced CMOS technology and (E^2PROM)-based architecture and offer densities from 32 to 512 macrocells with pin-to-pin propagation delays as small as 3.5 ns. MAX-7000 devices support in-system programmability and are available with 5.0, 3.3 and 2.5 V core operating voltages. There are three types of device in the MAX-7000 series. These include MAX-7000S, MAX-7000AE and MAX-7000B. Three types are pin-to-pin compatible when used in the same package. Figure 9.38 shows the basic architecture of the MAX-7000 series of CPLDs.

AMD offers the Mach-1 to Mach-5 series of CPLDs. While Mach-1 and Mach-2 are configured around 22V10 PALs, Mach-3 and Mach-4 use 34V16 PALs. Mach-5 is similar to the Mach-4 CPLD except that it offers higher speed performance. All Mach devices use E^2PROM technology.

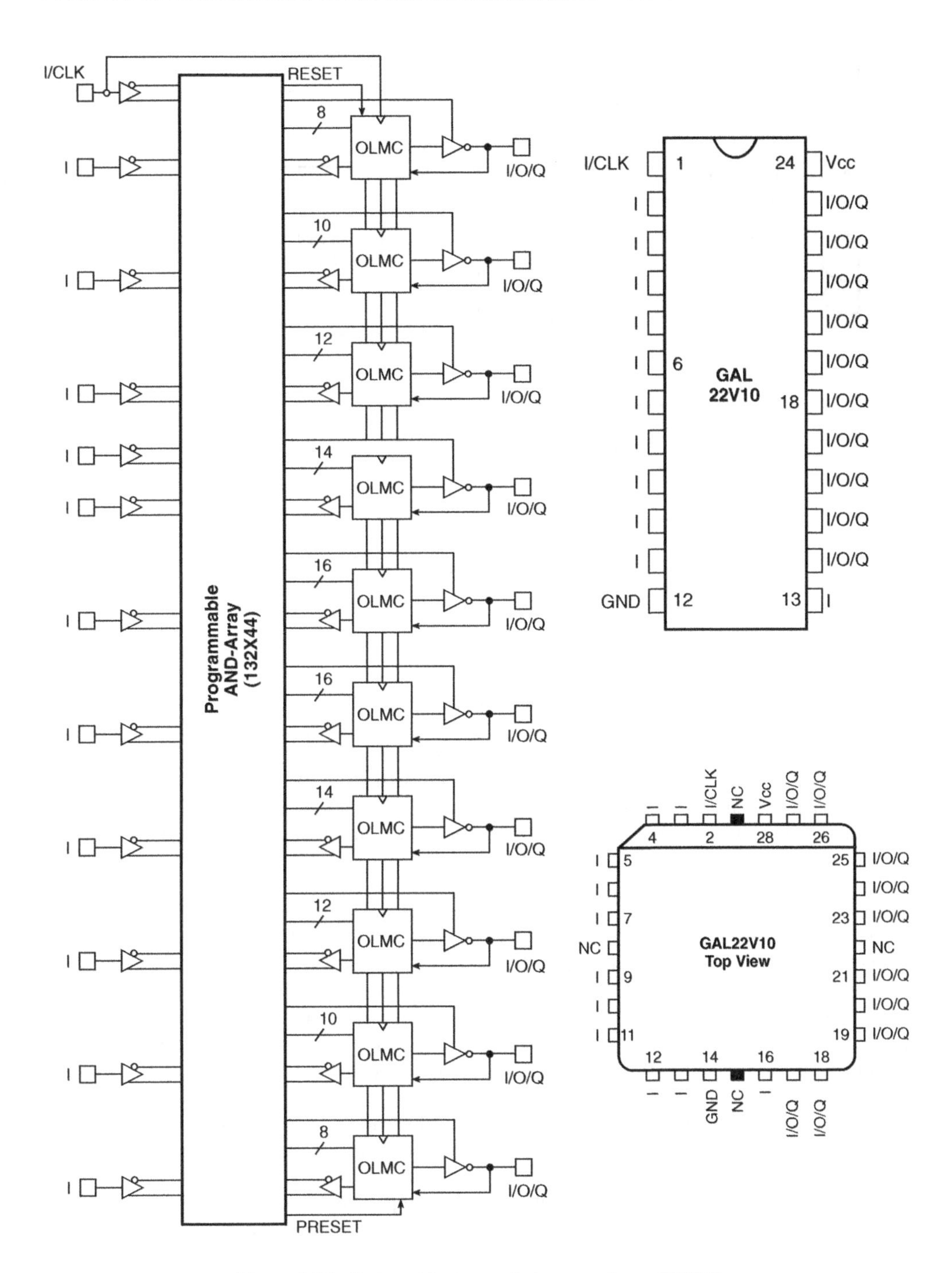

Figure 9.36 Basic architecture and pin connections of 22V10.

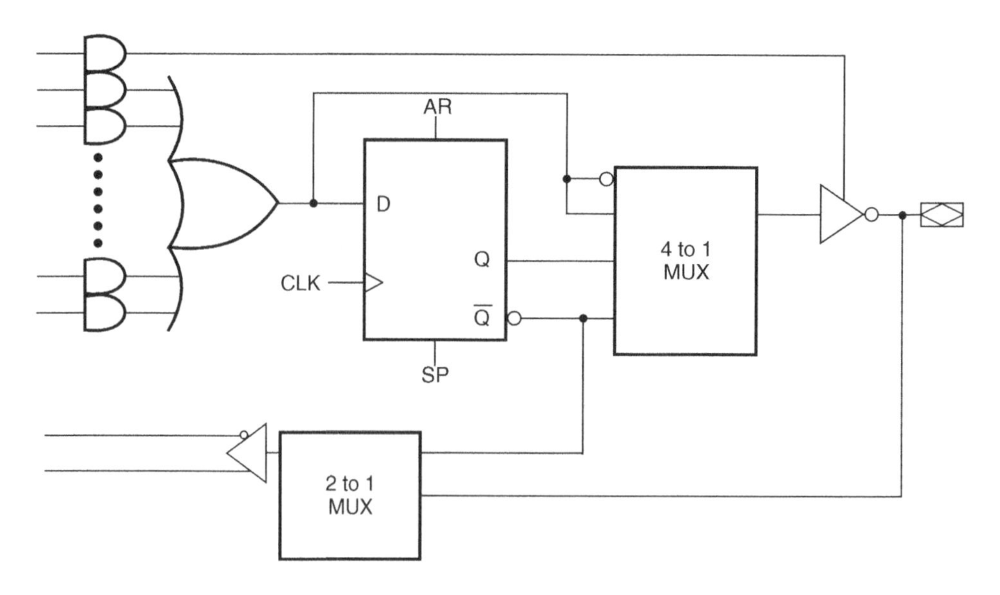

Figure 9.37 Architecture of an output GAL 22V10 logic macrocell.

Figure 9.39 shows the basic architecture of Mach-4 CPLDs. The number of 34V16-like PALs used varies from 6 to 16. Each 34V16-like PAL block consists of a maximum of 34 inputs and 16 outputs. The 34 inputs include 16 outputs that are fed back. All connections in the case of Mach-4 CPLDs, from one PAL block to another and also from a PAL block back to itself, are routed through a central switching matrix, on account of which all connections travel through the same path. This feature gives more predictable time delays in circuits implemented on Mach-4 devices.

Lattice offers the pLSI and ispLSI 1000-series, 2000-series and 3000-series of CPLDs. ispLSI devices are similar to pLSI devices except that they are in-system programmable. The three series of devices differ mainly in logic capacities and speed performance. The logic capacity in the case of the 1000-series CPLDs ranges from about 1200 to 4000 gates, and the pin-to-pin propagation delay is of the order of 10 ns. The ispLSI-1016 CPLD is one such device from the 1000-series of devices. It has a logic capacity of 2000 PLD gates and a pin-to-pin propagation delay of 7.5 ns. The device has four dedicated inputs, 32 universal I/O pins and 96 registers. It uses high-performance E^2CMOS technology, because of which it offers reprogrammability of the logic as well as the interconnects to provide truly reconfigurable systems.

The 2000-series devices have a logic capacity of 600–2000 equivalent gates that offer a higher ratio of macrocells to I/O pins. With a pin-to-pin propagation delay of 5.5 ns, they offer a higher speed performance compared with 1000-series devices. Of the three device families, the 3000-series has the highest logic capacity (up to 5000 equivalent gates). The propagation delay is in the range 10–15 ns. The 3000-series of devices offers some enhancements over the other two series of CPLDs to support more recent design approaches.

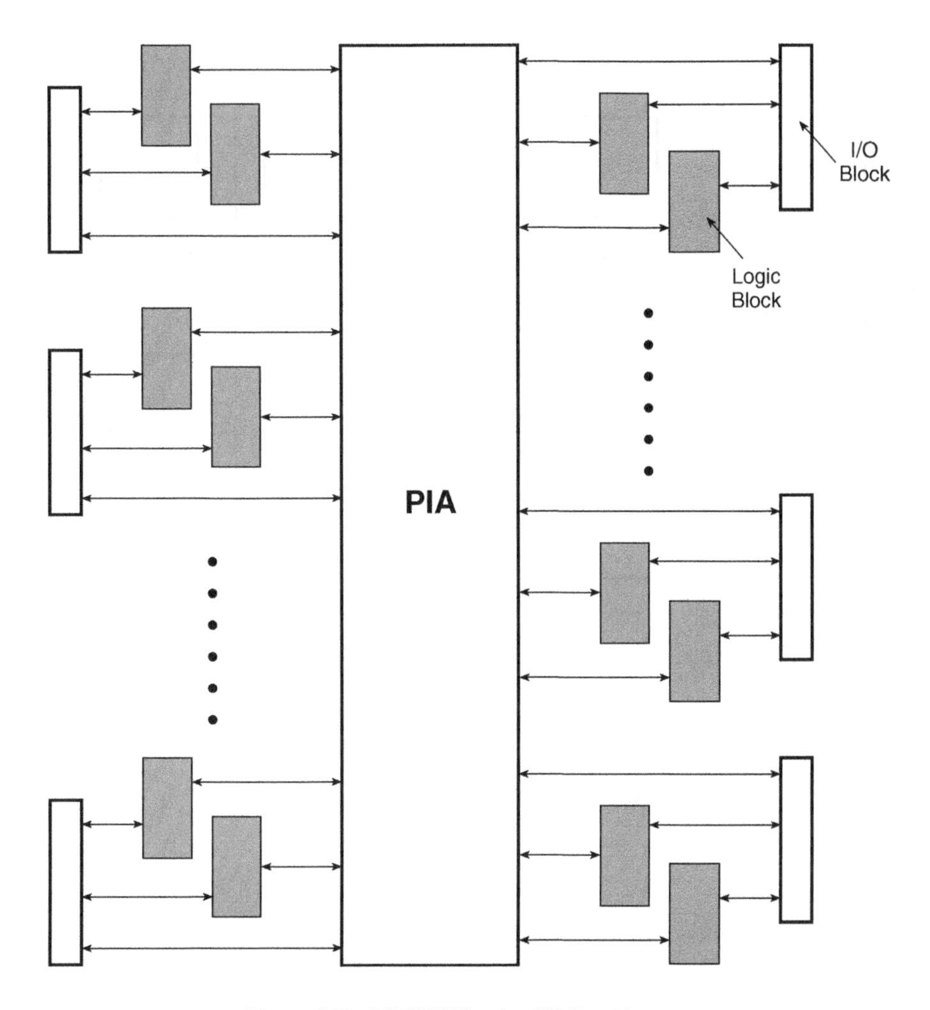

Figure 9.38 MAX-7000 series CPLD architecture.

FLASH-370 from Cypress is yet another popular class of CPLDs. FLASH-370 CPLDs use FLASH E^2PROM technology. Devices are not in-system programmable. One of the salient features of these devices is that they provide more inputs/outputs than the competing products featuring a linear relationship between the number of macrocells and the number of bidirectional I/O pins. FLASH-370 has a typical CPLD architecture as shown in Fig. 9.40, with multiple PAL-like blocks and a programmable interconnect matrix to connect them.

Xilinx, although mainly known for their range of FPGAs, offer CPLDs too. Major families of CPLDs from Xilinx include the XC-7000, CoolRunner and XC-9500 in-system programmable family of devices. The XC-7000 family of CPLDs further comprises two major series, namely XC-7200 and XC-7300. XC-7300 is an enhanced version of XC-7200 in terms of gate capacity and speed

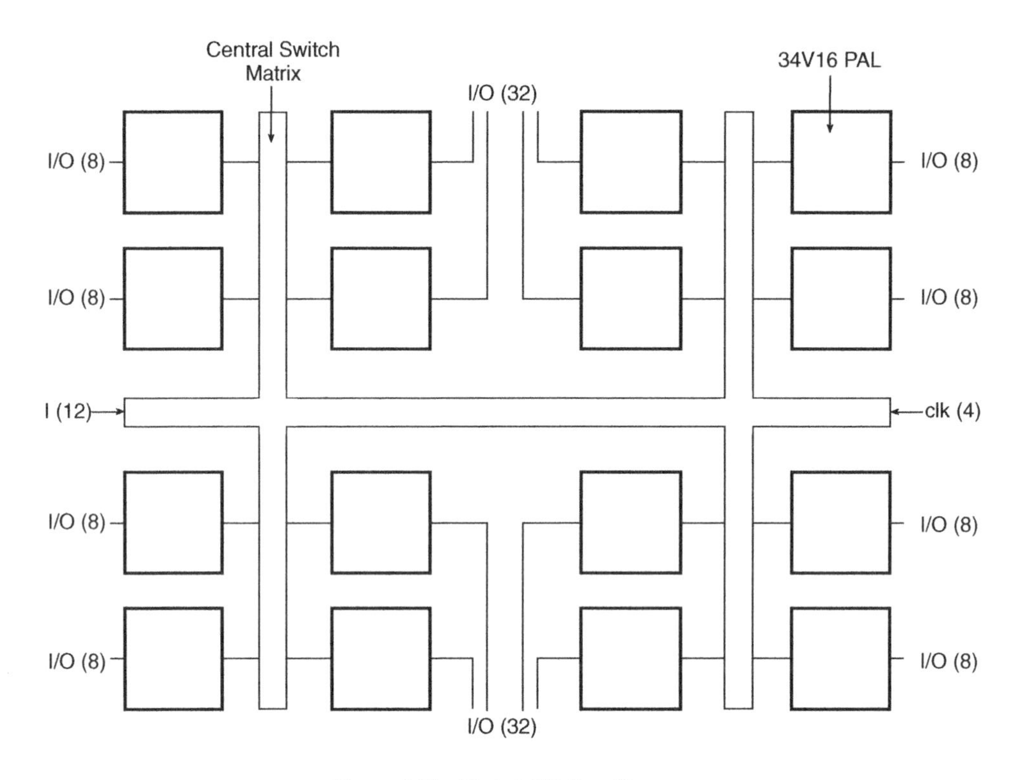

Figure 9.39 Mach-4 CPLD architecture.

performance. XC-7200 offers a logic capacity of 600–1500 gates with a speed performance of 25 ns pin-to-pin propagation delay. XC-7300 offers a gate capacity of up to 3000 gates. Each device in the XC-7000 family contains SPLD-like logic blocks, with each block having nine macrocells. A notable difference between the XC-7000 family of CPLDs and their counterparts from other manufacturers is that each macrocell has two OR gates whose outputs feed a two-bit arithmetic logic unit (ALU), which in turn can generate any function of its two inputs. The ALU output feeds a configurable flip-flop.

The CoolRunner family of CPLDs is characterized by high speed (5 ns pin-to-pin propagation delay) and low power consumption (100 μA of standby current). The family includes the XPLAE series of devices, available in 32, 64 and 128 macrocell versions, the XPLA2-series, which is SRAM-based and available in 320 and 920 macrocell capacities, and the XPLA3 series, available in 32, 64, 128, 256 and 384 macrocell versions.

The XC-9500 family of devices comprises the XC-9536, XC-9572, XC-95108, XC-95144, XC-95216 and XC-95288 series of CPLDs. The family offers a logic capacity ranging from 800 gates (in the case of XC-9536) to 6400 gates (in the case of XC-95288), with a propagation delay varying from 5 ns (in the case of XC-9536) to 15 ns (in the case of XC-95288). Architectural features of the XC-9500 family of CPLDs provide in-system programmability with a minimum of 10 000 program/erase cycles. Other features include output slew rate control and user-programmable ground pins, which help reduce system noise.

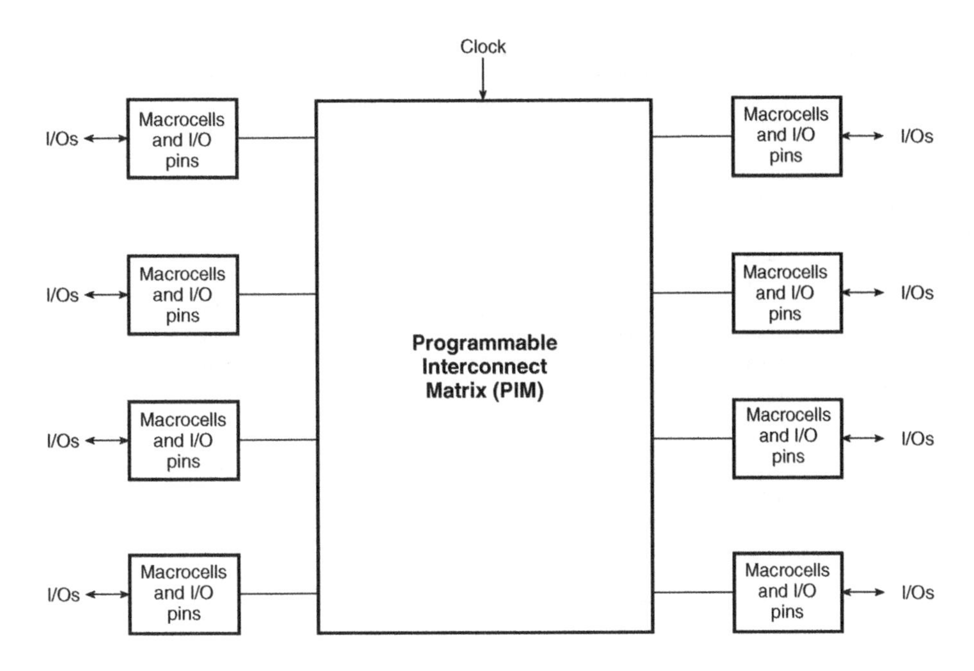

Figure 9.40 FLASH-370 CPLD architecture.

9.12.3 FPGAs

There are two broad categories of FPGAs, namely SRAM-based FPGAs and antifuse-based FPGAs. While Xilinx and Altera are the major players in the former category, antifuse-based devices are offered mainly by Xilinx, Actel, Quicklogic and Cypress. FPGAs were introduced by Xilinx with the XC-2000 series of devices, which have been subsequently followed up by the XC-3000 series, XC-4000 series and XC-5000 series of devices. Of all these, the XC-4000 series is the most widely used one. These are all SRAM-based. Xilinx has also introduced an antifuse-based FPGA family of FPGAs called XC-8100.

The basic architecture of the XC-4000 family is built around a two-dimensional array of configurable logic blocks (CLBs) that can be interconnected by horizontal and vertical routing channels and are surrounded by a perimeter of programmable input/output blocks (IOBs). CLBs provide the functional elements for constructing the user-desired logic function, and IOBs provide the interface between the package pins and internal signal lines. These devices are reconfigurable and are in-system programmable. Table 9.6 gives salient features of the XC-4000X and XC-4000E series of FPGAs.

Altera offers the FLEX-8000 and FLEX-10000 series of FPGAs. FLEX-8000 is SRAM-based. It combines the fine-grained architecture and high register count characteristics of FPGAs with the high speed and predictable interconnect timing delays of CPLDs. The basic logic element comprises a four-input look-up table (LUT) that provides combinational capability and a programmable register that provides sequential capability. Table 9.7 outlines salient features of the FLEX-8000 series of devices.

The FLEX-10000 series offers all the features of FLEX-8000 series devices, with the addition of variable-sized blocks of SRAM called embedded array blocks (EABs). Each of the EABs can be

Table 9.6 Salient features of the XC-4000X and XC-4000E series of FPGAs.

Device number	Logic cells	Maximum logic gates (no RAM)	CLB matrix	Number of CLBs	Number of flip-flops	Maximum user I/Os
XC4002XL	152	1 600	8 × 8	64	256	64
XC4003E	238	3 000	10 × 10	100	360	80
XC4005E/XL	466	5 000	14 × 14	196	616	112
XC4006E	608	6 000	16 × 16	256	768	128
XC4008E	770	8 000	18 × 18	324	936	144
XC4010E/XL	950	10 000	20 × 20	400	1120	160
XC4013E/XL	1368	13 000	24 × 24	576	1536	192
XC4020E/XL	1862	20 000	28 × 28	784	2016	224
XC4025E	2432	25 000	32 × 32	1024	2560	256
XC4028EX/	3078	28 000	32 × 32	1024	2560	256
XC4036EX/XL	3078	36 000	36 × 36	1296	3168	288
XC4044XL	3800	44 000	40 × 40	1600	3840	320
XC4052	4598	52 000	44 × 44	1936	4576	352
XC4062XL	5472	62 000	48 × 48	2304	5376	384
XC4085	7448	85 000	56 × 56	3136	7168	448

Table 9.7 Salient features of the FLEX-8000 series of devices.

Device number	Usable Gates	Flip-flops	Logic Array Blocks (LAB)	Logic Elements (LE)	Maximum User I/O PIns
EPF 8282A/AV	2500	282	26	208	78
EPF 8452A	4000	452	42	336	120
EPF 8636A	6000	636	63	504	136
EPF 8820A	8000	820	84	672	152
EPF 81188A	12000	1188	126	1008	184
EPF 81500A	16000	1500	162	1296	208

configured to serve as an SRAM block with a variable aspect ratio of 256 × 8, 512 × 4, 1K × 2 or 2K × 1.

AT&T offers SRAM-based FPGAs that are similar in architecture to those offered by Xilinx. The overall structure is called an optimized reconfigurable cell array (ORCA). The basic logic block is referred to as a programmable function unit (PFU). Similarities with the Xilinx-4000 series FPGAs include arithmetic circuitry being a part of the PFU and PFU configurability as a RAM. The PFU can be configured as either four four-input LUTs or as two five-input LUTs or as one six-input LUT. When configured as four-input LUTs, it is essential that the various LUT inputs come from the same PFU input. Although on the one hand this reduces the functionality of the PFU, on the other hand it significantly reduces the associated wiring cost.

Actel FPGAs use antifuse technology. Actel offers three main families of FPGA devices, namely Act-1, Act-2 and Act-3. All three series of devices have similar features. The structure is similar to that

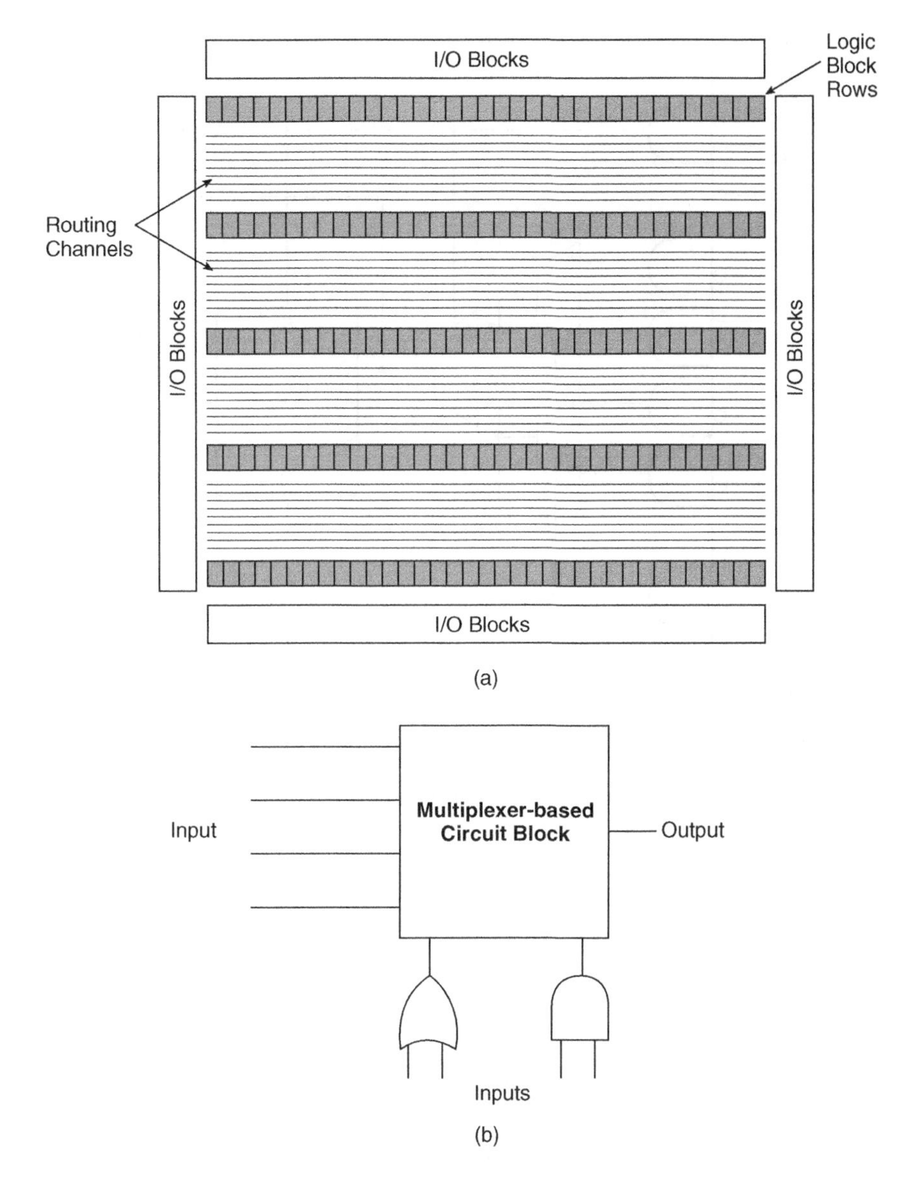

Figure 9.41 Actel FPGA.

of traditional gate arrays comprising logic blocks arranged in horizontal rows with horizontal routing channels between adjacent rows, as shown in Fig. 9.41(a). Actel chips also have vertical wires that overlay the logic blocks to provide signal paths that span multiple rows. These are not shown in Fig. 9.41(a). The logic block is not LUT based. Instead, it comprises an AND gate and an OR gate feeding a multiplexer circuit block, as shown in Fig. 9.41(b). The multiplexer circuit, along with the two gates, can realize a large range of logic functions. In the case of Act-3 FPGAs, 50 % of the logic blocks also contain a flip-flop.

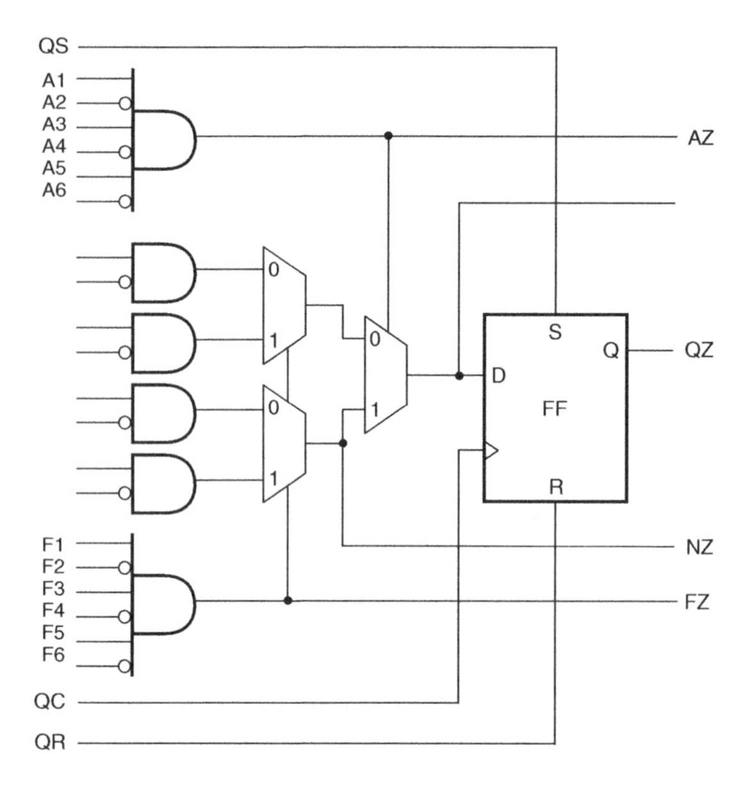

Figure 9.42 Quicklogic FPGA logic block.

Quicklogic also offers antifuse-based FPGAs, like Actel. They offer two families of devices, namely pASIC and pASIC-2. pASIC-2 is an enhanced version of pASIC. The overall structure is array based like the Xilinx FPGAs. The logic blocks are similar to those used in the Actel FPGAs, although more complex than their Actel counterparts. Also, each logic block contains a flip-flop. Figure 9.42 shows the architecture.

Review Questions

1. How does a programmable logic device differ from a fixed logic device? What are the primary advantages of using programmable logic devices?
2. Distinguish between a programmable logic array (PLA) device and a programmable array logic (PAL) device in terms of architecture and capability to implement Boolean functions.
3. How does a generic array logic (GAL) device differ from its PAL counterpart? Do they differ in their internal architecture? If yes, then how?
4. What are complex programmable logic devices (CPLDs)? Briefly outline salient features of these devices and application areas where these devices fit the best.
5. How does the architecture of a typical FPGA device differ from that of a CPLD? In what way does the architecture affect the timing performance in the two cases?

6. What are the various interconnect technologies used for the purpose of programming PLDs? Briefly describe each one of them.
7. What is a hardware description language? What are the requirements of a good HDL? Briefly describe the salient features of VHDL and Verilog.
8. What do you understand by the following as regards programmable logic devices?

   (a) combinational and registered outputs;
   (b) configurable output logic cell;
   (c) reprogrammable PLD;
   (d) in-system programmability.

Problems

1. Figure 9.43 shows a portion of the internal logic diagram of a certain PAL device that uses antifuse interconnect technology. In the diagram shown, a cross (×) represents an unprogrammed interconnect and the absence of a cross (×) at an intersection of input and product lines represents programmed interconnects; a dot (•) represents a hard-wired interconnect. Write (a) the Boolean expression for Y and (b) the Boolean expression for Y if the interconnect technology were fuse based.

   (a) $Y = \overline{A}.B + A.\overline{B}$*; (b) the same as in the case of (a)*

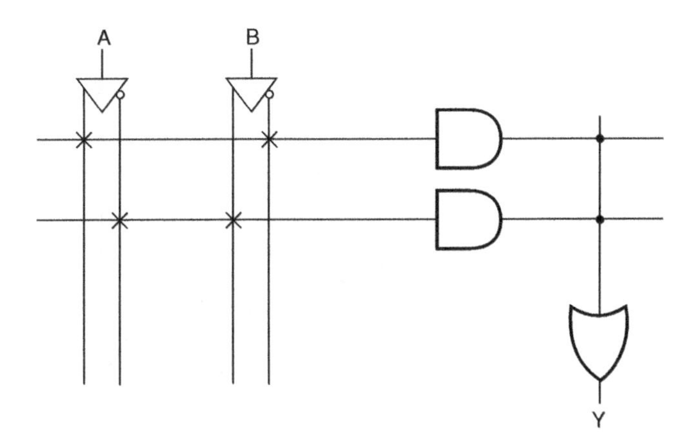

Figure 9.43 Problem 1.

2. Determine the size of PROM required for implementing the following logic circuits.

   (a) 16-to-1 multiplexer;
   (b) four-bit binary adder.

   (a) 1M×1; (b) 512×5

3. Determine the number of programmable interconnections in the following programmable logic devices.

   (a) 1K × 4 PROM;
   (b) PLA device with four input variables, 32 AND gates and four OR gates;
   (c) PAL device with eight input variables, 16 AND gates and four OR gates.

   (a) 4096; (b) 384; (c) 256

4. *A* and *B* are two binary variables. The objective is to design a magnitude comparator to produce $A = B$, $A < B$ and $A > B$ outputs. Design a suitable PLD with a PAL-like architecture using anti-fuse based interconnects.

   Fig. 9.44

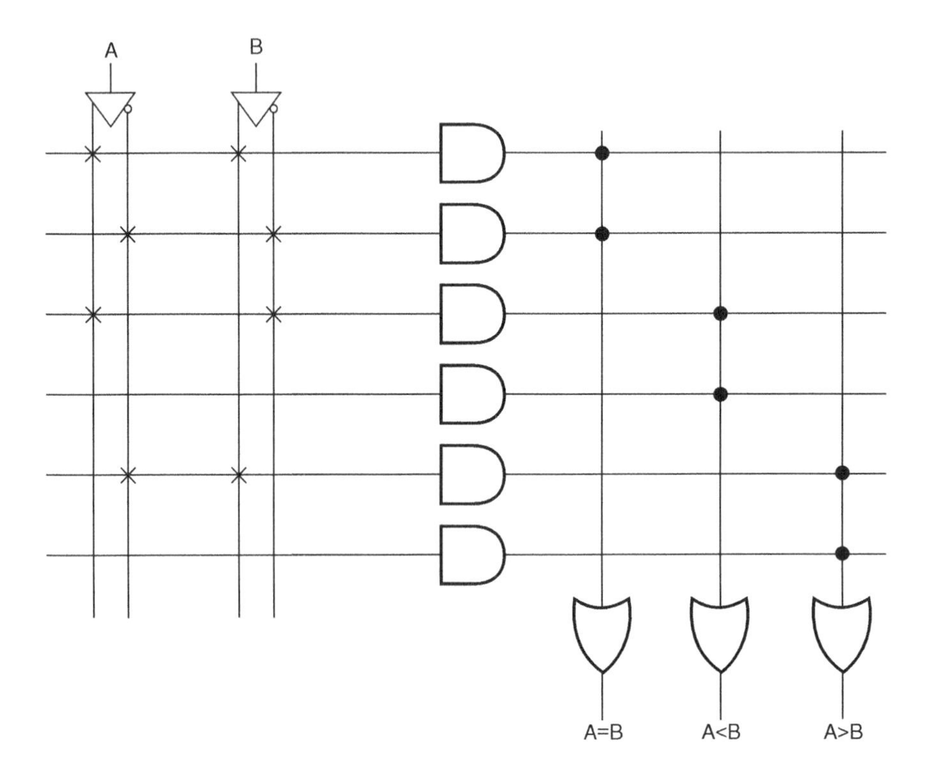

Figure 9.44 Answer to problem 4.

5. Figure 9.45 shows a programmed PAL device using fuse-based interconnects. Examine the logic diagram and determine the logic block implemented by the PLD. A cross (×) represents an unprogrammed interconnection and a dot (•) represents a hard-wired interconnection.

   Full adder

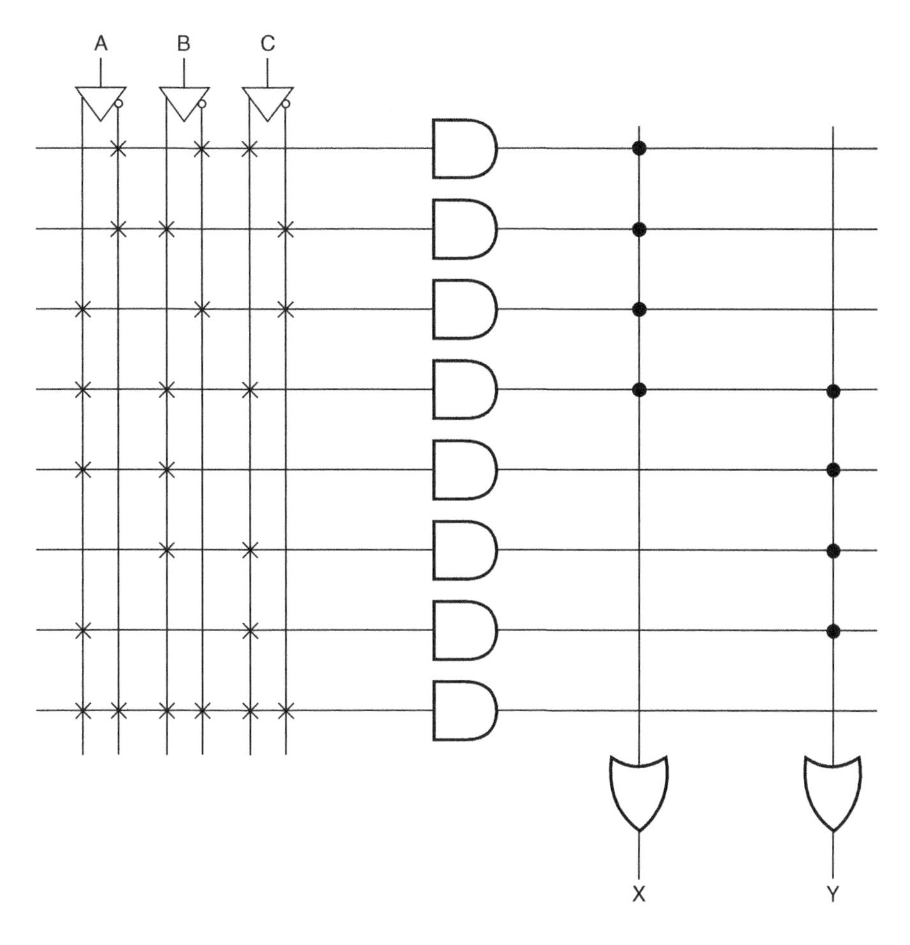

Figure 9.45 Problem 5.

Further Reading

1. Seals, R. C. and Whapshott, G. F. (1997) *Programmable Logic: PLDs and FPGAs*, McGraw-Hill, USA.
2. Dueck, R. (2003) *Digital Design with CPLD Applications and VHDL*, Thomson Delmar Learning, New York, USA.
3. Chartrand, L. (2003) *Digital Fundamentals: Experiments and Concepts with CPLD*, Thomson Delmar Learning, New York, USA.
4. Oldfield, J. and Dorf, R. (1995) *Field Programmable Gate Arrays*, John Wiley & Sons, Inc., New York, USA.
5. Trimberger, S. (Ed.) (1994) *Field Programmable Gate Array Technology*, Kluwer Academic Publishers, MA, USA.
6. Brown, S., Francis, R., Rose, J. and Vranesic, Z. (1992) *Field Programmable Gate Arrays*, Kluwer Academic Publishers, MA, USA.

10

Flip-Flops and Related Devices

Having discussed combinational logic circuits at length in previous chapters, the focus in the present chapter and in Chapter 11 will be on sequential logic circuits. While a logic gate is the most basic building block of combinational logic, its counterpart in sequential logic is the flip-flop. The chapter begins with a brief introduction to different types of multivibrator, including the bistable multivibrator, which is the complete technical name for a flip-flop, the monostable multivibrator and the astable multivibrator. The flip-flop is not only used individually for a variety of applications; it also forms the basis of many more complex logic functions. Counters and registers, to be covered in Chapter 11 are typical examples. There is a large variety of flip-flops having varying functional tables, input clocking requirements and other features. In this chapter, we will discuss all these basic types of flip-flop in terms of their functional aspects, truth tables, salient features and application aspects. The text is suitably illustrated with a large number of solved examples. Application-relevant information, including a comprehensive index of flip-flops and related devices belonging to different logic families, is given towards the end of the chapter. Pin connection diagrams and functional tables are given in the companion website.

10.1 Multivibrator

Multivibrators, like the familiar sinusoidal oscillators, are circuits with regenerative feedback, with the difference that they produce pulsed output. There are three basic types of multivibrator, namely the bistable multivibrator, the monostable multivibrator and the astable multivibrator.

10.1.1 Bistable Multivibrator

A *bistable multivibrator* circuit is one in which both LOW and HIGH output states are stable. Irrespective of the logic status of the output, LOW or HIGH, it stays in that state unless a change is

Digital Electronics: Principles, Devices and Applications Anil Kumar Maini

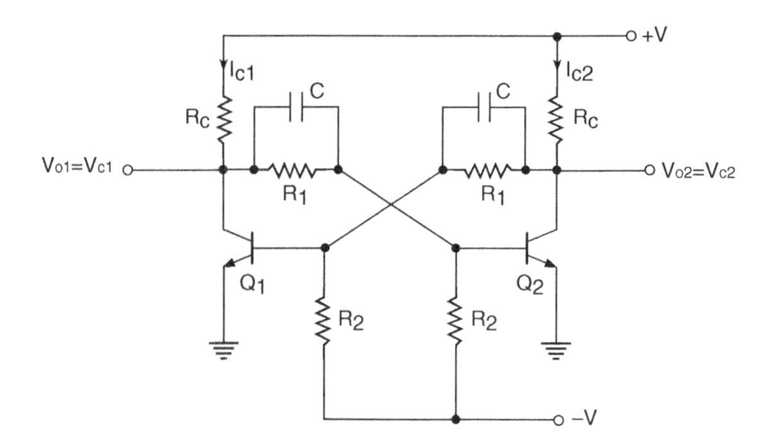

Figure 10.1 Bistable multivibrator.

induced by applying an appropriate trigger pulse. As we will see in the subsequent pages, the operation of a bistable multivibrator is identical to that of a flip-flop. Figure 10.1 shows the basic bistable multivibrator circuit. This is the fixed-bias type of bistable multivibrator. Other configurations are the self-bias type and the emitter-coupled type. However, the operational principle of all types is the same. The multivibrator circuit of Fig. 10.1 functions as follows.

In the circuit arrangement of Fig. 10.1 it can be proved that both transistors Q_1 and Q_2 cannot be simultaneously ON or OFF. If Q_1 is ON, the regenerative feedback ensures that Q_2 is OFF, and when Q_1 is OFF, the feedback drives transistor Q_2 to the ON state. In order to vindicate this statement, let us assume that both Q_1 and Q_2 are conducting simultaneously. Owing to slight circuit imbalance, which is always there, the collector current in one transistor will always be greater than that in the other. Let us assume that $I_{c2} > I_{c1}$. Lesser I_{c1} means a higher V_{c1}. Since V_{c1} is coupled to the Q_2 base, a rise in V_{c1} leads to an increase in the Q_2 base voltage. Increase in the Q_2 base voltage results in an increase in I_{c2} and an associated reduction in V_{c2}. Reduction in V_{c2} leads to a reduction in Q_1 base voltage and an associated fall in I_{c1}, with the result thatV_{c1} increases further. Thus, a slight circuit imbalance has initiated a regenerative action that culminates in transistor Q_1 going to cut-off and transistor Q_2 getting driven to saturation. To sum up, whenever there is a tendency of one of the transistors to conduct more than the other, it will end up with that transistor going to saturation and driving the other transistor to cut-off. Now, if we take the output from the Q_1 collector, it will be LOW ($= V_{CE1}$ sat.) if Q_1 was initially in saturation. If we apply a negative-going trigger to the Q_1 base to cause a decrease in its collector current, a regenerative action would set in that would drive Q_2 to saturation and Q_1 to cut-off. As a result, the output goes to a HIGH ($= +V_{CC}$) state. The output will stay HIGH until we apply another appropriate trigger to initiate a transition. Thus, both of the output states, when the output is LOW and also when the output is HIGH, are stable and undergo a change only when a transition is induced by means of an appropriate trigger pulse. That is why it is called a bistable multivibrator.

10.1.2 Schmitt Trigger

A Schmitt trigger circuit is a slight variation of the bistable multivibrator circuit of Fig. 10.1. Figure 10.2 shows the basic Schmitt trigger circuit. If we compare the bistable multivibrator circuit of Fig. 10.1

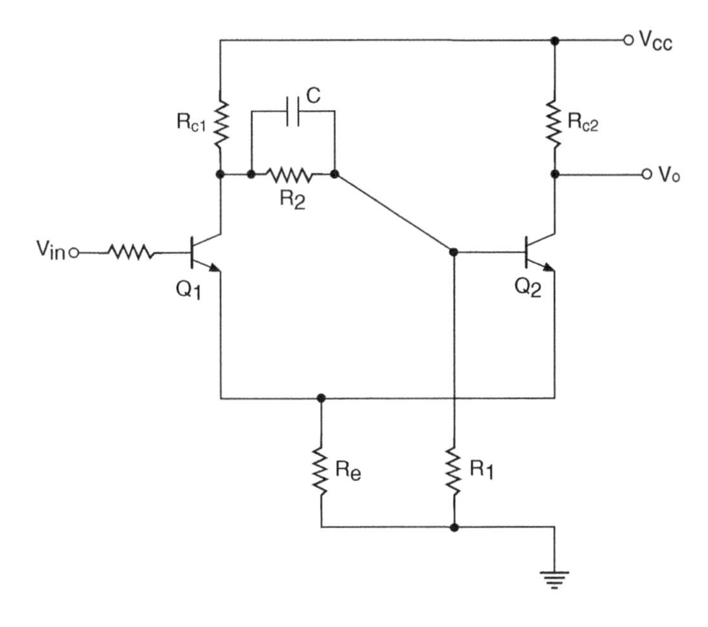

Figure 10.2 Schmitt trigger circuit.

with the Schmitt trigger circuit of Fig. 10.2, we find that coupling from Q_2 collector to Q_1 base in the case of a bistable circuit is absent in the case of a Schmitt trigger circuit. Instead, the resistance R_e provides the coupling. The circuit functions as follows.

When V_{in} is zero, transistor Q_1 is in cut-off. Coupling from Q_1 collector to Q_2 base drives transistor Q_2 to saturation, with the result that V_o is LOW. If we assume that V_{CE2} (sat.) is zero, then the voltage across R_e is given by the equation

$$\text{Voltage across } R_e = [V_{CC}.R_e/(R_e + R_{c2})] \tag{10.1}$$

This is also the emitter voltage of transistor Q_1. In order to make transistor Q_1 conduct, V_{in} must be at least 0.7 V more than the voltage across R_e. That is,

$$V_{in}(\text{min.}) = [V_{CC}.R_e/(R_e + R_{c2})] + 0.7 \tag{10.2}$$

When V_{in} exceeds this voltage, Q_1 starts conducting. The regenerative action again drives Q_2 to cut-off. The output goes to the HIGH state. Voltage across R_e changes to a new value given by the equation

$$\text{Voltage across } R_e = [V_{CC}.R_e/(R_e + R_{c1})] \tag{10.3}$$

$$V_{in} = [V_{CC}.R_e/(R_e + R_{c1})] + 0.7 \tag{10.4}$$

Transistor Q_1 will continue to conduct as long as V_{in} is equal to or greater than the value given by Equation (10.4). If V_{in} falls below this value, Q_1 tends to come out of saturation and conduct less heavily. The regenerative action does the rest, with the process culminating in Q_1 going to cut-off and Q_2 to saturation. Thus, the state of output (HIGH or LOW) depends upon the input voltage level.

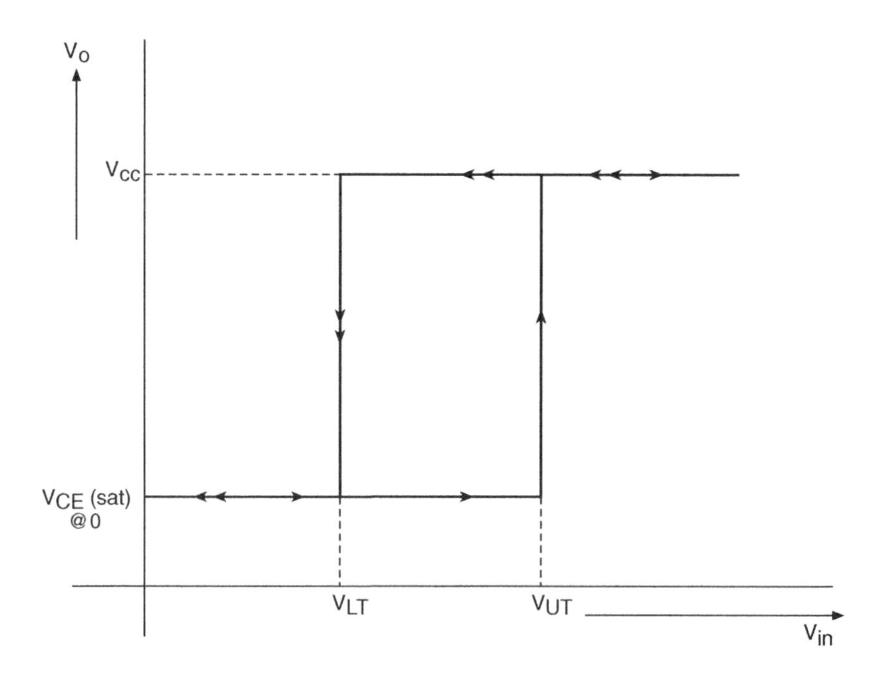

Figure 10.3 Transfer characteristics of a Schmitt trigger.

The HIGH and LOW states of the output correspond to two distinct input levels given by Equations (10.2) and (10.4) and therefore the values of R_{c1},R_{c2},R_e and V_{CC}. The Schmitt trigger circuit of Fig. 10.2 therefore exhibits hysteresis. Figure 10.3 shows the transfer characteristics of the Schmitt trigger circuit. The lower trip point V_{LT} and the upper trip point V_{UT} of these characteristics are respectively given by the equations

$$V_{LT} = [V_{CC}.R_e/(R_e + R_{c1})] + 0.7 \tag{10.5}$$

$$V_{UT} = [V_{CC}.R_e/(R_e + R_{c2})] + 0.7 \tag{10.6}$$

10.1.3 Monostable Multivibrator

A *monostable multivibrator*, also known as a *monoshot*, is one in which one of the states is stable and the other is quasi-stable. The circuit is initially in the stable state. It goes to the quasi-stable state when appropriately triggered. It stays in the quasi-stable state for a certain time period, after which it comes back to the stable state. Figure 10.4 shows the basic monostable multivibrator circuit. The circuit functions as follows. Initially, transistor Q_2 is in saturation as it gets its base bias from $+V_{CC}$ through R. Coupling from Q_2 collector to Q_1 base ensures that Q_1 is in cut-off. Now, if an appropriate trigger pulse induces a transition in Q_2 from saturation to cut-off, the output goes to the HIGH state. This HIGH output when coupled to the Q_1 base turns Q_1 ON. Since there is no direct coupling from Q_1 collector to Q_2 base, which is necessary for a regenerative process to set in, Q_1 is not necessarily

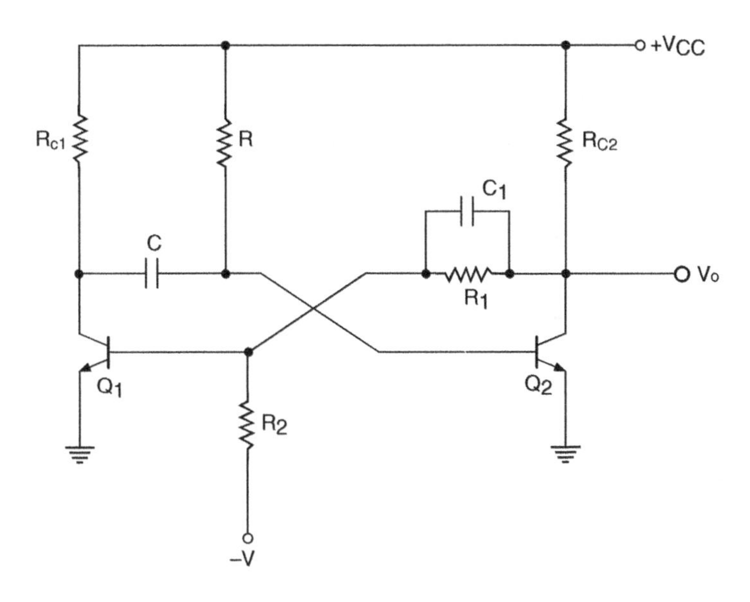

Figure 10.4 Monostable multivibrator.

in saturation. However, it conducts some current. The Q_1 collector voltage falls by $I_{c1}R_{c1}$ and the Q_2 base voltage falls by the same amount, as the voltage across a capacitor (C in this case) cannot change instantaneously. To sum up, the moment we applied the trigger, Q_2 went to cut-off and Q_1 started conducting. But now there is a path for capacitor C to charge from V_{CC} through R and the conducting transistor. The polarity of voltage across C is such that the Q_2 base potential rises. The moment the Q_2 base voltage exceeds the cut-in voltage, it turns Q_2 ON, which, owing to coupling through R_1, turns Q_1 OFF. And we are back to the original state, the stable state. Whenever we trigger the circuit into the other state, it does not stay there permanently and returns back after a time period that depends upon R and C. The greater the time constant RC, the longer is the time for which it stays in the other state, called the quasi-stable state.

10.1.3.1 Retriggerable Monostable Multivibrator

In a conventional monostable multivibrator, once the output is triggered to the quasi-stable state by applying a suitable trigger pulse, the circuit does not respond to subsequent trigger pulses as long as the output is in quasi-stable state. After the output returns to its original state, it is ready to respond to the next trigger pulse. There is another class of monostable multivibrators, called *retriggerable monostable multivibrators*. These respond to trigger pulses even when the output is in the quasi-stable state. In this class of monostable multivibrators, if n trigger pulses with a time period of T_t are applied to the circuit, the output pulse width, that is, the time period of the quasi-stable state, equals $(n-1)T_t + T$, where T is the output pulse width for the single trigger pulse and $T_t < T$. Figure 10.5 shows the output pulse width in the case of a retriggerable monostable multivibrator for repetitive trigger pulses.

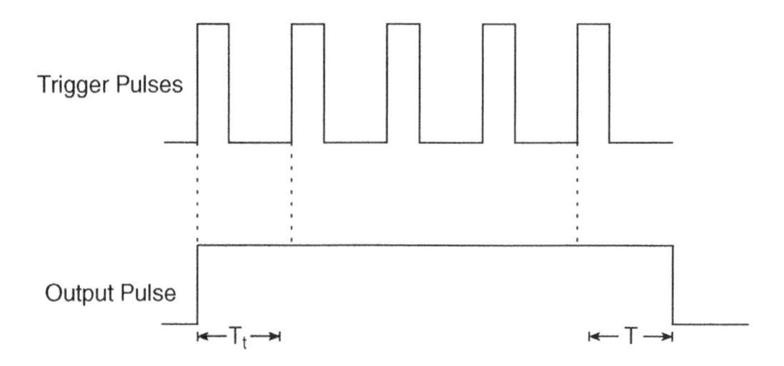

Figure 10.5 Retriggerable monostable multivibrator output for repetitive trigger pulses.

10.1.4 Astable Multivibrator

In the case of an astable multivibrator, neither of the two states is stable. Both output states are quasi-stable. The output switches from one state to the other and the circuit functions like a free-running square-wave oscillator. Figure 10.6 shows the basic astable multivibrator circuit. It can be proved that, in this type of circuit, neither of the output states is stable. Both states, LOW as well as HIGH, are quasi-stable. The time periods for which the output remains LOW and HIGH depends upon R_2C_2 and R_1C_1 time constants respectively. For $R_1C_1 = R_2C_2$, the output is a symmetrical square waveform. The circuit functions as follows. Let us assume that transistor Q_2 is initially conducting, that is, the output is LOW. Capacitor C_2 in this case charges through R_2 and the conducting transistor from V_{CC}, and, the moment the Q_1 base potential exceeds its cut-in voltage, it is turned ON. A fall in Q_1 collector

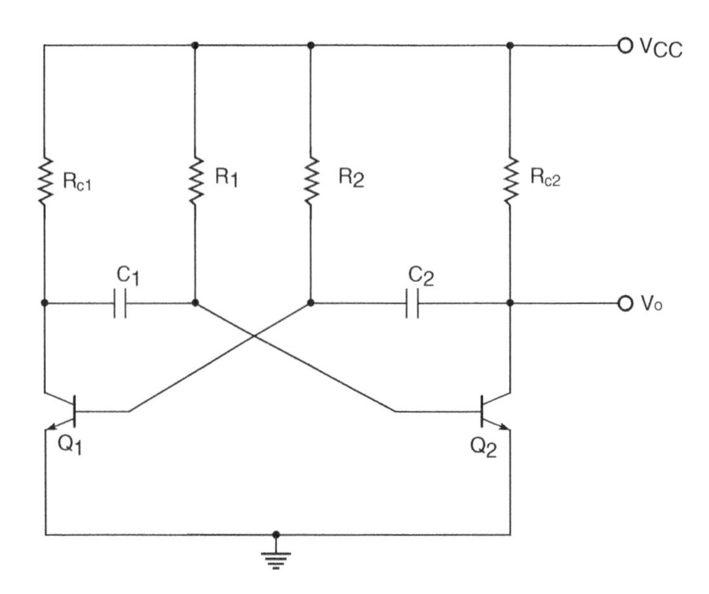

Figure 10.6 Astable multivibrator.

potential manifests itself at the Q_2 base as voltage across a capacitor cannot change instantaneously. The output goes to the HIGH state as Q_2 is driven to cut-off. However, C_1 has now started charging through R_1 and the conducting transistor Q_1 from V_{CC}. The moment the Q_2 base potential exceeds the cut-in voltage, it is again turned ON, with the result that the output goes to the LOW state. This process continues and, owing to both the couplings (Q_1 collector to Q_2 base and Q_2 collector to Q_1 base) being capacitive, neither of the states is stable. The circuit produces a square-wave output.

10.2 Integrated Circuit (IC) Multivibrators

In this section, we will discuss monostable and astable multivibrator circuits that can be configured around some of the popular digital and linear integrated circuits. The bistable multivibrator, which is functionally the same as a flip-flop, will not be discussed here. Flip-flops are discussed at length from Section 10.3 onwards.

10.2.1 Digital IC-Based Monostable Multivibrator

Some of the commonly used digital ICs that can be used as monostable multivibrators include 74121 (single monostable multivibrator), 74221 (dual monostable multivibrator), 74122 (single retriggerable monostable multivibrator) and 74123 (dual retriggerable monostable multivibrator), all belonging to the TTL family, and 4098B (dual retriggerable monostable multivibrator) belonging to the CMOS family. Figure 10.7 shows the use of IC 74121 as a monostable multivibrator along with a trigger input. The IC provides features for triggering on either LOW-to-HIGH or HIGH-to-LOW edges of the trigger pulses. Figure 10.7(a) shows one of the possible application circuits for HIGH-to-LOW edge triggering, and Fig. 10.7(b) shows one of the possible application circuits for LOW-to-HIGH edge triggering. The output pulse width depends on external R and C. The output pulse width can be computed from $T = 0.7RC$. Recommended ranges of values for R and C are 4–40 K Ω and 10 pf to 1000 μF respectively. The IC provides complementary outputs. That is, we have a stable LOW or HIGH state and the corresponding quasi-stable HIGH or LOW state available on Q and $\overline{Q}$ outputs.

Figure 10.8 shows the use of 74123, a retriggerable monostable multivibrator. Like 74121, this IC, too, provides features for triggering on either LOW-to-HIGH or HIGH-to-LOW edges of the trigger pulses. The output pulse width depends on external R and C. It can be computed from $T = 0.28RC\times$ $[1 + (0.7/R)]$, where R and C are respectively in kiloohms and picofarads and T is in nanoseconds. This formula is valid for $C > 1000$ pF. The recommended range of values for R is 5–50 KΩ . Figures 10.8(a) and (b) give application circuits for HIGH-to-LOW and LOW-to-HIGH triggering respectively. It may be mentioned here that there can be other triggering circuit options for both LOW-to-HIGH and HIGH-to-LOW edge triggering of monoshot.

10.2.2 IC Timer-Based Multivibrators

IC timer 555 is one of the most commonly used general-purpose linear integrated circuits. The simplicity with which monostable and astable multivibrator circuits can be configured around this IC is one of the main reasons for its wide use. Figure 10.9 shows the internal schematic of timer IC 555. It comprises two opamp comparators, a flip-flop, a discharge transistor, three identical resistors and an output stage. The resistors set the reference voltage levels at the noninverting input of the lower comparator and the inverting input of the upper comparator at $(+V_{CC}/3)$ and $(+2V_{CC}/3)$. The outputs of the two comparators feed the SET and RESET inputs of the flip-flop and thus decide the logic status

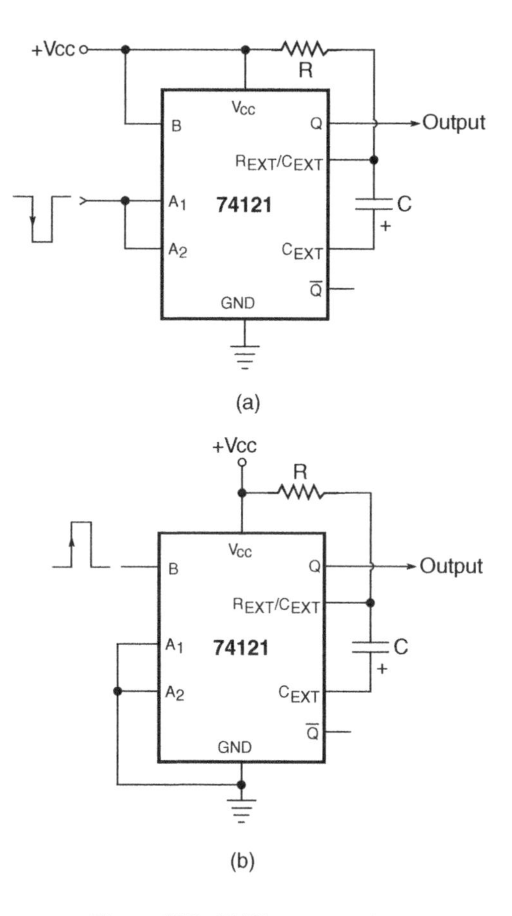

Figure 10.7 74121 as a monoshot.

of its output and subsequently the final output. The flip-flop complementary outputs feed the output stage and the base of the discharge transistor. This ensures that when the output is HIGH the discharge transistor is OFF, and when the output is LOW the discharge transistor is ON. Different terminals of timer 555 are designated as *ground* (terminal 1), *trigger* (terminal 2), *output* (terminal 3), *reset* (terminal 4), *control* (terminal 5), *threshold* (terminal 6), *discharge* (terminal 7) and $+V_{CC}$ (terminal 8). With this background, we will now describe the astable and monostable circuits configured around timer 555.

10.2.2.1 Astable Multivibrator Using Timer IC 555

Figure 10.10(a) shows the basic 555 timer based astable multivibrator circuit. Initially, capacitor C is fully discharged, which forces the output to go to the HIGH state. An open discharge transistor allows the capacitor C to charge from $+V_{CC}$ through R_1 and R_2. When the voltage across C exceeds $+2V_{CC}/3$, the output goes to the LOW state and the discharge transistor is switched ON at the same time.

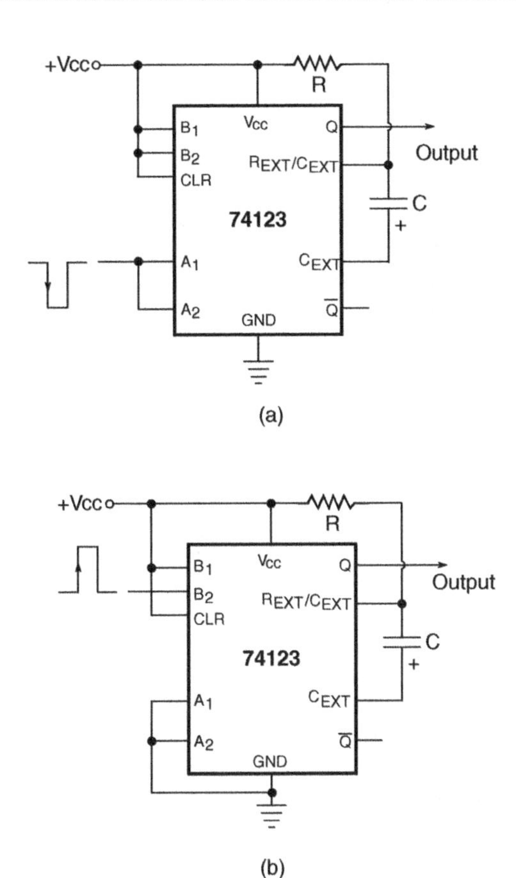

Figure 10.8 74123 as a retriggerable monoshot.

Capacitor C begins to discharge through R_2 and the discharge transistor inside the IC. When the voltage across C falls below $+V_{CC}/3$, the output goes back to the HIGH state. The charge and discharge cycles repeat and the circuit behaves like a free-running multivibrator. Terminal 4 of the IC is the RESET terminal. usually, it is connected to $+V_{CC}$. If the voltage at this terminal is driven below 0.4 V, the output is forced to the LOW state, overriding command pulses at terminal 2 of the IC. The HIGH-state and LOW-state time periods are governed by the charge ($+V_{CC}/3$ to $+2V_{CC}/3$) and discharge ($+2V_{CC}/3$ to $+V_{CC}/3$) timings. these are given by the equations

$$\text{HIGH-state time period } T_{HIGH} = 0.69(R_1 + R_2).C \tag{10.7}$$

$$\text{LOW-state time period } T_{LOW} = 0.69R_2.C \tag{10.8}$$

The relevant waveforms are shown in Fig. 10.10(b). The time period T and frequency f of the output waveform are respectively given by the equations

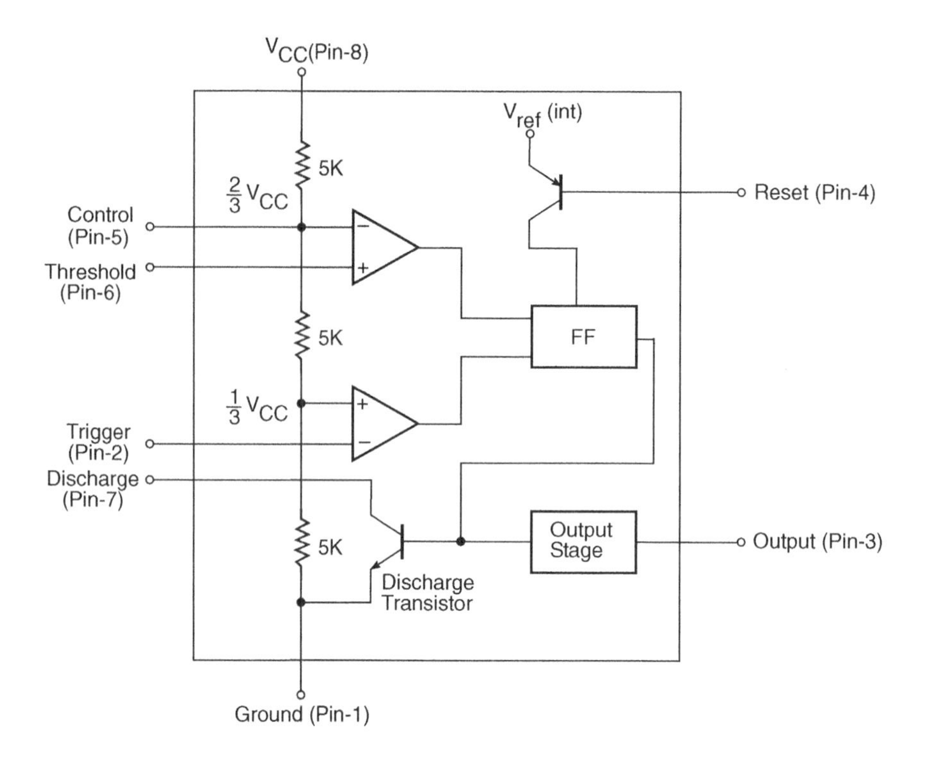

Figure 10.9 Internal schematic of timer IC 555.

$$\text{Time period } T = 0.69(R_1 + 2R_2).C \tag{10.9}$$

$$\text{Frequency } F = 1/[0.69(R_1 + 2R_2).C] \tag{10.10}$$

Remember that, when the astable multivibrator is powered, the first-cycle HIGH-state time period is about 30 % longer, as the capacitor is initially discharged and it charges from 0 (rather than $+V_{CC}/3$) to $+2V_{CC}/3$.

In the case of the astable multivibrator circuit in Fig. 10.10(a), the HIGH-state time period is always greater than the LOW-state time period. Figures 10.10(c) and (d) show two modified circuits where the HIGH-state and LOW-state time periods can be chosen independently. For the astable multivibrator circuits in Fig. 10.10(c) and (d), the two time periods are given by the equations

$$\text{HIGH-state time period} = 0.69R_1.C \tag{10.11}$$

$$\text{LOW-state time period} = 0.69R_2.C \tag{10.12}$$

For $R_1 = R_2 = R$

$$T = 1.38RC \text{ and } f = 1/1.38RC \tag{10.13}$$

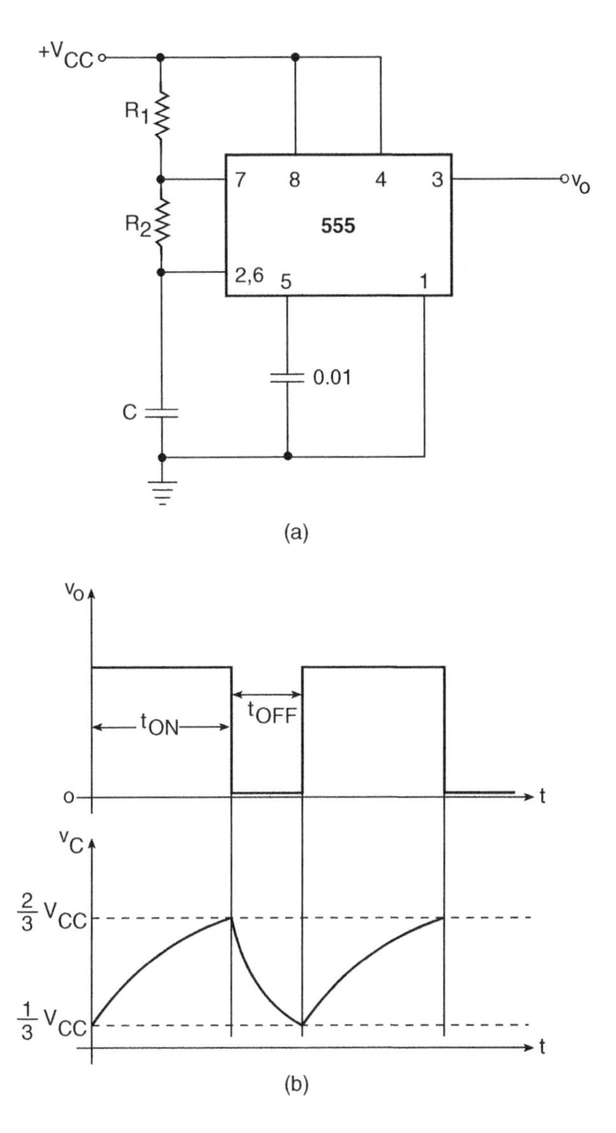

Figure 10.10 (a) Astable multivibrator using timer IC 555, (b) astable multivibrator relevant waveforms and (c, d) modified versions of the astable multivibrator using timer IC 555.

10.2.2.2 Monostable Multivibrator Using Timer IC 555

Figure 10.11(a) shows the basic monostable multivibrator circuit configured around timer 555. A trigger pulse is applied to terminal 2 of the IC, which should initially be kept at $+V_{CC}$. A HIGH at terminal 2 forces the output to the LOW state. A HIGH-to-LOW trigger pulse at terminal 2 holds the output in the HIGH state and simultaneously allows the capacitor to charge from $+V_{CC}$ through R. Remember that a LOW level of the trigger pulse needs to go at least below $+V_{CC}/3$. When the capacitor voltage exceeds $+2V_{CC}/3$, the output goes back to the LOW state. We will need to apply another trigger pulse to

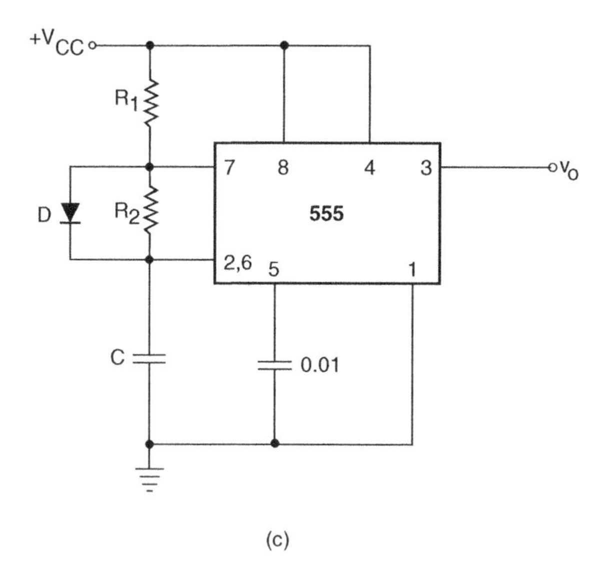

(c)

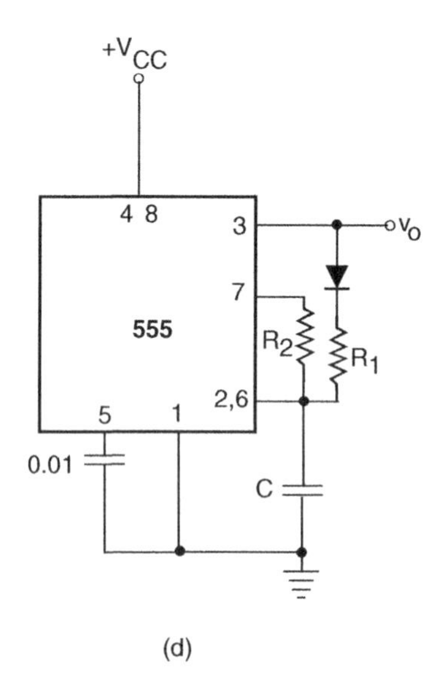

(d)

Figure 10.10 (*continued*).

terminal 2 to make the output go to the HIGH state again. Every time the timer is appropriately triggered, the output goes to the HIGH state and stays there for the time it takes the capacitor to charge from 0 to $+2V_{CC}/3$. This time period, which equals the monoshot output pulse width, is given by the equation

$$T = 1.1RC \tag{10.14}$$

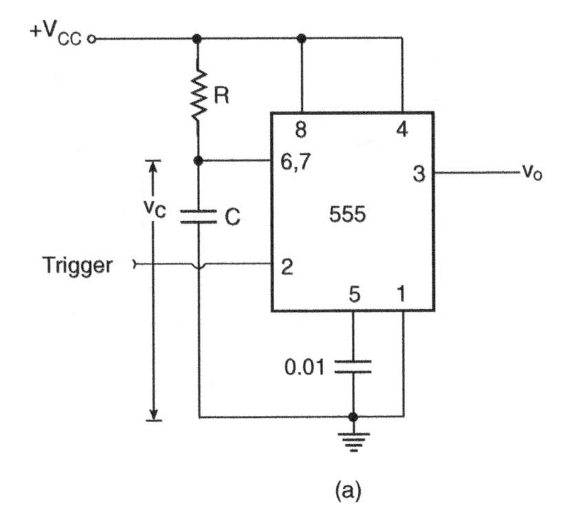

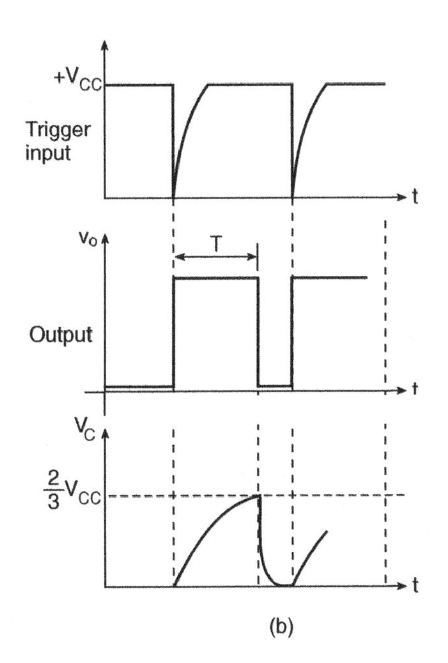

Figure 10.11 (a) Monostable multivibrator using timer 555 and (b) monostable multivibrator relevant waveforms.

Figure 10.11(b) shows the relevant waveforms for the circuit of Fig. 10.11(a).

It is often desirable to trigger a monostable multivibrator either on the trailing (HIGH-to-LOW) or leading (LOW-to-HIGH) edges of the trigger waveform. In order to achieve that, we will need an external circuit between the trigger waveform input and terminal 2 of timer 555. The external circuit ensures that terminal 2 of the IC gets the required trigger pulse corresponding to the desired edge of

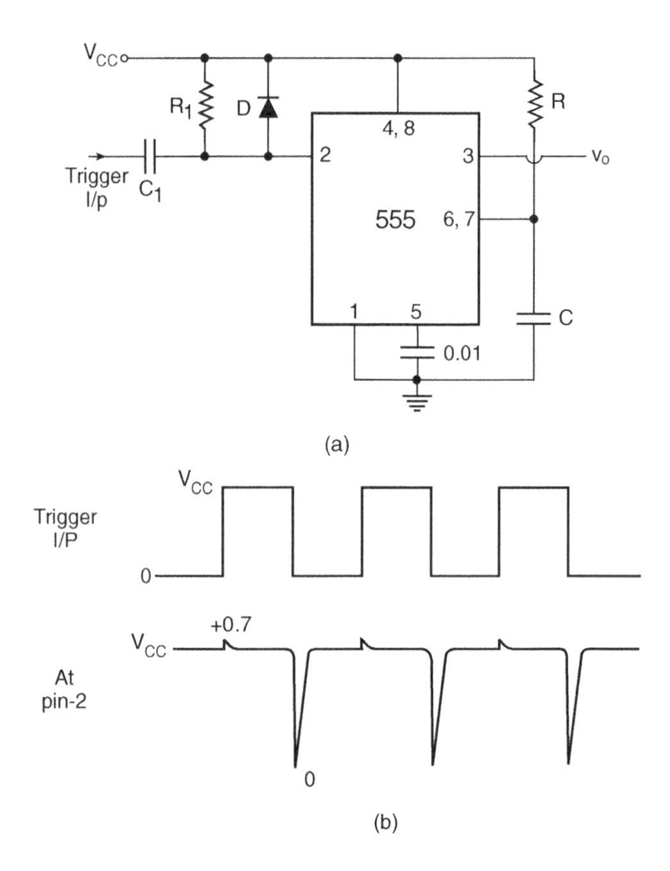

Figure 10.12 555 monoshot triggering on trailing edges.

the trigger waveform. Figure 10.12(a) shows the monoshot configuration that can be triggered on the trailing edges of the trigger waveform. R_1–C_1 constitutes a differentiator circuit. One of the terminals of resistor R_1 is tied to $+V_{CC}$, with the result that the amplitudes of differentiated pulses are $+V_{CC}$ to $+2V_{CC}$ and $+V_{CC}$ to ground, corresponding to the leading and trailing edges of the trigger waveform respectively. Diode D clamps the positive-going differentiated pulses to about +0.7 V. The net result is that the trigger terminal of timer 555 gets the required trigger pulses corresponding to HIGH-to-LOW edges of the trigger waveform. Figure 10.12(b) shows the relevant waveforms.

Figure 10.13(a) shows the monoshot configuration that can be triggered on the leading edges of the trigger waveform. The R_1–C_1 combination constitutes the differentiator producing positive and negative pulses corresponding to LOW-to-HIGH and HIGH-to-LOW transitions of the trigger waveform. Negative pulses are clamped by the diode, and the positive pulses are applied to the base of a transistor switch. The collector terminal of the transistor feeds the required trigger pulses to terminal 2 of the IC. Figure 10.13(b) shows the relevant waveforms.

For the circuits shown in Figs 10.12 and 10.13 to function properly, the values of R_1 and C_1 for the differentiator should be chosen carefully. Firstly, the differentiator time constant should be much smaller than the HIGH time of the trigger waveform for proper differentiation. Secondly, the differentiated pulse width should be less than the expected HIGH time of the monoshot output.

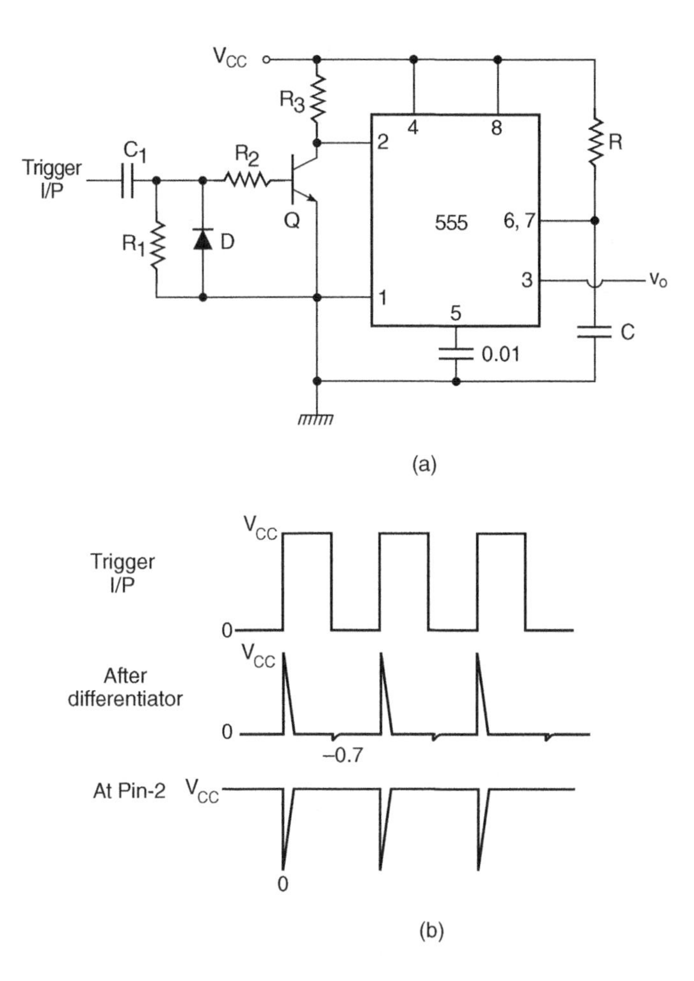

Figure 10.13 555 monoshot triggering on leading edges.

Example 10.1

The pulsed waveform of Fig. 10.14(b) is applied to the RESET terminal of the astable multivibrator circuit of Fig. 10.14(a). Draw the output waveform.

Solution

The circuit shown in Fig. 10.14(a) is an astable multivibrator with a 500 Hz symmetrical waveform applied to its RESET terminal. The RESET terminal is alternately HIGH and LOW for 1.0 ms. When the RESET input is LOW, the output is forced to the LOW state. When the RESET input is HIGH, an astable waveform appears at the output. The HIGH and LOW time periods of the astable multivibrator are determined as follows:

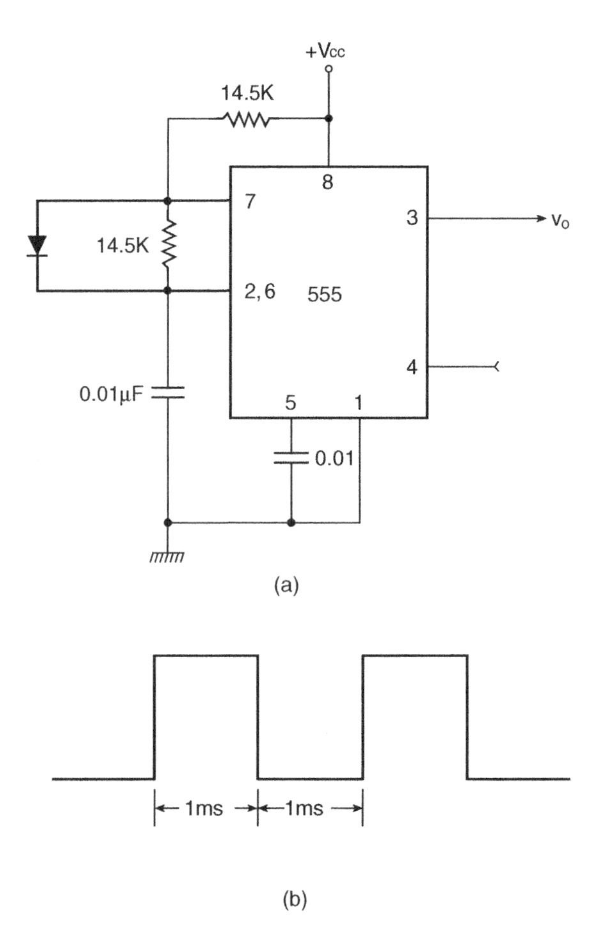

Figure 10.14 Example 10.1.

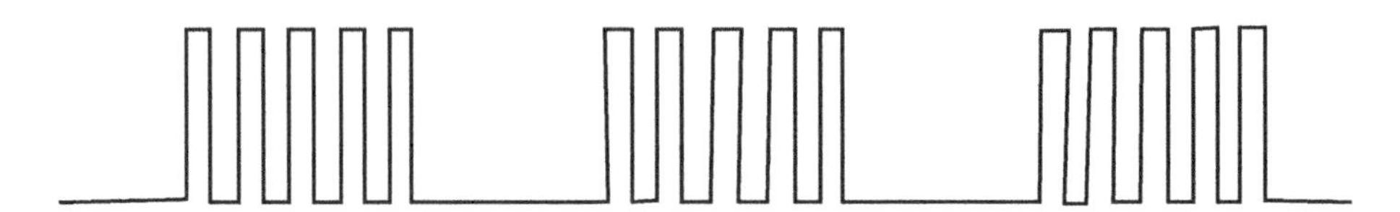

Figure 10.15 Solution to example 10.1.

$$\text{HIGH time} = 0.69 \times 14.5 \times 10^3 \times 0.01 \times 10^{-6} = 100\,\mu\text{s}$$

$$\text{LOW time} = 0.69 \times 14.5 \times 10^3 \times 0.01 \times 10^{-6} = 100\,\mu\text{s}$$

The astable output is thus a 5 kHz symmetrical waveform. Every time the RESET terminal goes to HIGH for 1.0 ms, five cycles of 5 kHz waveform appear at the output. Figure 10.15 shows the output waveform appearing at terminal 3 of the timer IC.

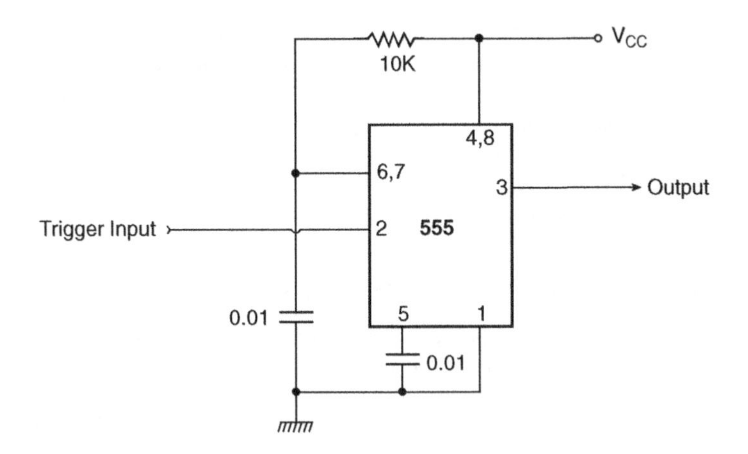

Figure 10.16 Example 10.2.

Example 10.2

Refer to the monostable multivibrator circuit in Fig. 10.16. The trigger terminal (pin 2 of the IC) is driven by a symmetrical pulsed waveform of 10 kHz. Determine the frequency and duty cycle of the output waveform.

Solution

- The frequency of the trigger waveform = 10 kHz.
- The time period between two successive leading or trailing edges = 100 μs.
- The expected pulse width of the monoshot output = $1.1RC = 1.1 \times 10^4 \times 10^{-8} = 110$ μs.
- The trigger waveform is a symmetrical one; it has HIGH and LOW time periods of 50 μs each. Since the LOW-state time period of the trigger waveform is less than the expected output pulse width, it can successfully trigger the monoshot on its trailing edges.
- Since the time period between two successive trailing edges is 100 μs and the expected output pulse width is 110 μs, only alternate trailing edges of the trigger waveform will trigger the monoshot.
- The frequency of the output waveform = 10/2 = 5 kHz.
- The time period of the output waveform = $1/(5 \times 10^3) = 200$ μs.
- Therefore, the duty cycle of the output waveform = 110/200 = 0.55.

10.3 *R-S* Flip-Flop

A flip-flop, as stated earlier, is a bistable circuit. Both of its output states are stable. The circuit remains in a particular output state indefinitely until something is done to change that output status. Referring to the bistable multivibrator circuit discussed earlier, these two states were those of the output transistor in saturation (representing a LOW output) and in cut-off (representing a HIGH output). If the LOW and HIGH outputs are respectively regarded as '0' and '1', then the output can either be a '0' or a '1'. Since either a '0' or a '1' can be held indefinitely until the circuit is appropriately triggered to go to the other state, the circuit is said to have memory. It is capable of storing one binary digit or one bit of digital information. Also, if we recall the functioning of the bistable multivibrator circuit, we find that, when one of the transistors was in saturation, the other was in cut-off. This implies that, if we had taken outputs from the collectors of both transistors, then the two outputs would be complementary. In the flip-flops of various types that are available in IC form, we will see that all these devices offer complementary outputs usually designated as Q and $\overline{Q}$.

The *R-S* flip-flop is the most basic of all flip-flops. The letters '*R*' and '*S*' here stand for RESET and SET. When the flip-flop is SET, its Q output goes to a '1' state, and when it is RESET it goes to a '0' state. The $\overline{Q}$ output is the complement of the Q output at all times.

10.3.1 R-S Flip-Flop with Active LOW Inputs

Figure 10.17(a) shows a NAND gate implementation of an *R-S* flip-flop with active LOW inputs. The two NAND gates are cross-coupled. That is, the output of NAND 1 is fed back to one of the inputs of NAND 2, and the output of NAND 2 is fed back to one of the inputs of NAND 1. The remaining inputs of NAND 1 and NAND 2 are the S and R inputs. The outputs of NAND 1 and NAND 2 are respectively Q and $\overline{Q}$ outputs.

The fact that this configuration follows the function table of Fig. 10.17(c) can be explained. We will look at different entries of the function table, one at a time.

Let us take the case of $R = S = 1$ (the first entry in the function table). We will prove that, for $R = S = 1$, the Q output remains in its existing state. In the truth table, Q_n represents the existing state and Q_{n+1} represents the state of the flip-flop after it has been triggered by an appropriate pulse at the R or S input. Let us assume that $Q = 0$ initially. This '0' state fed back to one of the inputs of gate 2 ensures that $\overline{Q} = 1$. The '1' state of $\overline{Q}$ fed back to one of the inputs of gate 1 along with $S = 1$ ensures that $Q = 0$. Thus, $R = S = 1$ holds the existing stage. Now, if Q was initially in the '1' state and not the '0' state, this '1' fed back to one of the inputs of gate 2 along with $R = 1$ forces $\overline{Q}$ to be in the '0' state. The '0' state, when fed back to one of the inputs of gate 1, ensures that Q remains in its existing state of logic '1'. Thus, whatever the state of Q, $R = S = 1$ holds the existing state.

Let us now look at the second entry of the function table where $S = 0$ and $R = 1$. We can see that such an input combination forces the Q output to the '1' state. On similar lines, the input combination $S = 1$ and $R = 0$ (third entry of the truth table) forces the Q output to the '0' state. It would be interesting to analyse what happens when $S = R = 0$. This implies that both Q and $\overline{Q}$ outputs should go to the '1' state, as one of the inputs of a NAND gate being a logic '0' should force its output to the logic '1' state irrespective of the status of the other input. This is an undesired state as Q and $\overline{Q}$ outputs are to be the complement of each other. The input condition (i.e. $R = S = 0$) that causes such a situation is therefore considered to be an invalid condition and is forbidden. Figure 10.17(b) shows the logic symbol of such a flip-flop. The R and S inputs here have been shown as active LOW inputs, which is obvious as this flip-flop of Fig. 10.17(a) is SET (that is, $Q = 1$) when $S = 0$ and RESET (that is, $Q = 0$) when $R = 0$. Thus, R and S are active when LOW. The term CLEAR input is also used sometimes in place of RESET. The operation of the *R-S* flip-flop of Fig. 10.17(a) can be summarized as follows:

1. SET = RESET = 1 is the normal resting condition of the flip-flop. It has no effect on the output state of the flip-flop. Both Q and $\overline{Q}$ outputs remain in the logic state they were in prior to this input condition.
2. SET = 0 and RESET = 1 sets the flip-flop. Q and $\overline{Q}$ respectively go to the '1' and '0' state.
3. SET = 1 and RESET = 0 resets or clears the flip-flop. Q and $\overline{Q}$ respectively go to the '0' and '1' state.
4. SET = RESET = 0 is forbidden as such a condition tries to set (that is, $Q = 1$) and reset (that is, $\overline{Q} = 1$) the flip-flop at the same time. To be more precise, SET and RESET inputs in the R-S flip-flop cannot be active at the same time.

The *R-S* flip-flop of Fig. 10.17(a) is also referred to as an *R-S* latch. This is because any combination at the inputs immediately manifests itself at the output as per the truth table.

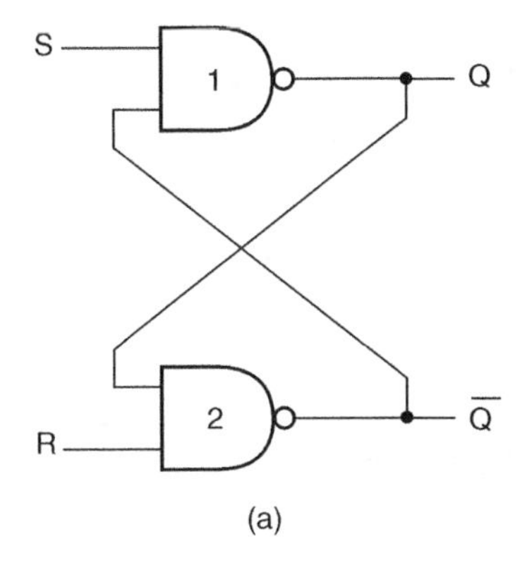

(a)

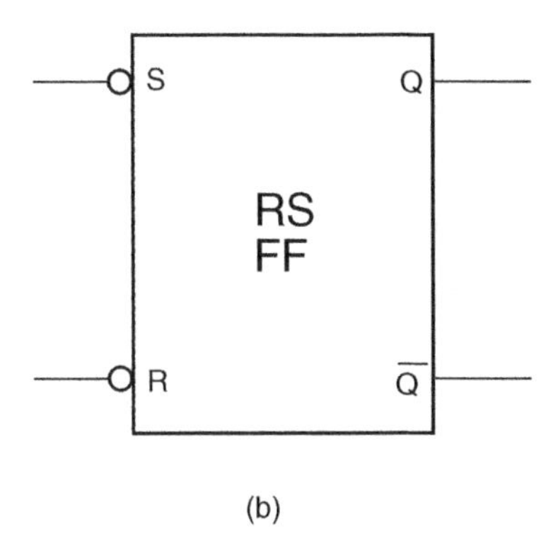

(b)

Operation Mode	S	R	Q_{n+1}
No change	1	1	Qn
SET	0	1	1
RESET	1	0	0
Forbidden	0	0	—

(c)

Figure 10.17 *R-S* flip-flop with active LOW inputs.

10.3.2 R-S Flip-Flop with Active HIGH Inputs

Figure 10.18(a) shows another NAND gate implementation of the *R-S* flip-flop. Figures 10.18(b) and (c) respectively show its circuit symbol and function table. Such a circuit would have active HIGH inputs. The input combination $R = S = 1$ would be forbidden as SET and RESET inputs in an *R-S* flip-flop cannot be active at the same time.

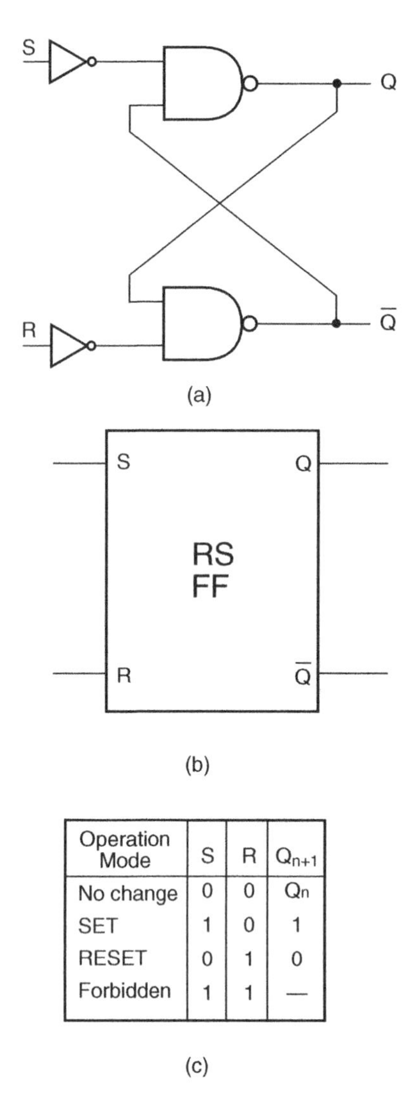

Operation Mode	S	R	Q_{n+1}
No change	0	0	Q_n
SET	1	0	1
RESET	0	1	0
Forbidden	1	1	—

Figure 10.18 *R-S* flip-flop with active HIGH inputs.

The *R-S* flip-flops (or latches) of Figs 10.17(a) and 10.18 (a) may also be implemented with NOR gates. The NOR gate counterparts of Fig. 10.17(a) and Fig. 10.18(a) are respectively shown in Figs 10.19(a) and (b).

So far we have discussed the operation of an *R-S* flip-flop with the help of its logic diagram and the function table on lines similar to the case of combinational circuits. We do, however, appreciate that a sequential circuit would be better explained if we expressed its output (immediately after it was clocked) in terms of its present output and its inputs. The function tables of Figs 10.17(c) and 10.18(c) may be redrawn as shown in Figs 10.20(a) and (b) respectively. This new form of representation is known as the characteristic table. Having done this, we could even write simplified Boolean expressions,

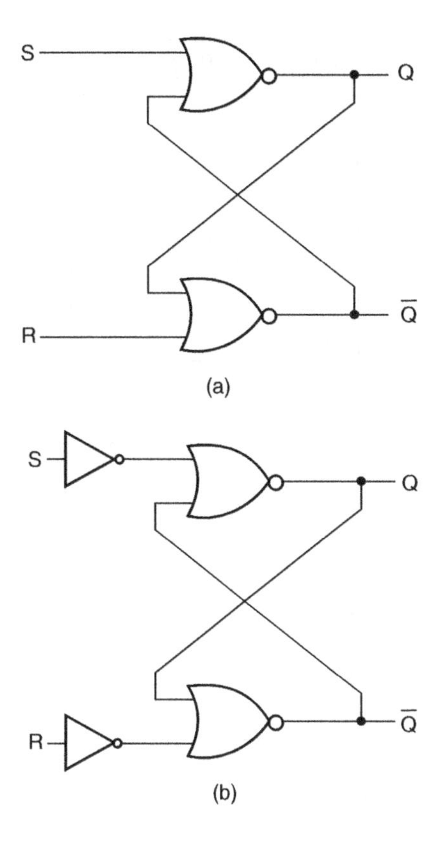

Figure 10.19 NOR implementation of an *R-S* flip-flop.

called characteristic equations, using any of the minimization techniques, such as Karnaugh mapping. The K-maps for the characteristic tables of Figs 10.20(a) and (b) are given in Figs 10.20(c) and (d) respectively. Characteristic equations for *R-S* flip-flops with active LOW and active HIGH inputs are given by the equations

$$Q_{n+1} = \overline{S} + R.Q_n \quad \text{and} \quad S + R = 1 \tag{10.15}$$

$$Q_{n+1} = S + \overline{R}.Q_n \quad \text{and} \quad S.R = 0 \tag{10.16}$$

$S+R=1$ indicates that $R=S=0$ is a prohibited entry. Similarly, $S.R=0$ only indicates that $R=S=1$ is a prohibited entry.

10.3.3 Clocked R-S Flip-Flop

In the case of a clocked *R-S* flip-flop, or for that matter any clocked flip-flop, the outputs change states as per the inputs only on the occurrence of a clock pulse. The clocked flip-flop could be a level-triggered one or an edge-triggered one. The two types are discussed in the next section. For the

Qn	S	R	Q_{n+1}
0	0	0	Indeter
0	0	1	1
0	1	0	0
0	1	1	0
1	0	0	Indeter
1	0	1	1
1	1	0	0
1	1	1	1

(a)

Qn	S	R	Q_{n+1}
0	0	0	0
0	0	1	0
0	1	0	1
0	1	1	Indeter
1	0	0	1
1	0	1	0
1	1	0	1
1	1	1	Indeter

(b)

Q_n \ SR	00	01	11	10
0	X	1		
1	X	1	1	

(c)

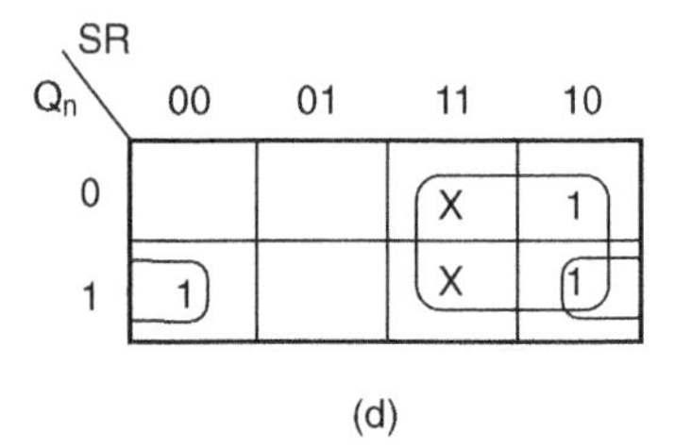

(d)

Figure 10.20 (a) Characteristic table of an *R-S* flip-flop with active LOW inputs, (b) the characteristic table of an *R-S* flip-flop with active HIGH inputs, (c) the K-map solution of an *R-S* flip-flop with active LOW inputs and (d) the K-map solution of an *R-S* flip-flop with active HIGH inputs.

time being, let us first see how the flip-flop of the previous section can be transformed into a clocked flip-flop. Figure 10.21(a) shows the logic implementation of a clocked flip-flop that has active HIGH inputs. The function table for the same is shown in Fig. 10.21(b) and is self-explanatory.

The basic flip-flop is the same as that shown in Fig. 10.17(a). The two NAND gates at the input have been used to couple the R and S inputs to the flip-flop inputs under the control of the clock signal. When the clock signal is HIGH, the two NAND gates are enabled and the S and R inputs are passed on to flip-flop inputs with their status complemented. The outputs can now change states as per the status of R and S at the flip-flop inputs. For instance, when $S = 1$ and $R = 0$ it will be passed on as 0 and 1 respectively when the clock is HIGH. When the clock is LOW, the two NAND gates produce a '1' at their outputs, irrespective of the S and R status. This produces a logic '1' at both inputs of the flip-flop, with the result that there is no effect on the output states. Figure 10.22(a) shows the clocked R-S flip-flop with active LOW R and S inputs. The logic implementation here is a modification of the basic R-S flip-flop in Fig. 10.18(a). The truth table of this flip-flop, as given in Fig. 10.22(b), is self-explanatory.

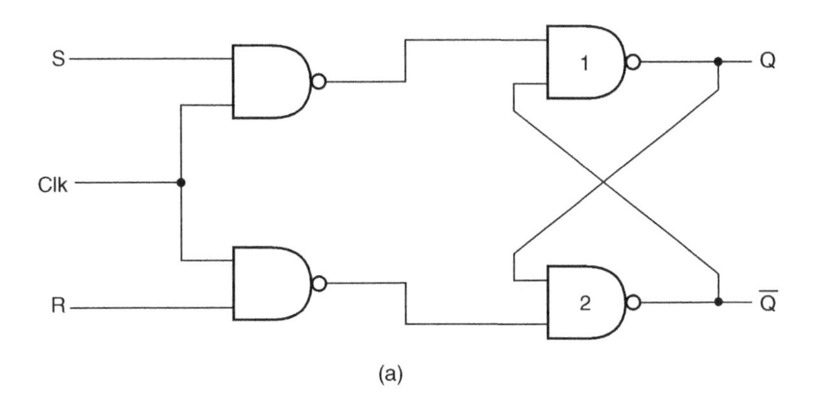

(a)

S	R	Clk	Q_{n+1}
0	0	0	Q_n
0	0	1	Q_n
0	1	0	Q_n
0	1	1	0
1	0	0	Q_n
1	0	1	1
1	1	0	Q_n
1	1	1	Invalid

(b)

Figure 10.21 Clocked R-S flip-flop with active HIGH inputs.

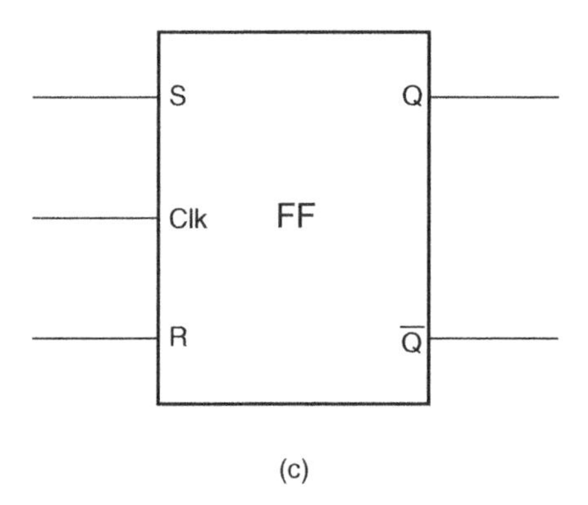

(c)

Figure 10.21 (*continued*).

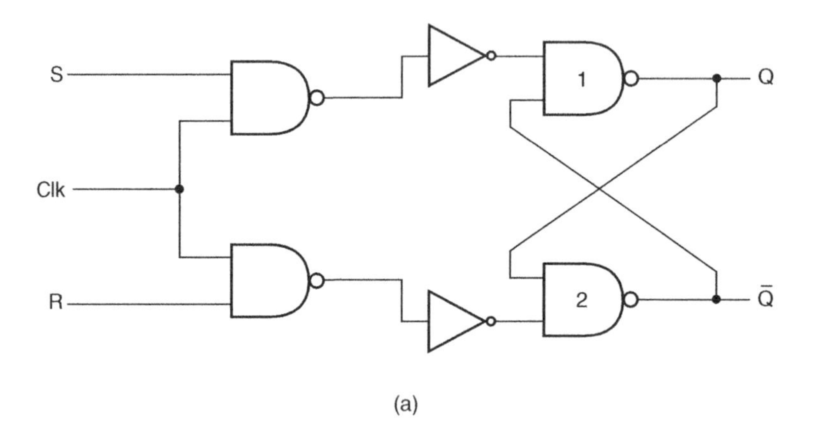

(a)

S	R	Clk	Q_{n+1}
0	0	0	Q_n
0	0	1	Invalid
0	1	0	Q_n
0	1	1	1
1	0	0	Q_n
1	0	1	0
1	1	0	Q_n
1	1	1	Q_n

(b)

Figure 10.22 Clocked *R-S* flip-flop with active LOW inputs.

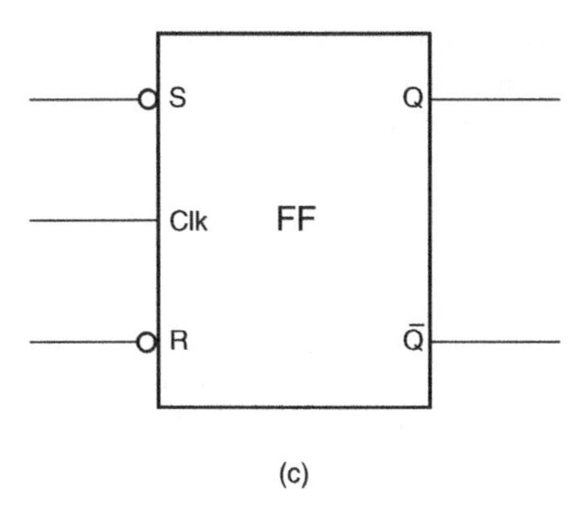

(c)

Figure 10.22 (*continued*).

10.4 Level-Triggered and Edge-Triggered Flip-Flops

In a *level-triggered* flip-flop, the output responds to the data present at the inputs during the time the clock pulse level is HIGH (or LOW). That is, any changes at the input during the time the clock is active (HIGH or LOW) are reflected at the output as per its function table. The clocked *R-S* flip-flop described in the preceding paragraphs is a level-triggered flip-flop that is active when the clock is HIGH.

In an *edge-triggered* flip-flop, the output responds to the data at the inputs only on LOW-to-HIGH or HIGH-to-LOW transition of the clock signal. The flip-flop in the two cases is referred to as positive edge triggered and negative edge triggered respectively. Any changes in the input during the time the clock pulse is HIGH (or LOW) do not have any effect on the output. In the case of an edge-triggered flip-flop, an edge detector circuit transforms the clock input into a very narrow pulse that is a few nanoseconds wide. This narrow pulse coincides with either LOW-to-HIGH or HIGH-to-LOW transition of the clock input, depending upon whether it is a positive edge-triggered flip-flop or a negative edge-triggered flip-flop. This pulse is so narrow that the operation of the flip–flop can be considered to have occurred on the edge itself.

Figure 10.23 shows the clocked *R-S* flip-flop of Fig. 10.21 with the edge detector block incorporated in the clock circuit. Figures 10.24 (a) and (b) respectively show typical edge detector circuits for positive

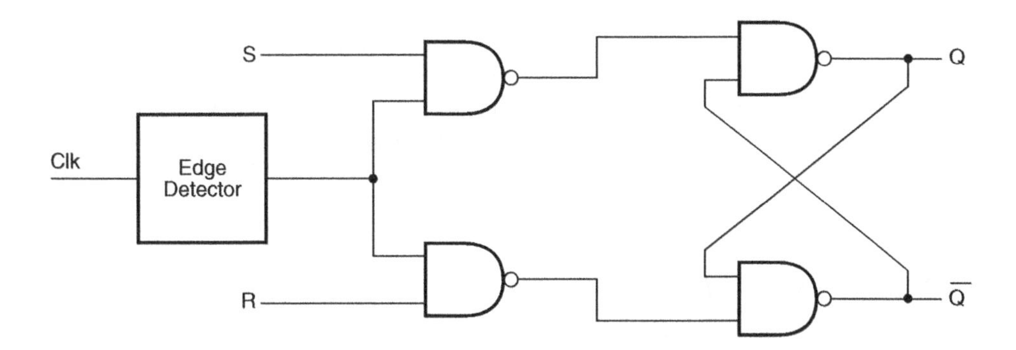

Figure 10.23 Edge-triggered *R-S* flip-flop.

and negative edge triggering. The width of the narrow pulse generated by this edge detector circuit is equal to the propagation delay of the inverter. Figure 10.25 shows the circuit symbol for the flip-flop of Fig. 10.23 for the positive edge-triggered mode [Fig. 10.25(a)] and the negative edge-triggered mode [Fig. 10.25(b)].

10.5 *J-K* Flip-Flop

A *J-K* flip-flop behaves in the same fashion as an *R-S* flip-flop except for one of the entries in the function table. In the case of an *R-S* flip-flop, the input combination $S = R = 1$ (in the case of a flip-flop with active HIGH inputs) and the input combination $S = R = 0$ (in the case of a flip-flop with active LOW inputs) are prohibited. In the case of a *J-K* flip-flop with active HIGH inputs, the output of the flip-flop toggles, that is, it goes to the other state, for $J = K = 1$. The output toggles for $J = K = 0$ in the case of the flip-flop having active LOW inputs. Thus, a *J-K* flip-flop overcomes the problem of a forbidden input combination of the *R-S* flip-flop. Figures 10.26(a) and (b) respectively show the circuit symbol of level-triggered *J-K* flip-flops with active HIGH and active LOW inputs, along with their function tables. Figure 10.27 shows the realization of a *J-K* flip-flop with an *R-S* flip-flop.

The characteristic tables for a *J-K* flip-flop with active HIGH *J* and *K* inputs and a *J-K* flip-flop with active LOW *J* and *K* inputs are respectively shown in Figs 10.28(a) and (b). The corresponding Karnaugh maps are shown in Fig. 10.28(c) for the characteristics table of Fig. 10.28(a) and in Fig. 10.28(d) for the characteristic table of Fig. 10.28(b). The characteristic equations for the Karnaugh maps of Figs 10.28(c) and (d) are respectively

$$Q_{n+1} = J.\overline{Q_n} + \overline{K}.Q_n \quad (10.17)$$

$$Q_{n+1} = \overline{J}.\overline{Q_n} + K.Q_n \quad (10.18)$$

10.5.1 J-K Flip-Flop with PRESET and CLEAR Inputs

It is often necessary to clear a flip-flop to a logic '0' state ($Q_n = 0$) or preset it to a logic '1' state ($Q_n = 1$). An example of how this is realized is shown in Fig. 10.29(a). The flip-flop is cleared (that is, $Q_n = 0$) whenever the CLEAR input is '0' and the PRESET input is '1'. The flip-flop is preset to the logic '1' state whenever the PRESET input is '0' and the CLEAR input is '1'. Here, the CLEAR and PRESET inputs are active when LOW. Figure 10.29(b) shows the circuit symbol of this presettable, clearable, clocked *J-K* flip-flop. Figure 10.29(c) shows the function table of such a flip-flop. It is evident from the function table that, whenever the PRESET input is active, the output goes to the '1' state irrespective of the status of the clock, *J* and *K* inputs. Similarly, when the flip-flop is cleared, that is, the CLEAR input is active, the output goes to the '0' state irrespective of the status of the clock, *J* and *K* inputs. In a flip-flop of this type, both PRESET and CLEAR inputs should not be made active at the same time.

10.5.2 Master–Slave Flip-Flops

Whenever the width of the pulse clocking the flip-flop is greater than the propagation delay of the flip-flop, the change in state at the output is not reliable. In the case of edge-triggered flip-flops, this pulse width would be the trigger pulse width generated by the edge detector portion of the flip-flop

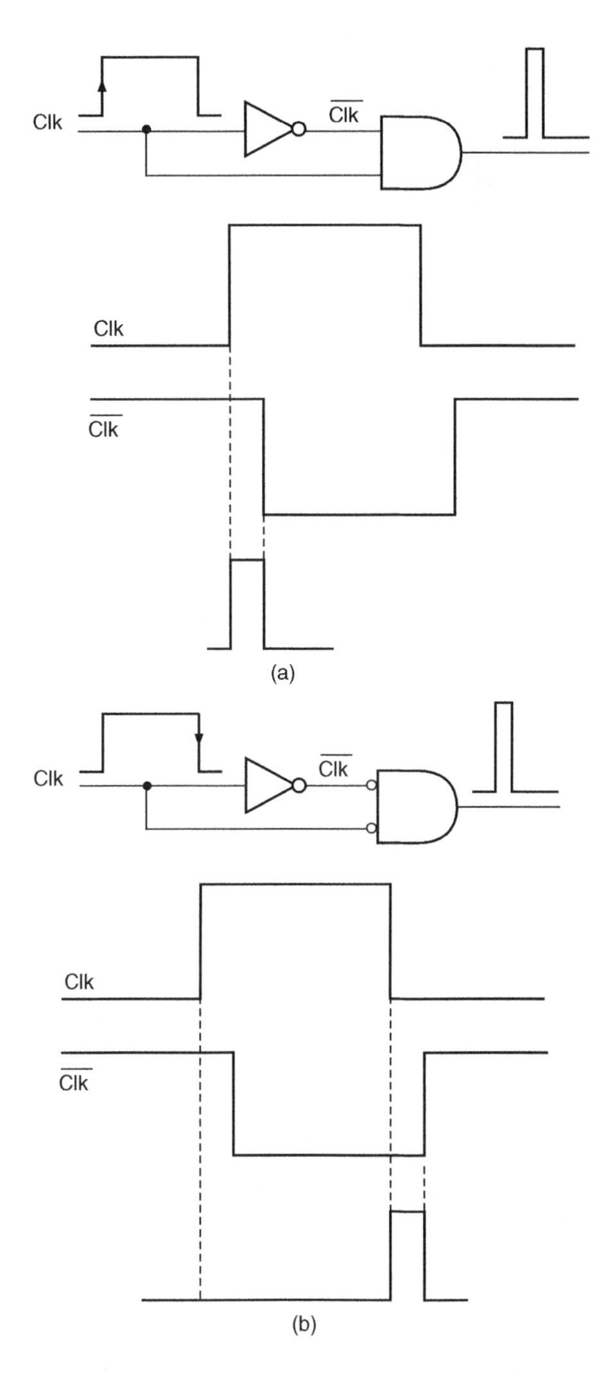

Figure 10.24 (a) Positive edge-triggered edge detector circuits and (b) negative edge-triggered edge detector circuits.

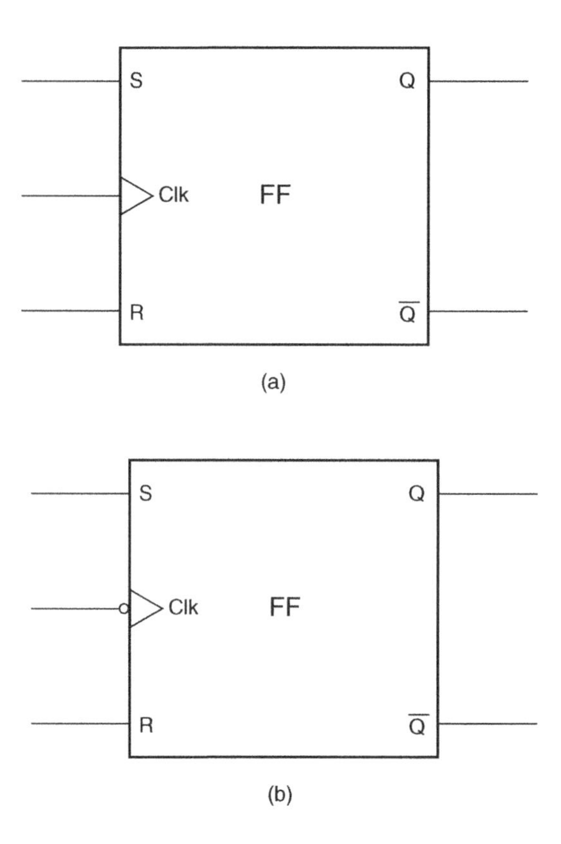

Figure 10.25 (a) Circuit symbol of a positive edge-triggered *R-S* flip-flop and (b) the circuit symbol of a negative edge-triggered *R-S* flip-flop.

and not the pulse width of the input clock signal. This phenomenon is referred to as the *race problem*. As the propagation delays are normally very small, the likelihood of the occurrence of a race condition is reasonably high. One way to get over this problem is to use a *master–slave* configuration. Figure 10.30(a) shows a master–slave flip-flop constructed with two *J-K* flip-flops. The first flip-flop is called the master flip-flop and the second is called the slave. The clock to the slave flip-flop is the complement of the clock to the master flip-flop. When the clock pulse is present, the master flip-flop is enabled while the slave flip-flop is disabled. As a result, the master flip-flop can change state while the slave flip-flop cannot. When the clock goes LOW, the master flip-flop gets disabled while the slave flip-flop is enabled. Therefore, the slave *J-K* flip-flop changes state as per the logic states at its *J* and *K* inputs. The contents of the master flip-flop are therefore transferred to the slave flip-flop, and the master flip-flop, being disabled, can acquire new inputs without affecting the output. As would be clear from the description above, a master–slave flip-flop is a pulse-triggered flip-flop and not an edge-triggered one. Figure 10.30(b) shows the truth table of a master–slave *J-K* flip-flop with active LOW PRESET and CLEAR inputs and active HIGH *J* and *K* inputs. The master–slave configuration has become obsolete. The newer IC technologies such as 74LS, 74AS, 74ALS, 74HC and 74HCT do not have master–slave flip-flops in their series.

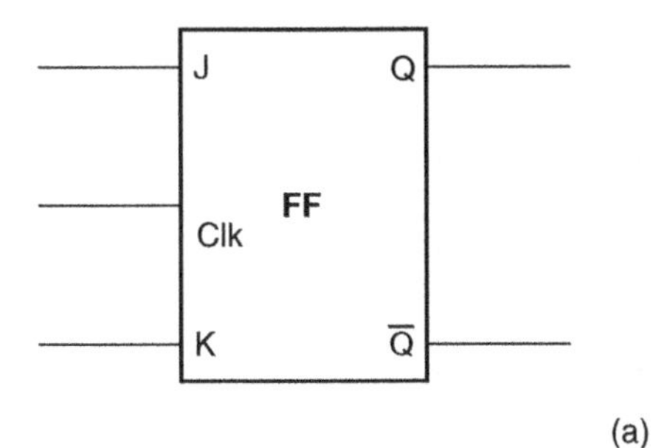

Operation Mode	J	K	Clk	Q_{n+1}
SET	1	0	1	1
RESET	0	1	1	0
NO CHANGE	0	0	1	Q_n
TOGGLE	1	1	1	$\overline{Q_n}$

(a)

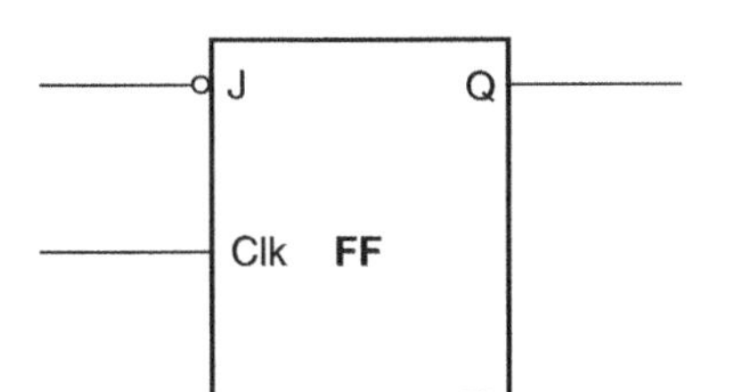

Operation Mode	J	K	Clk	Q_{n+1}
SET	0	1	1	1
RESET	1	0	1	0
NO CHANGE	1	1	1	Q_n
TOGGLE	0	0	1	$\overline{Q_n}$

(b)

Figure 10.26 (a) *J-K* flip-flop active HIGH inputs and (b) *J-K* flip-flop active LOW inputs.

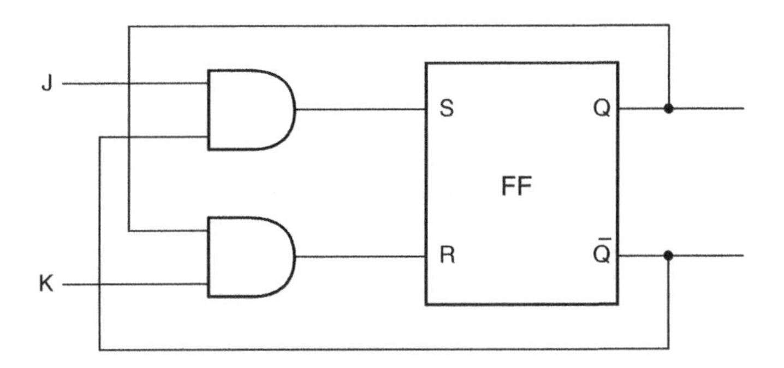

Figure 10.27 Realization of a *J-K* flip-flop using an *R-S* flip-flop.

Example 10.3

Draw the circuit symbol of the flip-flop represented by the function table of Fig. 10.31(a).

Solution

The first three entries of the function table indicate that the *J-K* flip-flop has active HIGH PRESET and CLEAR inputs. Referring to the fourth and fifth entries of the function table, it has active LOW *J* and *K* inputs. The seventh row of the function table confirms this. The output responds to positive (LOW-to-HIGH) edges of the clock input. Thus, the flip-flop represented by the given function table is a presettable, clearable, positive edge-triggered flip-flop with active HIGH PRESET and CLEAR

Q_n	J	K	Q_{n+1}
0	0	0	0
0	0	1	0
0	1	0	1
0	1	1	1
1	0	0	1
1	0	1	0
1	1	0	1
1	1	1	0

(a)

Q_n	J	K	Q_{n+1}
0	0	0	1
0	0	1	1
0	1	0	0
0	1	1	0
1	0	0	0
1	0	1	1
1	1	0	0
1	1	1	1

(b)

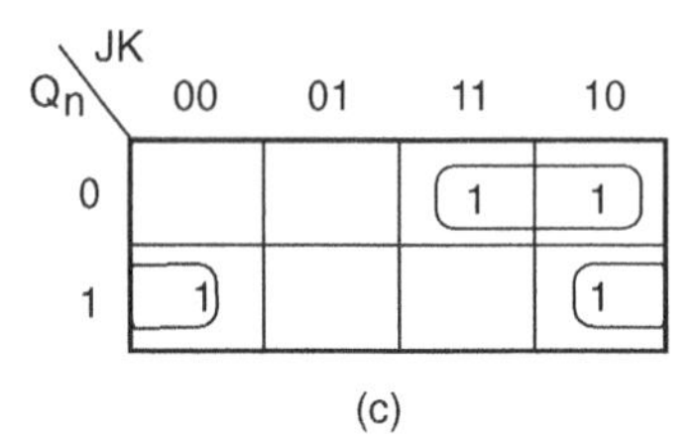

(c)

Q_n \ JK	00	01	11	10
0	1	1		
1		1	1	

(d)

Figure 10.28 (a) Characteristic table of a *J*-*K* flip-flop with active HIGH inputs, (b) the characteristic table of a *J*-*K* flip-flop with active LOW inputs, (c) the K-map solution of a *J*-*K* flip-flop with active HIGH inputs and (d) the K-map solution of a *J*-*K* flip-flop with active LOW inputs.

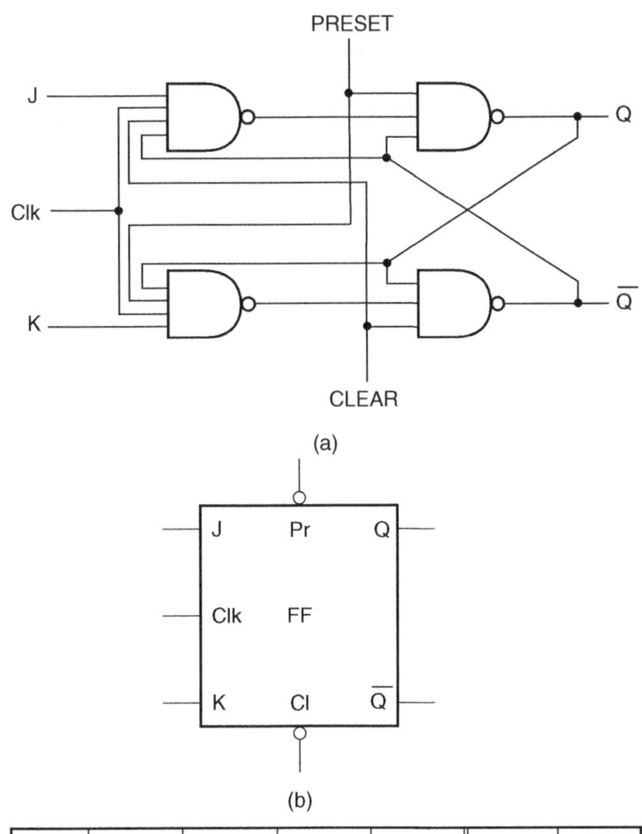

PR	CL	CLK	J	K	Q_{n+1}	$\overline{Q_{n+1}}$
0	1	X	X	X	1	0
1	0	X	X	X	0	1
0	0	X	X	X	---	---
1	1	1	0	0	Q_n	$\overline{Q_n}$
1	1	1	1	0	1	0
1	1	1	0	1	0	1
1	1	1	1	1	Toggle	
1	1	0	X	X	Q_n	$\overline{Q_n}$

(c)

Figure 10.29 *J-K* flip-flop with PRESET and CLEAR inputs.

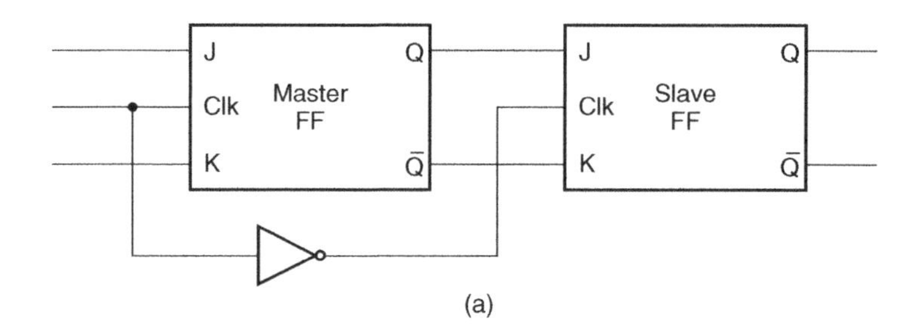

(a)

PR	CLR	CLK	J	K	Q_{n+1}	$\overline{Q_{n+1}}$
0	1	X	X	X	1	0
1	0	X	X	X	0	1
0	0	X	X	X	Unstable	
1	1	⎍	0	0	Q_n	$\overline{Q_n}$
1	1	⎍	1	0	1	0
1	1	⎍	0	1	0	1
1	1	⎍	1	1	Toggle	

(b)

Figure 10.30 Master–slave flip-flop.

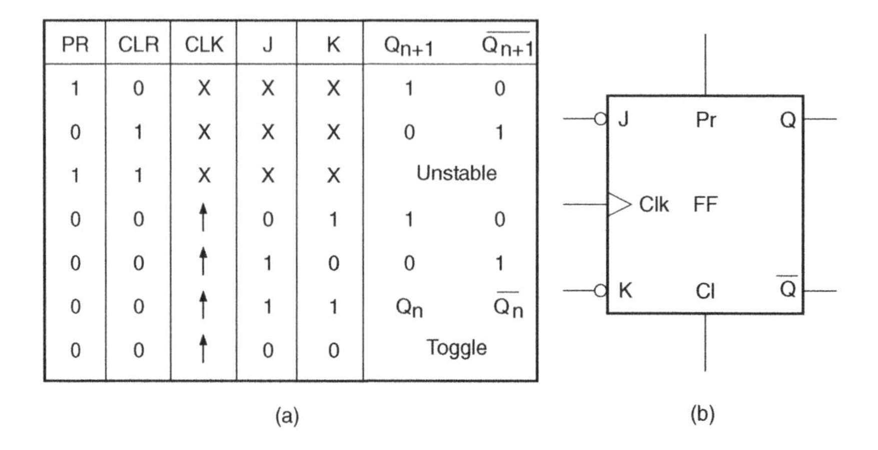

PR	CLR	CLK	J	K	Q_{n+1}	$\overline{Q_{n+1}}$
1	0	X	X	X	1	0
0	1	X	X	X	0	1
1	1	X	X	X	Unstable	
0	0	↑	0	1	1	0
0	0	↑	1	0	0	1
0	0	↑	1	1	Q_n	$\overline{Q_n}$
0	0	↑	0	0	Toggle	

(a) (b)

Figure 10.31 Example 10.3.

and active LOW J and K inputs. Figure 10.31(b) shows the circuit symbol of the flip-flop represented by this truth table.

Example 10.4

The 100 kHz square waveform of Fig. 10.32(a) is applied to the clock input of the flip-flops shown in Figs. 10.32(b) and (c). If the Q output is initially '0', draw the Q output waveform in the two cases. Also, determine the frequency of the Q output in the two cases.

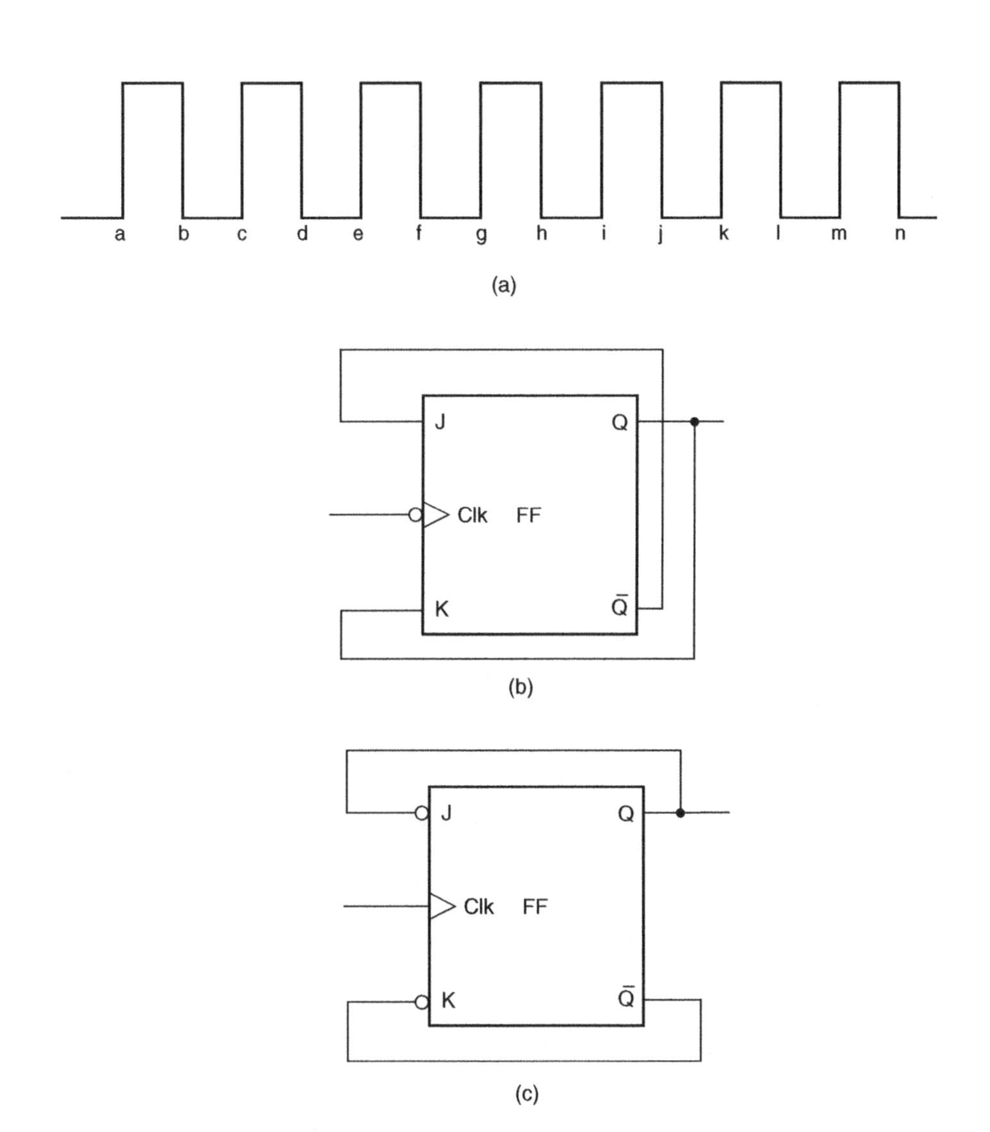

Figure 10.32 Example 10.4.

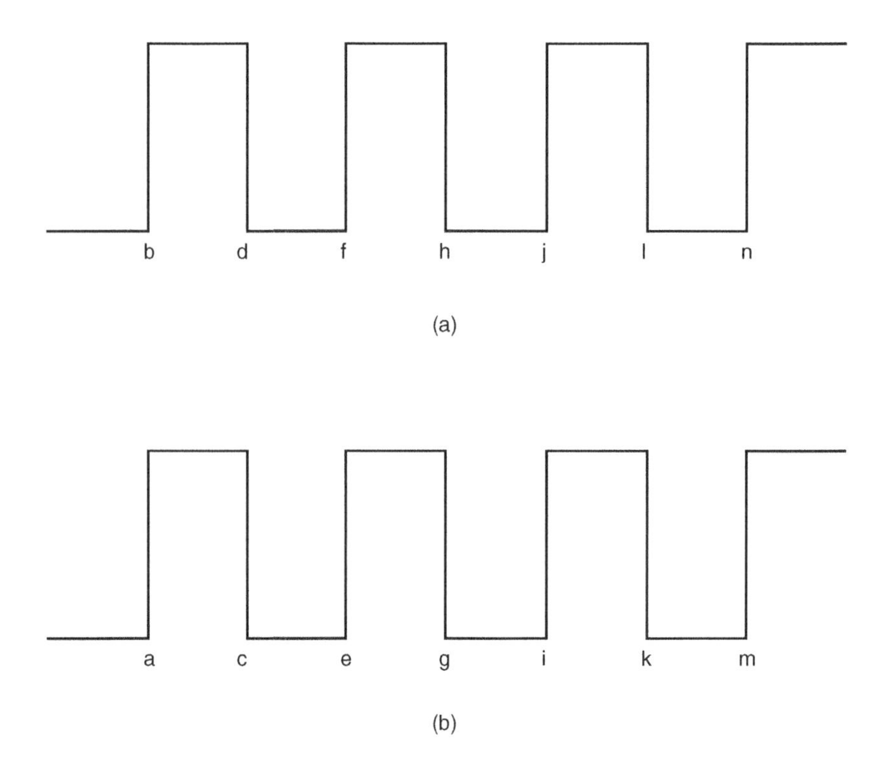

Figure 10.33 Solution to example 10.4.

Solution

Refer to the flip-flop of Fig. 10.32(b). Q is initially '0'. This makes the J and K inputs be initially '1' and '0' respectively. With the first trailing edge of the clock input, Q goes to the '1' state. Thus, J and K acquire a logic status of '0' and '1' respectively. With the next trailing edge of the clock input, Q goes to logic '0'. This process continues, and Q alternately becomes '1' and '0'. The Q output waveform for this case is shown in Fig. 10.33(a). In the case of the flip-flop of Fig. 10.32(c), J and K are initially '0' and '1' respectively. Thus, J is active. With the first leading edge of the clock input, Q and therefore J go to the logic '1' state. The second leading edge forces Q to go to the logic '0' state as now it is the K input that is in the logic '0' state and active. This circuit also behaves in the same way as the flip-flop of Fig. 10.32(b). The output goes alternately to the logic '0' and '1' state. However, the transitions occur on the leading edge of the clock input. Figure 10.33(b) shows the Q output waveform for this case. The frequency of the Q output waveform in the two cases is equal to half the frequency of the clock input, for obvious reasons, and is therefore 50 kHz.

10.6 Toggle Flip-Flop (*T* Flip-Flop)

The output of a *toggle flip-flop*, also called a T flip-flop, changes state every time it is triggered at its T input, called the toggle input. That is, the output becomes '1' if it was '0' and '0' if it was '1'.

Figures 10.34(a) and (b) respectively show the circuit symbols of positive edge-triggered and negative edge-triggered T flip-flops, along with their function tables.

If we consider the T input as active when HIGH, the characteristic table of such a flip-flop is shown in Fig. 10.34(c). If the T input were active when LOW, then the characteristic table would be as shown in Fig. 10.34(d). The Karnaugh maps for the characteristic tables of Figs 10.34(c) and (d) are shown in Figs 10.34(e) and (f) respectively. The characteristic equations as written from the Karnaugh maps are as follows:

$$Q_{n+1} = T.\overline{Q_n} + \overline{T}.Q_n \quad (10.19)$$

$$Q_{n+1} = \overline{T}.\overline{Q_n} + T.Q_n \quad (10.20)$$

It is obvious from the operational principle of the T flip-flop that the frequency of the signal at the Q output is half the frequency of the signal applied at the T input. A cascaded arrangement of nT flip-flops, where the output of one flip-flop is connected to the T input of the following flip-flop, can be used to divide the input signal frequency by a factor of 2^n. Figure 10.35 shows a divide-by-16 circuit built around a cascaded arrangement of four T flip-flops.

10.6.1 J-K Flip-Flop as a Toggle Flip-Flop

If we recall the function table of a J-K flip-flop, we will see that, when both J and K inputs of the flip-flop are tied to their active level ('1' level if J and K are active when HIGH, and '0' level when J and K are active when LOW), the flip-flop behaves like a toggle flip-flop, with its clock input serving as the T input. In fact, the J-K flip-flop can be used to construct any other flip-flop. That is why it is also sometimes referred to as a *universal flip-flop*. Figure 10.36 shows the use of a J-K flip-flop as a T flip-flop.

Example 10.5

Refer to the cascaded arrangement of two T flip-flops in Fig. 10.37(a). Draw the Q output waveform for the given input signal. If the time period of the input signal is 10 ms, find the frequency of the output signal? If, in the flip-flop arrangement of Fig. 10.37(a), FF-2 were positive edge triggered, draw the Q output waveform.

Solution

The Q output waveform is shown in Fig. 10.37(b) along with the Q output of FF-1. The output of the first T flip-flop changes state for every negative-going edge of the input clock waveform. Its frequency is therefore half the input signal frequency. The output of the first flip-flop acts as the clock input for the second T flip-flop in the cascade arrangement. The second flip-flop, too, toggles for every negative-going edge of the waveform appearing at its input. The final output thus has a frequency that is one-fourth of the input signal frequency:

- Now the time period of the input signal = 10 ms.
- Therefore, the frequency = 100 kHz.
- The frequency of the output signal = 25 kHz.

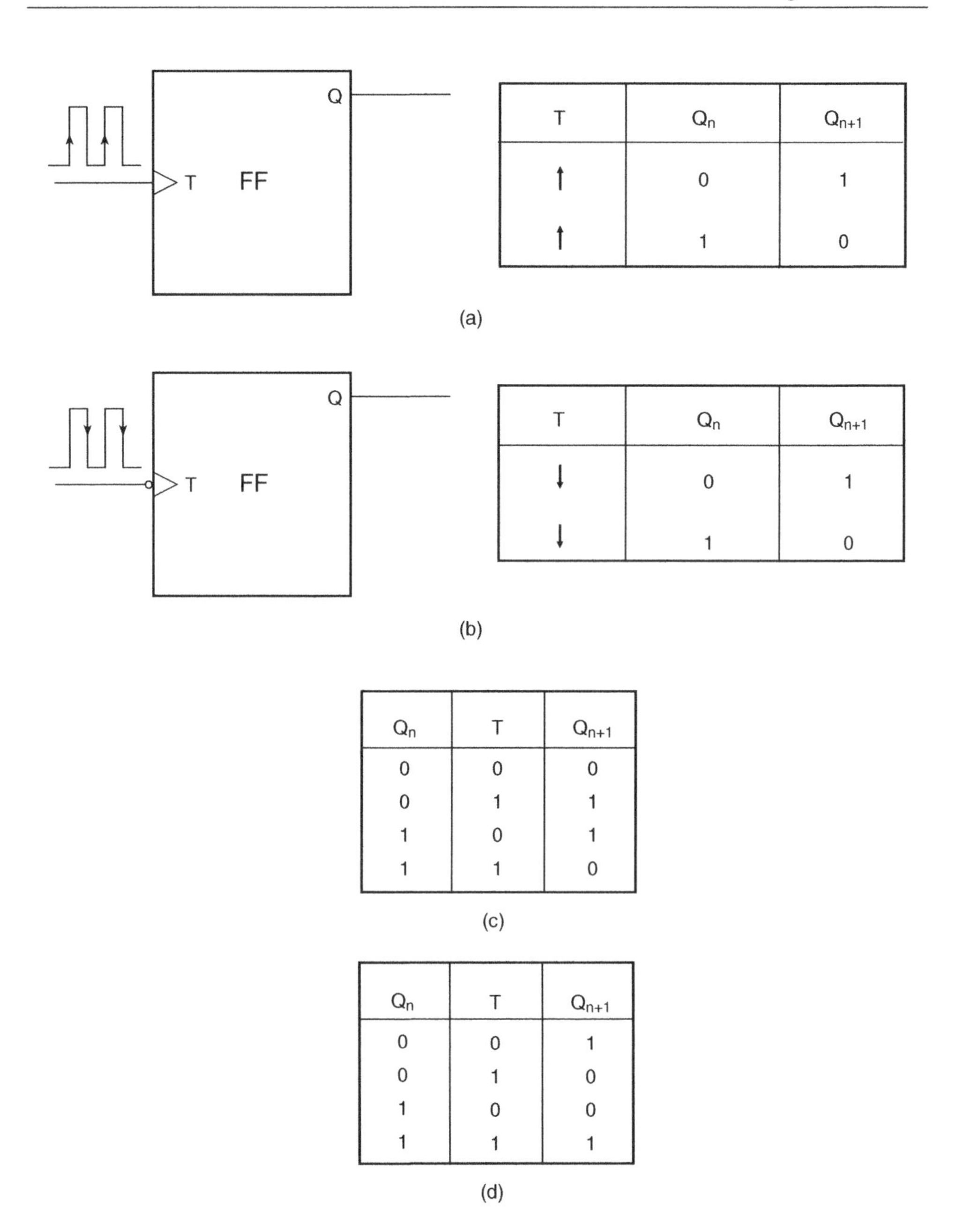

(a)

T	Q_n	Q_{n+1}
↑	0	1
↑	1	0

(b)

T	Q_n	Q_{n+1}
↓	0	1
↓	1	0

(c)

Q_n	T	Q_{n+1}
0	0	0
0	1	1
1	0	1
1	1	0

(d)

Q_n	T	Q_{n+1}
0	0	1
0	1	0
1	0	0
1	1	1

Figure 10.34 (a) Positive edge-triggered toggle flip-flop, (b) a negative edge-triggered toggle flip-flop, (c, d) characteristic tables of level-triggered toggle flip-flops and (e, f) Karnaugh maps for characteristic tables (c, d).

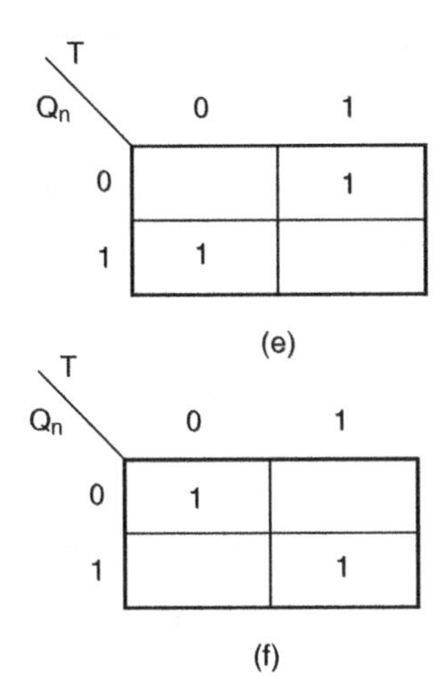

Figure 10.34 (*continued*).

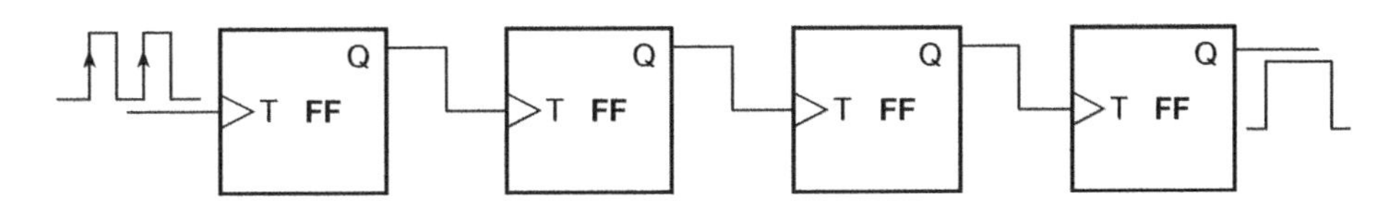

Figure 10.35 Cascade arrangement of T flip-flops.

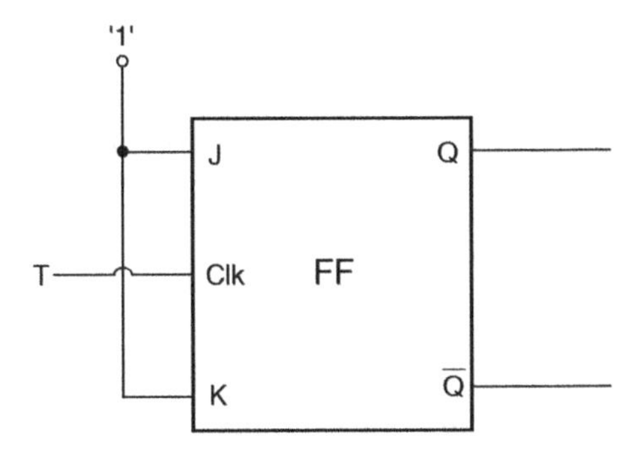

Figure 10.36 J-K flip-flop as a T flip-flop.

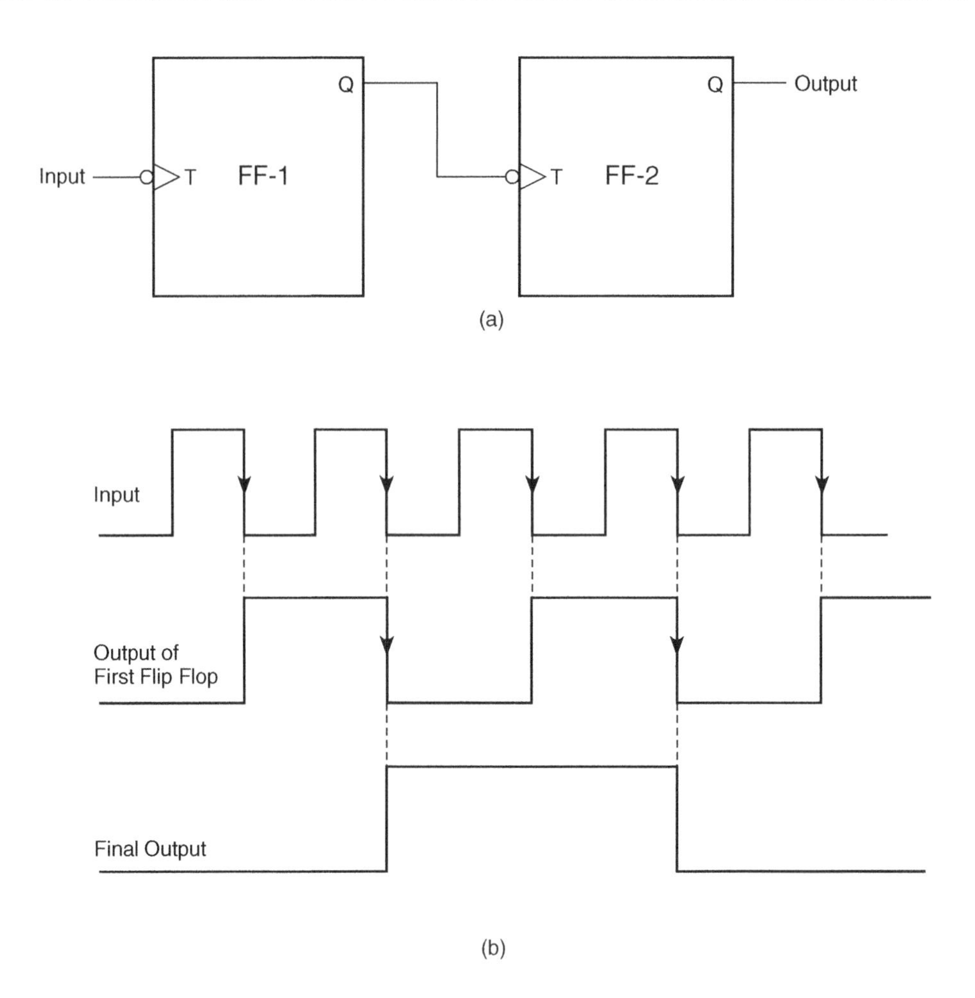

Figure 10.37 Example 10.5.

When the second flip-flop (FF-2) is a positive edge-triggered one, it will respond to the LOW-to-HIGH edges of the waveform appearing at its T input, which is the waveform appearing at the Q output of FF-1. The relevant waveforms in this case are shown in Fig. 10.38.

10.7 *D* Flip-Flop

A D flip-flop, also called a *delay flip-flop*, can be used to provide temporary storage of one bit of information. Figure 10.39(a) shows the circuit symbol and function table of a negative edge-triggered D flip-flop. When the clock is active, the data bit (0 or 1) present at the D input is transferred to the output. In the D flip-flop of Fig. 10.39, the data transfer from D input to Q output occurs on the negative-going (HIGH-to-LOW) transition of the clock input. The D input can acquire new status

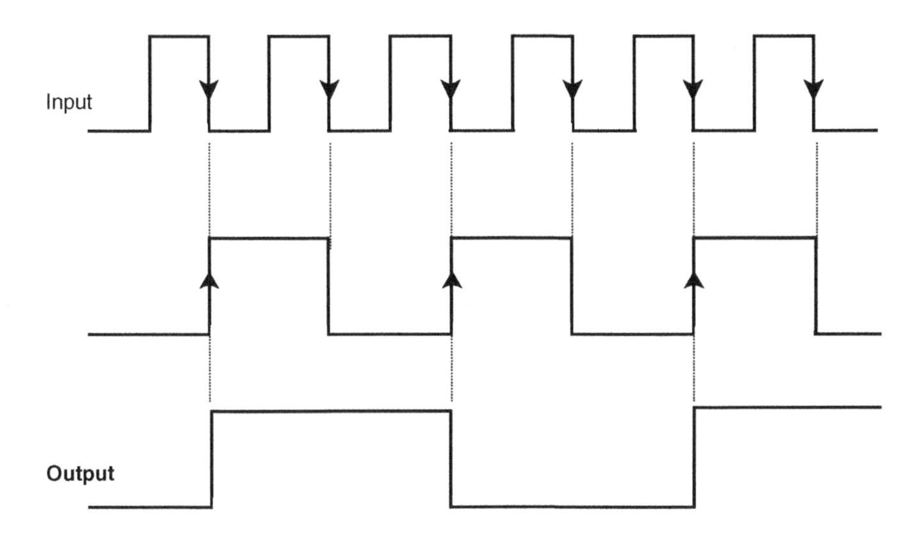

Figure 10.38 Example 10.5.

when the clock is inactive, which is the time period between successive HIGH-to-LOW transitions. The *D* flip-flop can provide a maximum delay of one clock period.

The characteristic table and the corresponding Karnaugh map for the *D* flip-flop of Fig. 10.39(a) are shown in Figs 10.39(c) and (d) respectively. The characteristic equation is as follows:

$$Q_{n+1} = D \tag{10.21}$$

10.7.1 J-K Flip-Flop as D Flip-Flop

Figure 10.40 shows how a *J-K* flip-flop can be used as a *D* flip-flop. When the *D* input is a logic '1', the *J* and *K* inputs are a logic '1' and '0' respectively. According to the function table of the *J-K* flip-flop, under these input conditions, the *Q* output will go to the logic '1' state when clocked. Also, when the *D* input is a logic '0', the *J* and *K* inputs are a logic '0' and '1' respectively. Again, according to the function table of the *J-K* flip-flop, under these input conditions, the *Q* output will go to the logic '0' state when clocked. Thus, in both cases, the *D* input is passed on to the output when the flip-flop is clocked.

10.7.2 D Latch

In a *D* latch, the output *Q* follows the *D* input as long as the clock input (also called the ENABLE input) is HIGH or LOW, depending upon the clock level to which it responds. When the ENABLE input goes to the inactive level, the output holds on to the logic state it was in just prior to the ENABLE input becoming inactive during the entire time period the ENABLE input is inactive.

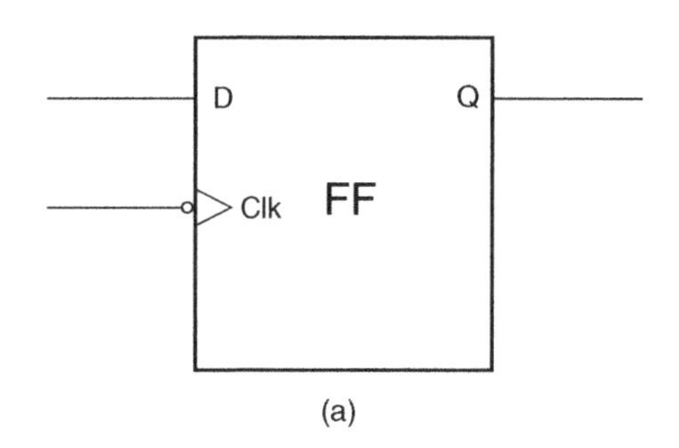

(a)

D	Clk	Q
0	↓	0
1	↓	1

(b)

Q_n	D	Q_{n+1}
0	0	0
0	1	1
1	0	0
1	1	1

(c)

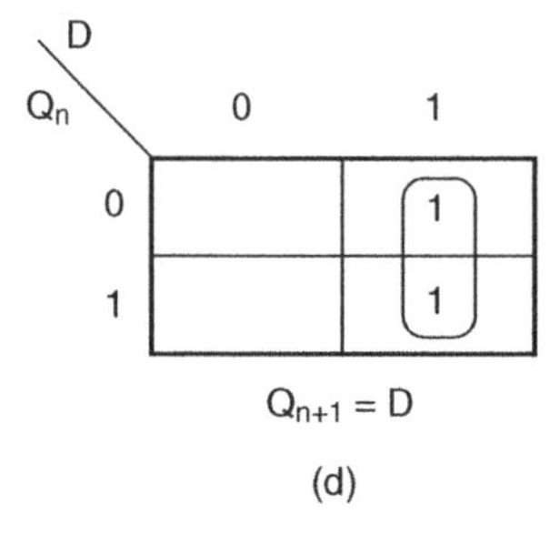

$Q_{n+1} = D$

(d)

Figure 10.39 *D* flip-flop.

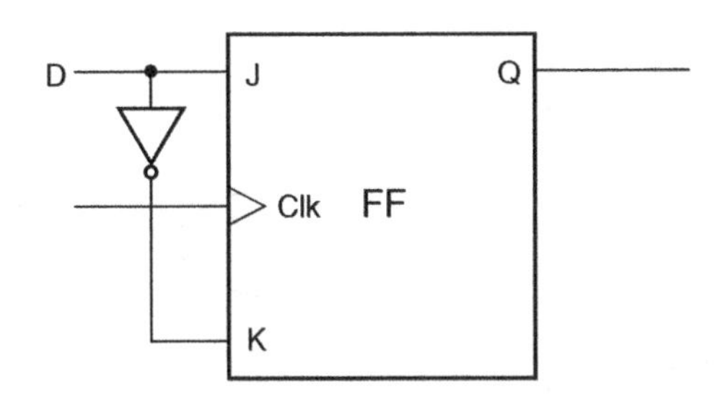

Figure 10.40 *J-K* flip-flop as a *D* flip-flop.

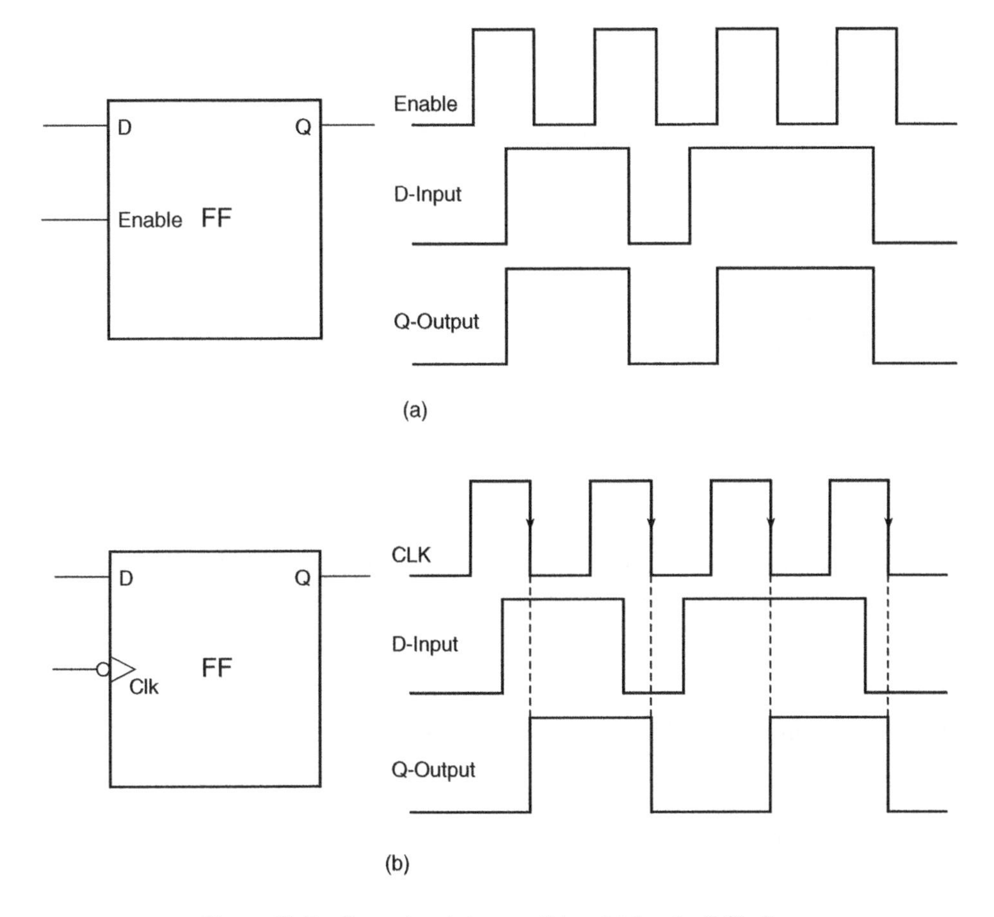

Figure 10.41 Comparison between a *D*-type latch and a *D* flip-flop.

A *D* flip-flop should not be confused with a *D* latch. In a *D* flip-flop, the data on the *D* input are transferred to the *Q* output on the positive- or negative-going transition of the clock signal, depending upon the flip-flop, and this logic state is held at the output until we get the next effective clock transition. The difference between the two is further illustrated in Figs 10.41(a) and (b) depicting the functioning of a *D* latch and a *D* flip-flop respectively.

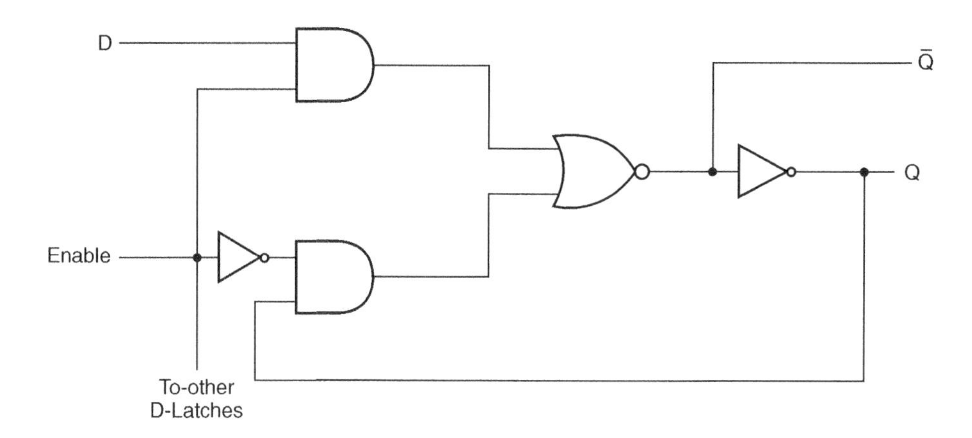

Figure 10.42 Example 10.6.

Example 10.6

Figure 10.42 shows the internal logic circuit diagram of one of the four D latches of a four-bit D latch in IC 7475. (a) Give an argument to prove that the Q output will track the D input only when the ENABLE input is HIGH. (b) Also, prove that the Q output holds the value it had just before the ENABLE input went LOW during the time the ENABLE input is LOW.

Solution

(a) When the ENABLE input is HIGH, the upper AND gate is enabled while the lower AND gate is disabled. The outputs of the upper and lower AND gates are D and logic '0' respectively. They constitute inputs of the NOR gate whose output is $\overline{D}$. The Q output is therefore D.

(b) When the ENABLE input goes LOW, the upper AND gate is disabled (with its output going to logic '0') and the lower AND gate is enabled (with its output becoming the same as the Q output owing to the feedback). The NOR gate output in this case is $\overline{Q}$, which means that the Q output holds its state as long as the ENABLE input is LOW.

10.8 Synchronous and Asynchronous Inputs

Most flip-flops have both synchronous and asynchronous inputs. Synchronous inputs are those whose effect on the flip-flop output is synchronized with the clock input. R, S, J, K and D inputs are all synchronous inputs. Asynchronous inputs are those that operate independently of the synchronous inputs and the input clock signal. These are in fact override inputs as their status overrides the status of all synchronous inputs and also the clock input. They force the flip-flop output to go to a predefined state irrespective of the logic status of the synchronous inputs. PRESET and CLEAR inputs are examples of asynchronous inputs. When active, the PRESET and CLEAR inputs place the flip-flop Q output in the '1' and '0' state respectively. Usually, these are active LOW inputs. When it is desired that the flip-flop functions as per the status of its synchronous inputs, the asynchronous inputs are kept in their inactive state. Also, both asynchronous inputs, if available on a given flip-flop, are not made active simultaneously.

10.9 Flip-Flop Timing Parameters

Certain timing parameters would be listed in the specification sheet of a flip-flop. Some of these parameters, as we will see in the paragraphs to follow, are specific to the logic family to which the flip-flop belongs. There are some parameters that have different values for different flip-flops belonging to the same broad logic family. It is therefore important that one considers these timing parameters before using a certain flip-flop in a given application. Some of the important ones are set-up and hold times, propagation delay, clock pulse HIGH and LOW times, asynchronous input active pulse width, clock transition time and maximum clock frequency.

10.9.1 Set-Up and Hold Times

The *set-up time* is the minimum time period for which the synchronous inputs (for example, *R*, *S*, *J*, *K* and *D*) and asynchronous inputs (for example, PRESET and CLEAR) must be stable prior to the active clock transition for the flip-flop output to respond reliably at the clock transition. It is usually denoted by t_s (min) and is usually defined separately for synchronous and asynchronous inputs. As an example, if in a *J-K* flip-flop the *J* and *K* inputs were to go to '1' and '0' respectively, and if the flip-flop were negative edge triggered, the set-up time would be as shown in Fig. 10.43(a). The set-up time in the case of 74ALS109A, which is a dual *J-K* positive edge-triggered flip-flop belonging to the advanced low-power Schottky TTL logic family, is 15 ns. Also, the asynchronous inputs, such as PRESET and CLEAR, if there, should be inactive prior to the clock transition for a certain minimum time period if the outputs have to respond as per synchronous inputs. In the case of 74ALS109A, the asynchronous input set-up time is 10 ns. The asynchronous input set-up time for active low PRESET and CLEAR inputs is shown in Fig. 10.43(b), assuming a positive edge-triggered flip-flop.

The *hold time* t_H (min) is the minimum time period for which the synchronous inputs (*R*, *S*, *J*, *K*, *D*) must remain stable in the desired logic state after the active clock transition for the flip-flop to respond reliably. The same is depicted in Fig. 10.43(a) if the desired logic status for *J* and *K* inputs is '1' and '0' respectively and the flip-flop is negative edge triggered. The hold time for flip-flop 74ALS109A is specified to be zero. To sum up, for a flip-flop to respond properly and reliably at the active clock transition, the synchronous inputs must be stable in their intended logic states and the asynchronous inputs must be stable in their inactive states for at least a time period equal to the specified minimum set-up times prior to the clock transition, and the synchronous inputs must be stable for a time period equal to at least the specified minimum hold time after the clock transition.

10.9.2 Propagation Delay

There is always a time delay, known as the *propagation delay*, from the time instant the signal is applied to the time the output makes the intended change. The flip-flop data sheet usually specifies propagation delays for both HIGH-to-LOW (t_{pHL}) and for LOW-to-HIGH (t_{pLH}) output transitions. The propagation delay is measured between 50 % points on input and output waveforms and is usually specified for all types of input including synchronous and asynchronous inputs. The propagation delays for LOW-to-HIGH and HIGH-to-LOW output transitions for a positive edge-triggered flip-flop are shown in Fig. 10.44. For flip-flop 74ALS109A, t_{pHL} and t_{pLH} for clock input to output are respectively 18 and 16 ns. The same for the asynchronous input to output for this flip-flop are 15 and 13 ns respectively.

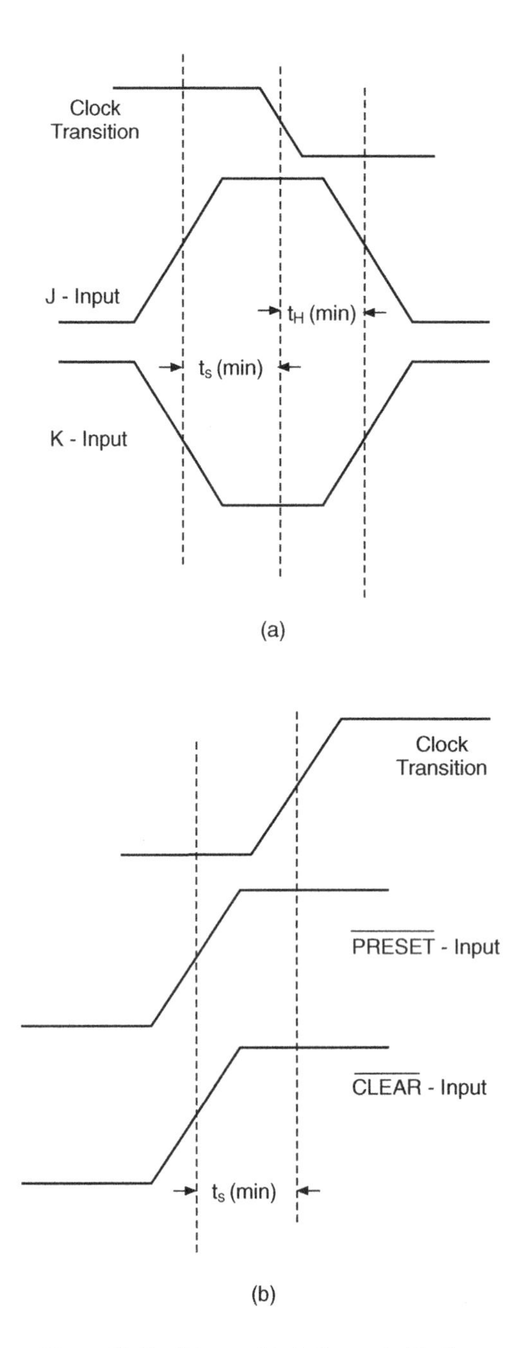

Figure 10.43 Set-up and hold times of a flip-flop.

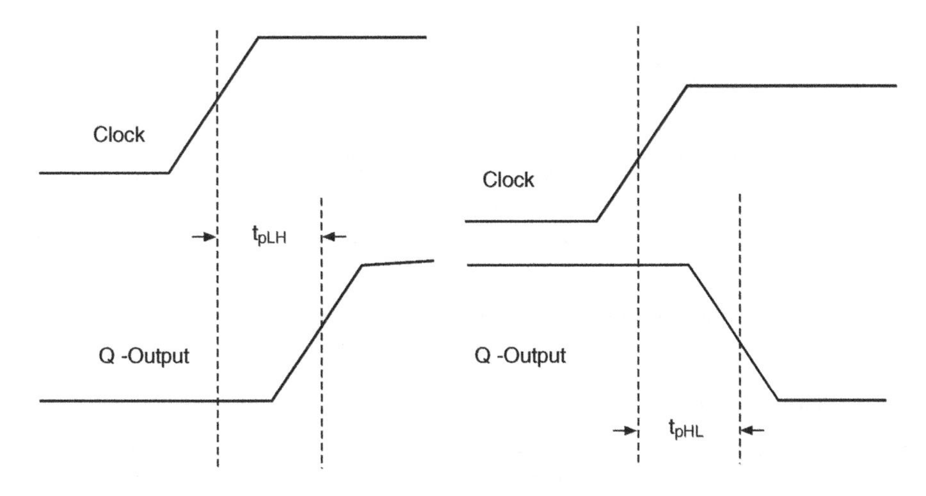

Figure 10.44 Propagation delay.

10.9.3 Clock Pulse HIGH and LOW Times

The clock pulse HIGH time t_W (H) and clock pulse LOW time, t_W (L) are respectively the minimum time durations for which the clock signal should remain HIGH and LOW. Failure to meet these requirements can lead to unreliable triggering. Figure 10.45 depicts these timing parameters. t_W (H) and t_W (L) for 74ALS109A are 4 and 5.5 ns respectively.

10.9.4 Asynchronous Input Active Pulse Width

This is the minimum time duration for which the asynchronous input (PRESET or CLEAR) must be kept in its active state, usually LOW, for the output to respond properly. It is 4 ns in the case of flip-flop 74ALS109A. Figure 10.46 shows this timing parameter.

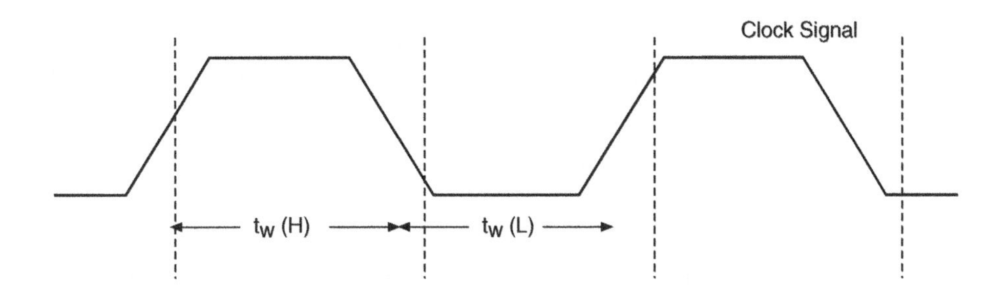

Figure 10.45 Clock pulse HIGH and LOW times.

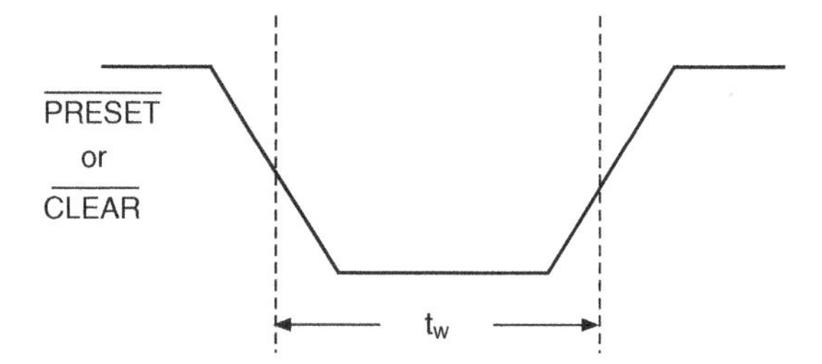

Figure 10.46 Asynchronous input active pulse width.

10.9.5 Clock Transition Times

The manufacturers specify the maximum transition times (rise time and fall time) for the output to respond properly. If these specified figures are exceeded, the flip-flop may respond erratically or even may not respond at all. This parameter is logic family specific and is not specified for individual devices. The allowed maximum transition time for TTL devices is much smaller than that for CMOS devices. Also, within the broad TTL family, it varies from one subfamily to another.

10.9.6 Maximum Clock Frequency

This is the highest frequency that can be applied to the clock input. If this figure is exceeded, there is no guarantee that the device will work reliably and properly. This figure may vary slightly from device to device of even the same type number. The manufacturer usually specifies a safe value. If this specified value is not exceeded, the manufacturer guarantees that the device will trigger reliably. It is 34 MHz for 74ALS109A.

10.10 Flip-Flop Applications

Flip-flops are used in a variety of application circuits, the most common among these being the *frequency division and counting* circuits and *data storage and transfer* circuits. These application areas are discussed at length in Chapter 11 on counters and registers. Both these applications use a cascaded arrangement of flip-flops with or without some additional combinational logic to perform the desired function. Counters and registers are available in IC form for a variety of digital circuit applications.

Other applications of flip-flops include their use for switch debouncing, where even an unclocked flip-flop (such as a NAND or a NOR latch) can be used, for synchronizing asynchronous inputs with the clock input and for identification of edges of synchronous inputs. These are briefly described in the following paragraphs.

10.10.1 Switch Debouncing

Owing to the switch bounce phenomenon, the mechanical switch cannot be used as such to produce a clean voltage transition. Refer to Fig. 10.47(a). When the switch is moved from position 1 to position 2, what is desired at the output is a clean voltage transition from 0 to $+V$ volts, as shown in Fig. 10.47(b). What actually happens is shown in Fig. 10.47(c). The output makes several transitions between 0 and

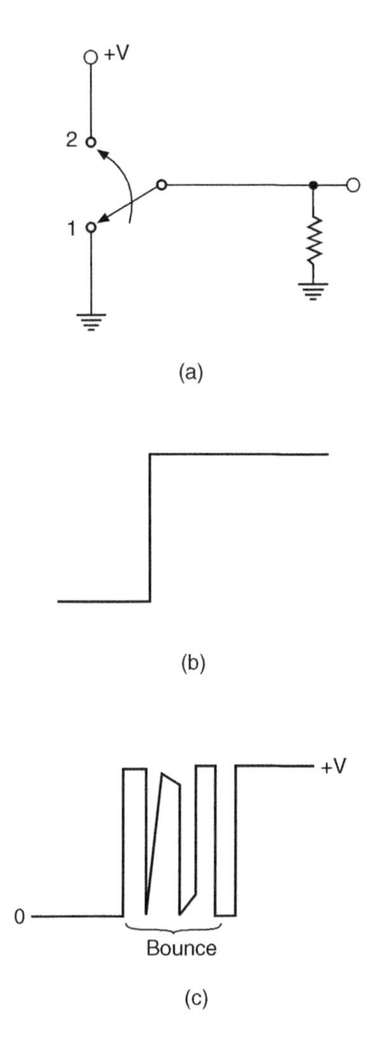

Figure 10.47 Switch bounce phenomenon.

$+V$ volts for a few milliseconds owing to contact bounce before it finally settles at $+V$ volts. Similarly, when it is moved from position 2 back to position 1, it makes several transitions before coming to rest at 0 V. Although this random behaviour lasts only for a few milliseconds, it is unacceptable for many digital circuit applications. A NAND or a NOR latch can solve this problem and provide a clean output transition. Figure 10.48 shows a typical switch debounce circuit built around a NAND latch. The circuit functions as follows.

When the switch is in position 1, the output is at a '0' level. When it is moved to position 2, the output goes to a '1' level within a few nanoseconds (depending upon the propagation delay of the NAND gate) after its first contact with position 2. When the switch contact bounces, it makes and breaks contact with position 2 before it finally settles at the intended position. Making of contact

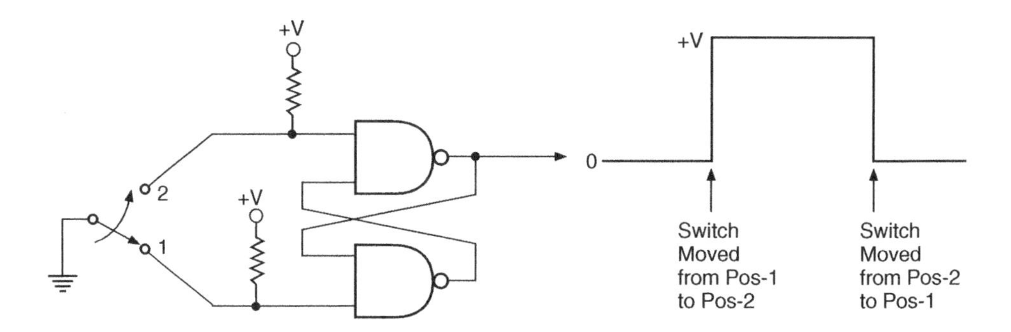

Figure 10.48 Switch debounce circuit.

always leads to a '1' level at the output, and breaking of contact also leads to a '1' level at the output owing to the fact that the contact break produces a '1' level at both inputs of the latch which forces the output to hold its existing logic state. The fact that when the switch is brought back to position 1 the output makes a neat transition to a '0' level can be explained on similar lines.

10.10.2 Flip-Flop Synchronization

Consider a situation where a certain clock input, which works in conjunction with various synchronous inputs, is to be gated with an asynchronously generated gating pulse, as shown in Fig. 10.49. The output in this case has the clock pulses at one or both ends shortened in width, as shown in Fig. 10.49. This problem can be overcome and the gating operation synchronized with the help of a flip-flop, as shown in Fig. 10.50.

10.10.3 Detecting the Sequence of Edges

Flip-flops can also be used to detect the sequence of occurrence of rising and falling edges. Figure 10.51 shows how a flip-flop can be used to detect whether a positive-going edge A follows or precedes another positive-going edge B. The two edges are respectively applied to *D* and clock inputs of a

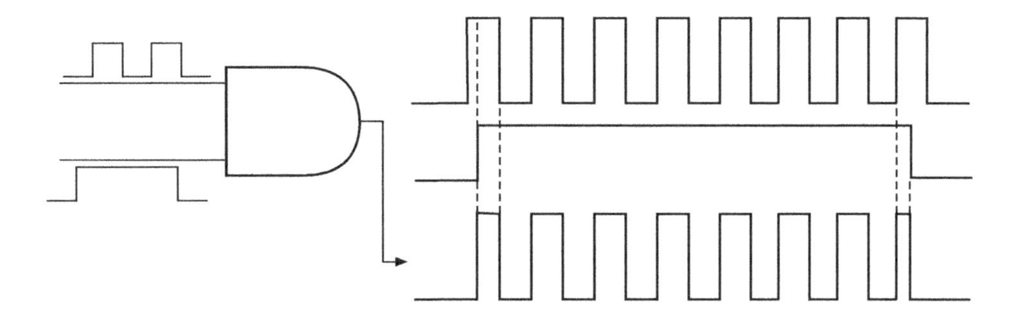

Figure 10.49 Gating of a clock signal.

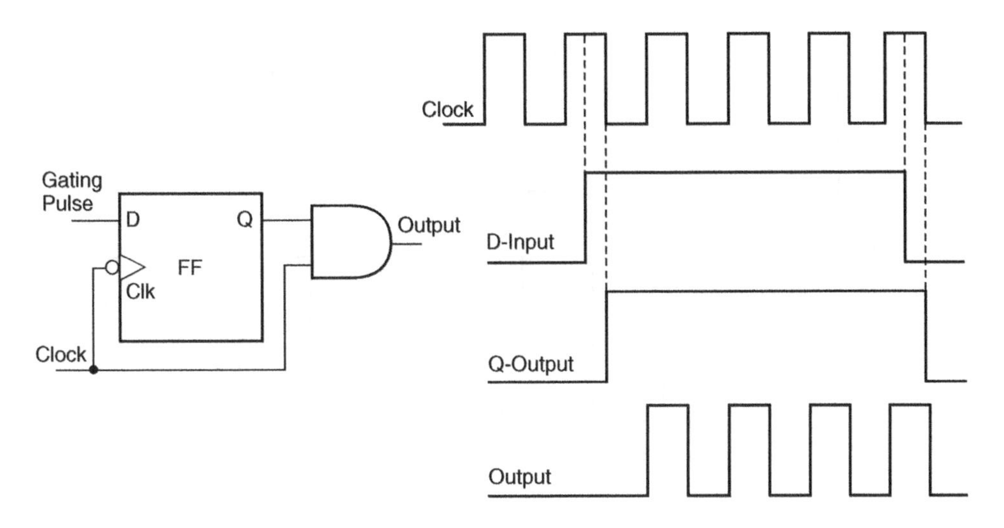

Figure 10.50 Flip-flop synchronization.

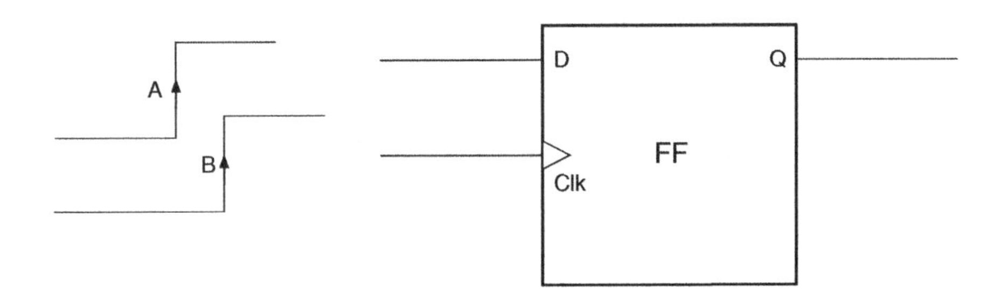

Figure 10.51 Detection of the sequence of edges.

positively edge-triggered D flip-flop. If edge A arrives first, then, on arrival of edge B, the output goes from 0 to 1. If it is otherwise, it stays at a '0' level.

Example 10.7

Figure 10.52 shows two pulsed waveforms A and B, with waveform A leading waveform B in phase, as shown in the figure. Suggest a flip-flop circuit to detect this condition by producing (a) a logic '1' Q output and (b) a logic '0' Q output.

Solution

(a) A positive edge-triggered D flip-flop, as shown in Fig. 10.53(a), can be used for the purpose. Waveform A is applied to the D input, and waveform B is applied to the clock input. If we examine the two waveforms, we will find that, on every occurrence of the leading edge of waveform B,

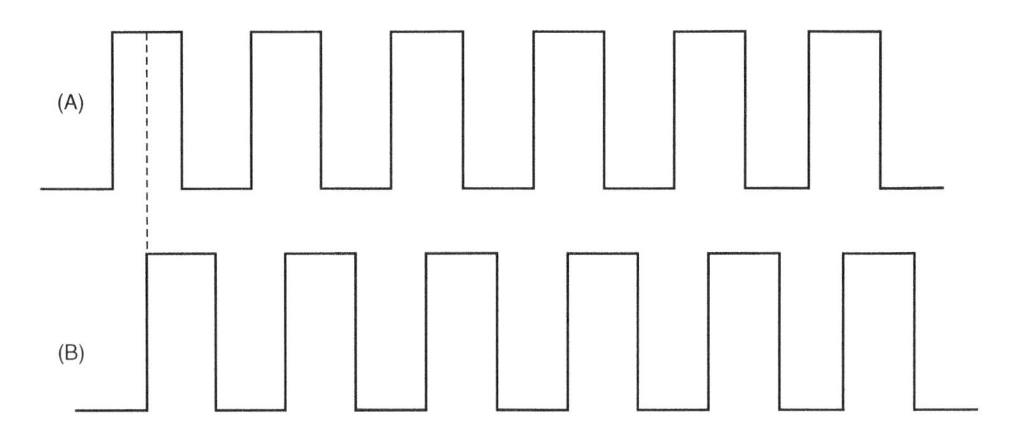

Figure 10.52 Example 10.7.

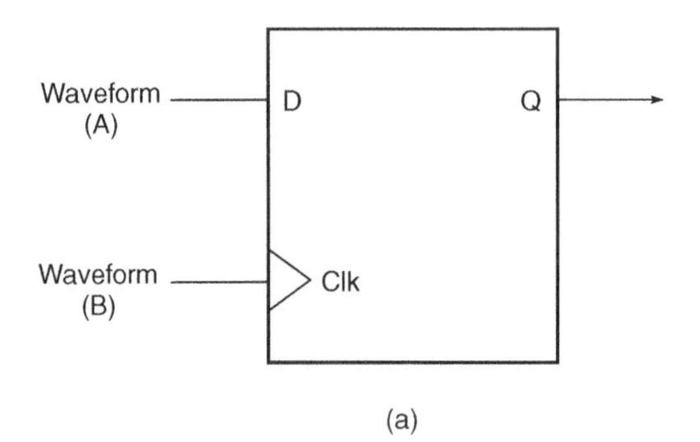

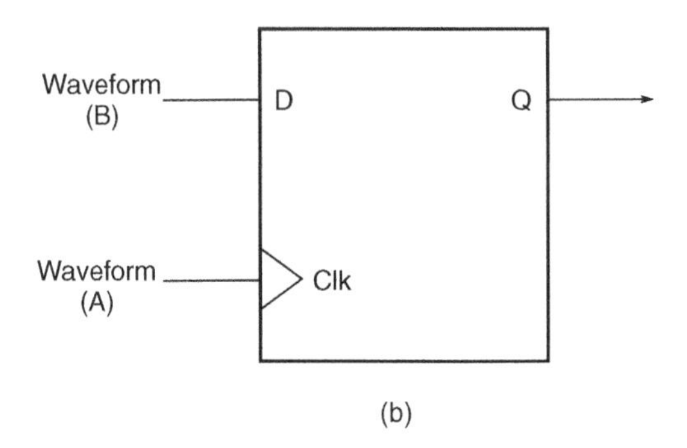

Figure 10.53 Solution to example 10.7.

waveform A is in a logic '1' state. Thus, the Q output in this case will always be in a logic '1' state.

(b) By interchanging the connections of waveforms A and B as shown in Fig. 10.53(b), the Q output will be in a logic '0' state as long as waveform A leads waveform B in phase. In this case, on every occurrence of the leading edge of waveform A (clock input), waveform B (D input) is in a logic '0' state.

10.11 Application-Relevant Data

Table 10.1 lists popular type numbers of flip-flops belonging to TTL, CMOS and ECL logic families. Application-relevant information of some of the popular type numbers is given in the companion website. The information given includes the pin connection diagram, package style and function table.

Table 10.1 Popular type numbers of flip-flops belonging to the TTL, CMOS and ECL logic families.

IC type number	Function	Logic family
54/7473	Dual *J-K* negative edge-triggered flip-flop with CLEAR	TTL
54/7474	Dual *D*-type positive edge-triggered flip-flop with PRESET and CLEAR	TTL
54/7475	Four-bit *D*-type latch	TTL
54/7476	Dual *J-K* flip-flop with PRESET and CLEAR	TTL
54/7478	Dual *J-K* flip-flop with PRESET and CLEAR	TTL
54/74107	Dual *J-K* flip-flop with CLEAR	TTL
54/74109	Dual *J-K* positive edge-triggered flip-flop with PRESET and CLEAR	TTL
54/74112	Dual *J-K* negative edge-triggered flip-flop with PRESET and CLEAR	TTL
54/74113	Dual *J-K* negative edge-triggered flip-flop with PRESET	TTL
54/74114	Dual *J-K* negative edge-triggered flip-flop with PRESET and CLEAR	TTL
54/74121	Monostable multivibrator	TTL
54/74122	Retriggerable monostable multivibrator	TTL
54/74123	Dual retriggerable monostable multivibrator	TTL
54/74174	Hex *D*-type flip-flop with CLEAR	TTL
54/74175	Quad edge triggered *D*-type flip-flop with CLEAR	TTL
54/74221	Dual monostable multivibrator	TTL
54/74256	Dual four-bit addressable latch	TTL
54/74259	Eight-bit addressable latch	TTL
54/74273	Octal *D*-type flip-flop with MASTER RESET	TTL
54/74279	Quad SET/RESET latch	TTL
54/74373	Octal transparent latch (three-state)	TTL
54/74374	Octal *D*-type flip-flop (three-state)	TTL
54/74377	Octal *D*-type flip-flop with common ENABLE	TTL
54/74378	Hex *D*-type flip-flop with ENABLE	TTL
54/74379	Four-bit *D*-type flip-flop with ENABLE	TTL
54/74533	Octal transparent latch (three-state)	TTL
54/74534	Octal *D*-type flip-flop (three-state)	TTL
54/74573	Octal *D*-type latch (three-state)	TTL
54/74574	Octal *D*-type flip-flop (three-state)	TTL

(continued overleaf)

Table 10.1 *(continued)*.

IC type number	Function	Logic family
4013	Dual *D*-type flip-flop	CMOS
4027	Dual *J-K* flip-flop	CMOS
4042	Quad *D*-type latch	CMOS
4044	Quad *R-S* latch with three-state output	CMOS
4047	Low-power monostable/astable multivibrator	CMOS
4076	Quad *D*-type flip-flop with three-state output	CMOS
40174	Hex *D*-type flip-flop	CMOS
40175	Quad *D*-type flip-flop	CMOS
4511	BCD to seven-segment latch/decoder/driver	CMOS
4528	Dual retriggerable resettable monostable multivibrator	CMOS
4543	BCD to seven-segment latch/decoder/driver for LCD	CMOS
4723	Dual four-bit addressable latch	CMOS
4724	Eight-bit addressable latch	CMOS
MC10130	Quad *D*-type latch	ECL
MC10131	Dual *D*-type master/slave flip-flop	ECL
MC10133	Quad *D*-type latch (negative transition)	ECL
MC10135	Dual *J-K* master/slave flip-flop	ECL
MC10153	Quad latch (positive transition)	ECL
MC10168	Quad *D*-type latch	ECL
MC10175	Quint latch	ECL
MC10176	Hex *D*-type master/slave flip-flop	ECL
MC10198	Monostable multivibrator	ECL
MC10231	High-Speed dual *D*-type M/S flip-flop	ECL
MC1666	Dual clocked *R-S* flip-flop	ECL
MC1668	Dual clocked latch	ECL
MC1670	*D*-type master/slave flip-flop	ECL
MC1658	Voltage-controlled multivibrator	ECL

Review Questions

1. Briefly describe the operational aspects of bistable, monostable and astable multivibrators. Which multivibrator closely resembles a flip-flop?
2. What is a flip-flop? Show the logic implementation of an *R-S* flip-flop having active HIGH *R* and *S* inputs. Draw its truth table and mark the invalid entry.
3. With the help of the logic diagram, describe the operation of a clocked *R-S* flip-flop with active LOW *R* and *S* inputs. Draw the truth table of this flip-flop if it were negatively edge triggered.
4. What is a clocked *J-K* flip-flop? What improvement does it have over a clocked *R-S* flip-flop?
5. Differentiate between:

   (a) synchronous and asynchronous inputs;
   (b) level-triggered and edge-triggered flip-flops;
   (c) active LOW and active HIGH inputs.

6. Briefly describe the following flip-flop timing parameters:

(a) set-up time and hold time;
(b) propagation delay;
(c) maximum clock frequency.

7. Draw the truth table for the following types of flip-flop:

   (a) a positive edge-triggered J-K flip-flop with active HIGH J and K inputs and active LOW PRESET and CLEAR inputs;
   (b) a negative edge-triggered J-K flip-flop with active LOW J and K inputs and active LOW PRESET and CLEAR inputs.

8. What is meant by the race problem in flip-flops? How does a master–slave configuration help in solving this problem?
9. Differentiate between a D flip-flop and a D latch.
10. Draw the function table for (a) a negative edge-triggered D flip-flop and (b) a D latch with an active LOW ENABLE input.
11. With the help of a schematic arrangement, explain how a J-K flip-flop can be used as a (a) a D flip-flop and (b) a T flip-flop.
12. With the help of a suitable circuit, briefly explain how a D flip-flop can be used to detect the sequence of occurrence of edges of synchronous inputs.

Problems

1. A 100 kHz clock signal is applied to a J-K flip-flop with $J = K = 1$.

   (a) If the flip-flop has active HIGH J and K inputs and is negative edge triggered, determine the frequency of the Q and $\overline{Q}$ outputs.
   (b) If the flip-flop has active LOW J and K inputs and is positive edge triggered, what should the frequency of the Q and $\overline{Q}$ outputs be? Assume that Q is initially '0'.

   (a) Q output = 50 kHz, $\overline{Q}$ output = 50 kHz;
   (b) both outputs remain in a logic '0' state

2. In a Schmitt trigger inverter circuit, the two trip points are observed to occur at 1.8 and 2.8 V. At what input voltage levels will this device make (a) HIGH-to-LOW transition and (b) LOW-to-HIGH transition?

   (a) 2.8 V; (b) 1.8 V

3. In the case of a presettable, clearable J-K flip-flop with active HIGH J and K inputs and active LOW PRESET and CLEAR inputs, what would the Q output logic status be for the following input conditions, assuming that Q is initially '0', immediately after it is clocked?

   (a) $J = 1$, $K = 0$, PRESET = 1, CLEAR = 1;
   (b) $J = 1$, $K = 1$, PRESET = 0, CLEAR = 1;
   (c) $J = 0$, $K = 1$, PRESET = 1, CLEAR = 0;
   (d) $J = K = 0$, PRESET = 0, CLEAR = 1.

   (a) 1; (b) 1; (c) 0; (d) 1

4. Figure 10.54 shows the function table of a certain flip-flop. Identify the flip-flop.

Negative edge-triggered J-K flip-flop with active HIGH J and K inputs and active LOW PRESET and CLEAR inputs

Pr	Cl	Clk	J	K	Q_n+1	$\overline{Q_n+1}$
1	0	X	X	X	0	1
0	1	X	X	X	1	0
0	0	X	X	X	Unstable	
0	0	↓	1	1	1	0
0	0	↓	0	1	0	1
0	0	↓	1	1	Toggle	
0	0	↓	0	0	Qn	Qn

Figure 10.54 Problem 4.

5. Derive the expression for Q_{n+1} in terms of Q_n and J and K inputs for a clocked J-K flip-flop with active LOW J and K inputs. Q_n and Q_{n+1} have the usual meaning.

$$Q_{n+1} = \overline{J}.\overline{Q_n} + K.Q_n$$

6. Consider a J-K flip-flop (J-$\overline{K}$ flip-flop to be more precise) where an inverter has been wired between the external $\overline{K}$ input and the internal K input as shown in Fig. 10.55. With the help of a characteristic table, write the characteristic equation for this flip-flop.

$$Q_{n+1} = J.\overline{Q_n} + K.Q_n$$

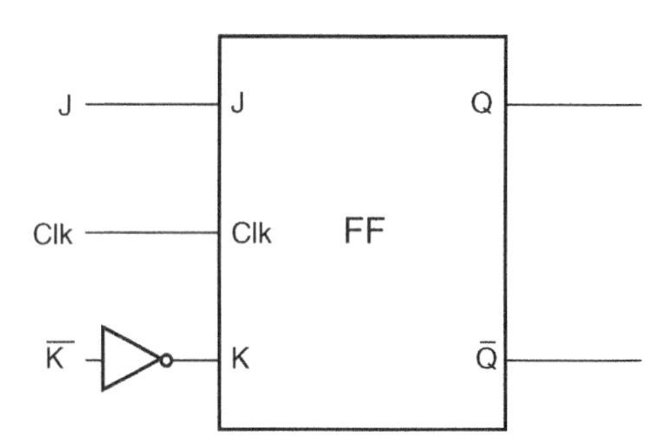

Figure 10.55 Problem 6.

Further Reading

1. Cook, N. P. (2003), *Practical Digital Electronics*, Prentice-Hall, NJ, USA.
2. Whitaker, C. (1996) *The Electronics Handbook*, CRC Press (in cooperation with IEEE Press), FL, USA.
3. Tokheim, R. L. (1994) *Schaum's Outline Series of Digital Principles*, McGraw-Hill Companies Inc., USA.
4. Tocci, R. J. (2006), *Digital Systems – Principles and Applications*, Prentice-Hall Inc., NJ, USA.
5. Malvino, A. P. and Leach, D. P. (1994) *Digital Principles and Applications*, McGraw-Hill, USA.

11

Counters and Registers

Counters and *registers* belong to the category of MSI sequential logic circuits. They have similar architecture, as both counters and registers comprise a cascaded arrangement of more than one flip-flop with or without combinational logic devices. Both constitute very important building blocks of sequential logic, and different types of counter and register available in integrated circuit (IC) form are used in a wide range of digital systems. While counters are mainly used in counting applications, where they either measure the time interval between two unknown time instants or measure the frequency of a given signal, registers are primarily used for the temporary storage of data present at the output of a digital circuit before they are fed to another digital circuit. We are all familiar with the role of different types of register used inside a microprocessor, and also their use in microprocessor-based applications. Because of the very nature of operation of registers, they form the basis of a very important class of counters called *shift counters*. In this chapter, we will discuss different types of counter and register as regards their operational basics, design methodology and application-relevant aspects. Design aspects have been adequately illustrated with the help of a large number of solved examples. A comprehensive functional index of a large number of integrated circuit counters and registers is given towards the end of the chapter.

11.1 Ripple (Asynchronous) Counter

A *ripple counter* is a cascaded arrangement of flip-flops where the output of one flip-flop drives the clock input of the following flip-flop. The number of flip-flops in the cascaded arrangement depends upon the number of different logic states that it goes through before it repeats the sequence, a parameter known as the modulus of the counter.

In a ripple counter, also called an *asynchronous counter* or a *serial counter*, the clock input is applied only to the first flip-flop, also called the input flip-flop, in the cascaded arrangement. The clock input to any subsequent flip-flop comes from the output of its immediately preceding flip-flop. For instance, the output of the first flip-flop acts as the clock input to the second flip-flop, the output of the second flip-flop feeds the clock input of the third flip-flop and so on. In general, in an arrangement of n

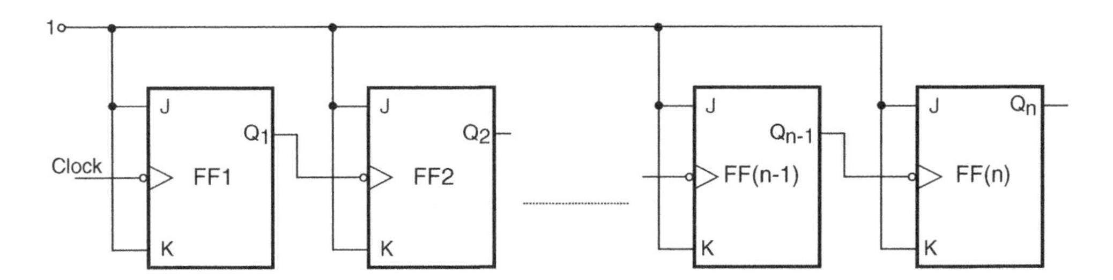

Figure 11.1 Generalized block schematic of n-bit binary ripple counter.

flip-flops, the clock input to the nth flip-flop comes from the output of the $(n-1)$th flip-flop for $n > 1$. Figure 11.1 shows the generalized block schematic arrangement of an n-bit binary ripple counter.

As a natural consequence of this, not all flip-flops change state at the same time. The second flip-flop can change state only after the output of the first flip-flop has changed its state. That is, the second flip-flop would change state a certain time delay after the occurrence of the input clock pulse owing to the fact that it gets its own clock input from the output of the first flip-flop and not from the input clock. This time delay here equals the sum of propagation delays of two flip-flops, the first and the second flip-flops. In general, the nth flip-flop will change state only after a delay equal to n times the propagation delay of one flip-flop. The term 'ripple counter' comes from the mode in which the clock information ripples through the counter. It is also called an 'asynchronous counter' as different flip-flops comprising the counter do not change state in synchronization with the input clock.

In a counter like this, after the occurrence of each clock input pulse, the counter has to wait for a time period equal to the sum of propagation delays of all flip-flops before the next clock pulse can be applied. The propagation delay of each flip-flop, of course, will depend upon the logic family to which it belongs.

11.1.1 Propagation Delay in Ripple Counters

A major problem with ripple counters arises from the propagation delay of the flip-flops constituting the counter. As mentioned in the preceding paragraphs, the effective propagation delay in a ripple counter is equal to the sum of propagation delays due to different flip-flops. The situation becomes worse with increase in the number of flip-flops used to construct the counter, which is the case in larger bit counters. Coming back to the ripple counter, an increased propagation delay puts a limit on the maximum frequency used as clock input to the counter. We can appreciate that the clock signal time period must be equal to or greater than the total propagation delay. The maximum clock frequency therefore corresponds to a time period that equals the total propagation delay. If t_{pd} is the propagation delay in each flip-flop, then, in a counter with N flip-flops having a modulus of less than or equal to 2^N, the maximum usable clock frequency is given by $f_{max} = 1/(N \times t_{pd})$. Often, two propagation delay times are specified in the case of flip-flops, one for LOW-to-HIGH transition (t_{pLH}) and the other for HIGH-to-LOW transition (t_{pHL}) at the output. In such a case, the larger of the two should be considered for computing the maximum clock frequency.

As an example, in the case of a ripple counter IC belonging to the low-power Schottky TTL (LSTTL) family, the propagation delay per flip-flop typically is of the order of 25 ns. This implies that a four-bit ripple counter from this logic family can not be clocked faster than 10 MHz. The upper limit on the clock frequency further decreases with increase in the number of bits to be handled by the counter.

11.2 Synchronous Counter

In a *synchronous counter*, also known as a *parallel counter*, all the flip-flops in the counter change state at the same time in synchronism with the input clock signal. The clock signal in this case is simultaneously applied to the clock inputs of all the flip-flops. The delay involved in this case is equal to the propagation delay of one flip-flop only, irrespective of the number of flip-flops used to construct the counter. In other words, the delay is independent of the size of the counter.

11.3 Modulus of a Counter

The *modulus* (MOD number) of a counter is the number of different logic states it goes through before it comes back to the initial state to repeat the count sequence. An n-bit counter that counts through all its natural states and does not skip any of the states has a modulus of 2^n. We can see that such counters have a modulus that is an integral power of 2, that is, 2, 4, 8, 16 and so on. These can be modified with the help of additional combinational logic to get a modulus of less than 2^n.

To determine the number of flip-flops required to build a counter having a given modulus, identify the smallest integer m that is either equal to or greater than the desired modulus and is also equal to an integral power of 2. For instance, if the desired modulus is 10, which is the case in a decade counter, the smallest integer greater than or equal to 10 and which is also an integral power of 2 is 16. The number of flip-flops in this case would be 4, as $16 = 2^4$. On the same lines, the number of flip-flops required to construct counters with MOD numbers of 3, 6, 14, 28 and 63 would be 2, 3, 4, 5 and 6 respectively. In general, the arrangement of a minimum number of N flip-flops can be used to construct any counter with a modulus given by the equation

$$(2^{N-1} + 1) \leq modulus \leq 2^N \quad (11.1)$$

11.4 Binary Ripple Counter – Operational Basics

The operation of a binary ripple counter can be best explained with the help of a typical counter of this type. Figure 11.2(a) shows a four-bit ripple counter implemented with negative edge-triggered *J-K* flip-flops wired as toggle flip-flops. The output of the first flip-flop feeds the clock input of the second, and the output of the second flip-flop feeds the clock input of the third, the output of which in turn feeds the clock input of the fourth flip-flop. The outputs of the four flip-flops are designated as Q_0 (LSB flip-flop), Q_1, Q_2 and Q_3 (MSB flip-flop). Figure 11.2(b) shows the waveforms appearing at Q_0, Q_1, Q_2 and Q_3 outputs as the clock signal goes through successive cycles of trigger pulses. The counter functions as follows.

Let us assume that all the flip-flops are initially cleared to the '0' state. On HIGH-to-LOW transition of the first clock pulse, Q_0 goes from '0' to '1' owing to the toggling action. As the flip-flops used are negative edge-triggered ones, the '0' to '1' transition of Q_0 does not trigger flip-flop FF1. FF1, along with FF2 and FF3, remains in its '0' state. So, on the occurrence of the first negative-going clock transition, $Q_0 = 1$, $Q_1 = 0$, $Q_2 = 0$ and $Q_3 = 0$.

On the HIGH-to-LOW transition of the second clock pulse, Q_0 toggles again. That is, it goes from '1' to '0'. This '1' to '0' transition at the Q_0 output triggers FF1, the output Q_1 of which goes from '0'

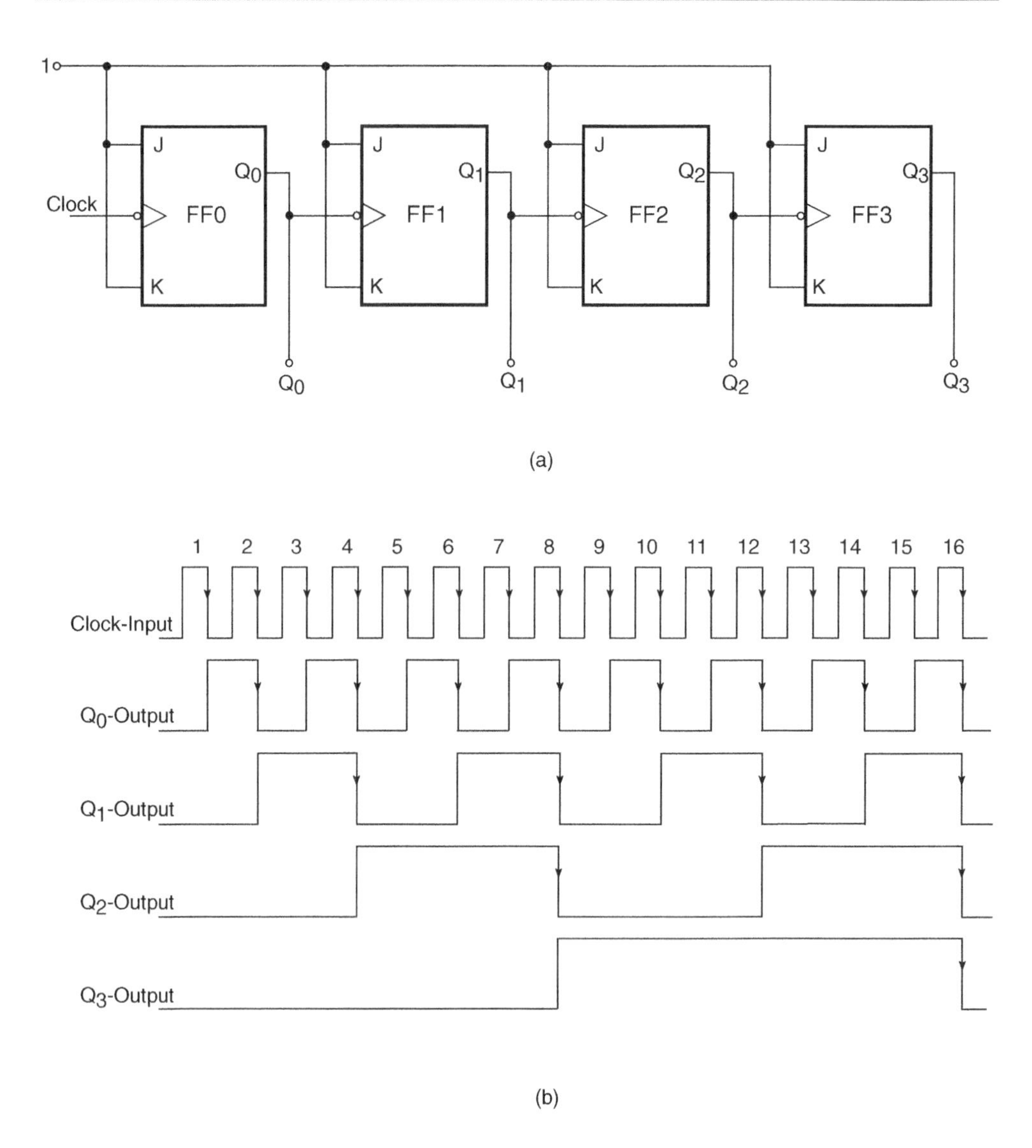

Figure 11.2 Four-bit binary ripple counter.

to '1'. The Q_2 and Q_3 outputs remain unaffected. Therefore, immediately after the occurrence of the second HIGH-to-LOW transition of the clock signal, $Q_0 = 0$, $Q_1 = 1$, $Q_2 = 0$ and $Q_3 = 0$. On similar lines, we can explain the logic status of Q_0, Q_1, Q_2 and Q_3 outputs immediately after subsequent clock transitions. The logic status of outputs for the first 16 relevant (HIGH-to-LOW in the present case) clock signal transitions is summarized in Table 11.1.

Thus, we see that the counter goes through 16 distinct states from 0000 to 1111 and then, on the occurrence of the desired transition of the sixteenth clock pulse, it resets to the original state of 0000 from where it had started. In general, if we had N flip-flops, we could count up to 2^N pulses before the counter resets to the initial state. We can also see from the Q_0, Q_1, Q_2 and Q_3 waveforms, as shown

Table 11.1 Output logic states for different clock signal transitions for a four-bit binary ripple counter.

Clock signal transition number	Q_0	Q_1	Q_2	Q_3
After first clock transition	1	0	0	0
After second clock transition	0	1	0	0
After third clock transition	1	1	0	0
After fourth clock transition	0	0	1	0
After fifth clock transition	1	0	1	0
After sixth clock transition	0	1	1	0
After seventh clock transition	1	1	1	0
After eighth clock transition	0	0	0	1
After ninth clock transition	1	0	0	1
After tenth clock transition	0	1	0	1
After eleventh clock transition	1	1	0	1
After twelfth clock transition	0	0	1	1
After thirteenth clock transition	1	0	1	1
After fourteenth clock transition	0	1	1	1
After fifteenth clock transition	1	1	1	1
After sixteenth clock transition	0	0	0	0

in Fig. 11.2(b), that the frequencies of the Q_0, Q_1, Q_2 and Q_3 waveforms are $f/2, f/4$, $f/8$ and $f/16$ respectively. Here, f is the frequency of the clock input. This implies that a counter of this type can be used as a divide-by-2^N circuit, where N is the number of flip-flops in the counter chain. In fact, such a counter provides frequency-divided outputs of $f/2^N$, $f/2^{N-1}$, $f/2^{N-2}$, $f/2^{N-3}$, . . . , $f/2$ at the outputs of the Nth, ($N-$ 1)th, ($N-$ 2)th, ($N-$ 3)th, . . . , first flip-flops. In the case of a four-bit counter of the type shown in Fig. 11.2(a), outputs are available at $f/2$ from the Q_0 output, at $f/4$ from the Q_1 output, at $f/8$ from the Q_2 output and at $f/16$ from the Q_3 output. It may be noted that frequency division is one of the major applications of counters.

Example 11.1

A four-bit binary ripple counter of the type shown in Fig. 11.2(a) is initially in the 0000 state before the clock input is applied to the counter. The clock pulses are applied to the counter at some time instant t_1 and then again removed some time later at another time instant t_2. The counter is observed to read 0011. How many negative-going clock transitions have occurred during the time the clock was active at the counter input?

Solution

It is not possible to determine the number of clock edges – it could have been 3, 19, 35, 51, 67, 83 . . . – as there is no means of finding out whether the counter has recycled or not from the given data. Remember that this counter would come back to the 0000 state after every 16 clock pulses.

Example 11.2

It is desired to design a binary ripple counter of the type shown in Fig. 11.1 that is capable of counting the number of items passing on a conveyor belt. Each time an item passes a given point, a pulse is generated that can be used as a clock input. If the maximum number of items to be counted is 6000, determine the number of flip-flops required.

Solution

- The counter should be able to count a maximum of 6000 items.
- An N-flip-flop would be able to count up to a maximum of $2^N - 1$ counts.
- On the 2^Nth clock pulse, it will get reset to all 0s.
- Now, $2^N - 1$ should be greater than or equal to 6000.
- That is, $2^N - 1 \geq 6000$, which gives $N \geq \log 6001/\log 2 \geq 3.778/0.3010 \geq 12.55$.
- The smallest integer that satisfies this condition is 13.
- Therefore, the minimum number of flip-flops required = 13

11.4.1 Binary Ripple Counters with a Modulus of Less than 2^N

An N-flip-flop binary ripple counter can be modified, as we will see in the following paragraphs, to have any other modulus less than 2^N with the help of simple externally connected combinational logic. We will illustrate this simple concept with the help of an example.

Consider the four-flip-flop binary ripple counter arrangement of Fig. 11.3(a). It uses *J-K* flip-flops with an active LOW asynchronous CLEAR input. The NAND gate in the figure has its output connected to the CLEAR inputs of all four flip-flops. The inputs to this three-input NAND gate are from the Q outputs of flip-flops FF0, FF1 and FF2. If we disregard the NAND gate for some time, this counter will go through its natural binary sequence from 0000 to 1111. But that is not to happen in the present arrangement. The counter does start counting from 0000 towards its final count of 1111. The counter keeps counting as long as the asynchronous CLEAR inputs of the different flip-flops are inactive. That is, the NAND gate output is HIGH. This is the case until the counter reaches 0110. With the seventh clock pulse it tends to go to 0111, which makes all NAND gate inputs HIGH, forcing its output to LOW. This HIGH-to-LOW transition at the NAND gate output clears all flip-flop outputs to the logic '0' state, thus disallowing the counter to settle at 0111. From the eighth clock pulse onwards, the counter repeats the sequence. The counter thus always counts from 0000 to 0110 and resets back to 0000. The remaining nine states, which include 0111, 1000, 1001, 1010, 1011, 1100, 1101, 1110 and 1111, are skipped, with the result that we get an MOD-7 counter.

Figure 11.3(b) shows the timing waveforms for this counter. By suitably choosing NAND inputs, one can get a counter with any MOD number less than 16. Examination of timing waveforms also reveals that the frequency of the Q_2 output is one-seventh of the input clock frequency.

The waveform at the Q_2 output is, however, not symmetrical as it would be if the counter were to go through its full binary sequence. The Q_3 output stays in the logic LOW state. It is expected to be so because an MOD-7 counter needs a minimum of three flip-flops. That is why the fourth flip-flop, which was supposed to toggle on the HIGH-to-LOW transition of the eighth clock pulse, and on every successive eighth pulse thereafter, never gets to that stage. The counter is cleared on the seventh clock pulse and every successive seventh clock pulse thereafter.

As another illustration, if the NAND gate used in the counter arrangement of Fig. 11.3(a) is a two-input NAND and its inputs are from the Q_1 and Q_3 outputs, the counter will go through 0000 to 1001 and then reset to 0000 again, as, the moment the counter tends to switch from the 1001 to the 1010 state, the NAND gate goes from the '1' to the '0' state, clearing all flip-flops to the '0' state.

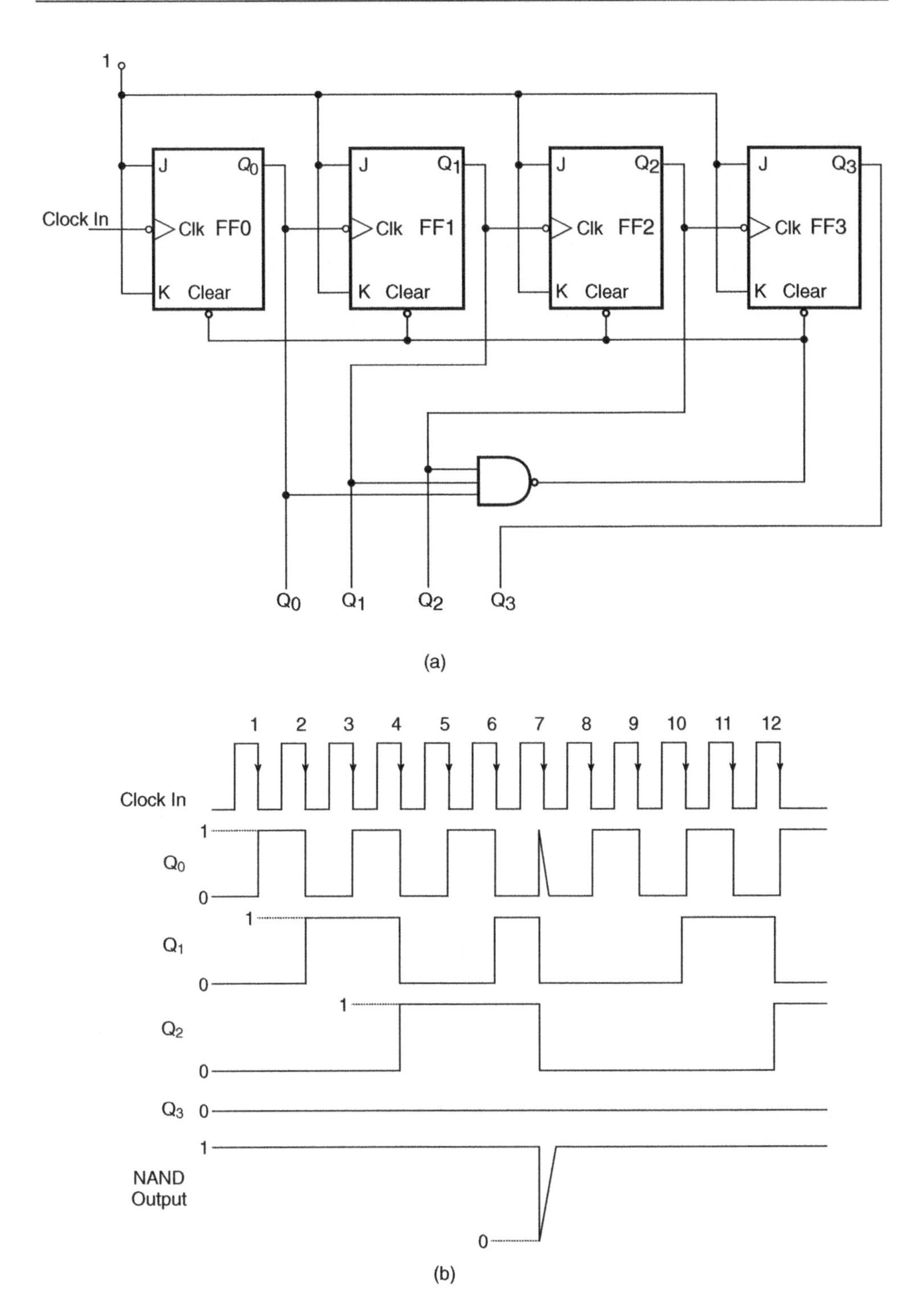

Figure 11.3 Binary ripple counter with a modulus of less than 2^N.

Steps to be followed to design any binary ripple counter that starts from 0000 and has a modulus of X are summarized as follows:

1. Determine the minimum number of flip-flops N so that $2^N \geq X$. Connect these flip-flops as a binary ripple counter. If $2^N = X$, do not go to steps 2 and 3.
2. Identify the flip-flops that will be in the logic HIGH state at the count whose decimal equivalent is X. Choose a NAND gate with the number of inputs equal to the number of flip-flops that would be in the logic HIGH state. As an example, if the objective were to design an MOD-12 counter, then, in the corresponding count, that is, 1100, two flip-flops would be in the logic HIGH state. The desired NAND gate would therefore be a two-input gate.
3. Connect the Q outputs of the identified flip-flops to the inputs of the NAND gate and the NAND gate output to asynchronous clear inputs of all flip-flops.

11.4.2 Ripple Counters in IC Form

In this section, we will look at the internal logic diagram of a typical binary ripple counter and see how close its architecture is to the ripple counter described in the previous section. Let us consider binary ripple counter type number 74293. It is a four-bit binary ripple counter containing four master–slave-type *J-K* flip-flops with additional gating to provide a divide-by-2 counter and a three-stage MOD-8 counter. Figure 11.4 shows the internal logic diagram of this counter. To get the full binary sequence of 16 states, the Q output of the LSB flip-flop is connected to the B input, which is the clock input of the next higher flip-flop. The arrangement then becomes the same as that shown in Fig. 11.2(a), with the exception of the two-input NAND gate of Fig. 11.4, which has been included here for providing the clearing features. The counter can be cleared to the 0000 logic state by driving both RESET inputs to the logic HIGH state. Tables 11.2 and 11.3 respectively give the functional table and the count sequence.

Example 11.3

Refer to the binary ripple counter of Fig. 11.5. Determine the modulus of the counter and also the frequency of the flip-flop Q_3 output.

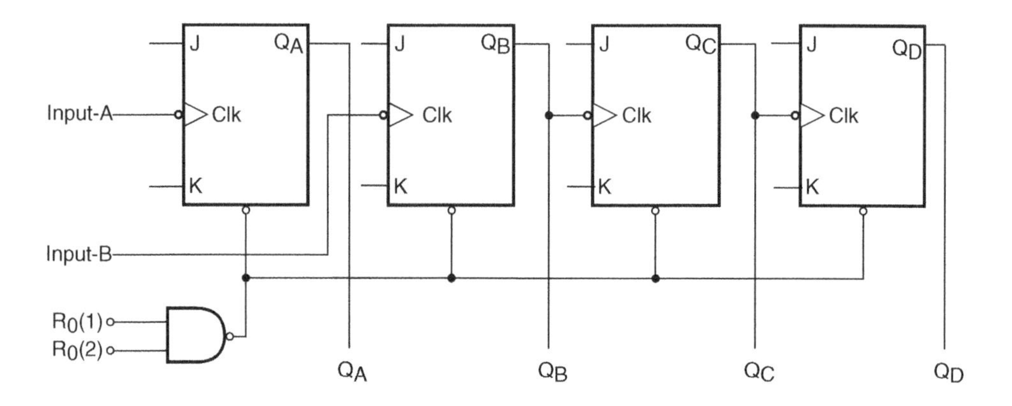

Figure 11.4 Logic diagram of IC 74293.

Table 11.2 Functional table for binary ripple counter, type number 74293.

RESET inputs		Outputs			
$R_0(1)$	$R_0(2)$	Q_D	Q_C	Q_B	Q_A
H	H	L	L	L	L
L	X		Count		
X	L		Count		

Table 11.3 Count sequence for binary ripple counter, type number 74293.

Count	Outputs			
	Q_D	Q_C	Q_B	Q_A
0	L	L	L	L
1	L	L	L	H
2	L	L	H	L
3	L	L	H	H
4	L	H	L	L
5	L	H	L	H
6	L	H	H	L
7	L	H	H	H
8	H	L	L	L
9	H	L	L	H
10	H	L	H	L
11	H	L	H	H
12	H	H	L	L
13	H	H	L	H
14	H	H	H	L
15	H	H	H	H

Solution

- The counter counts in the natural sequence from 0000 to 1011.
- The moment the counter goes to 1100, the NAND output goes to the logic '0' state and immediately clears the counter to the 0000 state.
- Thus, the counter is not able to stay in the 1100 state. It has only 12 stable states from 0000 to 1011.
- Therefore, the modulus of the counter = 12.
- The Q_3 output is the input clock frequency divided by 12.
- Therefore, the frequency of the Q_3 output waveform $= 1.2 \times 10^3/12 = 100$ kHz.

Example 11.4

Design a binary ripple counter that counts 000 and 111 and skips the remaining six states, that is, 001, 010, 011, 100, 101 and 110. Use presentable, clearable negative edge-triggered J-K flip-flops with active LOW PRESET and CLEAR inputs. Also, draw the timing waveforms and determine the frequency of different flip-flop outputs for a given clock frequency, f_c.

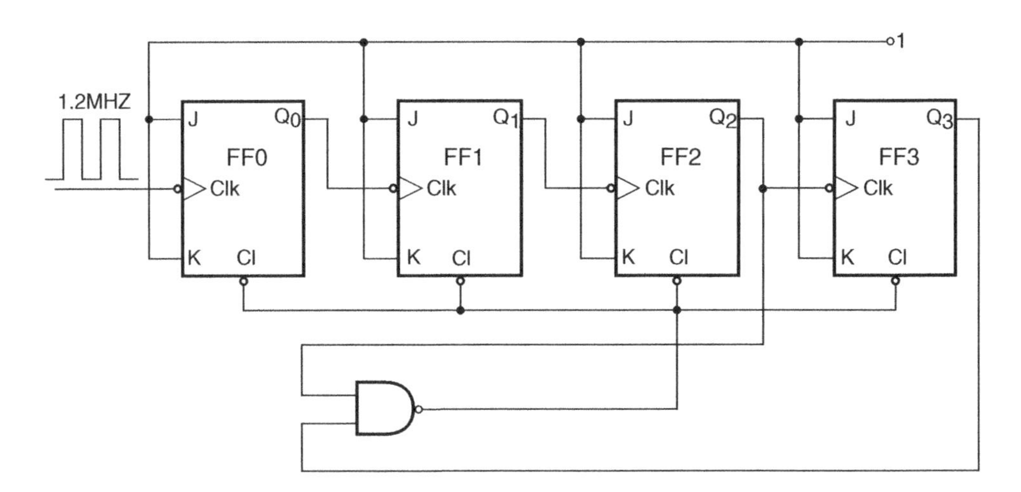

Figure 11.5 Example 11.3.

Solution

The counter is required to go to the 111 state from the 000 state with the first relevant clock transition. The second transition brings it back to the 000 state. That is, the three flip-flops toggle from logic '0' state to logic '1' state with every odd-numbered clock transition, and also the three flip-flops toggle from logic '1' state to logic '0' state with every even-numbered clock transition. Figure 11.6(a) shows the arrangement. The PRESET inputs of the three flip-flops have been tied to the NAND output whose inputs are Q_A, $\overline{Q_B}$ and $\overline{Q_C}$. Every time the counter is in the 000 state and is clocked, the NAND output momentarily goes from logic '1' state to logic '0' state, thus presetting the Q_A, Q_B and Q_C outputs to the logic '1' state. The timing waveforms as shown in Fig. 11.6(b) are self-explanatory.

The Q_A, $Q_B$and $Q_C$waveforms are identical, and each of them has a frequency of $f_c/2$, where f_c is the clock frequency.

Example 11.5

Refer to the binary ripple counter arrangement of Fig. 11.7. Write its count sequence if it is initially in the 0000 state. Also draw the timing waveforms.

Solution

The counter is initially in the 0000 state. With the first clock pulse, Q_0 toggles from the '0' to the '1' state, which means $\overline{Q_0}$ toggles from '1' to '0'. Since $\overline{Q_0}$ here feeds the clock input of next flip-flop, flip-flop FF1 also toggles. Thus, Q_1 goes from '0' to '1'. Since flip-flops FF2 and FF3 are also clocked from complementary outputs of their immediately preceding flip-flops, they also toggle. Thus, the counter moves from the 0000 state to the 1111 state with the first clock pulse.

With the second clock pulse, Q_0 toggles again, but the other flip-flops remain unaffected for obvious reasons and the counter is in the 1110 state. With subsequent clock pulses, the counter keeps counting downwards by one LSB at a time until it reaches 0000 again, after which the process repeats. The count sequence is given as 0000, 1111, 1110, 1101,1100, 1011, 1010, 1001, 1000,

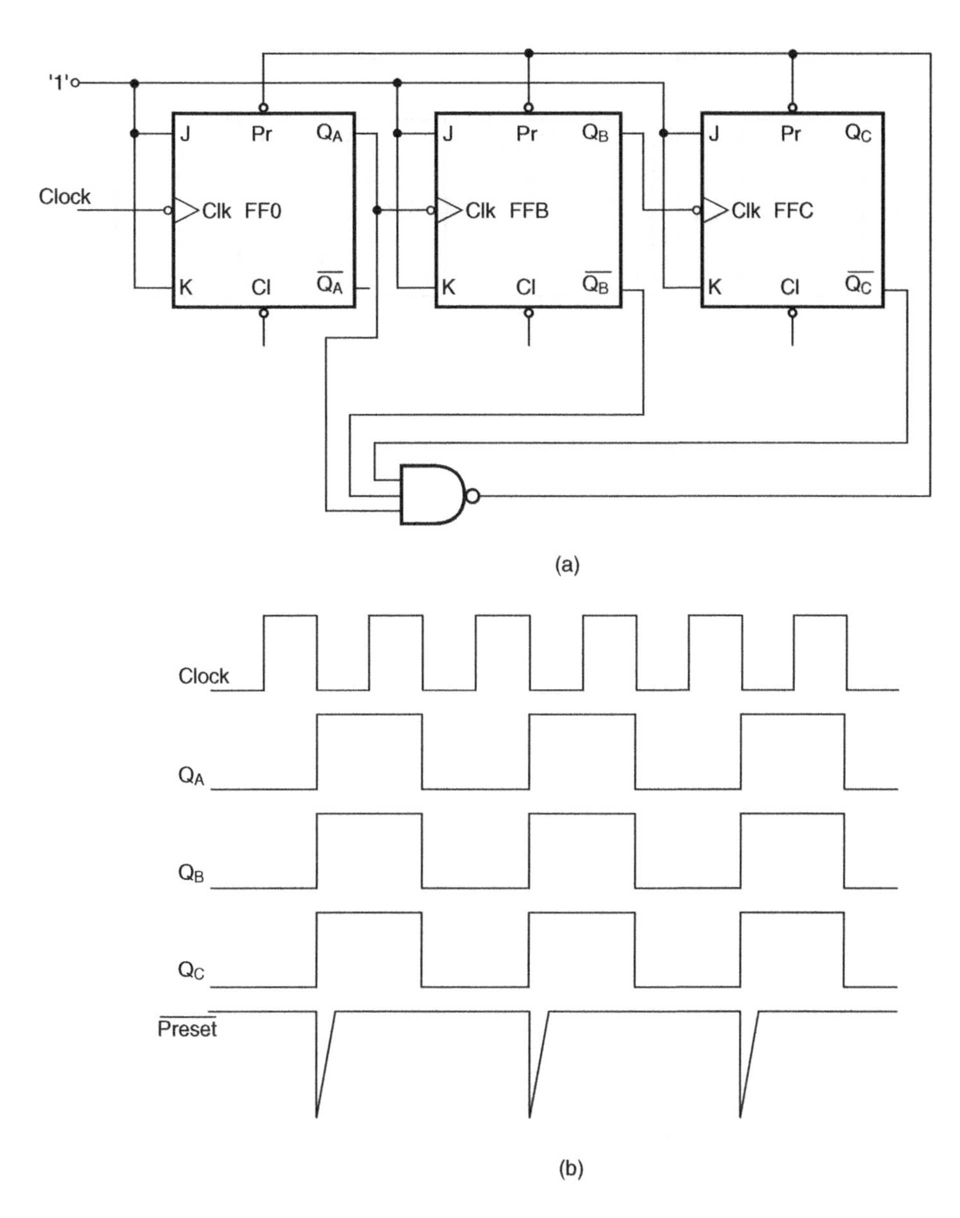

Figure 11.6 Example 11.4.

0111, 0110, 0101, 0100, 0011, 0010, 0001 and 0000. The timing waveforms are shown in Fig. 11.8. Thus, we have a four-bit counter that counts in the reverse sequence, beginning with the maximum count. This is a DOWN counter. This type of counter is discussed further in the subsequent paragraphs.

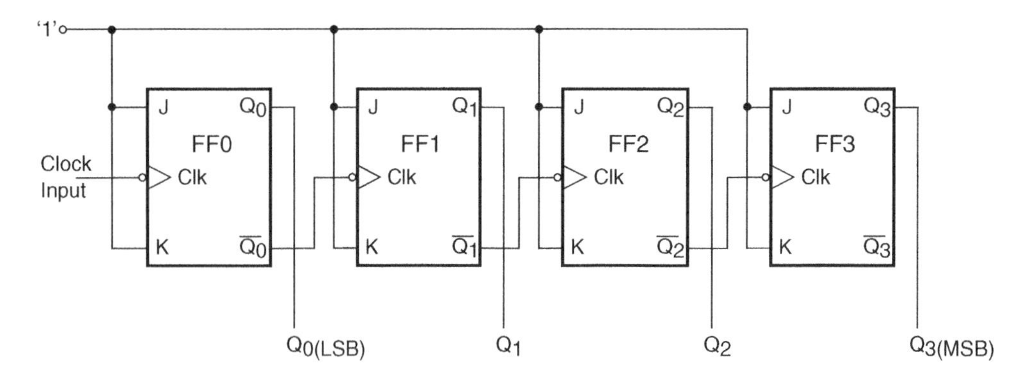

Figure 11.7 Counter schematic, example 11.5.

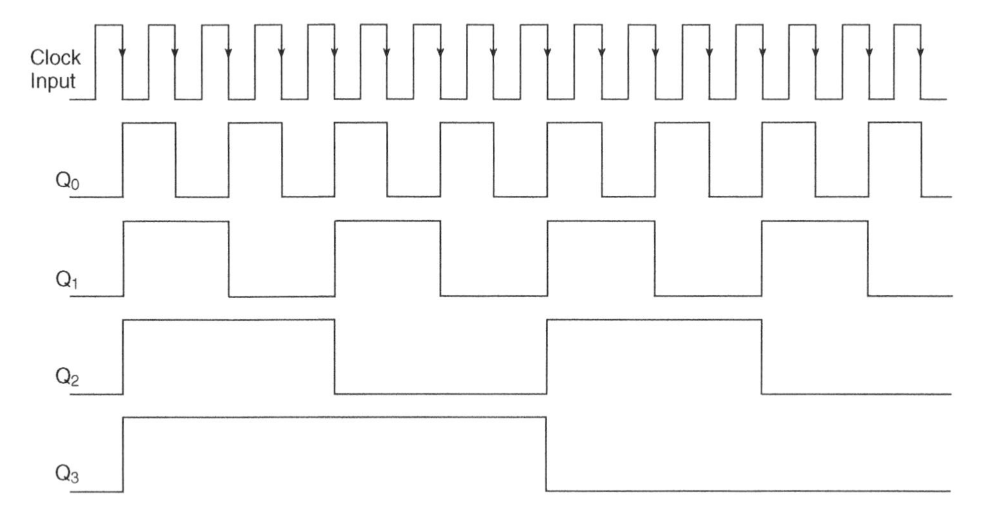

Figure 11.8 Timing waveforms, example 11.5.

From what we have discussed for a binary ripple counter, including the solved examples given to supplement the text, we can make the following observations:

1. If the flip-flops used to construct the counter are negative (HIGH-to-LOW) edge triggered and the clock inputs are fed from Q outputs, the counter counts in the normal upward count sequence.
2. If the flip-flops used to construct the counter are negative edge triggered and the clock inputs are fed from $\overline{Q}$ outputs, the counter counts in the reverse or downward count sequence.
3. If the flip-flops used to construct the counter are positive (LOW-to-HIGH) edge triggered and the clock inputs are fed from Q outputs, the counter counts in the reverse or downward count sequence.
4. If the flip-flops used to construct the counter are positive edge triggered and the clock inputs are fed from the $\overline{Q}$ outputs, the counter counts in the normal upward count sequence.

11.5 Synchronous (or Parallel) Counters

Ripple counters discussed thus far in this chapter are asynchronous in nature as the different flip-flops comprising the counter are not clocked simultaneously and in synchronism with the clock pulses. The total propagation delay in such a counter, as explained earlier, is equal to the sum of propagation delays due to different flip-flops. The propagation delay becomes prohibitively large in a ripple counter with a large count. On the other hand, in a synchronous counter, all flip-flops in the counter are clocked simultaneously in synchronism with the clock, and as a consequence all flip-flops change state at the same time. The propagation delay in this case is independent of the number of flip-flops used.

Since the different flip-flops in a synchronous counter are clocked at the same time, there needs to be additional logic circuitry to ensure that the various flip-flops toggle at the right time. For instance, if we look at the count sequence of a four-bit binary counter shown in Table 11.4, we find that flip-flop FF0 toggles with every clock pulse, flip-flop FF1 toggles only when the output of FF0 is in the '1' state, flip-flop FF2 toggles only with those clock pulses when the outputs of FF0 and FF1 are both in the logic '1' state and flip-flop FF3 toggles only with those clock pulses when Q_0, Q_1 and Q_2 are all in the logic '1' state. Such logic can be easily implemented with AND gates. Figure 11.9(a) shows the schematic arrangement of a four-bit synchronous counter. The timing waveforms are shown in Fig. 11.9(b). The diagram is self-explanatory. As an example, ICs 74162 and 74163 are four-bit synchronous counters, with the former being a decade counter and the latter a binary counter.

A synchronous counter that counts in the reverse or downward sequence can be constructed in a similar manner by using complementary outputs of the flip-flops to drive the J and K inputs of the following flip-flops. Refer to the reverse or downward count sequence as given in Table 11.5. As is evident from the table, FF0 toggles with every clock pulse, FF1 toggles only when Q_0 is logic '0', FF2 toggles only when both Q_0 and Q_1 are in the logic '0' state and FF3 toggles only when Q_0, Q_1 and Q_2 are in the logic '0' state.

Referring to the four-bit synchronous UP counter of Fig. 11.9(a), if the J and K inputs of flip-flop FF1 are fed from the $\overline{Q_0}$ output instead of the Q_0 output, the inputs to the two-input AND gate are $\overline{Q_0}$ and $\overline{Q_1}$ instead of Q_0 and Q_1, and the inputs to the three-input AND gate are $\overline{Q_0}$, $\overline{Q_1}$ and $\overline{Q_2}$ instead of Q_0, Q_1 and Q_2, we get a counter that counts in reverse order. In that case it becomes a four-bit synchronous DOWN counter.

Table 11.4 Count sequence of a four-bit binary counter.

Count	Q_3	Q_2	Q_1	Q_0	Count	Q_3	Q_2	Q_1	Q_0
0	0	0	0	0	8	1	0	0	0
1	0	0	0	1	9	1	0	0	1
2	0	0	1	0	10	1	0	1	0
3	0	0	1	1	11	1	0	1	1
4	0	1	0	0	12	1	1	0	0
5	0	1	0	1	13	1	1	0	1
6	0	1	1	0	14	1	1	1	0
7	0	1	1	1	15	1	1	1	1

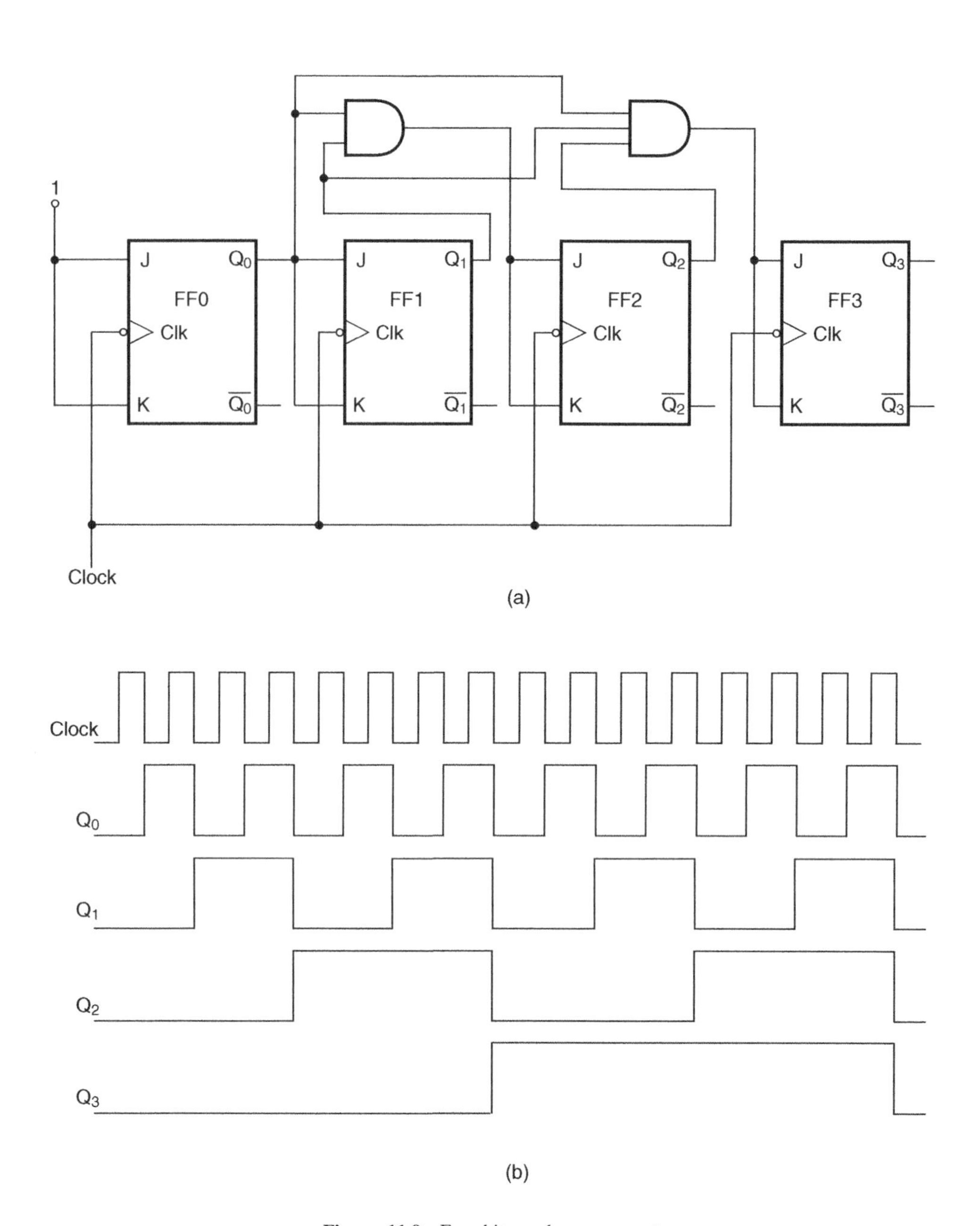

Figure 11.9 Four-bit synchronous counter.

Table 11.5 Reverse or downward count sequence synchronous counter.

Count	Q_3	Q_2	Q_1	Q_0	Count	Q_3	Q_2	Q_1	Q_0
0	0	0	0	0	8	1	0	0	0
1	1	1	1	1	9	0	1	1	1
2	1	1	1	0	10	0	1	1	0
3	1	1	0	1	11	0	1	0	1
4	1	1	0	0	12	0	1	0	0
5	1	0	1	1	13	0	0	1	1
6	1	0	1	0	14	0	0	1	0
7	1	0	0	1	15	0	0	0	1

11.6 UP/DOWN Counters

Counters are also available in integrated circuit form as UP/DOWN counters, which can be made to operate as either UP or DOWN counters. As outlined in Section 11.5, an UP counter is one that counts upwards or in the forward direction by one LSB every time it is clocked. A four-bit binary UP counter will count as 0000, 0001, 0010, 0011, 0100, 0101, 0110, 0111, 1000, 1001, 1010, 1011, 1100, 1101, 1110, 1111, 0000, 0001, . . . and so on. A DOWN counter counts in the reverse direction or downwards by one LSB every time it is clocked. The four-bit binary DOWN counter will count as 0000, 1111, 1110, 1101, 1100, 1011, 1010, 1001, 1000, 0111, 0110, 0101, 0100, 0011, 0010, 0001, 0000, 1111, . . . and so on.

Some counter ICs have separate clock inputs for UP and DOWN counts, while others have a single clock input and an UP/DOWN control pin. The logic status of this control pin decides the counting mode. As an example, ICs 74190 and 74191 are four-bit UP/DOWN counters in the TTL family with a single clock input and an UP/DOWN control pin. While IC 74190 is a BCD decade counter, IC 74191 is a binary counter. Also, ICs 74192 and 74193 are four-bit UP/DOWN counters in the TTL family, with separate clock input terminals for UP and DOWN counts. While IC 74192 is a BCD decade counter, IC 74193 is a binary counter.

Figure 11.10 shows a three-bit binary UP/DOWN counter. This is only one possible logic arrangement. As we can see, the counter counts upwards when UP control is logic '1' and DOWN

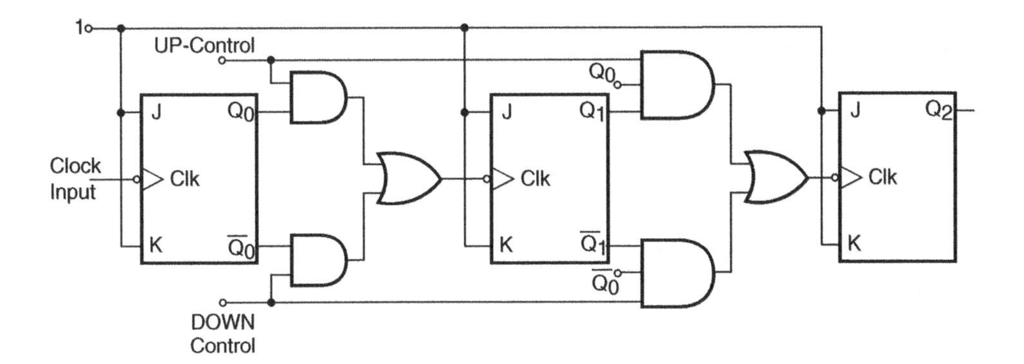

Figure 11.10 Four-bit UP/DOWN counter.

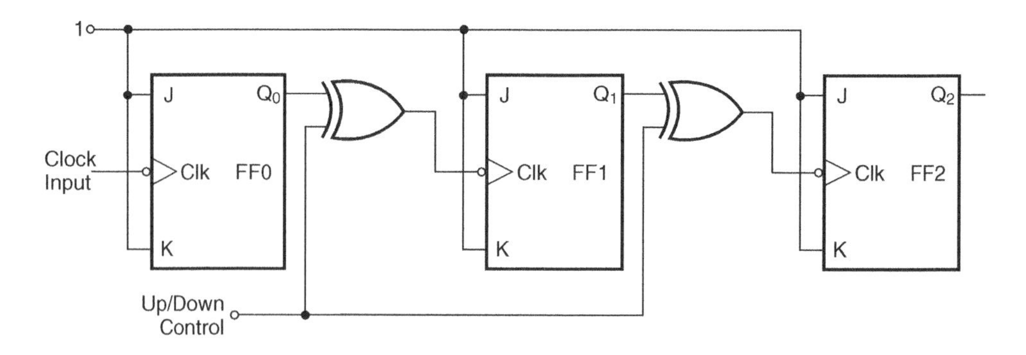

Figure 11.11 Three-bit UP/DOWN counter with a common clock input.

control is logic '0'. In this case the clock input of each flip-flop other than the LSB flip-flop is fed from the normal output of the immediately preceding flip-flop. The counter counts downwards when the UP control input is logic '0' and DOWN control is logic '1'. In this case, the clock input of each flip-flop other than the LSB flip-flop is fed from the complemented output of the immediately preceding flip-flop. Figure 11.11 shows another possible configuration for a three-bit binary ripple UP/DOWN counter. It has a common control input. When this input is in logic '1' state the counter counts downwards, and when it is in logic '0' state it counts upwards.

11.7 Decade and BCD Counters

A *decade counter* is one that goes through 10 unique output combinations and then resets as the clock proceeds further. Since it is an MOD-10 counter, it can be constructed with a minimum of four flip-flops. A four-bit counter would have 16 states. By skipping any of the six states by using some kind of feedback or some kind of additional logic, we can convert a normal four-bit binary counter into a decade counter. A decade counter does not necessarily count from 0000 to 1001. It could even count as 0000, 0001, 0010, 0101, 0110, 1001, 1010, 1100, 1101, 1111, 0000, ... In this count sequence, we have skipped 0011, 0100, 0111, 1000, 1011 and 1110.

A *BCD counter* is a special case of a decade counter in which the counter counts from 0000 to 1001 and then resets. The output weights of flip-flops in these counters are in accordance with 8421-code. For instance, at the end of the seventh clock pulse, the counter output will be 0111, which is the binary equivalent of decimal 7. In other words, different counter states in this counter are binary equivalents of the decimal numbers 0 to 9. These are different from other decade counters that provide the same count by using some kind of forced feedback to skip six of the natural binary counts.

11.8 Presettable Counters

Presettable counters are those that can be preset to any starting count either asynchronously (independently of the clock signal) or synchronously (with the active transition of the clock signal). The presetting operation is achieved with the help of PRESET and CLEAR (or MASTER RESET) inputs available on the flip-flops. The presetting operation is also known as the 'preloading' or simply the 'loading' operation.

Presettable counters can be UP counters, DOWN counters or UP/DOWN counters. Additional inputs/outputs available on a presettable UP/DOWN counter usually include PRESET inputs, from where any desired count can be loaded, parallel load (*PL*) inputs, which when active allow the PRESET inputs to be loaded onto the counter outputs, and terminal count (*TC*) outputs, which become active when the counter reaches the terminal count.

Figure 11.12 shows the logic diagram of a four-bit presettable synchronous UP counter. The data available on P_3, P_2, P_1 and P_0 inputs are loaded onto the counter when the parallel load ($\overline{PL}$) input goes LOW.

When the $\overline{PL}$ input goes LOW, one of the inputs of all NAND gates, including the four NAND gates connected to the PRESET inputs and the four NAND gates connected to the CLEAR inputs, goes to the logic '1' state. What reaches the PRESET inputs of FF3, FF2, FF1 and FF0 is $\overline{P_3}$, $\overline{P_2}$, $\overline{P_1}$ and $\overline{P_0}$ respectively, and what reaches their CLEAR inputs is P_3, P_2, P_1 and P_0 respectively. Since PRESET and CLEAR are active LOW inputs, the counter flip-flops FF3, FF2, FF1 and FF0 will respectively be loaded with P_3, P_2, P_1 and P_0. For example, if $P_3 = 1$, the PRESET and CLEAR inputs of FF3 will be in the '0' and '1' logic states respectively. This implies that the Q_3 output will go to the logic '1' state. Thus, FF3 has been loaded with P_3. Similarly, if $P_3 = 0$, the PRESET and CLEAR inputs of flip-flop FF3 will be in the '1' and '0' states respectively. The flip-flop output (Q_3 output) will be cleared to the '0' state. Again, the flip-flop is loaded with P_3 logic status when the $\overline{PL}$ input becomes active.

Counter ICs 74190, 74191, 74192 and 74193 are asynchronously presettable synchronous UP/DOWN counters. Many synchronous counters use synchronous presetting whereby the counter is preset or loaded with the data on the active transition of the same clock signal that is used for counting. Presettable counters also have terminal count ($\overline{TC}$) outputs, which allow them to be cascaded together to get counters with higher MOD numbers. In the cascade arrangement, the terminal count output of the lower-order counter feeds the clock input of the next higher-order counter. Cascading of counters is discussed in Section 11.10.

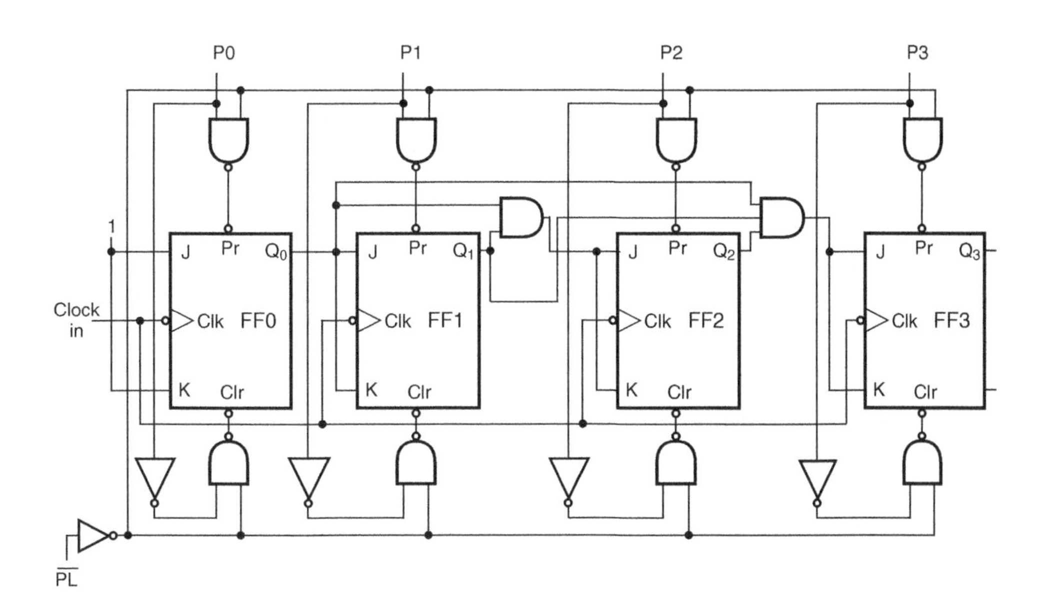

Figure 11.12 Four-bit presettable, clearable counter.

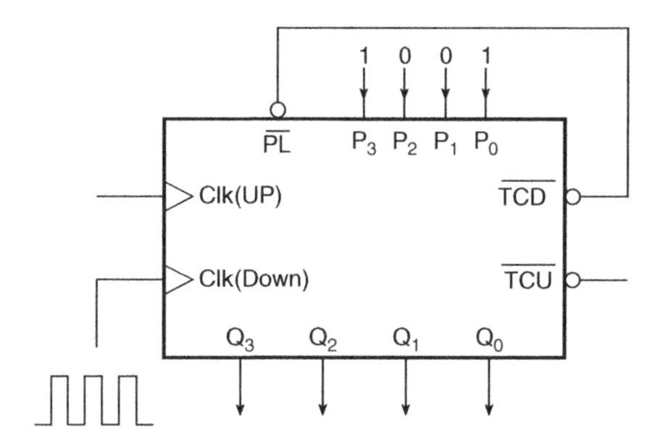

Figure 11.13 Presettable four-bit counter.

11.8.1 Variable Modulus with Presettable Counters

Presettable counters can be wired as counters with a modulus of less than 2^N without the need for any additional logic circuitry. When a presettable counter is preset with a binary number whose decimal equivalent is some number 'X', and if this counter is wired as a DOWN counter, with its terminal count (DOWN mode) output, also called borrow-out (B_o), fed back to the parallel load (PL) input, it works like an MOD-X counter.

We will illustrate this with the help of an example. Refer to Fig. 11.13. It shows a presettable four-bit synchronous UP/DOWN binary counter having separate clock inputs for UP and DOWN counting (both positive edge triggered), an active LOW parallel load input ($\overline{PL}$) and active LOW terminal count UP ($\overline{TCU}$) and terminal count DOWN ($\overline{TCD}$) outputs. This description is representative of IC counter type 74193. Let us assume that the counter is counting down and is presently in the 1001 state at time instant t_0. The $\overline{TCD}$ output is in the logic '1' state, and so is the $\overline{PL}$ input. That is, both are inactive. The counter counts down by one LSB at every positive-going edge of the clock input. Immediately after the ninth positive-going trigger (at time instant t_9), the counter is in the 0000 state, which is the terminal count. Coinciding with the negative-going edge of the same clock pulse, the $\overline{TCD}$ output goes to the logic '0' state, and so does the $\overline{PL}$ input. This loads the counter with 1001 at time instant t_{10}, as shown in the timing waveforms of Fig. 11.14. With the positive-going edges of the tenth clock pulse and thereafter, the counter repeats its DOWN count sequence. Examination of the Q_3 output waveform tells that its frequency is one-ninth of the input clock frequency. Thus, it is an MOD-9 counter. The modulus of the counter can be varied by varying the data loaded onto the parallel PRESET/LOAD inputs.

11.9 Decoding a Counter

The output state of a counter at any time instant, as it is being clocked, is in the form of a sequence of binary digits. For a large number of applications, it is important to detect or decode different states of the counter whose number equals the modulus of the counter. One typical application could be a need to initiate or trigger some action after the counter reaches a specific state. The decoding network therefore is going to be a logic circuit that takes its inputs from the outputs of the different flip-flops constituting the counter and then makes use of those data to generate outputs equal to the modulus or MOD-number of the counter.

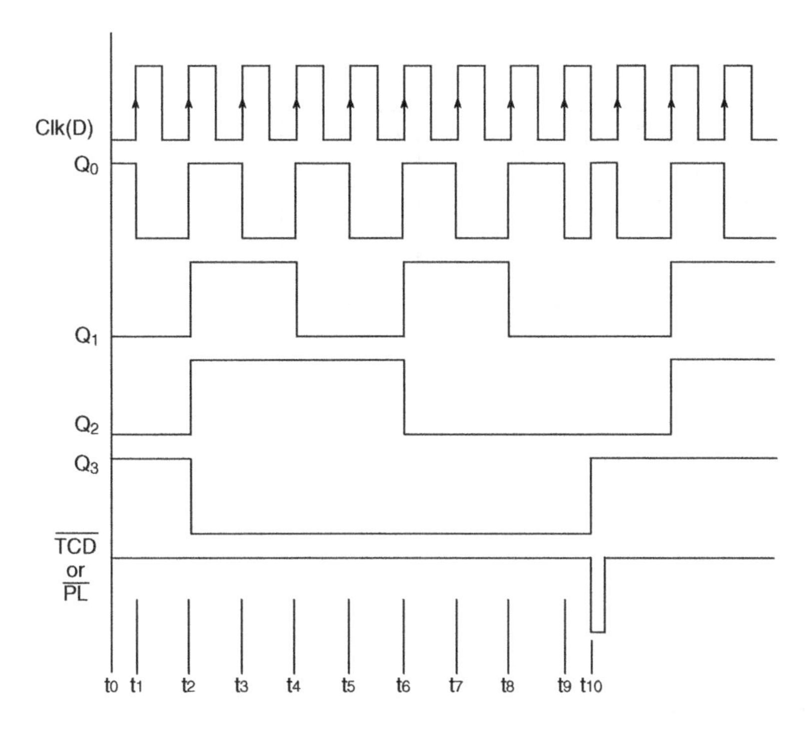

Figure 11.14 Timing waveforms for the counter of Fig. 11.13.

Depending upon the logic status of the decoded output, there are two basic types of decoding, namely *active HIGH* decoding and *active LOW* decoding. In the case of the former the decoder outputs are normally LOW, and for a given counter state the corresponding decoder output goes to the logic HIGH state. In the case of active LOW decoding, the decoder outputs are normally HIGH and the decoded output representing the counter state goes to the logic LOW state.

We will further illustrate the concept of decoding a counter with the help of an example. Consider the two-stage MOD-4 ripple counter of Fig. 11.15(a). This counter has four possible logic states, which need to be decoded. These include 00, 01, 10 and 11. Let us now consider the arrangement of four two-input AND gates as shown in Fig. 11.15(b) and what their outputs look like as the counter clock goes through the first four pulses. Before we proceed further, we have two important observations to make. Firstly, the number of AND gates used in the decoder network equals the number of logic states to be decoded, which further equals the modulus of the counter. Secondly, the number of inputs to each AND gate equals the number of flip-flops used in the counter. We can see from the waveforms of Fig. 11.15(b) that, when the counter is in the 00 state, the AND gate designated '0' is in the logic HIGH state and the outputs of the other gates designated '1', '2' and '3' are in the logic LOW state. Similarly, for 01, 10 and 11 states of the counter, the outputs of gates 1, 2 and 3 are respectively in the logic HIGH state. This is incidentally active HIGH decoding. We can visualize that, if the AND gates were replaced with NAND gates, with the inputs to the gates remaining the same, we would get an active LOW decoder. For a counter that uses N flip-flops and has a modulus of 'X', the decoder will have 'X' number of N-input AND or NAND gates, depending upon whether we want an active HIGH or active LOW decoder.

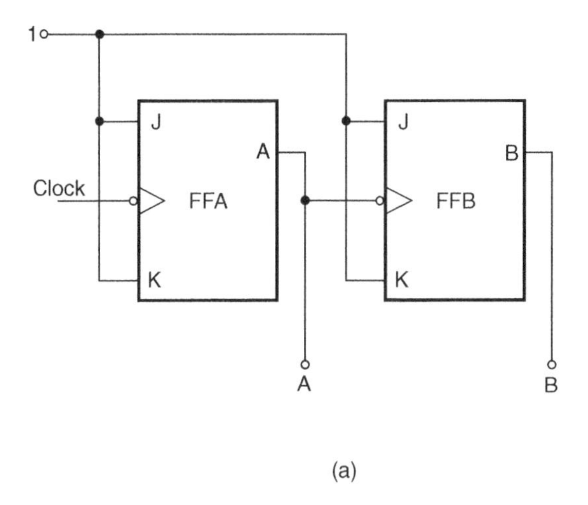

(a)

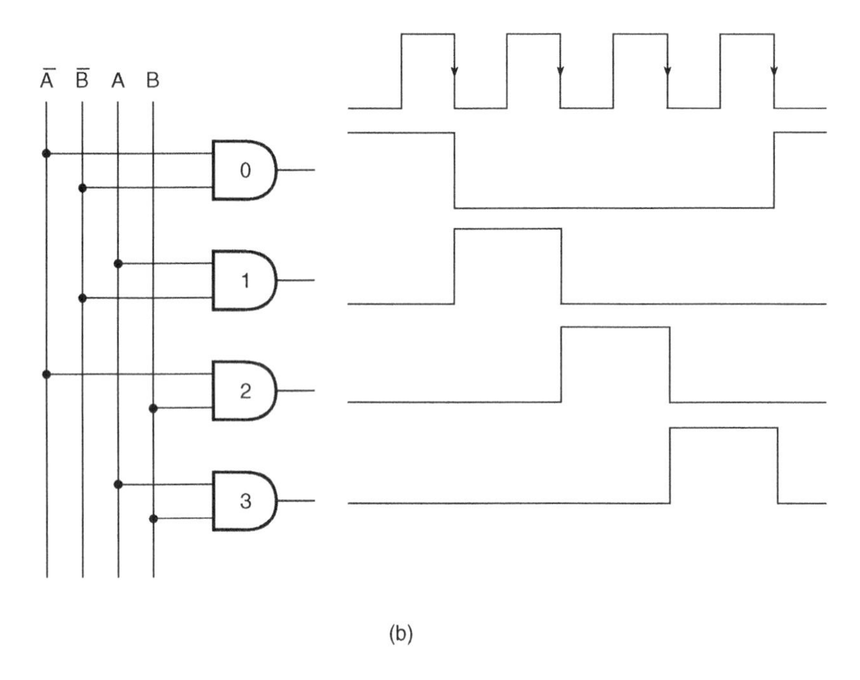

(b)

Figure 11.15 MOD-4 ripple counter with decoding logic.

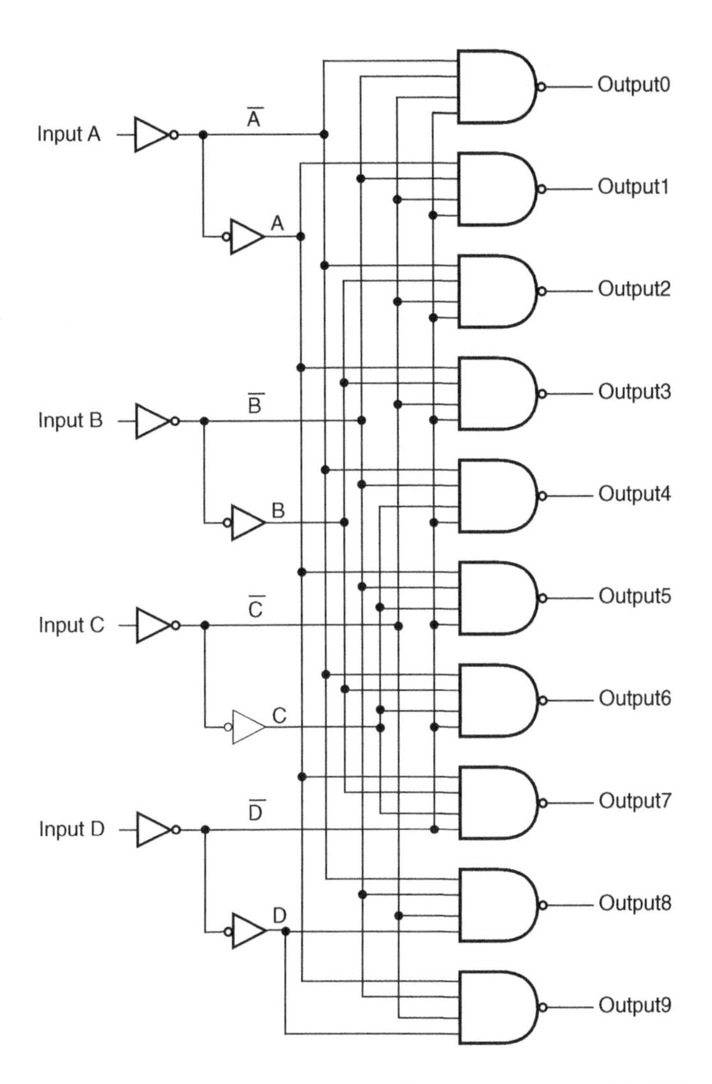

Figure 11.16 Logic diagram of four-line BCD-to-decimal decoder (IC 7442).

Figure 11.16 shows the logic diagram of a four-line BCD to decimal decoder with active low outputs. Full decoding of valid input logic states ensures that all outputs remain off or inactive for all invalid input conditions. Table 11.6 gives the functional table of the decoder of Fig. 11.16. The logic diagram shown in Fig. 11.16 is the actual logic diagram of IC 7442, which is a four-line BCD to decimal decoder in the TTL family.

The decoding gates used to decode the states of a ripple counter produce glitches (or spikes) in the decoded waveforms. These glitches basically result from the cumulative propagation delay as we move from one flip-flop to the next in a ripple counter. It can be best illustrated with the help of the MOD-4 counter shown in Fig. 11.17. The timing waveforms are shown in Fig. 11.18 and are self-explanatory.

Table 11.6 Functional table of the decoder of Fig. 11.16.

Decimal number	BCD input				Decimal output									
	D	*C*	*B*	*A*	0	1	2	3	4	5	6	7	8	9
0	L	L	L	L	L	H	H	H	H	H	H	H	H	H
1	L	L	L	H	H	L	H	H	H	H	H	H	H	H
2	L	L	H	L	H	H	L	H	H	H	H	H	H	H
3	L	L	H	H	H	H	H	L	H	H	H	H	H	H
4	L	H	L	L	H	H	H	H	L	H	H	H	H	H
5	L	H	L	H	H	H	H	H	H	L	H	H	H	H
6	L	H	H	L	H	H	H	H	H	H	L	H	H	H
7	L	H	H	H	H	H	H	H	H	H	H	L	H	H
8	H	L	L	L	H	H	H	H	H	H	H	H	L	H
9	H	L	L	H	H	H	H	H	H	H	H	H	H	L
Invalid	H	L	H	L	H	H	H	H	H	H	H	H	H	H
Invalid	H	L	H	H	H	H	H	H	H	H	H	H	H	H
Invalid	H	H	L	L	H	H	H	H	H	H	H	H	H	H
Invalid	H	H	L	H	H	H	H	H	H	H	H	H	H	H
Invalid	H	H	H	L	H	H	H	H	H	H	H	H	H	H
Invalid	H	H	H	H	H	H	H	H	H	H	H	H	H	H

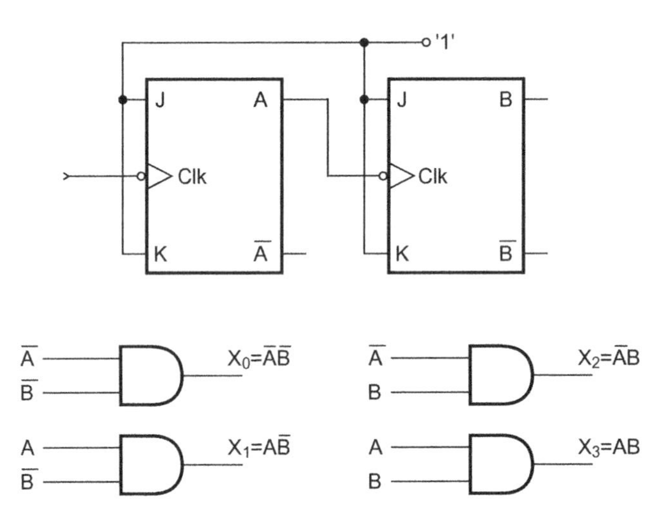

Figure 11.17 MOD-4 counter with decoding gates.

We can see the appearance of glitches at the output of decoding gates that decode X_0 and X_2 states.This problem for all practical purposes is absent in synchronous counters. Theoretically, it can even exist in a synchronous counter if the flip-flops used have different propagation delays.

One way to overcome this problem is to use a strobe signal which keeps the decoding gates disabled until all flip-flops have reached a stable state in response to the relevant clock transition. To implement

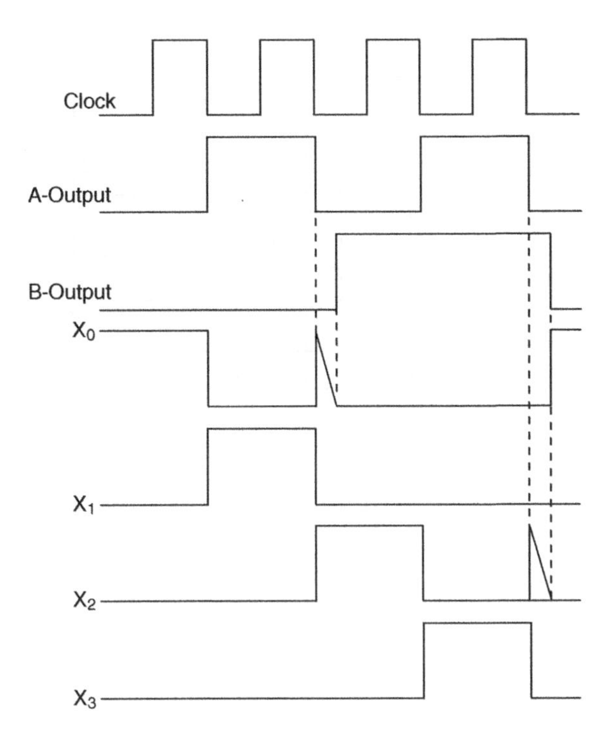

Figure 11.18 Glitch problem in decoders.

this, each of the decoding gates will have an additional input. This additional input of all decoding gates is tied together and the strobe signal applied to the common point.

One such decoder with additional strobe inputs to take care of glitch-related problems is IC 74154, which is a four-line to 16-line decoder in the TTL family. Figure 11.19 shows the internal logic diagram of IC 74154. We can see all NAND gates having an additional input line, which is controlled by strobe inputs $\overline{G}_1$ and $\overline{G}_2$.

11.10 Cascading Counters

A cascade arrangement allows us to build counters with a higher modulus than is possible with a single stage. The terminal count outputs allow more than one counter to be connected in a cascade arrangement. In the following paragraphs, we will examine some such cascade arrangements in the case of binary and BCD counters.

11.10.1 Cascading Binary Counters

In order to construct a multistage UP counter, all counter stages are connected in the count UP mode. The clock is applied to the clock input of a lowest-order counter, the terminal count UP (*TCU*), also called the carry-out (C_o), of this counter is applied to the clock input of the next higher counter stage

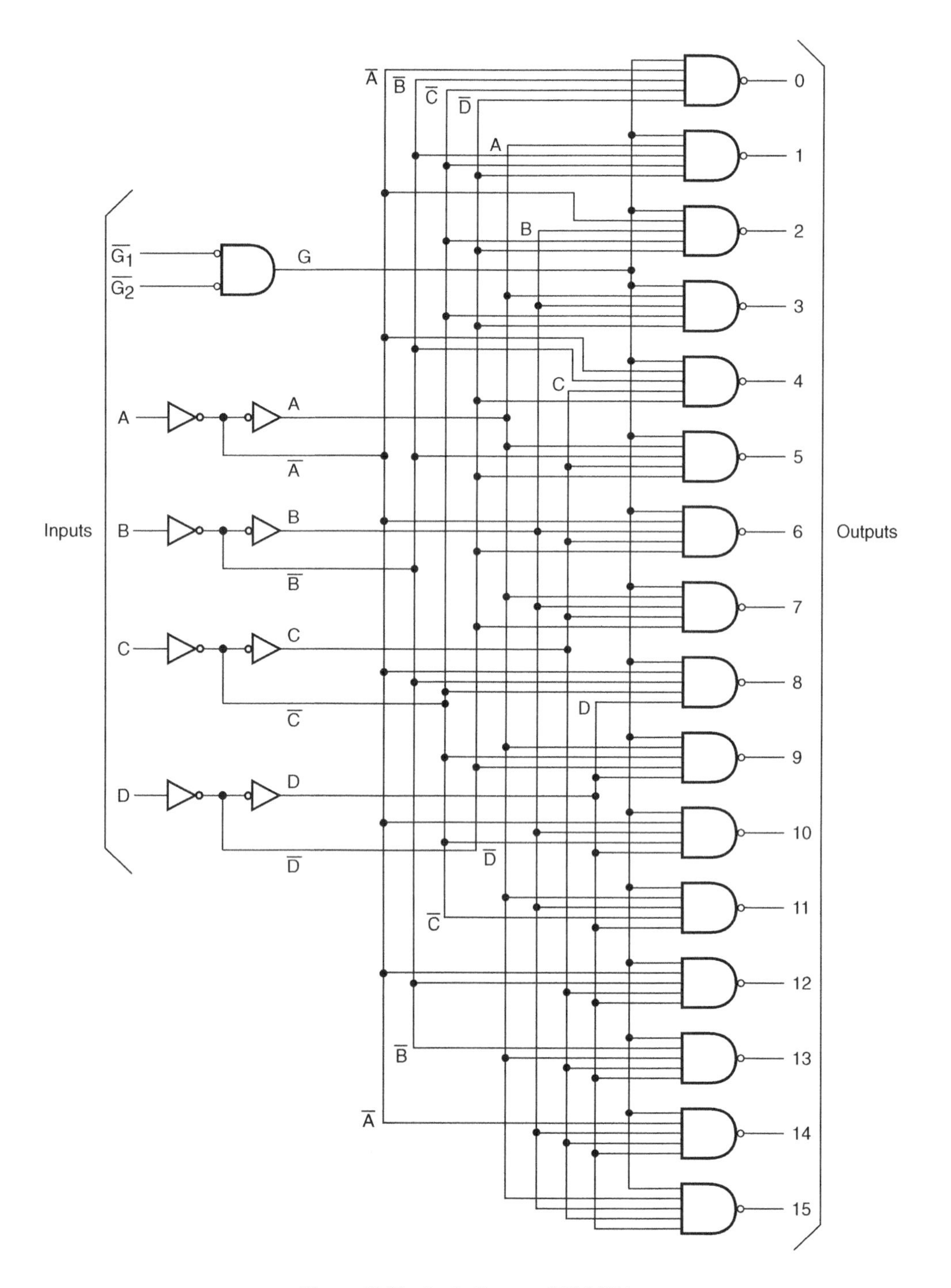

Figure 11.19 Logic diagram of IC 74154.

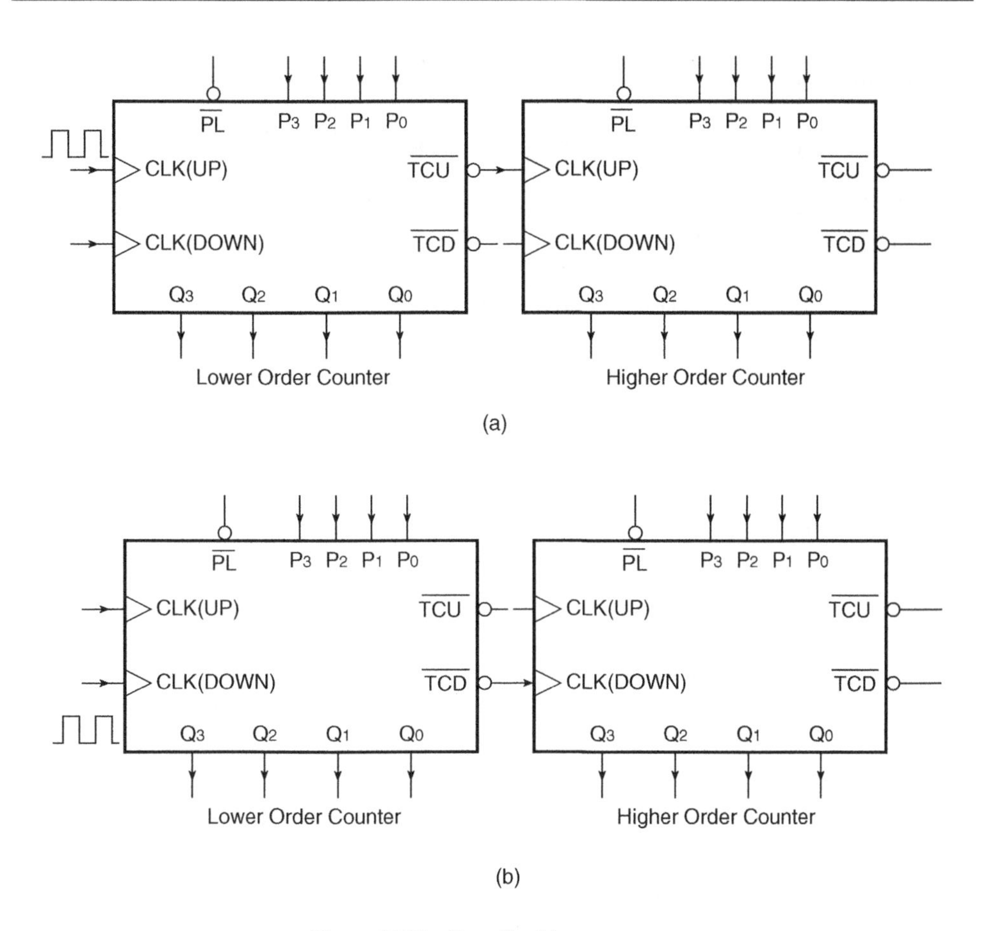

Figure 11.20 Cascading binary counters.

and the process continues. If it is desired to build a multistage DOWN counter, all counters are wired as DOWN counters, the clock is applied to the clock input of the lowest-order counter and the terminal count DOWN (*TCD*), also called the borrow-out (B_0), of the lowest-order counter is applied to the clock input of the next higher counter stage. The process continues in the same fashion, with the TCD output of the second stage feeding the clock input of the third stage and so on. The modulus of the multistage counter arrangement equals the product of the moduli of individual stages. Figures 11.20(a) and (b) respectively show two-stage arrangements of four-bit synchronous UP and DOWN counters respectively.

11.10.2 Cascading BCD Counters

BCD counters are used when the application involves the counting of pulses and the result of counting is to be displayed in decimal. A single-stage BCD counter counts from 0000 (decimal equivalent '0') to 1001 (decimal equivalent '9') and thus is capable of counting up to a maximum of nine pulses. The output in a BCD counter is in binary coded decimal (BCD) form. The BCD output needs

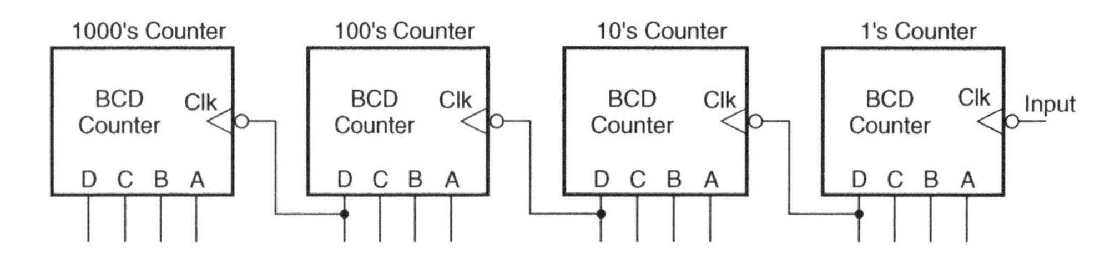

Figure 11.21 Cascading BCD counters.

to be decoded appropriately before it can be displayed. Decoding a counter has been discussed in the previous section. Coming back to the question of counting pulses, more than one BCD counter stage needs to be used in a cascade arrangement in order to be able to count up to a larger number of pulses. The number of BCD counter stages to be used equals the number of decimal digits in the maximum number of pulses we want to count up to. With a maximum count of 9999 or 3843, both would require a four-stage BCD counter arrangement with each stage representing one decimal digit.

Figure 11.21 shows a cascade arrangement of four BCD counter stages. The arrangement works as follows. Initially, all four counters are in the all 0s state. The counter representing the decimal digit of 1's place is clocked by the pulsed signal that needs to be counted. The successive flip-flops are clocked by the MSB of the immediately previous counter stage. The first nine pulses take 1's place counter to 1001. The tenth pulse resets it to 0000, and '1' to '0' transition at the MSB of 1's place counter clocks 10's place counter. 10's place counter gets clocked on every tenth input clock pulse. On the hundredth clock pulse, the MSB of 10's counter makes a '1' to '0' transition which clocks 100's place counter. This counter gets clocked on every successive hundredth input clock pulse. On the thousandth input clock pulse, the MSB of 100's counter makes '1' to '0' transition for the first time and clocks 1000's place counter. This counter is clocked thereafter on every successive thousandth input clock pulse. With this background, we can always tell the output state of the cascade arrangement. For example, immediately after the 7364th input clock pulse, the state of 1000's, 100's, 10's and 1's BCD counters would respectively be 0111, 0011, 0110 and 0100.

Example 11.6

Figure 11.22 shows a cascade arrangement of two 74190s. Both the UP/DOWN counters are wired as UP counters. What will be the logic status of outputs designated as A, B, C, D, E, F, G and H after the 34th clock pulse?

Solution

The cascade arrangement basically constitutes a two-stage BCD counter that can count from 0 to 99. The counter shown on the left forms 1's place counter, while the one on the right is 10's place counter. The ripple clock ($\overline{RC}$) output internally enabled by the terminal count ($\overline{\text{TC}}$) clocks 10's place counter on the tenth clock pulse and thereafter on every successive tenth clock pulse. At the end of the 34th clock pulse, 1's counter stores the binary equivalent of '4' and 10's counter stores the binary equivalent of '3'. Therefore, the logic status of A, B, C, D, E, F, G and H outputs will be 0, 0, 1, 0, 1,1, 0 and 0 respectively.

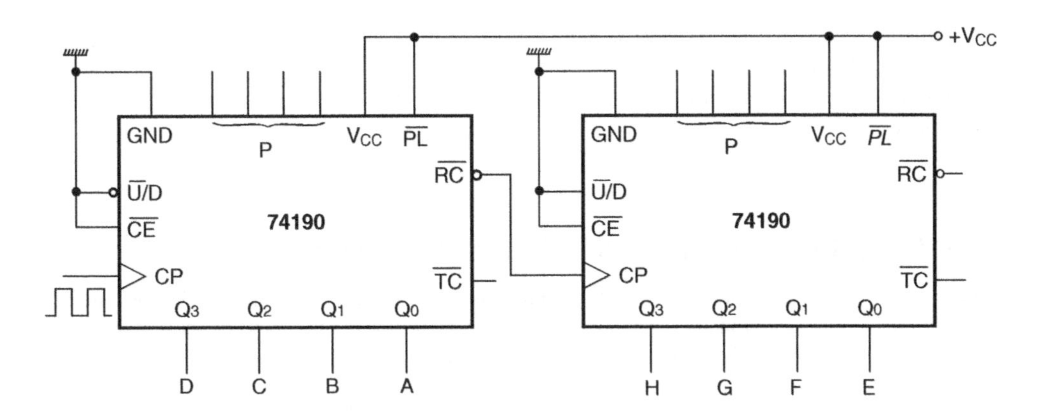

Figure 11.22 Cascade arrangement of two 74190s (example 11.6).

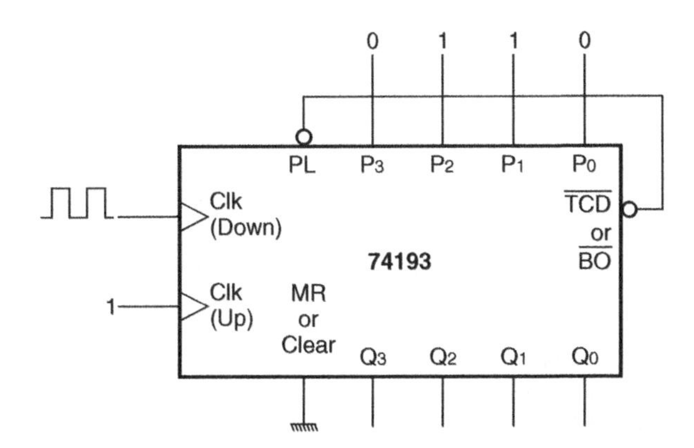

Figure 11.23 Presettable counter (example 11.7).

Example 11.7

Determine the modulus of the presettable counter shown in Fig. 11.23. If the counter were initially in the 0110 state, what would be the state of the counter immediately after the eighth clock pulse be?

Solution

- This presettable counter has been wired as a DOWN counter.
- The preset data input is 0110.
- Therefore, the modulus of the counter is 6 (the decimal equivalent of 0110).
- Now, the counter is initially in the 0110 state.
- Therefore, at the end of the sixth clock pulse, immediately after the leading edge of the sixth clock pulse, the counter will be in the 0000 state.

- A HIGH-to-LOW transition at the $\overline{TCD}$ output,coinciding with the trailing edge of the sixth clock pulse, loads 0110 to the counter output.
- Therefore, immediately after the leading edge of the eighth clock pulse, the counter will be in the 0100 state.

11.11 Designing Counters with Arbitrary Sequences

So far we have discussed different types of synchronous and asynchronous counters. A large variety of synchronous and asynchronous counters are available in IC form, and some of these have been mentioned and discussed in the previous sections. The counters discussed hitherto count in either the normal binary sequence with a modulus of 2^N or with slightly altered binary sequences where one or more of the states are skipped. The latter type of counter has a modulus of less than 2^N, N being the number of flip-flops used. Nevertheless, even these counters have a sequence that is either upwards or downwards and not arbitrary. There are applications where a counter is required to follow a sequence that is arbitrary and not binary. As an example, an MOD-10 counter may be required to follow the sequence 0000, 0010, 0101, 0001, 0111, 0011, 0100, 1010, 1000, 1111, 0000, 0010 and so on. In such cases, the simple and seemingly obvious feedback arrangement with a single NAND gate discussed in the earlier sections of this chapter for designing counters with a modulus of less than 2^N cannot be used.

There are several techniques for designing counters that follow a given arbitrary sequence. In the present section, we will discuss in detail a commonly used technique for designing synchronous counters using *J-K* flip-flops or *D* flip-flops. The design of the counters basically involves designing a suitable combinational logic circuit that takes its inputs from the normal and complemented outputs of the flip-flops used and decodes the different states of the counter to generate the correct logic states for the inputs of the flip-flops such as *J*, *K*, *D*, etc. But before we illustrate the design procedure with the help of an example, we will explain what we mean by the excitation table of a flip-flop and the state transition diagram of a counter. An excitation table in fact can be drawn for any sequential logic circuit, but, once we understand what it is in the case of a flip-flop, which is the basic building block of sequential logic, it would be much easier for us to draw the same for more complex sequential circuits such as counters, etc.

11.11.1 Excitation Table of a Flip-Flop

The excitation table is similar to the characteristic table that we discussed in the previous chapter on flip-flops. The excitation table lists the present state, the desired next state and the flip-flop inputs (*J*, *K*, *D*, etc.) required to achieve that. The same for a *J-K* flip-flop and a *D* flip-flop are shown in Tables 11.7 and 11.8 respectively. Referring to Table 11.7, if the output is in the logic '0' state and it is desired that it goes to the logic '1' state on occurrence of the clock pulse, the *J* input must be in the logic '1' state and the *K* input can be either in the logic '0' or logic '1' state. This is true as, for a '0' to '1' transition, there are two possible input conditions that can achieve this. These are $J = 1$, $K = 0$ (SET mode) and $J = K = 1$ (toggle mode), which further leads to $J = 1$, $K = X$ (either 0 or 1). The other entries of the excitation table can be explained on similar lines.

In the case of a *D* flip-flop, the *D* input is the same as the logic status of the desired next state. This is true as, in the case of a *D* flip-flop, the *D* input is transferred to the output on the occurrence of the clock pulse, irrespective of the present logic status of the *Q* output.

Table 11.7 Excitation table of a *J-K* flip-flop.

Present state (Q_n)	Next state (Q_{n+1})	J	K
0	0	0	X
0	1	1	X
1	0	X	1
1	1	X	0

Table 11.8 Excitation table of a *D* flip-flop.

Present state (Q_n)	Next state (Q_{n+1})	D
0	0	0
0	1	1
1	0	0
1	1	1

11.11.2 State Transition Diagram

The state transition diagram is a graphical representation of different states of a given sequential circuit and the sequence in which these states occur in response to a clock input. Different states are represented by circles, and the arrows joining them indicate the sequence in which different states occur. As an example, Fig. 11.24 shows the state transition diagram of an MOD-8 binary counter.

11.11.3 Design Procedure

We will illustrate the design procedure with the help of an example. We will do this for an MOD-6 synchronous counter design, which follows the count sequence 000, 010, 011, 001, 100, 110, 000, 010, . . . :

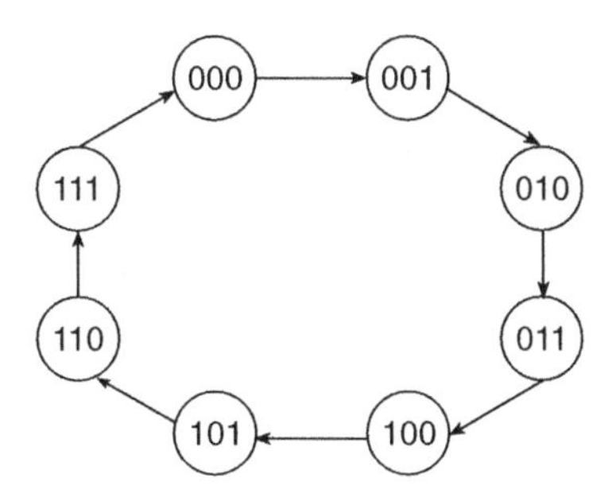

Figure 11.24 State transition diagram for an MOD-8 binary counter.

1. Determine the number of flip-flops required for the purpose. Identify the undesired states. In the present case, the number of flip-flops required is 3 and the undesired states are 101 and 111
2. Draw the state transition diagram showing all possible states including the ones that are not desired. The undesired states should be depicted to be transiting to any of the desired states. We have chosen the 000 state for this purpose. It is important to include the undesired states to ensure that, if the counter accidentally gets into any of these undesired states owing to noise or power-up, the counter will go to a desired state to resume the correct sequence on application of the next clock pulse. Figure 11.25 shows the state transition diagram

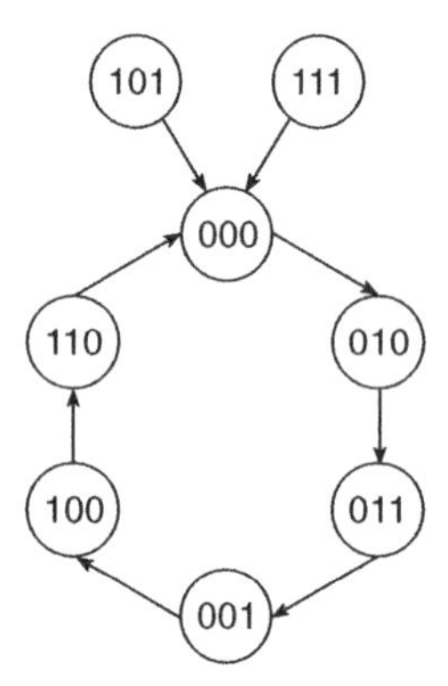

Figure 11.25 State transition diagram.

3. Draw the excitation table for the counter, listing the present states, the next states corresponding to the present states and the required logic status of the flip-flop inputs (the J and K inputs if the counter is to be implemented with J-K flip-flops). The excitation table is shown in Table 11.9

Table 11.9 Excitation table.

Present state			Next state			Inputs					
C	B	A	C	B	A	J_C	K_C	J_B	K_B	J_A	K_A
0	0	0	0	1	0	0	X	1	X	0	X
0	0	1	1	0	0	1	X	0	X	X	1
0	1	0	0	1	1	0	X	X	0	1	X
0	1	1	0	0	1	0	X	X	1	X	0
1	0	0	1	1	0	X	0	1	X	0	X
1	0	1	0	0	0	X	1	0	X	X	1
1	1	0	0	0	0	X	1	X	1	0	X
1	1	1	0	0	0	X	1	X	1	X	1

The circuit excitation table can be drawn very easily once we know the excitation table of the flip-flop to be used for building the counter. For instance, let us look at the first row of the excitation table (Table 11.9). The counter is in the 000 state and is to go to 010 on application of a clock pulse. That is, the normal outputs of C, B and A flip-flops have to undergo '0' to '0', '0' to '1' and '0' to '0' transitions respectively. Referring to the excitation table of a J-K flip-flop, the desired transitions can be realized if the logic status of J_A, K_A, J_B, K_B, J_C and K_C is as shown in the excitation table.

4. The next step is to design the logic circuits for generating J_A, K_A, J_B, K_B, J_C and K_C inputs from available A, $\overline{A}$, B, $\overline{B}$, C and $\overline{C}$ outputs. This can be done by drawing Karnaugh maps for each one of the inputs, minimizing them and then implementing the minimized Boolean expressions. The Karnaugh maps for J_A, K_A, J_B, K_B, J_C and K_C are respectively shown in Figs 11.26(a), (b), (c), (d), (e) and (f). The minimized Boolean expressions are as follows:

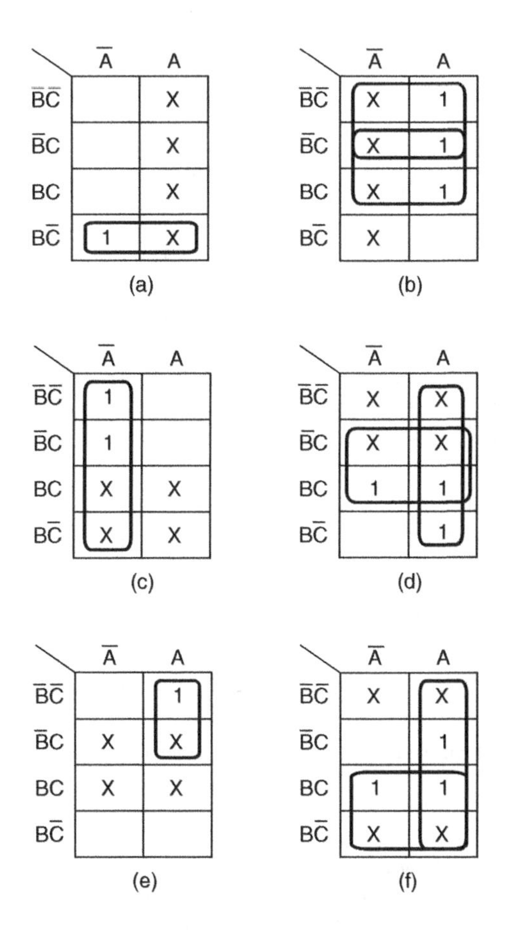

Figure 11.26 Karnaugh maps.

$$J_A = B.\overline{C} \tag{11.2}$$

$$K_A = \overline{B} + C \tag{11.3}$$

$$J_B = \overline{A} \tag{11.4}$$

$$K_B = A + C \tag{11.5}$$

$$J_C = A.\overline{B} \tag{11.6}$$

$$K_C = A + B \tag{11.7}$$

The above expressions can now be used to implement combinational circuits to generate J_A, K_A, J_B, K_B, J_C and K_C inputs. Figure 11.27 shows the complete counter circuit

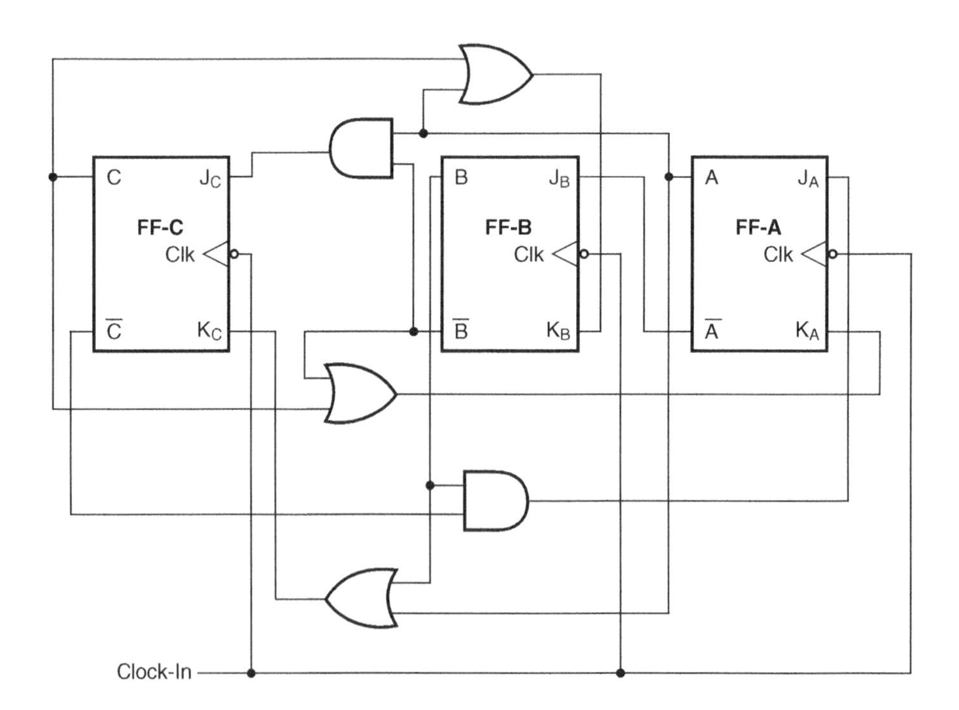

Figure 11.27 Counter with an arbitrary sequence.

The design procedure illustrated above can be used to design a synchronous counter for any given count sequence with the condition that no state occurs more than once in one complete cycle of the given count sequence as the design cannot handle a situation where a particular present state has more than one future state.

Table 11.10 Example 11.8.

Present state (Q_n)	Next state (Q_{n+1})	Inputs	
		X_1	X_2
0	0	0	0
0	1	0	1
1	0	1	X
1	1	X	1

X = don't care condition.

Example 11.8

Table 11.10 gives the excitation table of a certain flip-flop having X_1 and X_2 as its inputs. Draw the circuit excitation table of an MOD-5 synchronous counter using this flip-flop for the count sequence 000, 001, 011, 101, 110, 000,... If the present state is an undesired one, it should transit to 110 on application of a clock pulse. Design the counter circuit using the flip-flop whose excitation circuit is given in Table 11.10.

Solution

- The circuit excitation table is shown in Table 11.11.
- The number of flip-flops required is 3.
- X_1 (A) and X_2 (A) are the inputs of flip-flop A, which is also the LSB flip-flop.
- X_1 (B) and X_2 (B) represent the inputs to flip-flop B.
- X_1 (C) and X_2 (C) are the inputs to flip-flop C, which is also the MSB flip-flop.
- The next step is to draw Karnaugh maps, one each for different inputs to the three flip-flops.
- Figures 11.28(a) to (f) show the Karnaugh maps for X_1 (A), X_2 (A), X_1 (B), X_2 (B), X_1 (C) and X_2 (C) respectively.
- The minimized expressions are as follows:

$$X_1(A) = A \tag{11.8}$$

$$X_2(A) = A + \overline{B}.\overline{C} \tag{11.9}$$

$$X_1(B) = B \tag{11.10}$$

$$X_2(B) = A + B + C \tag{11.11}$$

$$X_1(C) = C \tag{11.12}$$

$$X_2(C) = B + C \tag{11.13}$$

- Figure 11.29 shows the circuit implementation.

Example 11.9

Design a synchronous counter that counts as 000, 010, 101, 110, 000, 010,... Ensure that the unused states of 001, 011, 100 and 111 go to 000 on the next clock pulse. Use J-K flip-flops. What will the counter hardware look like if the unused states are to be considered as 'don't care's.

Table 11.11 Example 11.8.

Present state			Next state			Inputs					
C	*B*	*A*	*C*	*B*	*A*	$X_1(A)$	$X_2(A)$	$X_1(B)$	$X_2(B)$	$X_1(C)$	$X_2(C)$
0	0	0	0	0	1	0	1	0	0	0	0
0	0	1	0	1	1	X	1	0	1	0	0
0	1	0	1	1	0	0	0	X	1	0	1
0	1	1	1	0	1	X	1	1	X	0	1
1	0	0	1	1	0	0	0	0	1	X	1
1	0	1	1	1	0	1	X	0	1	X	1
1	1	0	0	0	0	0	0	1	X	1	X
1	1	1	1	1	0	1	X	X	1	X	1

X = don't care condition.

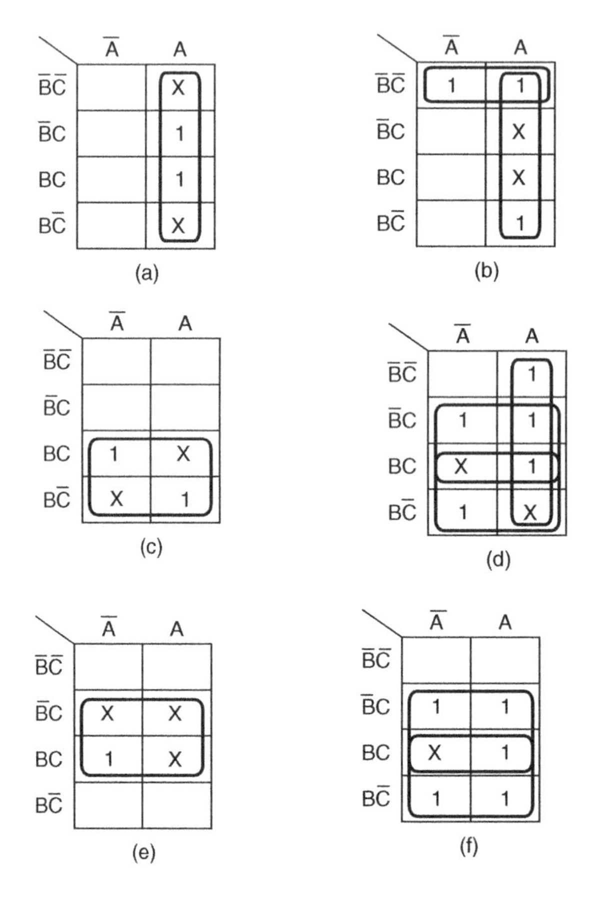

Figure 11.28 Karnaugh maps (example 11.8).

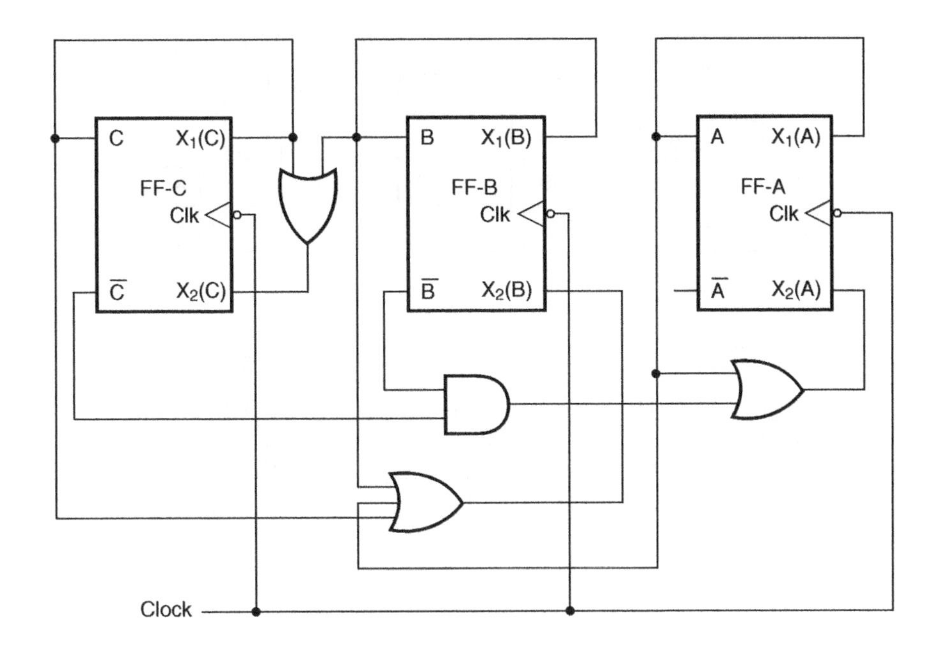

Figure 11.29 Counter circuit (example 11.8).

Table 11.12 Example 11.9.

Present state			Next state			Inputs					
C	B	A	C	B	A	J_A	K_A	J_B	K_B	J_C	K_C
0	0	0	0	1	0	0	X	1	X	0	X
0	0	1	0	0	0	X	1	0	X	0	X
0	1	0	1	0	1	1	X	X	1	1	X
0	1	1	0	0	0	X	1	X	1	0	X
1	0	0	0	0	0	0	X	0	X	X	1
1	0	1	1	1	0	X	1	1	X	X	0
1	1	0	0	0	0	0	X	X	1	X	1
1	1	1	0	0	0	X	1	X	1	X	1

Solution

- The number of flip-flops required is three.
- Table 11.12 shows the desired circuit excitation table.
- The Karnaugh maps for J_A, K_A, J_B, K_B, J_C and K_C are shown in Figs 11.30(a) to (f) respectively.
- The simplified Boolean expressions are as follows:

$$J_A = B.\overline{C} \tag{11.14}$$

$$K_A = 1 \tag{11.15}$$

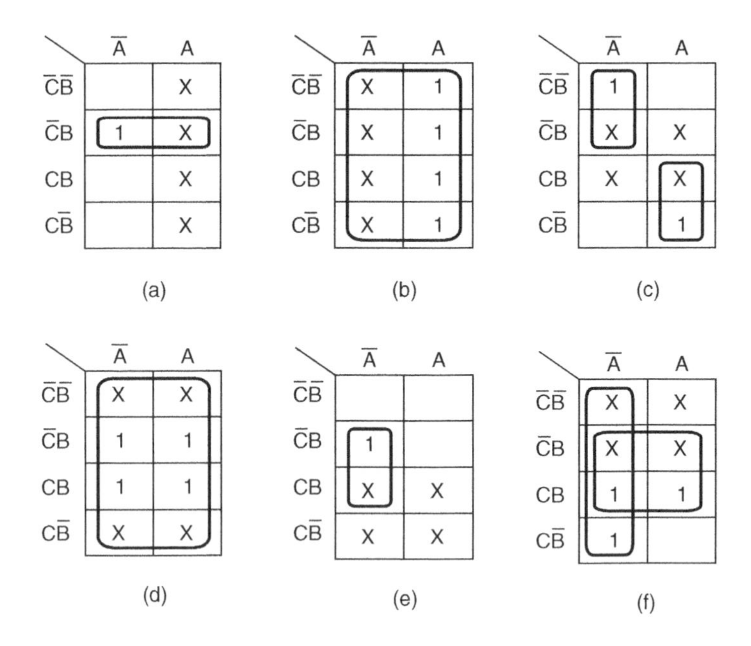

Figure 11.30 Karnaugh maps (example 11.9).

$$J_B = A.C + \overline{A}.\overline{C} \quad (11.16)$$

$$K_B = 1 \quad (11.17)$$

$$J_C = \overline{A}.B \quad (11.18)$$

$$K_C = \overline{A} + B \quad (11.19)$$

- The hardware implementation is shown in Fig. 11.31.
- In the case where the unused inputs are considered as 'don't cares', the circuit excitation table is modified to that shown in Table 11.13.
- Modified Karnaugh maps are shown in Fig. 11.32.
- The minimized Boolean expressions are derived from the Karnaugh maps of Figs 11.32(a) to (f).
- Minimized expressions for J_A, K_A, J_B, K_B, J_C and K_C respectively are as follows:

$$J_A = B.\overline{C} \quad (11.20)$$

$$K_A = 1 \quad (11.21)$$

$$J_B = 1 \quad (11.22)$$

$$K_B = 1 \quad (11.23)$$

$$J_C = B \quad (11.24)$$

$$K_C = \overline{A} \quad (11.25)$$

- Figure 11.33 shows the hardware implementation.

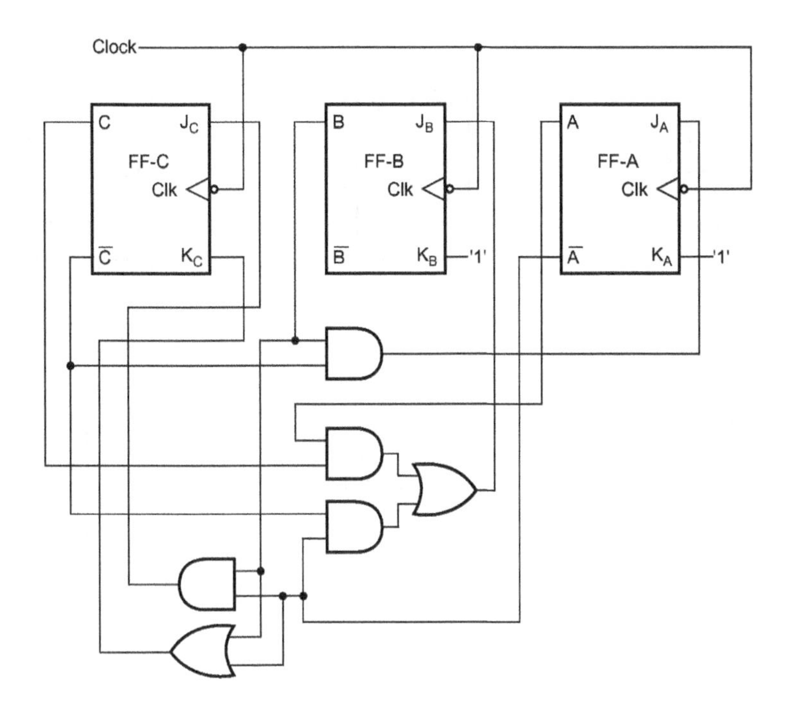

Figure 11.31 Hardware implementation of the counter circuit (example 11.9).

Table 11.13 Example 11.9.

Present state			Next state			Inputs					
C	B	A	C	B	A	J_A	K_A	J_B	K_B	J_C	K_C
0	0	0	0	1	0	0	X	1	X	0	X
0	0	1	X	X	X	X	X	X	X	X	X
0	1	0	1	0	1	1	X	X	1	1	X
0	1	1	X	X	X	X	X	X	X	X	X
1	0	0	X	X	X	X	X	X	X	X	X
1	0	1	1	1	0	X	1	1	X	X	0
1	1	0	0	0	0	0	X	X	1	X	1
1	1	1	X	X	X	X	X	X	X	X	X

11.12 Shift Register

A *shift register* is a digital device used for storage and transfer of data. The data to be stored could be the data appearing at the output of an encoding matrix before they are fed to the main digital system for processing or they might be the data present at the output of a microprocessor before they are fed

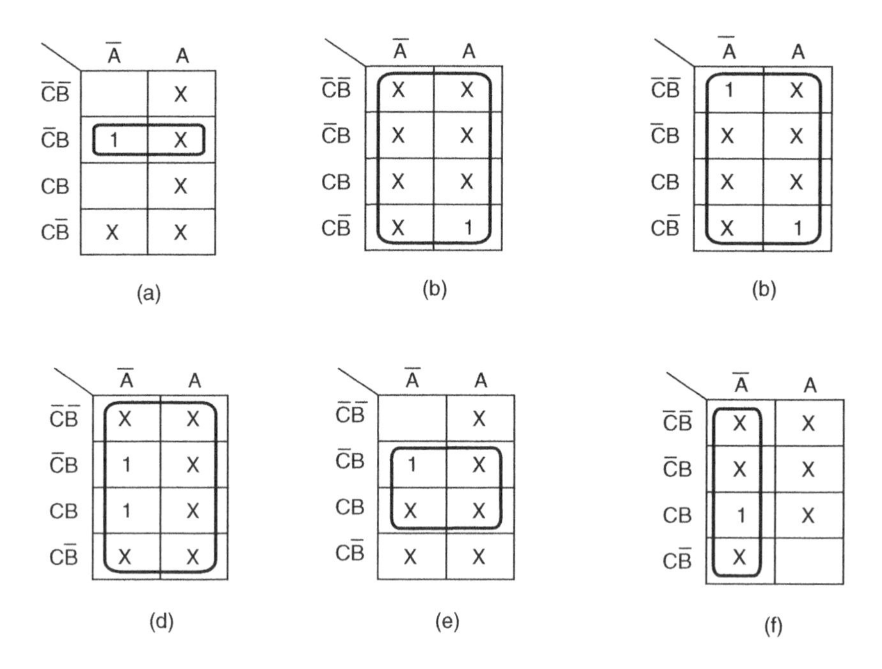

Figure 11.32 Modified Karnaugh maps (example 11.9).

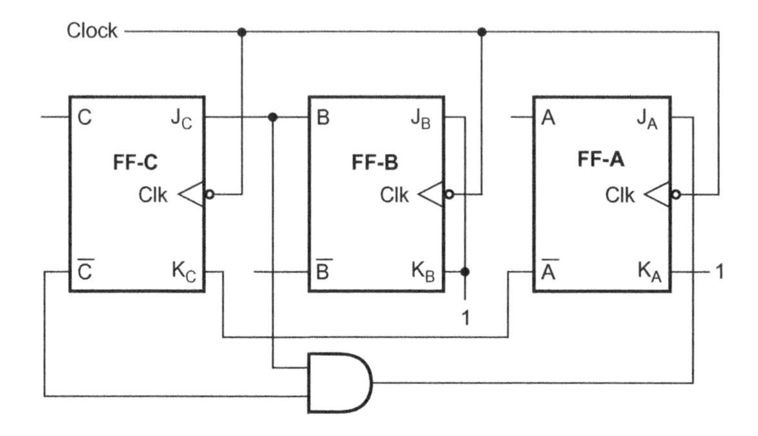

Figure 11.33 Hardware implementation of the counter circuit (example 11.9).

to the driver circuitry of the output devices. The shift register thus forms an important link between the main digital system and the input/output channels. The shift registers can also be configured to construct some special types of counter that can be used to perform a number of arithmetic operations such as subtraction, multiplication, division, complementation, etc. The basic building block in all shift registers is the flip-flop, mainly a D-type flip-flop. Although in many of the commercial shift register ICs their internal circuit diagram might indicate the use of *R-S* flip-flops, a careful examination will reveal that these *R-S* flip-flops have been wired as *D* flip-flops only.

The storage capacity of a shift register equals the total number of bits of digital data it can store, which in turn depends upon the number of flip-flops used to construct the shift register. Since each flip-flop can store one bit of data, the storage capacity of the shift register equals the number of flip-flops used. As an example, the internal architecture of an eight-bit shift register will have a cascade arrangement of eight flip-flops.

Based on the method used to load data onto and read data from shift registers, they are classified as serial-in serial-out (SISO) shift registers, serial-in parallel-out (SIPO) shift registers, parallel-in serial-out (PISO) shift registers and parallel-in parallel-out (PIPO) shift registers.

Figure 11.34 shows a circuit representation of the above-mentioned four types of shift register.

11.12.1 Serial-In Serial-Out Shift Register

Figure 11.35 shows the basic four-bit serial-in serial-out shift register implemented using *D* flip-flops. The circuit functions as follows. A reset applied to the CLEAR input of all the flip-flops resets their *Q* outputs to 0s. Refer to the timing waveforms of Fig. 11.36. The waveforms shown include the clock pulse train, the waveform representing the data to be loaded onto the shift register and the *Q* outputs of different flip-flops.

The flip-flops shown respond to the LOW-to-HIGH transition of the clock pulses as indicated by their logic symbols. During the first clock transition, the Q_A output goes from logic '0' to logic '1'.

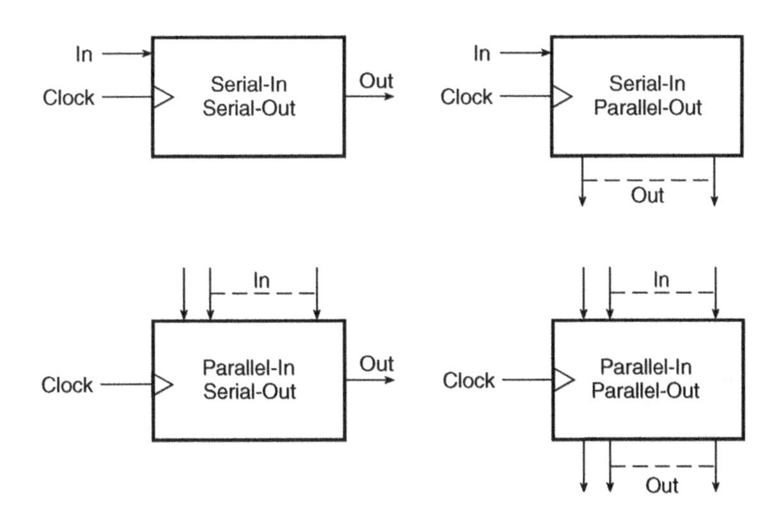

Figure 11.34 Circuit representation of shift registers.

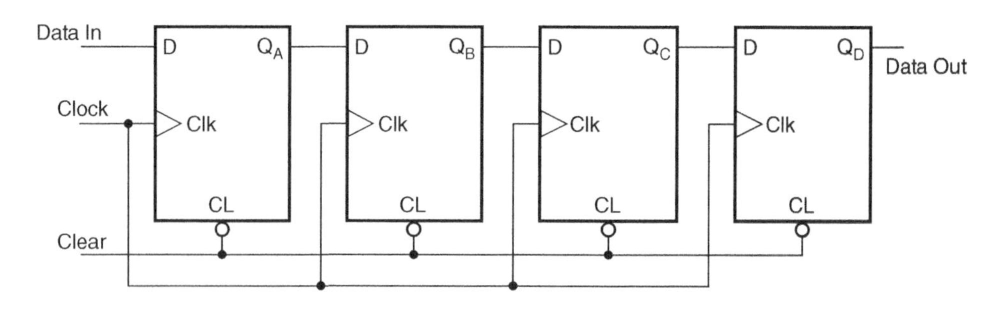

Figure 11.35 Serial-in, serial-out shift register.

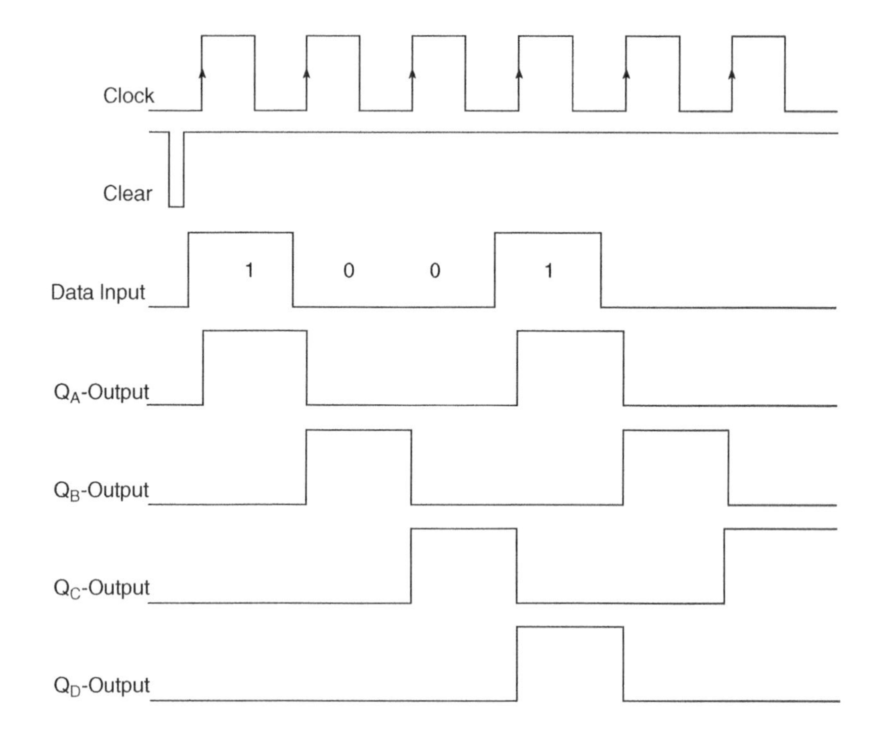

Figure 11.36 Timing waveforms for the shift register of Fig. 11.35.

The outputs of the other three flip-flops remain in the logic '0' state as their D inputs were in the logic '0' state at the time of clock transition. During the second clock transition, the Q_A output goes from logic '1' to logic '0' and the Q_B output goes from logic '0' to logic '1', again in accordance with the logic status of the D inputs at the time of relevant clock transition.

Thus, we have seen that a logic '1' that was present at the data input prior to the occurrence of the first clock transition has reached the Q_B output at the end of two clock transitions. This bit will reach the Q_D output at the end of four clock transitions. In general, in a four-bit shift register of the type

Table 11.14 Contents of four-bit serial-in serial-out shift register for the first eight clock cycles.

Clock	Q_A	Q_B	Q_C	Q_D
Initial contents	0	0	0	0
After first clock transition	1	0	0	0
After second clock transition	0	1	0	0
After third clock transition	0	0	1	0
After fourth clock transition	**1**	**0**	**0**	**1**
After fifth clock transition	0	1	0	0
After sixth clock transition	0	0	1	0
After seventh clock transition	0	0	0	1
After eighth clock transition	**0**	**0**	**0**	**0**

shown in Fig. 11.35, a data bit present at the data input terminal at the time of the nth clock transition reaches the Q_D output at the end of the $(n+4)$th clock transition. During the fifth and subsequent clock transitions, data bits continue to shift to the right, and at the end of the eighth clock transition the shift register is again reset to all 0s. Thus, in a four-bit serial-in serial-out shift register, it takes four clock cycles to load the data bits and another four cycles to read the data bits out of the register. The contents of the register for the first eight clock cycles are summarized in Table 11.14. We can see that the register is loaded with the four-bit data in four clock cycles, and also that the stored four-bit data are read out in the subsequent four clock cycles.

IC 7491 is a popular eight-bit serial-in serial-out shift register. Figure 11.37 shows its internal functional diagram, which is a cascade arrangement of eight R-S flip-flops. Owing to the inverter between the R and S inputs of the data input flip-flop, it is functionally the same as a D flip-flop. The data to be loaded into the register serially can be applied either at A or B input of the NAND gate. The other input is then kept in the logic HIGH state to enable the NAND gate. In that case, data present at A or B get complemented as they appear at the NAND output. Another inversion provided by the inverter, however, restores the original status so that for a logic '1' at the data input there is a logic '1' at the SET input of the flip-flop and a logic '0' at the RESET input of the flip-flop, and for a logic '0' at the data input there is a logic '0' at the SET input and a logic '1' at the RESET input of the flip-flop. The NAND gate provides only a gating function, and, if it is not required, the two inputs of the NAND can be shorted to have a single-line data input. The shift register responds to the LOW-to-HIGH transitions of the clock pulses.

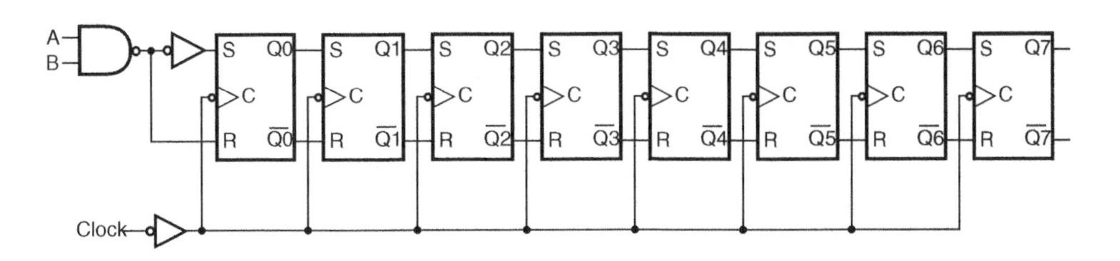

Figure 11.37 Logic diagram of IC 7491.

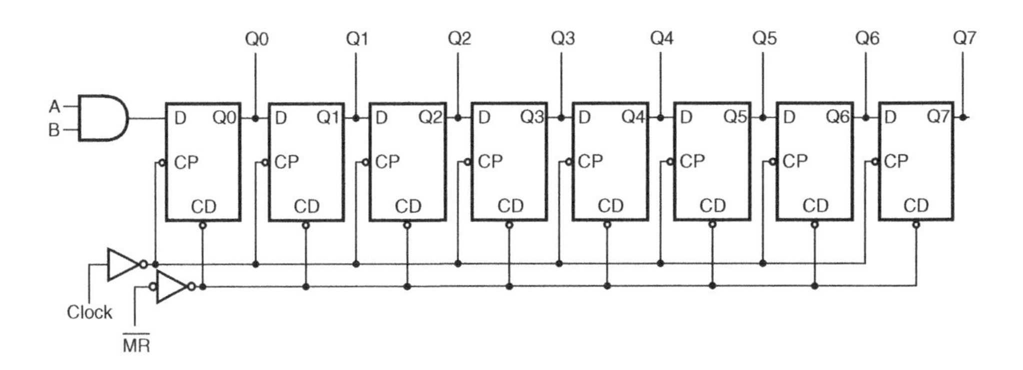

Figure 11.38 Logic diagram of IC 74164.

11.12.2 Serial-In Parallel-Out Shift Register

A serial-in parallel-out shift register is architecturally identical to a serial-in serial-out shift register except that in the case of the former all flip-flop outputs are also brought out on the IC terminals. Figure 11.38 shows the logic diagram of a typical serial-in parallel-out shift register. In fact, the logic diagram shown in Fig. 11.38 is that of IC 74164, a popular eight-bit serial-in parallel-out shift register. The gated serial inputs *A* and *B* control the incoming serial data, as a logic LOW at either of the inputs inhibits entry of new data and also resets the first flip-flop to the logic LOW level at the next clock pulse. Logic HIGH at either of the inputs enables the other input, which then determines the state of the first flip-flop.

Data at the serial inputs may be changed while the clock input is HIGH or LOW, and the register responds to LOW-to-HIGH transition of the clock. Figure 11.39 shows the relevant timing waveforms.

11.12.3 Parallel-In Serial-Out Shift Register

We will explain the operation of a parallel-in serial-out shift register with the help of the logic diagram of a practical device available in IC form. Figure 11.40 shows the logic diagram of one such shift register. The logic diagram is that of IC 74166, which is an eight-bit parallel/serial-in, serial-out shift register belonging to the TTL family of devices.

The parallel-in or serial-in modes are controlled by a SHIFT/LOAD input. When the SHIFT/LOAD input is held in the logic HIGH state, the serial data input AND gates are enabled and the circuit behaves like a serial-in serial-out shift register. When the SHIFT/LOAD input is held in the logic LOW state, parallel data input AND gates are enabled and data are loaded in parallel, in synchronism with the next clock pulse. Clocking is accomplished on the LOW-to-HIGH transition of the clock pulse via a two-input NOR gate. Holding one of the inputs of the NOR gate in the logic HIGH state inhibits the clock applied to the other input. Holding an input in the logic LOW state enables the clock to be applied to the other input. An active LOW CLEAR input overrides all the inputs, including the clock, and resets all flip-flops to the logic '0' state. The timing waveforms shown in Fig. 11.41 explain both serial-in, serial-out as well as parallel-in, serial-out operations.

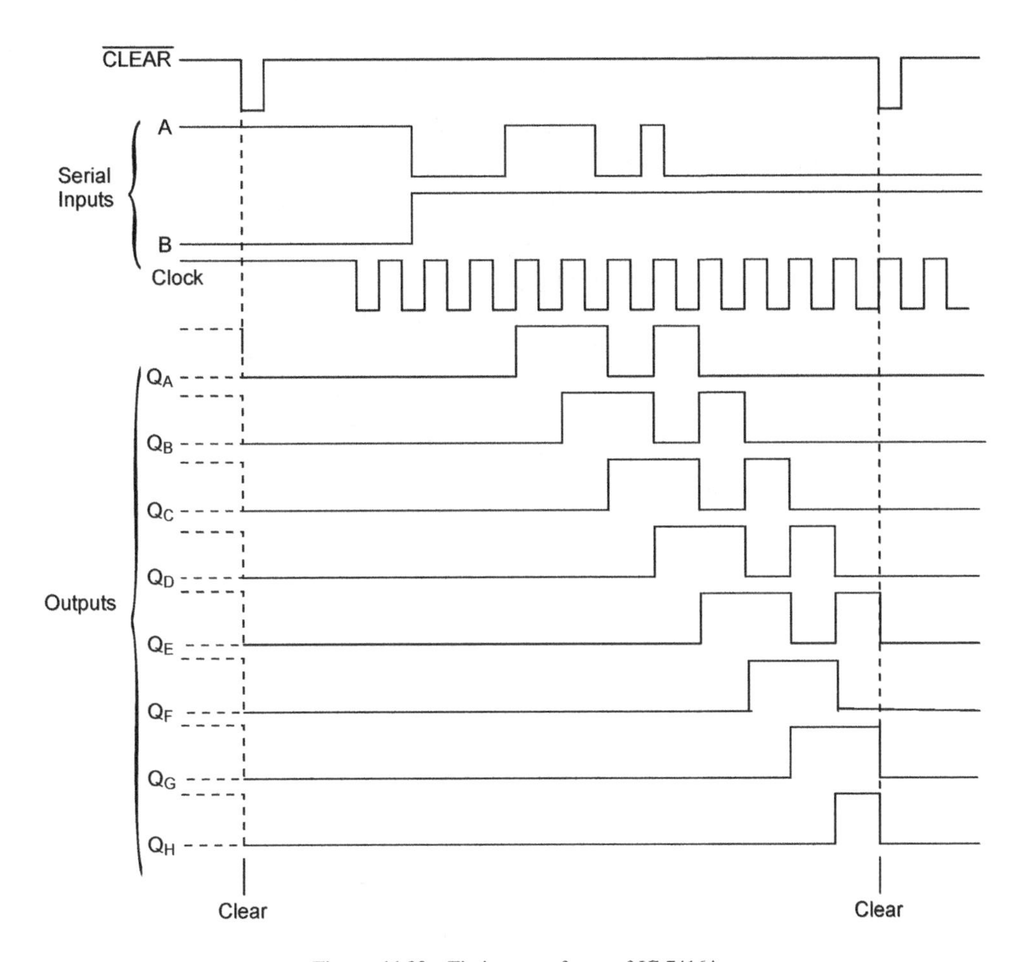

Figure 11.39 Timing waveforms of IC 74164.

11.12.4 Parallel-In Parallel-Out Shift Register

The hardware of a parallel-in parallel-out shift register is similar to that of a parallel-in serial-out shift register. If in a parallel-in serial-out shift register the outputs of different flip-flops are brought out, it becomes a parallel-in parallel-out shift register. In fact, the logic diagram of a parallel-in parallel-out shift register is similar to that of a parallel-in serial-out shift register. As an example, IC 74199 is an eight-bit parallel-in parallel-out shift register. Figure 11.42 shows its logic diagram. We can see that the logic diagram of IC 74199 is similar to that of IC 74166 mentioned in the previous section, except that in the case of the former the flip-flop outputs have been brought out on the IC terminals.

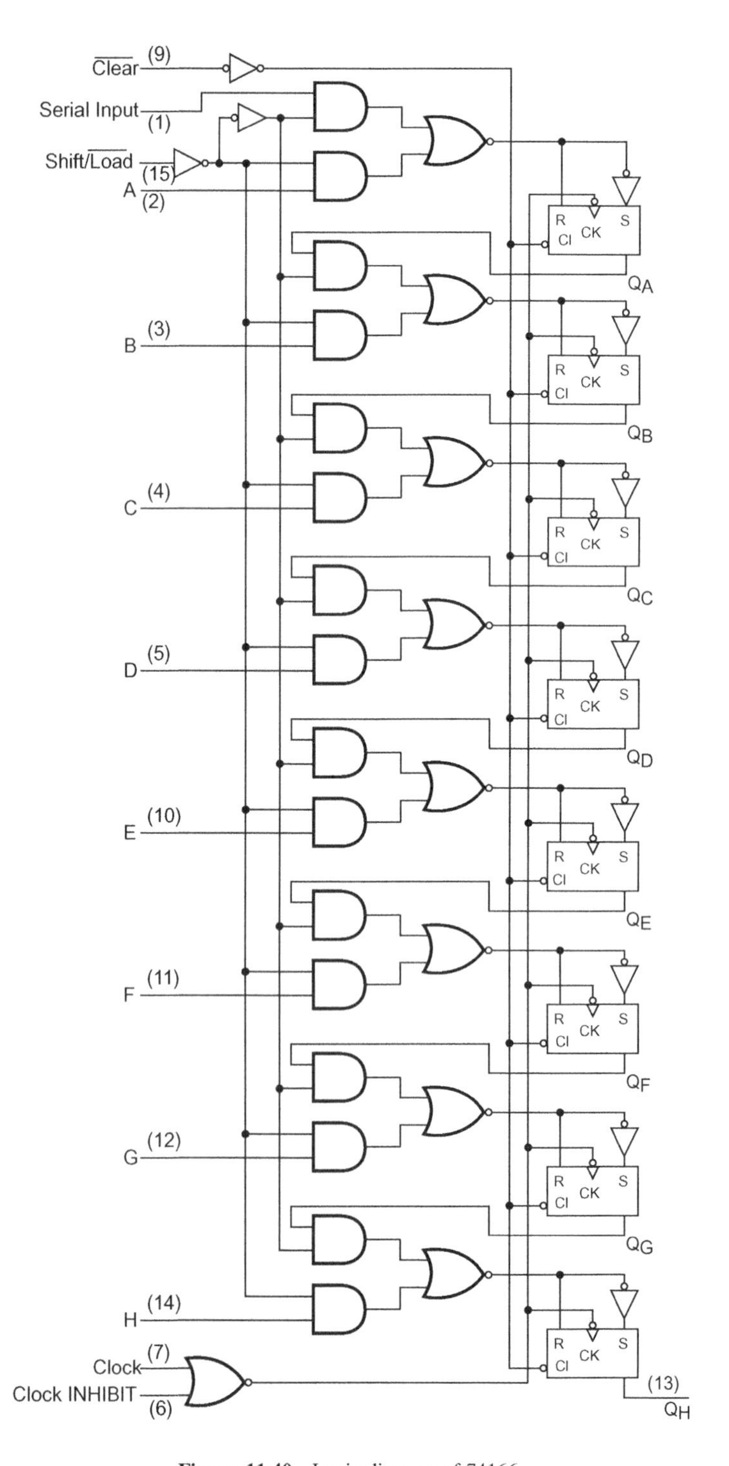

Figure 11.40 Logic diagram of 74166.

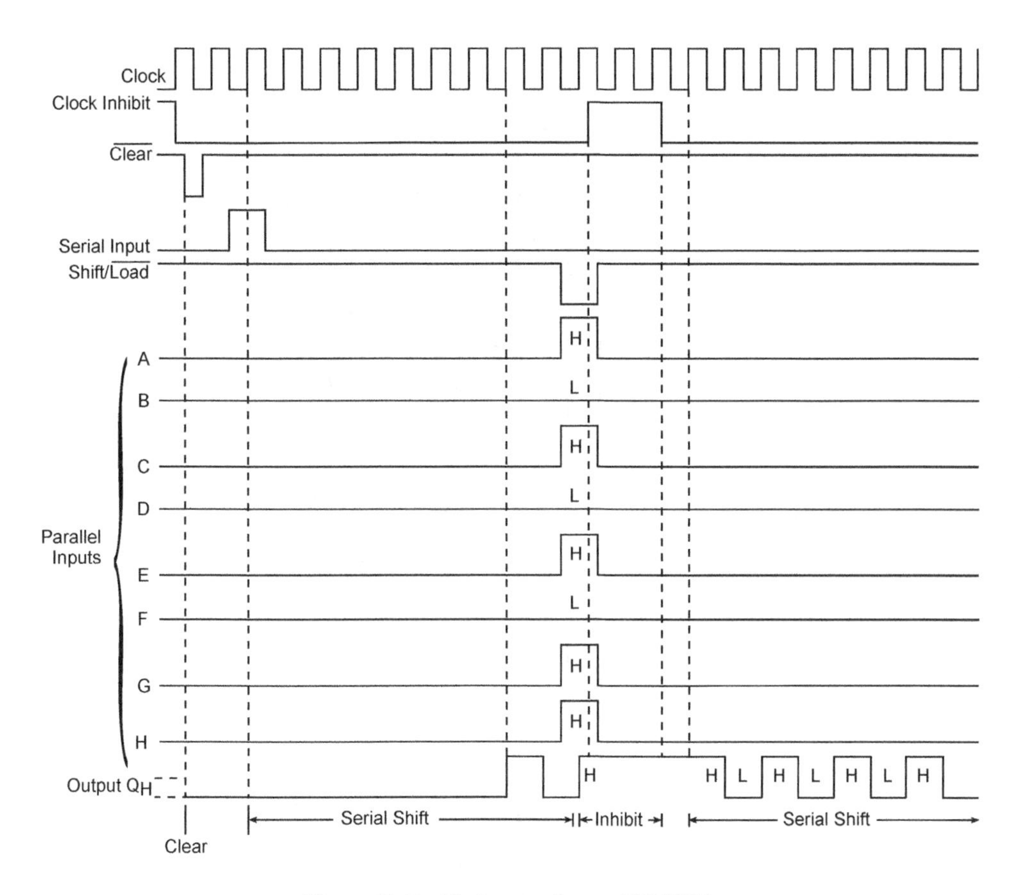

Figure 11.41 Timing waveforms of IC 74166.

11.12.5 Bidirectional Shift Register

A bidirectional shift register allows shifting of data either to the left or to the right. This is made possible with the inclusion of some gating logic having a control input. The control input allows shifting of data either to the left or to the right, depending upon its logic status.

11.12.6 Universal Shift Register

A universal shift register can be made to function as any of the four types of register discussed in previous sections. That is, it has serial/parallel data input and output capability, which means that it can function as serial-in serial-out, serial-in parallel-out, parallel-in serial out and parallel-in parallel-out shift registers.

IC 74194 is a common four-bit bidirectional universal shift register. Figure 11.43 shows the logic diagram of Ic 74194. the device offers four modes of operation, namely (a) inhibit clock, (b) shift right, (c) shift left and (d) parallel load. Clocking of the device is inhibited when both the mode control inputs S_1 and S_0 are in the logic LOW state. shift right and shift left operations are accomplished

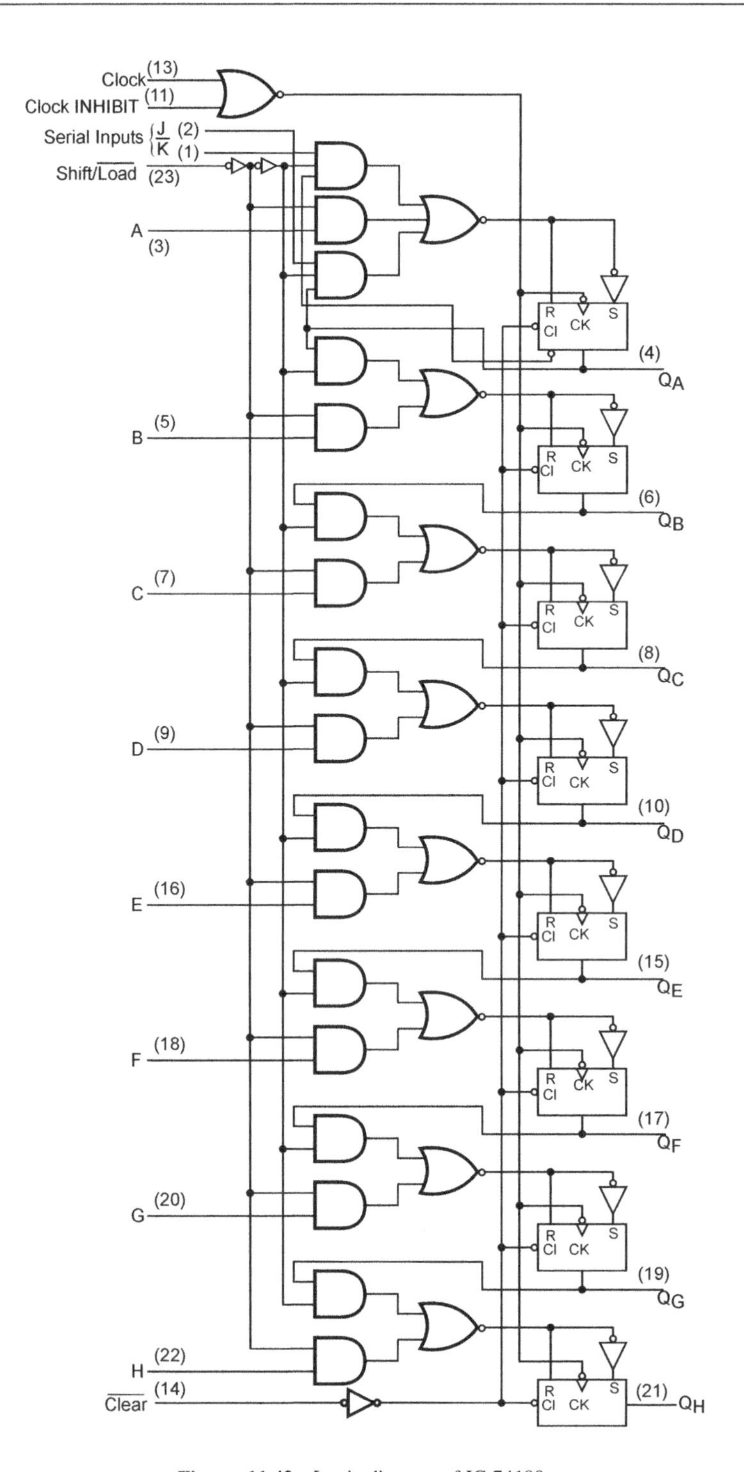

Figure 11.42 Logic diagram of IC 74199.

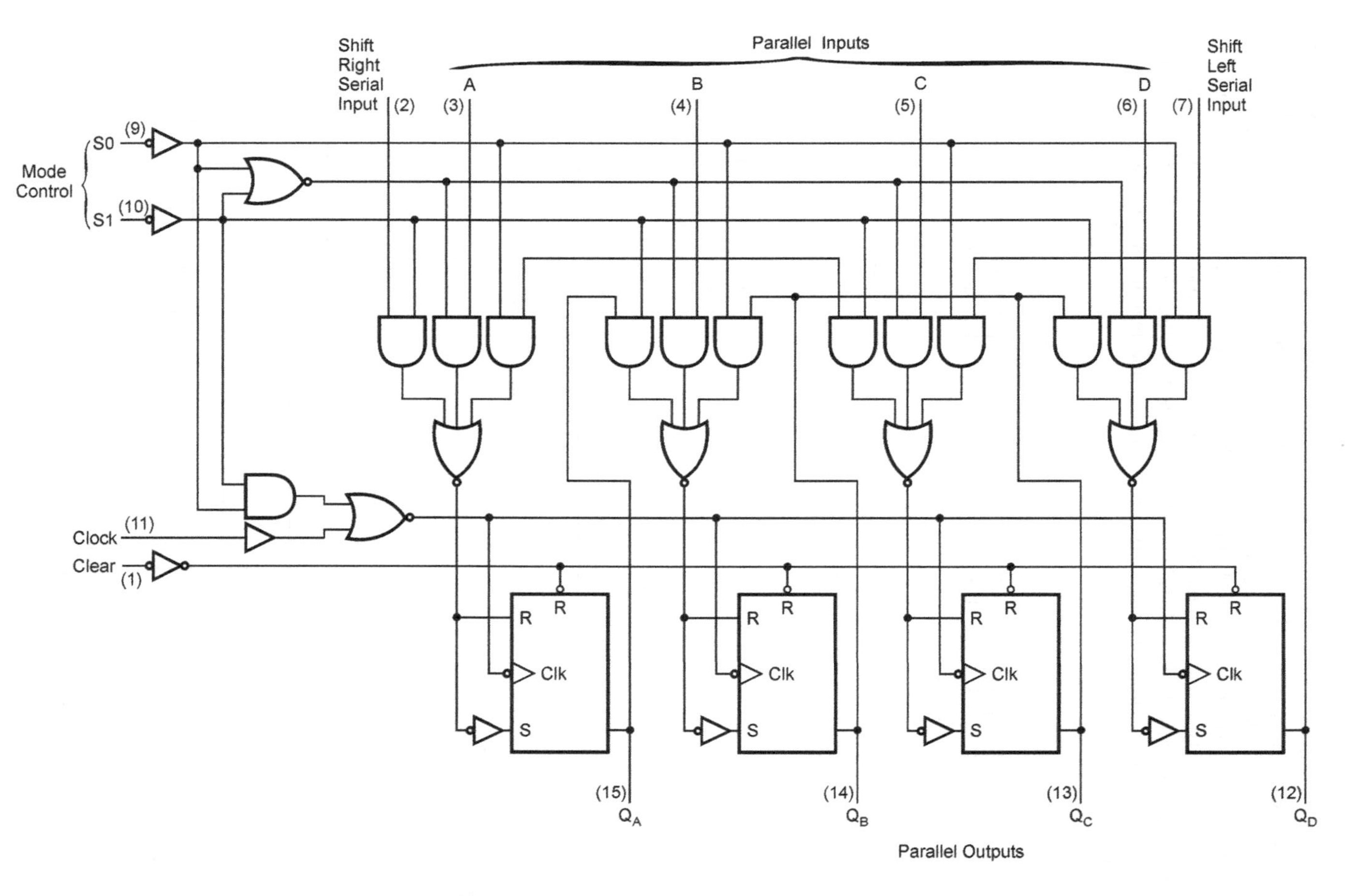

Figure 11.43 Logic diagram of IC 74194.

synchronously with LOW-to-HIGH transition of the clock with S_1 LOW and S_0 HIGH (for shift right) and S_1 HIGH and S_0 LOW (for shift left). Serial data are entered in the case of shift right and shift left operations at the corresponding data input terminals. Parallel loading is also accomplished synchronously with LOW-to-HIGH clock transitions by applying four bits of data and then driving the mode control inputs S_1 and S_0 to the logic HIGH state. Data are loaded into corresponding flip-flops and appear at the outputs with LOW-to-HIGH clock transition. Serial data flow is inhibited during parallel loading. Different modes of operation are apparent in the timing waveforms of Fig. 11.44.

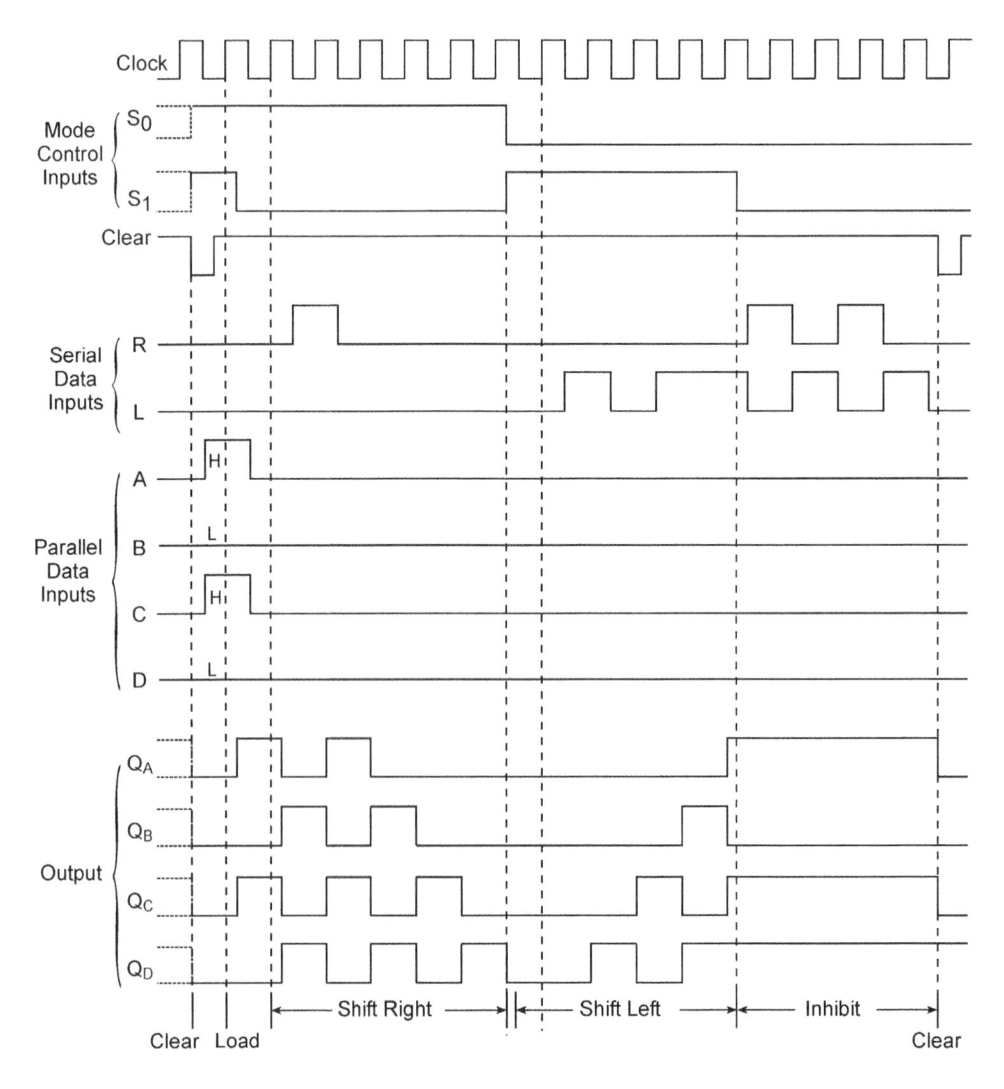

Figure 11.44 Timing waveforms of IC 74194.

11.13 Shift Register Counters

We have seen that both counters and shift registers are some kinds of cascade arrangement of flip-flops. A shift register, unlike a counter, has no specified sequence of states. However, if the serial output of the shift register is fed back to the serial input, we do get a circuit that exhibits a specified sequence of states. The resulting circuits are known as *shift register counters*. Depending upon the nature of the feedback, we have two types of shift register counter, namely the *ring counter* and the *shift counter*, also called the *Johnson counter*. These are briefly described in the following paragraphs.

11.13.1 Ring Counter

A *ring counter* is obtained from a shift register by directly feeding back the true output of the output flip-flop to the data input terminal of the input flip-flop. If D flip-flops are being used to construct the shift register, the ring counter, also called a circulating register, can be constructed by feeding back the Q output of the output flip-flop back to the D input of the input flip-flop. If J-K flip-flops are being used, the Q and $\overline{Q}$ outputs of the output flip-flop are respectively fed back to the J and K inputs of the input flip-flop. Figure 11.45 shows the logic diagram of a four-bit ring counter. Let us assume that flip-flop FF0 is initially set to the logic '1' state and all other flip-flops are reset to the logic '0' state. The counter output is therefore 1000. With the first clock pulse, this '1' gets shifted to the second flip-flop output and the counter output becomes 0100. Similarly, with the second and third clock pulses, the counter output will become 0010 and 0001. With the fourth clock pulse, the counter output will again become 1000. The count cycle repeats in the subsequent clock pulses. Circulating registers of this type find wide application in the control section of microprocessor-based systems where one event should follow the other. The timing waveforms for the circulating register of Figure 11.45, as shown in Fig. 11.46, further illustrate their utility as a control element in a digital system to generate control pulses that must occur one after the other sequentially.

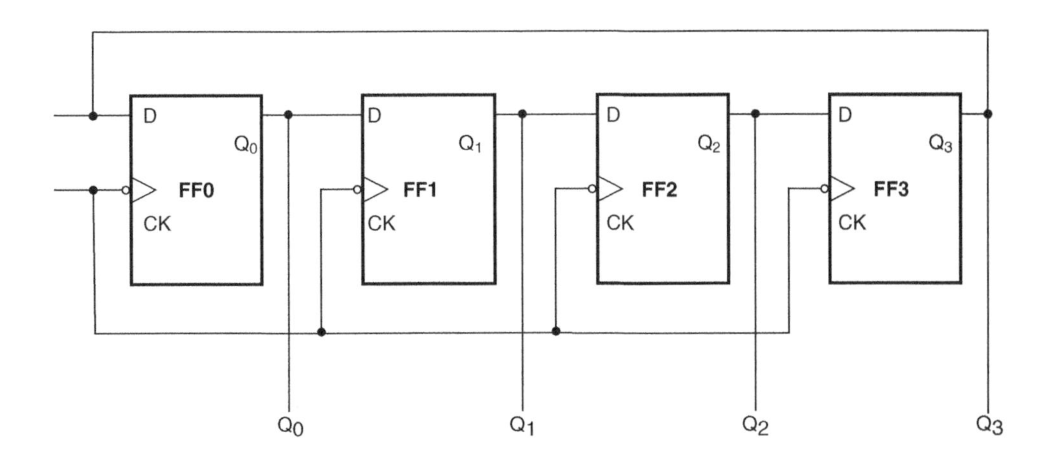

Figure 11.45 Four-bit ring counter.

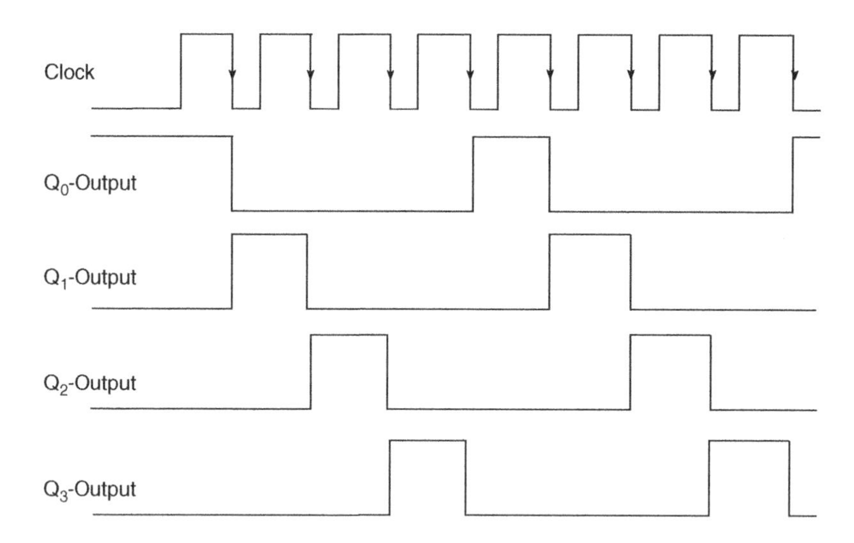

Figure 11.46 Timing waveforms of the four-bit ring counter.

11.13.2 Shift Counter

A *shift counter* on the other hand is constructed by having an inverse feedback in a shift register. For instance, if we connect the Q output of the output flip-flop back to the K input of the input flip-flop and the $\overline{Q}$ output of the output flip-flop to the J input of the input flip-flop in a serial shift register, the result is a shift counter, also called a *Johnson counter*. If the shift register employs D flip-flops, the $\overline{Q}$ output of the output flip-flop is fed back to the D input of the input flip-flop. If R-S flip-flops are used, the Q output goes to the R input and the $\overline{Q}$ output is connected to the S input. Figure 11.47 shows the logic diagram of a basic four-bit shift counter.

Let us assume that the counter is initially reset to all 0s. With the first clock cycle, the outputs will become 1000. With the second, third and fourth clock cycles, the outputs will respectively be 1100, 1110 and 1111. The fifth clock cycle will change the counter output to 0111. The sixth, seventh and eighth clock pulses successively change the outputs to 0011, 0001 and 0000. Thus, one count cycle

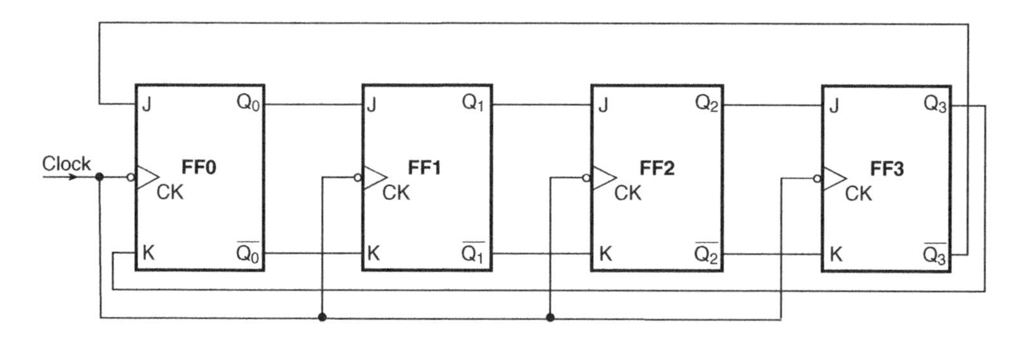

Figure 11.47 Four-bit shift counter.

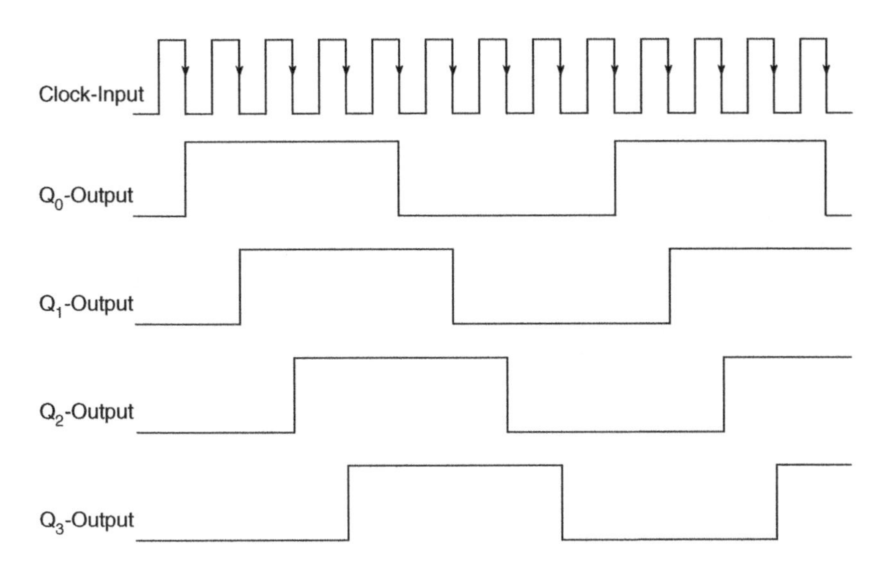

Figure 11.48 Timing waveforms of the shift counter.

is completed in eight cycles. Figure 11.48 shows the timing waveforms. Different output waveforms are identical except for the fact that they are shifted from the immediately preceding one by one clock cycle. Also, the time period of each of these waveforms is 8 times the period of the clock waveform. That is, this shift counter behaves as a divide-by-8 circuit.

In general, a shift counter comprising n flip-flops acts as a divide-by-$2n$ circuit. Shift counters can be used very conveniently to construct counters having a modulus other than the integral power of 2.

Example 11.10

Refer to Fig. 11.49, which shows an application circuit of eight-bit serial-in serial-out shift register type IC 7491 along with the waveform applied at the shorted A and B inputs:

(a) What will be the data bit present at the output at the end of the eleventh LOW-to-HIGH transition of the clock waveform?
(b) If there is a logic '1' at the end of the nth LOW-to-HIGH clock transition at the Q_3 output, what will the Q_5 output at the end of the $(n+2)$th transition be?

Solution

(a) At the end of the eighth LOW-to-HIGH clock transition, the data bits loaded into the register will be 10110010, with the '0' on the extreme right appearing at the Q_7 output (refer to the logic diagram of IC 7491 shown in Fig. 11.37). The ninth clock transition will shift this '0' out of the register, and the next adjacent bit (that is, '1') will take its place on the Q_7 output. Each subsequent clock pulse will shift the bits one step towards the right, with the result that at the end of the eleventh clock transition the Q_7 output will be a logic '0'.
(b) It will be a logic '1' only. The Q_3 output will be shifted two bit positions to the right by two clock transitions.

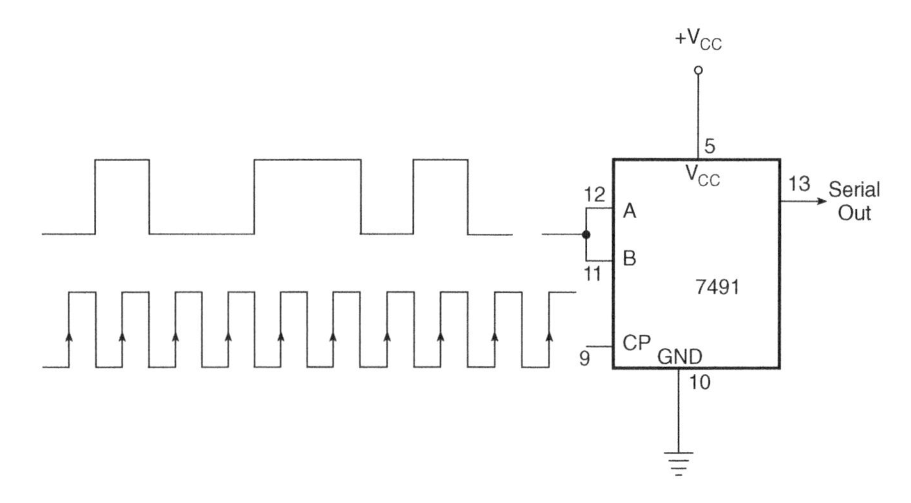

Figure 11.49 Example 11.10.

Example 11.11

Determine the number of flip-flops required to construct (a) a MOD-10 ring counter and (b) a MOD-10 Johnson counter. Also, write the count sequence in the two cases.

Solution

(a) The modulus of a ring counter is the same as the number of bits (or flip-flops). Therefore, the number of flip-flops required = 10. The count sequence is 1000000000, 0100000000, 0010000000, 0001000000, 0000100000, 0000010000, 0000001000, 0000000100, 0000000010, 0000000001 and back to 1000000000.

(b) The modulus of a Johnson counter is twice the number of flip-flops. Therefore, the number of flip-flops = 5. The count sequence is 00000, 10000, 11000, 11100, 11110, 11111, 01111, 00111, 00011, 00001 and back to 00000.

Example 11.12

Refer to the logic circuit of Fig. 11.50. Determine the modulus of this counter and write its counting sequence.

Solution

The LSB of the five-bit ring counter feeds the clock input of the *J-K* flip-flop that has been wired as a toggle flip-flop. The ring counter has a modulus of 5, and the *J-K* flip-flop works like a divide-by-2 circuit. The modulus of the counter circuit obtained by the cascade arrangement of the two is therefore 10. The counting sequence of this arrangement is given in Table 11.15.

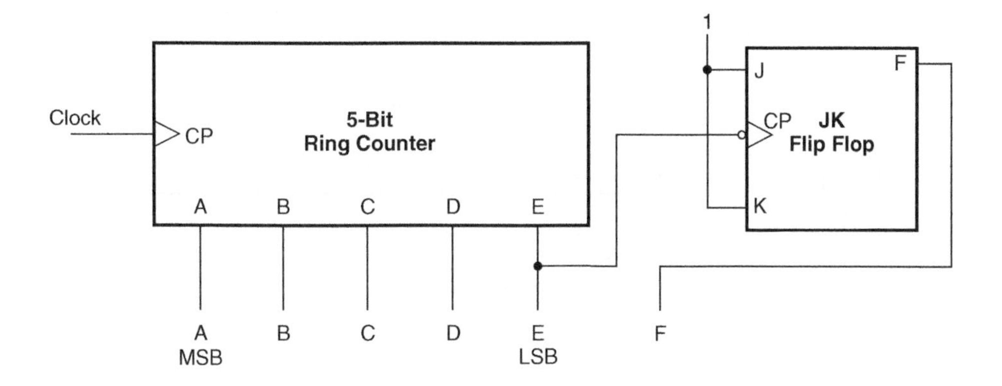

Figure 11.50 Example 11.12.

Table 11.15 Example 11.11.

Clock pulse	Outputs					
	A	*B*	*C*	*D*	*E*	*F*
1	1	0	0	0	0	0
2	0	1	0	0	0	0
3	0	0	1	0	0	0
4	0	0	0	1	0	0
5	0	0	0	0	1	0
6	1	0	0	0	0	1
7	0	1	0	0	0	1
8	0	0	1	0	0	1
9	0	0	0	1	0	1
10	0	0	0	0	1	1
11	1	0	0	0	0	0

It is very simple to write the count sequence. Firstly, we write the first 10 states of the ring counter output (designated by *A*, *B*, *C*, *D* and *E*). The logic status of *F* can be written by examining the logic status of *E*. *F* toggles whenever *E* undergoes '1' to '0' transition.

Example 11.13

Refer to the logic circuit arrangement of Fig. 11.51 built around an eight-bit serial-in/parallel-out shift register, type number 74164. A and B are the data inputs. The serial data feeding the register are obtained by an ANDing operation of A and B inputs inside the IC. $\overline{MR}$ *is an active LOW master reset. Write the logic status of register outputs for the first eight clock pulses.* Q_0 *represents the first flip-flop in this serial shift register.*

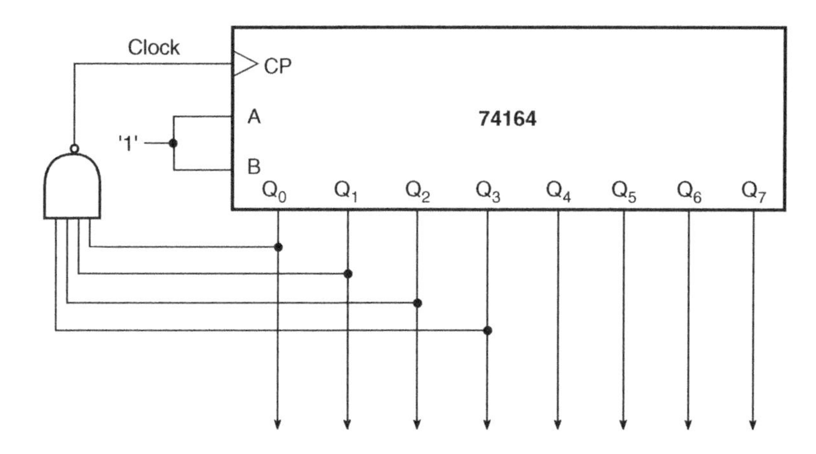

Figure 11.51 Example 11.13.

Solution

Initially, all outputs are in the logic '0' state. Since $A = B = 1$, the serial input to the shift register is a logic '1'. The $\overline{\text{MR}}$ input is initially inactive. For the first three clock pulses, the output status is 10000000, 11000000 and 11100000. With the fourth clock pulse, the output tends to go to 11110000, but it cannot be stable state as the NAND output goes from '1' to '0'. This resets the register to 00000000. Thus, the register transits from 11100000 to 00000000. With the fifth, sixth and seventh clock pulses, the circuit goes through 10000000, 11000000 and 11100000. The eight clock pulse again resets it to 00000000.

11.14 IEEE/ANSI Symbology for Registers and Counters

We introduced IEEE/ANSI symbology for digital integrated circuits as contained in IEEE/ANSI Standard 91-1984 in Section 4.22 of Chapter 4 on logic gates and related devices. A brief description of salient features of this symbology and its particular significance to sequential logic devices such as flip-flops, counters, registers, etc., was given, highlighting the use of dependency notation to provide almost complete functional information of the device. In this section, we will illustrate IEEE/ANSI symbology for counters and registers with the help of IEEE/ANSI symbols of some popular devices.

11.14.1 Counters

As an illustration, we will consider IEEE/ANSI symbols of a decade counter, type number 7490, and a presettable four-bit binary UP/DOWN counter, type number 74193. The IEEE/ANSI notation for IC 7490 and IC 74193 is shown in Figs 11.52(a) and (b) respectively.

The upper portion of the notation represents the common control block that affects all flip-flops constituting the counter. The lower portion represents individual flip-flops. Before we interpret different labels and inputs/outputs for the two counter ICs, we should know the following:

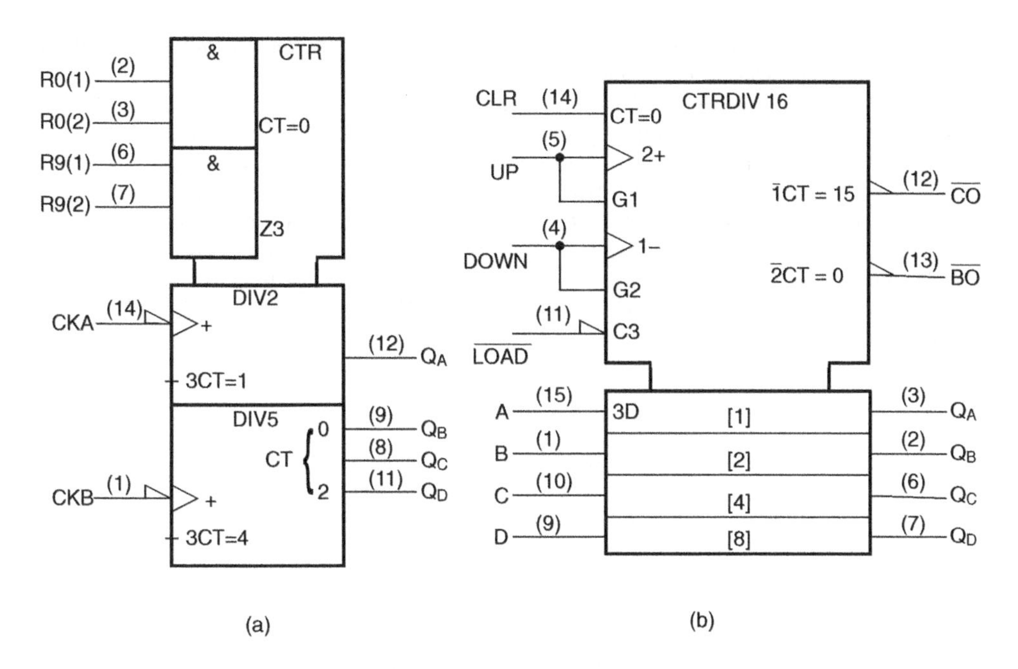

Figure 11.52 IEEE/ANSI notation for (a) IC 7490 and (b) IC 74193.

1. Letter 'C' represents control dependency. Use of the letter 'C' in the label of a certain input means that that particular input controls the entry of data into a storage element such as a flip-flop. The storage element or elements that are controlled by this input are indicated by a digit used as a suffix to the letter 'C'. The same digit appears as a prefix in the labels of all those storage elements that are controlled by this input.
2. Letter 'G' represents an AND dependency. The use of the letter 'G' followed by a digit in the label of an input means that this input is internally ANDed with another input or output and that the input or output will have the same digit as a prefix in its label.
3. Plus (+) and minus (−) signs in the labels indicate the count direction, with the former implying an UP count sequence and the latter implying a DOWN count sequence. These signs are used with clock inputs.

We will now interpret different inputs and outputs for the two counters. We will begin with IC 7490. Reset inputs R_0 (1) and R_0 (2) have an AND dependency, and when both of them are driven to the logic HIGH state the counter is reset to all 0s. Reset inputs R_9 (1) and R_9 (2) also have an AND dependency when both of them are driven to the logic HIGH state, the divide-by-2 portion of the counter is reset to count '1' (which is also the logic '1' state for the flip-flop true output) and the divide-by-5 portion of the counter is reset to count '4' (which is the 100 state for the counter outputs). If the two portions were used in cascade, the counter output would become 1001, which would mean that the counter is reset to count '9'. Clock A (CKA) and clock B (CKB) inputs allow the two portions of the counter to count in the upward sequence as indicated by the (+) sign.

We will now look at the IEEE/ANSI symbol of the other counter, that is, the counter IC type number 74193. Label CTR DIV16 means that IC 74193 is a divide-by-16 counter. Label CT = 0 with master reset (*MR*) input implies that the counter is reset to all 0s when the *MR* input is in the logic HIGH state. Label C3 with parallel load (*PL*) input means that the data on parallel load inputs P_0, P_1, P_2 and P_3 are loaded onto the corresponding flip-flops when the *PL* input is in the logic LOW state. We can see the prefix 3 in the labels of the flip-flops. The *CPU* input has an AND dependency with the *TCU* output and *CPD* input. In the case of the former, the *TCU* output goes to the logic LOW state when the *CPU* is LOW and the count reaches '15'. In the case of the latter, the *CPU* input should be in the logic HIGH state in order to allow the *CPD* to perform the count DOWN function. Similarly, the *CPD* input has an AND dependency with the *TCD* output and *CPU* input. In the case of the former, the *TCD* output goes to the logic LOW state when the *CPD* is LOW and the count reaches '0'. In the case of the latter, the *CPD* input should be in the logic HIGH state in order to allow the *CPU* to perform the count UP function.

11.14.2 Registers

As an illustration, we will consider IEEE/ANSI symbols of a serial-in serial-out shift register, type number 7491, and a serial-in parallel-out shift register, type number 74164. Figures 11.53(a) and (b) show the IEEE/ANSI notations for IC 7491 and IC 74164 respectively.

We will begin with shift register type number 7491. Label SRG8 stands for eight-bit shift register. Label C1/→ with the clock input means that the relevant clock transition performs two functions. Firstly, it loads data onto the data input as indicated by prefix '1' with the *D* input. Secondly, it performs a right shift operation. The *A* and *B* inputs have an AND dependency. When data are entered through either of the two inputs, the other input must be held in the logic HIGH state to allow the data bit to be loaded onto the data input terminal.

We will now consider shift register type number 74164. Label 'R' stands for reset operation. Whenever the *MR* input is driven to the logic LOW state, the shift register is reset to all 0s. The rest of the notations have already been explained in the case of register type number 7491.

11.15 Application-Relevant Information

Table 11.16 lists the commonly used IC counters and registers belonging to the TTL, CMOS and ECL logic families. Application-relevant information on more popular type numbers is given in the companion website. The information includes the pin configuration diagram, functional table and timing waveforms in some cases.

Review Questions

1. Differentiate between:

    (a) asynchronous and synchronous counters;
    (b) UP, DOWN and UP/DOWN counters;
    (c) presettable and clearable counters;
    (d) BCD and decade counters.

2. Indicate the difference between the counting sequences of:

    (a) a four-bit binary UP counter and a four-bit binary DOWN counter;
    (b) a four-bit ring counter and a four-bit Johnson counter.

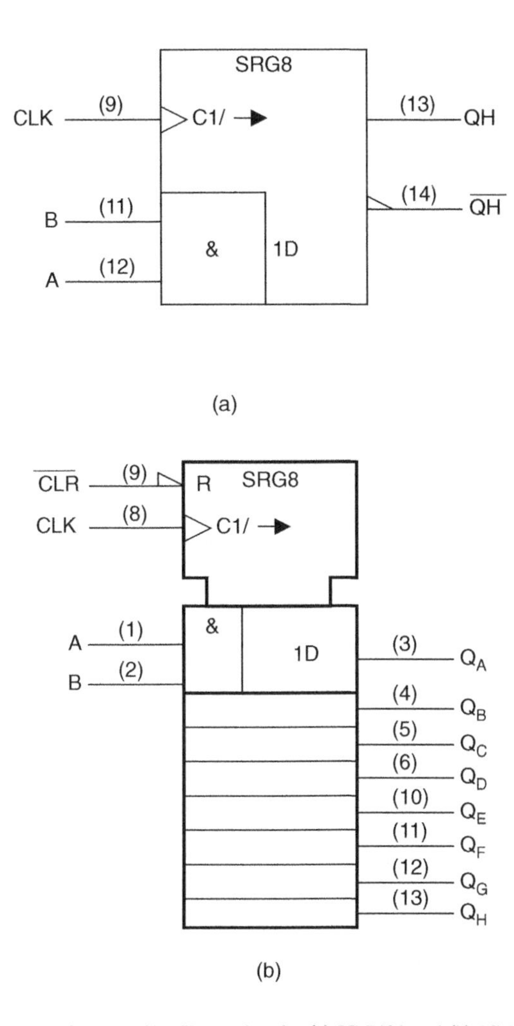

Figure 11.53 IEEE/ANSI notation for (a) IC 7491 and (b) IC 74164.

3. Briefly describe:

   (a) how the architecture of an asynchronous UP counter differs from that of a DOWN counter;
   (b) how the architecture of a ring counter differs from that of a shift counter.

4. Briefly explain why the maximum usable clock frequency of a ripple counter decreases as more flip-flops are added to the counter to increase its MOD-number.
5. Why is the maximum usable clock frequency in the case of a synchronous counter independent of the size of counter?
6. How can presettable counters be used to construct counters with variable modulus?

7. Indicate the type of shift register:

   (a) into which a complete binary number can be loaded in one operation and then shifted out one bit at a time;
   (b) into which data can be entered only one bit at a time but have all data bits available as outputs;
   (c) in which we have access to only the leftmost or rightmost flip-flop.

Table 11.16 Commonly used IC counters and registers belonging to the TTL, CMOS and ECL logic families.

Type number	Function	Logic family
7490	Decade counter	TTL
7491	Eight-bit shift register (serial-in/serial-out)	TTL
7493	Four-bit binary counter	TTL
74160	BCD decade counter with asynchronous CLEAR	TTL
74161	Four-bit binary counter with asynchronous CLEAR	TTL
74162	BCD decade counter with synchronous CLEAR	TTL
74163	Four-bit binary counter with synchronous CLEAR	TTL
74164	Eight-bit shift register (serial-in/parallel-out)	TTL
74165	Eight-bit shift register (parallel-in/serial-out)	
74166	Eight-bit shift register (parallel-in/serial-out)	TTL
74178	Four-bit parallel access shift register	TTL
74190	Presettable BCD decade UP/DOWN counter	TTL
74191	Presettable four-bit binary UP/DOWN counter	TTL
74192	Presettable BCD decade UP/DOWN counter	TTL
74193	Presettable four-bit binary UP/DOWN counter	TTL
74194	Four-bit right/left universal shift register	TTL
74198	Eight-bit universal shift register (parallel-in/parallel-out bidirectional)	TTL
74199	Eight-bit universal shift register (parallel-in/parallel-out bidirectional)	TTL
74290	Decade counter	TTL
74293	Four-bit binary counter	TTL
74390	Dual decade counter	TTL
74393	Dual four-bit binary counter	TTL
4014 B	Eight-bit static shift register (synchronous parallel or serial-in/serial-out)	CMOS
4015 B	Dual four-bit static shift register (serial-in/parallel-out)	CMOS
4017 B	Five-stage Johnson counter	CMOS
4021 B	Eght-bit static shift register (asynchronous parallel-in or synchronous serial-in/serial-out)	CMOS
4029 B	Synchronous presettable four-bit UP/DOWN counter	CMOS
4035 B	Four-bit universal shift register	CMOS
40160 B	Decade counter with asynchronous CLEAR	CMOS
40161 B	Binary counter with asynchronous CLEAR	CMOS
40162 B	Decade counter	CMOS
40163 B	Binary Counter	CMOS
40192 B	Presettable BCD UP/DOWN counter	CMOS
40193 B	Presettable Binary UP/DOWN counter	CMOS
4510 B	Presettable UP/DOWN BCD counter	CMOS

Table 11.16 *(continued).*

Type number	Function	Logic family
4518 B	Dual four-bit decade counter	CMOS
4520B	Dual four-bit binary counter	CMOS
4522 B	Four-bit BCD programmable divide-by-N counter	CMOS
4722 B	Programmable counter/timer	CMOS
4731 B	Quad 64-bit static shift register	CMOS
MC 10136	Universal hexadecimal counter	ECL
MC 10137	Universal decade counter	ECL
MC 10141	Four-bit universal shift register	ECL
MC 10154	Binary counter (four-bit)	ECL
MC 10178	Four-bit binary counter	ECL

8. What do you understand when the PRESET, CLEAR, UP/DOWN, master reset and parallel load functions of a counter are designated as $\overline{PR}$, CLR, $U/\overline{D}$, $\overline{MR}$ and PL respectively?
9. What are counters with arbitrary count sequences? Briefly describe the procedure for designing a counter with a given arbitrary count sequence.
10. Give at least one IC type number for:

    (a) a four-bit binary ripple counter;
    (b) a four-bit synchronous counter;
    (c) an eight-bit serial-in serial-out shift register;
    (d) a bidirectional universal shift register.

Problems

1. For the multistage counter arrangement of Fig.11.54, determine the frequency of the output signal.
   125 Hz

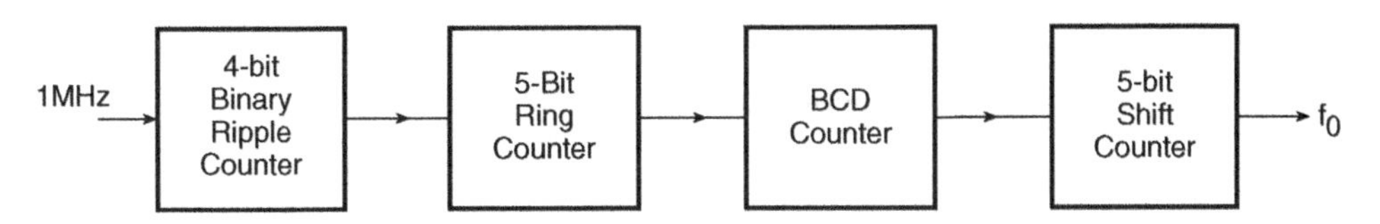

Figure 11.54 Problem 1.

2. A four-bit binary UP counter is initially in the 0000 state. Then the clock pulses are applied. Some time later the clock pulses are removed, and at that the counter is observed to be in the 0011 state. What is the minimum number of clock pulses that could possibly have occurred?
   3

3. An eight-bit binary ripple UP counter with a modulus of 256 is holding the count 01111111. What will be the count after 135 clock pulses be?
   00000110

4. Three four-bit BCD decade counters are connected in cascade. The MSB output of the first counter is fed to the clock input of the second counter, and the MSB output of the second counter is fed to the clock input of the third counter. If the counters are negatively edge triggered and the input clock frequency is 256 kHz, what is the frequency of the waveform available at the MSB of the third counter?

256 Hz

5. The flip-flops used in a four-bit binary ripple counter have a HIGH-to-LOW and LOW-to-HIGH propagation delay of 25 and 10 ns respectively. Determine the maximum usable clock frequency of this counter.

10 MHz

6. Refer to the counter schematic shown in Fig. 11.55. Determine the count sequence of this counter.

000, 001, 010, 011, 100, 101, 110, 000, . . .

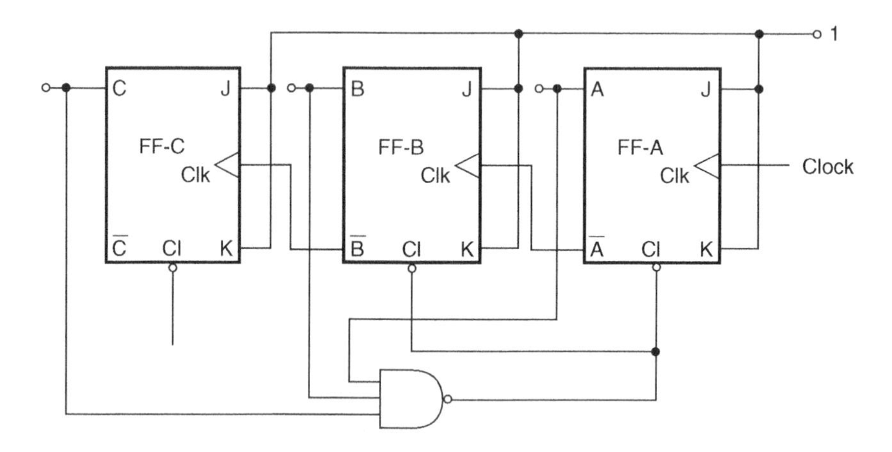

Figure 11.55 Problem 6.

7. Refer to the counter arrangement of Fig. 11.56. Determine the modulus of the counter and also the frequency of the *B* output and the duty cycle of the *C* output if the clock frequency is 600 kHz.

3; 200 kHz; 0 %

8. A four-bit ring counter and a four-bit Johnson counter are in turn clocked by a 10 MHz clock signal. Determine the frequency and duty cycle of the output of the output flip-flop in the two cases.

Ring counter: 2.5 MHz, 25 %; Johnson counter: 1.25 MHz, 50 %

9. A 100-stage serial-in/serial-out shift register is clocked at 100 kHz. How long will the data be delayed in passing through this register?

1 ms

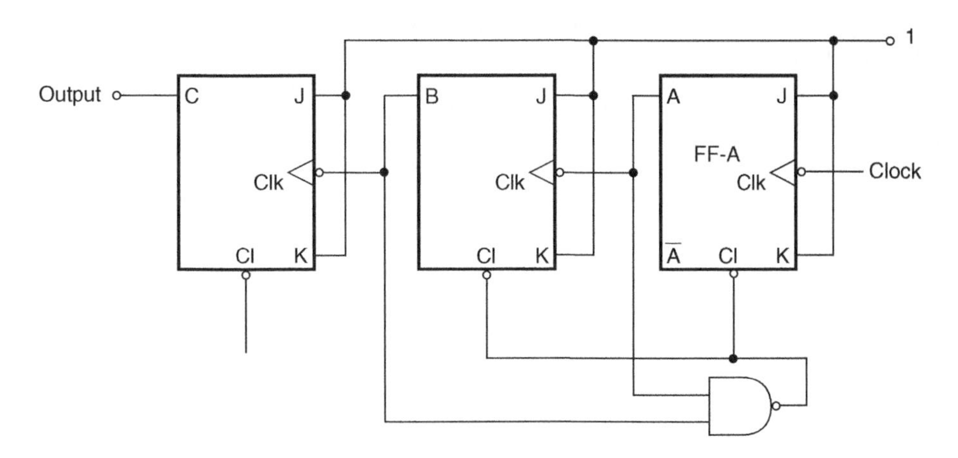

Figure 11.56 Problem 7.

10. Refer to the three-bit counter arrangement of Fig. 11.57. Determine its count sequence and also determine whether the counter is self-starting. (A counter is self-starting if it automatically goes to one of the desired states with subsequent clock pulse in case it lands itself accidentally into any of the undesired states.)

000, 001, 010, 011, 100, 000, . . . ; not self starting

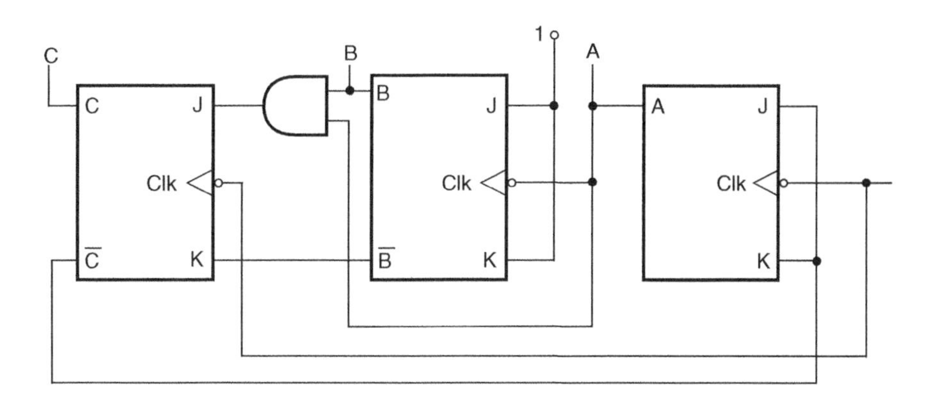

Figure 11.57 Problem 10.

Further Reading

1. Langholz, G., Mott, J. L. and Kandel, A (1998) *Foundations of Digital Logic Design*, World Scientific Publ. Co. Inc., NJ, USA.
2. Cook, N. P. (2003) *Practical Digital Electronics*, Prentice-Hall, NJ, USA.

3. Floyd, T. L. (2005) *Digital Fundamentals*, Prentice-Hall Inc., USA.
4. Tokheim, R. L. (1994) *Schaum's Outline Series of Digital Principles*, McGraw-Hill Companies Inc., USA.
5. Tocci, R. J. (2006) *Digital Systems – Principles and Applications*, Prentice-Hall Inc., NJ, USA.
6. Malvino, A. P. and Leach, D. P. (1994) *Digital Principles and Applications*, McGraw-Hill Book Company, USA.

12

Data Conversion Circuits – D/A and A/D Converters

Digital-to-analogue (D/A) and analogue-to-digital (A/D) converters constitute an essential link when digital devices interface with analogue devices, and vice versa. They are important building blocks of any digital system, including both communication and noncommunication systems, besides having other applications. A D/A converter is important not only because it is needed at the output of most digital systems, where it converts a digital signal into an analogue voltage or current so that it can be fed to a chart recorder, for instance, for measurement purposes, or a servo motor in a control application; it is also important because it forms an indispensable part of the majority of A/D converter types. An A/D converter, too, has numerous applications. When it comes to transmitting analogue data, it forms an essential interface with a digital communication system where the analogue signal to be transmitted is digitized at the sending end with an A/D converter. It is invariably used in all digital read-out test and measuring equipment. Whether it is a digital multimeter or a digital storage oscilloscope or even a pH meter, an A/D converter is an important and essential component of all of them. In this chapter, we will discuss the operational fundamentals, the major performance specifications, along with their significance, and different types and applications of digital-to-analogue and analogue-to-digital converters, in addition to application-relevant information of some of the popular devices. A large number of solved examples is also included to illustrate the concepts.

12.1 Digital-to-Analogue Converters

A D/A converter takes digital data at its input and converts them into analogue voltage or current that is proportional to the weighted sum of digital inputs. In the following paragraphs it is briefly explained

Digital Electronics: Principles, Devices and Applications Anil Kumar Maini

how different bits in the digital input data contribute a different quantum to the overall output analogue voltage or current, and also that the LSB has the least and the MSB the highest weight.

12.1.1 Simple Resistive Divider Network for D/A Conversion

Simple resistive networks can be used to convert a digital input into an equivalent analogue output. Figure 12.1 shows one such resistive network that can convert a three-bit digital input into an analogue output. This network, however, can be extended further to enable it to perform digital-to-analogue conversion of digital data with a larger number of bits. In the network of Fig. 12.1, if R_L is much larger than R, it can be proved with the help of simple network theorems that the output analogue voltage is given by

$$V_A = \frac{[V_1/R]+[V_2/(R/2)]+[V_3/(R/4)]}{[1/R]+[1/(R/2)]+[1/(R/4)]} \tag{12.1}$$

$$= \frac{[V_1/R]+[2V_2/R]+[4V_3/R]}{[1/R]+[2/R]+[4/R]} \tag{12.2}$$

$$= \frac{V_1+2V_2+4V_3}{7} \tag{12.3}$$

which can be further expressed as

$$V_A = \frac{V_1 \times 2^0 + V_2 \times 2^1 + V_3 \times 2^2}{2^3 - 1} \tag{12.4}$$

The generalized expression of Equation (12.4) can be extended further to an n-bit D/A converter to get the following expression:

$$V_A = \frac{V_1 \times 2^0 + V_2 \times 2^1 + V_3 \times 2^2 + \cdots + V_n \times 2^{n-1}}{2^n - 1} \tag{12.5}$$

In expression (12.5), if $V_1 = V_2 = \ldots = V_n = V$, then a logic '1' at the LSB position would contribute $V/(2^n - 1)$ to the analogue output, and a logic '1' in the next adjacent higher bit position would

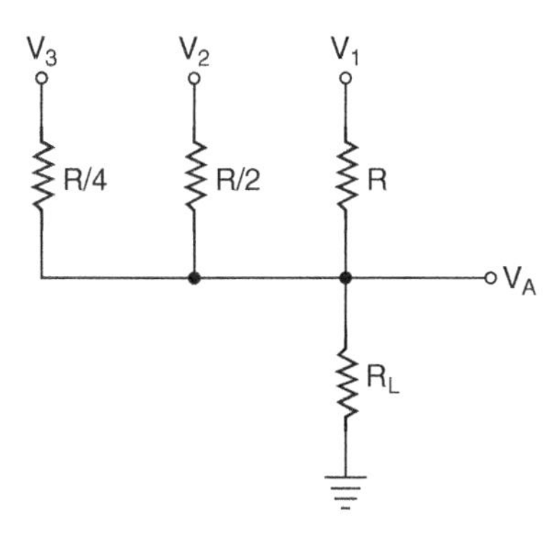

Figure 12.1 Simple resistive network for D/A conversion.

contribute $2V/(2^n - 1)$ to the output. The contributions of successive higher bit positions in the case of a logic '1' would be $4V/(2^n - 1)$, $8V/(2^n - 1)$, $16V/(2^n - 1)$ and so on. That is, the contribution of any given bit position owing to the presence of a logic '1' is twice the contribution of the adjacent lower bit position and half that of the adjacent higher bit position. When all input bit positions have a logic '1', the analogue output is given by

$$V_A = \frac{V(2^0 + 2^1 + 2^2 + \cdots + 2^{n-1})}{2^n - 1} = V \tag{12.6}$$

In the case of all inputs being in the logic '0' state, $V_A = 0$. Therefore, the analogue output varies from 0 to V volts as the digital input varies from an all 0s to an all 1s input.

12.1.2 Binary Ladder Network for D/A Conversion

The simple resistive divider network of Fig. 12.1 has two serious drawbacks. One, each resistor in the network is of a different value. Since these networks use precision resistors, the added expense becomes unattractive. Two, the resistor used for the most significant bit (MSB) is required to handle a much larger current than the LSB resistor. For example, in a 10-bit network, the current through the MSB resistor will be about 500 times the current through the LSB resistor.

To overcome these drawbacks, a second type of resistive network called the *binary ladder* (or *R*/2*R* ladder) is used in practice. The binary ladder, too, is a resistive network that produces an analogue output equal to the weighted sum of digital inputs. Figure 12.2 shows the binary ladder network for a four-bit D/A converter. As is clear from the figure, the ladder is made up of only two different values of resistor. This overcomes one of the drawbacks of the resistive divider network. It can be proved with the help of simple mathematics that the analogue output voltage V_A in the case of binary ladder network of Fig. 12.2 is given by

$$V_A = \frac{V_1 \times 2^0 + V_2 \times 2^1 + V_3 \times 2^2 + V_4 \times 2^3}{2^4} \tag{12.7}$$

In general, for an n-bit D/A converter using a binary ladder network

$$V_A = \frac{V_1 \times 2^0 + V_2 \times 2^1 + V_3 \times 2^2 + \cdots + V_n \times 2^{n-1}}{2^n} \tag{12.8}$$

For $V_1 = V_2 = V_3 = \cdots = V_n = V$, $V_A = [(2^n - 1)/2^n]V$. For $V_1 = V_2 = V_3 = \cdots = V_n = 0$, $V_A = 0$.

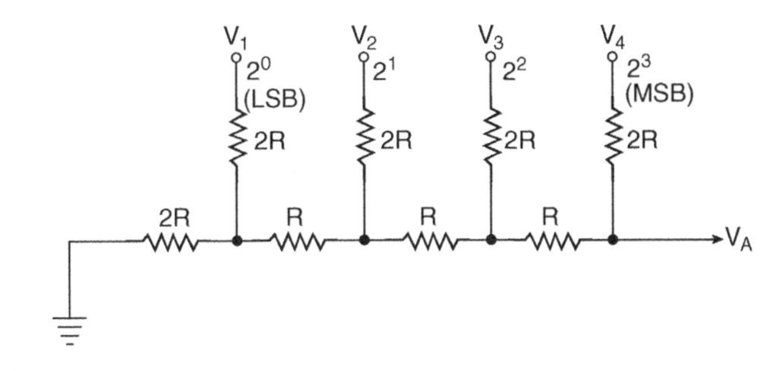

Figure 12.2 Binary ladder network for D/A conversion.

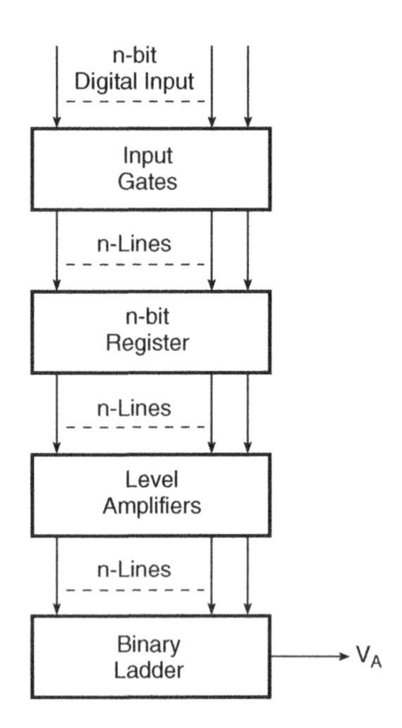

Figure 12.3 Block schematic representation of a D/A converter.

The analogue output voltage in this case varies from 0 (for an all 0s input) to $[(2^n - 1)/2^n]V$ (for an all 1s input).

Also, in the case of a resistive divider network, the LSB contribution to the analogue output is $[1/(2^n - 1)]V$. This is also the minimum possible incremental change in the analogue output voltage. The same in the case of a binary ladder network would be $(1/2^n)V$.

A binary ladder network is the most widely used network for digital-to-analogue conversion, for obvious reasons. Although actual D/A conversion takes place in this network, a practical D/A converter device has additional circuitry such as a register for temporary storage of input digital data and level amplifiers to ensure that the digital signals presented to the resistive network are all of the same level. Figure 12.3 shows a block schematic representation of a complete n-bit D/A converter. D/A converters of different sizes (eight-bit, 12-bit, 16-bit, etc.) are available in the form of integrated circuits.

12.2 D/A Converter Specifications

The major performance specifications of a D/A converter include resolution, accuracy, conversion speed, dynamic range, nonlinearity (NL) and differential nonlinearity (DNL) and monotonocity.

12.2.1 Resolution

The *resolution* of a D/A converter is the number of states (2^n) into which the full-scale range is divided or resolved. Here, n is the number of bits in the input digital word. The higher the number of bits, the better is the resolution. An eight-bit D/A converter has 255 resolvable levels. It is said to have a percentage resolution of (1/255) × 100 = 0.39 % or simply an eight-bit resolution. A 12-bit D/A converter would have a percentage resolution of (1/4095) × 100 = 0.0244 %. In general, for an n-bit D/A converter, the percentage resolution is given by $(1/2^n - 1) \times 100$. The resolution in millivolts for the two cases for a full-scale output of 5 V is approximately 20 mV (for an eight-bit converter) and 1.2 mV (for a 12-bit converter).

12.2.2 Accuracy

The *accuracy* of a D/A converter is the difference between the actual analogue output and the ideal expected output when a given digital input is applied. Sources of error include the *gain error* (or full-scale error), the *offset error* (or zero-scale error), *nonlinearity errors* and a drift of all these factors. The gain error [Fig. 12.4(a)] is the difference between the actual and ideal output voltage, expressed as a percentage of full-scale output. It is also expressed in terms of LSB. As an example, an accuracy of ±0.1 % implies that the analogue output voltage may be off by as much as ±5 mV for a full-scale output of 5 V throughout the analogue output voltage range. The offset error is the error at analogue zero [Fig. 12.4(b)].

12.2.3 Conversion Speed or Settling Time

The *conversion speed* of a D/A converter is expressed in terms of its settling time. The *settling time* is the time period that has elapsed for the analogue output to reach its final value within a specified error band after a digital input code change has been effected. General-purpose D/A converters have a settling time of several microseconds, while some of the high-speed D/A converters have a settling

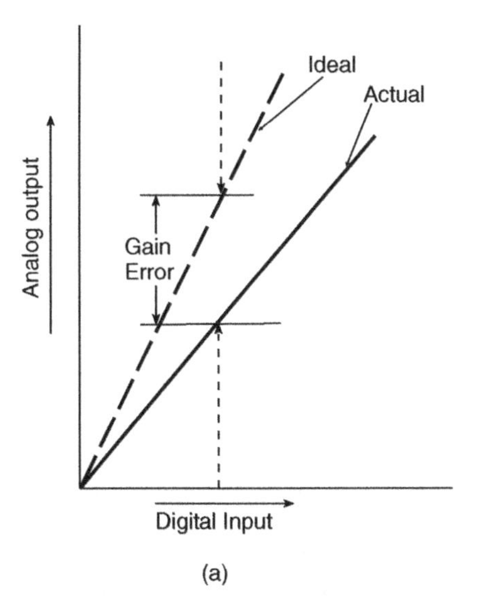

Figure 12.4 (a) Gain error and (b) offset error.

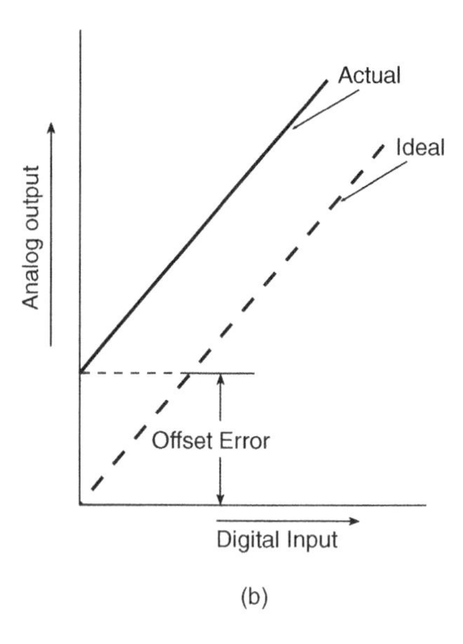

Figure 12.4 (*continued*).

time of a few nanoseconds. The settling time specification for D/A converter type number AD 9768 from Analog Devices USA, for instance, is 5 ns.

12.2.4 Dynamic Range

This is the ratio of the largest output to the smallest output, excluding zero, expressed in dB. For linear D/A converters it is $20 \times \log 2^n$, which is approximately equal to $6n$. For companding-type D/A converters, discussed in Section 12.3, it is typically 66 or 72 dB.

12.2.5 Nonlinearity and Differential Nonlinearity

Nonlinearity (NL) is the maximum deviation of analogue output voltage from a straight line drawn between the end points, expressed as a percentage of the full-scale range or in terms of LSBs. *Differential nonlinearity* (DNL) is the worst-case deviation of any adjacent analogue outputs from the ideal one-LSB step size.

12.2.6 Monotonocity

In an ideal D/A converter, the analogue output should increase by an identical step size for every one-LSB increment in the digital input word. When the input of such a converter is fed from the output of a counter, the converter output will be a perfect staircase waveform, as shown in Fig. 12.5. In such cases, the converter is said to be exhibiting perfect monotonocity. A D/A converter is considered as monotonic if its analogue output either increases or remains the same but does not decrease as the digital input code advances in one-LSB steps. If the DNL error of the converter is less than or equal to twice its worst-case nonlinearity error, it guarantees monotonocity.

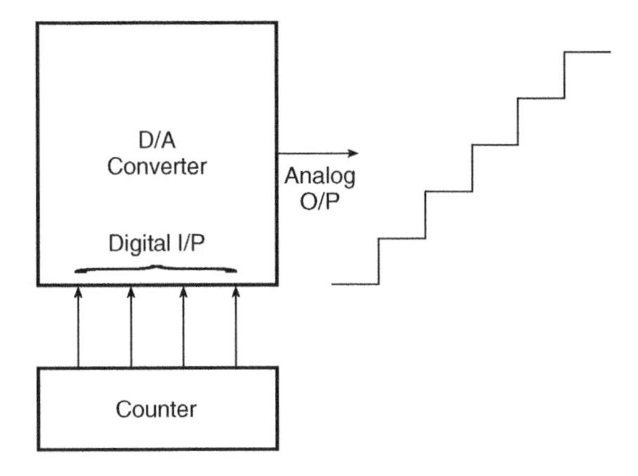

Figure 12.5 Monotonocity in a D/A converter.

12.3 Types of D/A Converter

The D/A converters discussed in this section include the following:

1. Multiplying-type D/A converters.
2. Bipolar-output D/A converters.
3. Companding D/A converters.

12.3.1 Multiplying D/A Converters

In a *multiplying-type D/A converter*, the converter multiplies an analogue reference by the digital input. Figure 12.6 shows the circuit representation. Some D/A converters can multiply only positive digital words by a positive reference. This is known as single quadrant (QUAD-I) operation. Two-quadrant operation (QUAD-I and QUAD-III) can be achieved in a D/A converter by configuring the output for bipolar operation. This is accomplished by offsetting the output by a negative MSB (equal to the analogue output of 1/2 of the full-scale range) so that the MSB becomes the sign bit.

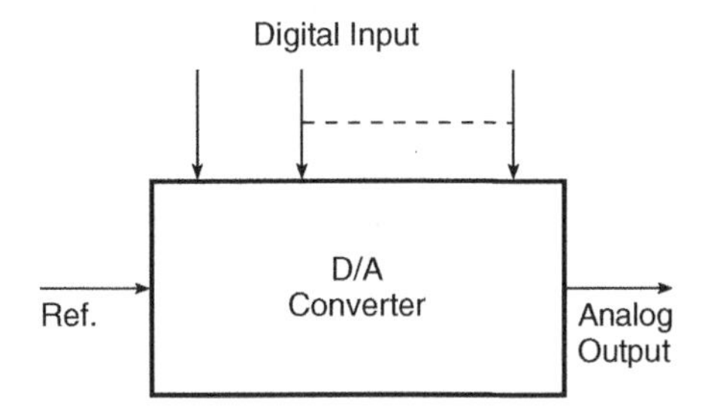

Figure 12.6 Multiplying-type D/A converter.

Some D/A converters even provide four-quadrant operation by allowing the use of both positive and negative reference. Multiplying D/A converters are particularly useful when we are looking for digitally programmable attenuation of an analogue input signal.

12.3.2 Bipolar-Output D/A Converters

In *bipolar-output D/A converters* the analogue output signal range includes both positive and negative values. The transfer characteristics of an ideal two-quadrant bipolar-output D/A converter are shown in Fig. 12.7.

12.3.3 Companding D/A Converters

Companding-type D/A converters are so constructed that the more significant bits of the digital input have a larger than binary relationship to the less significant bits. This decreases the resolution of the more significant bits, which in turn increases the analogue signal range. The effect of this is to compress more data into more significant bits.

12.4 Modes of Operation

D/A converters are usually operated in either of the following two modes of operation:

1. Current steering mode.
2. Voltage switching mode.

12.4.1 Current Steering Mode of Operation

In the *current steering mode* of operation of a D/A converter, the analogue output is a current equal to the product of a reference voltage and a fractional binary value D of the input digital word. D is equal to the sum of fractional binary values of different bits in the digital word. Also, fractional binary values of different bits in an n-bit digital word starting from the LSB are $2^0/2^n, 2^1/2^n, 2^2/2^n, \ldots, 2^{n-1}/2^n$.

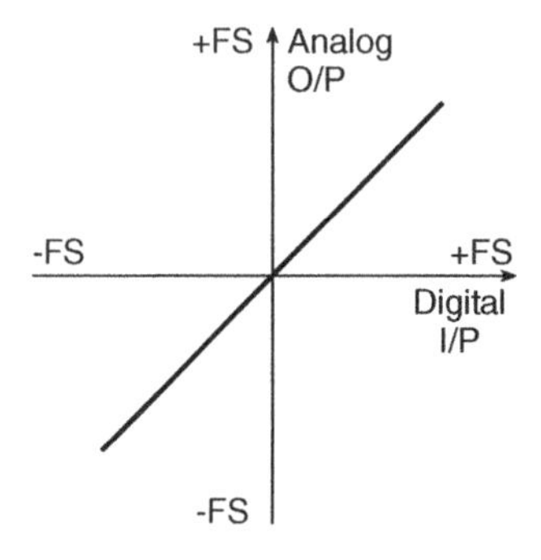

Figure 12.7 Bipolar-output D/A converter transfer characteristics.

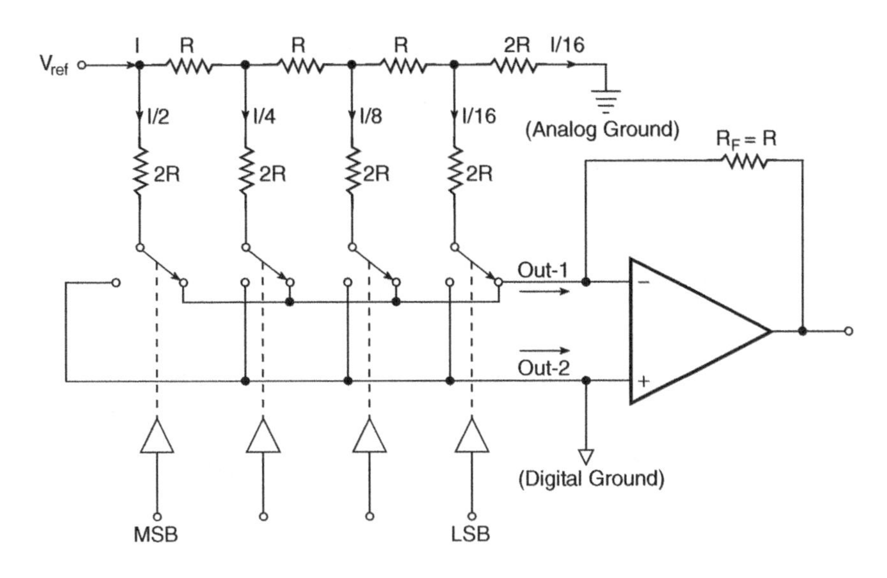

Figure 12.8 Current steering mode of operation of a D/A converter.

The output current is often converted into a corresponding voltage using an external opamp wired as a current-to-voltage converter. Figure 12.8 shows the circuit arrangement. The majority of D/A converters in IC form have an in-built opamp that can be used for current-to-voltage conversion. For the circuit arrangement of Fig. 12.8, if the feedback resistor R_F equals the ladder resistance R, the analogue output voltage at the opamp output is $-(D.V_{\text{ref}})$.

The arrangement of the four-bit D/A converter of Fig. 12.8 can be conveniently used to explain the operation of a D/A converter in the current steering mode. The $R/2R$ ladder network divides the input current I due to a reference voltage V_{ref} applied at the reference voltage input of the D/A converter into binary weighted currents, as shown. These currents are then steered to either the output designated Out-1 or Out-2 by the current steering switches. The positions of these current steering switches are controlled by the digital input word. A logic '1' steers the corresponding current to Out-1, whereas a logic '0' steers it to Out-2. For instance, a logic '1' in the MSB position will steer the current $I/2$ to Out-1. A logic '0' steers it to Out-2, which is the ground terminal. In the four-bit converter of Fig. 12.8, the analogue output current (or voltage) will be maximum for a digital input of 1111. The analogue output current in this case will be $I/2 + I/4 + I/8 + I/16 = (15/16)I$. The analogue output voltage will be $(-15/16)IR_F = (-15/16)IR$. Also, $I = V_{\text{ref}}/R$ as the equivalent resistance of the ladder network across V_{ref} is also R. The analogue output voltage is then $[(-15/16)(V_{\text{ref}})/R] \times R = (-15/16)V_{\text{ref}}$. Here, 15/16 is nothing but the fractional binary value of digital input 1111. In general, the maximum analogue output voltage is given by $-(1-2^{-n}) \times V_{\text{ref}}$, where n is the number of bits in the input digital word.

12.4.2 Voltage Switching Mode of Operation

In the *voltage switching mode* of operation of a $R/2R$ ladder type D/A converter, the reference voltage is applied to the Out-1 terminal and the output is taken from the reference voltage terminal. Out-2 is joined to analogue ground. Figure 12.9 shows a four-bit D/A converter of the $R/2R$ ladder type in

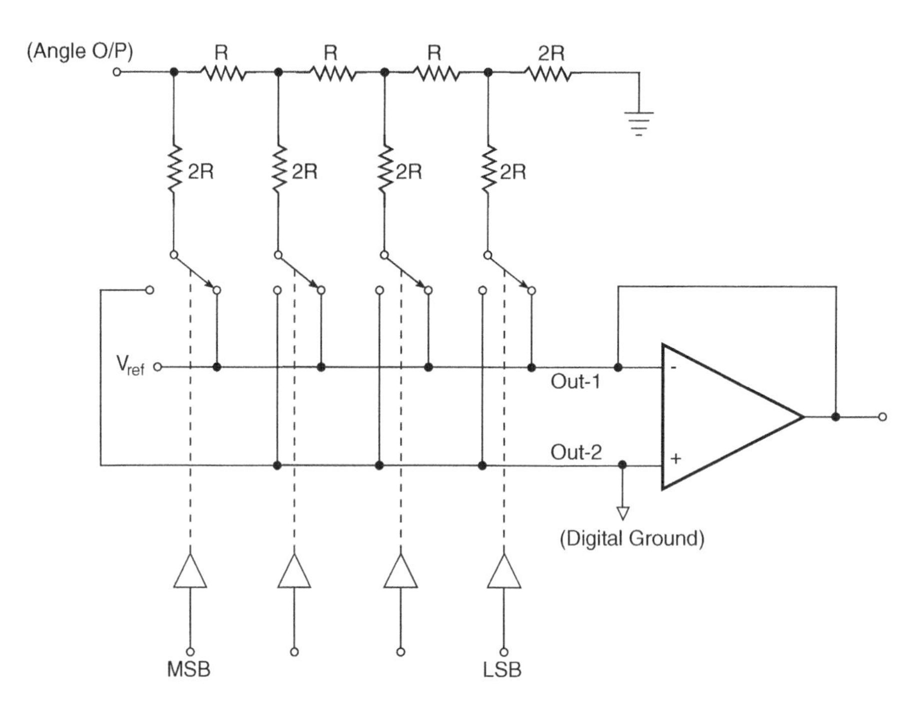

Figure 12.9 Voltage switching mode of operation of a D/A converter.

voltage switching mode of operation. The output voltage is the product of the fractional binary value of the digital input word and the reference voltage applied at the Out-1 terminal, i.e. $D.V_{\text{ref}}$. As the positive reference voltage produces a positive analogue output voltage, the voltage switching mode of operation is possible with a single supply. As the circuit produces analogue output voltage, it obviates the need for an opamp and the feedback resistor. However, the reference voltage applied to the Out-1 terminal in this case will see different input impedances for different digital inputs. For this reason, the source of the input is buffered.

12.5 BCD-Input D/A Converter

A BCD-input D/A converter accepts the BCD equivalent of decimal digits at its input. A two-digit BCD D/A converter for instance is an eight-bit D/A converter. Figure 12.10 shows the circuit representation of an eight-bit BCD-type D/A converter. Such a converter has 99 steps and accepts decimal digits 00 to 99 at its input. A 12-bit converter will have 999 steps. The weight of the different bits in the least significant digit (LSD) will be 1 (for A_0), 2 (for B_0), 4 (for C_0) and 8 (for D_0). The weights of the corresponding bits in the next higher digit will be 10 times the weights of corresponding bits in the lower adjacent digit. For the D/A converter shown in Fig. 12.10 the weight of the different bits in the most significant digit (MSD) will be 10 (for A_1), 20 (for B_1), 40 (for C_1) and 80 (for D_1). In general, an n-bit D/A converter of the BCD input type will have $(10^{n/4} - 1)$ steps. The percentage resolution of such a converter is given by $[1/(10^{n/4} - 1)] \times 100$.

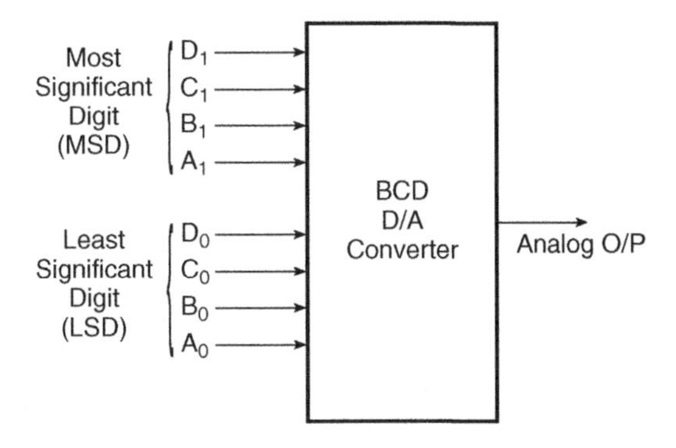

Figure 12.10 BCD-input D/A converter.

Example 12.1

An eight-bit D/A converter has a step size of 20 mV. Determine the full-scale output and percentage resolution.

Solution

- $(1/2^8) \times V = 20 \times 10^{-3}$, where V is the voltage corresponding to a logic '1'.
- This gives $V = 20 \times 10^{-3} \times 2^8 = 5.12$V.
- The full-scale output $= [(2^n - 1)/2^n] \times V = [(2^8 - 1)/2^8] \times 5.12 = (255/256) \times 5.12 = 5.1$V.
- The percentage resolution $= [1/(2^n - 1)] \times 100 = 100/255 = 0.392\%$.
- The percentage resolution can also be determined from: (Step size/full-scale output) $\times$ 100 = $(20 \times 10^{-3}/5.1) \times 100 = 0.392\,\%$.

Example 12.2

Refer to Fig. 12.11. This BCD D/A converter has a step size of 6.25 mV. Determine the full-scale output.

Solution

- A step size of 6.25 mV implies that A_0 has a weight of 6.25 mV.
- The weights of B_0, C_0 and D_0 would respectively be 12.5, 25 and 50 mV.
- Now, the weight of A_1 will be 10 times the weight of A_0, i.e. the weight of A_1 will be 62.5 mV.
- The weights of B_1, C_1 and D_1 will accordingly be 125, 250 and 500 mV respectively.
- On similar lines, the weights of A_2, B_2, C_2 and D_2 will respectively be 625 mV, 1.25 V, 2.5 V and 5 V.
- For full-scale output, the input will be decimal 999. Each of the three four-bit groups will be 1001.
- Therefore, the full-scale analogue output = 6.25 + 50 + 62.5 +500 + 625 + 5000 mV = 6.24375 V.
- The full-scale analogue output can also be determined from the product of the step size and number of steps. That is, the full-scale output = 6.25 × 999 = 6.24375 V.

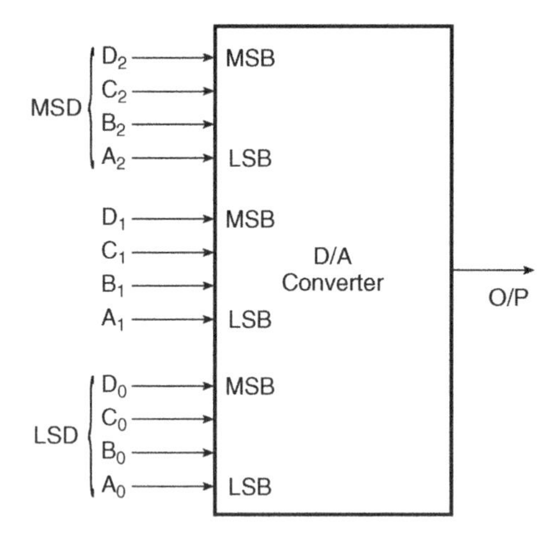

Figure 12.11 BCD-input D/A converter (example 12.2).

Example 12.3

A certain eight-bit D/A converter has a full-scale output of 5 mA and a full-scale error of ±0.25 % of full scale. Determine the range of expected analogue output for a digital input of 10000010.

Solution

- Step size $= \dfrac{\text{Full-scale output}}{\text{Number of steps}}$

$$= \frac{5 \times 10^{-3}}{2^8 - 1}$$

$$= 19.6\ \mu\text{A}$$

- For a digital input of 10000010 ($= 130_{10}$) the analogue output is given by $130 \times 19.6 = 2.548$ mA.
- Error $= \pm \dfrac{0.25 \times 5 \times 10^{-3}}{100}$

$$= \pm 12.5\ \mu\text{A}$$

- The expected analogue output will therefore be in the range 2.5355–2.5605 mA.

Example 12.4

An experimenter connects a four-bit ripple counter to a four-bit D/A converter to perform a staircase test using a 1 kHz clock as shown in Fig. 12.12. The output staircase waveform is shown in Fig. 12.13. The cause of the incorrect staircase signal is later determined to be a wrong connection between the counter output and the D/A converter input. What is it?

Solution

The correct staircase waveform would be generated at the output of the D/A converter if the counter outputs Q_0 (LSB), Q_1, Q_2 and Q_3 (MSB) were connected to the corresponding inputs

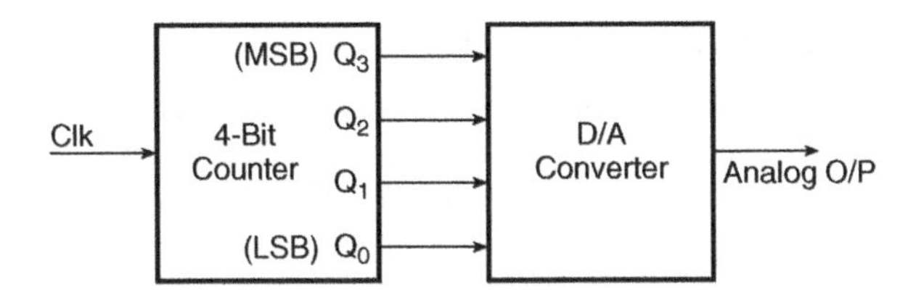

Figure 12.12 Example 12.4.

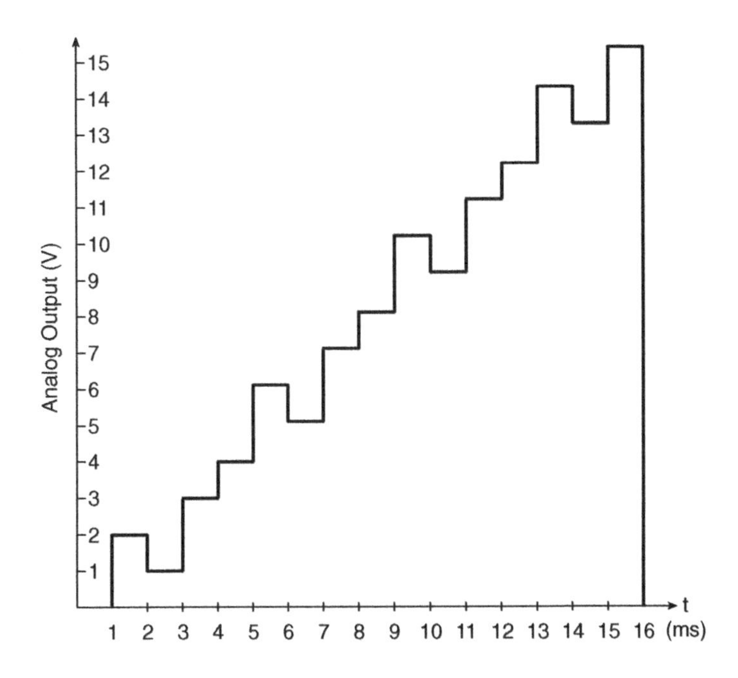

Figure 12.13 Staircase waveform (example 12.4).

of the D/A converter in the same order. If we carefully examine the given staircase waveform and recall the sequence in which the counter will advance, it can be visualized that the given staircase waveform would result if the interconnections of the LSB and the next adjacent higher bit of the counter output and the corresponding inputs of the D/A converter were interchanged. While in one complete cycle the counter counts as 0000, 0001, 0010, 0011, 0100, 0101, 0110, 0111, 1000, 1001, 1010, 1011, 1100, 1101, 1110 and 1111, the D/A converter, owing to interchanged connections, gets inputs as 0000, 0010, 0001, 0011, 0100, 0110, 0101, 0111, 1000, 1010, 1001, 1011, 1100, 1110, 1101 and 1111. The corresponding analogue outputs are 0, 2, 1, 3, 4, 6, 5, 7, 8, 10, 9, 11, 12, 14, 13 and 15 V, as shown in the staircase waveform of Fig. 12.13.

12.6 Integrated Circuit D/A Converters

This section presents application-relevant information on some of the commonly used D/A converter IC type numbers, as it is not possible to give a detailed description of each one of them. The type numbers included for this purpose are DAC-08/0800, DAC-80, DAC-0808, AD 7524 and DAC-1408A/1508A.

12.6.1 DAC-08

DAC-08 is an eight-bit monolithic D/A converter. Its major performance specifications include a settling time of 85 ns, a monotonic multiplying performance over a wide 20-to-1 reference current range, a direct interface to all popular logic families, high voltage compliance complementary current outputs, nonlinearities of ±0.1 % over the entire operating temperature range and a wide power supply range of ±4.5 V to ±18 V. Figures 12.14(a) and (b) respectively show the basic circuit configurations for positive low impedance output operation and negative low impedance output operation. DAC-08

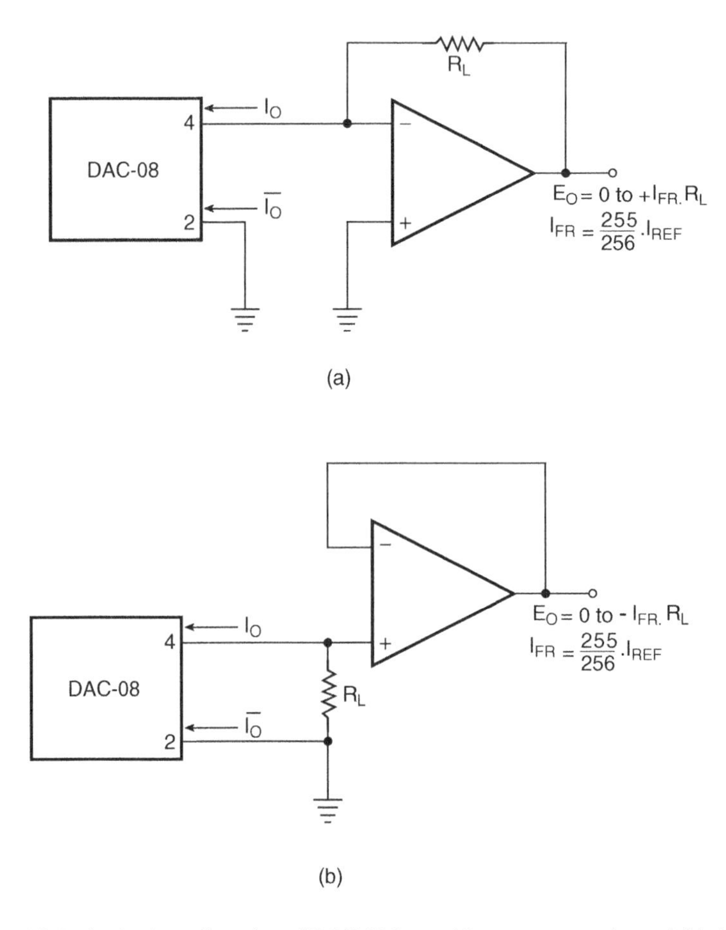

Figure 12.14 (a) Basic circuit configuration of DAC-08 for positive output operation and (b) the basic circuit configuration of DAC-08 for negative output operation.

applications include waveform generators, servomotor and pen drivers, audio encoders and digitally controlled attenuators, analogue meter drivers, programmable power supplies, high-speed modems, CRT display drivers, etc.

12.6.2 DAC-0808

DAC-0808 is an eight-bit D/A converter featuring a full-scale output current settling time of 150 ns while dissipating only 33 mW with ± 5 V supplies. Relative accuracies of better than ± 0.19 % ensure eight-bit monotonocity and linearity, while zero-level output current of less than 4 mA provides eight-bit zero accuracy for $I_{ref} \geq 2$ mA. It has a wide power supply voltage range of ± 4.5 V to ± 18 V. It can interface directly with popular TTL, DTL or CMOS logic families and is a direct replacement for the D/A converter MC 1508/MC 1408. Figure 12.15 shows the application circuit of DAC-0808 wired as a voltage-output D/A converter.

12.6.3 DAC-80

DAC-80 is a 12-bit D/A converter. Both current and voltage-output versions are available. Its salient features include low power dissipation (345 mW), full ± 10 V swing with ±12 V supplies, TTL and CMOS-compatible digital inputs, ±1/2 LSB maximum nonlinearity over 0–70 °C, guaranteed monotonocity over 0–70 °C and 4 ms settling time to ±0.01 % of full-scale and monolithic design. Figures 12.16 and 12.17 show the pin connection diagrams of current-output and voltage-output models of DAC-80.

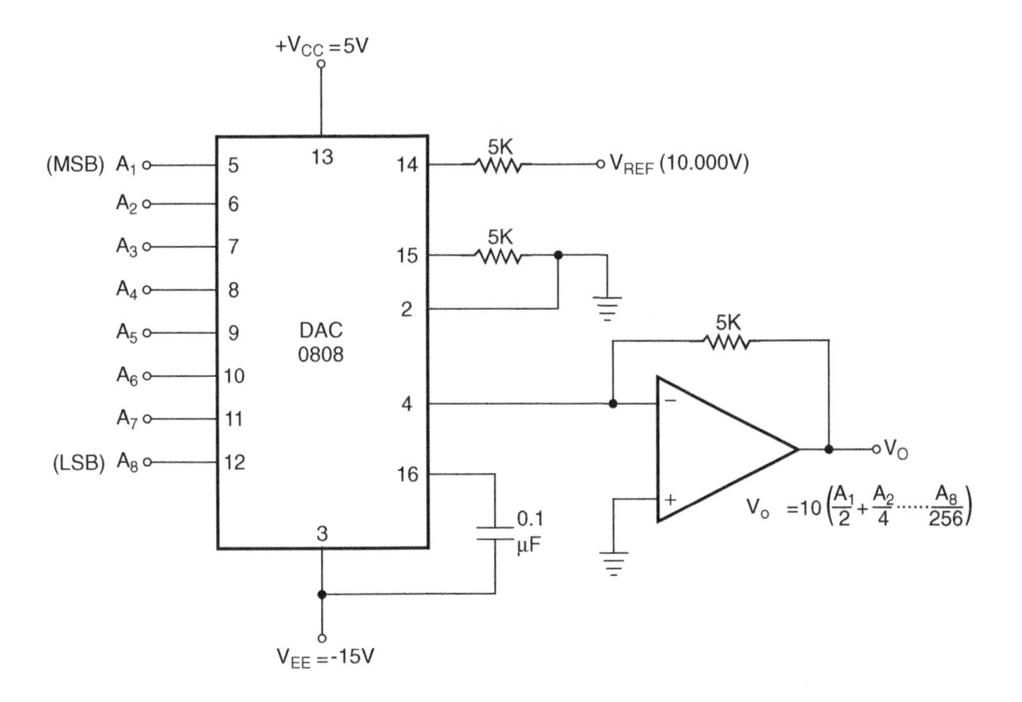

Figure 12.15 DAC-0808 wired as a voltage-output D/A converter.

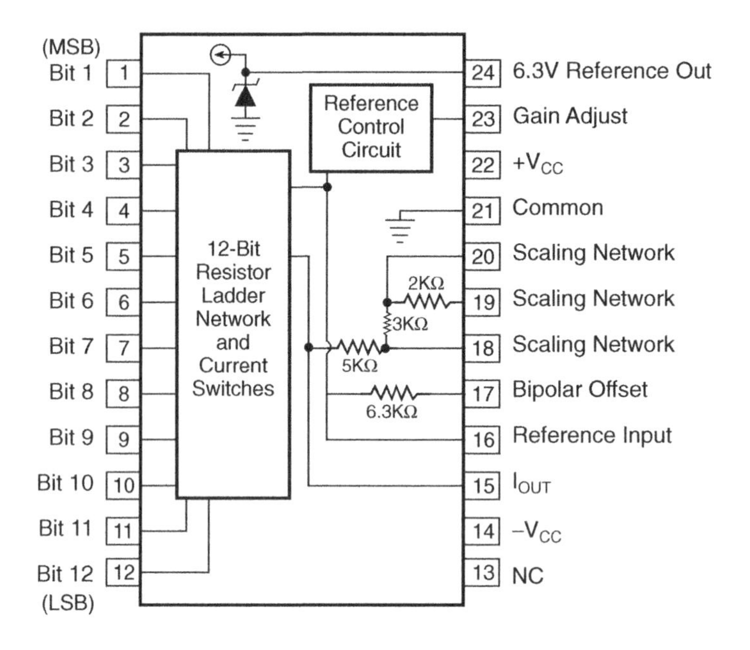

Figure 12.16 Pin connection diagram of DAC-80 (current-output version).

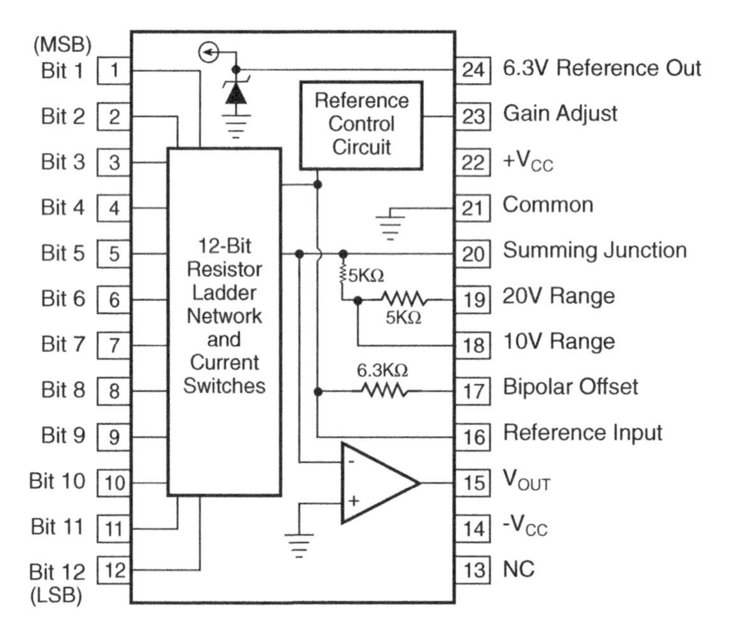

Figure 12.17 Pin connection diagram of DAC-80 (voltage-output version).

12.6.4 AD 7524

AD 7524 is an eight-bit monolithic CMOS DAC designed for direct interface to most microprocessors such as 6800, 8085, Z80, etc. It has an accuracy of 1/8 LSB, with a typical power dissipation of less than 10 mW. Monotonocity is guaranteed over full operation temperature range. It has a settling time of 250 ns (typical) for the output current to settle within 1/2 LSB for a supply voltage of +15 V. Its excellent multiplying characteristics (two or four-quadrant) make AD 7524 an ideal choice for many microprocessor-controlled gain setting and signal control applications. It has a wide power supply range of +5 V to +15 V. Figure 12.18 shows the functional diagram which resembles the functional diagram of any current-output multiplying D/A converter.

12.6.5 DAC-1408/DAC-1508

DAC-1508/1408 is a general-purpose, high-speed multiplying-type eight-bit D/A converter. DAC-1508 is identical to DAC-1408 except for the operational temperature range, which is −55°C to +125 °C in the case of DAC-1508, as against 0–70 °C for DAC-1408. It is pin and functionally compatible with DAC-0808.

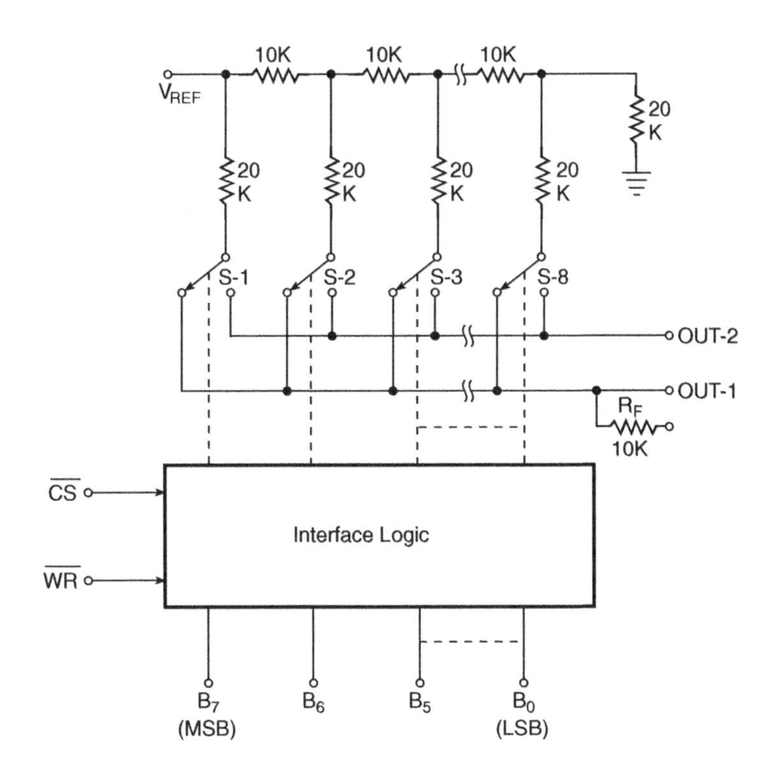

Figure 12.18 Functional diagram of AD 7524.

12.7 D/A Converter Applications

In addition to being an integral part of some of the architectures of popular varieties of A/D converters, D/A converters are extensively used in a variety of other application circuits. Some common applications include multipliers, digitally controlled dividers, programmable integrators, low-frequency function generators and digitally controlled filters.

12.7.1 D/A Converter as a Multiplier

The basic D/A converter operated in the current steering mode with the output opamp wired as a current-to-voltage converter works as a multiplier where the output voltage is the product of the analogue input applied at the V_{ref} terminal and the digital word input. CMOS D/A converters are much better suited to multiplying applications as the multiplying capabilities of other types of D/A converter are restricted to a limited range of input voltage. One such application circuit where the multiplying capability of the D/A converter is used is the digitally controlled audio signal attenuator. Figure 12.19 shows the circuit diagram. The audio signal is applied to the V_{ref} input and the attenuation code is applied to the digital input. The analogue output is the attenuated version of the input.

As audio attenuators, conventional D/A converters provide a limited range of attenuation which is 256:1 or 48 dB for an eight-bit converter and 4096:1 or 72 dB for a 12-bit converter. Logarithmic D/A converters, which give a logarithmic relationship between the digital fraction and the output signal matching the response of the human ear, are particularly suitable for this application. These are coded to give attenuation in equal decimal steps.

12.7.2 D/A converter as a Divider

If the feedback resistance is used as the input resistor and the D/A converter is connected as a feedback element, the circuit acts as a divider or a programmable gain element. Figure 12.20 shows the circuit configuration. The output is given by $V_o = -(V_{in}/D)$. For smaller values of digital fraction D the output increases, and the designer should ensure that the amplifier does not saturate under these conditions.

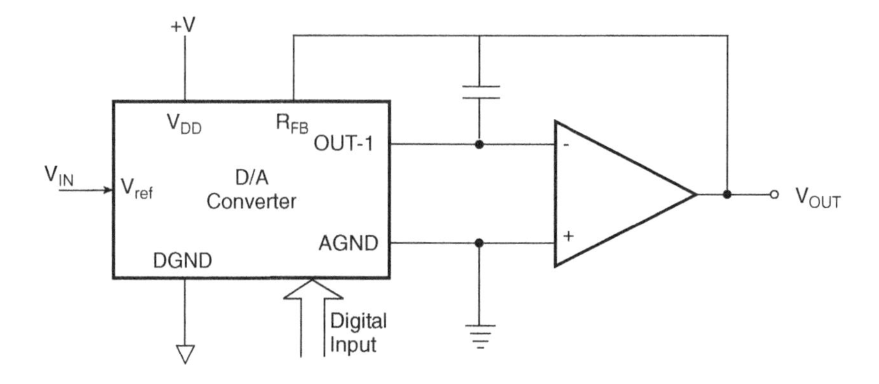

Figure 12.19 Digitally controlled audio signal attenuator.

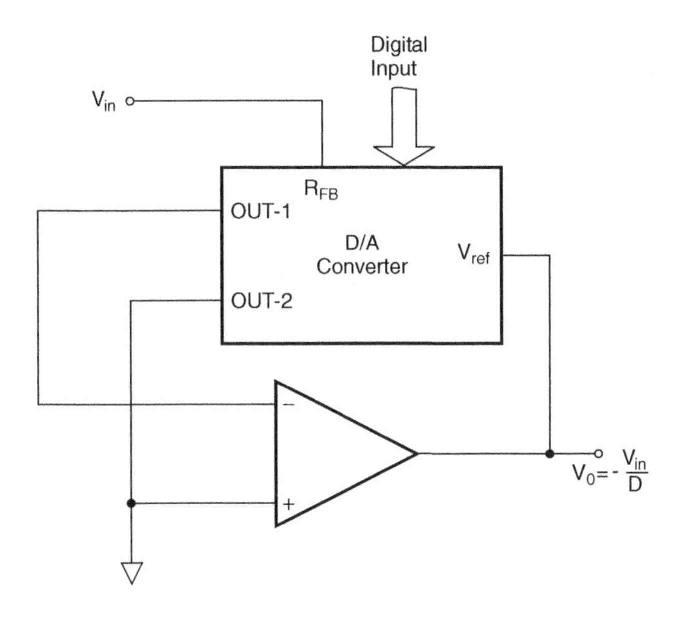

Figure 12.20 Digitally controlled divider.

12.7.3 Programmable Integrator

The programmable integrator forms the basis of a number of medium-frequency function generators. Figure 12.21 shows an inverting type of programmable integrator. The output is expressed by

$$V_o = [-1/C.(R_{DAC} + R_1)].D \int V_{in}.dt \tag{12.9}$$

where R_{DAC} isthe input resistance of the D/A converter at the V_{ref} terminal. Resistance R_1 has been used to get an appropriate value of the integrator time constant for the full-scale value of D. The integrator time constant given by $[C.(R_{DAC}+ R_1)/D]$ is largest when the input digital code is near zero and

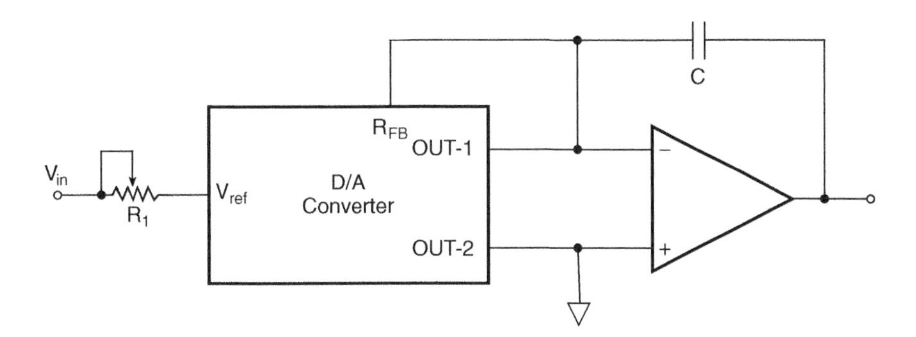

Figure 12.21 Inverting-type programmable integrator.

smallest when D has the full-scale value. Figure 12.22 shows the noninverting type of programmable integrator. The output in this case is given by

$$V_o = (D/CR_1)\int V_{in}.dt \qquad (12.10)$$

12.7.4 Low-Frequency Function Generator

Figure 12.23 shows one possible circuit configuration of a D/A converter based low-frequency function generator. There is no limit to the lowest frequency possible using this configuration. The upper limit

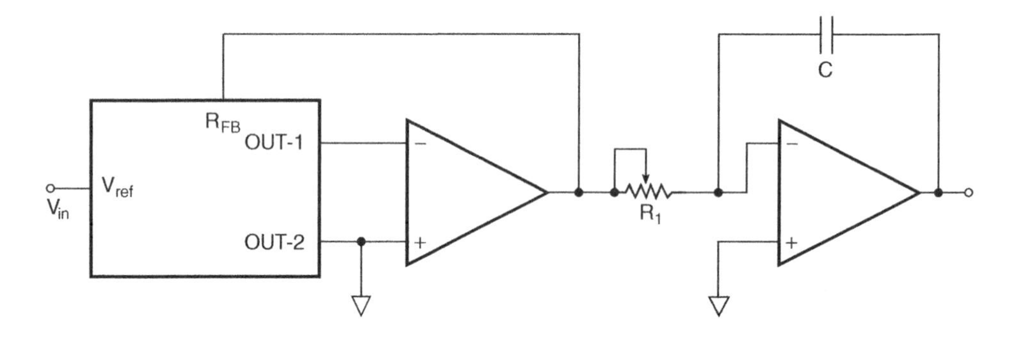

Figure 12.22 Non-inverting programmable integrator.

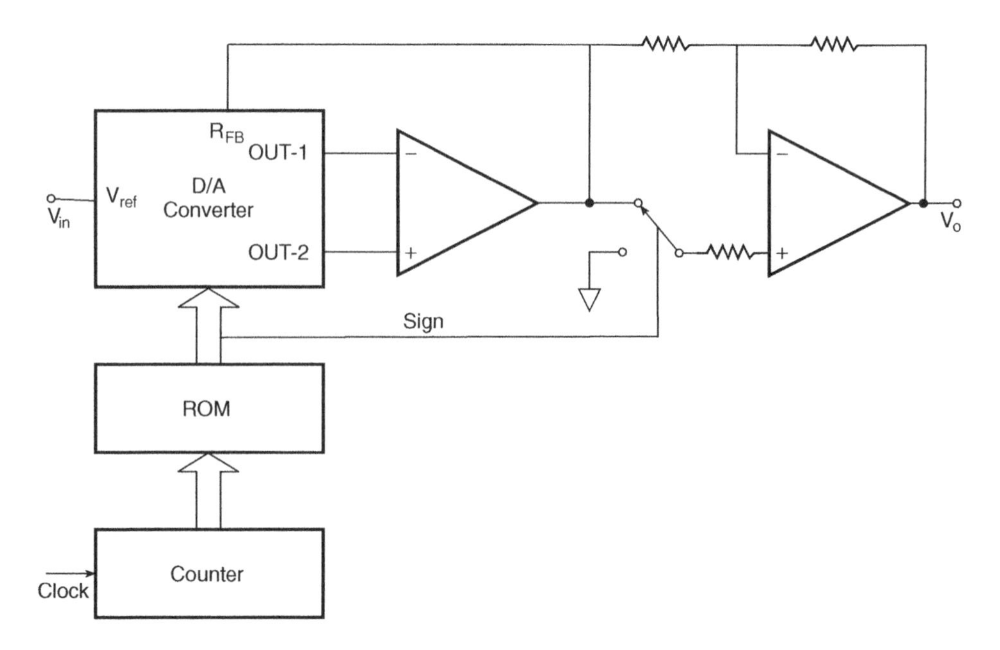

Figure 12.23 Low-frequency function generator.

is determined by the settling time of the D/A converter, the required resolution and the permissible quantization noise.

Since most of the functions are symmetric, it is usual to synthesize only half of the waveform and then invert it for the second half. This is true for pulse, triangular, ramp and trapezoidal waveforms. For sinusoidal waveforms it is necessary only to synthesize one-quarter of the waveform. In the arrangement of Fig. 12.23, the frequency is determined by the clock frequency and the waveform by the contents of the ROM.

12.7.5 Digitally Controlled Filters

Active filters having low noise and distortion with controllable gain, centre frequency and Q-factor can be constructed using multiplying-type D/A converters. Three basic types of first-order low-pass filter are shown in Figs 12.24, 12.25 and 12.26. The low-pass circuit of Fig. 12.24 has a R_{DAC}-dependent cut-off frequency given by

$$\omega = [R_1/(R_1 + R_2)] \times [D/C.R_{DAC}] \quad (12.11)$$

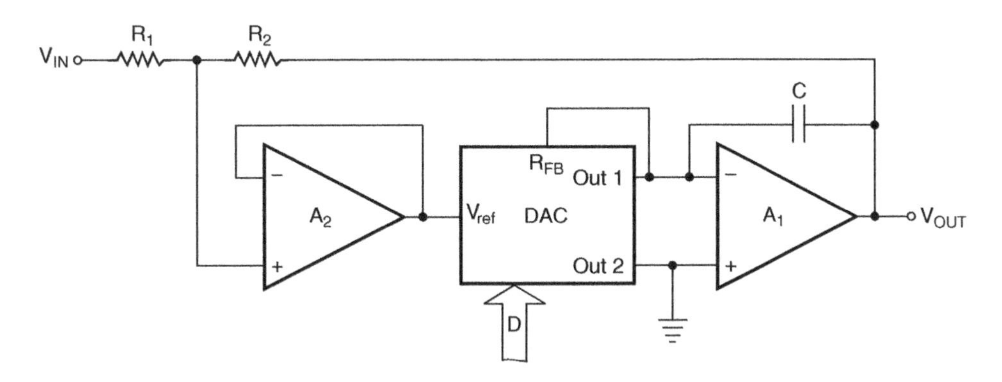

Figure 12.24 Low-pass filter with R_{DAC}-dependent cut-off frequency.

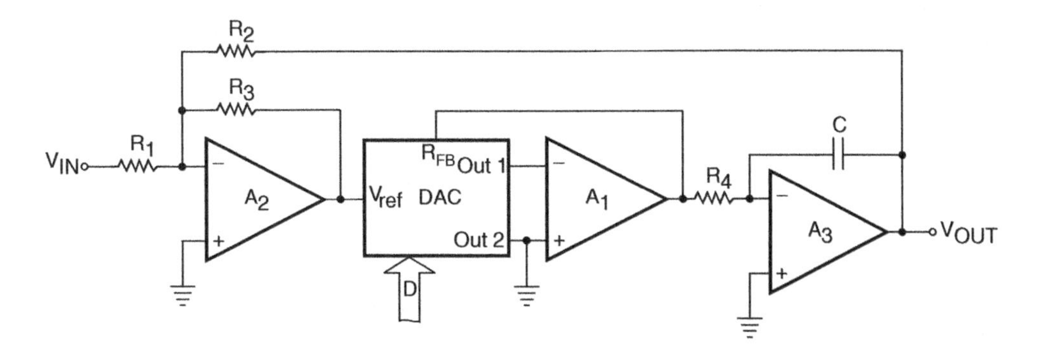

Figure 12.25 Low-pass filter with cut-off frequency independent of R_{DAC}.

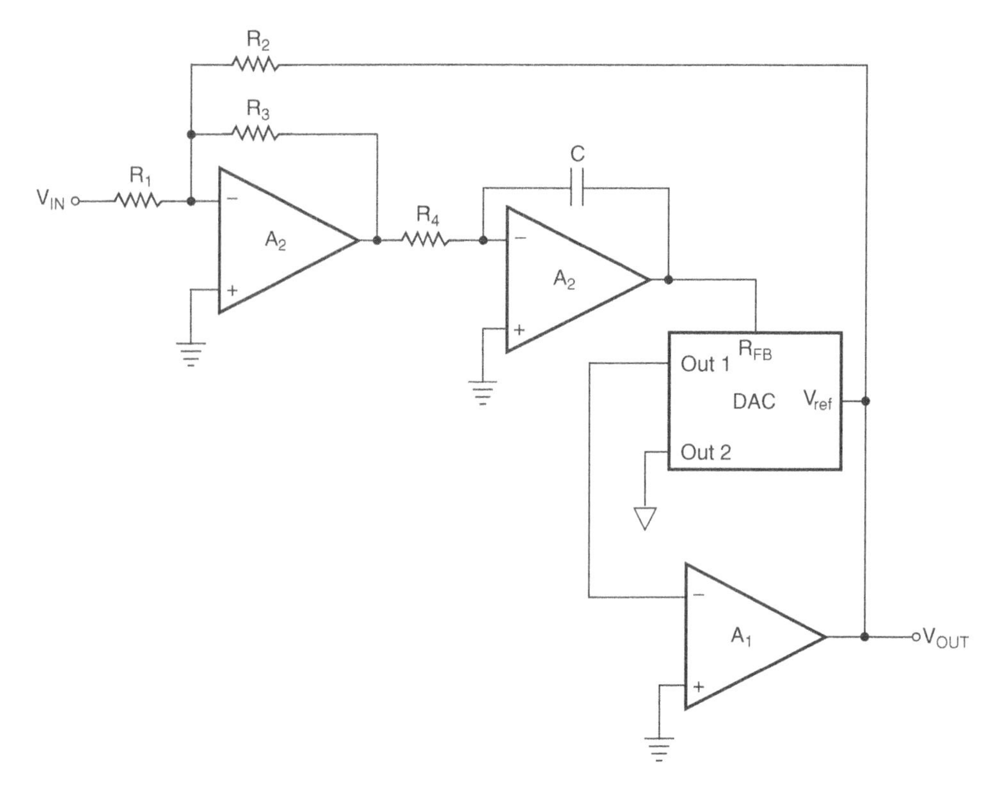

Figure 12.26 Low-pass filter with a programmable time constant.

Also, the transfer function for this low-pass filter is given by

$$V_{out}/V_{in} = (-R_2/R_1)\{1/[1+j\omega(R_1+R_2).R_{DAC}.C/R_1.D]\} \tag{12.12}$$

The cut-off frequency can be made independent of R_{DAC} by using the D/A converter as a programmable gain element, as shown in Fig. 12.25. In this case, the cut-off frequency ω is given by

$$\omega = R_3.D/R_2.R_4.C \tag{12.13}$$

and the transfer function is given by

$$V_{out}/V_{in} = (-R_2/R_1)\{1/[1+j\omega(R_2.R_4.C/R_3.D)]\} \tag{12.14}$$

If it is required to have a proportional adjustment of the filter time constant rather than its cut-off frequency, the circuit of Fig. 12.25 is rearranged and the D/A converter is connected in the divider configuration as shown in Fig. 12.26. The time constant is given by

$$Time\ constant = R_2.R_4.C.D/R_3 \tag{12.15}$$

and the transfer function is given by

$$V_{out}/V_{in} = (R_2/R_1)\{1/[1+j\omega(R_2.R_4.C.D/R_3)]\} \tag{12.16}$$

It may be mentioned here that other types of digitally controlled filter are also possible using D/A converters. One such possibility, for instance, is by using state variable techniques, which can be used to design D/A converter based programmable filters to get low-pass, high-pass and band-pass functions from the same circuit.

12.8 A/D Converters

After digital-to-analogue converters, the discussion in the following paragraphs is on another vital data conversion integrated circuit component known as the analogue-to-digital (A/D) converter. An A/D converter is a very important building block and has numerous applications. It forms an essential interface when it comes to analysing analogue data with a digital computer. It is an indispensable part of any digital communication system where the analogue signal to be transmitted is digitized at the sending end with the help of an A/D converter. It is invariably used in all digital read-out test and measuring equipment. Be it a digital voltmeter or a laser power meter, or for that matter even a pH meter, an A/D converter is the heart of all of them.

An A/D converter takes at its input an analogue voltage and after a certain amount of time produces a digital output code representing the analogue input. The A/D conversion process is generally more complex than the D/A conversion process. There are various techniques developed for the purpose of A/D conversion, and these techniques have different advantages and disadvantages with respect to one another, which have been utilized in the fabrication of different categories of A/D converter ICs. A D/A converter circuit, as we will see in the following paragraphs, forms a part of some of the types of A/D converter.

We begin with a brief interpretation of the terminology and the major specifications that are relevant to the understanding of A/D converters. The idea is to enable the designers to make a judicious choice of A/D converter suitable for their application. A brief comparative study of different types of A/D converter and the suitability of each one of these types for a given application requirement is also discussed. This is followed by application-relevant information on some of the more popular A/D converter IC type numbers.

12.9 A/D Converter Specifications

The major performance specifications of an A/D converter include resolution, accuracy, gain and offset errors, gain and offset drifts, the sampling frequency and aliasing phenomenon, quantization error, nonlinearity, differential nonlinearity, conversion time, aperture and acquisition times and code width.

Each one of these is briefly described in the following paragraphs.

12.9.1 Resolution

The *resolution* of an A/D converter is the quantum of the input analogue voltage change required to increment its digital output from one code to the next higher code. An n-bit A/D converter can resolve one part in $2^n - 1$. It may be expressed as a percentage of full scale or in bits. The resolution of an eight-bit A/D converter, for example, can be expressed as one part in 255 or as 0.4 % of full scale or simply as eight-bit resolution. If such a converter has a full-scale analogue input range of 10 V, it can resolve a 40 mV change in input.

12.9.2 Accuracy

The *accuracy* specification describes the maximum sum of all errors, both from analogue sources (mainly the comparator and the ladder resistors) and from the digital sources (quantization error) of the A/D converter. These errors mainly include the gain error, the offset error and the quantization error. The accuracy describes the actual analogue input and full-scale weighted equivalent of the output code corresponding to the actual analogue input. The accuracy specification is rarely provided on the datasheets, and quite often several sources of errors are listed separately.

12.9.3 Gain and Offset Errors

The *gain error* is the difference between the actual full-scale transition voltage and the ideal full-scale transition voltage. It is expressed either as a percentage of the full-scale range (% of FSR) or in LSBs. The *offset error* is the error at analogue zero for an A/D converter operating in bipolar mode. It is measured in % of FSR or in LSBs.

12.9.4 Gain and Offset Drifts

The *gain drift* is the change in the full-scale transition voltage measured over the entire operating temperature range. It is expressed in full scale per degree Celsius or ppm of full scale per degree Celsius or LSBs. The *offset drift* is the change with temperature in the analogue zero for an A/D converter operating in bipolar mode. It is generally expressed in ppm of full scale per degree Celsius or LSBs.

12.9.5 Sampling Frequency and Aliasing Phenomenon

If the rate at which the analogue signal to be digitized is sampled is at least twice the highest frequency in the analogue signal, which is what is embodied in the Shannon–Nyquist sampling theorem, then the analogue signal can be faithfully reproduced from its quantized values by using a suitable interpolation algorithm. The accuracy of the reproduced signal is, however, limited by the quantization error (discussed in Section 12.9.6). If the sampling rate is inadequate, i.e. if it is less than the Nyquist rate, then the reproduced signal is not a faithful reproduction of the original signal and these spurious signals, called aliases, are produced. The frequency of an aliased signal is the difference between the signal frequency and the sampling frequency. For example, if sampled at a 1.5 kHz rate, a 2 kHz sine wave would be reconstructed as a 500 Hz sine wave. This problem is called *aliasing* and, in order to avoid it, the analogue input signal is low-pass filtered to remove all frequency components above half the sampling rate. This filter, called an *anti-aliasing filter*, is used in all practical A/D converters.

12.9.6 Quantization Error

The *quantization error* is inherent to the digitizing process. For a given analogue input voltage range it can be reduced by increasing the number of digitized levels. An A/D converter having an n-bit output can only identify 2^n output codes while there are an infinite number of analogue input values adjacent to the LSB of the A/D converter that are assigned the same output code. For instance, if we are digitizing an analogue signal with a peak value of 7 V using three bits, then all analogue voltages equal to or greater than 5.5 V and less than or equal to 6.5 V will be represented by the same output code, i.e. 110 (if the output coding is in straight binary form). The error is ±0.5 V or ±1/2 LSB, as a one-LSB change in the output corresponds to an analogue change of 1 V in this case. The ±1/2 LSB limit to resolution is known as the fundamental quantization error. Expressed as a percentage, the quantization error in an eight-bit converter is one part in 255 or 0.4 %.

12.9.7 Nonlinearity

The *nonlinearity* specification [also referred to as the integral nonlinearity (INL) by some manufacturers] of an A/D converter describes its departure from a linear transfer curve. The nonlinearity error does not include gain, offset and quantization errors. It is expressed as a percentage of full scale or in LSBs.

12.9.8 Differential Nonlinearity

This indicates the worst-case difference between the actual analogue voltage change and the ideal one-LSB voltage change. The DNL specification is as important as the INL specification, as an A/D converter having a good INL specification may have a poor-quality transfer curve if the DNL specification is poor. DNL is also expressed as a percentage of full scale or in LSBs. DNL in fact explains the smoothness of the transfer characteristics and is thus of great importance to the user. Figure 12.27 shows the transfer curve for a three-bit A/D converter with a 7 V full-scale range, 1/4-LSB INL and one-LSB DNL. Figure 12.28 shows the same for a 7 V full-scale range, one-LSB INL and 1/4-LSB DNL. Although the former has a much better INL specification, the latter, with a better DNL specification, has a much better and smoother curve and may thus be preferred. Too high a value of DNL may even grossly degrade the converter resolution. In a four-bit converter

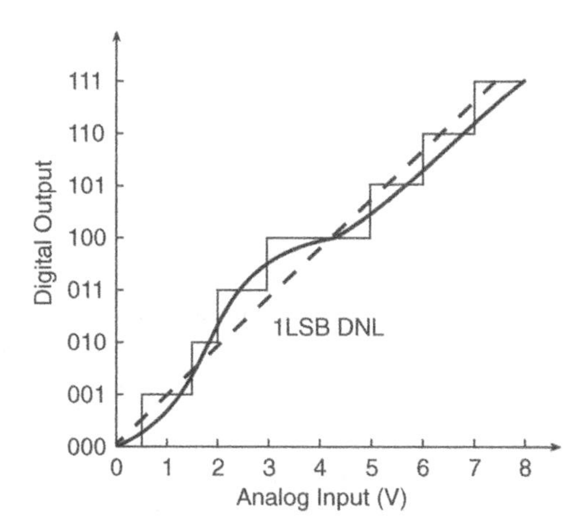

Figure 12.27 Transfer characteristics of a three-bit A/D converter (INL = one LSB, DNL = 1LSB).

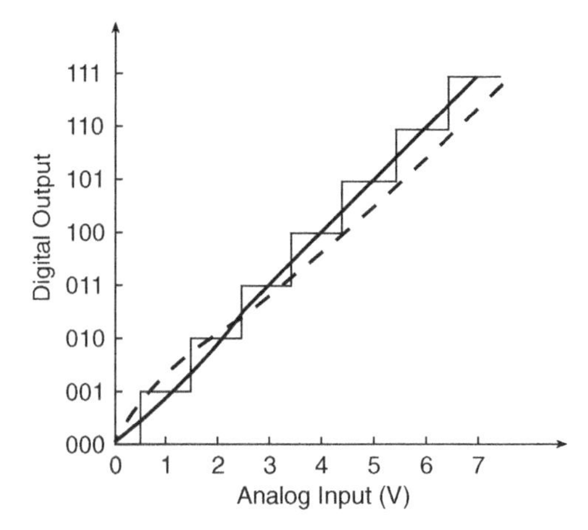

Figure 12.28 Transfer characteristics of a three-bit A/D converter (INL = one LSB, DNL = 1/4 LSB).

with a ±2 LSB DNL, the 16-step transfer curve may be reduced to a six-step curve. The DNL specification should in no case be ignored, unless the INL specification is tight enough to guarantee the desirable DNL.

12.9.9 Conversion Time

This is the time that elapses from the time instant of the start of the conversion signal until the conversion complete signal occurs. It ranges from a few nanoseconds for flash-type A/D converters to a few microseconds for successive approximation type A/D converters and may be as large as tens of milliseconds for dual-slope integrating A/D converters.

12.9.10 Aperture and Acquisition Times

When a rapidly changing signal is digitized, the input signal amplitude will have changed even before the conversion is complete, with the result that the output of the A/D converter does not represent the signal amplitude at the start. A *sample-and-hold* circuit with a buffer amplifier is used at the input of the A/D converter to overcome this problem. The aperture and acquisition times are the parameters of the sample-and-hold circuit. The signal to be digitized is sampled with an electronic switch that can be rapidly turned ON and OFF. The sampled amplitude is then stored on the hold capacitor. The A/D converter digitizes the stored voltage, and, after the conversion is complete, a new sample is taken and held for the next conversion. The *acquisition time* is the time required for the electronic switch to close and the hold capacitor to charge, while the *aperture time* is the time needed for the switch completely to open after the occurrence of the hold signal. Ideally, both times should be zero. The maximum sampling frequency is thus determined by the aperture and acquisition times in addition to the conversion time.

12.9.11 Code Width

The *code width* is the quantum of input voltage change that occurs between the output code transitions expressed in LSBs of full scale. *Code width uncertainty* is the dynamic variation or *jitter* in the code width owing to noise.

12.10 A/D Converter Terminology

Some of the more commonly used terms while interpreting the specifications and salient features of A/D converters are briefly described in the following paragraphs.

12.10.1 Unipolar Mode Operation

In the unipolar mode of operation, the analogue input to the A/D converter varies from 0 to full-scale voltage of one polarity only.

12.10.2 Bipolar Mode Operation

An A/D converter configured to convert both positive and negative analogue input voltages is said to be operating in bipolar mode.

12.10.3 Coding

Coding defines the nature of the A/D converter output data format. Commonly used formats include straight binary, offset binary, complementary binary, 2's complement, low byte and high byte.

12.10.4 Low Byte and High Byte

In A/D converters with a resolution greater than eight bits, some products are offered in high-byte or low-byte format to simplify their interface with eight-bit microprocessor systems. The low-byte output contains the least significant bit and some or all of the lower eight bits of the A/D converter output. In the high byte, the output contains the MSB and some or all of the upper eight bits.

12.10.5 Right-Justified Data, Left-Justified Data

Data bit sets shorter than eight bits are placed in byte-oriented data output format, starting with the right side of the data output transfer register. This could apply to the upper or lower byte. For example, a 12-bit ADC will have four extra bits which could be right justified. Data bit sets shorter than eight bits are placed in left-justified data, starting with the left side of the data output transfer register. This could apply to the lower or upper byte. For example, a 12-bit ADC will have four extra bits which could be left justified.

12.10.6 Command Register, Status Register

The command register is an internal register of the ADC that can be programmed by the user to select various modes of operation such as unipolar or bipolar mode selection, range selection, data output format selection, etc. The status register indicates the current status of the analogue-to-digital conversion with a 'busy' or 'conversion complete' signal.

12.10.7 Control Lines

Digital input/output pins that activate/monitor and control ADC operation are called control lines. Some examples are chip select, write, start convert, conversion complete, etc.

Example 12.5

Determine the resolution of a 12-bit A/D converter having a full-scale analogue input voltage of 5 V.

Solution

- A 12-bit A/D converter resolves the analogue input voltage into $(2^{12} - 1)$ levels.
- The resolution $= 5/(2^{12} - 1) = 5000/(4096 - 1) = (5000/4095) = 1.22mV$.

Example 12.6

The data sheet of a certain eight-bit A/D converter lists the following specifications: resolution eight bits; full-scale error 0.02 % of full scale; full-scale analogue input +5 V. Determine (a) the quantization error (in volts) and (b) the total possible error (in volts).

Solution

(a) The eight-bit A/D converter has $2^8 - 1 = 255$ steps. Therefore, the quantization error = 5/255 = 5000/255 = 19.607 mV.
(b) The full-scale error $= 0.02\% offullscale = 0.02 \times 5000/100 = 1$ mV. Therefore, the total possible error $= 19.607 + 1 = 20.607mV$.

12.11 Types of A/D Converter

Analogue-to-digital converters are often classified according to the conversion process or the conversion technique used to digitize the signal. Based on various conversion methodologies, common types of A/D converter include flash or simultaneous or direct-conversion A/D converters, half-flash A/D converters, counter-type A/D converters, tracking A/D converters, successive approximation type A/D converters, single-slope, dual-slope and multislope A/D converters and sigma-delta A/D converters.

Each of the above-mentioned types of A/D converter is described in the following paragraphs.

12.11.1 Simultaneous or Flash A/D Converters

The simultaneous method of A/D conversion is based on using a number of comparators. The number of comparators needed for n-bit A/D conversion is $2^n - 1$. One such system capable of converting an

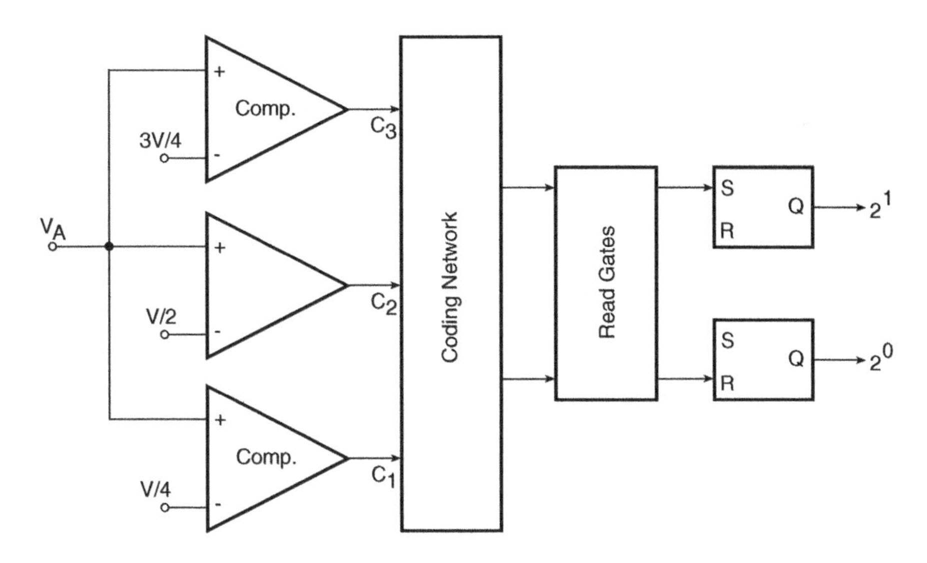

Figure 12.29 Two-bit simultaneous A/D converter.

analogue input signal into a two-bit digital output is shown in Fig. 12.29. The analogue signal to be digitized serves as one of the inputs to each of the comparators. The second input for each of the comparators is a reference input, different for each comparator. The reference voltages to be used for comparators are in general $V/2^n$, $2V/2^n$, $3V/2^n$, $4V/2^n$ and so on. Here, V is the maximum amplitude of the analogue signal that the A/D converter can digitize, and n is the number of bits in the digitized output. In the present case of a two-bit A/D converter, the reference voltages for the three comparators will be $V/4$, $V/2$ and $3V/4$. If we wanted a three-bit output, the reference voltages would have been $V/8$, $V/4$, $3V/8$, $V/2$, $5V/8$, $3V/4$ and $7V/8$. Referring to Fig. 12.29, the output status of various comparators depends upon the input analogue signal V_A. For instance, when the input V_A lies between $V/4$ and $V/2$, the C_1 output is HIGH whereas the C_2 and C_3 outputs are both LOW. The results are summarized in Table 12.1. The three comparator outputs can then be fed to a coding network (comprising logic gates, etc.) to provide two bits that are the digital equivalent of the input analogue voltage. The bits at the output of the coding network can then be entered into a flip-flop register for storage. Figure 12.30 shows the arrangement of a three-bit simultaneous-type A/D converter.

The construction of a simultaneous A/D converter is quite straightforward and relatively easy to understand. However, as the number of bits in the desired digital signal increases, the number of

Table 12.1 Simultaneous or Flash A/D converters.

Input analogue voltage(v_a)	C_1	C_2	C_3	2^1	2^2
0 to $V/4$	LOW	LOW	LOW	0	0
$V/4$ to $V/2$	HIGH	LOW	LOW	0	1
$V/2$ to $3V/4$	HIGH	HIGH	LOW	1	0
$3V/4$ to V	HIGH	HIGH	HIGH	1	1

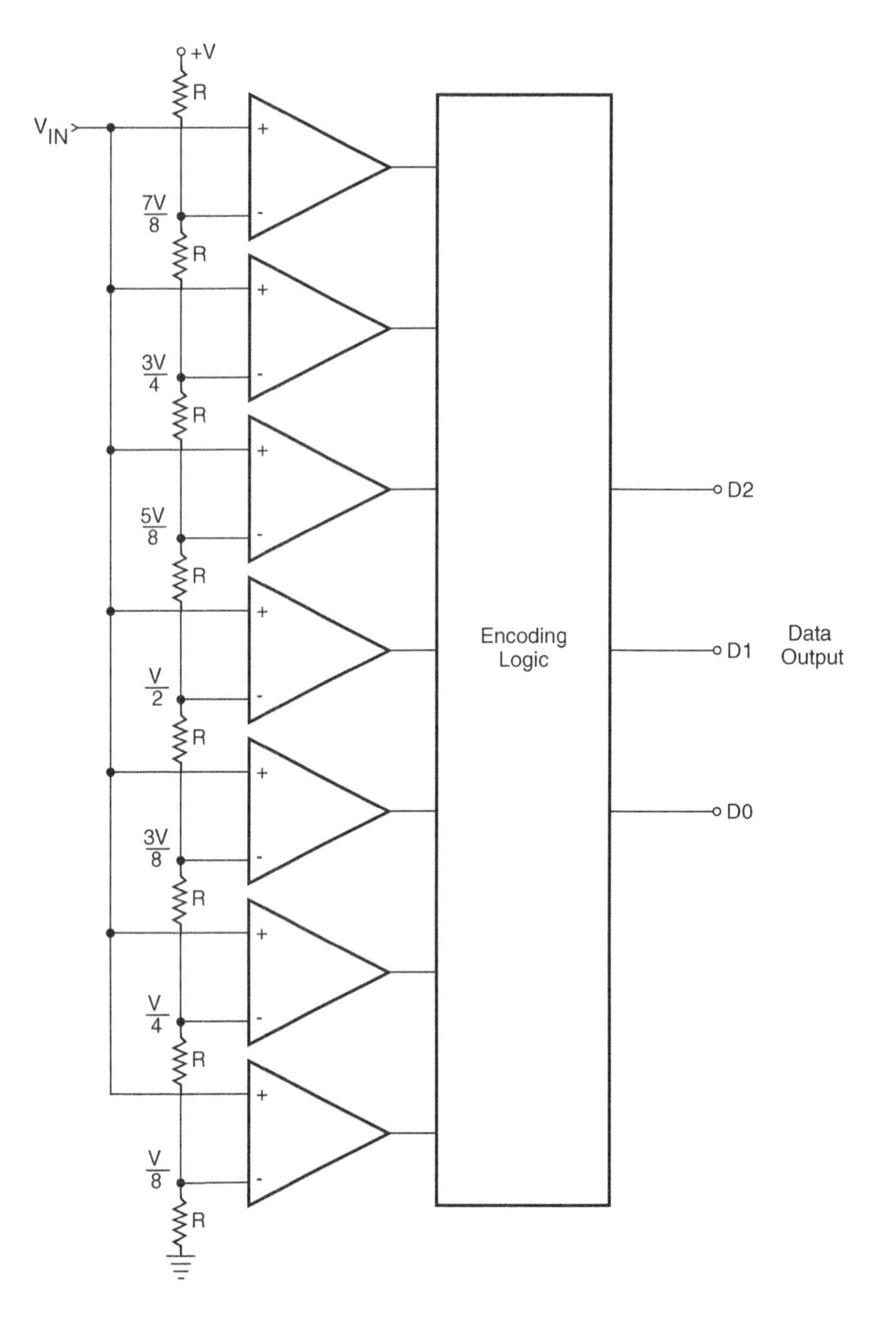

Figure 12.30 Three-bit simultaneous A/D converter.

comparators required to perform A/D conversion increases very rapidly, and it may not be feasible to use this approach once the number of bits exceeds six or so. The greatest advantage of this technique lies in its capability to execute extremely fast analogue-to-digital conversion.

12.11.2 Half-Flash A/D Converter

The *half-flash A/D converter*, also known as the *pipeline A/D converter*, is a variant of the flash-type converter that largely overcomes the primary disadvantage of the high-resolution full-flash converter, namely the prohibitively large number of comparators required, without significantly degrading its high-speed conversion performance. Compared with a full-flash converter of certain resolution, while the number of comparators and associated resistors is drastically reduced in a half-flash converter, the conversion time increases approximately by a factor of 2. For an n-bit flash converter the number of comparators required is $2^n[(2^n - 1)$ for encoding of amplitude and one comparator for polarity], while the same for an equivalent half-flash converter would be $2 \times 2^{n/2}$. In the case of an eight-bit converter, the number is 32 (for half-flash) against 256 (for full flash). How it is achieved is explained in the following paragraphs considering the example of an eight-bit half-flash converter.

A half-flash converter uses two full-flash converters, with each full-flash converter having a resolution equal to half the number of bits of the half-flash converter. That is, an eight-bit half-flash converter uses two four-bit flash converters. In addition, it uses a four-bit D/A converter and an eight-bit latch. Figure 12.31 shows the basic architecture of such a converter. The timing and control circuitry is omitted for the sake of simplicity. The circuit functions as follows.

The most significant four-bit A/D converter converts the input analogue signal into a corresponding four-bit digital code, which is stored in the most significant four bits of the output latch. This four-bit digital code, however, represents the low-resolution sample of the input. Simultaneously, it is converted back into an equivalent analogue signal with a four-bit D/A converter. The approximate value of the analogue signal so produced is then subtracted from the sampled value and the difference is converted

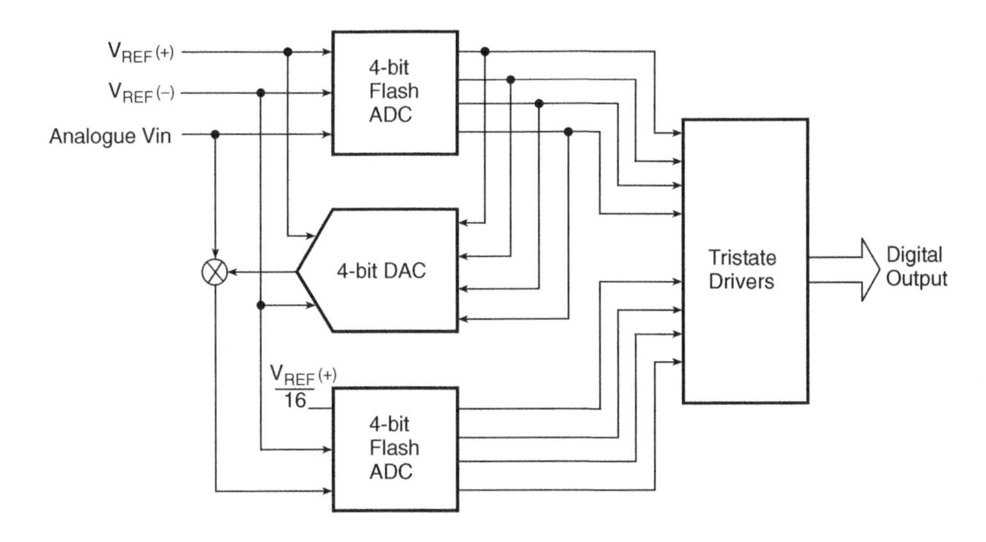

Figure 12.31 Eight-bit half-flash A/D converter.

into digital code using a least significant four-bit A/D converter. The least significant A/D converter is referenced to one-sixteenth (= $1/2^4$) of the reference voltage used by the most significant A/D converter. The new four-bit digital output is stored in the least significant four bits of the output latch. The latch now contains the eight-bit digital equivalent of the analogue input. The digitized output is the same as would be produced by an eight-bit full-flash converter. The only difference is that the conversion process takes a little longer. It may also be mentioned here that the eight-bit half-flash converter can be used either as a four-bit full-flash converter or as an eight-bit half-flash converter. Some half-flash converters use a single full-flash converter and reuse it for both conversions. This is achieved by using additional sample-and-hold circuitry.

12.11.3 Counter-Type A/D Converter

It is possible to construct higher-resolution A/D converters with a single comparator by using a variable reference voltage. One such A/D converter is the *counter-type A/D converter* represented by the block schematic of Fig. 12.32. The circuit functions as follows. To begin with, the counter is reset to all 0s. When a convert signal appears on the start line, the input gate is enabled and the clock pulses are applied to the clock input of the counter. The counter advances through its normal binary count sequence. The counter output feeds a D/A converter and the staircase waveform generated at the output of the D/A converter forms one of the inputs of the comparator. The other input to the comparator is the analogue input signal. Whenever the D/A converter output exceeds the analogue input voltage, the comparator changes state. The gate is disabled and the counter stops. The counter output at that instant of time is then the required digital output corresponding to the analogue input signal.

The counter-type A/D converter provides a very good method for digitizing to a high resolution. This method is much simpler than the simultaneous method for higher-resolution A/D converters. The drawback with this converter is that the required conversion time is longer. Since the counter always begins from the all 0s position and counts through its normal binary sequence, it may require as many as 2^n counts before conversion is complete. The average conversion time can be taken to be $2^n/2 = 2^{n-1}$ counts. One clock cycle gives one count. As an illustration, if we have a four-bit converter and a 1 MHz clock, the average conversion time would be 8 ms. It would be as large as 0.5 ms for a 10-bit converter of this type at a 1 MHz clock rate. In fact, the conversion time doubles for each bit

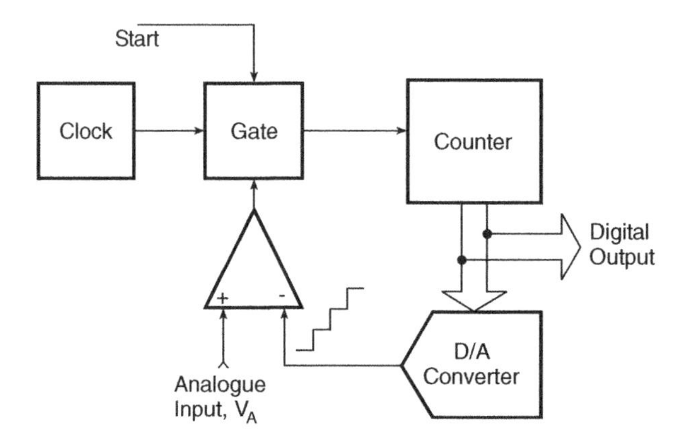

Figure 12.32 Counter-type A/D converter.

added to the converter. Thus, the resolution can be improved only at the cost of a longer conversion time. This makes the counter-type A/D converter unsuitable for digitizing rapidly changing analogue signals.

12.11.4 Tracking-Type A/D Converter

In the counter-type A/D converter described above, the counter is reset to zero at the start of each new conversion. The D/A converter output staircase waveform always begins at zero and increases in steps until it reaches a point where the analogue output of the D/A converter exceeds the analogue input to be digitized. As a result, the counter-type A/D converter of the type discussed above is slow. The *tracking-type A/D converter*, also called the *delta-encoded A/D converter*, is a modified form of counter-type converter that to some extent overcomes the shortcoming of the latter. In the modified arrangement, the counter, which is primarily an UP counter, is replaced with an UP/DOWN counter. It counts in upward sequence whenever the D/A converter output analogue voltage is less than the analogue input voltage to be digitized, and it counts in the downward sequence whenever the D/A converter output analogue voltage is greater than the analogue input voltage. In this type of converter, whenever a new conversion is to begin, the counter is not reset to zero; in fact it begins counting either up or down from its last value, depending upon the comparator output. The D/A converter output staircase waveform contains both positive-going and negative-going staircase signals that track the input analogue signal.

12.11.5 Successive Approximation Type A/D Converter

The development of A/D converters has progressed in a quest to reduce the conversion time. The successive approximation type A/D converter aims at approximating the analogue signal to be digitized by trying only one bit at a time. The process of A/D conversion by this technique can be illustrated with the help of an example. Let us take a four-bit successive approximation type A/D converter. Initially, the counter is reset to all 0s. The conversion process begins with the MSB being set by the start pulse. That is, the flip-flop representing the MSB is set. The counter output is converted into an equivalent analogue signal and then compared with the analogue signal to be digitized. A decision is then taken as to whether the MSB is to be left in (i.e. the flip-flop representing the MSB is to remain set) or whether it is to be taken out (i.e. the flip-flop is to be reset) when the first clock pulse sets the second MSB. Once the second MSB is set, again a comparison is made and a decision taken as to whether or not the second MSB is to remain set when the subsequent clock pulse sets the third MSB. The process continues until we go down to the LSB. Note that, every time we make a comparison, we tend to narrow down the difference between the analogue signal to be digitized and the analogue signal representing the counter count. Refer to the operational diagram of Fig. 12.33. It is clear from the diagram that, to reach any count from 0000 to 1111, the converter requires four clock cycles. In general, the number of clock cycles required for each conversion will be n for an n-bit A/D converter of this type.

Figure 12.34 shows a block schematic representation of a successive approximation type A/D converter. Since only one flip-flop (in the counter) is operated upon at one time, a ring counter, which is nothing but a circulating register (a serial shift register with the outputs Q and $\overline{Q}$ of the last flip-flop connected to the J and K inputs respectively of the first flip-flop), is used to do the job. Referring to Fig. 12.33, the dark lines show the sequence in which the counter arrives at the desired count, assuming that 1001 is the desired count. This type of A/D converter is much faster than the counter-type A/D converter previously discussed. In an n-bit converter, the counter-type A/D converter on average would

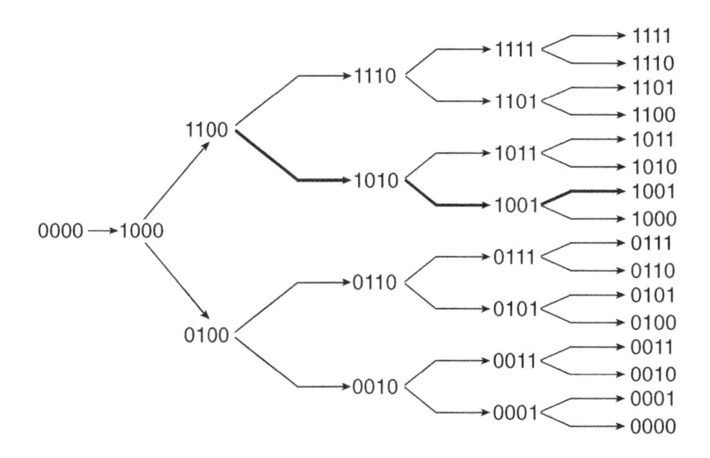

Figure 12.33 Conversion process in a successive approximation type A/D converter.

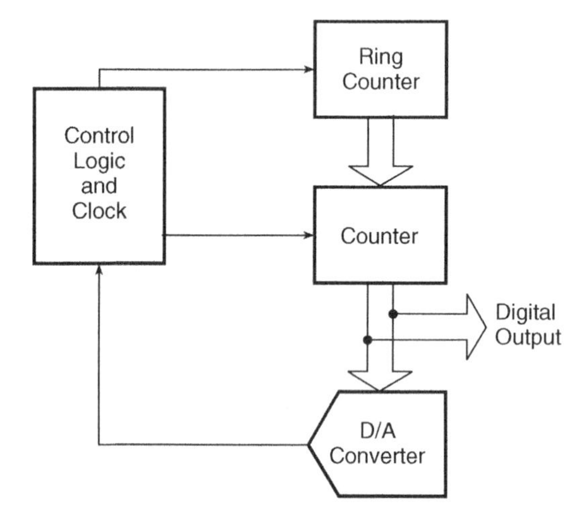

Figure 12.34 Block schematic representation of a successive-approximation A/D converter.

require 2^{n-1} clock cycles for each conversion, whereas a successive approximation type converter requires only n clock cycles. That is, an eight-bit A/D converter of this type operating on a 1 MHz clock has a conversion time of 8 μs.

12.11.6 Single-, Dual- and Multislope A/D Converters

Figure 12.35 shows a block schematic representation of a *single-slope A/D converter*. In this type of converter, one of the inputs to the comparator is a ramp of fixed slope, while the other input is the analogue input to be digitized. The counter and the ramp generator are initially reset to 0s. The

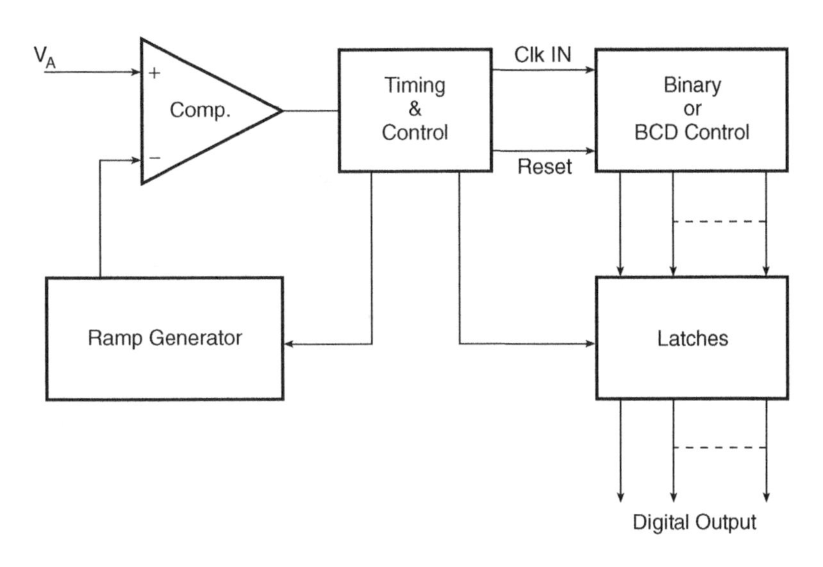

Figure 12.35 Block schematic representation of a single-slope A/D converter.

counter starts counting with the first clock cycle input. The ramp is also synchronized to start with the first clock input. The counter stops when the ramp amplitude equals the analogue input. In this case, the counter count is directly proportional to the analogue signal. It is a low-cost, reasonably high-accuracy converter but it suffers from the disadvantage of loss of accuracy owing to changes in the characteristics of the ramp generator. This shortcoming is overcome in a dual-slope integrating-type A/D converter.

Figure 12.36 shows a block schematic arrangement of a *dual-slope integrating A/D converter*. The converter works as follows. Initially, switch S is connected to the analogue input voltage V_A to be digitized. The output of the integrator is mathematically given by

$$v_o = (-1/RC) \int V_A.dt = (-V_A/RC).t \qquad (12.17)$$

The moment v_o tends to go below zero, clock pulses reach the clock input terminal of the counter which is initially cleared to all 0s. The counter begins counting from 0000. . . 0. At the (2^n)th clock pulse, the counter is again cleared, the '1' to '0' transition of the MSB of the counter sets a flip-flop that controls the state of switch S which now connects the integrator input to a reference voltage of polarity opposite to that of the analogue input. The integrator output moves in the positive direction; the counter has again started counting after being reset (at, say, $t=T_1$). The moment the integrator output tends to exceed zero, the counter stops as the clock pulses no longer reach the clock input of the counter. The counter output at this stage (say, at $t=T_2$) is proportional to the analogue input. Mathematically, it can be proved that $n=(V_A/V_R).2^n$, where n is the count recorded in the counter at $t=T_2$. Figure 12.37 illustrates the concept further with the help of relevant waveforms. This type of A/D converter is very popular in digital voltmeters owing to its good conversion accuracy and low cost. Also, the accuracy is independent of both the integrator capacitance and the clock frequency, as they affect the negative and positive slope in the same manner. Yet another advantage of the dual-slope integrator A/D converter is that the fixed analogue input integration period results in rejection of noise frequencies present

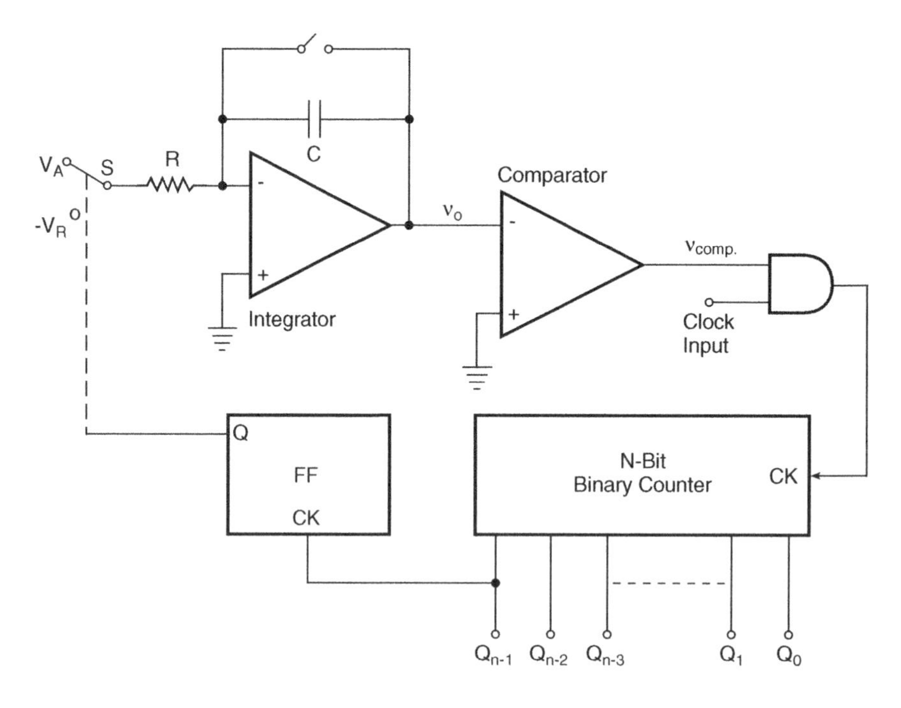

Figure 12.36 Block schematic representation of a dual-slope A/D converter.

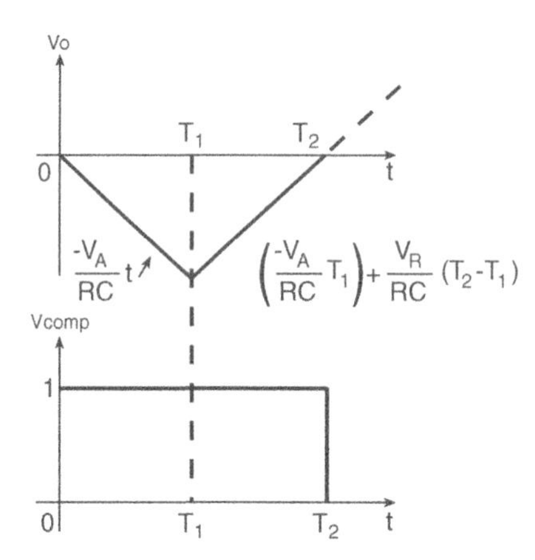

Figure 12.37 Relevant waveforms in a dual-slope A/D converter.

in the analogue input and having time periods that are equal to or submultiples of the integration time. The proper choice of integration time can therefore achieve excellent rejection of 50/60 Hz line ripple.

There are also multislope converter architectures that are aimed at further enhancing the performance of integrating A/D converters. For example, the *triple-slope architecture* is used to increase the conversion speed at the cost of added complexity. Increase in conversion speed is accomplished by carrying out integration from reference voltage at two distinct rates, a high-speed rate and a low-speed rate. The counter is also divided into two sections, one for MSB bits and the other for LSB bits. A properly designed triple-slope converter achieves increased conversion speed without compromising the inherent linearity, differential linearity and stability characteristics of the dual-slope converter.

Bias currents, offset voltages and gain errors associated with operational amplifiers used as integrators and comparators do introduce some errors. These can be cancelled by using additional charge/discharge cycles and then using the results to correct the initial measurement. One such A/D converter is the *quad-slope converter* which uses two charge/discharge cycles as compared with one charge/discharge cycle in the case of the dual-slope converter. Quad-slope A/D converters have a much higher accuracy than their dual-slope counterparts.

12.11.7 Sigma-Delta A/D Converter

The sigma-delta A/D converter employs a different concept from what has been discussed so far for the case of various types of A/D converter. While the A/D converters covered so far rely on sampling of the analogue signal at the Nyquist frequency and encode the absolute value of the sample, in the case of a sigma-delta converter, as explained in the following paragraphs, the analogue signal is oversampled by a large factor (i.e. the sampling frequency is much larger than the Nyquist value), and also it is not the absolute value of the sample but the difference between the analogue values of two successive samples that is encoded by the converter.

In the case of the A/D converters discussed prior to this and sampled at the Nyquist rate f_s, the RMS value of the quantization noise is uniformly distributed over the Nyquist band of DC to $f_s/2$, as shown in Fig. 12.38(a). The signal-to-noise ratio for a full-scale sine wave input in this case is given by $S/N = (6.02n + 1.76)$ dB, n being the number of bits. The only way to increase the signal-to-noise ratio is by increasing the number of bits. On the other hand, a sigma-delta converter attempts to enhance the signal-to-noise ratio by oversampling the analogue signal, which has the effect of spreading the noise spectrum over a much larger bandwidth and then filtering out the desired band. If the analogue signal were sampled at a rate of Kf_s, the quantization noise would be spread over DC to $Kf_s/2$, as shown in Fig. 12.38(b). K is a constant referred to as the oversampling ratio. The enhanced S/N ratio means higher resolution, which is achieved by other types of A/D converter by way of increasing the number of bits.

It may be mentioned here that, if we simply use oversampling to improve the resolution, it would be required to oversample by a factor of 2^{2N} to achieve an N-bit increase in resolution. The sigma-delta converter does not require to be oversampled by such a large factor because it not only limits the signal pass band but also shapes the quantization noise in such a way that most of it falls outside this pass band, as shown in Fig. 12.38(c). The following paragraphs explain the operational principle of the sigma-delta A/D converter.

The heart of the sigma-delta converter is the delta modulator. Figure 12.39 shows a block schematic representation of a delta modulator, which is basically a one-bit quantizer of the flash type (single comparator). The output of the delta modulator is a bit stream of 1s and 0s, with the number of 1s relative to the number of 0s over a given number of clock cycles indicating the amplitude of the analogue signal over that time interval. An all 1s sequence over a given interval corresponds to the maximum positive amplitude, and an all 0s sequence indicates the maximum negative amplitude. An equal number of 1s and 0s indicates a zero amplitude. Other values between the positive and negative maxima are indicated by a proportional number of 1s relative to the number of 0s. This is further illustrated in Fig. 12.40.

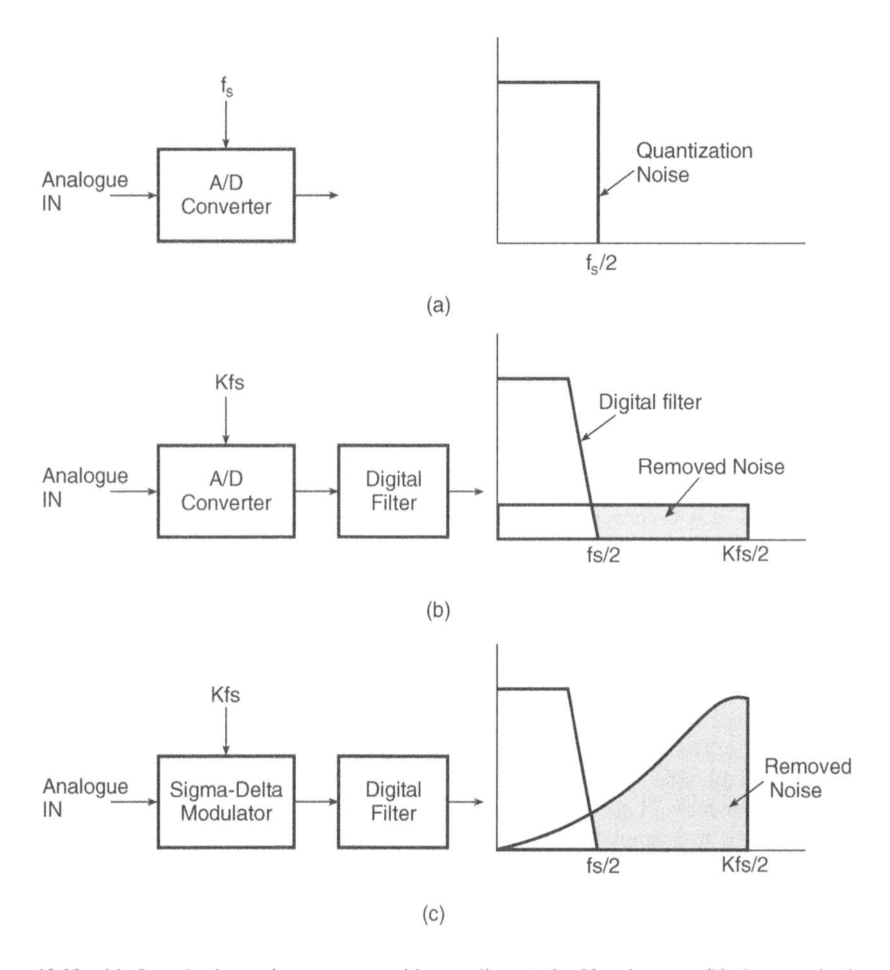

Figure 12.38 (a) Quantization noise spectrum with sampling at the Nyquist rate, (b) the quantization noise spectrum with oversampling and (c) the quantization noise spectrum with oversampling in a sigma-delta converter.

Coming back to the delta modulator (Fig. 12.39), the input to the one-bit quantizer, which is basically a comparator, is from the output of an integrator. The integrator in turn is fed from the difference between the analogue input signal and the analogue equivalent of the quantized output produced by a one-bit D/A converter. A one-bit D/A converter is nothing but a two-way switch that feeds either $+V_{ref}$ or $-V_{ref}$ to the summing point, depending upon the bit status at its input. The negative feedback loop ensures that the average value of the D/A converter output nearly equals the analogue input so as to produce a near-zero input to the integrator.

An increase in analogue signal amplitude produces a larger number of 1s at the quantizer output and consequently a higher average value of the analogue signal at the D/A converter output. This means that the number of 1s in the quantizer output bit stream over a given time interval represents the analogue signal amplitude. The single-bit data stream can then be encoded into the desired output format. One simple way to do this could be to use a counter to count the number of 1s in the data stream over fixed

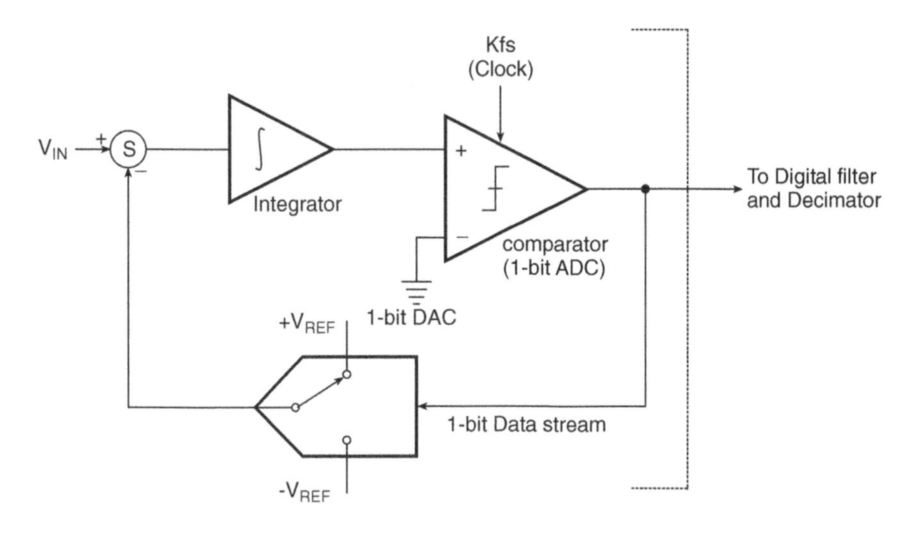

Figure 12.39 Block schematic representation of a delta modulator.

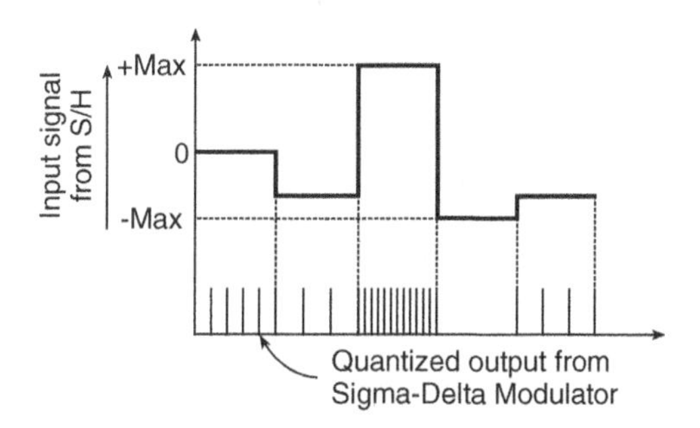

Figure 12.40 Generation of a one-bit data stream.

intervals of time, in which case the counter output would represent the digital equivalent of the analogue signal over those time intervals. Practical sigma-delta A/D converters use a digital decimation filter at the output of the delta modulator to process the one-bit data stream to produce an output in the desired format.

Sigma-delta A/D converters are widely used for contemporary voice-band, audio and high-resolution precision industrial measurement applications. Their highly digital architecture is ideally suited for such applications as it allows easy addition of digital functionality without significantly increasing the cost. AD 1871 from Analog Devices is one such high-performance A/D converter of sigma-delta architecture intended for digital audio applications.

Example 12.7

Determine the conversion time of a 12-bit A/D converter of the counter type shown earlier in Fig. 12.32 for an input clock frequency of 1 MHz.

Solution

- The counter-type A/D converter shown in Fig. 12.32 has a variable conversion time that is maximum when the input analogue voltage is just below the full-scale analogue input voltage.
- An average conversion time equal to half the maximum conversion time is usually defined in the case of such converters.
- The maximum conversion time equals the time taken by $2^{12} - 1 = 4095$ cycles of clock input.
- The clock time period $= 1/(1 \times 10^6) = 1\ \mu s$.
- Therefore, the maximum conversion time $= 4095 \times 1 = 4095\ \mu s = 4.095$ ms.
- The average conversion time $= (4.095/2) = 2.047$ ms.

Example 12.8

The D/A converter of a counter-type A/D converter (refer to Fig. 12.32) produces a staircase output having a step size of 10 mV. The A/D converter has a 10-bit resolution and is specified to have a quantization error of $\pm 1/2$ LSB. Determine the digital output for an analogue input of 4.012 V. Assume that the comparator has a comparison threshold of 1 mV.

Solution

- The comparator has a comparison threshold of 1 mV.
- With reference to Fig. 12.32, this implies that, for the comparator to change state, the voltage at the relevant input should be 1 mV more than the voltage at the other input.
- Now, one of the inputs to the comparator is the analogue input voltage (= 4.012 V in the present case).
- The other input to the comparator is a voltage that is equal to the sum of the D/A converter output voltage and a fixed voltage corresponding to 1/2 LSB.
- This is the case when the quantization error of the A/D converter is specified to be $\pm$ 1/2 LSB.
- In the case of a quantization error of one LSB, the D/A converter directly feeds the other input of the comparator.
- In the present case, one LSB corresponds to 10 mV.
- Therefore, 1/2 LSB corresponds to 5 mV.
- For an analogue input of 4.012 V, the voltage at the other input needs to be 4.013 V (owing to the comparator threshold of 1 mV).
- This implies that the D/A converter output needs to be 4.008 V.
- Therefore, the number of steps $= 4.008/(10 \times 10^{-3}) = 400.8 = 401$.
- The digital output is the binary equivalent of $(401)_{10}$,which equals 0110010001.

Example 12.9

A 10-bit A/D converter of the successive approximation type has a resolution (or quantization error) of 10 mV. Determine the digital output for an analogue input of 4.365 V.

Solution

- In the case of a successive approximation type A/D converter, the final analogue output of its D/A converter portion always settles at a value below the analogue input voltage to be digitized within the resolution of the converter.
- The analogue input voltage = 4.365 V.
- The resolution = 10 mV.
- The number of steps = $4.365/(10 \times 10^{-3}) = 436.5$.
- Step number 436 will produce a D/A converter output of $436 \times 10 = 4360$mV = 4.36V, and step number 437 will produce a D/Aconverter of 4.37 V.
- The A/D converter will settle at step 436.
- The digital output will be the binary equivalent of$(436)_{10}$ which is 0110110100.

Note. When this converter actually performs the conversion, in the tenth clock cycle, the LSB will be set to '1' initially. This would produce a D/A converter output of 4.37 V which exceeds the analogue input voltage of 4.365 V. The comparator changes state, which in turn resets the LSB to '0', bringing the D/A converter output to 4.36 V. This is how a converter of this type settles where a D/A converter output settles at a value that is one step below the value that makes it exceed the analogue input to be digitized.

Example 12.10

Compare the average conversion time of an eight-bit counter-type A/D converter with that of an eight-bit successive approximation type A/D converter if both are working at a 10 MHz clock frequency.

Solution

- The clock time period = 0.1 μs.
- The average conversion time in the case of a counter-type A/D converter is given by $[(2^8 - 1)/2] \times 0.1 = 12.75\mu s$.
- The conversion time in the case of a successive approximation type A/D converter is given by $8 \times 0.1 = 0.8$ μs.

12.12 Integrated Circuit A/D Converters

This section presents application-relevant information of some of the popular A/D converter IC type numbers, as it is not possible to give a detailed description of each one of them. The type numbers included for this purpose are ADC 0800, ADC 0808, ADC 80, ADC 84, ICL 7106/ICL 7107 and AD 7820.

12.12.1 ADC-0800

ADC-0800 is a successive approximation type eight-bit A/D converter. The internal architecture of ADC-0800 is shown in Fig. 12.41. The digital output is in complementary form and is also tristate to permit bussing on common data lines. Its salient features include ratiometric conversion, no missing codes, tristate outputs and a conversion time of 50 μs (typical), ±1-LSB linearity and a clock frequency range of 50–800 kHz.

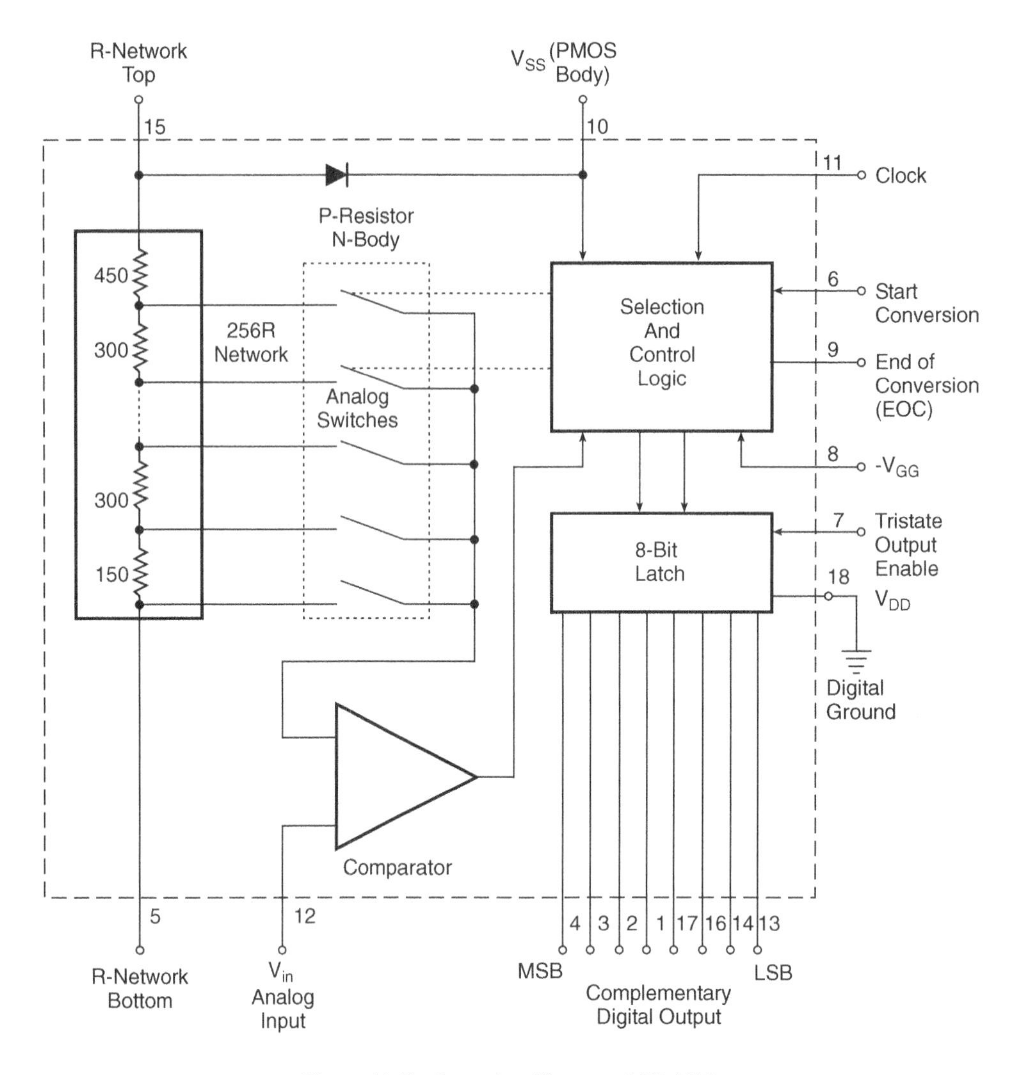

Figure 12.41 Internal architecture of AD 0800.

Figures 12.42(a) and (b) show application circuits using AD-0800. Figure 12.42(a) shows typical circuit connections for a ±5 V input voltage range and TTL-compatible output levels, whereas Fig. 12.42(b) shows the connections for a 0–10 V input range and 0–10 V output levels.

12.12.2 ADC-0808

ADC 0808 is an eight-bit CMOS successive approximation type A/D converter. The device has an eight-channel multiplexer and a microprocessor-compatible control logic. Salient features of the device include eight-bit resolution, no missing codes, a conversion time of 100 μs (typical),

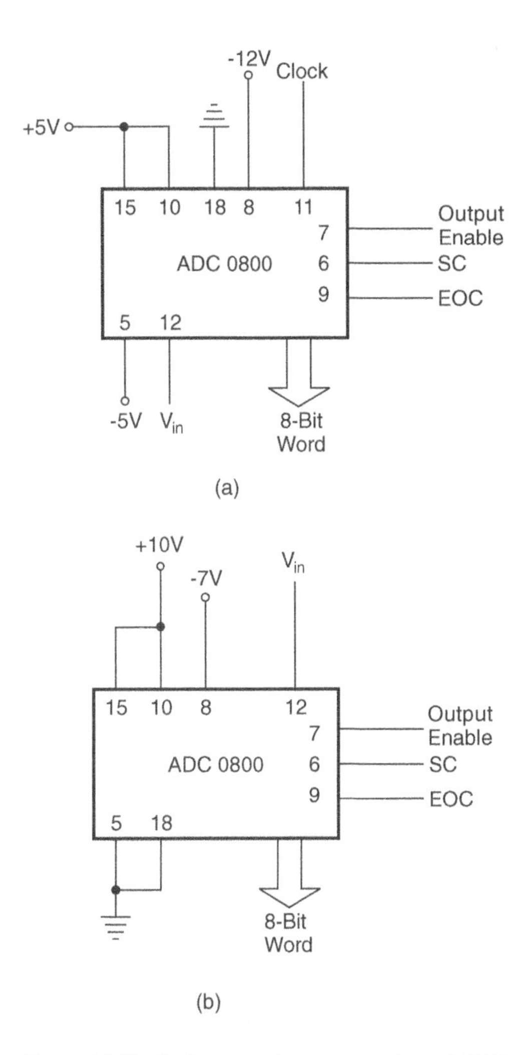

Figure 12.42 Basic application circuits using AD 0800.

stand-alone operation or easy interface to all microprocessors, a 0–5 V analogue input range with a single 5 V supply and latched tristate outputs. Figure 12.43 shows the internal architecture of the device.

12.12.3 ADC-80/AD ADC-80

ADC-80/AD ADC-80 is a 12-bit A/D converter of the successive approximation type. It has an on-chip clock generator, reference and comparator. AD ADC80 is pin-to-pin compatible with industry-standard ADC-80. Figure 12.44 shows the internal block schematic/pin connections of AD-ADC-80. The salient features of the device include low cost, ±0.012 % linearity, a conversion time of 25 μs (max.),

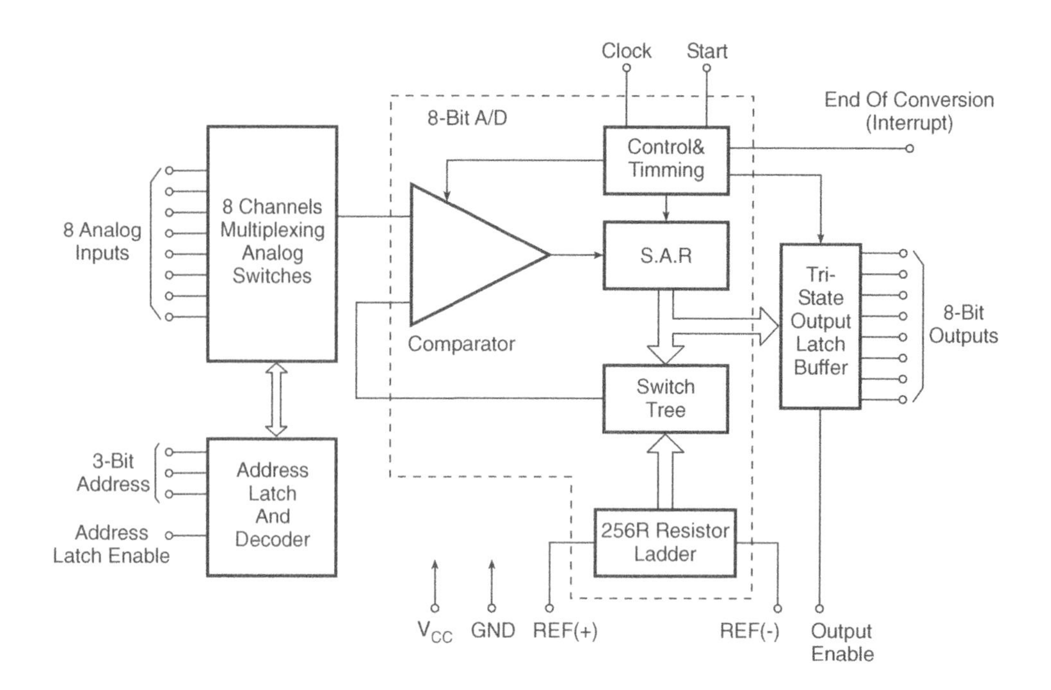

Figure 12.43 Internal architecture of AD 0808.

± 12 V or ± 15 V operation, guaranteed no missing codes over a temperature range from -25 °C to $+85$ °C and a maximum power dissipation of 595 mW.

12.12.4 ADC-84/ADC-85/AD ADC-84/AD ADC-85/AD-5240

ADC-84 and ADC-85 families of 10-bit (ADC 84-10 and ADC 85-10) and 12-bit (ADC 84-12 and ADC 85-12) converters are complete A/D converters like the industry-standard ADC-80, with an internal clock (1.9 MHz in the case of the 10-bit converters and 1.35 MHz in the case of the 12-bit converters), comparator, reference (6.3 V) and input buffer amplifier. These have a conversion time of 10 μs (for 12-bit operation) and 6 μs (for 10-bit operation). Figure 12.45 shows an internal block schematic/pin connection diagram of ADC-84/ADC-85/AD-5240.

12.12.5 AD 7820

AD 7820 is a μP-compatible, eight-bit A/D converter built around half-flash architecture. It incorporates internal sample-and-hold circuitry, which eliminates the need for an external sample-and-hold circuit for signals having slew rates of less than 100 mV/μs. Figure 12.46 shows the internal architecture/pin connection diagram of AD 7820. Other features include a 1.36 μs conversion time, a single +5 V supply and tristate buffered outputs.

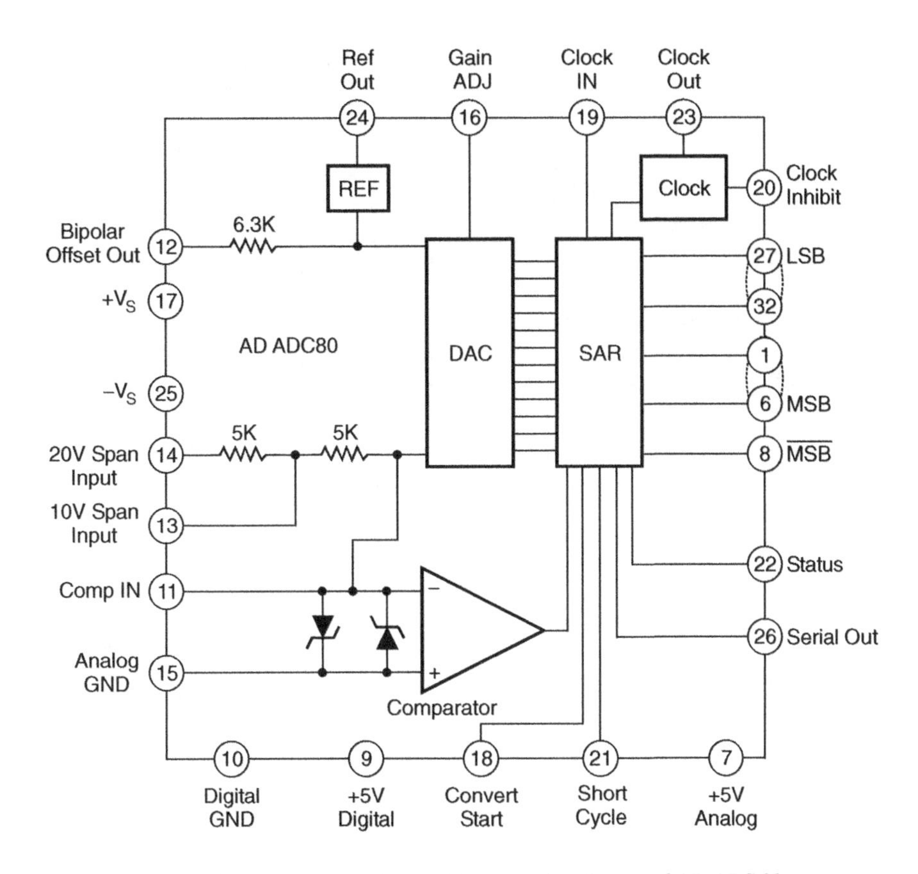

Figure 12.44 Internal architecture/pin connection diagram of AD ADC 80.

12.12.6 ICL 7106/ICL 7107

ICL 7106 and ICL 7107 are the most commonly used A/D converter ICs for digital panel meter (DPM) and digital voltmeter (DVM) applications, with the former used with LCD and the latter used with LED displays. The two types are high-performance, low-power A/D converter ICs of the dual-slope integrating type from Intersil, containing all the necessary building blocks such as a clock generator, a reference, seven segment decoders, display drivers, etc., for directly driving seven segment displays. Figure 12.47 shows the pin connection diagram of ICL 7106/7107 in a dual in-line package. Notice that pin-21 in the case of ICL 7106 is the back plane drive pin, whereas in the case of ICL 7107 it is the ground pin.

Salient features include low cost, low power consumption (typically less than 10 mW), low noise (less than 15 μV peak to peak), true polarity at zero for precise null detection, true differential input and reference, a rollover error of less than one count and so on. The reference voltage is set to be half the full-scale analogue input. For a maximum analogue input of more than what is acceptable at analogue input terminals (±4 V for ±5 V supplies), the input should be scaled down by a factor of 10. The scale-down factors are 100 and 1000 for $(20 < V_{in} \leq 200)$ V and $(200 < V_{in} \leq 2000)$ V respectively.

Figures 12.48 and 12.49 show the basic application circuits of ICL 7106 and ICL 7107 respectively. ICL 7106 operates from a single supply (9 V in the circuit shown), whereas ICL 7107 operates from

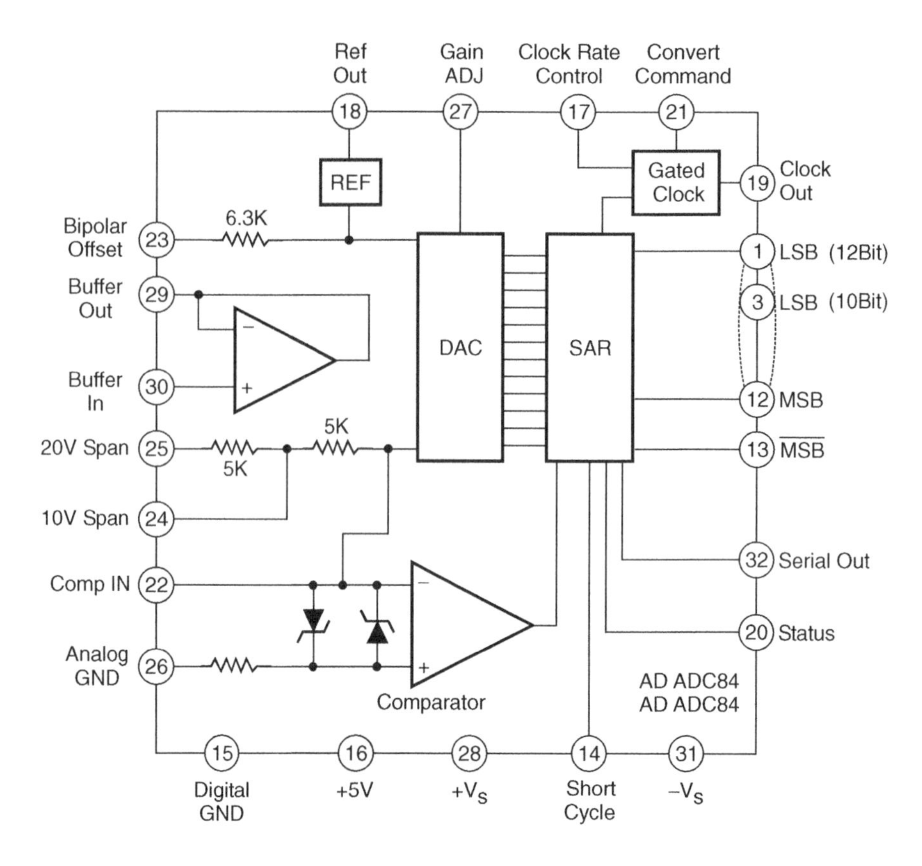

Figure 12.45 Internal architecture/pin connection diagram of AD ADC-84/ADC-85/AD-5240.

dual supplies of ±5 V. The following guidelines should be adhered to when selecting the values of the components for these circuits.

1. The integrating resistor R_1 should be large enough to remain in the linear region over the input voltage range but small enough for undue leakage current requirements not to be placed on the PC board. A value of 470 Ω is the optimum for a 2 V scale. For a 200 mV scale, 47 Ω should be used.
2. For a conversion rate of three readings per second (48 kHz clock), the nominal value of the integrating capacitor C_7 is 0.22 μF. A capacitor with low dielectric absorption should be used to prevent rollover errors. Polypropylene or polycarbonate capacitors should be preferred. If the oscillator frequency is different, C_7 should be changed in inverse proportion in order to maintain the same output swing.
3. Capacitor C_8, the auto zero capacitor, influences the noise of the system. For a 200 mV full scale, where the system noise is critical, a 0.47 μF capacitor is recommended for C_8. A smaller-value capacitor can be used on larger scales. For instance, 0.047 μF would do for a 2 V full scale. A smaller auto zero capacitor has the additional advantage of a faster recovery from overload condition.

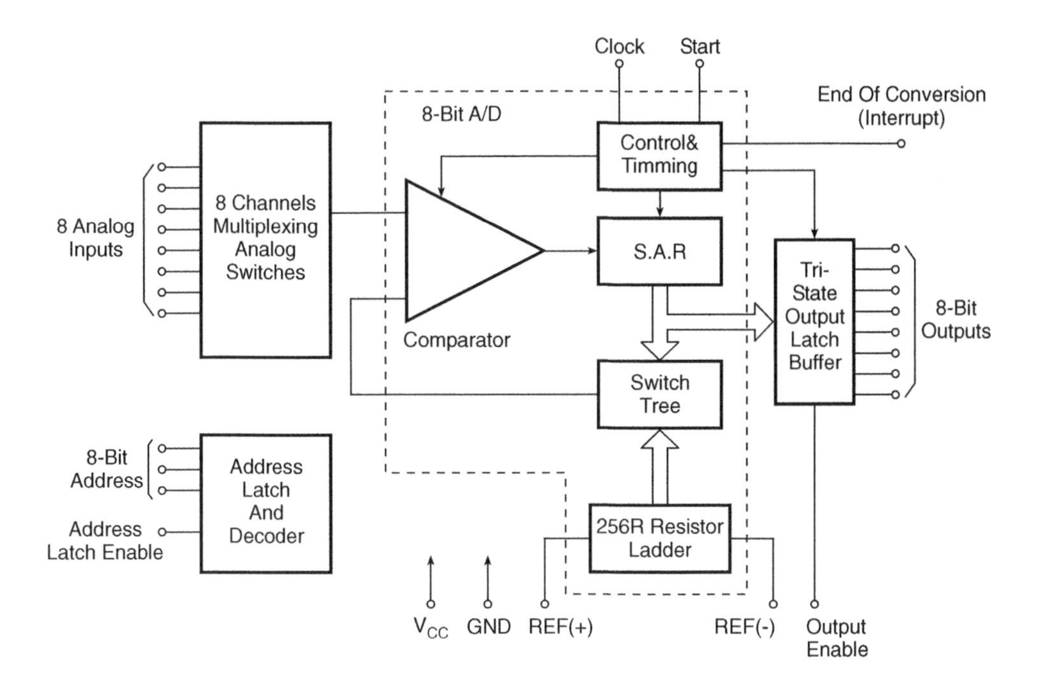

Figure 12.46 Internal architecture/pin connection diagram of AD 7820.

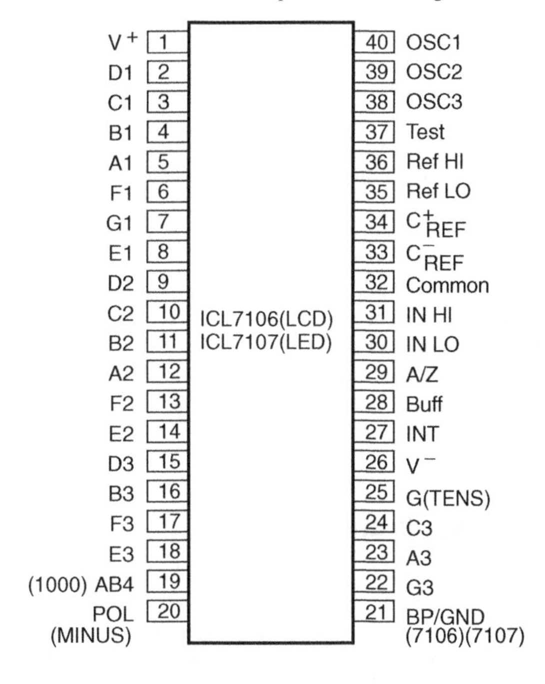

Figure 12.47 Pin connection diagram of ICL 7106/7107.

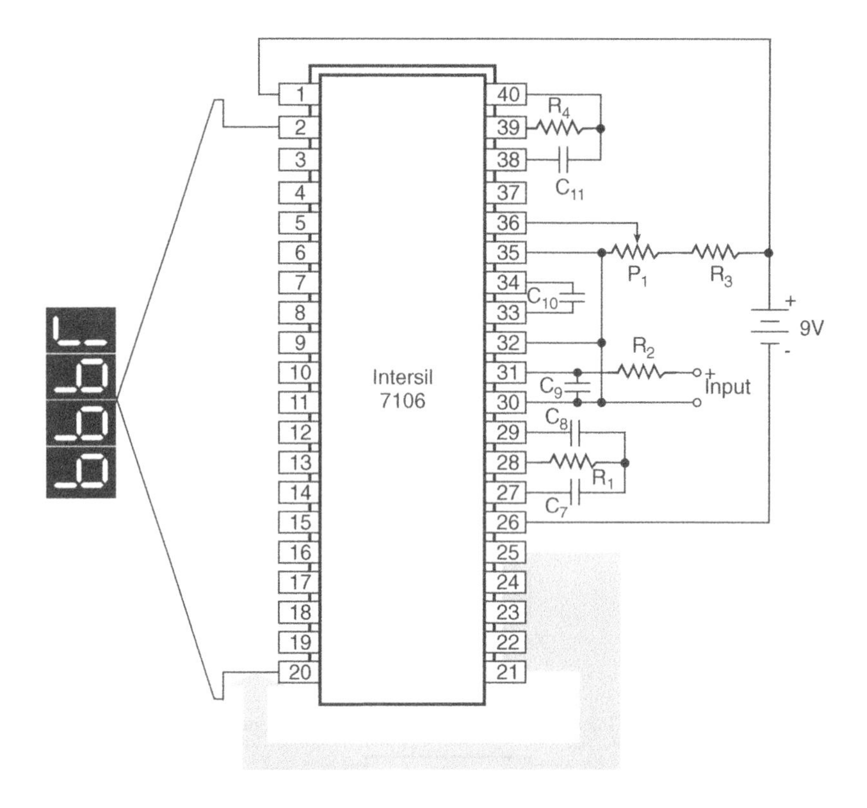

Figure 12.48 Application circuit using ICL 7106.

4. A 0.1 μF capacitor generally works well as the reference capacitor C_{10} connected between pins 33 and 34. However, if the REF/LO (pin 35) is not at analogue common (pin 30) and a 200 mV scale is being used, a larger value is generally required to prevent rollover error. A 1 μF capacitor will hold the rollover error in this case to 0.5 count.
5. The oscillator frequency is given by $f = [0.45/(R_4.C_{11})]$. R_4 is selected to be 100 Ω. C_{11} is computed from the equation for a known value of oscillator frequency. For $f = 48$ kHz (three readings per second), C_{11} turns out to be 100 pF.
6. The reference voltage V_{ref} is selected on the basis of the analogue input required to generate a full-scale output of 2000 counts and is $V_{in}/2$. It will be 100 mV for a 200 mV full scale and 1 V for a 2 V full scale.

12.13 A/D Converter Applications

Like D/A converters, A/D converters have numerous applications. A/D converters are used in virtually all those applications where the analogue signal is to be processed, stored or transported in digital form. They form an essential interface when it comes to analysing analogue data with a digital computer, the process being known as 'data acquisition'. They are an indispensable component of any digital communication system where the analogue signal to be transmitted is

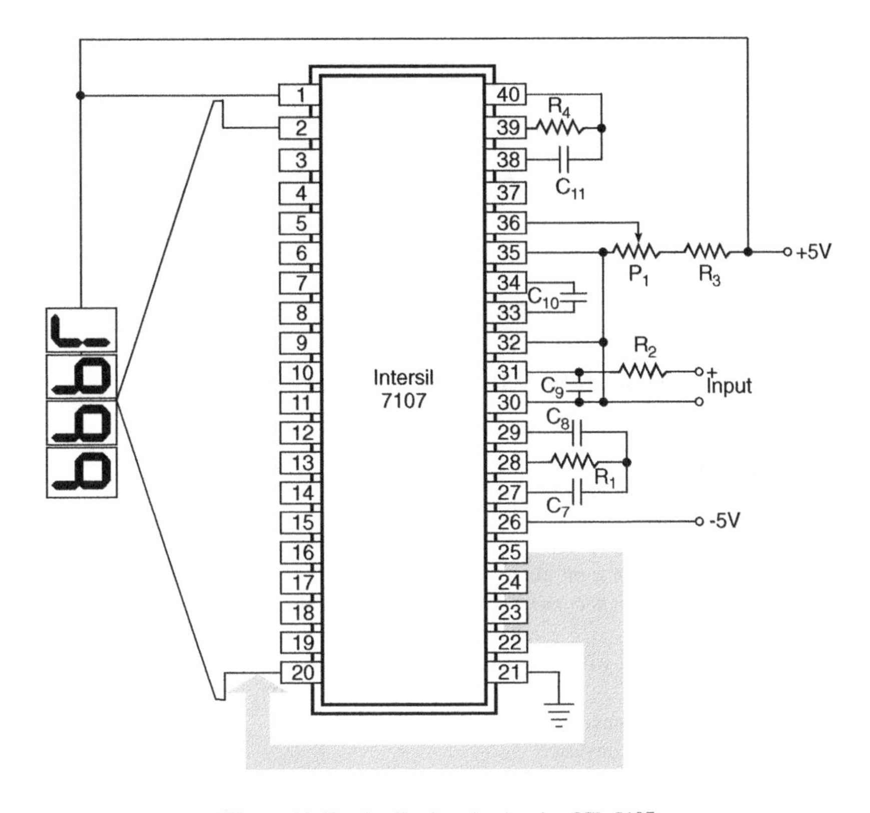

Figure 12.49 Application circuit using ICL 7107.

digitized at the sending end with an A/D converter. They are invariably used in all digital read-out test and measuring equipment such as digital multimeters (DMMs), digital storage oscilloscopes (DSOs), etc. Also, A/D converters are integral to contemporary music reproduction technology, as most of it is done on computers. In the case of analogue recording too, an A/D converter is needed to create the PCM data stream that goes onto a compact disc. While digital test and measurement instruments are discussed in detail in Chapter 16, the use of A/D converters for data acquisition, which forms the basis of most other applications, is discussed in the next section.

12.13.1 Data Acquisition

There are a large number of applications where an analogue signal is digitized to be subsequently stored or processed in a digital computer. The computer may store the data to be later passed on to a D/A converter to reconstruct the original signal, as in a digital-storage oscilloscope. It may process the digitized signal to generate the desired outputs in a process control application. Figure 12.50 shows the basic data acquisition building block. The computer generates a start-of-conversion signal. At the time instant of occurrence of the end-of-conversion signal generated by the A/D converter, the computer loads the digital output of the A/D converter onto its memory.

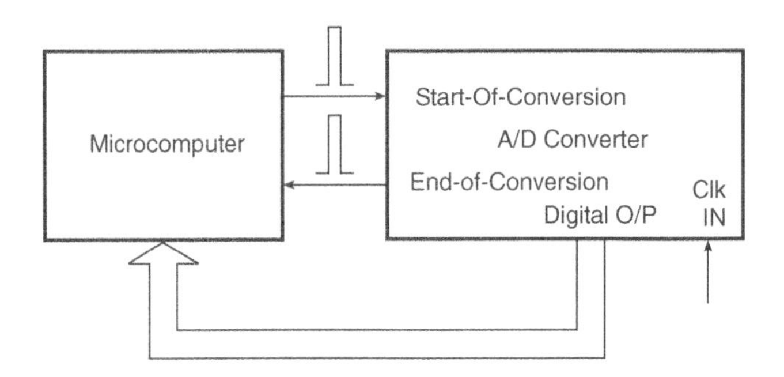

Figure 12.50 A/D converter for data acquisition.

Review Questions

1. Briefly describe the process of digital-to-analogue conversion in a binary ladder network. How does it differ from the simple resistive network used for the same purpose? Write an expression for the output analogue voltage for an n-bit binary ladder network.
2. Briefly describe the *resolution*, *accuracy*, *conversion time* and *monotonocity* specifications of a D/A converter.
3. Briefly describe the following with reference to D/A converters:

   (a) a multiplying-type D/A converter;
   (b) a companding-type D/A converter;
   (c) the current steering mode of operation;
   (d) the voltage switching mode of operation.

4. With reference to A/D converters, differentiate between:

   (a) resolution and accuracy;
   (b) nonlinearity (NL) and differential nonlinearity (DNL).

5. Briefly describe the principle of operation of a simultaneous or flash-type A/D converter. What are the merits and demerits of this type of converter? How does the architecture of a flash converter differ from that of a half-flash converter?
6. Describe with the help of a schematic diagram the operation of a tracking-type A/D converter. Explain how it overcomes the inherent disadvantage of a longer conversion time of the conventional counter-type A/D converter.
7. Describe with the help of a schematic diagram the principle of operation of a successive approximation type A/D converter. Explain the sequence of operation of conversion of an analogue signal to its digital equivalent when the expected digital output is 1010.
8. Explain the following:

   (a) why a tracking type A/D converter is particularly suitable for fast-changing analogue signals;
   (b) the use of a D/A converter as a programmable integrator;
   (c) why a dual-slope integrating-type A/D converter has a higher accuracy than a single-slope integrating-type A/D converter;
   (d) the use of a D/A converter as a digitally controlled voltage attenuator.

Problems

1. Determine the percentage resolution of (a) an eight-bit and (b) a 12-bit D/A converter.

   (a) 0.39 %; (b) 0.024 %

2. An eight-bit D/A converter produces an analogue output of 12.5 mV for a digital input of 00000010. Determine the analogue output for a digital input of 00000100.

   25 mV

3. A 12-bit D/A converter has a resolution of 2.44 mV. Determine its analogue output for a digital input of 111111111111.

   10 V

4. How many bits should a current-output D/A converter have for its full-scale output to be 20 mA and its resolution to be better than 25 mA?

   10 bits

5. Compare (a) the step size and (b) the percentage resolution of a D/A converter having an eight-bit binary input with those of a D/A converter having an eight-bit BCD input. Both have a full-scale output of 10 V.

   Binary input: (a) 39.2 mV, (b) 0.39 %; BCD input: (a) 101 mV, (b) 1 %

6. Compare the average conversion time of an eight-bit counter-type A/D converter with the conversion time of a 12-bit successive approximation type A/D converter. Assume a clock frequency of 10 MHz.

   Counter-type A/D converter 12.8 μs, successive approximation type 1.2 μs

7. A certain 12-bit successive approximation type A/D converter has a full-scale analogue input of 10 V. It operates at a clock frequency of 1MHz. Determine the conversion time for an analogue input of (a) 1.25 V, (b) 2.50 V, (c) 3.75 V, (d) 7.5 V and (e) 10 V.

   (a) 12 μs; (b) 12 μs; (c) 12 μs; (d) 12 μs; (e) 12 μs

Further Reading

1. Demler, M. (2006) *High Speed Analog-to-Digital Conversion*, Academic Press, CA, USA.
2. Jespers, P. G. A. (2001) *Integrated Converters: D to A and A to D Architectures, Analysis and Simulation* (Textbooks in Electrical and Electronic Engineering), Oxford University Press, New York, USA.
3. Razavi, B. (2001)*Principles of Data Conversion System Design*, Oxford University Press, New York, USA.
4. Coombs Jr, C. F. (1999) *Electronic Instrument Handbook*, McGraw-Hill Inc., USA
5. Webster, J. G. (1999) *The Measurement, Instrumentation and Sensors Handbook*, CRC Press (in cooperation with IEEE Press), FL, USA.

13

Microprocessors

The microprocessor is the heart of a microcomputer system. In fact, it forms the central processing unit of any microcomputer and has been rightly referred to as the *computer on a chip*. This chapter gives an introduction to microprocessor fundamentals, followed by application-relevant information, such as salient features, pin configuration, internal architecture, instruction set, etc., of popular brands of eight-bit, 16-bit, 32-bit and 64-bit microprocessors from international giants like INTEL, MOTOROLA and ZILOG.

13.1 Introduction to Microprocessors

A microprocessor is a programmable device that accepts binary data from an input device, processes the data according to the instructions stored in the memory and provides results as output. In other words, the microprocessor executes the program stored in the memory and transfers data to and from the outside world through I/O ports. Any microprocessor-based system essentially comprises three parts, namely the microprocessor, the memory and peripheral I/O devices. The microprocessor is generally referred to as the heart of the system as it performs all the operations and also controls the rest of the system. The three parts are interconnected by the data bus, the address bus and the control bus (Fig. 13.1).

The *memory* stores the binary instructions and data for the microprocessor. The memory can be classified as the primary or main memory and secondary memory. Read/write memory (R/WM) and read only memory (ROM) are examples of primary memory and are used for executing and storing programs. Magnetic disks and tapes are examples of secondary memory. They are used to store programs and results after the completion of program execution. Microprocessors do not execute programs stored in the secondary memory directly. Instead, they are first copied on to the R/W primary memory.

Digital Electronics: Principles, Devices and Applications Anil Kumar Maini

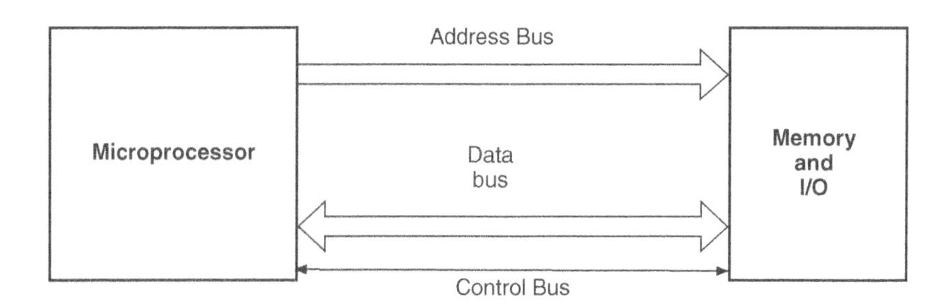

Figure 13.1 Microprocessor-based system.

Input/output devices are means through which the microprocessor interacts with the outside world. The commonly used input devices include keyboards, A/D converters, switches, cameras, scanners, microphones and so on. LEDs, seven-segment displays, LCD displays, printers and monitors are some of the commonly used output devices.

A *bus* is basically a communication link between the processing unit and the peripheral devices. It is a group of wires that carry information in the form of bits. The *address bus* is unidirectional and is used by the CPU to send out the address of the memory location to be accessed. It is also used by the CPU to select a particular input or output port. It may consist of 8, 16, 20 or an even greater number of parallel lines. The number of bits in the address bus determines the maximum number of data locations in the memory that can be accessed. A 16-bit address bus, for instance, can access 2^{16} data locations. It is labelled as $A_0, \ldots, A_{n-1}$, where n is the width (in bits) of the address bus.

The *data bus* is bidirectional, that is, data flow occurs both to and from the microprocessor and peripherals. Data bus size has a considerable influence on the computer architecture, as parameters such as the word length and the quantum of data that can be manipulated at a time are determined by the size of the data bus. There is an internal data bus, which may not be of the same width as the external data bus that connects the microprocessor to I/O and memory. The size of the internal data bus determines the largest number that can be processed by the microprocessor in a single operation. The largest number that can be processed, for instance, by a microprocessor having a 16-bit internal data bus is 65535. The data bus is labelled as $D_0, \ldots, D_{n-1}$, where n is the data bus width (in bits).

The *control bus* contains a number of individual lines carrying synchronizing signals. The term ‘bus’ would normally imply a group of lines working in unison. The control bus (if we call it a bus) sends out control signals to memory, I/O ports and other peripheral devices to ensure proper operation. It carries control signals such as memory read, memory write, read input port, write output port, hold, interrupt, etc. For instance, if it is desired to read the contents of a particular memory location, the CPU first sends out the address of that location on the address bus and a ‘memory read’ control signal on the control bus. The memory responds by outputting data stored in the addressed memory location onto the data bus. ‘Interrupt’ tells the CPU that an external device needs to be read or serviced. ‘Hold’ allows a device such as the direct memory access (DMA) controller to take over the address and data buses.

Figure 13.2 shows the bus interface between the microprocessor and its peripheral devices. The microprocessor considered in the diagram is an eight-bit microprocessor such as Intel’s 8085.

Microprocessor-based systems can be categorized as general-purpose reprogrammable systems and embedded systems. Reprogrammable systems include microcomputers and mainframe and miniframe computers where microprocessors are used for computing and data processing. In embedded systems, they perform a specific task and are not available for reprogramming to the end-user. Examples of these systems include mobile phones, washing machines, microwave ovens, dish washers and so on.

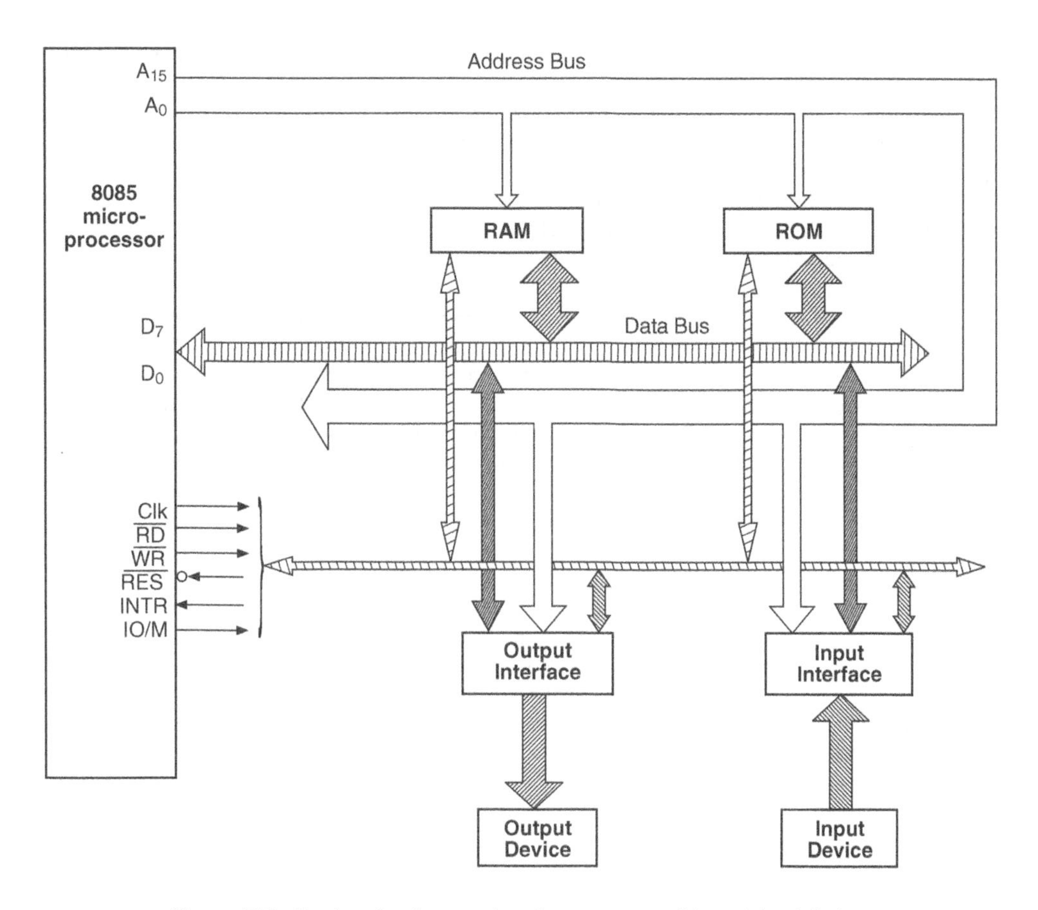

Figure 13.2 Bus interface between the microprocessor and its peripheral devices.

In most of these systems, the microprocessor, memory and I/O ports are combined onto one chip, known as the *microcontroller*. Microcontrollers are discussed in detail in Chapter 14.

13.2 Evolution of Microprocessors

The evolution of microprocessors has been known to follow Moore's law, which suggests that the complexity of an integrated circuit, with respect to the minimum component cost, doubles every 24 months. This rule has been generally followed, since the humble beginning of microprocessors as the drivers for calculators to the present-day scenario where every system, from the largest mainframes to the smallest handheld computers, uses a microprocessor at its core.

The first microprocessor was introduced in 1971 by the Intel Corporation. It was a four-bit microprocessor, Intel 4004. Other four-bit microprocessors developed were Intel 4040 by Intel, PPS-4 by Rockwell International, T3472 by Toshiba and so on. The first eight-bit microprocessor, named Intel 8008, was also developed by Intel in the year 1972. All these microprocessors were made using PMOS technology. The first microprocessor using NMOS technology was Intel 8080, developed by Intel in the year 1973. Intel 8080 was followed by Intel 8085 in the year 1975, which became very popular. Other popular eight-bit microprocessors were Zilog's Z80 (1976) and Z800, Motorola's MC6800 (1974) and MC6809 (1978), National Semiconductor's NSC 800, RCA's 1802 (1976) and so on.

The first multichip 16-bit microprocessor was National Semiconductor's IMP-16, introduced in 1973. The first 16-bit single-chip microprocessor was Texas Instrument's TMS 9900. Intel's first 16-bit microprocessor was Intel 8086 introduced in the year 1978. Other 16-bit microprocessors developed by Intel were Intel 80186 (1982), Intel 8088, Intel 80188 and Intel 80286 (1982). Other popular 16-bit microprocessors include Motorola's 68000 (1979), 68010 and 68012, Zilog's Z8000, Texas Instruments TMS 9900 series and so on.

32-bit microprocessors came into existence in the 1980s. The world's first single-chip 32-bit microprocessor was introduced by AT&T Bell Labs in the year 1980. It was named BELLMAC-32A. The first 32-bit processor introduced by Intel was iapx 432, introduced in 1981. The more popular 32-bit microprocessor was Intel 80386, introduced by Intel in 1985. It was widely used for desktop computers. The 32-bit microprocessor family of Intel includes Intel 486, Pentium, Pentium Pro, Pentium II, Pentium III and Pentium IV. AMD's K5, K6 and K7, Motorola's 68020 (1985), 68030 and 68040, National Semiconductor's 32032 and 32332 and Zilog's Z80000 are other popular 32-bit microprocessors. All these microprocessors are based on CISC (Complex Instruction Set Computers) architecture. The first commercial RISC (Reduced Instruction Set Computers) design was released by MIPS Technologies, the 32-bit R2000. Some of the popular RISC processors include Intel's 80860 and 80960, Motorola's 88100 and Motorola's, IBM and Apple's PowerPC series of microprocessors.

While 64-bit microprocessor designs have been in use in several markets since the early 1990s, the early 2000s have seen the introduction of 64-bit microchips targeted at the PC market. Some of the popular 64-bit microprocessors are AMD's AMD64 (2003) and Intel's x86-64 chips. Popular 64-bit RISC processors include SUN's ULTRASPARC, PowerPC 620, Intel's Itanium, MIPS R4000, R5000, R10000 and R12000 and so on.

13.3 Inside a Microprocessor

Figure 13.3 shows a simplified typical schematic arrangement of a microprocessor. The figure shown is a generalized one and is not the actual structure of any of the commercially available microprocessors. The important functional blocks include the arithmetic logic unit (ALU), the register file and the control unit.

These functional blocks are briefly described in the following paragraphs.

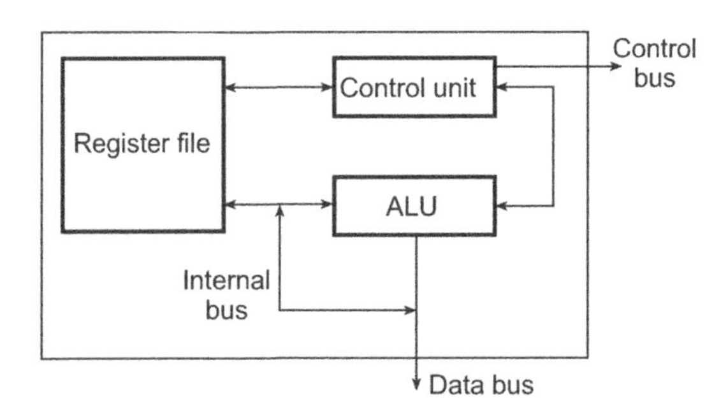

Figure 13.3 Typical schematic arrangement of a microprocessor.

13.3.1 Arithmetic Logic Unit (ALU)

The arithmetic logic unit (ALU) is the core component of all microprocessors. It performs the entire integer arithmetic and bit-wise logical operations of the microprocessor. ALU is a combinational logic circuit and has two data input lines, a data output line and a status line. It gets data from the registers of the microprocessor, processes the data according to the instructions from the control unit and stores the results in its output registers. All modern ALUs use binary data in 2's complement format.

The integer arithmetic operations performed by the ALU include addition and subtraction. It performs AND, OR, NOT and EXCLUSIVE-OR logical operations. Some 16-bit, 32-bit and 64-bit microprocessors also perform multiplication and division operations. In other microprocessors, the multiplication and division operations are performed by writing algorithms using addition and subtraction operations. Some such algorithms were outlined in Chapter 3 on digital arithmetic. ALU also performs the bit-shifting operations and the comparison of data operations.

13.3.2 Register File

The register file comprises various registers used primarily to store data, addresses and status information during the execution of a program. Registers are sequential logic devices built using flip-flops. Some of the commonly found registers in most of the microprocessors include the *program counter*, *instruction registers*, *buffer registers*, *the status register*, *the stack pointer*, *general-purpose registers* and *temporary registers*.

13.3.2.1 Program Counter

The *program counter* is a register that stores the address of the next instruction to be executed and hence plays a central role in controlling the sequence of machine instructions that the processor executes. After the instruction is read into the memory, the program counter is automatically incremented by '1'. This is of course on the assumption that the instructions are executed sequentially. Its contents are affected by jump and call instructions. In the case of a jump instruction, the program counter is first loaded with the new address and then incremented thereafter until another jump instruction is encountered. When the microprocessor receives an instruction to begin a subroutine, the contents of the program counter are incremented by '1' and are saved in the stack. The program counter is loaded with the address of the first instruction of the subroutine. Its contents are incremented by '1' until a return instruction is encountered. The saved stack contents are then loaded into the program counter and the program continues, executing each instruction sequentially until another jump instruction or a subroutine call is encountered. The interrupt process also alters the contents of the program counter.

13.3.2.2 Instruction Register

The *instruction register* stores the code of the instruction currently being executed. The control unit extracts the operation code from the instruction register, which determines the sequence of signals necessary to perform the processing required by the instruction.

13.3.2.3 Buffer Register

Buffer registers interface the microprocessor with its memory system. The two standard buffer registers are the *memory address register* (MAR) and the *memory buffer register* (MBR). The MAR is connected to the address pins of the microprocessor and holds the absolute memory address of the data or instruction to be accessed. The MBR, also known as the memory data register, is connected to the data pins of the microprocessor. It stores all data written to and read from memory.

13.3.2.4 Status Register

The *status register* stores the status outputs of the result of an operation and gives additional information about the result of an ALU operation. The status of bits stored in the status register tells about the occurrence or nonoccurrence of different conditions, and one or more bits may be updated at the end of an operation. Each bit is a Boolean flag representing a particular condition. The most common conditions are the carry, overflow, zero and negative. For instance, a '1' in the carry status bit position shows that the result of the operation generates a carry. The significance of the status register lies in the fact that the condition code set by the status of different bits in the status register forms the basis of decision-making by the microprocessor during the execution of a program.

13.3.2.5 Stack Pointer

The *stack pointer* is a register used to store the address of a memory location belonging to the most recent entry in the stack. In fact, a stack is a block of memory locations designated for temporary storage of data. It is used to save data of another general-purpose register during execution of a subroutine or when an interrupt is serviced. The data are moved from a general register to the stack by a PUSH instruction at the beginning of a subroutine call, and back to the general register by a POP instruction at the end of the subroutine call. Microprocessors use a stack because it is faster to move data using PUSH and POP instructions than to move data to/from memory using a MOVE instruction.

13.3.2.6 General-purpose Registers

There is a set of registers for general-purpose use, designated as *general-purpose registers*. They are used explicitly to store data and address information. Data registers are used for arithmetic operations, while the address registers are used for indexing and indirect addressing. These enhance the processing speed of the microprocessor by avoiding a large number of external memory read/write operations while an ALU operation is being performed, as it is much easier and faster to read from or write into an internal register than to read from or write into an external memory location. Earlier microprocessors had only one register called the *accumulator* for ALU operations. It needed at least four assembly language instructions to perform a simple addition, including carrying data from an external memory location to the accumulator, adding the contents of the accumulator to those of another memory location, storing the result in the accumulator and transferring the contents of the accumulator back to the external memory location. With the availability of a greater number of general-purpose registers, it would be possible to perform many ALU operations without even a need to store data in external memory.

13.3.2.7 Temporary Registers

These are used when data have to be stored during the execution of a machine instruction. They are completely hidden from the user of the microprocessor.

13.3.3 Control Unit

The *control unit* governs and coordinates the activities of different sections of the processor and I/O devices. It is responsible for controlling the cycle of fetching machine instructions from memory and executing them. It also coordinates the activities of input and output devices. It is undoubtedly the most complex of all functional blocks of the microprocessor and occupies most of the chip area. The control unit is a sequential logic circuit, which steps the processor through a sequence of synchronized operations. It sends a stream of control signals and timed pulses to the components and external pins of the microprocessor. As an illustration, to execute an instruction from the memory, the control unit sends out a 'read' command to the memory and reads the instruction (or data) that comes back on the data bus. The control unit then decodes the instruction and sends appropriate signals to the ALU, the general-purpose registers, the multiplexers, the demultiplexers, the program counter and so on. If the instruction was to store data in the memory, the control unit sends out the address of the memory location on the address bus, the data to be stored on the data bus and a 'write' command on a control line.

Control units are categorized into two types depending upon the way they are built. These include *hard-wired* and *microcoded* control units. Hard-wired controllers are sequential logic circuits, the states of which correspond to the phases of the instruction execution cycle. In the case of hard-wired controllers, there is an electronic circuitry in the control unit to generate control signals for each instruction. They are very compact and fast, but are difficult to design. This design is also known as RISC (Reduced Instruction Set Computer) design. Microcoded control units are easy to design, and execution of an instruction in this case involves executing a microprogram consisting of a sequence of microinstructions. This design is also known as CISC (Complex Instruction Set Computer) design. Microcoded control units offer more flexibility than do hard-wired control units but they are comparatively slower than the latter.

Figure 13.4 shows a more descriptive block diagram of a microprocessor. Multiplexers and demultiplexers do not represent primary functions and are there to facilitate the flow of data between different blocks and also between different blocks and the outside world.

13.4 Basic Microprocessor Instructions

Microprocessors perform various basic operations including data transfer instructions, arithmetic instructions, logic instructions, control transfer instructions and machine control instructions.

13.4.1 Data Transfer Instructions

Data transfer instructions transfer data from one location designated as the source location to another location designated as the destination. The data transfer could take place from one register to another, from one memory location to another memory location, from a memory location to a register or from a register to a memory location, and so on. In fact, they are more correctly referred to as data movement operations as the contents of the source are not transferred but are copied into the destination register without modifying the contents of the source. It may be mentioned here that these operations do not affect the flags. Data transfer operations of the 8085 microprocessor are of three types, namely MOVE, LOAD and STORE:

MOV destination, source	Copy data from the source to the destination location
LDA address	Copy the data byte at the memory location specified by the 16-bit address into the accumulator
STA address	Copy the data from the accumulator to the memory location specified by the 16-bit address

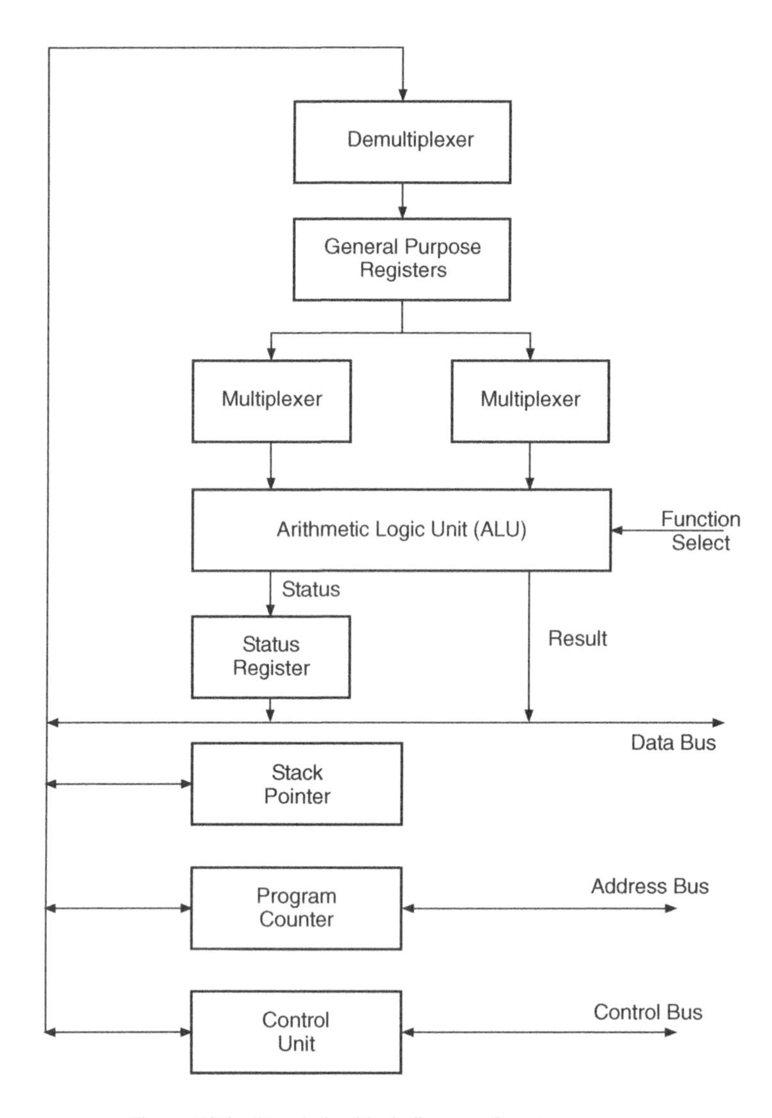

Figure 13.4 Descriptive block diagram of a microprocessor.

13.4.2 Arithmetic Instructions

Arithmetic instructions performed by microprocessors include addition, subtraction, multiplication, division, comparison, negation, increment and decrement. It may be mentioned here that most of the eight-bit microprocessors do not support multiplication and division operations. These operations are supported by the 16-bit and 32-bit microprocessors. The arithmetic operations supported by the 8085 microprocessor are addition, subtraction, increment and decrement operations. Examples are as follows:

ADD R Adds the contents of the register to the accumulator
ADI eight-bit Adds the eight-bit data to the accumulator

SUB R	Subtracts the contents of the register from the accumulator
SUI eight-bit	Subtracts eight-bit data from the contents of the accumulator
INR R	Increments the contents of the register
DCR R	Decrements the contents of the register

13.4.3 Logic Instructions

Microprocessors can perform all the logic functions of hard-wired logic. The basic logic operations performed by all microprocessors are AND, OR, NOT and EXCLUSIVE-OR. Other logic operations include 'shift' and 'rotate' operations. All these operations are performed on a bit-for-bit basis on bytes or words. For instance, 11111111 AND 10111010 equals 10111010, and 11111111 OR 10111010 equals 11111111. Some microprocessors also perform bit-level instructions such as 'set bit', 'clear bit' and 'complement bit' operations. It may be mentioned that logic operations always clear the carry and overflow flags, while the other flags change to reflect the condition of the result.

The basic shift operations are the 'shift left' and 'shift right' operations. In the *shift left* operation, also known as the arithmetic shift left, all bits are shifted one position to the left, with the rightmost bit set to '0' and the leftmost bit transferred to the carry position in the status register. In the *shift right* operation, also known as logic shift right, all bits are shifted one bit position to the right, with the leftmost bit set to '0' and the rightmost bit transferred to the carry position in the status register. If in the shift right operation the leftmost bit is left unchanged, it is called arithmetic shift right. In a 'rotate' operation, the bits are circulated back into the register. Carry may or may not be included. As an illustration, in a 'rotate left' operation without carry, the leftmost bit goes to the rightmost bit position, and, in a 'rotate right' with carry included, the rightmost bit goes to the carry position and the carry bit takes the position of the leftmost bit.

Examples of logic instructions performed by the 8085 microprocessor include the following:

ANA R/M	Logically AND the contents of the register/memory with the contents of the accumulator
ANI eight-bit	Logically AND the eight-bit data with the contents of the accumulator
ORA R/M	Logically OR the contents of the register/memory with the contents of the accumulator
ORI eight-bit	Logically OR the eight-bit data with the contents of the accumulator
XRA R/M	Logically EXCLUSIVE-OR the contents of the register memory with the contents of the accumulator
XRI eight-bit	Logically EXCLUSIVE-OR the eight-bit data with the contents of the accumulator
CMA	Complement the contents of the accumulator
RLC	Rotate each bit in the accumulator to the left position
RRC	Rotate each bit in the accumulator to the right position

13.4.4 Control Transfer or Branch or Program Control Instructions

Microprocessors execute machine codes from one memory location to the next, that is, they execute instructions in a sequential manner. Branch instructions change the flow of the program either unconditionally or under certain test conditions. Branch instructions include 'jump', 'call', 'return' and 'interrupt'.

'Jump' instructions are of two types, namely 'unconditional jump' instructions and 'conditional jump' instructions. If the microprocessor is so instructed as to load a new address in the program counter and start executing instructions at that address, it is termed an unconditional jump. In the case of a conditional jump, the program counter is loaded with a new instruction address only if and when certain conditions are established by the microprocessor after reading the appropriate status register bits. 'Call' instructions transfer the flow of the program to a subroutine. The 'call' instruction differs from the 'jump' instruction as 'call' saves a return address (the address of the program counter plus one) on the stack. The 'return' instruction returns control to the instruction whose address was stored in the stack when the 'call' instruction was encountered. 'Interrupt' is a hardware-generated call (externally driven from a hardware signal) or a software-generated call (internally derived from the execution of an instruction or by some internal event). Examples of transfer control instructions of the 8085 microprocessor are as follows:

JMP 16-bit address	Change the program sequence to the location specified by the 16-bit address
JZ 16-bit address	Change the program sequence to the location specified by the 16-bit address if a zero flag is set
JC 16-bit address	Change the program sequence to the location specified by the 16-bit address if a carry flag is set
CALL 16-bit address	Change the program sequence to the location of the subroutine specified by the 16-bit address
RET	Return to the calling program

13.4.5 Machine Control Instructions

Machine control instructions include HALT and NOP instructions. Machine control instructions performed by the 8085 microprocessor include the following:

HLT	Stop processing and wait
NOP	No operation

13.5 Addressing Modes

Microprocessors perform operations on data stored in the register or memory. These data are specified in the operand field of the instruction. The data can be specified in various ways as a direct data value or stored in some register or memory location, and so on. These are referred to as the *addressing modes* of the microprocessor. In other words, the addressing mode as expressed in the instruction tells us how and from where the microprocessor can get the data to act upon. Addressing modes are of direct relevance to compiler writers and to programmers writing the code in assembly language.

Different microprocessor architectures provide a variety of addressing modes. RISC microprocessors have far fewer addressing modes than CISC microprocessors. The most commonly used addressing modes are absolute, immediate, register direct, register indirect, indexed, program counter relative, implicit and relative addressing modes. They account for more than 90 % of the total addressing modes.

13.5.1 Absolute or Memory Direct Addressing Mode

In *absolute addressing mode,* the data are accessed by specifying their address in the memory [Fig. 13.5(a)]. This mode is useful for accessing fixed memory locations, such as memory mapped I/O devices. For example, the instruction MOV A, 30H in the 8085 microprocessor moves the contents of memory location 30H into the accumulator [Fig. 13.5(b)]. In this case the accumulator has the value 07H.

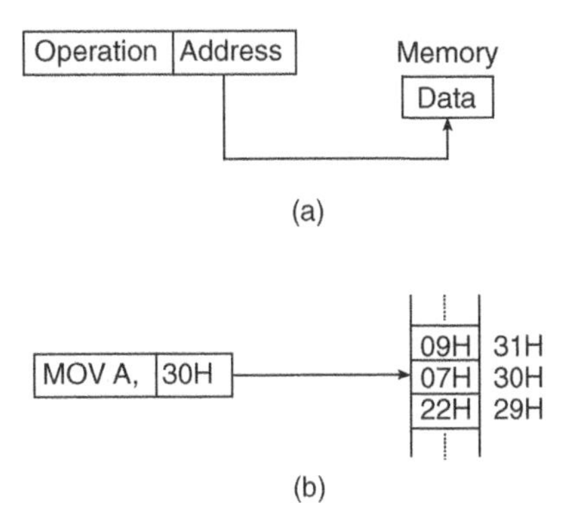

Figure 13.5 Absolute addressing mode.

Figure 13.6 Immediate addressing mode.

13.5.2 Immediate Addressing Mode

In *immediate addressing mode* the value of the operand is held within the instruction itself (Fig. 13.6). This mode is useful for accessing constant values in a program. It is faster than the absolute addressing mode and requires less memory space. For example, the instruction MVI A, #30H moves the data value 30H into the accumulator. The sign # in the instruction tells the assembler that the addressing mode used is immediate.

13.5.3 Register Direct Addressing Mode

In *register direct addressing mode,* data are accessed by specifying the register name in which they are stored [Fig. 13.7(a)]. Operations on registers are very fast, and hence instructions in this mode require less time than absolute addressing mode instructions. As an example, the instruction MOV A, R1 in the 8051 microprocessor moves the contents of register R1 into the accumulator [Fig. 13.7(b)]. The contents of the accumulator after the instruction are 06H.

13.5.4 Register Indirect Addressing Mode

In all the modes discussed so far, either the value of the data or their location is directly specified. The *indirect addressing mode* uses a register to hold the actual address where the data are stored. That is, in this case the memory location of the data is stored in a register [Fig. 13.8(a)]. In other words, in indirect addressing mode, the address is specified indirectly and has to be looked up. This addressing mode is useful when implementing the pointer data type of high-level language.

In the 8085 microprocessor, the R0 and R1 registers are used as an eight-bit index and the DPTR as a 16-bit index. The mnemonic symbol used for indirect addressing is @. As an example, the instruction

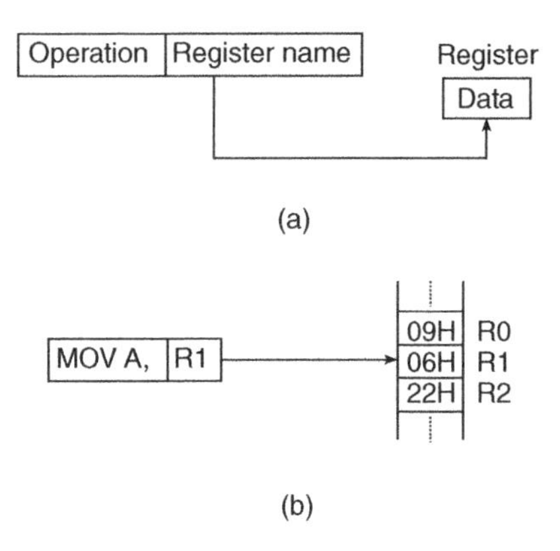

Figure 13.7 Register direct addressing mode.

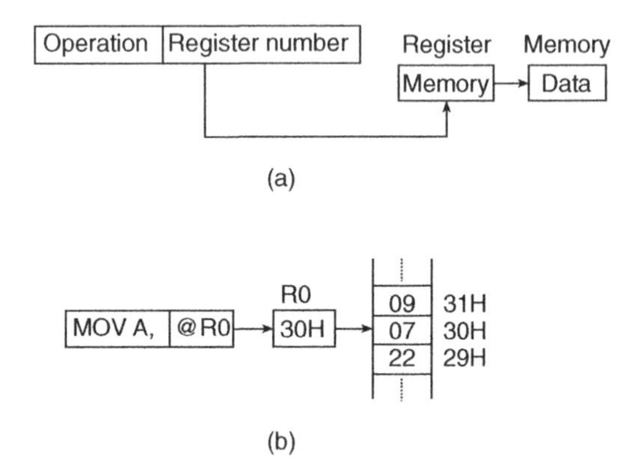

Figure 13.8 Register indirect addressing mode.

MOV A, @R0 moves the contents of the memory location whose address is stored in R0 into the accumulator. The value of the accumulator in this example is 07H [Fig. 13.8(b)]. This addressing mode can also be enhanced with an offset for accessing data structures in data space memory. This is referred to as *register indirect* with displacement. As an example, the instruction MOVC A, @A+DPTR copies the code byte at the memory address formed by adding the contents of A and DPTR to A.

13.5.5 Indexed Addressing Mode

In the indexed addressing mode, the address is obtained by adding the contents of a register to a constant (Fig. 13.9). The instruction 'move the contents of accumulator A to the memory location whose address is given by the contents of register 1 plus 5' is an example of indexed addressing. The indexed addressing mode is useful whenever the absolute location of the data is not known until the program is running. This addressing mode is used to access a continuous table or array of data items stored in memory. The content of the constant gives the starting address, while the contents of the

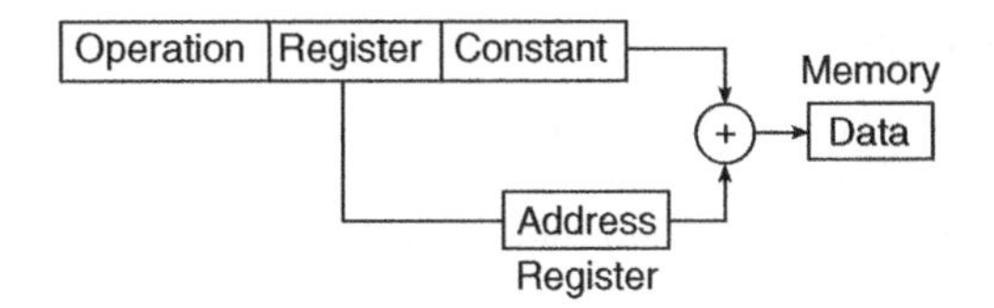

Figure 13.9 Indexed addressing mode.

register determine the element of the array or table to be accessed. If the program counter is used in the indexed addressing mode, it is known as the *program counter relative addressing mode.*

13.5.6 Implicit Addressing Mode and Relative Addressing Mode

In *implicit addressing mode,* no operand is used in the instruction and the location of the operand is obvious from the instruction itself. Examples include 'clear carry flag', 'return from subroutine' and so on.

The relative addressing mode is used for 'jump' and 'branch' instructions only. In this, a displacement is added to the address in the program counter and the next instruction is fetched from the new address in the program counter. This mode is particularly useful in connection with conditional jumps.

13.6 Microprocessor Selection

There are thousands of microprocessors available on the market. Selection of the right microprocessor for a given application is not an easy task and cannot take place in a vacuum; it must be done with the application in mind. Not only this, the quantity to be produced and the experience and capabilities of the designers must also be considered. The selection process begins with the definition of the application to be followed by matching a given processor with the well-defined application.

13.6.1 Selection Criteria

Sometimes it becomes difficult to extract microprocessor requirements from the application at the early stage of the project. This may be due to several factors, which include the following:

1. Speed compatibility of the microprocessor with peripherals.
2. The time-critical behaviour of the application.
3. The size of the program required to implement certain functions is not known in advance.

These ambiguities serve as a warning that perhaps the project is not adequately defined for the microprocessor selection to be made. Factors to be considered while selecting the microprocessor are price, power consumption, performance, availability, software support and code density.

13.6.1.1 Price

Price is one of the important factors that is considered by designers to evaluate a processor. It assumes more importance for those embedded systems that have price constraint.

13.6.1.2 Power consumption

Power consumption is an important factor for battery-operated systems. The power consumption of a microprocessor varies with the supply voltage (square of supply voltage), speed (linearly) and with the software the chip is running. The bus structure of the processor and its interconnection with the memory ICs should also be looked into.

13.6.1.3 Performance

Processors that are good for one task may not be suitable for another. It is therefore very important to define the processor requirements for the given application. These include the estimated size and complexity of the program, speed requirements (time-critical functions), the language to be used, the arithmetic functions needed, memory requirements (ROM, RAM and mass storage), I/O requirements and interrupt source and response time required.

After defining the application requirements, they should be matched with those that a processor can offer. Table 13.1 enumerates the main parameters of the processor to be considered while selecting it for a particular application.

13.6.1.4 Availability

Before zeroing onto a particular microprocessor, it is important to ensure that it is easily available.

Table 13.1 Microprocessor characteristics checklist.

Instruction set	Data types: bit operations, long words Arithmetic functions: multiply and divide Encoding efficiency: RISC or CISC
Register set	Number of registers Width of registers Number of special-purpose registers
Addressing	Number of modes: direct, indirect, etc. Segmented or linear addressing Memory and I/O address ranges (memory mapped/I/O mapped) Memory management
Bus and control signals	Bus timings Interrupts DMA/bus arbitration control signals Data and address bus width Clock speed and bus cycle time
Miscellaneous	Prefetch (instruction queue length), cache memory Coprocessor support: floating point, I/O processors Power requirements
Nontechnical considerations	Documentation quality and availability Development tools: emulators, debuggers and logic analysers Software support: OS, compiler, assembler, utilities

13.6.1.5 Software support

The associated software with the microprocessor, such as the debugger, compiler and operating system, constitutes one of the factors that needs to be considered.

13.6.1.6 Code density

The code density is the ratio between the size of the source code and the size of the object code. The smaller the object code, the better is the code density. Processors having high code densities require less memory to execute the code. RISC processors have poor code density compared with CISC processors.

Moreover, there is seldom one right microprocessor for a given task. There are several chips that can be used for a given task. Factors such as past experience, the market reputation of the processor and availability are considered before making the final decision.

13.6.2 Microprocessor Selection Table for Common Applications

Single-chip microcomputers are commonly used in control applications. In more complex control applications requiring large amounts of I/O, memory or high-speed processing, eight-bit or 16-bit microprocessors are used. Data processing applications, which require more memory and I/O, use a PC. The 32-bit and 64-bit microprocessors are used in systems that require high performance such as engineering workstations and in multi-user systems. Table 13.2 gives typical microprocessor types for various application classes.

Table 13.2 Microprocessor types for various application classes.

Application classes			Typical device types					
Type	Speed and complexity	Type example	Single-chip microcomputer			Microprocessor		
			four-bit	eight-bit	16-bit	eight-bit	16-bit	32/64-bit
Control	Low	Automatic thermostat	✓	✓				
	Medium	Digital multimeter		✓		✓		
	High	Engine control		✓	✓	✓	✓	
Data processing	Low	Home computer				✓		
	Medium	Mid-range PC				✓	✓	
	High	Engineering workstation, multiuser computer					✓	✓

13.7 Programming Microprocessors

Microprocessors execute programs stored in the memory in the form of a sequence of binary digits. Programmers do not write the program in binary form but write it either in the form of a text file containing an assembly-language source code or using a high-level language. Programs such as editor, assembler, linker and debugger enable the user to write the program in assembly language, convert it into binary code and debug the binary code. *Editor* is a program that allows the user to enter, modify and store a group of instructions or text under a file name. The assembly language source code is translated into an object code by a program called *assembler*. *Linker* converts the output of the assembler into a format that can be executed by the microprocessor. The *debugger* is a program that allows the user to test and debug the object file.

Programming in assembly language produces a code that is fast and takes up little memory. However, it is difficult to write large programs using assembly language. Another disadvantage of assembly language programming is that it is specific to a particular microprocessor. High-level language programming overcomes these problems. Some of the popular high-level languages used include C, C++, Pascal and so on. Compiler programs are primarily used to translate the source code from a high-level language to a lower-level language (e.g. assembly language or machine language). Figures 13.10(a) and (b) show the various steps involved in executing assembly language programs and programs written in high-level languages respectively.

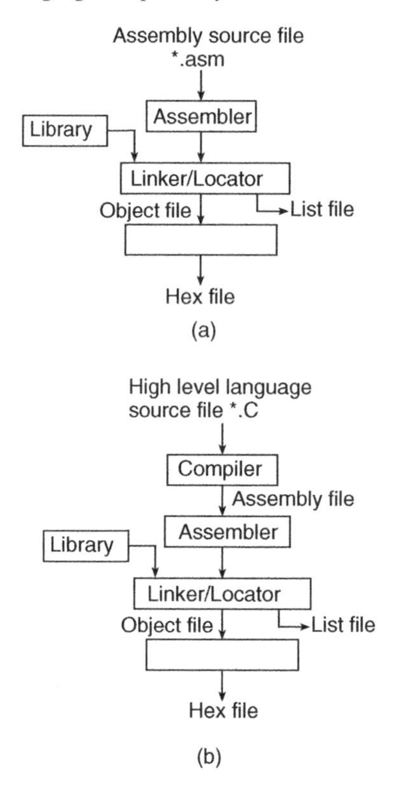

Figure 13.10 (a) Various steps involved in executing assembly language programs and (b) various steps involved in executing programs written in high-level languages.

13.8 RISC Versus CISC Processors

CISC is an acronym for Complex Instruction Set Computer. The primary goal of CISC architecture is to complete a task in as few lines of assembly as possible. This is achieved by building processor hardware that is capable of understanding and executing a series of complex operations. In this case, each instruction can execute several low-level instructions. One of the primary advantages of this system is that the compiler has to do very little work to translate a high-level language statement into assembly. Because the length of the code is relatively short, very little RAM is required to store instructions. In a nutshell, the emphasis is to build complex instructions directly into the hardware. Examples of CISC processors are the CDC 6600, System/360, VAX, PDP-11, the Motorola 68000 family, and Intel and AMD x86 CPUs.

RISC is an acronym for Reduced Instruction Set Computer. This type of microprocessor emphasizes simplicity and efficiency. RISC designs start with a necessary and sufficient instruction set. The objective of any RISC architecture is to maximize speed by reducing clock cycles per instruction. Almost all computations can be done from a few simple operations. The goal of RISC architecture is to maximize the effective speed of a design by performing infrequent operations in software and frequent functions in hardware, thus obtaining a net performance gain.

To understand this phenomenon, consider any assembly-level language program. It has been observed that it uses the MOV instruction much more frequently than the MUL instruction. Therefore, if the architectural design implements MOV in hardware and MUL in software, there will be a considerable gain in speed, which is the basic feature of RISC technology. Examples of RISC processors include Sun's SPARC, IBM and Motorola's PowerPCs, and ARM-based processors.

The salient features of a RISC processor are as follows:

1. The microprocessor is designed using hard-wired control. For example, one bit can be dedicated for one instruction. Generally, variable-length instruction formats require microcode design. All RISC instructions have fixed formats, so no microcode is required.
2. The RISC microprocessor executes most of the instructions in a single clock cycle. This is due to the fact that they are implemented in hardware.
3. The instruction set typically includes only register-to-register load and store.
4. The instructions have a simple format with few addressing modes.
5. The RISC microprocessor has several general-purpose registers and large cache memories, which support the very fast access of data.
6. The RISC microprocessor processes several instructions simultaneously and so includes pipelining.
7. The software can take advantage of more concurrency.

13.9 Eight-Bit Microprocessors

This section describes the block diagram, pin-out diagram, salient features and instruction set of the most popular eight-bit microprocessors, namely 8085 of Intel, Z80 of Zilog and 6800 of Motorola.

13.9.1 8085 Microprocessor

Figure 13.11 gives the pin-out configuration and Fig. 13.12 shows a block diagram of the 8085 microprocessor. Table 13.3 lists the pin details.

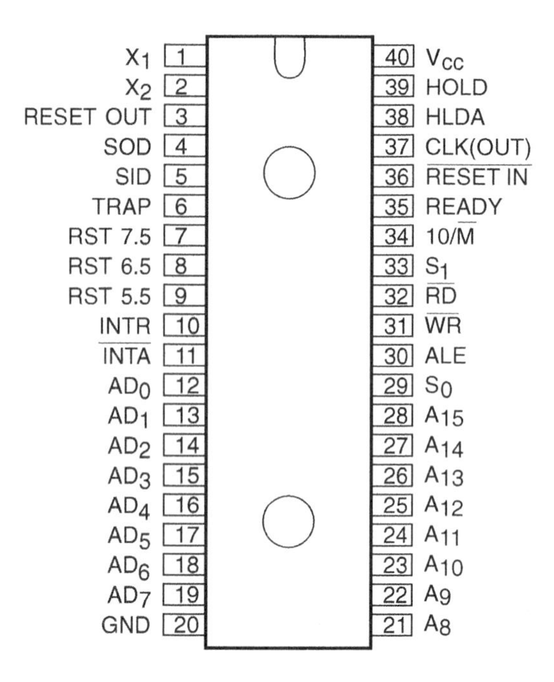

Figure 13.11 Pin-out configuration of 8085.

13.9.1.1 8085 Registers

The 8085 microprocessor registers include an eight-bit accumulator, an eight-bit flag register (five one-bit flags, namely sign, zero, auxiliary carry, parity and carry), eight-bit B and C registers (which can be used as one 16-bit BC register pair), eight-bit D and E registers (which can be used as one 16-bit DE register pair), eight-bit H and L registers (which can be used as one 16-bit HL register pair), a 16-bit stack pointer and a 16-bit program counter.

13.9.1.2 Addressing Modes

8085 has four addressing modes. These include register addressing, register indirect addressing, direct addressing mode and immediate addressing mode.

13.9.1.3 8085 Instructions

An instruction is a binary pattern designed inside a microprocessor to perform a specific function. The entire group of instructions a microprocessor can perform is referred to as its *instruction set*. An *instruction cycle* is defined as the time required to complete the execution of an instruction. An 8085 instruction cycle consists of 1–6 machine cycles. A *machine cycle* is defined as the time required to complete one operation of accessing memory, I/O and so on. This will comprise 3–6 T-states, which is defined as one subdivision of the operation performed in one clock period.

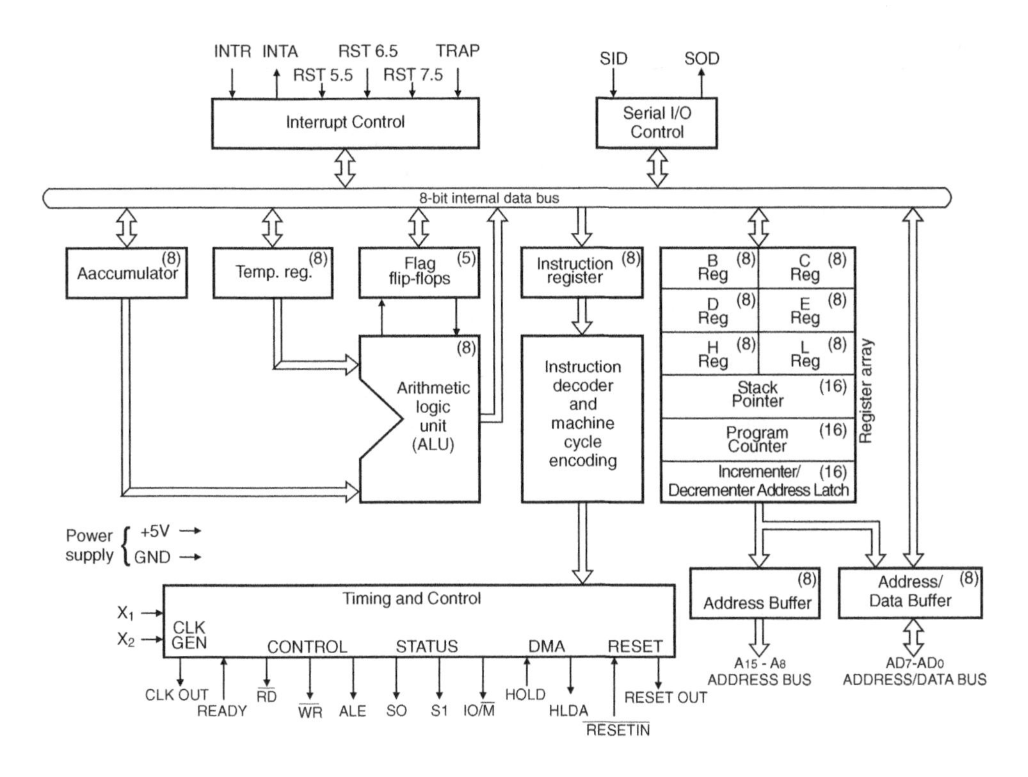

Figure 13.12 Block diagram of 8085.

Table 13.3 Pin details of 8085.

Signals	Description
Address bus (12–19, 21–29)	A 16-bit address bus. The lower eight bits are multiplexed with the data bus. The most significant eight bits of the memory address (or I/O address) are denoted by A_8–A_{15}. The lower eight bits of the memory address (or I/O address) appear on the multiplexed address/data bus (AD_0–AD_7) for the first clock cycle of the machine cycle. It then becomes the data bus during the second and third clock cycles
Data bus (12–19)	Eight-bit data bus is multiplexed with lower eight bits of the address bus (AD_0–AD_7)
Control and status signals	
ALE (Address Latch Enable) (30)	It is a positive-going pulse during the first clock state of the machine cycle that indicates that the bits on AD_7–AD_0 are address bits. It is used to latch the low-order address on the on-chip latch from the multiplexed bus
READ ($\overline{RD}$) (32)	A LOW on $\overline{RD}$ indicates that the selected memory or I/O device is ready to be read and the data bus is available for data transfer
WRITE ($\overline{WR}$) (31)	A LOW on $\overline{WR}$ indicates that data on the data bus are to be written into a selected memory or I/O location. Data are set up at the trailing edge of the $\overline{WR}$ signal

(*continued overleaf*)

Table 13.3 (*continued*).

Signals	Description
IO/$\overline{M}$(34)	This is a status signal that is used to differentiate between I/O and memory operations
S_1 and $S_{0(29,33)}$	These are status signals and can identify various operations
Power supply and clock frequency	
V_{CC} (40)	+ 5 V
V_{SS}	Ground
X_1, X_2 (20)	A crystal, LC or RC network is connected at these two pins to drive the internal clock generator. X_1 can also be an external clock input from a logic gate. The frequency is internally divided by 2 to give the internal operating frequency of the processor. The crystal frequency must be at least 1 MHz and must be twice the desired internal clock frequency
CLK OUT – clock output (37)	This output signal can be used as a system clock for devices on the board. The period of CLK is twice the X_1, X_2 input period
Interrupts and other operations: 8085 has five interrupt signals	
INTR: INTerrupt Request (10)	This is a general-purpose interrupt signal. The microprocessor issues an interrupt acknowledge signal (INTA) when the interrupt is requested
RST 7.5 (7) RST 6.5 (8) RST 5.5 (9)	These are restart interrupts. These are vectored interrupts and transfer the program control to specific memory locations
TRAP (6)	It is a nonmaskable interrupt and has the highest priority
In addition to these interrupts RESET, HOLD and READY pins accept externally initiated signals as inputs	
HOLD (39) HLDA (38)	A HOLD signal indicates that another master device is requesting the use of data and address buses. The microprocessor, upon receiving the HOLD request, will relinquish the use of the bus after completion of the current bus transfer. It sends the HOLD ACKNOWLEDGE (HLDA) signal, indicating that it will relinquish the bus in the next clock cycle
READY (35)	A READY signal is used to delay the microprocessor READ or WRITE cycles until a slow-responding peripheral is ready to send or accept data. If READY is HIGH during the READ or WRITE cycle, it indicates that the memory or peripheral is ready to send or receive data. If READY is LOW, the processor will wait for an integral number of clock cycles for READY to go to HIGH
$\overline{RSEST\ IN}$ (36) RESET OUT (3)	A LOW on the RESET IN pin causes the program counter to be set to zero, the buses are tristated and the microprocessor is reset. RESET OUT indicates that the microprocessor is being reset
Serial I/O parts	
SID (5)	Serial Input Data
SOD (4)	Serial Output Data

13.9.2 Motorola 6800 Microprocessor

This is an eight-bit microprocessor housed in a 40-pin dual in-line package (DIP) and released at the same time as Intel 8080. An important feature of 6800 is that it does not have I/O instructions, and therefore 6800-based systems had to use memory-mapped I/O for input/output capabilities. Motorola 6800 started a family of 680X microcontrollers and microprocessors, many of which are in use today. 6800 microprocessors can operate at a maximum frequency of 2 MHz. Figure 13.13 shows a block schematic representation of the internal architecture of the Motorola 6800 microprocessor.

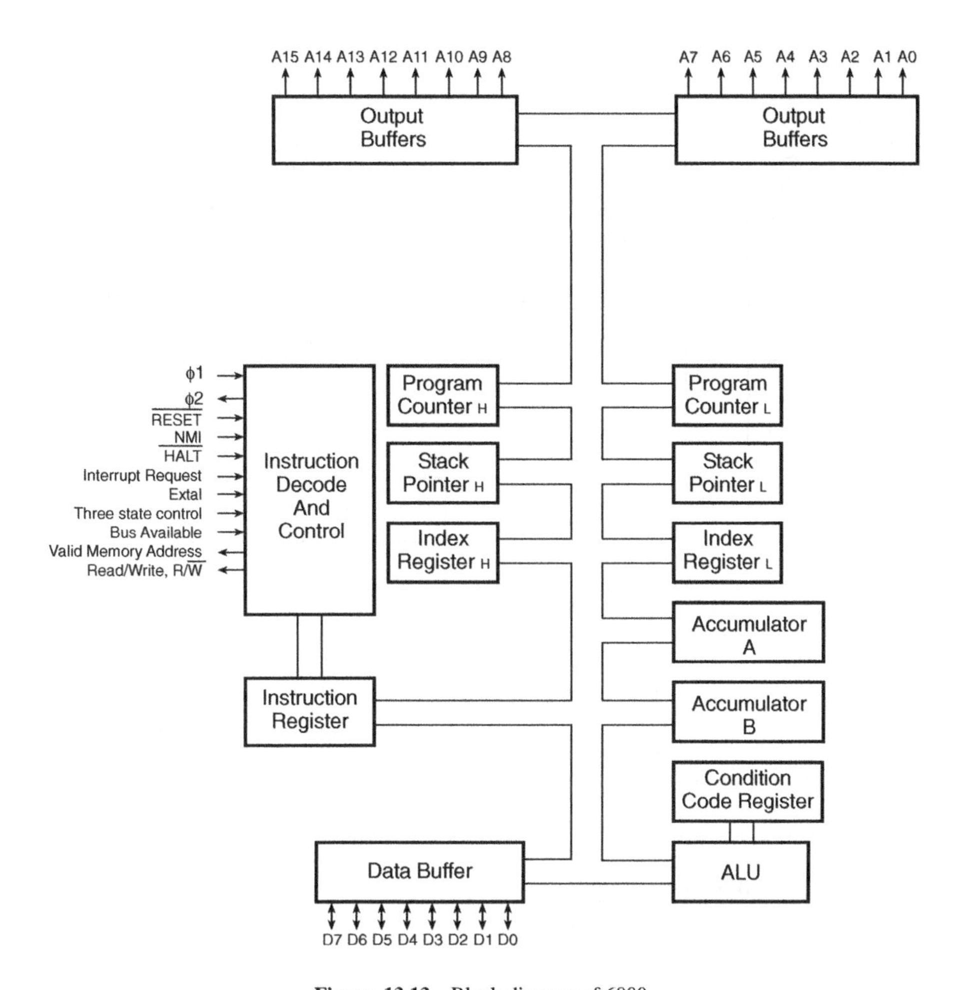

Figure 13.13 Block diagram of 6800.

13.9.2.1 6800 Registers

The 6800 microprocessors have six internal registers, namely accumulator A (ACCA), accumulator B (ACCB), an index (IX), a program counter, a stack pointer (SP) and a condition code register.

13.9.2.2 Addressing Modes

It has the implied addressing mode, accumulator addressing mode, immediate addressing mode, direct addressing mode, extended addressing mode, relative addressing mode and indexed addressing mode.

13.9.2.3 Instruction Set

The 6800 instruction set consists of 72 instructions. It supports data moving instructions, arithmetic instructions (add, subtract, negate, increment, decrement and compare), logic instructions (AND,

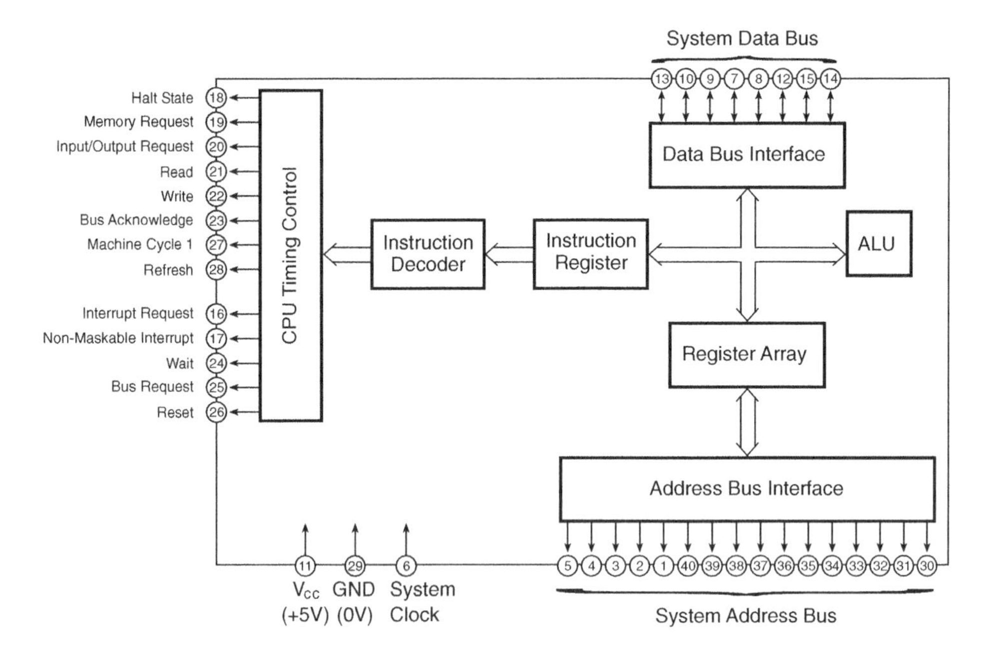

Figure 13.14 Block diagram of Z-80.

OR, EXCLUSIVE-OR, complement and shift/rotate), control transfer (conditional, unconditional, call subroutine and return from subroutine) and others – clear/set condition flags, bit test, stack operations, software interrupt, etc.

13.9.3 Zilog Z80 Microprocessor

The Zilog Z80 microprocessor is an eight-bit processor that is object-code compatible with Intel 8080. It is available in industry-standard 40-pin dual in-line and 44-pin chip carrier packages. The maximum operating frequency is 2.5 MHz. Figure 13.14 shows a block diagram of Z80.

13.9.3.1 Z80 registers

The Z80 microprocessor has registers compatible with the 8080 microprocessor as well as some other registers. The 8080-compatible registers include the accumulator, flag register (F), general-purpose registers (six programmable general-purpose registers designated B, C, D, E, H and L), stack pointer (SP) and program counter. The registers introduced with Z80 are the alternate accumulator register (A′), the alternate flag register (F′), the alternate B, C, D, E, H and L registers (represented as A′, B′, C′, D′, E′, H′ and L′), the index registers (IX and IY), the interrupt vector register (I) and the memory refresh register (R).

13.9.3.2 Instruction set

The Z80 microprocessor has 158 instructions. They perform data copy (transfer) or load operations, arithmetic, logic operations, bit manipulation, branch operations and machine control operations.

13.10 16-Bit Microprocessors

Eight-bit microprocessors are limited in their speed (the number of instructions that can be executed in 1 s), directly addressable memory, data handling capability, etc. Advances in semiconductor technology have made it possible for the manufacturers to develop 16-bit, 32-bit, 64-bit and even-larger-bit microprocessors. This section describes the block diagram, pin-out configuration and salient features of some of the most popular 16-bit microprocessors including 8086 of Intel and Motorola's MC68000.

13.10.1 8086 Microprocessor

This is a 16-bit microprocessor introduced by Intel. It was designed using HMOS technology and contains approximately 29 000 transistors. It has a maximum operating frequency of 10 MHz. The 8086, 8088, 80186 and 80286 microprocessors have the same basic set of registers and addressing modes. The 8086 microprocessor is available in DIP, CeraDIP and PLCC packages. Figure 13.15 shows a block diagram of 8086.

13.10.1.1 8086 registers

8086 has four segment registers and other general-purpose registers. The segment registers include code segment (CS), stack segment (SS), data segment (DS) and extra segment (ES). The general-purpose registers of 8086 include the accumulator register, base register, count register, data register, stack pointer (SP), base pointer (BP), source index (SI) and destination index (DI). The stack pointer, base pointer, source index and destination index registers are both general and index registers. Other registers include the instruction pointer (IP) and the flag register containing nine one-bit flags.

13.10.1.2 Addressing modes

The addressing modes of 8086 are implied addressing, register addressing, immediate addressing, direct addressing, register indirect addressing, base addressing, indexed addressing, base indexed addressing and base indexed with displacement addressing.

13.10.1.3 Internal Architecture and Pin-out Configuration

The internal functions of the 8086 processor are portioned logically into two processing units. The first is the bus interface unit (BIU) and the second is the execution unit (EU), as shown in Fig. 13.15. The BIU provides the functions related to instruction fetching and queuing, operand fetch and store and address relocation. It also provides the basic bus control. The EU receives prefetched instructions from the BIU queue and provides unrelocated operand addresses to the BIU.

13.10.1.4 Instruction set

The instruction set includes the following: data transfer operations, arithmetic operations, logical instructions, string manipulation instructions, control transfer instructions, processor control instructions and input/output operations.

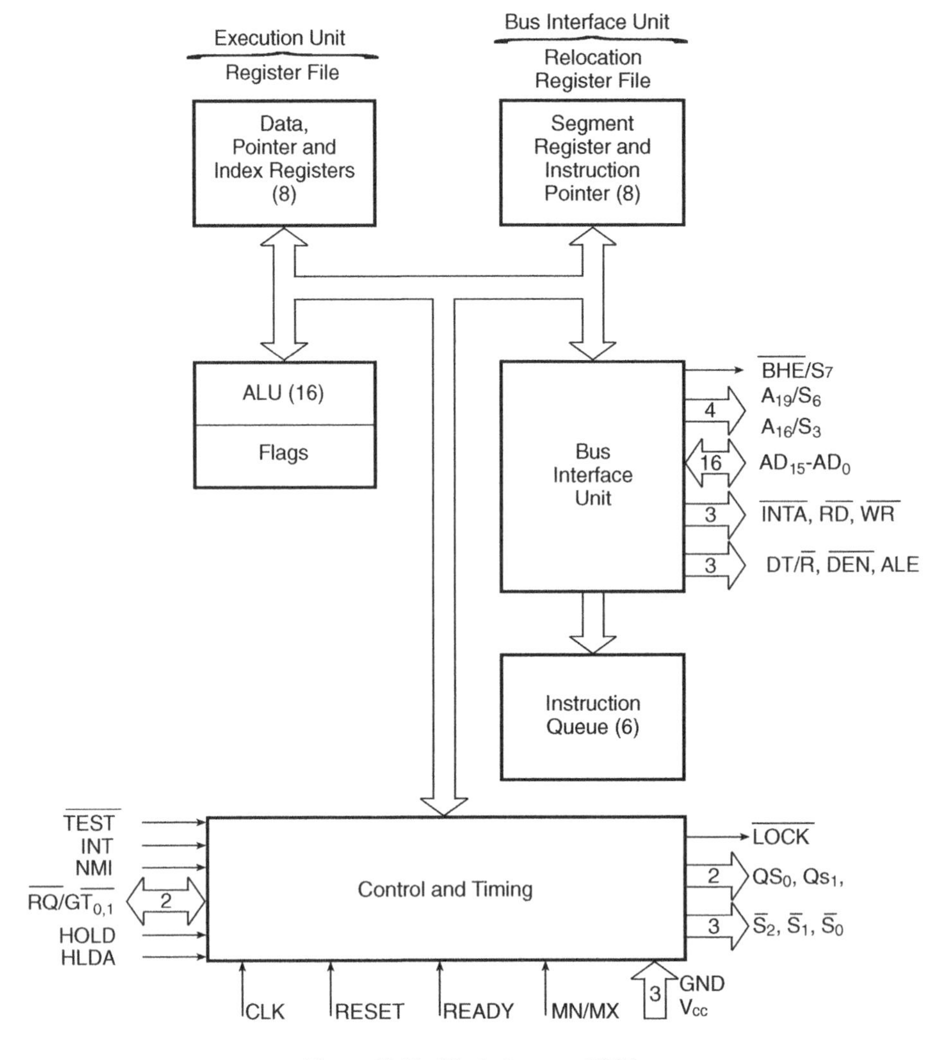

Figure 13.15 Block diagram of 8086.

13.10.2 80186 Microprocessor

The Intel 80186 is packaged in a 68-pin leadless package. It includes the Intel 8086 and several additional functional units on a single chip. The major on-chip circuits included are a clock generator, two independent DMA channels, a programmable interrupt controller, three programmable 16-bit timers and a chip select unit. It operates at a maximum frequency of 10 MHz.

13.10.3 80286 Microprocessor

The 80286 microprocessor is an advanced version of the 8086 microprocessor that was designed for multi-user and multitasking environments. It addresses 16 MB of physical memory and 1 GB of virtual memory by using its memory management system. The 80286 is packaged in a 68-pin ceramic flat package and PGA, CLCC and PLCC packages. The 80286 microprocessor can work at a maximum frequency of 12.5 MHz.

13.10.4 MC68000 Microprocessor

68000 is the first member of Motorola's family of 16-bit and 32-bit processors. It is a successor to the 6809 and was followed by the 68010. The 68000 has 32-bit registers but only a 16-bit ALU and external data bus. It has 24-bit addressing and a linear address space. Addresses are computed as 32-bit, but the top eight bits are cut to fit the address bus into a 64-pin package (address and data share a bus in the 40-pin packages of the 8086). It is available in several clock frequencies. These include 6, 8, 10, 12.5, 16.67 and 25 MHz. The 68000 microprocessor is available in two packages, namely the 64-pin ceramic DIP and the 68-pin ceramic LLCC package. Figure 13.16 shows a simplified block diagram of the 68000 microprocessor.

13.10.4.1 68000 registers

The 68000 microprocessor has 16 32-bit registers and a 32-bit program counter. There are eight data registers for byte (eight-bit), word (16-bit) and long-word (32-bit) operations. There are seven address registers. These seven registers and the user stack pointer (USP) may be used as software stack pointers and base address registers. They are also used for word and long-word operations. Data, address and USP registers may also be used as index registers. In supervisor mode, the upper byte of the status register and the supervisor stack pointer (SSP) are also available to the programmer. The status register contains the interrupt mask as well as the condition codes [extend (X), negative (N), zero (Z), overflow (V) and carry (C)]. It also has status bits to indicate whether the processor is in trace (T) mode or in supervisor (S) mode.

13.10.4.2 Instruction Set

68000 has the following instruction types: data movement operations, integer arithmetic operations, logical operations, shift and rotate operations, bit manipulation operations, program control operations and system control operations.

13.10.4.3 Addressing Modes

The 68000 microprocessor supports the following addressing modes:

1. Register direct addressing (data register direct and address register direct).
2. Absolute data addressing (absolute short and absolute long).
3. Program counter relative addressing (relative with offset, relative with index and offset).
4. Register indirect addressing (register indirect, post-increment register indirect, predecrement register indirect, register indirect with offset, indexed register indirect with offset).

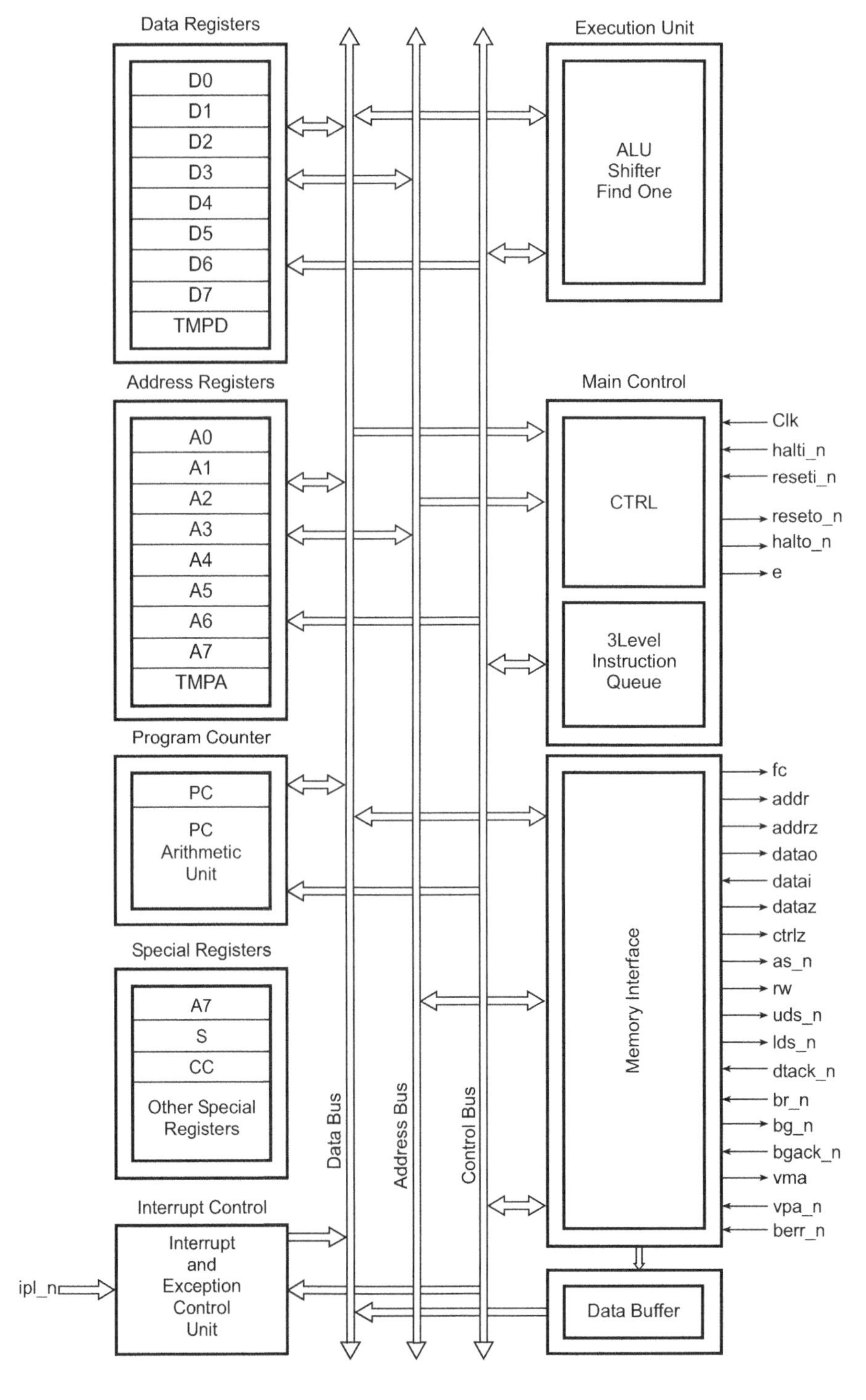

Figure 13.16 Block diagram of the 68000 microprocessor.

5. Immediate addressing (immediate and quick immediate).
6. Implied addressing (implied register).

13.11 32-Bit Microprocessors

This section describes the block diagram, internal architecture, salient features and instruction set of some of the most popular 32-bit microprocessors, namely 80386 of Intel and 68020 and 68030 of Motorola. It also gives an introduction to Intel's 80486 and Pentium series of processors.

13.11.1 80386 Microprocessor

80386 is a 32-bit microprocessor and is the logical extension of 80286. It provides multitasking support, memory management, pipeline architecture, address translation caches and a high-speed bus interface in a single chip. 80386 can be operated from a 12.5, 16, 20, 25 or 33 MHz clock. The 80386 has three processing modes, namely the protected mode, the real address mode and the virtual 8086 mode. The protected mode is the natural 32-bit environment of the 80386 processor. In this mode, all instructions and features are available. The real address mode is the mode of the processor immediately after RESET. In real mode, 80386 appears to programmers as a fast 8086 with some new instructions. Most applications of the 80386 will use the real mode for initialization only. The virtual 8086 mode (also called the V86 mode) is a dynamic mode in the sense that the processor can switch repeatedly and rapidly between V86 mode and protected mode.

Two versions of 80386, namely the 80386DX and the 80386SX, are commonly available. 80386SX is a reduced bus version of the 80386. The 80386DX addresses 4 GB of memory through its 32-bit data bus and 32-bit address bus. The 80386SX addresses 16 MB of memory with its 24-bit address bus. It was developed after the 80386DX for applications that did not require the full 32-bit bus version. A new version of 80386, named the 80386EX, incorporates the AT bus system, dynamic RAM controller, programmable chip selection guide, 26 address pins, 16 data pins and 24 I/O pins. Figure 13.17 shows the block diagram of the 80386 processor.

13.11.1.1 80386 DX registers

80386 DX contains a total of 32 registers. These registers may be grouped into general registers, segment registers, status and instruction registers, control registers, system address registers and debug and test registers.

13.11.1.2 Instruction Set

80386 DX executes the following instruction types:

1. Data movement instructions (general-purpose data movement instructions, stack manipulation instructions and type conversion instructions).
2. Binary arithmetic instructions (addition and subtraction instructions, comparison and size change instructions, multiplication instructions and division instructions).
3. Decimal arithmetic instructions (packed BCD adjustment and unpacked BCD adjustment instructions).

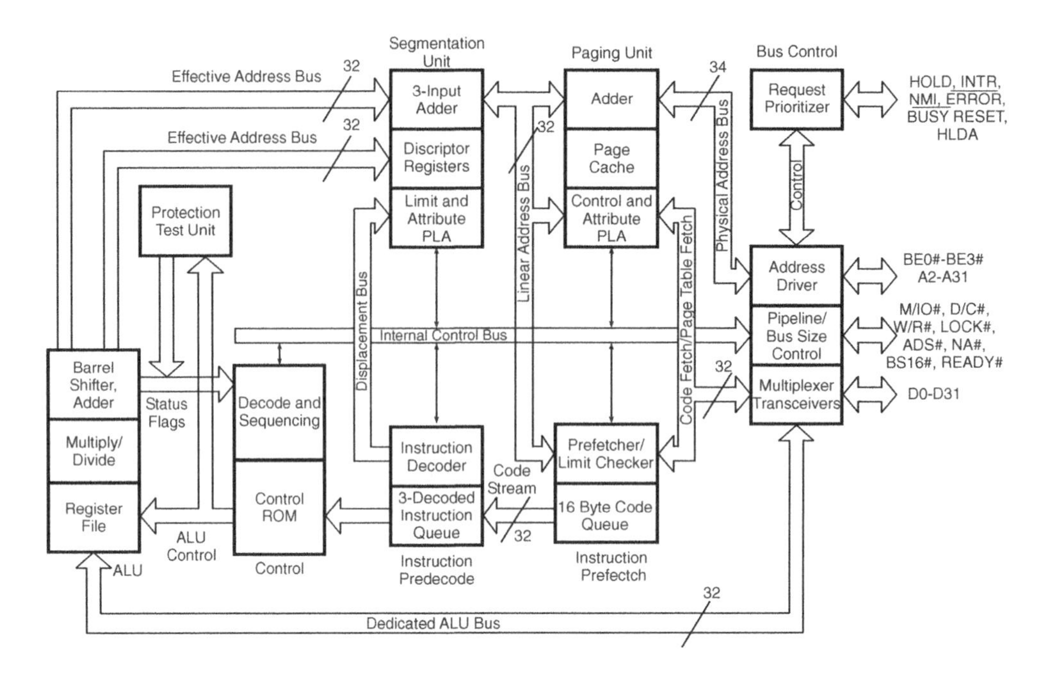

Figure 13.17 Block diagram of the 80386 microprocessor.

4. Logical instructions (Boolean operation, bit test and modify, bit scan, rotate and shift, byte set ON condition).
5. Control transfer instructions (unconditional transfer, conditional transfer, software-generated interrupts).
6. String and character translation instructions (repeat prefixes, indexing and direction flag control, string instructions).
7. Instructions for block-structured languages.
8. Flag control instructions (carry and direction flag control instructions and flag transfer instructions).
9. Coprocessor interface instructions.
10. Segment register instructions (segment register transfer, far-control transfer and data pointer instructions).
11. Miscellaneous instructions (address calculation, no-operation instruction and translate instruction).

13.11.1.3 Addressing Modes

80386 DX supports a total of 11 addressing modes as follows:

1. *Register and immediate modes.* These two modes provide for instructions that operate on register or immediate operands. These include register addressing mode and immediate addressing mode.
2. *32-bit memory addressing modes.* The remaining nine modes provide a mechanism for specifying the effective address of an operand. Here, the effective address is calculated by using combinations of displacement, base, index and scale address elements. The combination of these four elements

makes up the additional nine addressing modes. These include the direct mode, register indirect mode, based mode, index mode, scaled index mode, based index mode, based scaled index mode, based index mode with displacement and based scaled mode with displacement.

13.11.2 MC68020 Microprocessor

This is a 32-bit microprocessor introduced by Motorola. It can execute an object code written for MC68000, and therefore upward compatibility is maintained. It can operate at 12.5, 16.67, 20, 25 or 33 MHz. The MC68020 is supported by an array of peripheral devices and can directly be interfaced to coprocessor chips such as the MC68881/MC68882 floating-point and MC68851 memory management unit (MMU) coprocessor. It can directly address 4 GB of memory. The 68020 microprocessor also has an on-chip cache of size 128 words (16-bit). It is available in a PGA 114 ceramic-pin grid-array package and in CQFP 132 (Ceramic Quad Flat Package). Figure 13.18 shows the block diagram of Motorola's MC68020.

13.11.2.1 68020 Registers

68020 is a true 32-bit processor and it is object-code compatible with 68000. It has many more registers than 68000. Besides the eight data registers, seven address registers, one program counter and one status register (SR), there are three stack pointer (SP) registers instead of two. There is also one 16-bit vector-based register (VBR), two three-bit function code registers, one 32-bit cache address register (CAAR) and one 32-bit cache control register (CACR).

13.11.2.2 Instruction set

More than 20 new instructions have been added over MC68000. The new instructions include some minor improvements and extensions to the supervisor state, several instructions for software management of a multi-processing system, some support for high-level languages, bigger multiply (32×32) and divide (64/32) instructions and bit field manipulations.

13.11.2.3 Addressing modes

The 68020 microprocessor supports a total of 18 addressing modes with nine basic types:

1. Register direct (data register direct and address register direct).
2. Register indirect (address register indirect, address register indirect with post-increment, address register indirect with predecrement and address register indirect with displacement).
3. Register address indirect with index (register address indirect with index and register address indirect with index).
4. Memory indirect (memory indirect post-indexed and memory indirect pre-indexed).
5. Program counter indirect with displacement.
6. Program counter indirect with index (eight-bit displacement and base displacement).
7. Program counter memory indirect (post-indexed, pre-indexed).
8. Absolute data addressing (short and long).
9. Immediate addressing.

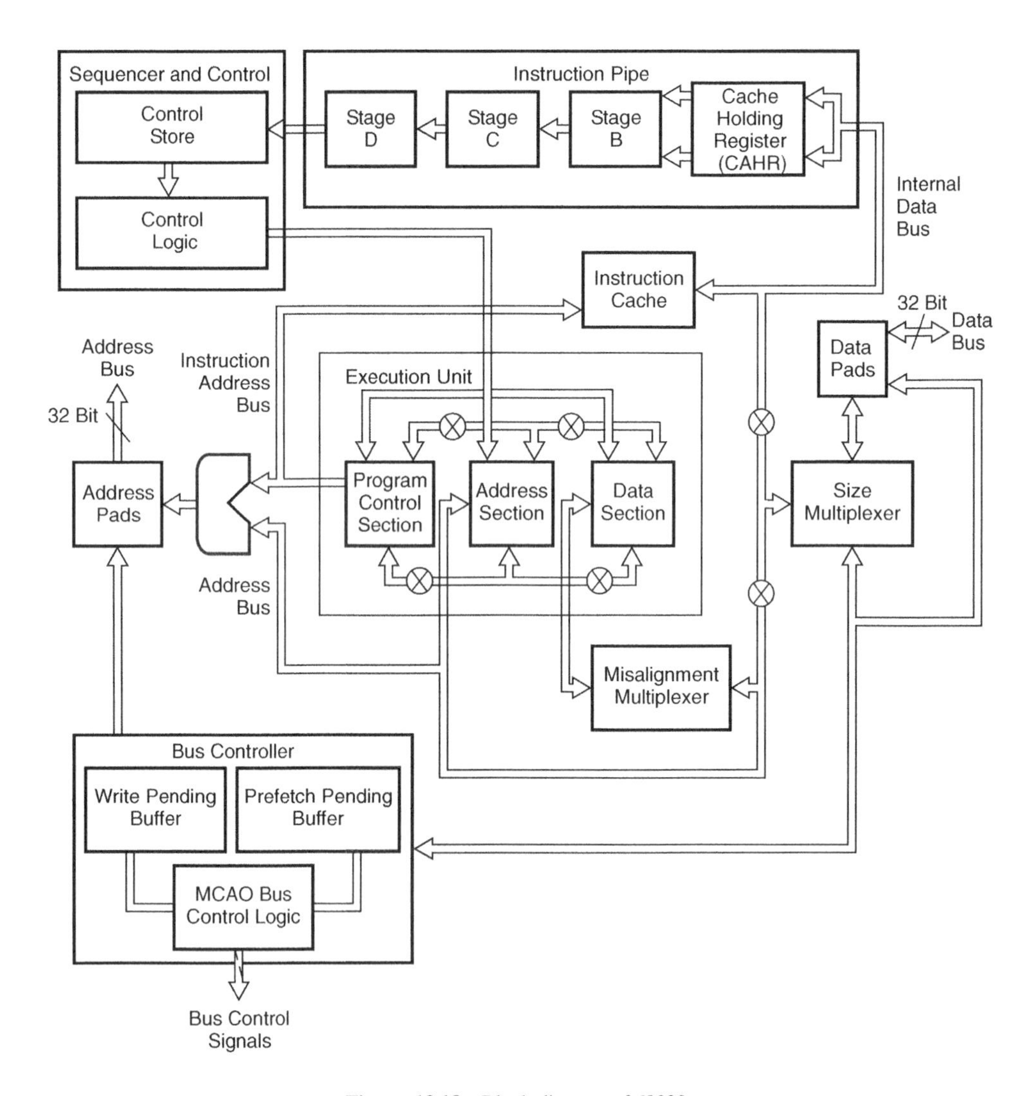

Figure 13.18 Block diagram of 68020.

13.11.3 MC68030 Microprocessor

The MC68030 is a second-generation full 32-bit virtual-memory microprocessor designed using HCMOS technology (Combining HMOS and CMOS on the same device) from Motorola. It is based on an MC68020 core with additional features. The MC68030 is a member of the M68000 family of devices that combines a central processing unit (CPU) core, a data cache, an instruction cache, an enhanced bus controller and a memory management unit (MMU) in a single VLSI device. It can be operated at 16.67, 20 and 33 MHz clocks. The MC68030 is upward-object-code compatible with the earlier members of the M68000 family and has the added features of an on-chip MMU, a data cache and an improved bus interface. The MC68030 fully supports the nonmultiplexed bus structure of the MC68020, with 32 bits of address and 32 bits of data. The MC68030 bus has an enhanced controller that supports both asynchronous and synchronous bus cycles and burst data transfers.

13.11.4 80486 Microprocessor

The 80486 (i486 is the trade name) offers high performance for DOS, OS/2, Windows and UNIX System V applications. It is 100 % compatible with 80386 DX and SX microprocessors. One million transistors integrate cache memory, floating-point hardware and a memory management unit on-chip while retaining binary compatibility with previous members of the x86 architectural family. Frequently used instructions execute in one cycle, resulting in RISC performance levels. An eight-byte unified code and data cache combined with an 80/106 MB/s burst bus at 25/33 MHz ensure high system throughput even with inexpensive DRAMs.

The 80486 microprocessor is currently available in versions operating at 25, 33, 50, 66 and 100 MHz frequency. It is available as 80486DX and 80486SX. The only difference between these two devices is that 80486SX does not contain the numeric coprocessor. The 80487SX numeric coprocessor is available as a separate component for the 80486SX microprocessor.

Salient features of the 80486 processor include:

1. Full binary compatibility with 386 DX CPU, 386 SX CPU, 376 embedded processor and 80286, 8086 and 8088 processors.
2. Execution unit designed to execute frequently used instructions in one clock cycle.
3. 32-bit integer processor for performing arithmetic and logical operations.
4. Internal floating-point arithmetic unit for supporting the 32-, 64- and 80-bit formats specified in IEEE standard 754 (object-code compatible with 80387 DX and 387 SX math coprocessors).
5. Internal 8 kB cache memory, which provides fast access to recently used instructions and data.
6. Bus control signals for maintaining cache consistency in multiprocessor systems.
7. Segmentation, a form of memory management for creating independent, protected address space.
8. Paging, a form of memory management that provides access to data structures larger than the available memory space by keeping them partly in memory and partly on disk.
9. Restartable instructions that allow a program to be restarted following an exception (necessary for supporting demand-paged virtual memory).
10. Pipelined instruction execution overlaps the interpretation of different instructions.
11. Debugging registers for hardware support of instruction and data breakpoints.

The 80486 is object-code compatible with three other 386 processors, namely the 386 DX processor, the 386 SX processor (16-bit data bus) and the 376 embedded processor (16-bit data bus). 80486SX is also available in the same package with a few differences, as mentioned below:

1. Pin B15 is NMI on the 80486DX and pin A15 is NMI on 80486SX.
2. Pin A15 is $\overline{IGNNE}$ on 80486DX. It is not present on 80486SX.
3. C14 is $\overline{FERR}$ on 80486DX and pins B15 and C14 on 80486SX are not connected.

The architecture of 80486DX is almost the same as that of 80386 except that it contains a math coprocessor and an 8K byte level 1 cache memory. 80486SX does have the math coprocessor. Figure 13.19 shows the internal architecture of 80486DX. The major difference between 80386 and 80486 is that almost half of the instructions of 80486 execute in one clocking period instead of the two clocking periods for the 80386 microprocessor for the same instructions.

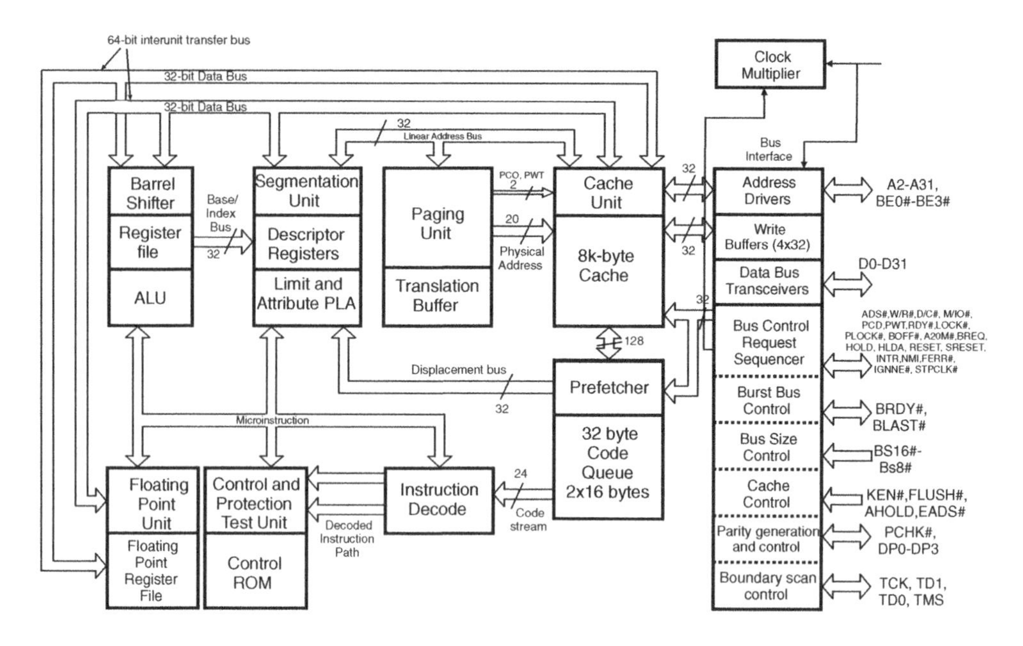

Figure 13.19 Internal architecture of 80486DX.

13.11.4.1 80486DX instruction set

The instruction set can be divided into 11 categories including data transfer operations, arithmetic operations, shift/rotate instructions, bit manipulation instructions, string manipulation instructions, control transfer instructions, high-level language support instructions, operating system support instructions, floating point processor control instructions and floating-point control instructions

13.11.4.2 80486DX registers

It contains all the registers of the 80386 microprocessor and 80386 math coprocessor. The register set is divided into the following categories: base architecture registers, general-purpose registers, instruction pointers, flag registers, segment registers, systems-level registers, control registers, system address registers, floating-point registers, data registers, tag word, status word, instruction and data pointers and control word and debug and test registers.

13.11.4.3 80486DX operating modes

The operating mode of the 80486 processor determines which instructions and architectural features are accessible. The 80486 has three modes for running programs. These are as follows:

1. The protected mode uses the native 32-bit instruction set of the processor. In this mode, all instructions and architectural features are available.

2. The real address mode (also called the 'real mode') emulates the programming environment of the 8086 processor, with a few extensions (such as the ability to break out of this mode). Reset initialization places the processor into real mode.
3. The virtual 8086 mode (also called the 'V86 mode') is another form of 8086 emulation mode. Unlike the real address mode, virtual 8086 mode is compatible with protection and memory management. The processor can enter virtual 8086 mode from protected mode to run a program written for the 8086 processor, then leave virtual 8086 mode and re-enter protected mode to run programs that use the 32-bit instruction set.

13.11.5 PowerPC RISC Microprocessors

The PowerPC family of microprocessors are high-performance superscalar RISC microprocessors developed jointly by MOTOROLA, IBM and Apple. They are used in personal computers, workstations and servers as well as for industrial and commercial embedded applications. Different versions of PowerPC microprocessors include PowerPC 601, 602, 603, ec603e, 604, 604e, 620, 740, 750, 745, 755, 750CX, 750CXE, 750FX and 750X microprocessors. PowerPC 601, 602, 603, ec603e and 604 are 32-bit microprocessors with a 32-bit address bus and a 64-bit data bus. They have 32 32-bit general-purpose registers and 32 64-bit floating-point registers. PowerPC 601 was introduced in the year 1993 for desktop PCs and low-end workstation applications. It uses 0.5 μm process technology and is available in 100 and 120 MHz clock frequency versions. PowerPC 602 was designed for graphical and multimedia applications. PowerPC 603 was introduced in the year 1993 and was used for applications where low power consumption was a critical requirement. It operates at 100 MHz. Its version 603e is an embedded microprocessor operating at 300 MHz. PowerPC 604 introduced in the year 1994 is available in different versions operating at 80, 100, 120, 133 and 250 MHz.

PowerPC 740/750 are 32-bit RISC microprocessors with special added features including a faster memory bus (66 MHz), larger L1 caches, enhanced integer and floating-point units and higher core frequency. PowerPC 750CX was developed by IBM using an 0.18 μm copper process. PowerPC 750FX was introduced in the year 2002 and had an operational frequency of up to 900 MHz. PowerPC 750GX introduced in the year 2004 is the latest and most powerful G3 processor from IBM.

13.12 Pentium Series of Microprocessors

The Pentium family of processors has its roots in the Intel 486 processor and has the same instruction set with a few additional instructions. Pentium processors have a 64-bit data bus and represent a major step forward in personal computer CPU design. The first Pentium processors (the P5 variety) were introduced in 1993. They were fabricated in 0.8 μm bipolar complementary metal oxide semiconductor (BiCMOS) technology. The P5 processor runs at a clock frequency of either 60 or 66 MHz and has 3.1 million transistors. The next version of the Pentium processor family was the P54 processor. The P54 processors were fabricated in 0.6 μm BiCMOS technology. The P54 was followed by P54C, introduced in 1994, which used a 0.35 μm CMOS process, as opposed to the bipolar CMOS process used for the earlier Pentiums. The P5 operated on 5 V supply and the P54 and P54C series operated on a 3.5 V supply voltage. All these processors had a problem in the floating-point unit. They were followed by the P55C processor, also referred to as the Pentium MMX. It was based on the P5 core and fabricated using the 0.35 μm process. The performance of the P55C was improved over the previous versions by doubling the level 1 CPU cache from 16 to 32 kB. Intel has retained the Pentium trademark for naming later generations of processor architectures, which are internally quite different from the Pentium itself. These include Pentium Pro, Pentium II, Pentium M, Pentium D and Pentium Extreme Edition.

The Pentium processor has two primary operating modes and a system management mode. The operating mode determines which instructions and architectural features are accessible. These modes are as follows:

1. *Protected mode.* This is the native state of the microprocessor. In this mode, all instructions and architectural features are available, providing the highest performance and capability.
2. *Real address mode.* This mode provides the programming environment of the Intel 8086 processor, with a few extensions. Reset initialization places the processor in real mode where, with a single instruction, it can switch to protected mode.
3. *System management mode.* It provides an operating system and application independent transparent mechanism to implement system power management and OEM differentiation features. SMM is entered through activation of an external interrupt pin (SMI#), which switches the CPU to a separate address space while saving the entire content of the CPU.

13.12.1 Salient Features

The Pentium series (P5, P54 and P54C) of microprocessors has the following advanced features:

1. *Superscalar execution.* The Intel 486 processor can execute only one instruction at a time. With superscalar execution, the Pentium processor can sometimes execute two instructions simultaneously.
2. *Pipeline architecture.* Like the Intel 486 processor, the Pentium processor executes instructions in five stages. This staging, or pipelining, allows the processor to overlap multiple instructions so that it takes less time to execute two instructions in a row. Because of its superscalar architecture, the Pentium processor has two independent processor pipelines.
3. *Branch target buffer.* The Pentium processor fetches the branch target instruction before it executes the branch instruction.
4. *Dual 8 kB on-chip caches.* The Pentium processor has two separate 8 kB caches on chip, one for instructions and the other for data. This allows the Pentium processor to fetch data and instructions from the cache simultaneously.
5. *Write-back cache.* When data are modified, only the data in the cache are changed. Memory data are changed only when the Pentium processor replaces the modified data in the cache with a different set of data.
6. *64-bit bus.* With its 64-bit wide external data bus (in contrast to the Intel 486 processor's 32-bit wide external bus), the Pentium processor can handle up to twice the data load of the Intel 486 processor at the same clock frequency.
7. *Instruction optimization.* The Pentium processor has been optimized to run critical instructions in fewer clock cycles than the Intel 486 processor.
8. *Floating-point optimization.* The Pentium processor executes individual instructions faster through execution pipelining, which allows multiple floating-point instructions to be executed at the same time.
9. *Pentium extensions.* The Pentium processor has fewer instruction set extensions than the Intel 486 processors. The Pentium processor also has a set of extensions for multiprocessor (MP) operation. This makes a computer with multiple Pentium processors possible.

13.12.2 Pentium Pro Microprocessor

Pentium Pro is a sixth-generation x86 architecture microprocessor (P6 core) from Intel. It was originally intended to replace the earlier Pentium series of microprocessors in a full range of applications, but was later reduced to a narrow role as a server and high-end desktop chip. The Pentium Pro was capable of both dual- and quad-processor configurations. The Pentium Pro achieves a performance approximately 50 % higher than that of a Pentium of the same clock speed. In addition to its new way of processing instructions, the Pentium Pro incorporates several other technical features including superpipelining, an integrated level 2 cache, 32-bit optimization, a wider address bus, greater multiprocessing, out-of-order completion of instructions, a superior branch prediction unit and speculative execution.

13.12.3 Pentium II Series

Pentium II is an x86 architecture microprocessor introduced by Intel in the year 1997. It was based on a modified version of the P6 core improved 16-bit performance and the addition of the MMX SIMD instruction set. The Pentium II series of processors are available in speeds of 233, 266, 300, 330, 350, 400 and 450 MHz. Some of the product highlights include the use of Intel's 0.25 μm manufacturing process for increased processor core frequencies and reduced power consumption, the use of MMX bus (DIB) architecture to increase bandwidth and performance over single-bus processors, a 32 kB nonblocking level 1 cache, a 512 kB unified, nonblocking level 2 cache and data integrity and reliability features.

13.12.4 Pentium III and Pentium IV Microprocessors

Pentium III is an x86 architecture microprocessor from Intel, introduced in the year 1999. Initial versions were very similar to the earlier Pentium II. The most notable difference is the addition of SSE instructions and the introduction of a serial number which was embedded in the chip during the manufacturing process. Pentium III processors are available in speeds of 650, 667, 700, 733, 750, 800, 850 and 866 MHz and 1 GHz. The Pentium III processor integrates PC dynamic execution microarchitecture, DIB architecture, a multitransaction system bus and Intel's MMX media enhancement technology. In addition to these features, it offers Internet streaming and single-instruction multiple-data (SIMD) extension. It has 70 new instructions to enable advanced imaging, 3D, streaming audio and video and speech recognition. Pentium III processors were superseded by Pentium IV.

Pentium IV is a seventh-generation x86 architecture microprocessor from Intel. It uses a new CPU design, called the netburst architecture. The netburst microarchitecture featured a very deep instruction pipeline, with the intention of scaling to very high frequencies. It also introduced the SSE2 instruction set for faster SIMD integer and 64-bit floating-point computation. It operates at frequencies of over 1 GHz.

13.12.5 Pentium M, D and Extreme Edition Processors

Pentium M is an x86 architecture microprocessor from Intel, introduced in the year 2003. It forms part of the Intel Centrino platform. The processor was originally designed for use in laptop personal computers (thus the 'M' for mobile).

Pentium D is a series of microprocessors from Intel introduced in the year 2005. Pentium D was the first multicore CPU along with the Pentium Extreme Edition. It is the final processor to carry the

Pentium brand name. The Pentium Extreme Edition series of microprocessors was introduced by Intel in the year 2005. It is based on the dual-core Pentium D processor.

13.12.6 Celeron and Xeon Processors

Celeron processors were introduced by Intel as a low-cost CPU alternative for the Pentium processors. They were basically Pentium II processors without any L2 cache at all. However, this reduced the performance of Celeron processors as compared with AMD and Cyrix chips. Hence, subsequent Celeron versions (300A and up) were provided with 128 kB of L2 cache. It was about one-fourth the size of the Pentium cache but operated at the full speed of the respective CPU, rather than at half-speed as in the Pentium processors. Later Celeron versions were based on the Pentium III, Pentium IV and Pentium M processors. These processors are suitable for most applications, but their performance is somewhat limited when it comes to running intense applications. Xeon are high-end processors having a full-speed L2 cache of the same size as the Pentium cache. These processors are used for high-performance servers and workstations.

13.13 Microprocessors for Embedded Applications

Embedded microprocessors are microprocessors designed for embedded applications and not for use in personal computers. They are mostly used for embedded data control applications such as data processing, data formatting, I/O control, DMA data transfer, etc. In other words, they are designed for specific applications rather than for general-purpose applications. Intel has developed a number of embedded microprocessors, namely Intel 80960, Intel 80376 and embedded versions of 80486, 80386 and 80186 microprocessors. Other embedded microprocessors include Motorola's Coldfire, Sun's Sparc, Hitachi's SuperH, Advanced RISC Machines' ARM, and MIPS Computer Systems Inc.'s MIPS processors.

The Intel 80960 and 80376 microprocessors are 32-bit microprocessors designed for sophisticated industrial control applications. Embedded versions of 80486 include 486GX, 486SX, 496DX2 and 486DX4 microprocessors. The embedded versions of 80386 include 386CXSA, 386CXSB, 386EX and 386SXSA microprocessors. Scalable processor architecture (SPARC) microprocessors are 32-bit and 64-bit CISC processors from Sun Microsystems. ARM microprocessors are 32-bit RISC microprocessors and are mostly used in the mobile electronics market, where low power is the most critical design requirement. MIPS (Microprocessor without Interlocked Pipeline Stages) is a RISC microprocessor from MIPS Computer Systems Inc. They are available in 32-bit and 64-bit versions.

13.14 Peripheral Devices

Microprocessors and peripheral devices provide a complete solution in increasingly complex application environments. A peripheral device typically belongs to the category of MSI logic devices. This section gives an introduction to the popular peripheral devices that are used along with the microprocessor in a microcomputer system. The different peripheral devices used in a microcomputer system include a programmable counter/timer, a programmable peripheral interface (PPI), EPROM, RAM, a programmable interrupt controller (PIC), a direct memory access (DMA) controller, a programmable communication interface – a universal synchronous/asynchronous receiver/transmitter (USART), a math coprocessor, a programmable keyboard/display interface, a CRT controller, a floppy disk controller and clock generators and transceivers.

13.14.1 Programmable Timer/Counter

The programmable timer/counter is used for the generation of an accurate time delay for event counting, rate generation, complex waveform generation applications and so on. Examples of programmable timer/counter devices include Intel's 8254 and 8253 family of devices. Intel 8254 contains three 16-bit counters that can be programmed to operate in several different modes. Some of the functions common to microcomputers and implementable with 8254 are a real-time clock, an event counter, a digital one-shot, programmable rate generator, a square-wave generator, a binary rate multiplier, a complex waveform generator and a complex motor controller. It is available in 24-pin CERDIP and plastic DIP packages.

13.14.2 Programmable Peripheral Interface

Programmable peripheral interface (PPI) devices are used to interface the peripheral devices with the microprocessors. 8255 PPI is a widely used programmable parallel I/O device. It is available in PDIP, CerDIP, PLCC and MQPF packages. 8255 can be programmed to transfer data under various conditions, from simple I/O to interrupt I/O. It can function in bit reset (BSR) mode or I/O mode. In I/O mode it has three ports, namely port A, port B and port C. The I/O mode is further divided into three different modes, namely mode 0, mode 1 and mode 2. In mode 0, all ports function as simple I/O ports. Mode 1 is a handshake mode whereby port A and/or B use bits from port C as handshake signals. In mode 2, port A can be set up for bidirectional data transfer using handshake signals from port C, and port B can be set up either in mode 0 or in mode 1. In BSR mode, individual bits in port C can be set or reset.

13.14.3 Programmable Interrupt Controller

A programmable interrupt controller (PIC) is a device that allows priority levels to be assigned to its interrupt outputs. It functions as an overall manager in an interrupt-driven system environment. When the device has multiple interrupt outputs, it will assert them in the order of their relative priority. Common modes of a PIC include hard priorities, rotating priorities and cascading priorities. Intel 8259 is a family of programmable interrupt controllers (PICs) designed and developed for use with the Intel 8085 and Intel 8086 microprocessors. The family originally consisted of the 8259, 8259A, and 8259B PICs, although a number of manufacturers make a wide range of compatible chips today.

It handles up to eight vectored priority interrupts for the CPU. It is designed to minimize the software and real-time overhead in handling multi-level priority interrupts. It accepts requests from peripheral equipment, determines which of the incoming requests is of the highest priority, ascertains whether an incoming request has a higher priority value than the level currently being serviced and issues an interrupt to the CPU on the basis of this determination.

13.14.4 DMA Controller

In a direct memory access (DMA) data transfer scheme, data are transferred directly from an I/O device to memory, or vice versa, without going through the CPU. The DMA controller is used to control the process of data transfer. Its primary function is to generate, upon a peripheral request, a sequential memory address that will allow the peripheral to read or write data directly to or from memory. One of the popular known programmable DMA controllers is Intel's 8257. It is a four-channel direct memory access (DMA) controller. It is specifically designed to simplify the transfer of data at high speeds for microcomputer systems. It has a priority logic that resolves the peripheral requests and issues a composite hold request to the CPU. It maintains the DMA cycle count for each channel and outputs a control signal to notify the peripheral that the programmed number of DMA cycles is completed.

13.14.5 Programmable Communication Interface

Programmable communication interfaces (PCIs) are interface devices that are used for data communication applications with microprocessors. They basically convert the data from the microprocessor into a format acceptable for communication and also convert the incoming data into a format understood by the microprocessor.

8251 is a PCI device designed for Intel's 8085, 8086 and 8088 microprocessors and is used in serial communication applications. It is a 28-pin chip available in DIP and PLCC packages. It is basically a universal synchronous/asynchronous receiver/transmitter (USART) that accepts data characters from the CPU in parallel format and then converts them into a continuous serial data stream for transmission. Simultaneously, it can receive a serial data stream and convert it into parallel data characters for the CPU. The USART will signal the CPU whenever it can accept a new character for transmission or whenever it has received a character for the CPU.

13.14.6 Math Coprocessor

Math coprocessors are special-purpose processing units that assist the microprocessor in performing certain mathematical operations. The arithmetic operations performed by the coprocessor are floating-point operations, trigonometric, logarithmic and exponential functions and so on. Examples include Intel's 8087, 80287, etc. The 8087 numeric coprocessor provides the instructions and data types needed for high-performance numeric application, providing up to 100 times the performance of a CPU alone. Another widely used math coprocessor is 80287. The 80287 numeric processor extension (NPX) provides arithmetic instructions for a variety of numeric data types in 80286 systems. It also executes numerous built-in transcendental functions (e.g. tangent and log functions).

13.14.7 Programmable Keyboard/Display Interface

Programmable keyboard/display interfaces are devices used for interfacing the keyboard and the display to the microprocessor. The keyboard section of the device debounces the keyboard entries and provides data to the microprocessor in the desired format. The display section converts the data output of the microprocessor into the form desired by the display device in use.

8279 is a general-purpose programmable keyboard and display I/O interface device designed for use with Intel microprocessors. The keyboard portion can provide a scanned interface to a 64-contact key matrix. Keyboard entries are debounced and strobbed in eight-character FIFO. If more than eight characters are entered, overrun status is set. Key entries set the interrupt output line to the CPU. The display portion provides a scanned display interface for LED, incandescent and other popular display technologies. Both numeric and alphanumeric segment displays may be used. The 8279 has a 16×8 display RAM.

13.14.8 Programmable CRT Controller

The programmable CRT controller is a device to interface CRT raster scan displays with the microprocessor system. Its primary function is to refresh the display by buffering the information from the main memory and keeping track of the display position of the screen. One of the commonly used programmable CRT controllers is Intel's 8275H. It allows a simple interface to almost any raster scan CRT display with minimum external hardware and software overheads. The number of display characters per row and the number of character rows per frame are software programmable.

13.14.9 Floppy Disk Controller

The floppy disk controller is used for disk drive selection, head loading, the issue of read/write commands, data separation and serial-to-parallel and parallel-to-serial conversion of data. Examples of floppy disk controllers include Intel's 82078, 82077 and 8272.

13.14.10 Clock Generator

The clock generator is a circuit that produces a timing signal for synchronization of the circuit's operation. Examples of clock generators used in microprocessor systems include 8284 and 82284. 8284 generates the system clock for the 8086 and 8088 processors. It requires a crystal or a TTL signal source for producing clock waveforms. It provides local READY and MULTIBUS READY synchronization.

82284 is a clock generator/driver that provides clock signals for the 80286 processor and support components. It also contains logic to supply READY to the CPU from either asynchronous or synchronous sources and synchronous RESET from an asynchronous input with hysteresis. The 82284 is packaged in 18-pin DIP and contains a crystal-controlled oscillator, an MOS clock generator, a peripheral clock generator, multibus ready synchronization logic and system reset generation logic.

13.14.11 Octal Bus Transceiver

Bus transceivers are devices with a high-output drive capability for interconnection with data buses. In a microprocessor-based system they provide an interface between the microprocessor bus and the system data bus. 8286 is an eight-bit bipolar transceiver with a three-state output that is used in a wide variety of buffering applications in microcomputer systems. It comes in a 20-pin DIP package.

Review Questions

1. Briefly describe the difference between a microprocessor and a microcomputer. What are the three main constituents of a microprocessor and what is the basic function performed by each one of them.
2. What are the different types of register found in a typical microprocessor? Briefly describe the function of each one of them.
3. Distinguish between the following

    (a) address bus and data bus;
    (b) direct addressing mode and indirect addressing mode;
    (c) programmable timer and clock generator;
    (d) programmable interrupt controller and DMA controller;
    (e) RISC and CISC microprocessors.

4. Briefly describe the parameters that you would consider while choosing the right microprocessor for your application, emphasizing the significance of each parameter.
5. With the help of a labelled diagram, briefly describe the operational role of the three types of bus in a microcomputer system.
6. Briefly describe the primary functions of the following peripheral devices. Also, give at least one device type number for each of them:

   (a) programmable timer;
   (b) clock generator;
   (c) programmable peripheral interface;
   (d) DMA controller;
   (e) programmable interrupt controller.

7. Briefly describe salient features of the Pentium series of microprocessors.
8. Compare and contrast:

   (a) eight-bit microprocessors,
   (b) 16-bit microprocessors,
   (c) 32-bit microprocessors and
   (d) 64-bit microprocessors
   from Intel and Motorola.

Further Reading

1. Brey, B. B. (2000) *The Intel Microprocessors 8086/8088, 80186/80188, 80286, 80326, 80486, Pentium, and Pentium Pro Processor Architecture, Programming, and Interfacing*, Prentice-Hall, NJ, USA.
2. Floyd, T. L. (2005) *Digital Fundamentals*, Prentice-Hall Inc., USA.
3. Crisp, J. (2004) *Introduction to Microprocessors and Microcontrollers*, Newnes, Oxford, UK.
4. Tocci, R. J. and Ambrosio, F. J. (2002) *Microprocessors and Microcomputers: Hardware and Software*, Prentice-Hall, NJ, USA.
5. Rafiquzzaman, M. (1990) *Microprocessor and Microcomputer-based System Design*, CRC Press, FL, USA.

14

Microcontrollers

Microcontrollers are hidden inside almost every product or device with which its user can interact. In fact, any device that has a remote control or has an LED/LCD screen and a keypad has an embedded microcontroller. Some common products where one is sure to find the use of a microcontroller include automobiles, microwave ovens, TVs, VCRs, high-end stereo systems, camcorders, digital cameras, washing machines, laser printers, telephone sets with caller ID facility, mobile phones, refrigerators and so on. This chapter focuses on microcontroller fundamentals and the application-related aspects of it. Beginning with an introductory description of the device, with particular reference to its comparison with a microprocessor, the chapter covers the general architecture and the criteria to be followed to choose the right device for a given application. This is followed by application-relevant information, such as salient features, pin configuration, internal architecture, etc., of popular brands of eight-bit, 16-bit, 32-bit and 64-bit microcontrollers from major international manufacturers. Intel's 8051 family of microcontrollers is described in more detail.

14.1 Introduction to the Microcontroller

The microcontroller may be considered as a specialized computer-on-a-chip or a single-chip computer. The word 'micro' suggests that the device is small, and the word 'controller' suggests that the device may be used to control one or more functions of objects, processes or events. It is also called an embedded controller as microcontrollers are often embedded in the device or system that they control.

The microcontroller contains a simplified processor, some memory (RAM and ROM), I/O ports and peripheral devices such as counters/timers, analogue-to-digital converters, etc., all integrated on a single chip. It is this feature of the processor and peripheral components available on a single chip that distinguishes it from a microprocessor-based system. A microprocessor is nothing but a processing

Digital Electronics: Principles, Devices and Applications Anil Kumar Maini

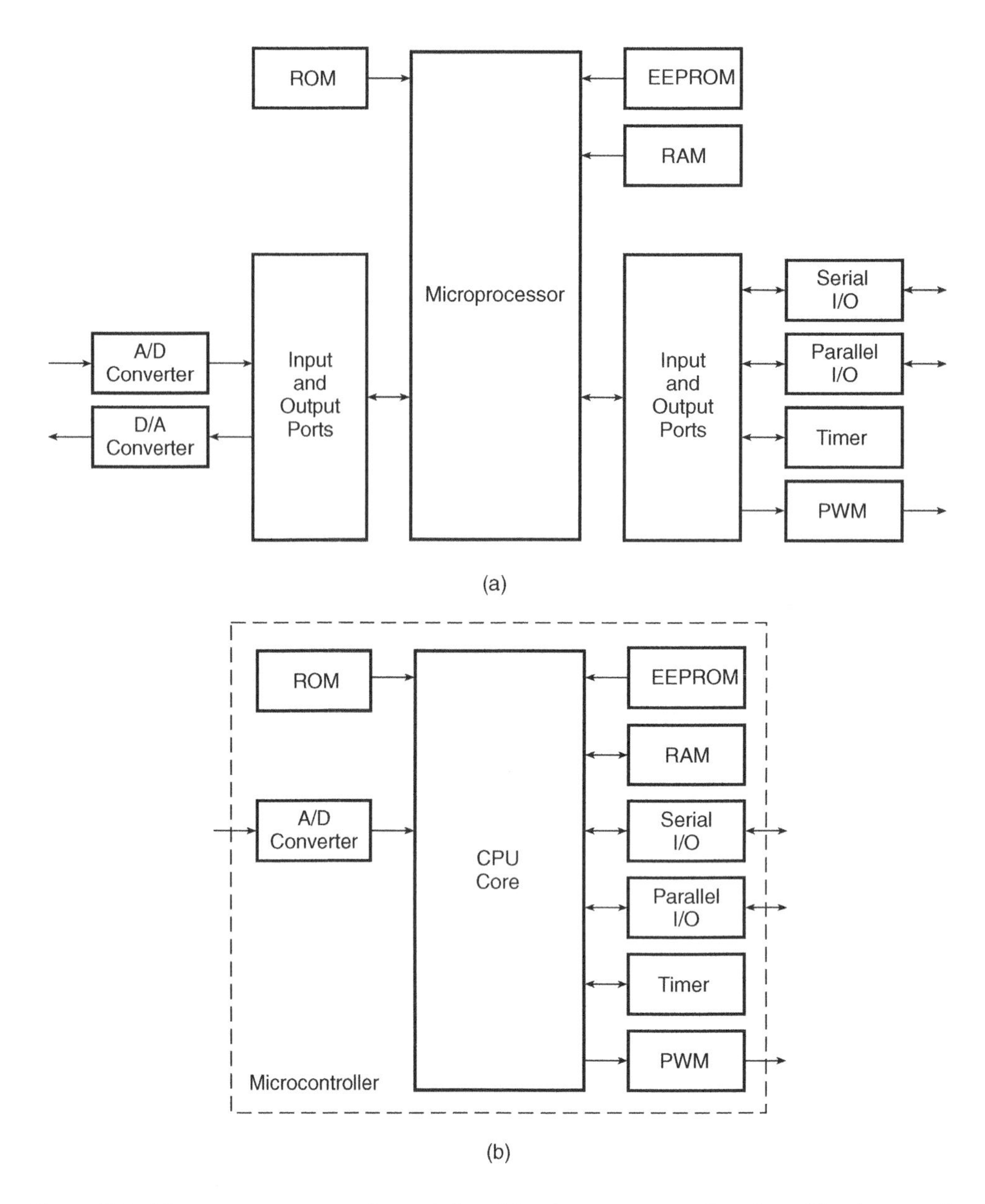

Figure 14.1 Microprocessor versus microcontroller: (a) microprocessor configuration; (b) microcontroller configuration.

unit with some general-purpose registers. A microprocessor-based system also has RAM, ROM, I/O ports and other peripheral devices to make it a complete functional unit, but all these components are external to the microprocessor chip. While a microprocessor-based system is a general-purpose system that may be programmed to do any of the large number of functions it is capable of doing, microcontrollers are dedicated to one task and run one specific program. This program is stored in ROM and generally does not change.

Figure 14.1 further illustrates the basic difference between a microprocessor-based system and a microcontroller. As is evident from the two block schematics shown in the figure, while a microprocessor-based system needs additional chips to make it a functional unit, in a microcontroller the functions of all these additional chips are integrated on the same chip.

14.1.1 Applications

Microcontrollers are embedded inside a surprisingly large number of product categories including automobiles, entertainment and consumer products, test and measurement equipment and desktop computers, to name some prominent ones.

Any device or system that measures, stores, controls, calculates or displays information is sure to have an embedded microcontroller as a part of the device or system. In automobiles, one or more microcontrollers may be used for engine control, car cruise control (Fig. 14.2), antilock brakes and so on. Test and measurement equipment such as signal generators, multimeters, frequency counters, oscilloscopes, etc., make use of microcontrollers to add features such as the ability to store measurements, to display messages and waveforms and to create and store user routines. In desktop computers, microcontrollers are used in peripheral devices such as keyboards, printers, modems, etc. Consumer and entertainment products such as TVs, video recorders, camcorders, microwave ovens, washing machines, telephones with caller ID facility, cellular phones, air conditioners, refrigerators and many more products make extensive use of microcontrollers to add new control and functional features.

14.2 Inside the Microcontroller

Figure 14.3 shows the block schematic arrangement of various components of a microcontroller. As outlined earlier, a microcontroller is an integrated chip with an on-chip CPU, memory, I/O ports and some peripheral devices to make a complete functional unit. A typical microcontroller as depicted in Fig. 14.4 has the following components: a central processing unit (CPU), a random access memory (RAM), a read only memory (ROM), special-function registers and peripheral components including serial and/or parallel ports, timers and counters, analogue-to-digital (A/D) converters and digital-to-analogue (D/A) converters.

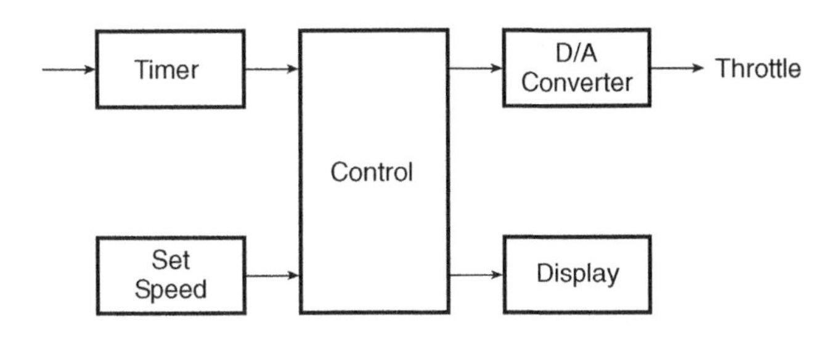

Figure 14.2 Microcontroller-based car cruise control.

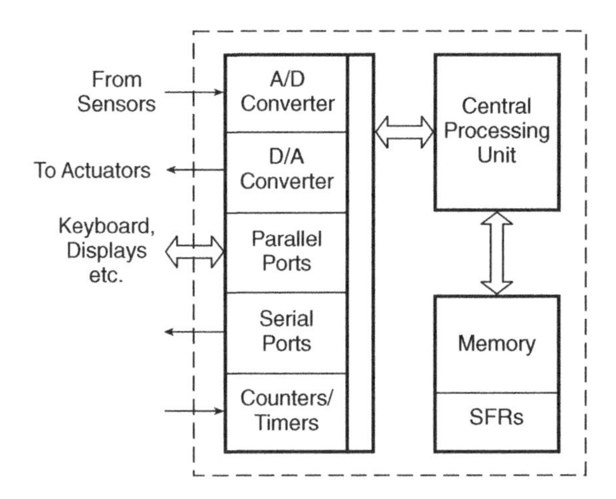

Figure 14.3 Inside the microcontroller.

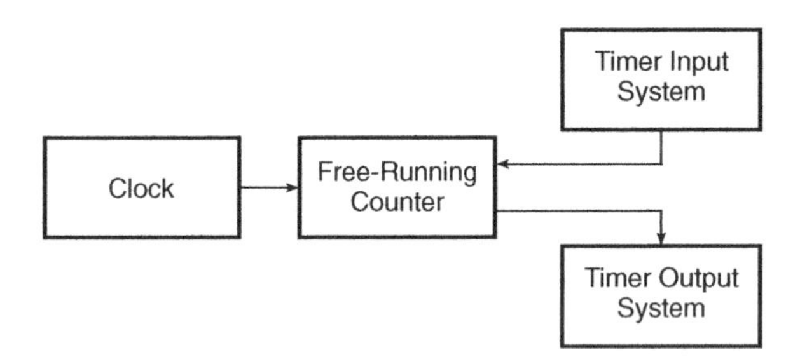

Figure 14.4 Timer subsystem.

14.2.1 Central Processing Unit (CPU)

The central processing unit processes the program. It executes the instructions stored in the program memory pointed to by the program counter in synchronization with the clock signal. The processor complexity could vary from simple eight-bit processors to sophisticated 32-bit or even 64-bit processors. Some common microcontrollers using eight-bit processors include 68HC11 (Freescale Semiconductor – earlier part of Motorola), the 80C51 family of microcontrollers (Intel and Dallas Semiconductor), Zilog-eZ8 and Zilog-eZ80 (Zilog) and XC800 (Infineon). Examples of microcontrollers using 16-bit processors include the 8096 family (Intel), 68HC12 and 68HC16 (Freescale Semiconductor), the F2MC family (Fujitsu) and the XC166 family (Infineon). Examples of microcontrollers using 32-bit processors include 683XX, MPC 860 (PowerQUICC), MPC 8240/8250 (PowerQUICC-II) and MPC 8540/8555/8560 (PowerQUICC-III) (all from Freescale Semiconductor), the TRICORE family (Infineon) and the FR/FR-V family (Fujitsu).

14.2.2 Random Access Memory (RAM)

RAM is used to hold intermediate results and other temporary data during the execution of the program. Typically, microcontrollers have a few hundreds of bytes of RAM. As an example, microcontroller type numbers 8XC51/80C31, 8XC52/80C32 and 68HC12 respectively have 128, 256 and 1024 bytes of RAM.

14.2.3 Read Only Memory (ROM)

ROM holds the program instructions and the constant data. Microcontrollers use one or more of the following memory types for this purpose: ROM (mask-programmed ROM), PROM (one-time programmable ROM, which is not field programmable), EPROM (field programmable and usually UV erasable), EEPROM (field programmable, electrically erasable, byte erasable) and flash (similar to EEPROM technology). Microcontroller type numbers 8XC51, 8XC51FA and 8XC52 have 4K, 8K and 16K of ROM. As another example, the 68HC12 16-bit microcontroller has 32K of flash EEPROM, 768 bytes of EEPROM and 2K of erase-protected boot block.

14.2.4 Special-Function Registers

Special-function registers control various functions of a microcontroller. There are two categories of these registers. The first type includes those registers that are wired into the CPU and do not necessarily form part of addressable memory. These registers are used to control program flow and arithmetic functions. Examples include status register, program counter, stack pointer, etc. These registers are, however, taken care of by compilers of high-level languages, and therefore programmers of high-level languages such as C, Pascal, etc., do not need to worry about them. The other category of registers is the one that is required by peripheral components. The contents of these registers could, for instance, set a timer or enable serial communication and so on. As an example, special-function registers available on the 80C51 family of microcontrollers (80C51, 87C51, 80C31) include a program counter, stack pointer, RAM address register, program address register and PC incrementer.

14.2.5 Peripheral Components

Peripheral components such as analogue-to-digital converters, I/O ports, timers and counters, etc., are available on the majority of microcontrollers. These components perform functions as suggested by their respective names. In addition to these, microcontrollers intended for some specific or relatively more complex functions come with many more on-chip peripherals. Some of the common ones include the pulse width modulator, serial communication interface (SCI), serial peripheral interface (SPI), interintegrated circuit (I^2C) two-wire communication interface, RS 232 (UART) port, infrared port (IrDA), USB port, controller area network (CAN) and local interconnect network (LIN). These peripheral devices are briefly described in the following paragraphs.

14.2.5.1 Analogue-to-Digital Converters

Analogue-to-digital and digital-to-analogue converters provide an interface with analogue devices. For example, the analogue-to-digital converter provides an interface between the microcontroller and the sensors that produce analogue electrical equivalents of the actual physical parameters to be controlled.

The digital-to-analogue converter, on the other hand, provides an interface between the microcontroller and the actuators that provide the control function. As an example, both 68HC11 and 68HC12 from Freescale Semiconductor have eight-channel, eight-bit analogue-to-digital converters. The digital-to-analogue converter function in microcontrollers is provided by a combination of pulse width modulator (PWM) followed by a filter. As an example, 68HC12 has an on-chip 16-bit/two-channel PWM. Analogue-to-digital and digital-to-analogue converters are discussed at length in Chapter 12.

14.2.5.2 I/O Ports

I/O ports provide an interface between the microcontroller and the peripheral I/O devices such as the keyboard, display, etc. The 80C51 family of microcontrollers has four eight-bit I/O ports. Microcontroller 68HC11 offers 38 general-purpose I/O pins including 16 bidirectional I/O pins, 11 input-only pins and 11 output-only pins.

14.2.5.3 Counters/Timers

Counters/timers usually perform the following three functions. They are used to keep time and/or measure the time interval between events, count the number of events and generate baud rates for the serial ports. Microcontroller 68HC11 has a 16-bit timer system comprising three input capture channels, four output compare channels and one additional channel that can be configured as either an input or an output channel. Another popular microcontroller type number, PIC 16F84, has an eight-bit timer/counter with an eight-bit prescaler.

Figure 14.4 shows a generalized block schematic representation of the timer subsystem of a microcontroller. The clock signal controls all timing activities of the microcontroller. The counter is used both to capture external timing events (accomplished by the timer input block) and to generate timing events for external devices (accomplished by the timer output block). While the former process is typically used to measure the frequency and time interval of periodic signals, the latter generates control signals for external devices.

It may be mentioned here that a timing event to be captured or generated is nothing but a change in logic status on one of the microcontroller I/O pins configured as an input pin if the event is to be captured and as an output pin if it is to be generated. Figure 14.5 shows a block schematic arrangement of the timer input block of Fig. 14.4. As shown in the figure, the counter captures the input time event in the form of its contents at the time of occurrence of the event. In fact, the counter captures the relative time of the event as the counter is free running. Absolute timing values can be computed from the relative system clock values. As an example, consider a microcontroller with a 10 MHz clock and a 16-bit counter/timer subsystem. This counter will take 6.5536 ms to count from 0000 to FFFF (hex notation). Let us assume that it is desired to find the frequency of a periodic signal whose successive rising or falling edges are observed to occur at 0010 and 0150. 0010 and 0150 respectively correspond to 16 and 336 in decimal. Therefore, the time interval between two successive edges equals $320 \times 0.1 = 32$ μs. The signal frequency is therefore (1/32) MHz = 31.25 kHz.

Figure 14.6 shows a block schematic arrangement of the timer output block of Fig. 14.4. The diagram is self-explanatory. Again, free-running counter values can be used to synchronize the time of the desired logic state changes on the output pin. This feature can also be used to generate an aperiodic pulse or a periodic signal of any desired duty cycle.

For timer input and output operations, the microcontroller needs to set up some special registers. For timer input operation, as shown in Fig. 14.5, registers are required to program the event (logic HIGH or logic LOW), configure the physical I/O pin as an input pin and also set up parameters for the

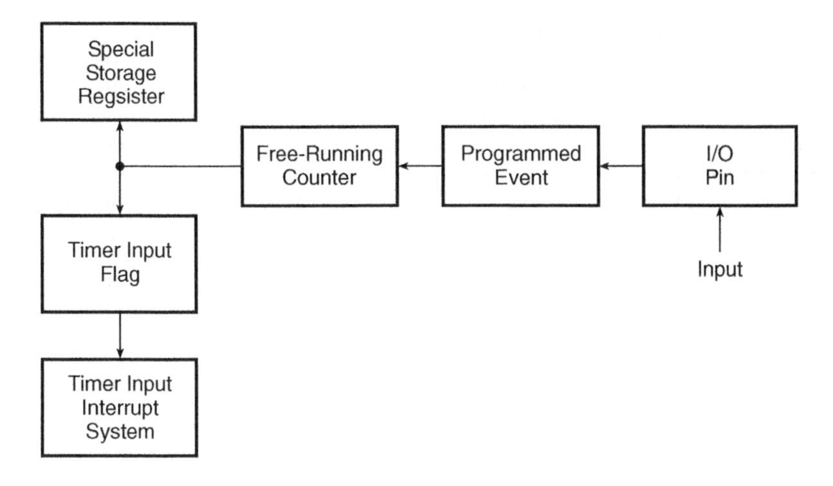

Figure 14.5 Timer input subsystem.

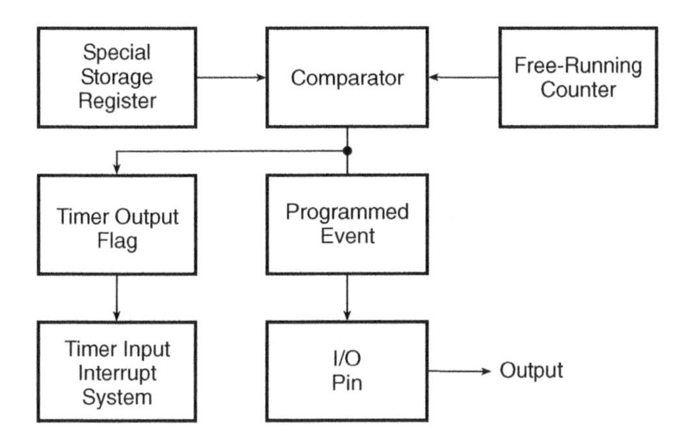

Figure 14.6 Timer output subsystem.

related interrupt, if used. Another register is used to capture the counter value at the time of occurrence of the event. For time output operation, as shown in Fig. 14.6, the physical I/O pin is to be configured as an output pin, the event is to be programmed and the timing value is to be set in the special register to tell when the programmed event should appear on the output pin. The output timer system also has an associated interrupt that can be utilized if needed.

14.2.5.4 Serial Communication Interfaces

There are two types of serial communication interface, namely the asynchronous communication interface and the synchronous communication interface. The asynchronous communication interface uses a start and stop bit protocol to synchronize the transmitter and receiver. Start and stop bits are embedded in each data byte. Compared with the synchronous communication interface, it offers lower data transmission rates. It is also referred to as the universal asynchronous receiver/transmitter (UART) or the serial communication interface (SCI). The synchronous communication interface uses a synchronized clock to transmit and receive each bit. Synchronization of transmitter and receiver clocks is usually accomplished by using an additional clock line linking the transmitter and the receiver. It is not recommended for long distance communication. It is also referred to as the serial peripheral interface (SPI). Microcontroller 68HC11 offers an asynchronous non-return-to-zero serial communication interface and also a synchronous serial peripheral interface. The 80C51 family of microcontrollers offers a full duplex-enhanced UART interface.

Since a large number of peripheral devices are equipped to communicate with an RS-232-compatible interface, which is a serial interface standard that specifies the different aspects, including electrical, mechanical, functional and procedural specifications, a variety of chips are available to translate microcontroller signals to RS-232-compatible signals. These chips are equipped to provide interfacing for a two-way communication system.

14.2.5.5 Interintegrated Circuit (I^2C) Bus

The interintegrated circuit (I^2C) bus is a two-wire, low- to-medium-speed serial communication interface developed by Philips Semiconductors in the early 1980s for chip-to-chip communications. The two wires in the I^2C bus are called clock (SCL) and data (SDA). The SDA wire carries data, while the SCL wire synchronizes the transmitter and receiver during data transfer.

It is a proven industry-standard communication protocol used in a variety of electronic products, which is particularly facilitated by its low cost and powerful features. It is supported by a large number of semiconductor and system manufacturers who offer a variety of electronic products including input and output devices, different types of sensor, memory devices, displays, data entry devices, etc. Some of the important features offered by I^2C devices are briefly described in the following paragraphs.

I^2C devices offer master–slave hierarchy. These are classified as either master (the device that initiates the message) or slave (the device that responds to the message). The device can be either master only or slave only or can be switched between master and slave depending upon the application requirement. One possible master–slave configuration is the one where one master (e.g. a microcontroller) is connected to many chips configured as slaves, as shown in Fig. 14.7. Each of the I^2C slave devices is

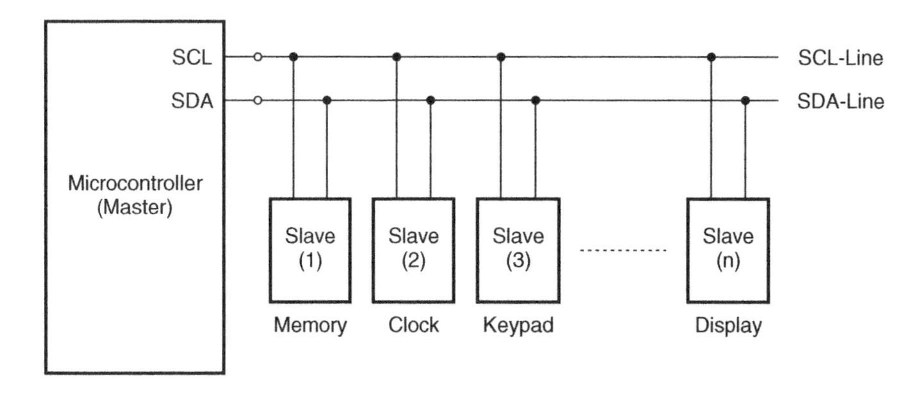

Figure 14.7 Master–slave configuration – one I^2C master and multiple slaves.

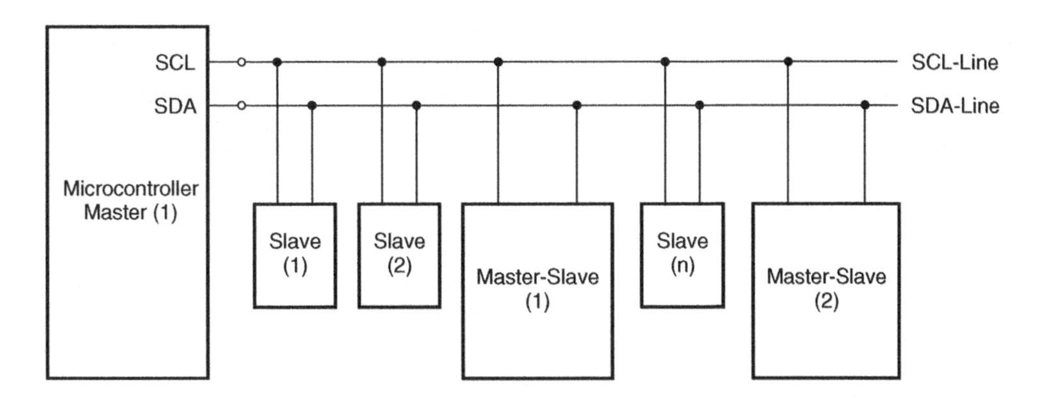

Figure 14.8 Master–slave configuration – multiple-master support arrangement.

identifiable by a unique address. When the master device sends a message, it includes the address of the intended slave device at the beginning of the message.

The I^2C interface also supports multiple master devices at the same time. The bus has a special feature that allows it to resolve signal conflicts should two or more master devices try to talk on the bus at the same time. A master I^2C device that detects the conflict, called arbitration loss, terminates its use of the bus, thus allowing the message sent by another master to cross the bus unharmed. Figure 14.8 shows one such multimaster support arrangement.

14.2.5.6 Controller Area Network (CAN) Bus

The controller area network (CAN) bus is a rugged serial communication interface used in a broad range of embedded as well as automation control applications. It was introduced by Bosch in 1986 for in-vehicle networks in automobiles. The CAN protocol was internationally standardized in 1993 as ISO-11898-1 and comprises the data link layer of the seven-layer ISO/OSI reference model. The protocol provides two communication services, namely data frame transmission (sending of a message) and remote transmission request (requesting of a message). All other services such as error signalling, automatic retransmission of erroneous frames, etc., are performed by CAN chips. Some of the important features of the CAN protocol include the following. It provides a multimaster hierarchy. This allows the user to build intelligent and redundant systems. It uses the broadcast communication method. The sender of a message transmits to all devices connected to the bus. All devices read the message and decode it if it is intended for them. This feature guarantees data integrity. Data integrity is also ensured by error detection mechanisms and automatic retransmission of faulty messages.

CAN protocol provides low-speed fault-tolerant transmission at a rate of 125 kbps up to a distance of 40 m, which can function over one wire if there is a short. Transmission without fault tolerance is provided at a rate of 1 Mbps up to a distance of 40 m. Transmission rates of 50 kbps are achievable up to a distance of 1 km.

14.2.5.7 Local Interconnect Network (LIN) Bus

The local interconnect network (LIN) bus is a broadcast serial network that is used as a low-cost subnetwork of a CAN bus to integrate intelligent sensors or actuators in modern automobiles. It comprises one master (typically a moderately powerful microcontroller) and up to 16 slaves (less powerful, cheaper microcontrollers or ASICs). It does not offer a collision detection feature and therefore all messages are initiated by the master with at the most one slave replying to a given message identifier. Multiple such LIN networks may all be linked to a CAN upper layer network through their respective masters.

Example 14.1

A certain microcontroller has an on-chip 16-bit counter/timer system. It is used to measure the width of an input pulse. The microcontroller has been programmed to measure the time of occurrence of rising and falling edges of an input pulse on a certain I/O pin. If the microcontroller uses an 8 MHz clock and the count values observed at the time of occurrence of rising and falling edges of the input pulse are 001F and 00F1 (in hex), determine the pulse width as measured by the microcontroller.

Solution

- Since the microcontroller uses a 16-bit counter, it counts from 0000 to FFFF (in hex) or 0 to 65536 in decimal.
- The rising edge of the input pulse occurs at 001F, the decimal equivalent of which is 31.
- The falling edge occurs at 00F1, the decimal equivalent of which is 241.
- Therefore, the input pulse width accounts for $241 - 31 = 210$ clock cycles.
- The clock signal time period $= 1/8 = 0.125$ μs.
- Therefore, the time period corresponding to 210 cycles $= 210 \times 0.125 = 26.25$ μs.
- The pulse width measured by the microcontroller $= 26.25$ μs.

Example 14.2

It is desired to design a microcontroller-based periodic signal generator with minimum and maximum time period specifications of 125 ns and 100 ms. What should the system clock frequency be?

Solution

- The minimum time period that can be generated by the microcontroller equals the time period corresponding to one clock cycle.
- Therefore, one clock cycle time period = 125 ns.
- The clock frequency = 1/125 GHz = 1000/125 MHz = 8 MHz.

14.3 Microcontroller Architecture

Microcontroller architecture may be defined in several ways. These include architecture used by the processor to access memory, architecture used for mapping special-function registers into memory space and the processor architecture itself.

14.3.1 Architecture to Access Memory

There are two fundamental architectures used by the processing units to access memory, namely Von Neumann architecture and Harvard architecture.

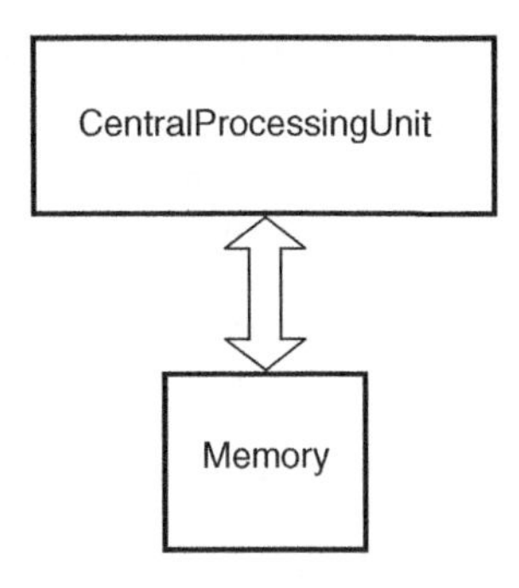

Figure 14.9 Von Neumann architecture.

Von Neumann architecture uses a single memory to hold both program instructions and data. There is one common data and address bus between processor and memory (Fig. 14.9). Instructions and data are fetched in sequential order, thus limiting the operation data transfer rate or the throughput. This phenomenon is commonly referred to as the Von Neumann bottleneck. The throughput is very small compared with the size of the memory. In present-day machines, the throughput is also very small compared with the rate at which the processor itself can work. In the condition where the processor is required to perform minimal processing on large amounts of data, the processor is forced to wait for vital data to be transferred from or to memory. Microcontroller type number 68HC11 uses Von Neumann architecture.

Harvard architecture uses physically separate memories for program instructions and data. It therefore requires separate buses for program and data, as shown in Fig. 14.10. In such architecture, instructions and operands can be fetched simultaneously, which makes microcontrollers using this architecture much faster compared with the ones using Von Neumann architecture. Also, different data and program bus widths are possible, which allows the program and data memory to be better optimized to architectural requirements. In fact, the word width, timing, implementation technology and memory address structure can be different in the two cases. Program memory is usually much larger than data memory, which implies that the address bus for the program memory is wider than the address bus for the data memory.

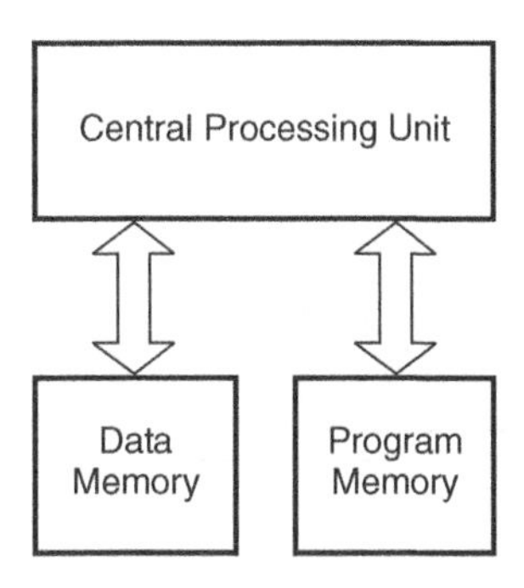

Figure 14.10 Harvard architecture.

14.3.2 Mapping Special-Function Registers into Memory Space

There are again two fundamental architectures used for mapping special-function registers into memory space. In the first type of arrangement, I/O space and memory space are separated as shown in Fig. 14.11(a). I/O devices have a separate address space, which is accomplished by either an extra I/O pin on the CPU physical interface or through a dedicated I/O bus. As a result of this, access to I/O control registers requires special instructions. It is particularly attractive in CPUs having a limited addressing capability. It is generally found on Intel microprocessors.

In the second arrangement, called the memory-mapped I/O, I/O control registers are mapped into memory address space as shown in Fig. 14.11(b). Read and write operations to the control registers are done via absolute memory addresses, which could be variables at absolute addresses or pointers to absolute addresses in high-level languages. In this case, no special instructions are needed to access I/O control registers. The memory-mapped I/O uses the same bus to address both memory and I/O devices. CPU instructions used to read from or write to memory are also used in accessing I/O devices.

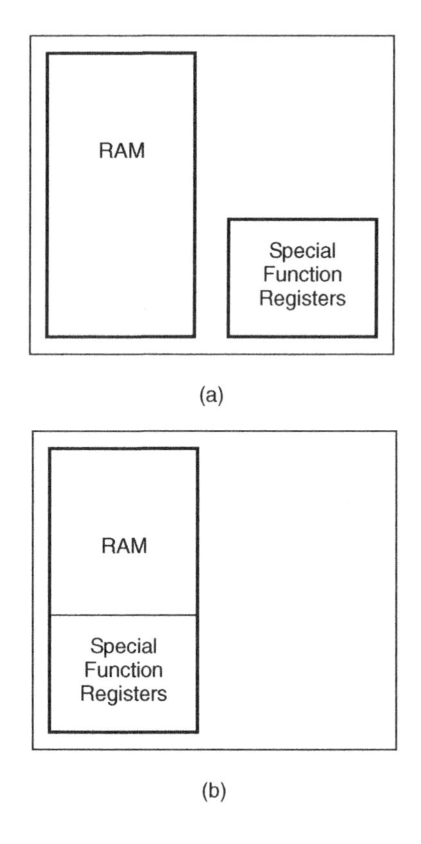

Figure 14.11 (a) Separate I/O and memory spaces and (b) memory-mapped I/O.

14.3.3 Processor Architecture

Processor architecture can be categorized as one of the following four architectures: accumulator-based architecture, register-based architecture, stack-based architecture and pipeline architecture.

14.3.3.1 Accumulator-based Architecture

In accumulator-based architecture, as shown in Fig. 14.12, instructions begin and end in accumulators (Acc A and Acc B in Fig. 14.12), which are specially designated registers. In a typical operation, one of the operands is found in the accumulator and the other is to be fetched from memory. The result of the operation is placed in the accumulator. As one of the operands needs to be continually fetched from memory, this architecture is slower than the register-based and stack-based architectures. However, accumulator-based architecture has the ability to run fairly complicated instructions.

14.3.3.2 Register-based Architecture

In register-based architecture, as shown in Fig. 14.13, both operands are stored in registers and the result of operation is also stored in a register. The registers are typically colocated with the processor. Since the processor and registers operate at the same speed, this architecture is much faster than the previously discussed accumulator-based architecture. The contents of the register are read from and written to memory using background operation.

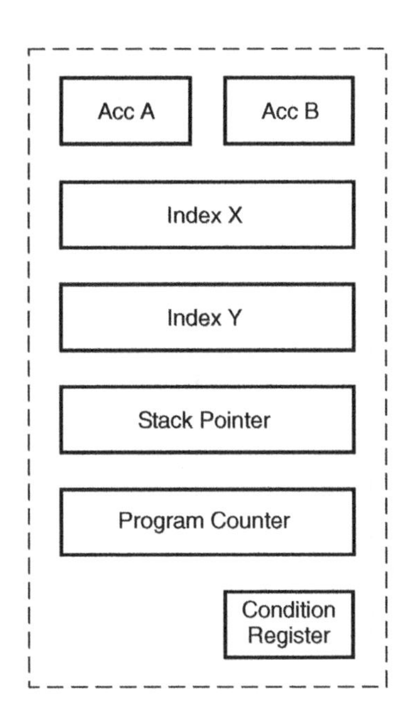

Figure 14.12 Accumulator-based processor architecture.

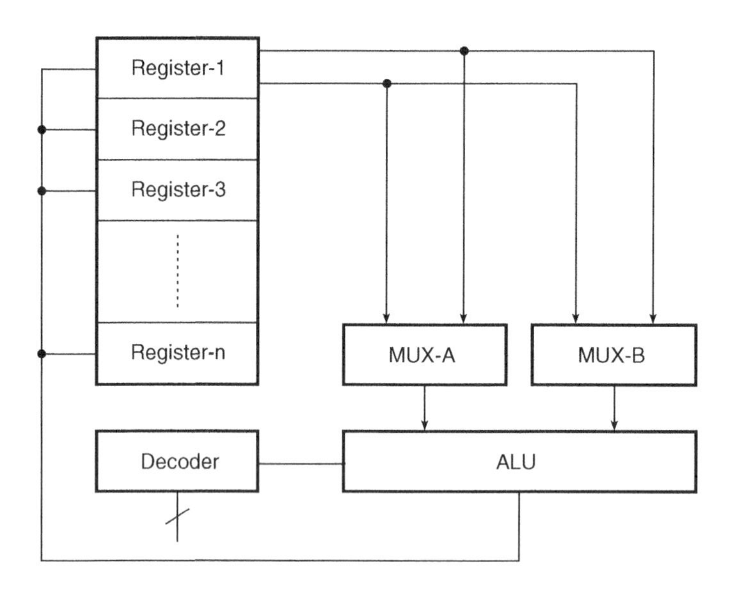

Figure 14.13 Register-based processor architecture.

14.3.3.3 Stack-based Architecture

In stack-based architecture, both operands and the operation to be performed are stored on the stack, which could be configured around dedicated registers or a special portion of RAM. The result of operation is placed back on the stack. Figure 14.14 shows typical block schematic arrangement of this type of architecture.

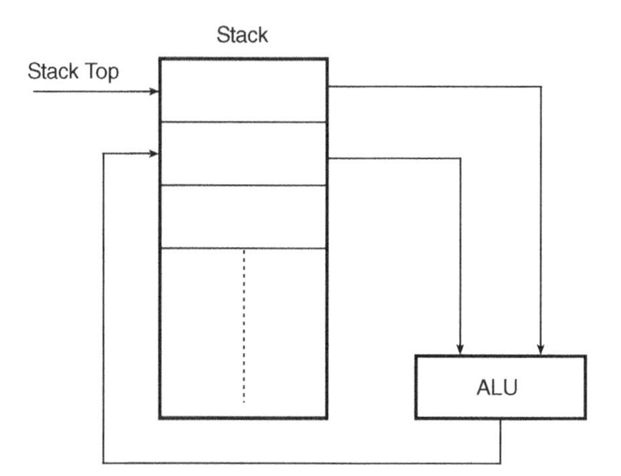

Figure 14.14 Stack-based processor architecture.

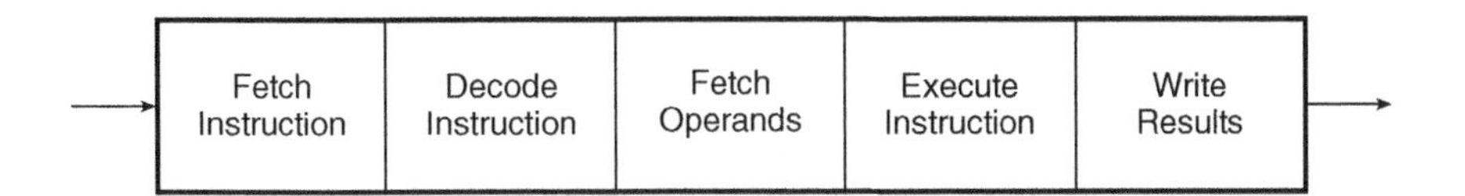

Figure 14.15 Pipelined architecture.

14.3.3.4 Pipeline Architecture

In pipelined architecture, as shown in Fig. 14.15, there are separate hardware stages for execution of different steps involved in execution of an instruction. These different steps include fetching an instruction from memory, decoding the instruction, fetching instruction operands from memory or registers, executing the instruction and then finally placing the result back on the memory. Pipelining allows these stages to overlap and perform with parallelism. The overall result is that the execution of an instruction is completed on every clock cycle. Instructions in a pipelined architecture are usually simple instructions that can be implemented within a single stage. These simple instructions act as building blocks for more complex instructions.

14.4 Power-Saving Modes

Power consumption is one of the important issues in battery-powered devices. Most microcontrollers come with various power-saving features. For a given application requirement, designers use these features to keep the power consumption down to an optimum value without compromising the operational requirements of the device. It may be mentioned here that not all modes are for power saving. Some microcontrollers support in-circuit debugging. As an example, some of the power-saving modes available with the 80C51 family of microcontrollers are briefly outlined in the following paragraphs.

The *stop clock mode* allows the clock oscillator to be stopped or the clock speed to be reduced to as low as 0 MHz. When the oscillator is stopped, the special-function registers and RAM retain their values. This mode allows reduced power consumption by lowering the clock frequency to any value.

The *idle mode* is another power-saving mode available with the 80C51 family of microcontrollers. In this mode, the processor puts itself to sleep while all on-chip peripheral components stay active. The processor contents, the on-chip RAM and all special-function registers remain intact during the idle mode. The instruction that invokes this mode is the last instruction executed in the normal operating mode before the idle mode is activated. The idle mode can be terminated by either an enabled interrupt or by a hardware reset. By an enabled interrupt, the process is picked up at the interrupt service routine and continued. Hardware reset starts the processor in the same manner as it does on a power-on reset.

The *power down mode* is recommended for the lowest power consumption. When this mode is enabled, the oscillator stops and the instruction that invokes the power down mode is the last instruction executed. Special-function registers and on-chip RAM retain their values down to a V_{CC} amplitude of 2.0 V. V_{CC} must be brought to the minimum specified operating voltage before this mode is deactivated. Either a hardware reset or an external interrupt can be used to terminate the power down mode. While a hardware reset redefines all the special-function registers and retains on-chip RAM values, an external interrupt allows both special-function registers and the on-chip RAM to retain their values. For proper termination of the power-down mode, a reset or external interrupt should not be executed unless $V_{CC}$is restored to its normal operating level and also has been held active long enough for the oscillator to start and stabilize.

Yet another mode available with the 80C51 family of microcontrollers that helps in power saving is the LPEP. The EPROM array contains some analogue circuits that are not required for a V_{CC} of less than 4.0 V. This feature can be used to save power by setting the LPEP bit, resulting in reduced supply current. This mode should be used only for applications that require a V_{CC} of less than 4.0 V.

14.5 Application-Relevant Information

This section briefly presents application-relevant information in terms of general specifications, microcontroller-related features and peripheral features on some of the common types of microcontroller from well-known international manufacturers including Intel, Freescale Semiconductor, Microchip Technology, Altera, Atmel, Zilog, Lattice Semiconductor, National Semiconductor, Applied Micro Circuits Corporation (AMCC), Fujitsu, Infineon, Dallas Semiconductor, Philips Semiconductors, Texas Instruments, Xilinx, NEC, Toshiba and so on. Some of the more widely used type numbers, including the 80C51 family of microcontrollers (Intel and many more manufacturers), the 89C51 microcontroller (Intel and many more manufacturers), the 68HC11 family of microcontrollers (Freescale Semiconductor) and the PIC 16X84 family of microcontrollers (Microchip Technology), are discussed in a little more detail. For these type numbers, information such as architecture, pin connection diagrams, functional description of different pins, addressing modes, etc., is also presented.

14.5.1 Eight-Bit Microcontrollers

This subsection outlines salient features of popular eight-bit microcontrollers. For most of the type numbers, the information is contained under two headings, namely microcontroller-related features and peripheral-related features.

14.5.1.1 80C51/87C51/80C31 (Dallas Semiconductor and Other Manufacturers)

Microcontroller-related Features

MCS-51 architecture, CMOS technology, 4K × 8 ROM (no ROM in 80C31), 128 × 8 RAM, memory addressing capability of 64K (ROM and RAM), special-function registers, six interrupt sources, three power control modes including STOP CLOCK, IDLE and POWER DOWN modes, two clock speed ranges of 0–16 MHz and 0–33 MHz, low EMI (inhibit ALE) and three package style options (40-pin dual in-line, 44-pin plastic leaded chip carrier and 44-pin plastic quad flat pack).

Peripheral-related Features

Two 16-bit counters/timers, four eight-bit I/O ports and full duplex-enhanced UART.

Architecture and Pin Connection Diagram

Figure 14.16 shows the architecture and Fig. 14.17 shows the pin connection diagram in the 40-pin dual in-line package.

Registers

Registers are categorized as general-purpose registers and special-function registers. The 80C51 family of microcontrollers has an accumulator, B-register and four register banks, each having eight-bit wide registers R0 to R7. Registers R0 through R7 are used as scratch-pad registers. In addition, there is

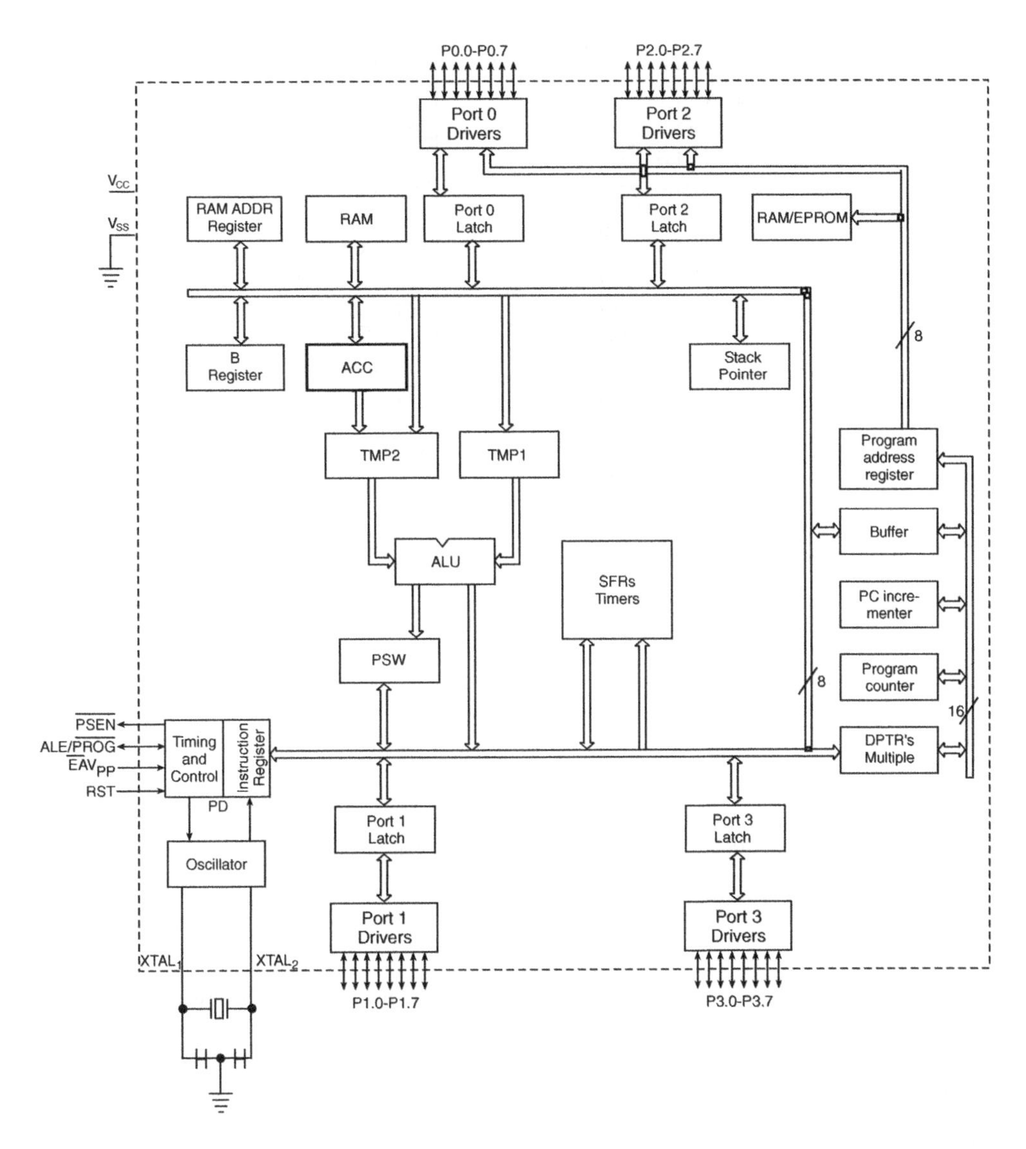

Figure 14.16 Architecture of the 80C51 microcontroller family.

an eight-bit wide stack pointer and a 16-bit wide program counter. Special-function registers include program status word (PSW), data pointer (DPTR), timer registers, control registers and capture registers.

Addressing Modes

The 80C51 family of microcontrollers supports five addressing modes including register addressing, direct addressing, register indirect addressing, immediate addressing and base register plus index register addressing.

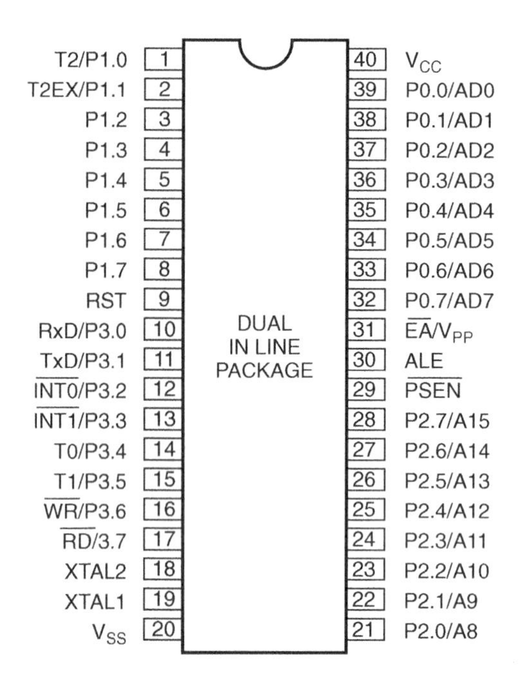

Figure 14.17 PIN connection diagram in the 40-pin DIP package.

Instruction Set

The instruction set of the 80C51 family of microcontrollers consists of 111 instructions divided into five categories, namely data transfer instructions, arithmetic instructions, logical instructions, Boolean variable manipulation instructions and control transfer instructions.

Interrupts

The 80C51 family of microcontrollers supports five vectored interrupts. These include external interrupt 0, external interrupt 1, timer/counter 0 interrupt, timer/counter 1 interrupt and serial port interrupts.

Power Modes

The 80C51 family of microcontrollers offers various operational modes that can be used to reduce power consumption. These include STOP CLOCK MODE which enables the clock speed to be reduced down to 0 MHz, IDLE MODE when the CPU puts itself to sleep while all of the on-chip peripherals stay active and POWER DOWN MODE in which the oscillator is stopped. In addition to the power-saving operational modes, it also offers ONCE™ (On-Circuit Emulation) MODE which facilitates in-circuit testing and debugging.

14.5.1.2 80C31FA/8XC51FA/FB/FC (Dallas Semiconductor and Other Manufacturers)

The same as 80C51 except for the size of ROM and RAM, which is 0K/8K/16K/32K (ROM) and 256 bytes (RAM).

14.5.1.3 80C31RA$^+$/8XC51RA$^+$/RB$^+$/RC$^+$(Dallas Semiconductor and Other manufacturers)

The same as 80C51 except for the size of ROM and RAM, which is 0K/8K/16K/32K (ROM) and 512 bytes (RAM).

14.5.1.4 8XC51RD$^+$(Dallas Semiconductor and Other Manufacturers)

The same as 80C51 except for the size of ROM and RAM, which is 64K (ROM) and 1024 bytes (RAM).

14.5.1.5 80C32/8XC52/54/58 (Dallas Semiconductor and Other Manufacturers)

The same as 80C51 except for the size of ROM and RAM, which is 0K/8K/16K/32K (ROM) and 256 bytes (RAM).

14.5.1.6 89C51 (ATMEL and Other Manufacturers)

Microcontroller-related Features

MCS-51 architecture, CMOS technology, 4K $\times$ 8 of in-system reprogrammable ROM, 128 $\times$ 8 internal RAM, memory addressing capability of 64K (ROM and RAM), special-function registers, six interrupt sources, two power-saving modes (IDLE and POWER DOWN modes), a clock speed range of 0–24 MHz, low EMI (inhibit ALE), three package style options (40-pin dual in-line, 44-pin plastic leaded chip carrier and 44-pin plastic quad flat pack) and compatible with the industry-standard MCS-51 instruction set and pin-out.

Peripheral-related Features

Two 16-bit counters/timers, 32 programmable I/O lines and a programmable serial channel.

Architecture and Pin Connection Diagram

The architecture and pin connection diagram are the same as those given earlier for the case of the 80C51 family of microcontrollers in Fig. 14.16 (architecture) and Fig. 14.17 (pin connection diagram).

14.5.1.7 68HC05 Family of Microcontrollers (Freescale Semiconductor)

Microcontroller-related Features

Fully static chip design using a standard eight-bit M68HC05 core, a clock speed of 4 MHz, 920 bytes of on-chip RAM, 32K of ROM, 7932 bytes of EEPROM (maximum values across the family of devices), power-saving WAIT mode and available in 40-pin DIP and 42-pin SDIP package styles.

Peripheral-related Features

Two serial interface channels, a multifunction timer with periodic interrupt, eight A/D converter channels, three PWM channels and 80 I/O lines (maximum values across the family of devices).

14.5.1.8 68HC11 Family of Microcontrollers (Freescale Semiconductor)

Microcontroller-related Features

Fully static chip design using an eight-bit M68HC11 core, a clock speed of 5 MHz, 0/256/512/768/1024 bytes of on-chip RAM (in different variants), 0/12/20 kB of on-chip ROM or EPROM (in different variants), 0/512/2048 bytes of on-chip EEPROM (in different variants), power-saving STOP and WAIT modes and available in six different package styles.

Peripheral-related Features

Asynchronous non-return-to-zero (NRZ) serial communication interface (SCI), synchronous serial peripheral interface (SPI), eight-channel, eight-bit analogue-to-digital converter, 16-bit timer system including three input capture channels, four output compare channels and an additional channel configurable as an input or an output channel, eight-bit pulse accumulator and 38 general-purpose I/O pins including 16 bidirectional I/O pins, 11 input-only pins and 11 output-only pins.

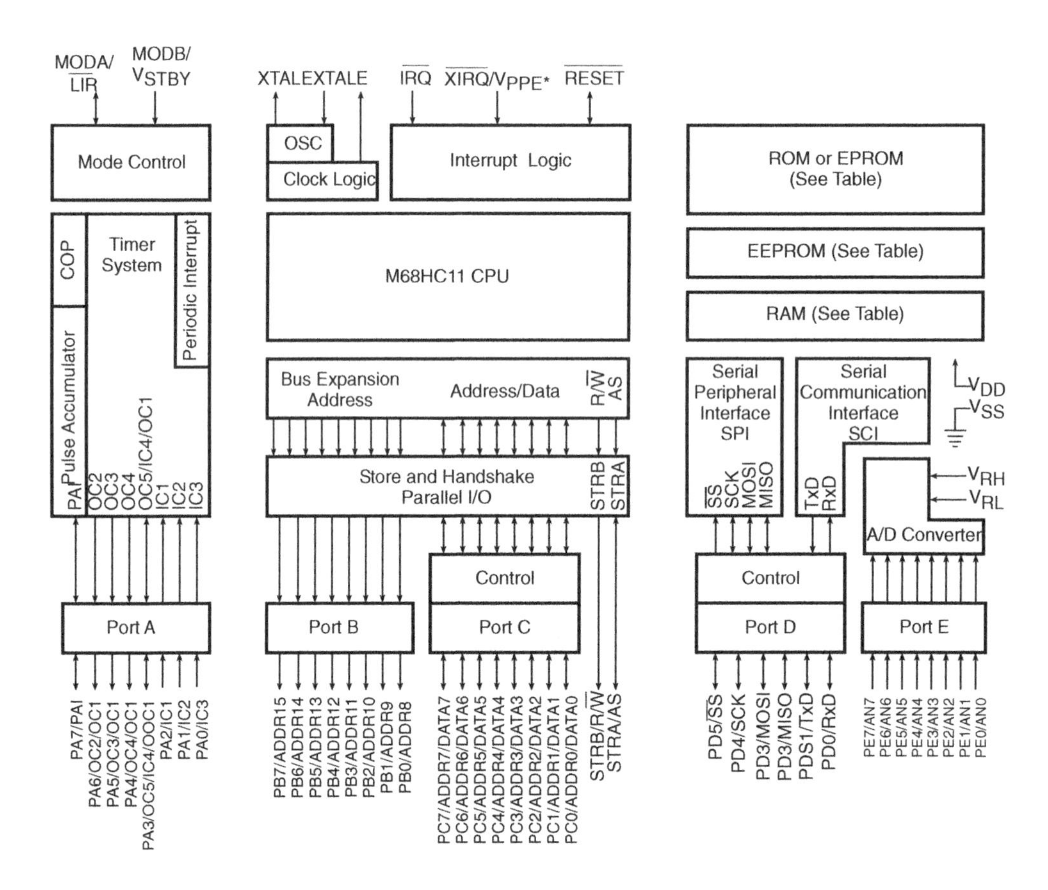

Figure 14.18 Architecture of the 68HC11 family of microcontrollers.

Architecture and Pin Connection Diagram

Figure 14.18 shows the architecture of the 68HC11 family of microcontrollers. Pin connection diagrams are shown in Fig. 14.19 (56-pin SDIP package) and Fig. 14.20 (48-pin DIP package). DIP and SDIP respectively stand for dual in-line package and shrink dual in-line package.

14.5.1.9 PIC 16X84 Family of Microcontrollers (Microchip Technology)

PIC 16C84 and PIC 16F84 are the two microcontrollers in the PIC 16X84 family of microcontrollers from Microchip Technology. PIC 16F84 is an improved version of PIC 16C84.

Microcontroller-related Features

High-performance RISC CPU, 14-bit wide instructions, eight-bit wide data path, 1024 × 14 EEPROM program memory, 64 bytes of on-chip data EEPROM, 36 × 8 general-purpose registers (16C84), 68 bytes of data RAM (16F84), 15 special-function hardware registers (16F84), a clock speed of 10/20

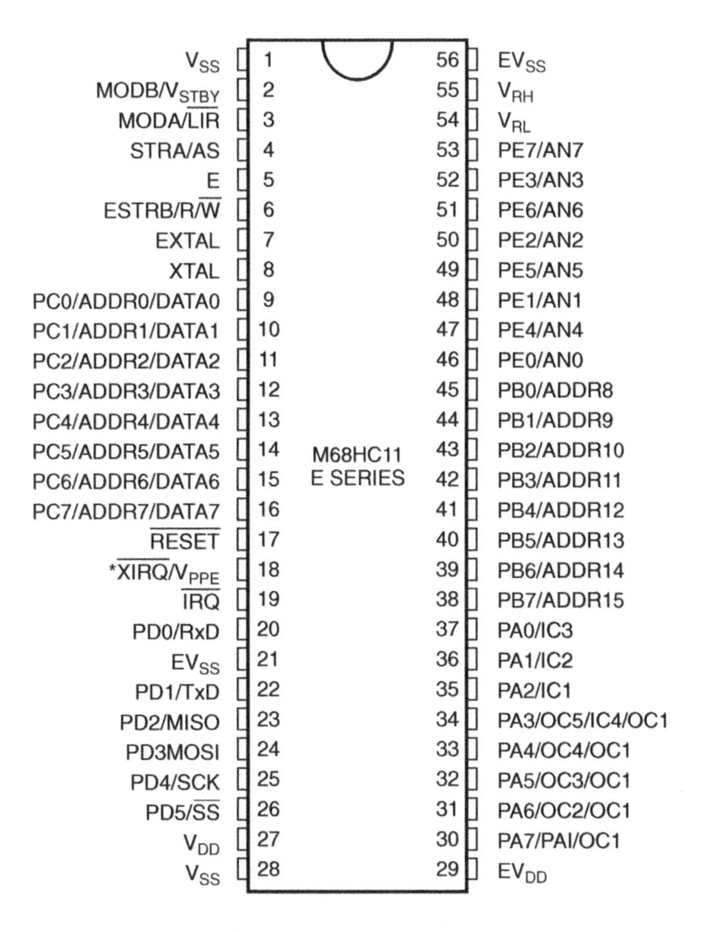

Figure 14.19 68HC11 in the 56-pin SDIP package.

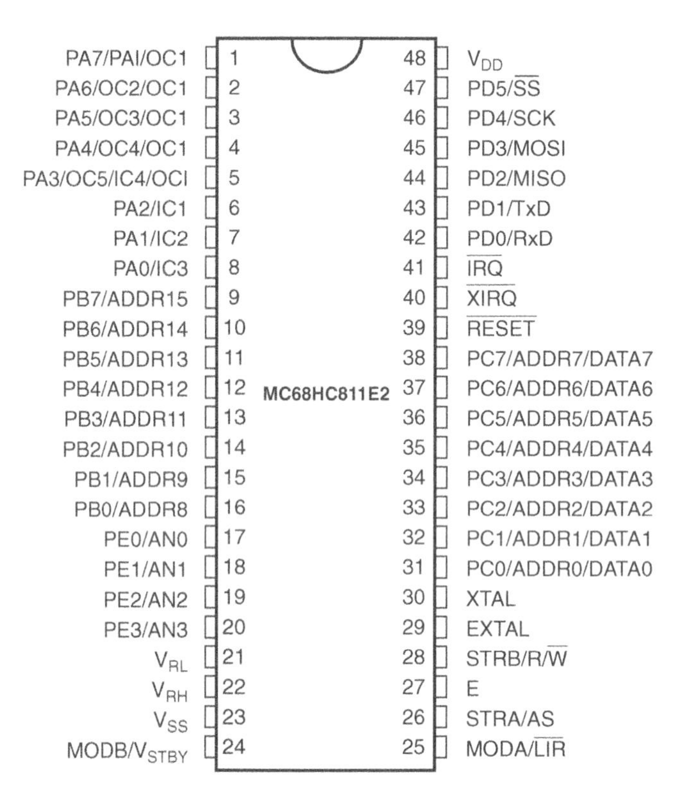

Figure 14.20 68HC11 in the 48-pin DIP package.

MHz (16C84/16F84), direct, indirect and relative addressing modes, power-saving SLEEP mode and four interrupt sources.

Peripheral-related Features

Thirteen I/O pins with individual direction control, high current sink/source for direct LED drive and eight-bit timer/counter with an eight-bit programmable prescaler.

Architecture and Pin Connection Diagram

Figure 14.21 shows the architecture. Figure 14.22 shows the pin connection diagram in the 18-pin DIP package.

14.5.1.10 XC-800 Family of Microcontrollers (Infineon)

The XC-800 family of microcontrollers offers high-performance eight-bit microcontrollers, with some of the members providing advanced networking capabilities by integrating both a CAN controller and LIN support on a single chip. Salient features of two of its members, i.e. XC-886/888 and XC-866, are briefly outlined in the following paragraphs.

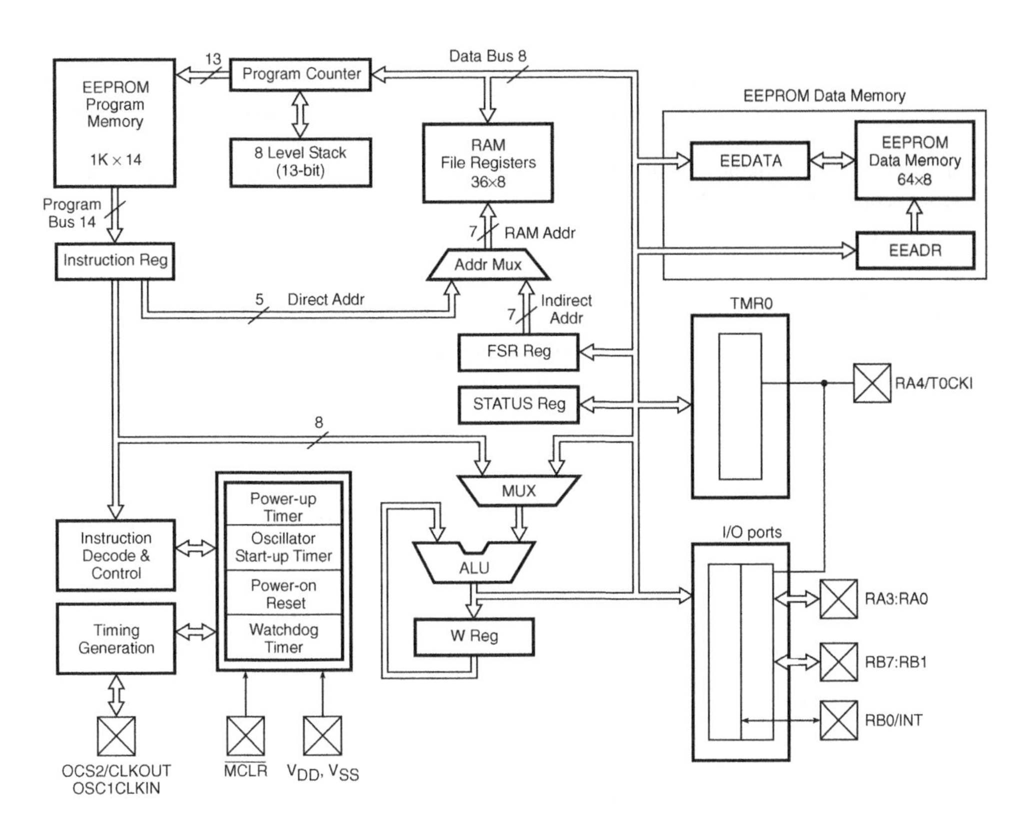

Figure 14.21 Architecture of the PIC 16X84 microcontroller family

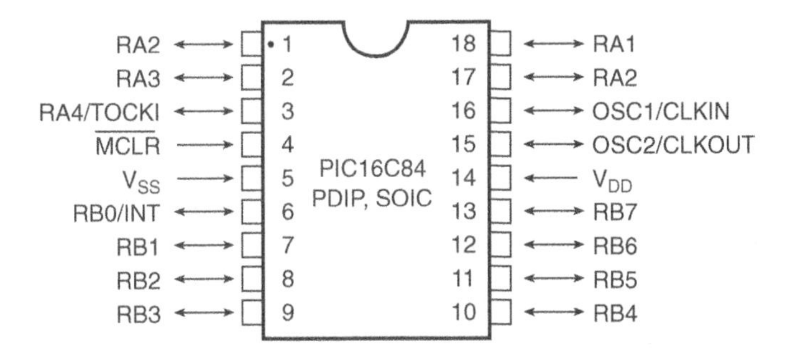

Figure 14.22 Pin connection diagram in the 18-pin DIP package.

Microcontroller-related Features

High-performance XC-800 core based on industry-standard 8051 architecture, a clock speed of 24 MHz, 24K or 32K of flash memory (XC-886/888), 256 bytes of RAM and 1536 bytes of XRAM (XC-886/888), 256 bytes of RAM and 512 bytes of XRAM (XC-866) and four power-saving modes including SLOW DOWN mode, IDLE mode, POWER DOWN mode and clock gating control.

Peripheral-related Features
Total of 34/48 general-purpose I/O ports, including eight analogue ports (XC-886/888) and 27 general-purpose I/O ports (XC-866), eight-channel, 10-bit analogue-to-digital converter, four 16-bit general-purpose timers (XC-886/888) and three 16-bit timers (XC-866), programmable 16-bit watchdog timer (WDT), two UARTs, including one for LIN simulation (XC-886/888) and one for LIN simulation, and one serial peripheral interface (XC-866).

14.5.2 16-Bit Microcontrollers

This subsection outlines salient features of some of the popular 16-bit microcontrollers. Again, the information is mainly contained under the headings microcontroller-related features and peripheral-related features.

14.5.2.1 68HC12 Family of Microcontrollers (Freescale Semiconductor)

Microcontroller-related Features
High-performance 16-bit CPU12 core having a 20-bit ALU, upward compatibility with the 68HC11 microcontroller instruction set, enhanced indexed addressing and fuzzy logic instructions, 1024 bytes of RAM, 32K of flash EEPROM and 768 bytes of EEPROM, a clock speed of 8 MHz, slow-mode clock divider, computer operating properly (COP) watchdog timer and available in 80-pin QFP and 112-pin TQFP packages.

Peripheral-related Features
Eight-channel, 10-bit analogue-to-digital converter, eight-channel, 16-bit input capture or output compare channels, up to 63 I/O lines, 16-bit pulse accumulator, eight-bit/four-channel or 16-bit/two-channel pulse width modulator, asynchronous serial communication interface (SCI) and synchronous serial peripheral interface (SPI).

Architecture and Pin Connection Diagram
Figure 14.23 shows the architecture of the 68HC12 family of microcontrollers. The pin connection diagram is shown in Fig. 14.24 (112-pin TQFP).

14.5.2.2 68HC16 Family of Microcontrollers (Freescale Semiconductor)

The 68HC16 family of microcontrollers is the 16-bit enhancement of the eight-bit 68HC11 family of microcontrollers. This family of microcontrollers has been designed to provide many powerful features without the need for CPU intervention.

Microcontroller-related Features
8K of ROM, 4K of RAM, clock speeds of 16, 20 and 25 MHz and available in 132-pin PQFP and 144-pin LQFP packages.

Peripheral-related Features
Twenty-four I/O lines, general-purpose timer, asynchronous serial communication interface (SCI) and synchronous serial peripheral interface (SPI).

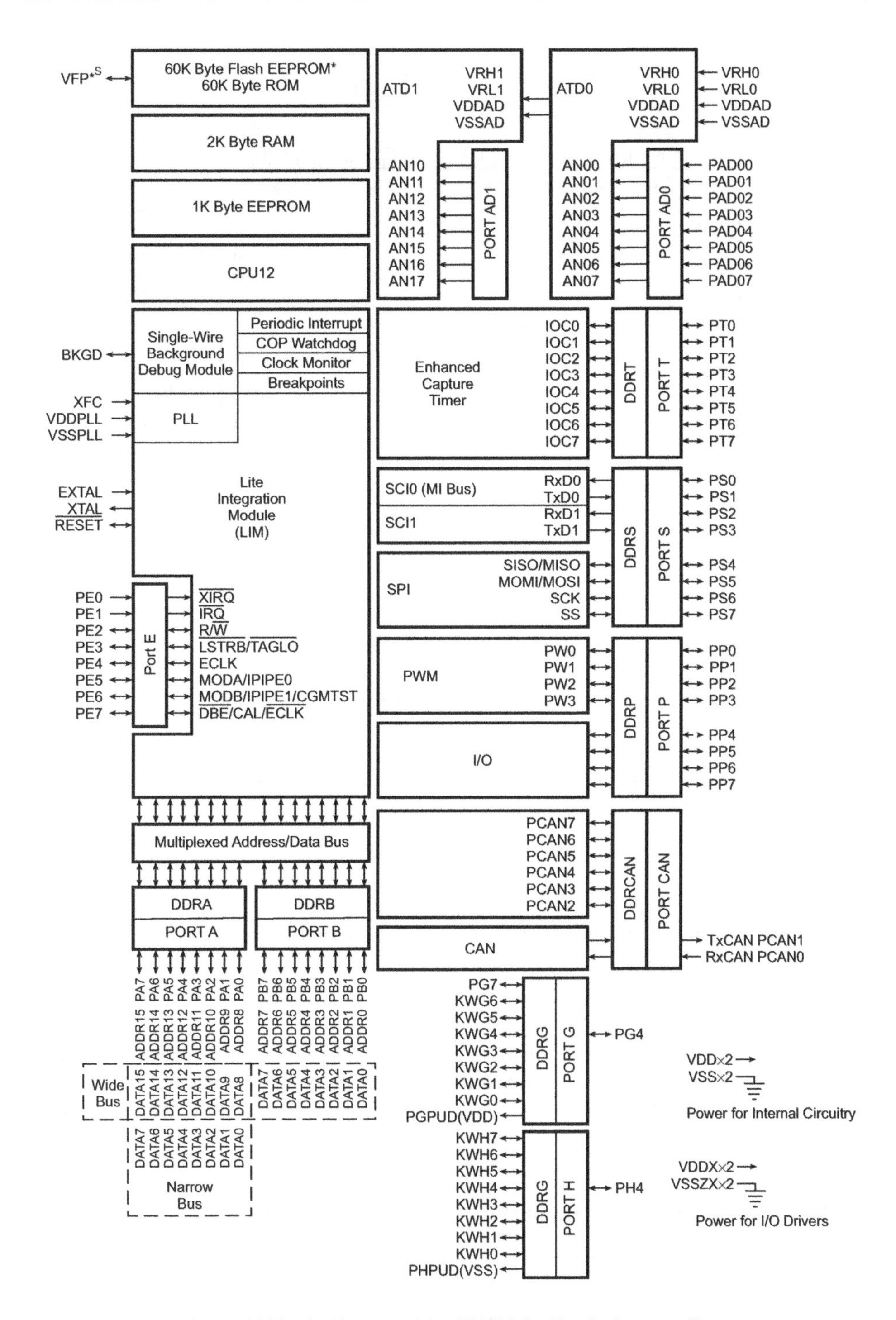

Figure 14.23 Architecture of the 68HC12 family of microcontrollers.

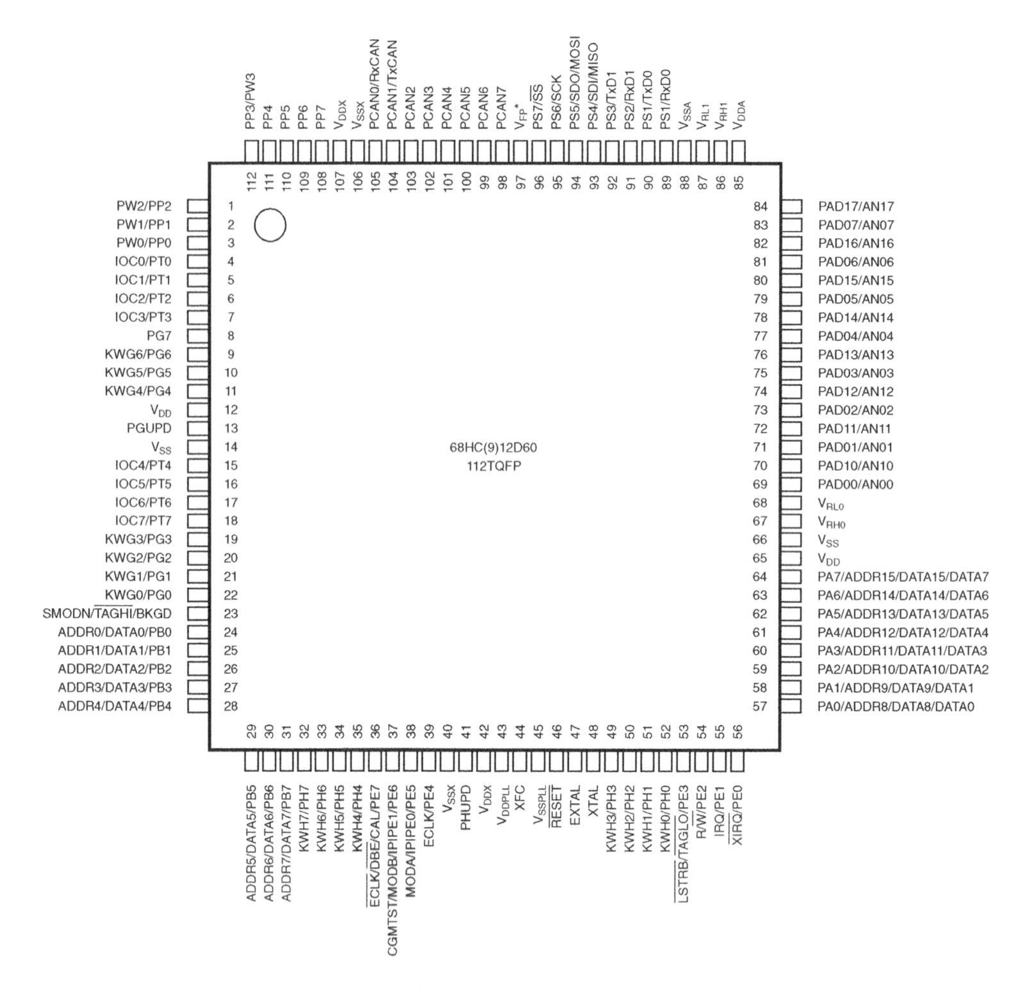

Figure 14.24 Pin connection diagram in the 112-pin TQFP package.

14.5.3 32-Bit Microcontrollers

This subsection outlines salient features of some of the popular 32-bit microcontrollers. The families of microcontrollers that are briefly described in the following paragraphs include 683XX, MCORE, MPC500 and MCFXXX families (Freescale), the LPC-3000 family (Philips Semiconductors) and the TRICORE family (Infineon).

14.5.3.1 683XX Family of Microcontrollers (Freescale Semiconductors)

Different members of this family include 68302, 68306, 68331/332/336, 68340, 68360 and 68375/376. 68302 uses an integrated multiprotocol processor. All other members of the family use a CPU32 core. The CPU32 core is a 32-bit processing unit based on the 68000 software model and instruction set with some additional features from 68010 and 68020. It also has some new features added to the core for control operations. Salient features of this family of microcontrollers are as follows. The family offers 10K of RAM, 256K of flash, a clock speed of 33 MHz, 48 I/O lines, a 16-bit timer, a 16-channel/10-bit analogue-to-digital converter and four serial communication channels. It may be mentioned here that the above-mentioned values are the maximum available ones across the family of devices.

14.5.3.2 MCORE Family of Microcontrollers (Freescale Semiconductors)

This family of microcontrollers is built around a processing core known as the MCORE microRISC engine. The design of the core combines high performance with low power consumption, which makes the MCORE family of microcontrollers particularly suitable for battery-operated and mobile applications. Salient features of this family of microcontrollers are as follows. The family offers 32-bit wide load/store architecture, 16-bit wide instructions for fast instruction throughput between the core and the memory, 32 general-purpose registers and a four-stage instruction pipeline that facilitates most instructions to be completed in one clock cycle. Other features include 32K of RAM, 256K of flash, 33 MHz of clock speed, two serial communication channels, 104 I/O lines, an eight-channel analogue-to-digital converter and two timers. Again, the above-mentioned values are the maximum available ones across the family of devices.

14.5.3.3 MPC500 Family of Microcontrollers (Freescale Semiconductors)

The MPC500 family of microcontrollers is configured around a 32-bit PowerPC core. Different members of the family include MPC555, MPC556, MPC561, MPC562, MPC563, MPC564, MPC565 and MPC566. PowerPC architecture based design provides compatibility with the PowerPC instruction set, including floating-point operations. Salient features include 36K of RAM, 1024K of flash, a 66 MHz clock, three serial communication channels, 101 I/O lines, 40 channels of analogue-to-digital conversion and 70 timer channels. These microcontrollers are particularly suitable for scientific applications requiring complex operations.

14.5.3.4 MCFXXX Family of Microcontrollers (Freescale Semiconductors)

The MCFXXX family of microcontrollers is configured around a ColdFire Version 2 core. Different members of the family include MCF5206, MPC5207, MPC5208, MPC5211, MPC5212, MPC5213, MPC5214, MPC5216, MPC5232, MPC5233, MPC5234, MPC5235, MPC5249, MPC5270, MPC5271, MPC5272, MPC5274, MPC5275, MPC5280, MPC5281, MPC5282, MPC5327, MPC5328 and MPC5329. The core uses variable-instruction-length RISC architecture. ColdFire instructions, which are similar to those in the 680X0 instruction set, are processed in a pipelined architecture of fetch and decode/execute units. The core also contains an enhanced multiply-and-accumulate (eMAC) unit, which has been designed to support DSP applications. Other features include 64K of RAM, 66 MHz of clock, 5 serial communication channels, including an I^2C bus and CAN support, 150 I/O lines and four timer channels. Again, the above-mentioned values are the maximum available ones across the family of devices.

14.5.3.5 LPC3000 Family of Microcontrollers (Philips Semiconductors)

The LPC-3000 family of 32-bit microcontrollers is based on Philips' Nexperia platform. It is configured around an ARM926EJ core with the VFP9 floating-point coprocessor. The family offers enhanced signal-processing performance with the 926EJ core equipped with features such as single-cycle multiply-accumulate packed data and saturating arithmetic. The vector coprocessor is a high-speed floating-point unit and is IEEE754 compliant.

The LPC3000 family of microcontrollers incorporates 32K of instruction cache and 32K of data cache, which operate concurrently owing to the use of Harvard architecture. The family combines high performance with low power dissipation, which is made possible by its low-voltage operation at 1.2 V. It operates at clock speeds in excess of 200 MHz and supports a wide range of peripherals. As an example, LPC3180 (the first member of the LPC3000 family of microcontrollers) has multiple serial interfaces including seven UARTs, two single master I^2C interfaces and two SPI controllers, USB on-the-go, a 32-bit general-purpose timer with a 16-bit prescaler with capture and compare capability, a watchdog timer, PWM blocks with an output rate of up to 50 kHz and up to 55 general-purpose I/O pins.

14.5.3.6 TRICORE™ Family of Microcontrollers (Infineon)

The TRICORE family of 32-bit microcontrollers uses a unified, single-core 32-bit microcontroller–DSP architecture optimized for real-time embedded systems. The architecture combines the real-time capability of a microcontroller with the computational power of a DSP and the high performance features of RISC load/store architecture. The TRICORE family of microcontrollers offers various subfamilies, which include the AUDO-NextGeneration family, the AUDO1 family, the TC116X family and the TC1130 family. The family offers clock speeds ranging from 40 MHz (AUDO1 family) to 150 MHz (AUDO NextGeneration family) and is equipped with almost every microcontroller-related and peripheral-related features in terms of on-chip memory, power-saving modes, serial interfaces, counters/timers, PWM blocks, I/O ports, A/D converters and so on.

14.6 Interfacing Peripheral Devices with a Microcontroller

This section briefly describes the interfacing of some common external peripheral devices with the microcontroller. The peripheral devices discussed in this section include LEDs, electromechanical relays, seven-segment displays, keypads, LCD displays and analogue-to-digital and digital-to-analogue converters. Only the basic fundamentals are discussed here. A detailed description of the software routines is beyond the scope of this book.

14.6.1 Interfacing LEDs

The commonly used configuration to connect an LED to a microcontroller is shown in Fig. 14.25(a). The LED glows when the microcontroller pin is driven LOW and is OFF when the pin is set HIGH. The LEDs are connected in this fashion as the current-sinking capability of microcontrollers is of the order of a few tens of milliamperes and the current-sourcing capability is of the order of microamperes. The resistor is used to limit the current through the LED.

The value of the resistance is chosen according to the equation

$$R = (V_{CC} - V_{LED})/I \qquad (14.1)$$

where V_{LED} is the voltage across the LED and I is the current.

Typical values of V_{LED} and I are 1.5 V and 20 mA respectively. If the current-sourcing capability of the microcontroller is sufficient to drive the LED directly, then the LED is connected to the

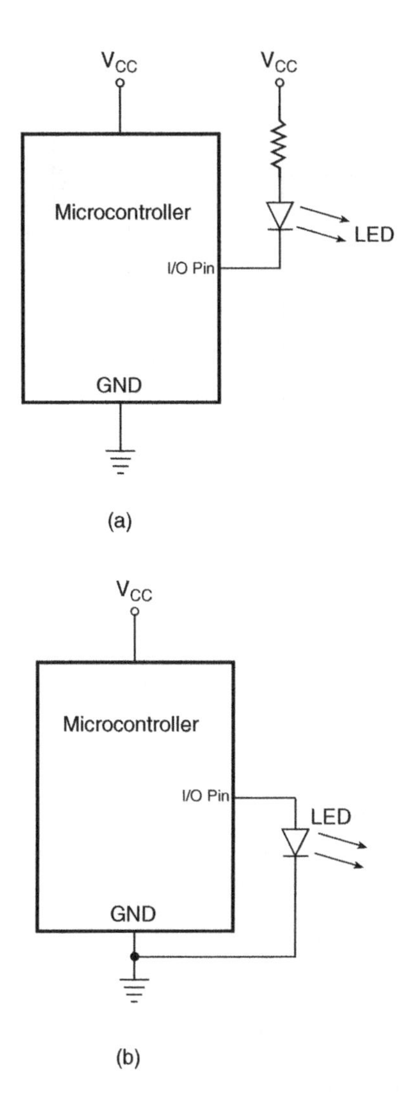

Figure 14.25 Interfacing LEDs to a microcontroller.

microcontroller as shown in Fig.14.25(b). The LED in this case glows when the microcontroller pin is set HIGH.

14.6.2 Interfacing Electromechanical Relays

Figure 14.26 shows the typical connection diagram for interfacing an electromechanical relay to a microcontroller. The NPN transistor is used to provide the desired current to the relay coil as the microcontroller cannot drive the relay directly. The freewheeling diode is required as the current

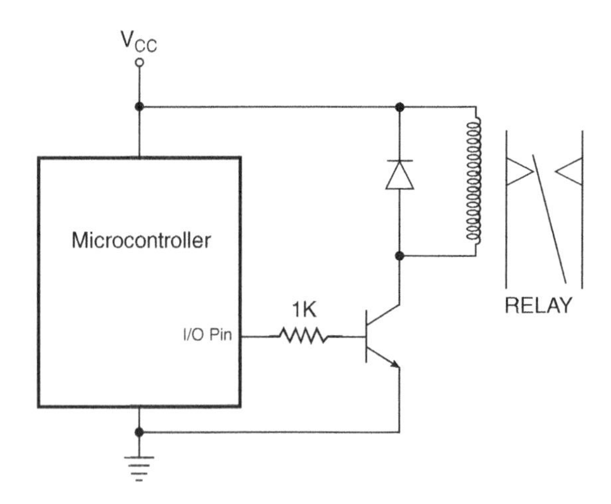

Figure 14.26 Interfacing an electromechanical relay to a microcontroller.

through the inductor cannot be instantaneously reduced to zero. When the microcontroller pin is set HIGH, the transistor is switched on. Current flows through the relay coil and the contact is closed. When the microcontroller pin is LOW, the transistor is switched off and the inductor current now flows through the freewheeling diode and slowly decays to zero value.

14.6.3 Interfacing Keyboards

Keyboards are used to enter data, values, etc., into the microcontroller system. They are generally available in three configurations, namely the lead-per-key keyboard, the matrix keyboard and the coded keyboard. Lead-per-key or linear keyboards are used when very few keys have to be sensed. Coded keypads are generally used in telephonic applications. They are high-quality durable keyboards and permit a multiple key press to be detected easily. They are used when the number of keys is 16 or less, as they are very expensive. The most commonly used keyboard is the matrix keyboard where the keys are arranged in a matrix, with keys in the same row and column sharing the same access lines. When the number of keys exceeds 10, more often than not matrix keyboards are used. Interfacing a matrix keyboard with the microcontroller is discussed in the following paragraphs.

When the keyboards are connected to a microcontroller, following factors must be considered:

1. *Contact bounce.* Contact bounce refers to multiple 'make' and 'break' oscillations of contact during the key-pressing operation (Fig. 14.27). Good-quality keyboards have bounce periods of 1–5 ms, whereas low-cost keyboards have bounce periods of tens of milliseconds. If the bounce is not taken into consideration, the microprocessor responds as if the key has been pressed and released several times when in fact it has been pressed only once. Contact debouncing through either a hardware or a software routine is done to avoid the undesirable multiple-contact effects during a key closure, so that it appears as a single ON or OFF operation. Hardware debouncing is done using an RC circuit [Fig. 14.28(a)] or a Schmitt trigger circuit [Fig. 14.28(b)]. If debouncing is done by a software routine, a delay of 20–50 ms is given after a key press before the routine for that key is executed.

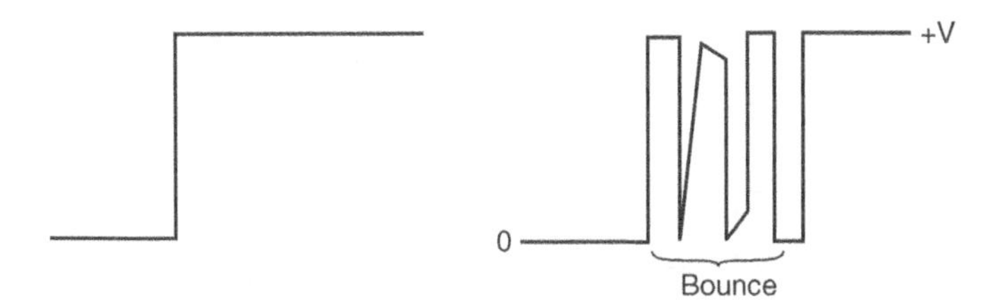

Figure 14.27 Contact bounce.

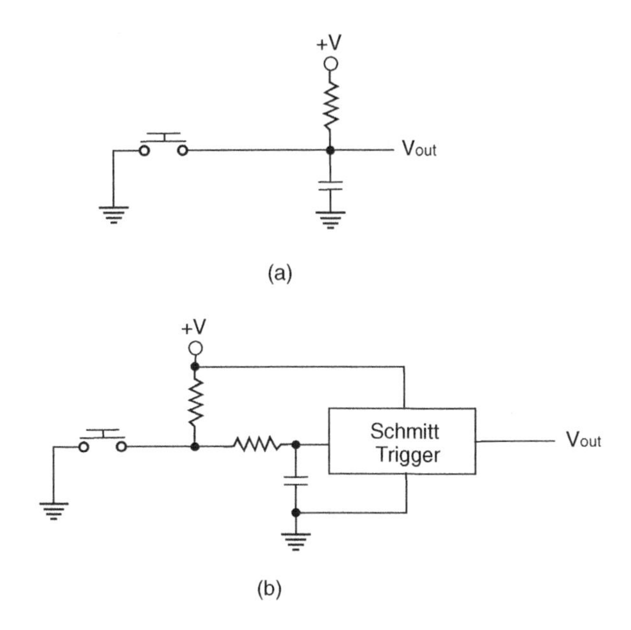

Figure 14.28 (a) Hardware debouncing using an RC circuit and (b) hardware debouncing using a Schmitt trigger circuit.

2. *Multiple keys.* If more than one key is pressed, then only routines corresponding to valid multiple-key presses should be executed. Also, the first valid key press pattern is executed.
3. *Key hold.* There are two types of keyboard actuation, namely the two-key lock-out and the N-key rollover. The two-key lock-out takes into account only one key pressed. An additional key pressed and released does not generate any codes. The system is simple to implement and most often used. The N-key rollover will ignore all keys pressed until only one remains down.

Figure 14.29 shows the connection of a 16-key matrix keypad with a microcontroller. Here, each column and row access line is connected to the microcontroller pin. The columns are generally at a HIGH level. The row lines are configured as output lines and the column lines are used as scan lines. The key actuation is sensed by sending a LOW to each row one at a time through a software routine via the row 1, row 2, row 3 and row 4 lines. The column lines are checked for each row to see

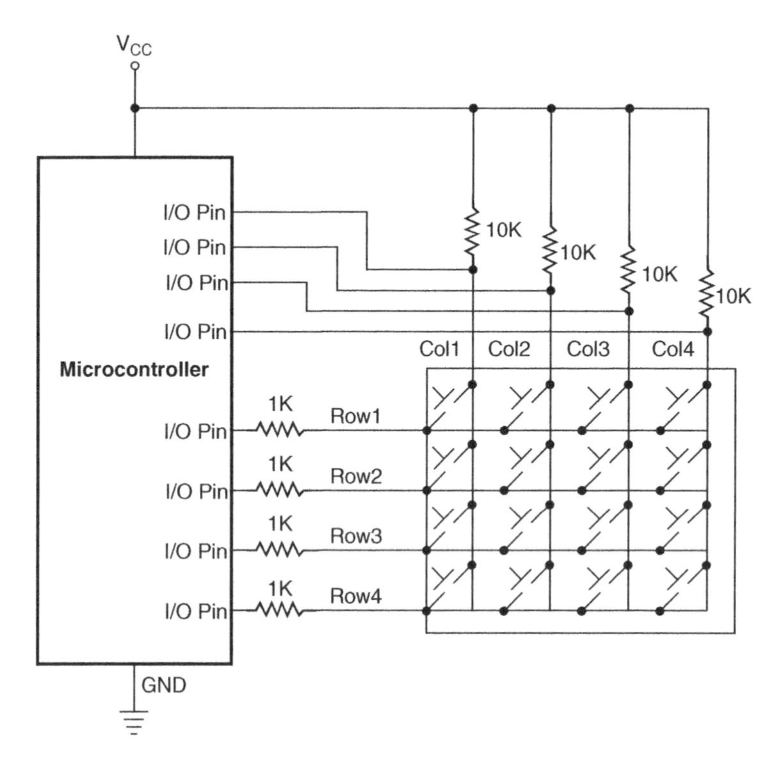

Figure 14.29 Connection of a 16-key matrix keypad with a microcontroller.

whether any of the normally HIGH column lines are pulled LOW. If a column is driven LOW, then, by determining which row and column line is LOW, the key is identified and the routine corresponding to that key press is executed.

14.6.4 Interfacing Seven-Segment Displays

Seven-segment displays commonly contain LED segments arranged as a figure-of-eight pattern, with one common lead (anode or cathode) and seven individual leads for each segment. When the common lead is the anode it is referred to as the common anode (CA), and when the common lead is the cathode it is referred to as the common cathode (CC). Figure 14.30 shows one of the possible configurations of interfacing a CC display with the microcontroller. The IC CD4511 is a BCD to seven-segment decoder/driver. The microcontroller feeds the BCD equivalent of the digit to be displayed to the 4511 IC.

Seven-segment displays can also be connected directly without the use of a BCD to seven-segment decoder. In this case the seven-segment code of the digit is generated by the microcontroller program itself. Figure 14.31 shows the direct circuit connection for CA display.

If more than one display is to be used, the displays are time multiplexed. The human eye cannot detect the blinking display if each display is relit every 10 ms or so. The 10 ms time is divided by the number of displays used to find the interval between updating each display. In the case of CC displays

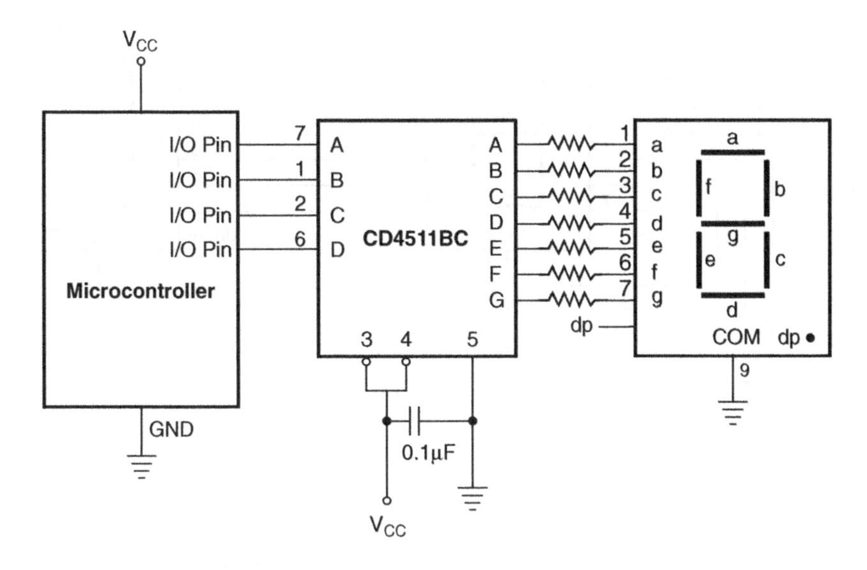

Figure 14.30 Possible configurations of interfacing a CC display with a microcontroller.

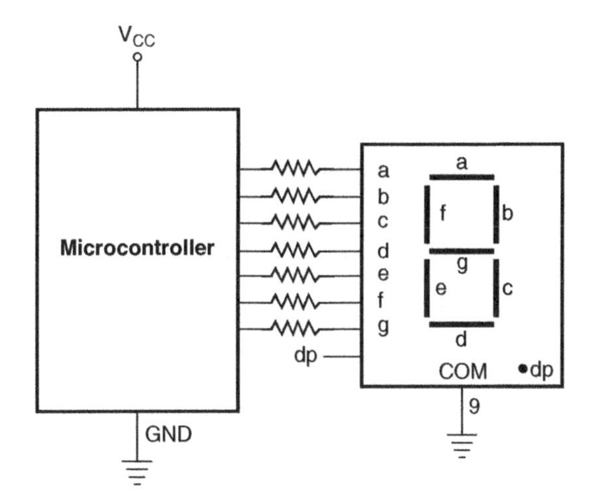

Figure 14.31 Direct circuit connection for CA display.

the display is selected by driving the common cathode to logic LOW, and in the case of CA displays the display is selected by driving the common anode to logic HIGH. Figure 14.32 shows the multiplexed circuit for two CC displays. The IC 74138 is a 3-to-8 line decoder used for selecting the display. Figure 14.33 shows the multiplexing in the case of CA displays for direct connection without the use of a BCD to seven-segment driver.

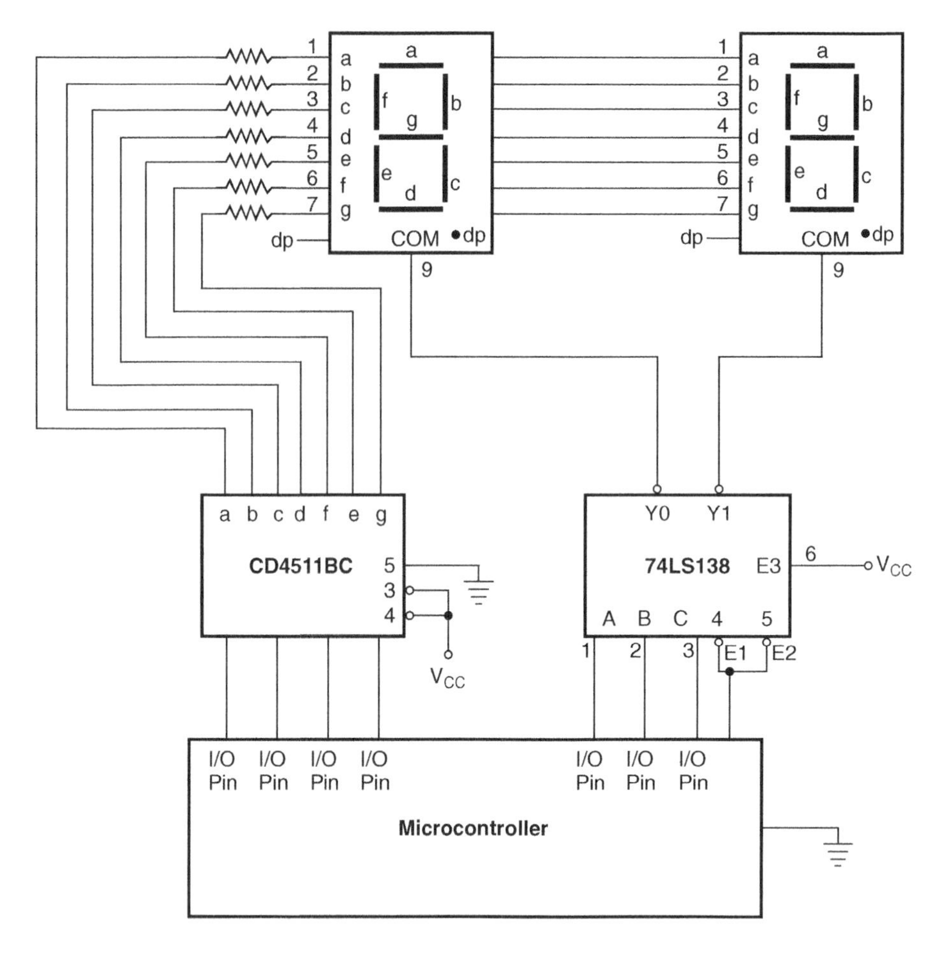

Figure 14.32 Multiplexed circuit for two CC displays.

14.6.5 Interfacing LCD Displays

Liquid crystal displays allow a better user interface compared with LED displays as it is much easier to display text messages in LCD displays. They also consume much less power than LED displays. However, LED displays have better intensity than LCD displays.

LCD displays are available typically in 8×2, 16×2, 20×2 or 20×4 formats. 20×2 means two lines of 20 characters each. These displays come with an LCD controller that drives the display. Figure 14.34 shows the interface of an LCD display with a microcontroller. There are three control lines, namely EN (enable), RS (register select) and RW (read/write). The EN line is used to instruct the LCD that the microcontroller is sending the data. When the RS line is HIGH, the data comprise text data to be displayed on the LCD. When the RS is LOW, the data are treated as a command or instruction to

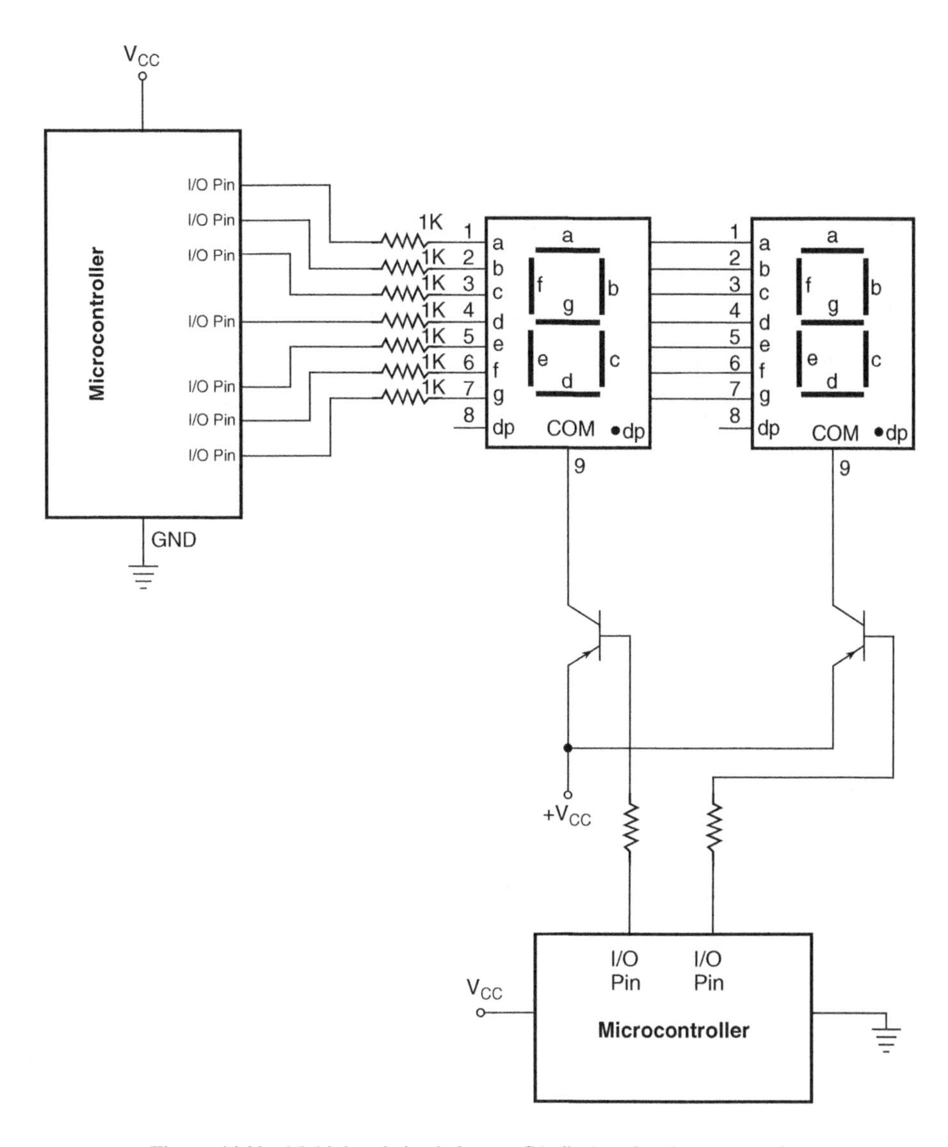

Figure 14.33 Multiplexed circuit for two CA displays for direct connection.

the LCD module. When the RW line is LOW, the instruction on the data bus is written on the LCD. When the RW line is HIGH, the data are being read from the LCD.

The software routine initializes the LCD firstly by setting the width of the data bus, selecting the character, font, etc., clearing the LCD, turning on the LCD module and the cursor, setting the cursor position and so on. Then the data to be displayed are sent on the data lines, and the three control signals are made use of to ensure proper LCD operation.

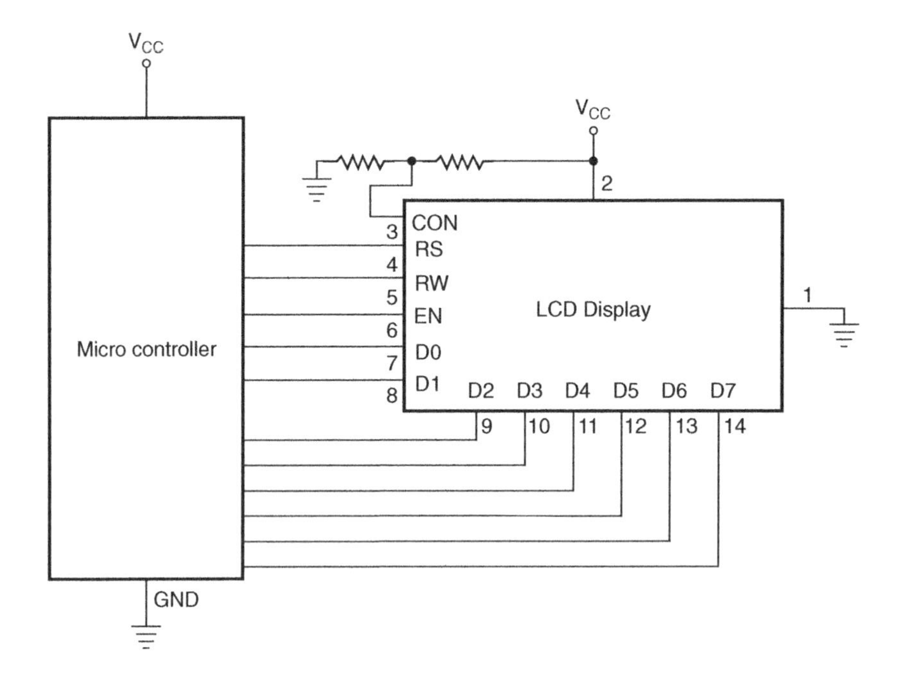

Figure 14.34 LCD display interface with a microcontroller.

14.6.6 Interfacing A/D Converters

A/D converters are used to interface the microcontroller with the analogue world. Figure 14.35 shows the interface of A/D converter type number AD571 with the microcontroller. AD571 is an eight-bit A/D converter. As can be seen from the figure, the data output lines and the control lines of the A/D converter are connected to the microcontroller I/O pins. The microcontroller sends commands such as the start of conversion, selection of the input channel if the A/D converter has more than one input channel, etc. It also senses signals from the A/D converter such as the end of conversion to store the digital bits. In the present case, the microcontroller sends a LOW on the $BLANK/\overline{DR}$ line to start the conversion process. It then waits for the data ready ($\overline{DR}$) signal to go to LOW. After that, the digital output bits are received by the microcontroller and processed according to the software routine.

14.6.7 Interfacing D/A Converters

When interfacing a D/A converter to the microcontroller, the digital data lines and the control lines, such as the start of conversion and chip select lines, are connected to the microcontroller I/O pins. The software routine generates the required signals to start the conversion process. Figure 14.36 shows the interface of D/A converter type number DAC 809 with the microcontroller. DAC-809 is a eight-bit D/A converter. Here, the output is current, so a current-to-voltage converter is required at the output.

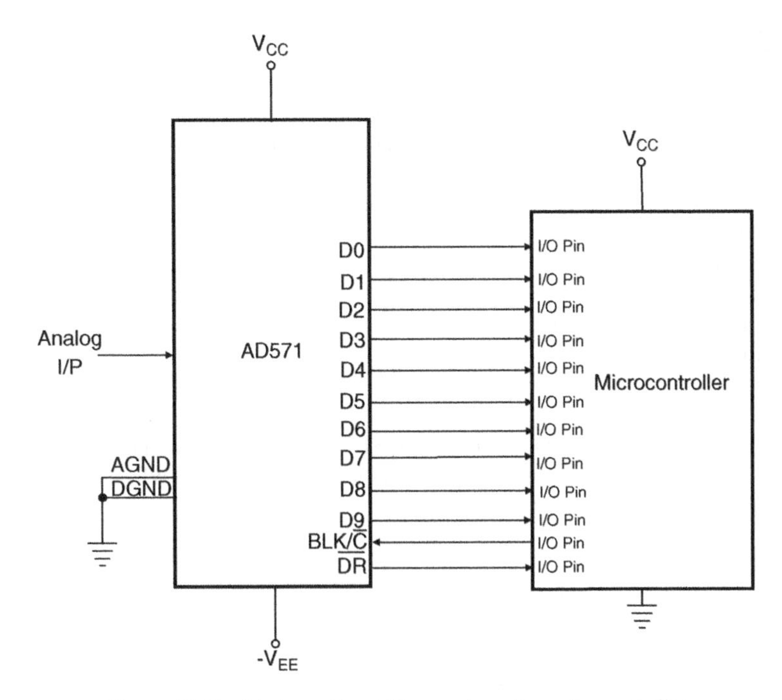

Figure 14.35 Interface of an A/D converter with a microcontroller.

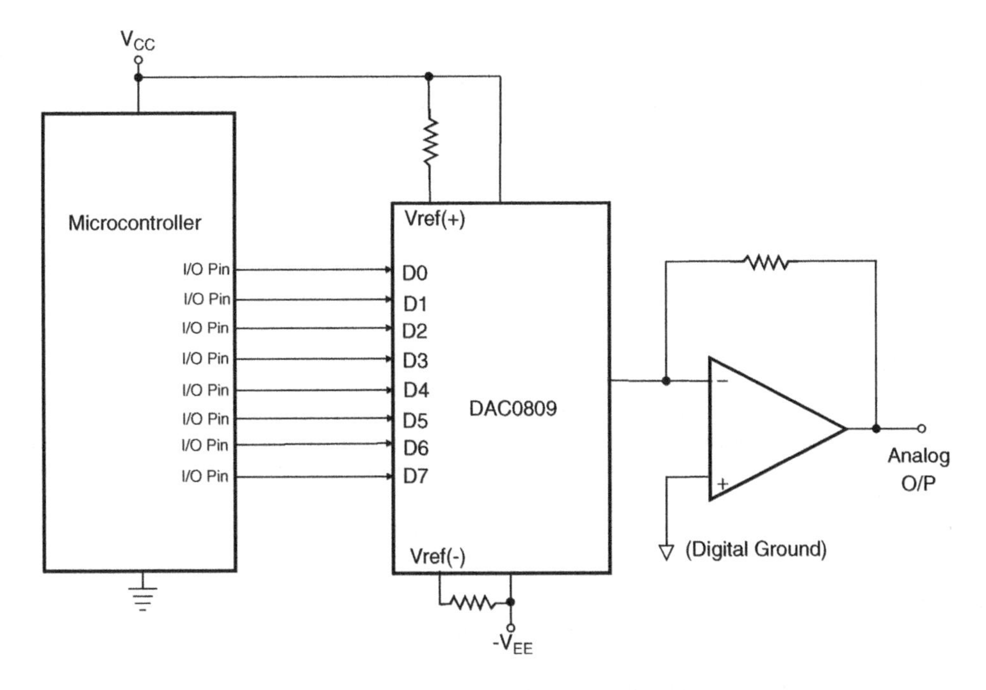

Figure 14.36 Interface of a D/A with a microcontroller.

Review Questions

1. What are the fundamental differences between a microprocessor and a microcontroller? Briefly describe some of the major application areas of microcontrollers.
2. What hardware components are likely to be found inside a typical microcontroller? Briefly describe the function of each one of them.
3. Name any three serial interfaces that are usually found on microcontrollers. Briefly describe where these are generally used.
4. What are the salient features of:

   (a) an interintegrated circuit (I^2C) bus;
   (b) Harvard architecture;
   (c) a memory-mapped I/O.

5. Briefly describe the salient features of the 80C51 family of eight-bit microcontrollers. Which microcontroller from Freescale Semiconductor does it closely resemble? Why and when would one like to choose a microcontroller other than 8051?
6. With reference to internal architecture, how do you compare eight-bit, 16-bit and 32-bit microcontrollers?
7. What are the basic differences between

   (a) the 80C51 and 89C51 families of microcontrollers;
   (b) the 68HC11 and 68HC16 families of microcontrollers;
   (c) the 80C51 and 16C84 families of microcontrollers.

8. With the help of relevant diagrams, briefly explain the difference between interfacing an LED type of display and an LCD type of display to a given microcontroller.
9. What are the interface requirements on the part of the microcontroller if it were to be interfaced with:

   (a) a keypad;
   (b) an LED;
   (c) another microcontroller.

Problems

1. A microcontroller with an eight-bit counter/timer system is used to measure the width of an input pulse. The microcontroller has been programmed to measure the time of occurrence of rising and falling edges of an input pulse on a certain I/O pin. If the microcontroller uses a 10 MHz clock and the count values observed at the time of occurrence of rising and falling edges of the input pulse are FE and 9A (in hex), determine the pulse width as measured by the microcontroller.

   10 μs

2. A microcontroller with a 16-bit counter/timer system is used to measure the frequency of an input pulse train. The microcontroller has been programmed to measure the time of occurrence of two successive leading edges of the input pulse signal on a certain I/O pin. If the microcontroller uses an 8 MHz clock and the count values observed at the time of occurrence of two successive rising

edges are observed to be FEEA and FE86 (in hex), determine the pulse width as measured by the microcontroller.

80 kHz

3. It is desired to design a microcontroller-based periodic signal generator with minimum and maximum time period specifications of 50 ns and 150 ms. Determine the minimum clock speed requirement of the microcontroller.

20 MHz

Further Reading

1. Susnea, L. and Mitescu, M. (2005) *Microcontrollers in Practice*, Springer Series, Springer, Germany.
2. Predko, M. (1999) *Programming and Customizing the 8051*, McGraw-Hill Professional, USA.
3. Van Sickle, T. (2000) *Programming Microcontrollers in C*, Elsevier Science, MA, USA.
4. Predko, M. (1998) *Handbook of Microcontrollers*, McGraw-Hill/Tab Electronics, USA.

15

Computer Fundamentals

This chapter focuses mainly on computer hardware fundamentals, with a brief introduction to some of the relevant software-related topics. The chapter begins with a brief description of different types of computer system, from giant supercomputers to tiny digital assistants, which is then followed up by anatomical description of a generalized computer system, with particular reference to microcomputer systems. Other hardware-related topics that are extensively covered include input/output devices and memory devices.

15.1 Anatomy of a Computer

The basic functional blocks of a computer comprise the central processing unit (CPU), memory and input and output ports. These functional blocks are depicted in the block schematic arrangement of Fig. 15.1. As is clear from the figure, these functional blocks are connected to each other by internal buses. The CPU is the brain of the computer. It is basically a microprocessor with associated circuits. Ports are physical interfaces on the computer, through which the computer interacts with the input and output devices. Memories are storage devices used for storing data and instructions. The CPU fetches the data and instructions by sending the address of the memory location on the address bus. The data and the instructions are then transferred to the CPU by the data bus. The CPU then executes the instructions and stores the processed data in the memory or sends them to an output device via the data bus. It may be mentioned here that in most cases the instructions modify the data stored in the memory or obtained from an input device.

15.1.1 Central Processing Unit

As mentioned above, the CPU is the brain of the computer. The fundamental operation of the CPU is to execute a sequence of stored instructions called a program. In other words, it controls the execution

Digital Electronics: Principles, Devices and Applications Anil Kumar Maini

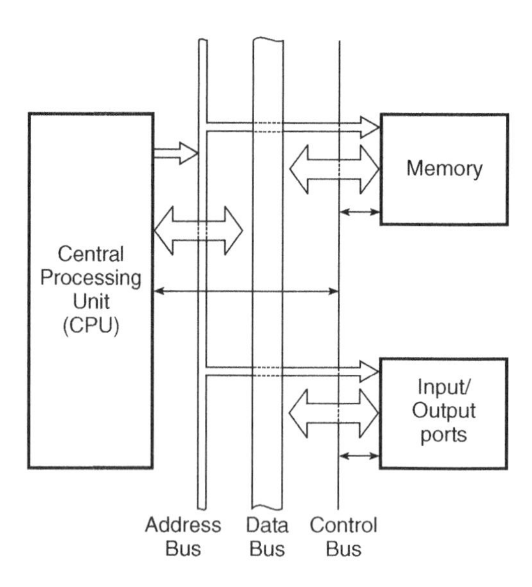

Figure 15.1 Block schematic of a typical computer.

of the computer software programs. It fetches and executes the instructions from the memory in a sequential manner. It may be mentioned here that the CPU can operate on more than one instruction at a time. Early CPUs were custom designed for a particular type of computer. But they have given way to a standardized class of processors that are used for generic applications. Since the advent of microprocessors in the 1970s, single-chip microprocessors have totally replaced all other types of CPU, and today the term 'CPU' refers to a microprocessor.

A microprocessor is a programmable device that accepts binary data from an input device, processes the data according to the instructions stored in the memory and provides results as output. The important functional blocks of a microprocessor are the arithmetic logic unit, the control unit and the register file. Microprocessors were discussed at length in Chapter 13.

15.1.2 Memory

There are several types of memory used in a computer. They can be classified as primary memory and secondary memory. Primary memory is directly connected to the CPU and is accessible to the CPU without the use of input/output channels. Primary memory can be classified into process registers, main memory, cache memory and read only memory (ROM). Process registers are present inside the CPU and store information to carry out the current instruction. Main memory is a random access memory (RAM) that stores the programs that are currently being run and the data related to these programs. It is a volatile memory and is used for temporary storage of data and programs. Cache memory is a special type of internal memory that can be accessed much faster than the main RAM. It is used by the CPU to enhance its performance. ROM is a nonvolatile memory that stores the system programs including the basic input/output system (BIOS), start-up programs and so on.

Secondary or auxiliary memory cannot be accessed by the CPU directly. It is accessed by the CPU through its input/output channels. Secondary memory has a much greater capacity than primary memory, but it is much slower than the primary memory. It is used to store programs and data for future use. Most commonly used secondary memory devices include the hard disk, floppy disks, compact disks (CDs), USB disks and so on. The hard disk is used for storing the high-level operating systems, application software and the user data files. Floppy disks have a limited capacity of 1.44 MB and have been replaced by CDs and USB drives. Floppy disks, CDs and USB drives are also referred to as off-line storage devices as they can be easily removed from the computer. Different types of memory are covered in Section 15.4.

15.1.3 Input/Output Ports

A port is a physical interface on the computer through which the input and output devices are connected to and interact with the computer. Ports are also used as an interface to connect two computers to each other. The ports on the computer can be configured as input and output ports through software. These ports are of two types, namely serial ports and parallel ports. Serial ports send and receive one bit at a time through a single wire pair. Parallel ports send multiple bits at the same time over a set of wires. Serial ports are used to connect devices such as modems, digital cameras, etc., to the computer. Parallel ports are used to connect printers, scanners, CD burners, external hard drives, etc., to the computer. Serial and parallel ports are discussed in detail in Section 15.8.

15.2 A Computer System

Figure 15.2 shows the block diagram of a typical computer system. The diagram basically shows the interconnection of the computer with the commonly used input/output devices. Input devices convert the raw data to be processed into a computer-understandable format. Some of the commonly used input devices include the keyboard, mouse, scanner and so on. Output devices convert the processed data into a format understandable by the user. Commonly used output devices include the monitor, printer, cameras, and so on. Input and output devices are discussed at length in Section 15.7.

15.3 Types of Computer System

Computers can be classified into various types, depending upon the technology used or the size and capacity or the applications for which they are designed.

15.3.1 Classification of Computers on the Basis of Applications

Based on the application or the purpose, computers are often classified as general-purpose computers and special-purpose or dedicated computers. *General-purpose computers* are comparatively more flexible and thus can be used to work on a large variety of problems including business and scientific problems. For instance, banking applications such as financial accounting, pay-roll processing, etc., at the head-office level would require the services of a general-purpose computer. The size and capacity of a general-purpose computer could of course vary, depending upon the quantum of data and the

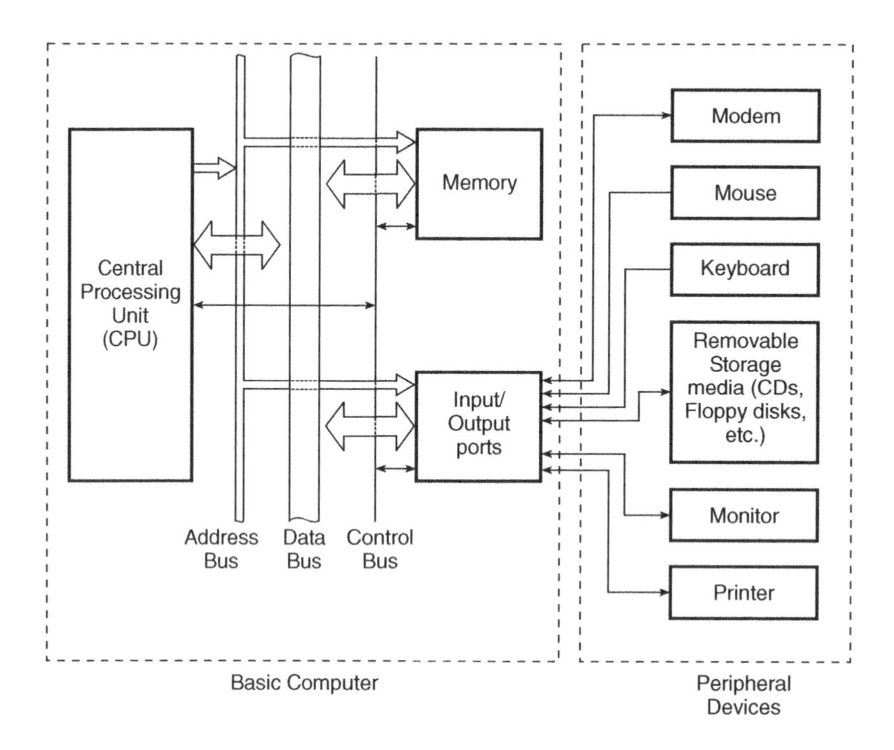

Figure 15.2 Block diagram of a typical computer system.

complexity of data processing to be done. *Special-purpose computers*, on the other hand, are designed for a dedicated application. These computers perform a certain predecided and fixed sequence of operations. Typical applications include the computers used for weather forecasting, aircraft control systems, missile and other weapon guidance systems, etc.

15.3.2 Classification of Computers on the Basis of the Technology Used

Based on the technology used, the computers are classified as analogue computers, digital computers and hybrid computers. In analogue computers, the input data comprise continuously changing electrical or nonelectrical (temperature, pressure, speed, volume, etc.) information. There are numerous examples of analogue computational devices. One such device is the speedometer of an automobile. The input data to this device or machine are the continuously varying rotational speed of its driveshaft. The rotational motion is converted into a linear movement of a needle pointer that indicates the speed in km/h. A tachometer used to measure the rotational speed is another device of the same type. The input data in the case of a digital computer are discrete in nature. They are represented by a binary notation in the form of 0s and 1s. A hybrid computer is a mixture of the two. It attempts to combine the good points of both analogue and digital computers. In a typical hybrid computer, the measuring functions are performed the analogue way while the control and logic functions are digital in nature.

15.3.3 Classification of Computers on the Basis of Size and Capacity

Based on their size and capacity, computers are classified as mainframe computers, minicomputers, microcomputers and supercomputers.

15.3.3.1 Mainframe Computers

A mainframe computer is the largest, fastest and perhaps one of the most expensive computer systems of general use. Before the advent of minicomputers and microcomputers respectively in the third- and fourth-generation periods, all data processing was done on mainframe systems only. Thousands of such machines are still in use in medium- and large-size business houses, universities, hospitals, etc.

These machines have a very large primary storage capability and have a very high processing speed. Because of their size and speed, mainframe systems must be placed on special platforms that allow wiring and cooling systems. These machines are useful not only because they have an enormous storage capacity but also because of their capability to support a large number of terminals. Modern-day mainframe computers are defined by their high-quality internal engineering, reliability, technical support and security features, along with their performance qualities. Their applications include the processing of a huge amount of different kinds of data such as census, industry/consumer statistics, financial transactions processing, etc., in large private and public enterprises, government agencies, etc. Examples of mainframe computers include IBM's zSeries and System z9 servers, Unisys's ClearPath mainframes, the zSeries 800 from Hitachi and IBM, the Nonstop systems from HP and so on.

15.3.3.2 Minicomputers

A minicomputer more or less resembles a mainframe system except that it is comparatively smaller and less expensive. They represent a class of multi-user computers that are used for middle-range computing applications, in between the mainframe systems and the microcomputers. Minicomputers were developed during the third-generation period. PDP-8 and PDP-11 from Digital Equipment Corporation (DEC) are examples of the popular minicomputers developed in the late 1960s. Minicomputers gave way to microcomputers in the mid-1980s and early 1990s.

15.3.3.3 Microcomputers

The microcomputer, the development of which was made possible largely owing to the development of the microprocessor, is a compact, relatively inexpensive and complete computer. The most obvious, though not the only difference between a microcomputer and a mainframe is the physical size. While a mainframe system may fill a room, a microcomputer may be put on a desktop or may even fit into a brief case. Although microcomputers can be distinguished from mainframe and minicomputers on the basis of size, technology used, applications and so on, these dividing lines are hazy and these categories almost overlap with each other owing to brisk advances in technology. Like mainframes and minis, today's microcomputers do data processing, manipulate lists, store, retrieve and sort information. Unlike mainframes and minis, microcomputers do not require any specialized environment for operation and can be effectively made use of by people who do not have any comprehensive formal training in computer techniques. In fact, these machines are designed to be used both at the workplace and at home. The concept of office automation has become feasible only with the advent of microcomputers.

15.3.3.4 Personal Computers

A personal computer, popularly known as a PC, is a stand-alone microcomputer that is used in a varied range of applications, from writing letters to the present-day desktop publishing, from playing video games to enquiring about railway and air schedules, from simple graphics to designing an advertisement, from simple financial accounting to preparing spread sheets and so on.

With the development of microprocessors and related peripherals, the personal computer of today is as powerful as a minicomputer of yesteryears. The processing speed has touched GHz and the hard disk capacity has reached tens of GBs. The contemporary microprocessors for the PC platform offer applications including internet audio and streaming video, image processing, video content creation, speech, computer-aided simulation and design, games, multimedia and multitasking user environments. Depending upon their size and portability, they can be classified as desktops, laptops and palmtops.

Desktops are personal computers for use on a desk in an office or at home. They are currently the most popular type of computer in use. Laptops, also referred to as notebooks, are mobile personal computers that can be carried in a briefcase. They do not always require an external power source and run on rechargeable batteries for 4–5 h. Some of the famous manufacturers of laptops include IBM, Compaq, Acer, Dell, HP and so on.

15.3.3.5 Workstations

Workstations are high-end technical computing desktop microcomputers designed primarily to be used by one person at a time, but they can also be connected remotely to other users if needed. They offer high performance compared with a personal computer, especially with respect to graphics, processing power and multitasking ability. Today, workstations use many technologies common to the personal computers.

15.3.3.6 Supercomputers

Supercomputers are the fastest and most powerful of all computer systems. They are typically 200 times faster than the mainframes. Supercomputers are mainly used for calculation-intensive applications requiring enormous amounts of data to be processed in a very short time. These include weather forecasting, weapons research, breaking secret codes, designing aircraft, molecular modelling, physical simulations and so on. Supercomputers are mainly used in universities, military agencies and scientific research laboratories. Supercomputers are highly parallel systems, i.e. they perform many tasks simultaneously. They generate a lot of heat and need a proper cooling mechanism. Some of the popular supercomputers include Cray-1, Cray X-MP/4, Cray-2, Intel's ASCI Red/9152 and ASCI Red/9632 and IBM's Blue Gene/L.

15.4 Computer Memory

Computer memory refers to components, devices, chips and recording media that are used for temporary, semi-permanent and permanent storage of data. As mentioned in the previous section, there are several types of memory device used in a computer. These include RAM, ROM, cache, flash memory, hard disk, floppy disk, CDs and so on. Memory devices can be broadly classified into two types, namely primary memory and secondary storage. Figure 15.3 shows the various types of memory device present in a typical computer system. It may be mentioned here that, in computer terminology, 'memory' usually refers to RAM and ROM and the term 'storage' refers to hard disks, floppy disks

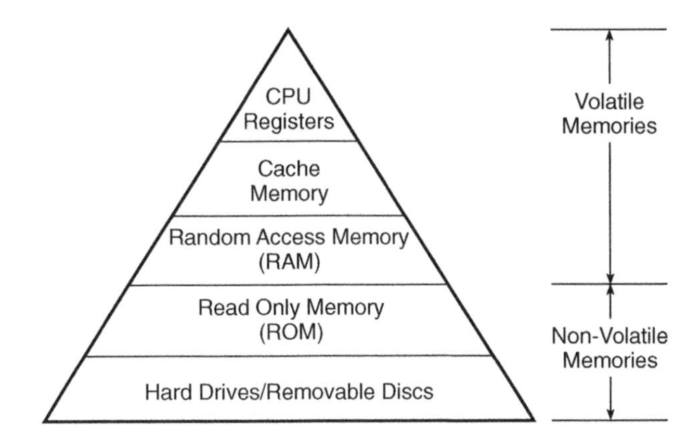

Figure 15.3 Various types of memory present in a typical computer system.

and CDs. Primary memory is described in this section, and secondary storage media are discussed in Section 15.10.

15.4.1 Primary Memory

The primary memory holds the program instructions for the program to be executed, the input data to be processed and the intermediate results of any calculations when processing is being done. Primary memory is also used for storing BIOS and start-up programs.

When a program and data are entered into a computer, the control unit directs them to the primary memory. Each program instruction and each data item is stored in a memory location that has a unique address. These data and instructions are held till new data items and instructions are written over them. Thus, the same data can be accessed repeatedly if so desired and the same instructions can be executed repeatedly if so required. This is what is known as the stored program concept. The primary memory of a computer further comprises process registers, random access memory (RAM), cache memory and read only memory (ROM). Process registers are memory cells built into the CPU that contain the specific data needed by the CPU. Cache memory is basically a type of RAM memory.

15.4.1.1 Random Access Memory

RAM is a read/write memory where the data can be read from or written into any of the memory locations regardless of the order in which they are arranged. Therefore, all the memory locations in a RAM can be accessed at the same speed. RAM is used to store data, program instructions and the results of any intermediate calculations during the execution of a program. Also, the same data can be read any number of times and different data can be written into the same memory location, with every fresh data item overwriting the existing one. It is typically used for short-term data storage as it cannot retain data when the power is turned off.

RAM is available in the form of ICs as well as in the form of plug-in modules. The plug-in modules are small circuit boards containing memory ICs and having input and output lines connected to an edge connector. They are available as single in-line memory modules (SIMMs) and dual in-line memory modules (DIMMs). More than one memory IC (or chip) can be used to build the RAM for larger systems. The capacity or size of a RAM is measured in bytes. RAM chips are available in the memory capacities ranging from 2 kB to as much as 32 MB. 1 kB of memory equals $2^{10} = 1024$ bytes and 1 MB of memory equals 2^{20} bytes. The terms 'kilo' (k) and 'mega' (M) have been used, as 2^{10} and 2^{20} are approximately equal to 1000 and 1 000 000 respectively. As an illustration, a microcomputer with a 64 kB of RAM has $64\times2^{10} = 2^6\times2^{10} = 2^{16} = 65\ 536$ bytes of memory. The two categories of RAM are static RAM (SRAM) and dynamic RAM (DRAM). RAM is discussed in detail in Section 15.5.

15.4.1.2 Read Only Memory

In the case of ROM, instructions can be written into the memory only once at the manufacturer's premises. These instructions can, however, be read from a ROM as many times as desired. Once it is written, a ROM cannot be written into again. The contents of a ROM can thus be accessed by a CPU but cannot be changed by it. The instructions stored on a ROM vary with the type of application for which it is made. The ROM for a general-purpose microcomputer, for instance, would contain system programs such as those designed to handle operating system instructions.

In the case of some special types of ROM, it is possible for users to have their own instructions stored on the ROM as per their requirements. Such ROM chips are called PROMs (Programmable Read Only Memory). PROM contents, once programmed, cannot be changed. But then there are some special types of PROMs whose contents can be erased and then reprogrammed. These are known as EPROMs (Erasable Programmable Read Only Memory). ROM memories are discussed in detail in Section 15.6.

15.5 Random Access Memory

In this section we will discuss at length the types of RAM and their basic construction, properties, applications and so on.

RAM has three basic building blocks, namely an array of memory cells arranged in rows and columns with each memory cell capable of storing either a '0' or a '1', an address decoder and a read/write control logic. Depending upon the nature of the memory cell used, there are two types of RAM, namely static RAM (SRAM) and dynamic RAM (DRAM). In SRAM, the memory cell is essentially a latch and can store data indefinitely as long as the DC power is supplied. DRAM on the other hand, has a memory cell that stores data in the form of charge on a capacitor. Therefore, DRAM cannot retain data for long and hence needs to be refreshed periodically. SRAM has a higher speed of operation than DRAM but has a smaller storage capacity.

15.5.1 Static RAM

As mentioned before, the basic element of SRAM is a latch memory cell. Figure 15.4 shows a basic SRAM memory cell. The memory cell is selected by setting the 'select' line active. The data bit is written into the cell by placing it on the 'data in' line and is read from the 'data out' line.

SRAMs can be broadly classified as asynchronous SRAM and synchronous SRAM. Asynchronous SRAMs are those whose operations are not synchronized with the system clock, i.e. they operate independently of the clock frequency. 'Data in' and 'data out' in these RAMs are controlled by address transition. Synchronous SRAMs are those whose timings are initiated by clock edges. 'Address', 'data in', 'data out' and all other control signals are synchronized with the clock signal. Synchronous

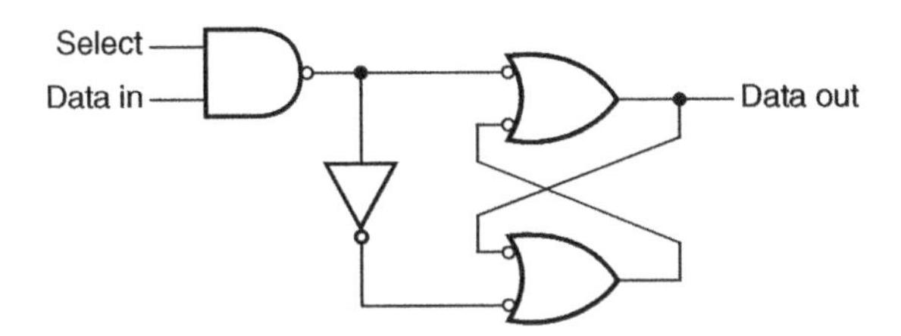

Figure 15.4 Basic SRAM memory cell.

SRAMs normally have an address burst feature, which allows the memory to read and write at more than one location using a single address. Both synchronous and asynchronous SRAMs are available in bipolar, MOS and BiCMOS technologies. While bipolar SRAM offers a relatively higher speed of operation, MOS technology offers a higher capacity and reduced power consumption. Figures 15.5(a) and (b) show the basic bipolar memory cell and the MOS (NMOS more specifically) memory cell respectively.

15.5.1.1 Asynchronous SRAM

Figure 15.6 shows the typical architecture of a 64×8 asynchronous SRAM. It is capable of storing 64 words of eight bits each. The main blocks include a 6-to-64 line address decoder, I/O buffers, 64 memory cells and control logic for read/write operations. The memory cells in a row are represented as a register. Each register is an eight-bit register and can be read from as well as written into. As can be seen from the figure, all the cells inside the same register share the same decoder output line, also referred to as 'row line'. The control functions are provided by $R/\overline{W}$ (read/write) and $\overline{CS}$ (chip select) inputs. $R/\overline{W}$ and $\overline{CS}$ inputs are also referred to as $\overline{WE}$ (write enable) and $\overline{CE}$ (chip enable) inputs respectively. The 'data input' and 'data output' lines are usually combined by using common input/output lines in order to conserve the number of pins on the IC package.

The memory is selected by making $\overline{CS}=0$. During the 'read' operation the status of the $R/\overline{W}$ and $\overline{CS}$ pins is '1' and '0' respectively, while during the 'write' operation it is '0' and '0' respectively. During the 'read' operation the input buffers are disabled and the contents of the selected register appear at the output. During the 'write' operation the input buffers are enabled and the output buffers are disabled. The contents of the input buffers are loaded into the selected register, the previous data of which are overwritten by the new data. The output buffers, being tristate, are in the high-impedance state during the write operation. $\overline{CS}=1$ deselects the chip, and both the input and the output data buffers get disabled and go to the high-impedance state. The contents of the memory in this case remain unaffected. 'Chip select' inputs are particularly important when more than one RAM memory chip is combined to get a larger memory capacity.

In the case of larger SRAM memories, there are two address decoders, one for rows and one for columns. They are referred to as row decoders and column decoders respectively. Some of the address lines are fed to the row decoder and the rest of the address lines are fed to the column decoder. Figure 15.7 shows the architecture of a typical 16K×8 asynchronous SRAM. The memory cells are arranged in eight arrays of 128 rows and 128 columns each. Memories with a single address decoder are referred

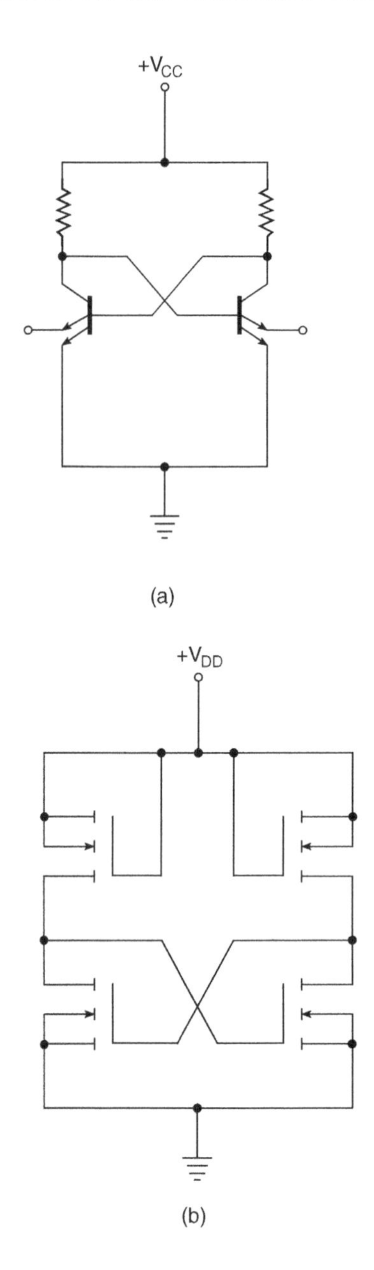

Figure 15.5 (a) Basic bipolar memory cell and (b) a basic MOS memory cell.

to as two-dimensional memories, and those with two decoders are referred to as three-dimensional memories.

Figures 15.8(a) and (b) show the timing diagrams during 'read' and 'write' operations respectively. The diagrams are self-explanatory. Read and write cycle time intervals of a few nanoseconds to a few tens of nanoseconds are common in the case of asynchronous SRAMs.

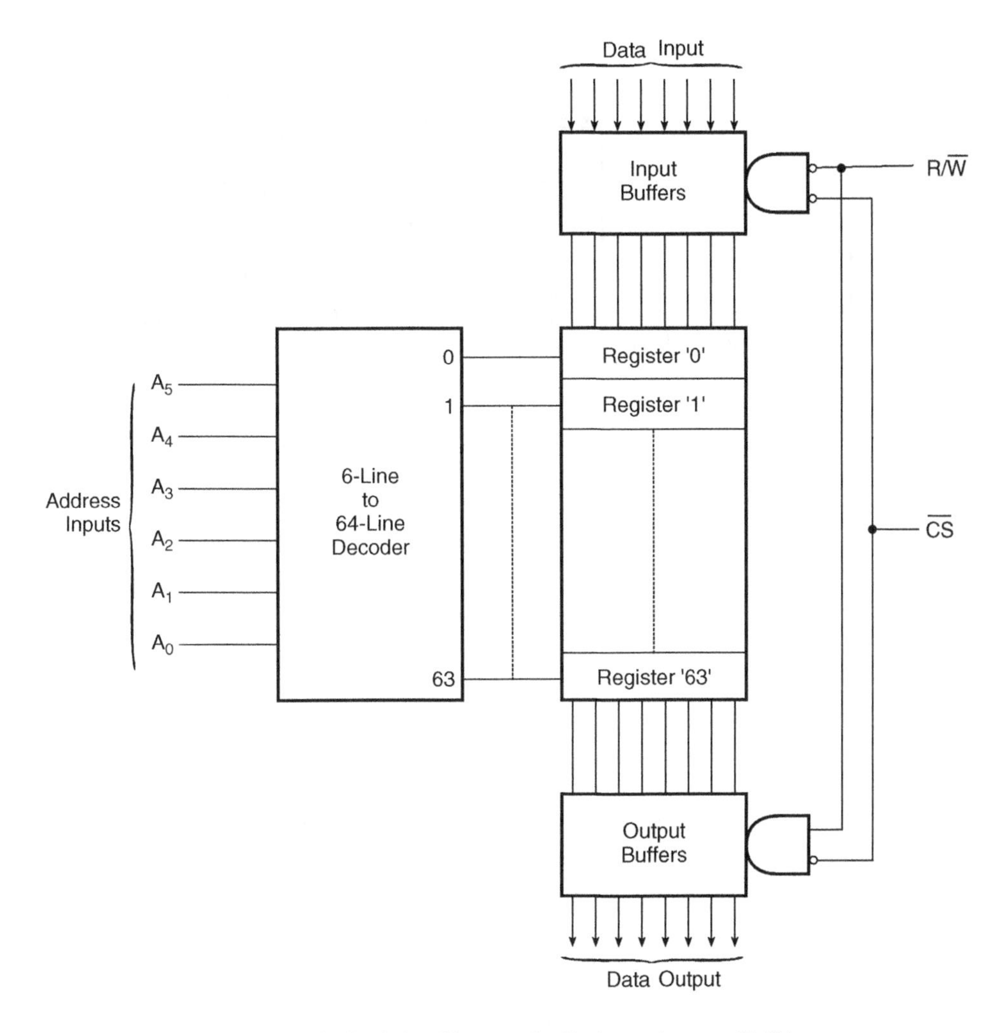

Figure 15.6 Typical architecture of a 64×8 asynchronous SRAM.

The different timing intervals shown in the diagram are defined as follows:

- Complete read cycle time t_{RC}. This is defined as the time interval for which a valid address code is applied to the address lines during the 'read' operation.
- RAM access time t_{ACC}. This is defined as the time lapse between the application of a new address input and the appearance of valid output data.
- Chip enable access time t_{CO}. This is defined as the time taken by the RAM output to go from the Hi-Z state to a valid data level once $\overline{CS}$ is activated.
- Chip disable access time t_{OD}. This is defined as the time taken by the RAM to return to the Hi-Z state after $\overline{CS}$ is deactivated.

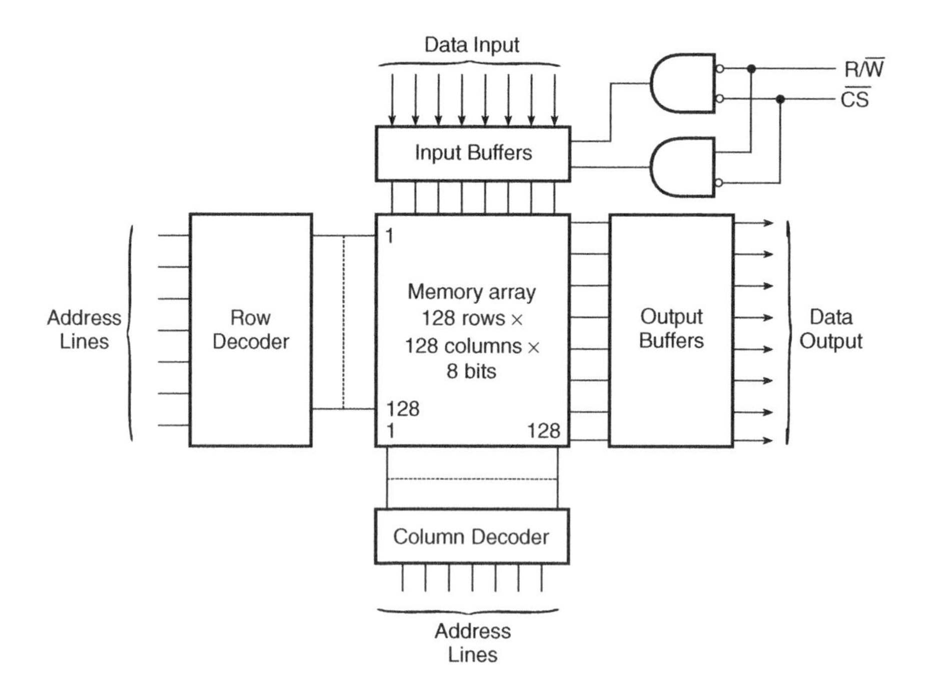

Figure 15.7 Typical architecture of a 16K×8 asynchronous SRAM.

- Complete write cycle time t_{WC}. This is defined as the time interval for which a valid address code is applied to the address lines during the 'write' operation.
- Write pulse width t_W. This is the time for which $R/\overline{W}$ is held LOW during the 'write' operation.
- Address set-up time t_{AS}.This is the time interval between the appearance of a new address and $R/\overline{W}$ going LOW.
- Data set-up time t_{DS}. This is defined as the time interval for which the $R/\overline{W}$ must remain LOW after valid data are applied to the data inputs.
- Data hold time t_{DH}. This is defined as the time interval for which valid input data must remain on the data lines after the $R/\overline{W}$ input goes HIGH.
- Address hold time interval t_{AH}. This is defined as the time interval for which the valid address must remain on the address lines after the $R/\overline{W}$ input goes HIGH.

15.5.1.2 Synchronous SRAM

Synchronous SRAM, as mentioned before, is synchronized with the system clock. In the case of a computer system it operates at the same clock frequency as the microprocessor. This synchronization of microprocessor and memory ensures faster execution speeds. The basic difference between the architecture of synchronous and asynchronous SRAMs is that the synchronous SRAM makes use of clocked registers to synchronize 'address', $R/\overline{W}$, $\overline{CS}$ and 'data in' lines to the system clock. Figure 15.9 shows the basic architecture of a 32K × 8 synchronous SRAM with a burst feature. As we can see from the figure, the memory array block, the address decoder block and $R/\overline{W}$ and $\overline{CS}$ are the same

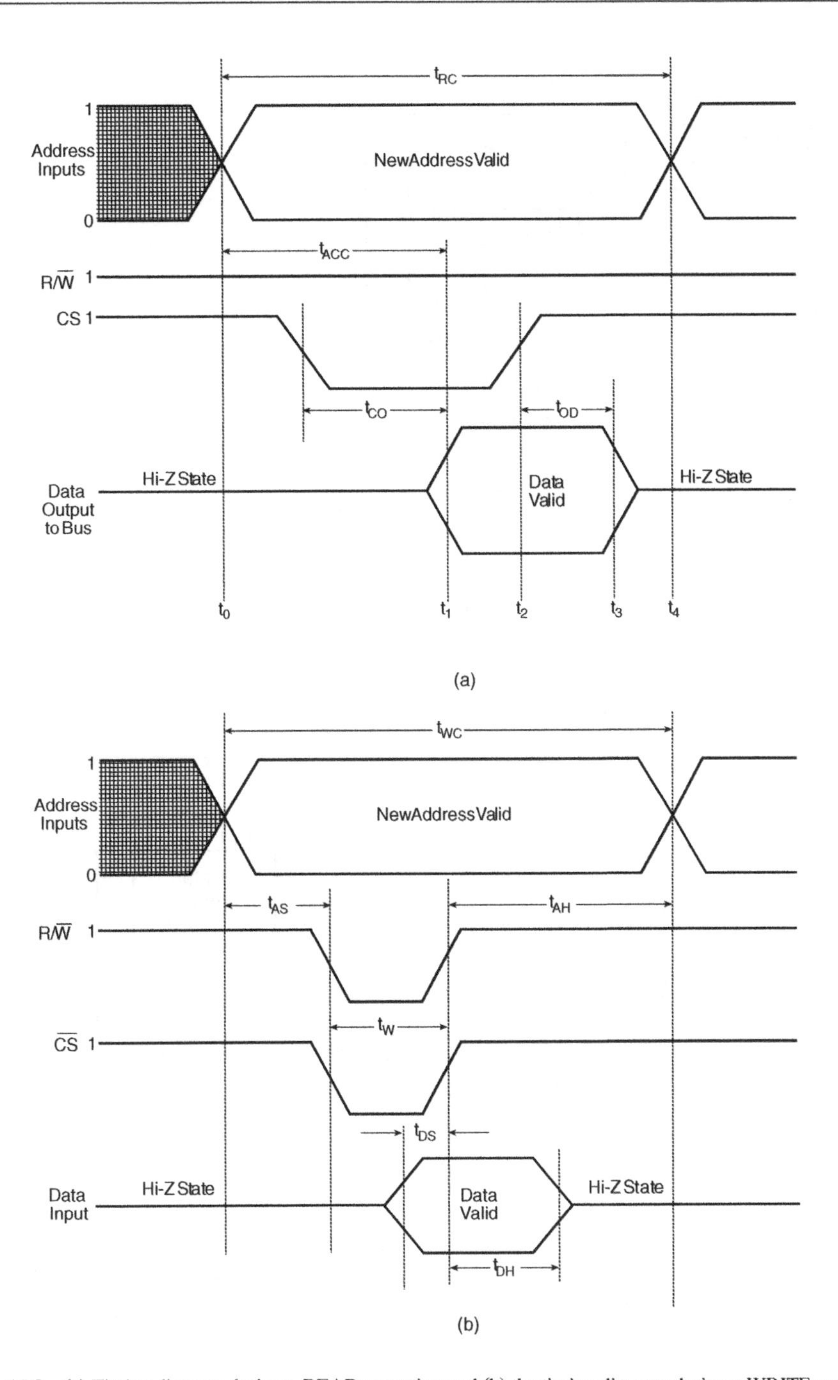

Figure 15.8 (a) Timing diagram during a READ operation and (b) the timing diagram during a WRITE operation.

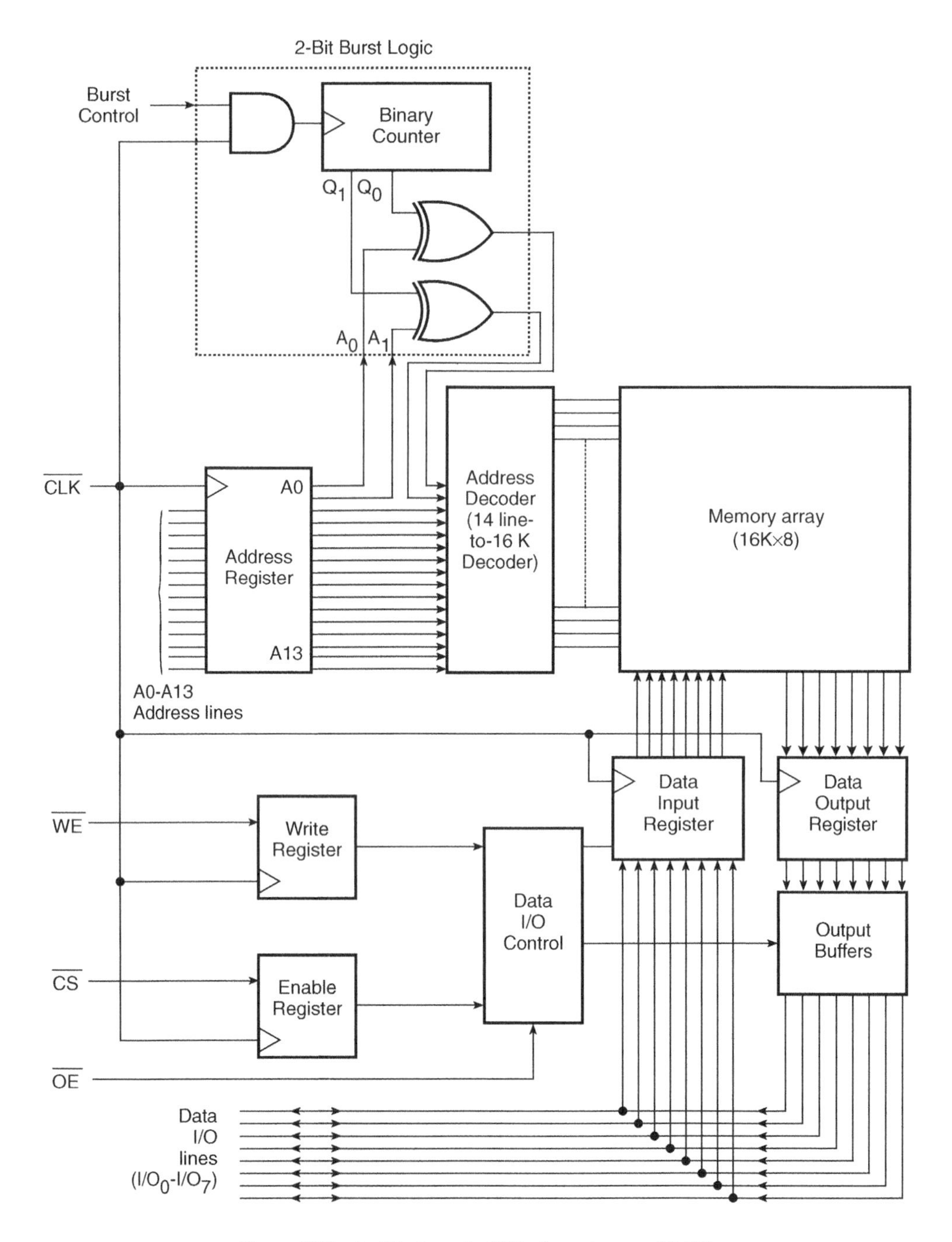

Figure 15.9 Architecture of a 16K ×8 synchronous SRAM.

as in the case of an asynchronous SRAM. As mentioned before, most synchronous SRAMs have an address burst feature. In this case, when an external address is latched to the address register, a certain number of lowest address bits are applied to the burst logic. Burst logic comprises a binary counter and EXCLUSIVE-OR gates. The output of the burst logic, which basically produces a sequence of internal addresses, is fed to the address bus decoder. In the case of a two-bit burst logic, the internal address sequence generated is given by A_1A_0, $A_1\overline{A_0}$, $\overline{A_1}A_0$, $\overline{A_1}\ \overline{A_0}$, where A_0 and A_1 are the address bits applied to the burst logic. The burst logic shown in Fig. 15.9 is also a two-bit logic.

15.5.2 Dynamic RAM

The memory cell in the case of a DRAM comprises a capacitor and a MOSFET. The cell holds a value of '1' when the capacitor is charged and '0' when it is discharged. The main advantage of this type of memory is its higher density, or more bits per package, compared with SRAM. This is because the memory cell is very simple compared with that of SRAM. Also, the cost per bit is less in the case of a DRAM. The disadvantage of this type of memory is the leakage of charge stored on the capacitors of various memory cells when they are storing a '1'. To prevent this from happening, each memory cell in a DRAM needs to be periodically read, its charge (or voltage) compared with a reference value and then the charge restored to the capacitor. This process is known as 'memory refresh' and is done approximately every 5–10 ms.

Figure 15.10 shows the basic memory cell of a DRAM and its principle of operation. The MOSFET acts like a switch. When in the 'write' mode ($R/\overline{W} = 0$), the input buffers are enabled while the output buffers are disabled. When '1' is to be stored in the memory, the 'data in' line must be in the HIGH state and the corresponding 'row line' should also be in the HIGH state so that the MOSFET is switched ON. This connects the MOSFET to the 'data in' line, and it charges the capacitor to a positive voltage level. When '0' needs to be stored, the 'data in' line is LOW and the capacitor also acquires the same level. When the 'row line' is taken to the LOW state, the MOSFET is switched OFF and is disconnected from the bit line. This traps the charge on the capacitor. In 'read' mode ($R/\overline{W} = 1$), the output buffers are enabled while the input buffers are disabled. When the 'row line' is taken to HIGH logic, the MOSFET is switched ON and connects the capacitor to the 'data out' line through the output

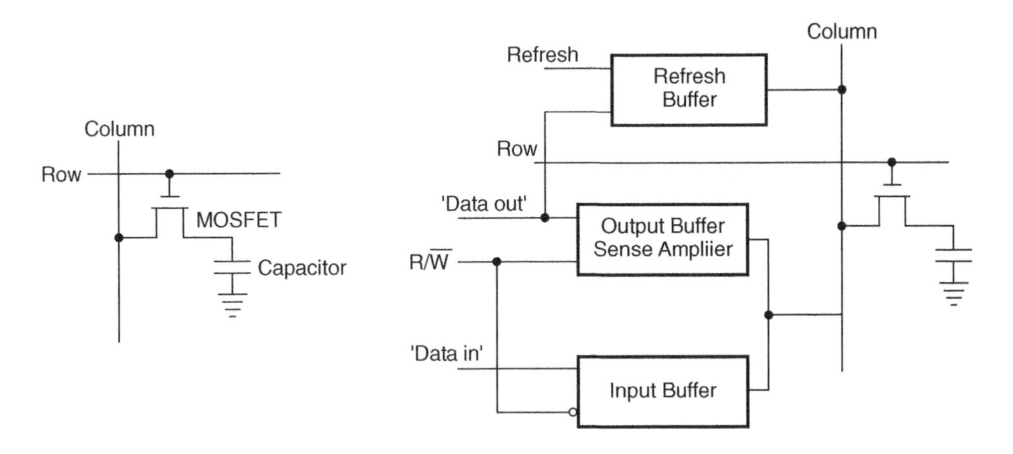

Figure 15.10 Basic memory cell of a DRAM.

buffer. The refresh operation is performed by setting $R/\overline{W} = 1$ and by enabling the refresh buffer. There are two basic modes of refreshing the memory, namely the burst refresh and distributed refresh modes. In burst refresh mode, all rows in the memory array are refreshed consecutively during the refresh burst cycle. In distributed refresh mode, each row is refreshed at intervals interspaced between 'read' and 'write' operations.

15.5.2.1 DRAM Architecture

The architecture of DRAM memory is somewhat different from that of SRAM memory. Row and column address lines are usually multiplexed in a DRAM. This is done to reduce the number of pins on the package. Row address select (RAS) and column address select (CAS) inputs are used to indicate whether a row or a column is to be addressed. Address multiplexing is particularly attractive for higher-capacity DRAMs. A 4 MB RAM, for instance, would require 22 address inputs ($2^{22} = 4M$).

Figure 15.11 shows the architecture of a 16K × 1 DRAM. The heart of a DRAM is an array of single-bit memory cells. Each cell has a unique position as regards row and column. Other important blocks include address decoders (row decoder and column decoder) and refresh control and address latches (row address latch and column address latch). As can be seen from the figure, seven address lines are time multiplexed at the beginning of the memory cycle by the RAS and CAS lines. Firstly, the seven-bit address (A_0–A_6) is latched into the row address latch, and then the seven-bit address is latched into the column address latch (A_7–A_{13}). They are then decoded to select the particular memory location. Larger word sizes can be achieved by combining more than one chip. This is discussed in the next section. Figures 15.12(a) and (b) respectively show the timing diagrams for read and write operations. A DRAM is relatively slower than a SRAM. The typical access time is in the range 100–250 ns.

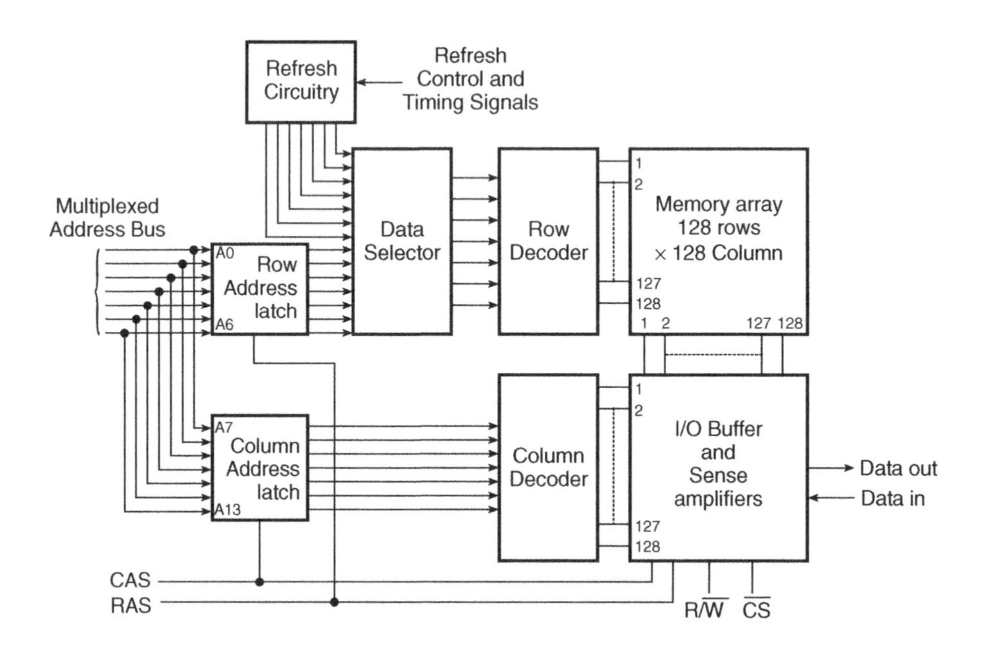

Figure 15.11 Architecture of a 16K × 1 DRAM.

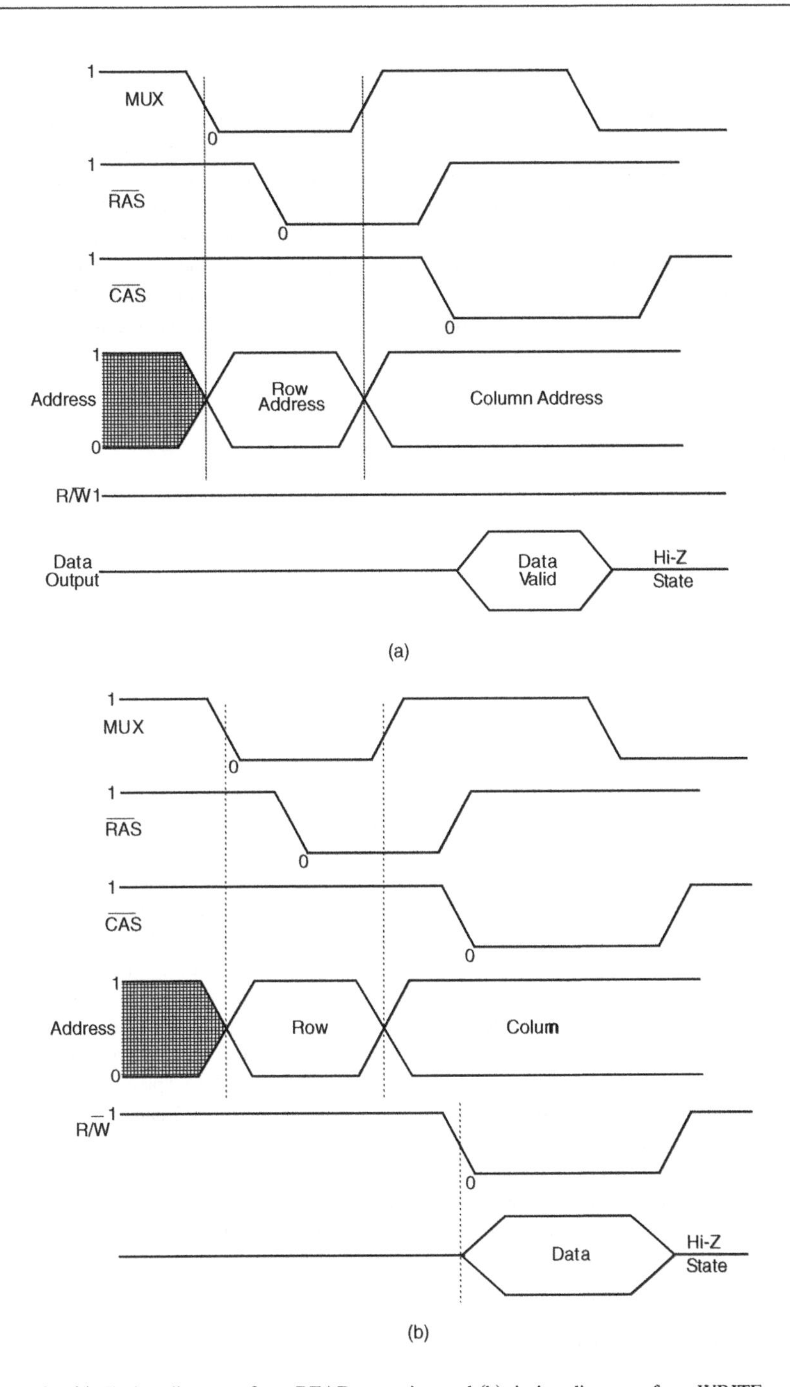

Figure 15.12 (a) Timing diagrams for a READ operation and (b) timing diagrams for a WRITE operation.

15.5.2.2 Types of DRAM

DRAM memories can be further classified as fast page mode (FPM) DRAM, extended data output (EDO) DRAM, burst extended data output (BEDO) DRAM and synchronous (S) DRAM. In FPM DRAM, the row address is specified only once for access to several successive column addresses. Hence, the read and write times are reduced. EDO DRAM is similar to FPM DRAM, with the additional feature that a new access cycle can be started while keeping the data output of the previous cycle active. BEDO DRAM is an EDO DRAM with address burst capability. All the types of DRAM discussed hitherto are asynchronous DRAMs, and their operation is not synchronized with the system clock. SDRAM, as the name suggests, is a synchronous DRAM whose operation is synchronized with the system clock.

15.5.3 RAM Applications

One of the major applications of RAM is its use in cache memories. It is also used as main memory to store temporary data and instructions in a computer.

15.5.3.1 Cache Memory

Advances in microprocessor technology and also the software have greatly enhanced the application potential of present-day computers. These enhanced performance features and increased speed can be optimally utilized to the maximum only if the computer has the required capacity of main (or internal) memory. The computer's main memory, as we know, stores program instructions and data that the CPU needs during normal operation. In order to get the maximum out of the system, this would normally require all of the system's main memory to have a speed comparable with that of the CPU. It is not economical for all the main memory to be high speed. This is where the cache memory comes in.

Cache memory is a block of high-speed memory located between the main memory and the CPU. The cache memory block is the one that communicates directly with the CPU at high speed. It stores the most recently used instructions or data. When the processor needs data, it checks in the high-speed cache to see if the data are there. If they are there, called a 'cache hit', the CPU accesses the data from the cache. If they are not there, called a 'cache miss', then the CPU retrieves them from the relatively slower main memory. Cache memory mostly uses SRAM chips, but it can also use DRAM.

There are two levels of cache memory. The first is the level 1 cache (L1 or primary or internal cache). It is physically a part of the microprocessor chip. The second is the level 2 cache (L2 or secondary or external cache). It is in the form of memory chips mounted external to the microprocessor. It is larger than the L1 cache. The L1 and L2 cache memories range from 2 to 64 kB and from 256 kB to 2 MB in size respectively. Some systems have higher-level caches (L3, L4, etc.), but L1 and L2 are the most common. Figure 15.13 shows the use of L1 and L2 cache memories in a computer system.

15.6 Read Only Memory

ROM is a nonvolatile memory that is used for permanent or semi-permanent storage of data. The contents of ROM are retained even after the power is turned off. In this section we will be discussing at length the ROM architecture, types of ROM and typical applications.

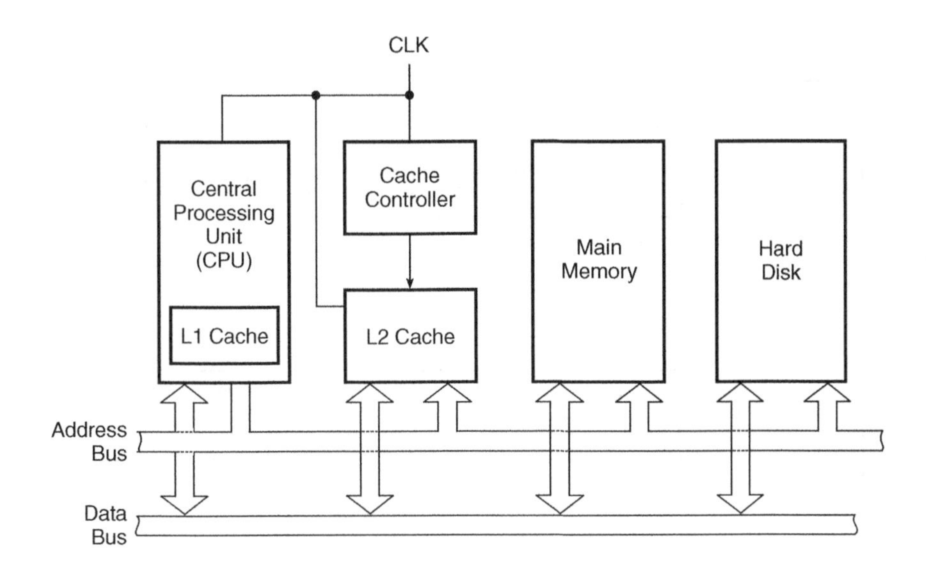

Figure 15.13 Cache memory in a computer system.

15.6.1 ROM Architecture

The internal structure or architecture of a ROM comprises three basic parts, namely the array of memory cells, the address decoder and the output buffers. The address decoder comprises a single decoder in the case of small memories. In the case of large memories it comprises two decoders referred to as row and column decoders. The operation of a ROM can be best explained with the help of the simplified representation of a 32 × 8 ROM, as shown in Fig. 15.14.

The array of memory cells stores the data to be programmed into the ROM. The number of memory cells in a row equals the word size, and the number of memory cells in a column equals the number of such words to be stored. In the memory shown in Fig. 15.14, the word size is eight bits and the number of words is 32. The data outputs of each of the memory cells in the array are connected to an internal data bus that runs through the entire circuit. The address decoder, a 1-of-32 decoder in this case, sets the corresponding 'row line' HIGH when a binary address is applied at its input lines. A five-bit address code ($A_4A_3A_2A_1A_0$) is needed to address 32 memory cells. As an illustration, an address code of 10011 will identify the nineteenth row.The output is read from the column lines. The data placed on the internal data bus by the memory cells are fed to the output buffers. $\overline{CS}$ is an active LOW input used to select the memory device. In the case of larger memories, the address decoder comprises row as well as column decoders. Let us consider a 2K-bit ROM device with 256 × 8 organization. The memory is arranged in the format of a 32 × 64 matrix instead of a 256 × 8 matrix. Five of the address lines are connected to the row decoder, and the remaining three lines are connected to the column decoder. The row decoder is a 1-of-32 decoder, and it selects one of the 32 rows. The column decoder comprises eight 1-of-8 decoders. It selects eight of the total 64 columns. Thus, an eight-bit word appears on the data output when the address is applied and $\overline{CS} = 0$.

Figure 15.15 shows the typical timing diagram of a ROM read operation. It shows that there is a time delay that occurs between the application of an address input and the availability of corresponding data at the output. It is this time delay that determines the ROM operating speed. This time delay is

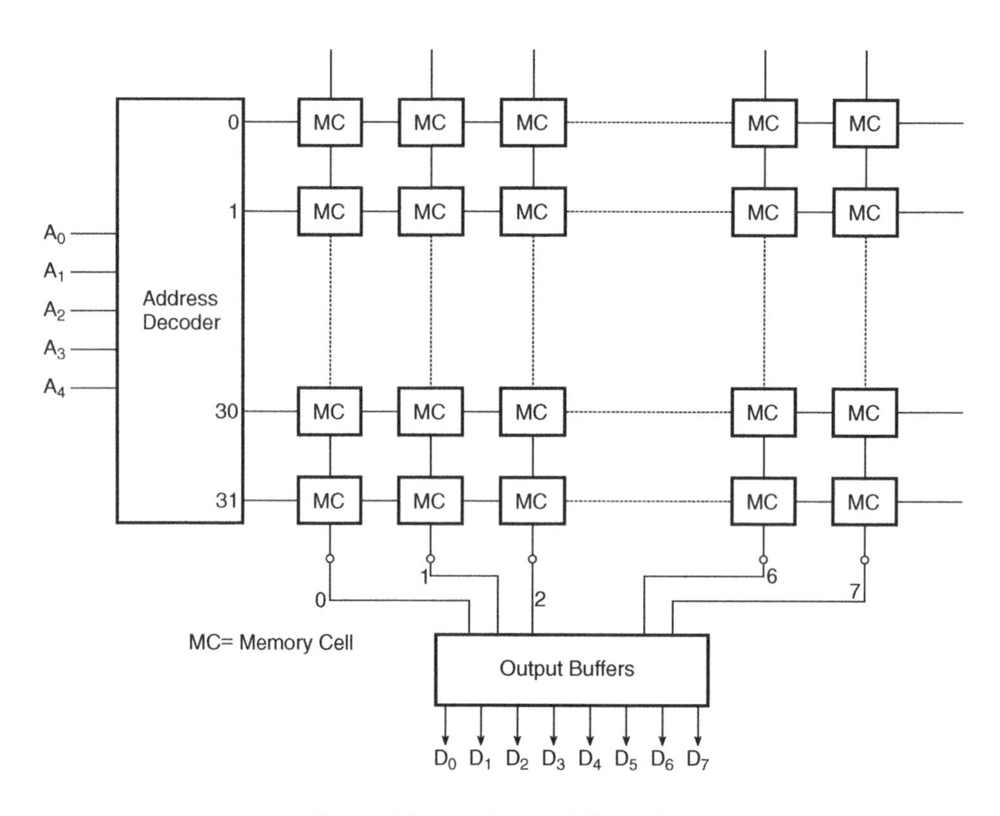

Figure 15.14 Architecture of 32 × 8 ROM.

known as the access time, t_{ACC}. Another useful timing parameter is the output enable time, t_{OE}, which is the time delay between application of input and appearance of valid data output.

Typical bipolar ROMs have access times of 30–90 ns. In the case of NMOS devices, the access times range from 35 to 500 ns. The output enable time, t_{OE}, in the case of bipolar ROMs is in the range 10–20 ns. For MOS-based ROMs, t_{OE} is in the range 25–100 ns.

15.6.2 Types of ROM

Depending upon the methodology of programming, erasing and reprogramming information into ROMs, they are classified as mask-programmed ROMs, programmable ROMs (PROMs) and erasable programmable ROMs (EPROMs) [ultraviolet-erasable programmable ROMs (UV EPROMs) and electrically erasable programmable ROMs (EEPROMs)].

15.6.2.1 Mask-programmed ROM

In the case of a mask-programmed ROM, the ROM is programmed at the manufacturer's site according to the specifications of the customer. A photographic negative, called a mask, is used to store the required data on the ROM chip. A different mask would be needed for storing each different set

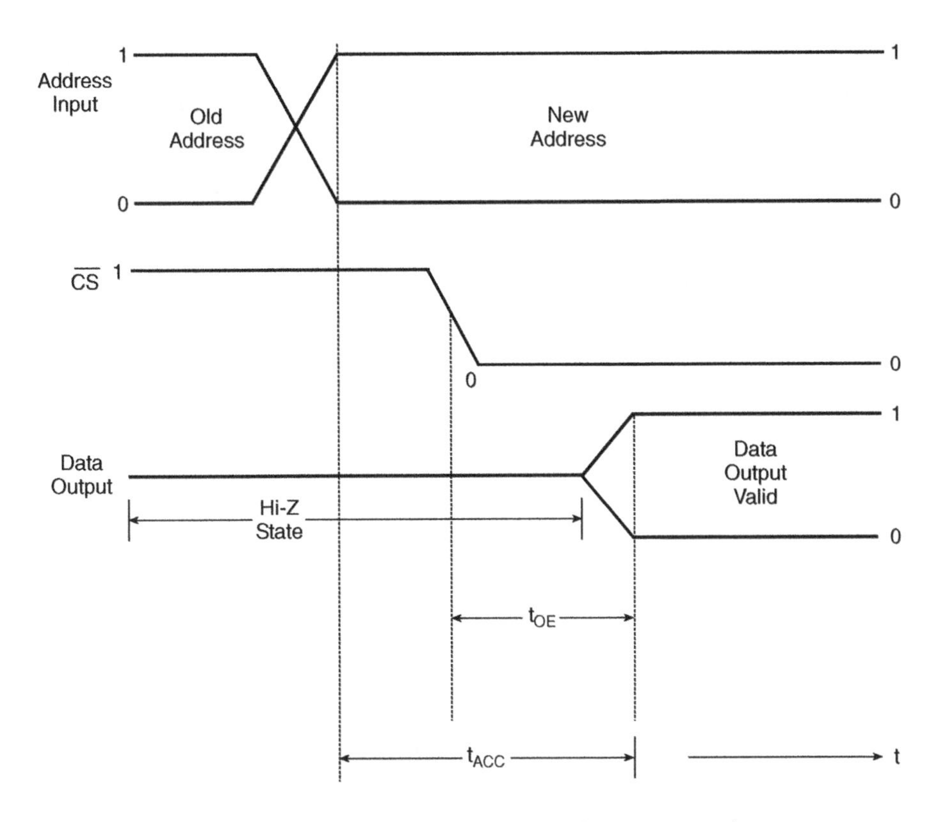

Figure 15.15 Typical timing diagram of a ROM READ operation.

of information. As preparation of a mask is an expensive proposition, mask-programmed ROM is economical only when manufactured in large quantities. The limitation of such a ROM is that, once programmed, it cannot be reprogrammed.

The basic storage element is an NPN bipolar transistor, connected in common-collector configuration, or a MOSFET in common drain configuration. Figures 15.16(a) and (b) show a MOSFET-based basic cell connection when storing a '1' and '0' respectively. As is clear from the figure, the connection of the 'row line' to the gate of the MOSFET stores '1' at the location when the 'row line' is set to level '1'. A floating-gate connection is used to store '0'. Figures 15.16(c) and (d) show the basic bipolar memory cell connection when storing a '1' and '0' respectively.

Figure 15.17 shows the internal structure of a 4×4 bipolar mask-programmed ROM. The data programmed into the ROM are given in the adjoining truth table. The transistors with an open base store a '0', whereas those with their bases connected to the corresponding decoder output store a '1'. As an illustration, transistors Q_{30}, Q_{20}, Q_{10} and Q_{00} in row 0 store '1', '0', '1' and '0' respectively. The stored information in a given row is available at the output when the corresponding decoder is enabled, and that 'row line' is set to level '1'. The output of the memory cells appears at the column lines. For example, when the address input is '11', row 3 is enabled and the data item at the output is 0110.

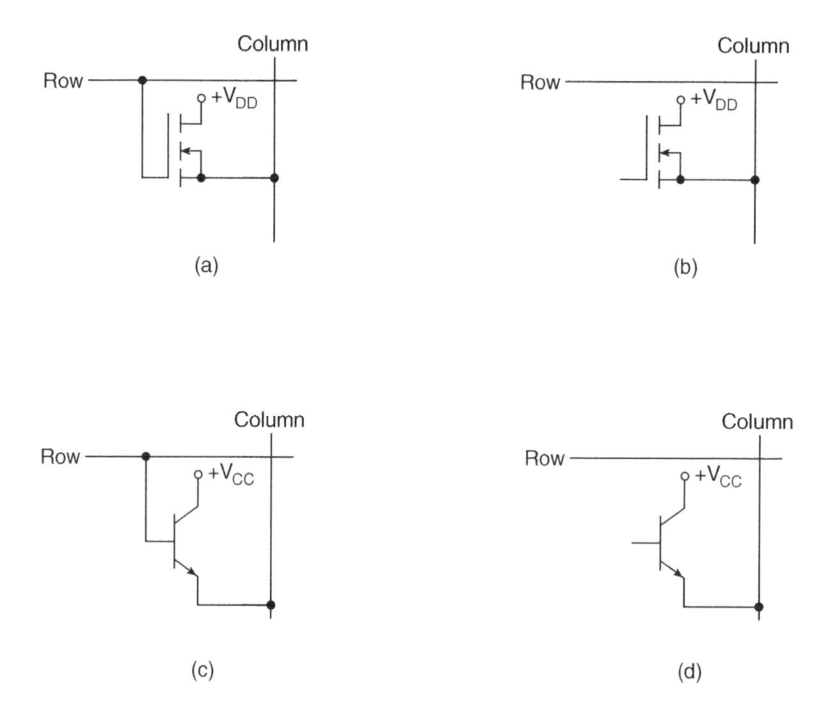

Figure 15.16 Basic cell connection of a mask-programmed ROM.

In the ROM architecture shown in Fig. 15.17, the number of memory cells in a row represents the word size. The four memory cells in a row here constitute a four-bit register. There are four such registers in this ROM. In a 16×8 ROM of this type there will be 16 rows of such transistor cells, with each row having eight memory cells. The decoder in that case would be a 1-of-16 decoder.

15.6.2.2 Programmable ROM

In the case of PROMs, instead of being done at the manufacturer's premises during the manufacturing process, the programming is done by the customer with the help of a special gadget called a PROM programmer. Since the data, once programmed, cannot be erased and reprogrammed, these devices are also referred to as one-time programmable ROMs.

The basic memory cell of a PROM is similar to that of a mask-programmed ROM. Figures 15.18(a) and (b) show a MOSFET-based memory cell and bipolar memory cell respectively. In the case of a PROM, each of the connections that were left either intact or open in the case of a mask-programmed ROM are made with a thin fusible link, as shown in Fig. 15.18. The different interconnect technologies used in programmable logic devices are comprehensively covered in Chapter 9. Basic fuse technologies used in PROMs are metal links, silicon links and PN junctions. These fusible links can be selectively blown off to store desired data. A sufficient current is injected through the fusible link to burn it open to store '0'. The programming operation, as said earlier, is done with a PROM programmer. The PROM chip is plugged into the socket meant for the purpose. The programmer circuitry selects each address of the PROM one by one, burns in the required data and then verifies the correctness of the

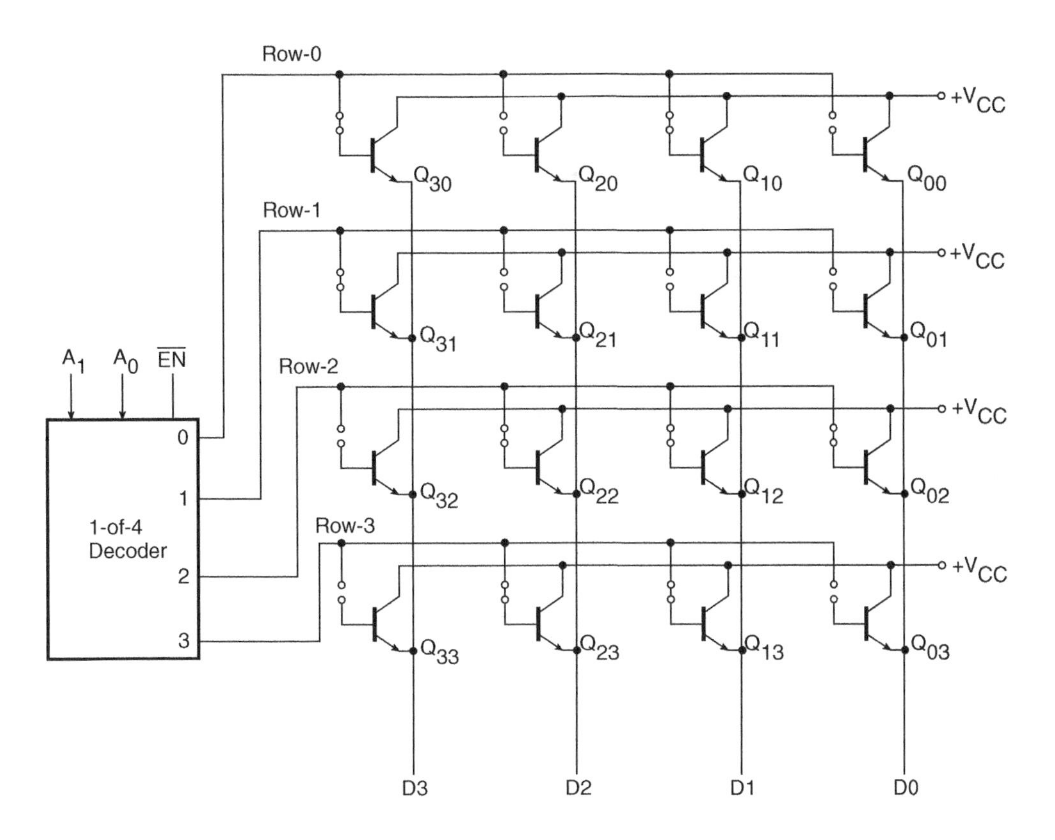

Truth Table

Address		Data			
A_1	A_0	D_3	D_2	D_1	D_0
0	0	1	0	1	0
0	1	1	0	0	0
1	0	0	1	1	1
1	1	0	1	1	0

Figure 15.17 Internal structure of a 4×4 bipolar mask-programmed ROM.

data before proceeding to the next address. The data are fed to the programmer from a keyboard or a disk drive or from a computer.

PROM chips are available in various word sizes and capacities. 27LS19, 27S21, 28L22, 27S15, 24S41, 27S35, 24S81, 27S45, 27S43 and 27S49 are respectively 32×8, 256×4, 256×8, 512×8, $1K \times 4$, $1K \times 8$, $2K \times 4$, $2K \times 8$, $4K \times 8$ and $8K \times 8$ PROMS. The typical access time in the case of these devices is in the range 50–70 ns. MOS PROMs are available with much greater capacities than bipolar PROMs. Also, the power dissipation is much lower in MOS PROMs than it is in the case of bipolar PROMs with similar capacities.

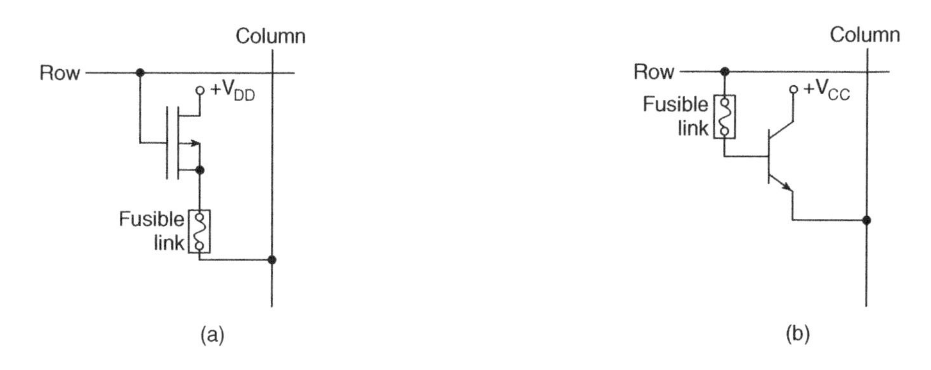

Figure 15.18 Basic memory cell of a PROM.

15.6.2.3 Erasable PROM

EPROM can be erased and reprogrammed as many times as desired. Once programmed, it is nonvolatile, i.e. it holds the stored data indefinitely. There are two types of EPROM, namely the ultraviolet-erasable PROM (UV EPROM) and electrically erasable PROM (EEPROM).

The memory cell in a UV EPROM is a MOS transistor with a floating gate. In the normal condition, the MOS transistor is OFF. It can be turned ON by applying a programming pulse (in the range 10–25 V) that injects electrons into the floating-gate region. These electrons remain trapped in the gate region even after removal of the programming pulse. This keeps the transistor ON once it is programmed to be in that state even after the removal of power. The stored information can, however, be erased by exposing the chip to ultraviolet radiation through a transparent window on the top of the chip meant for the purpose. The photocurrent thus produced removes the stored charge in the floating-gate region and brings the transistor back to the OFF state. The erasing operation takes around 15–20 min, and the process erases information on all cells of the chip. It is not possible to carry out any selective erasure of memory cells. Intel's 2732 is 4K × 8 UV EPROM hardware implemented with NMOS devices. Type numbers 2764, 27128, 27256 and 27512 have capacities of 8K × 8, 16K × 8, 32K × 8 and 64K × 8 respectively. The access time is in the range 150–250 ns. UV EPROMs suffer from disadvantages such as the need to remove the chip from the circuit if it is to be reprogrammed, the nonfeasibility of carrying out selective erasure and the reprogramming process taking several tens of minutes. These are overcome in the EEPROMs and flash memories discussed in the following paragraphs.

The memory cell of an EEPROM is also a floating-gate MOS structure with the slight modification that there is a thin oxide layer above the drain of the MOS memory cell. Application of a high-voltage programming pulse between gate and drain induces charge in the floating-gate region which can be erased by reversing the polarity of the pulse. Since the charge transport mechanism requires very low current, erasing and programming operations can be carried out without removing the chip from the circuit. EEPROMs have another advantage – it is possible to erase and rewrite data in the individual bytes in the memory array. The EEPROMs, however, have lower density (bit capacity per square mm of silicon) and higher cost compared with UV EPROMs.

15.6.2.4 Flash Memory

Flash memories are high-density nonvolatile read/write memories with high density. Flash memory combines the low cost and high density features of an UV EPROM and the in-circuit electrical erasability feature of EEPROM without compromising the high-speed access of both. Structurally, the memory cell of a flash memory is like that of an EPROM. The basic memory cell of a flash memory is shown in Fig. 15.19. It is a stacked-gate MOSFET with a control gate and floating gate in addition to drain and source. The floating gate stores charge when sufficient voltage is applied to the control gate. A '0' is stored when there is more charge, and a '1' when there is less charge. The amount of charge stored on the floating gate determines whether or not the MOSFET is turned ON.

It is called a flash memory because of its rapid erase and write times. Most flash memory devices use a 'bulk erase' operation in which all the memory cells on the chip are erased simultaneously. Some flash memory devices offer a 'sector erase' mode in which specific sectors of the memory device can be erased at a time. This mode comes in handy when only a portion of the memory needs to be updated.

Figure 15.20 shows the basic array of a 4×4 flash memory. As in the case of earlier memories, there is an address decoder that selects the row. During the read operation, for a cell containing a '1' there is current through the bit line which produces a voltage drop across the active load. This is compared with the reference voltage, and the output bit is '1'. If the memory cell has a '0', there is very little current in the bit line. Memory sticks are flash memories. They are available in 4, 8, 16, 32, 64 and 128 MB sizes.

To sum up, while PROMs are least complex and low cost, they cannot be erased and reprogrammed. UV EPROMs are a little more complex and costly, but then they can be erased and reprogrammed by being taken out of the circuit. Flash memories are in-circuit electrically erasable either sectorwise or in bulk mode. The most complex and most expensive are the EEPROMs, but then they offer byte-by-byte electrical erasability in circuit.

15.6.3 Applications of ROMs

The majority of ROM applications originate from the need for nonvolatile storage of data or program codes. Some of the common application areas include firmware, bootstrap memory, look-up tables, function generators and auxiliary memory.

The most common application of ROM chips is in the storage of data and program codes that must be made available to microprocessor-based systems such as microcomputers on power-up. This component of the software is referred to as firmware as it comes embedded in the hardware with the machine. Even consumer products such as CD players, microwave ovens, washing machines, etc., have embedded microcontrollers that have a microprocessor to control and monitor the operation according to the information stored on the ROM.

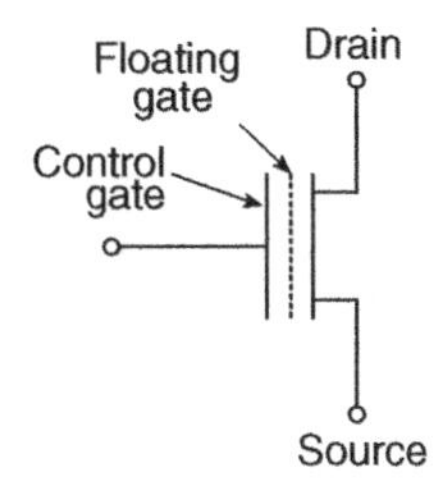

Figure 15.19 Basic cell of flash memory.

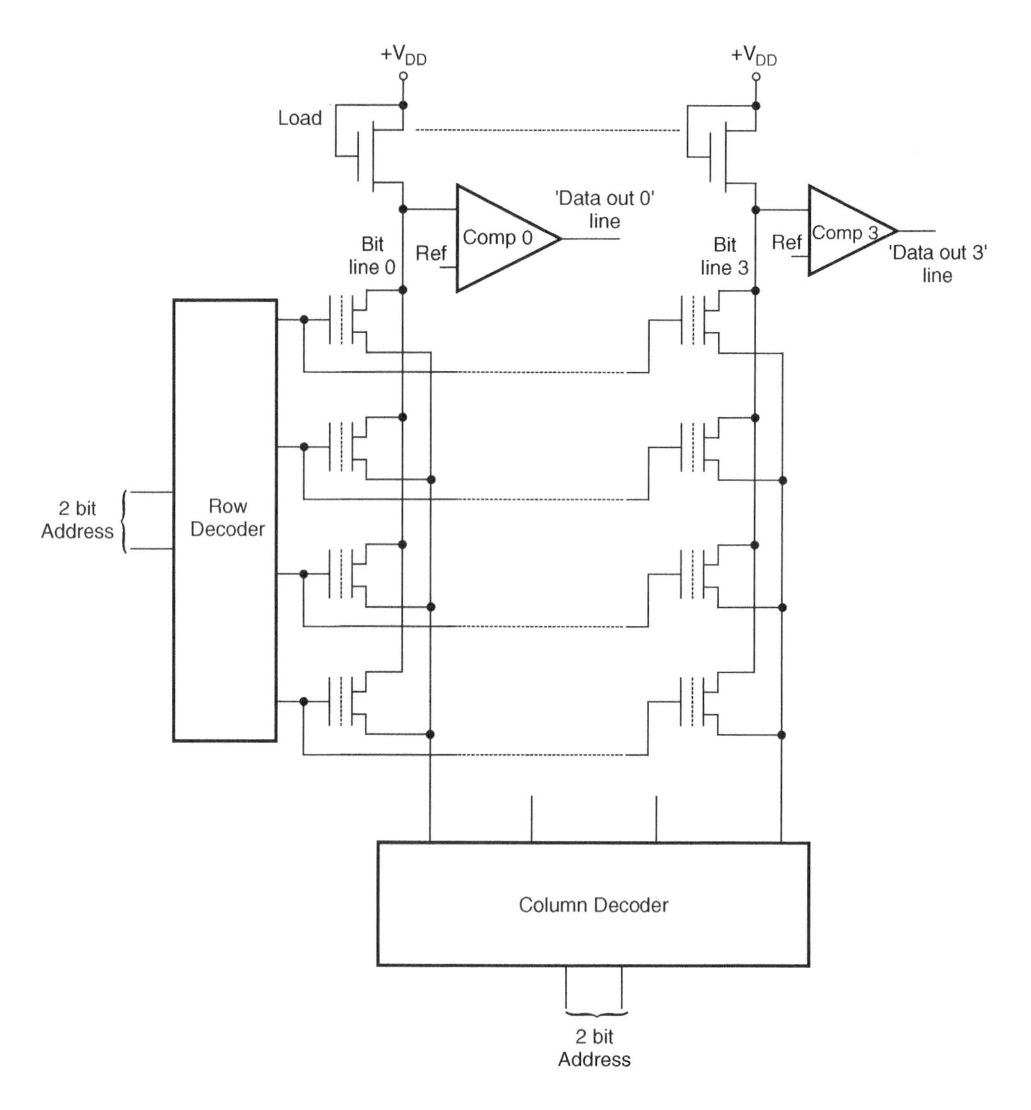

Figure 15.20 Basic array of 4 × 4 flash memory.

ROMs are also used to store the 'bootstrap program' in computers. It is a relatively small program containing instructions that will cause the CPU to initialize the system hardware after it is powered on. The bootstrap program then loads the operating system programs stored in the secondary memory into its main internal memory. The computer then begins to execute the operating system program. This start-up operation is also called the 'booting operation'.

ROMs are frequently used as 'look-up tables'. There are two sets of data, one constituting the address and the other corresponding to the data stored in various memory locations of the ROM. Corresponding to each address input, there is a unique data ouput. One typical application is that of code conversion. As an illustration, a ROM can be used to build a binary-to-BCD converter where each memory location stores the BCD equivalent of the corresponding address code expressed in binary.

A ROM can be an important building block in a waveform generator. In a typical waveform generation set-up, ROM is used as a look-up table, with each of its memory locations storing a unique digital code corresponding to a different amplitude of the waveform to be generated. The address inputs of the ROM are fed from the output of a counter. The data outputs of ROM feed a D/A converter whose output constitutes the desired analogue waveform. This concept is also utilized in speech synthesizers, where the digital equivalent of speech waveform values are stored in the ROM.

Today, ROMs have become a viable alternative to the use of magnetic disks for auxiliary storage, more so for lower-capacity requirements. The low power consumption of flash memories, for instance, makes them particularly attractive for notebook computers.

Example 15.1

A certain ROM is capable of storing 16 kB of data. If the internal architecture of the ROM uses a square matrix of registers, determine (a) the number of registers in each row, (b) the number of registers in each column, (c) the total number of address inputs, (d) the type of row decoder and (e) the type of column decoder.

Solution

(a) The ROM capacity = 16K = 16 × 1024 = 16 384 bytes. Therefore, the total number of registers = 16 384. Since the registers are arranged in a square matrix, the number of rows equals the number of columns. The number of registers in each row = 128.
(b) The number of registers in each column = 128.
(c) The total number of memory locations = 16 384 = 2^{14}. Therefore, the total number of address inputs = 14.
(d) 1-of-7 decoder.
(e) 1-of-7 decoder.

Example 15.2

Determine the minimum size of a ROM required to convert a four-bit straight binary code into a Gray code equivalent. Also, write data to be programmed in various memory locations of the ROM.

Solution

- Table 15.1 shows the four-bit straight binary numbers and their Gray code equivalents.
- It is clear from the table that the MSB of the straight binary number is the same as the MSB of the Gray code equivalent.
- This can therefore be passed on as such to the output.
- In that case, each memory location of the ROM needs to store only three-bit data as the fourth bit is available as such from the input.
- The required size of the ROM is therefore 16 × 3.
- The three-bit data to be programmed into 16 different memory locations of the ROM corresponding to address inputs of 0000 to 1111 in the same order would be 000, 001, 011, 010, 110, 111, 101, 100, 100, 101, 111, 110, 010, 011, 001 and 000.
- Figure 15.21 shows this in ROM representation.

Table 15.1 Example 15.2.

Binary code				Gray code			
A_3	A_2	A_1	A_0	D_3	D_2	D_1	D_0
0	0	0	0	0	0	0	0
0	0	0	1	0	0	0	1
0	0	1	0	0	0	1	1
0	0	1	1	0	0	1	0
0	1	0	0	0	1	1	0
0	1	0	1	0	1	1	1
0	1	1	0	0	1	0	1
0	1	1	1	0	1	0	0
1	0	0	0	1	1	0	0
1	0	0	1	1	1	0	1
1	0	1	0	1	1	1	1
1	0	1	1	1	1	1	0
1	1	0	0	1	0	1	0
1	1	0	1	1	0	1	1
1	1	1	0	1	0	0	1
1	1	1	1	1	0	0	0

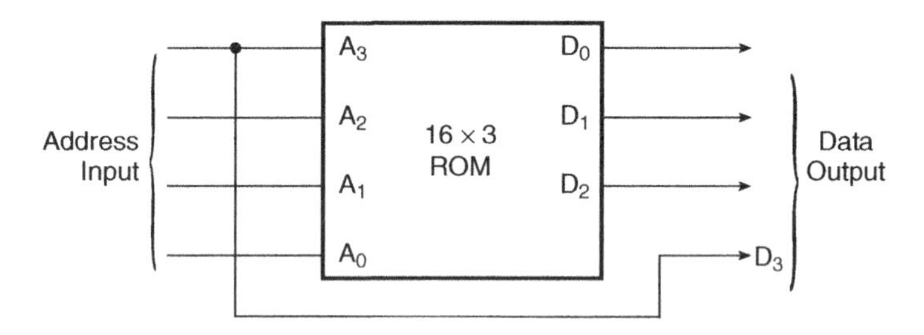

Figure 15.21 Solution to problem 15.2.

15.7 Expanding Memory Capacity

When a given application requires a RAM or ROM with a capacity that is larger than what is available on a single chip, more than one such chip can be used to achieve the objective. The required enhancement in capacity could be either in terms of increasing the word size or increasing the number of memory locations. How this can be achieved is illustrated in the following paragraphs with the help of examples.

15.7.1 Word Size Expansion

Let us take up the task of expanding the word size of an available 16 × 4 RAM chip from four bits to eight bits. Figure 15.22 shows a diagram where two such RAM chips have been used to achieve the

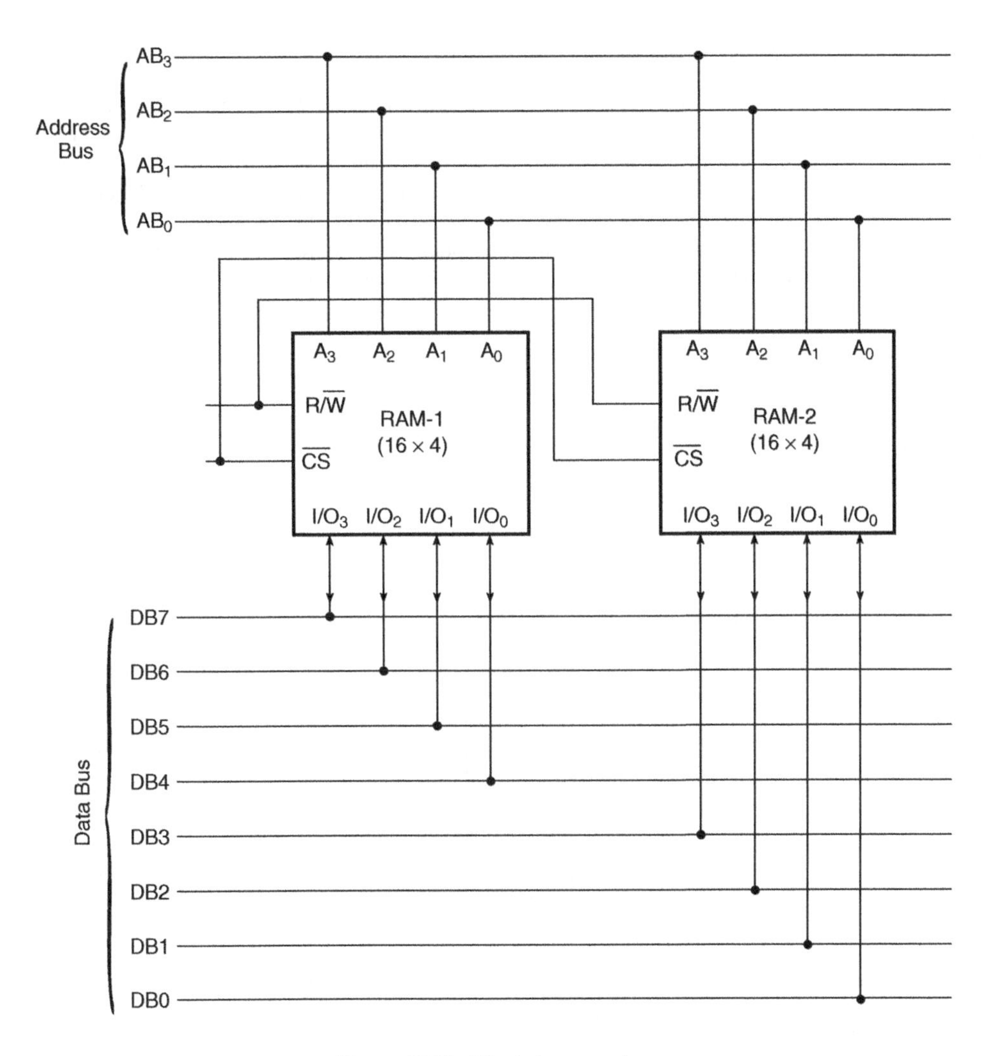

Figure 15.22 Word size expansion.

desired effect. The arrangement is straightforward. Both chips are selected or deselected together. Also, the input that determines whether it is a 'read' or 'write' operation is common to both chips. That is, both chips are selected for 'read' or 'write' operation together. The address inputs to the two chips are also common. The memory locations corresponding to various address inputs store four higher-order bits in the case of RAM-1 and four lower-order bits in the case of RAM-2. In essence, each of the RAM chips stores half of the word. Since the address inputs are common, the same location in each chip is accessed at the same time.

15.7.2 Memory Location Expansion

Figure 15.23 shows how more than one memory chip can be used to expand the number of memory locations. Let us consider the use of two 16×8 chips to get a 32×8 chip. A 32×8 chip would need five address input lines. Four of the five address inputs, other than the MSB address bit, are common to both 16×8 chips. The MSB bit feeds the input of one chip directly and the input of the other chip after inversion. The inputs to the two chips are common.

Now, for first half of the memory locations corresponding to address inputs 00000 to 01111 (a total of 16 locations), the MSB bit of the address is '0', with the result that RAM-1 is selected

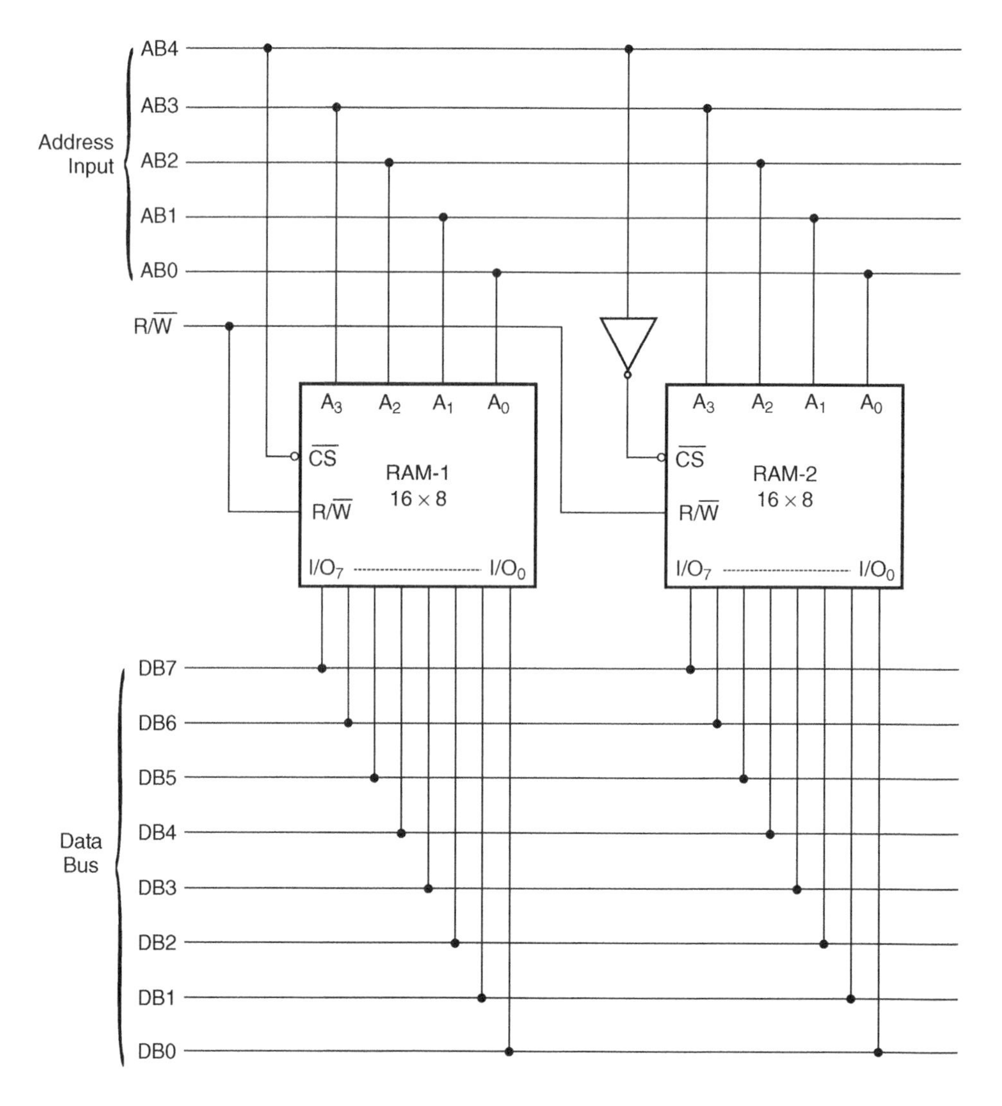

Figure 15.23 Memory location expansion.

and RAM-2 is deselected. For the remaining address inputs of 10000 to 11111 (again, a total of 16 locations), RAM-1 is deselected while RAM-2 is selected. Thus, the overall arrangement offers a total of 32 locations, 16 provided by RAM-1 and 16 provided by RAM-2. The overall capacity is thus 32×8.

Example 15.3

Two 16 MB RAMs are used to build a RAM capacity of 32 MB. Show the configuration and also state the address inputs for which the two RAMs will be active. The two RAMs have common I/O pins, a WRITE ENABLE input that is active LOW and a CHIP SELECT input that is active HIGH.

Solution

Figure 15.24 shows the arrangement. Since the overall RAM capacity is 32 MB, it will have 25 address inputs (AB0 to AB24) as $32\text{M} = 2^{25}$. For address inputs $(0000000)_{hex}$ to $(0FFFFFF)_{hex}$, which account for 16M ($=2^{24}$) memory locations, RAM-1 is enabled and 16 M locations of RAM-1 are available. RAM-2 is deselected for these address inputs. For address inputs $(1000000)_{hex}$ to $(1FFFFFF)_{hex}$, the total number of addresses in this group again being equal to 16M, RAM-2 is selected and RAM-1 is deselected. 16M locations of RAM-2 are available. Thus, out of 32 MB, 16 MB is stored in RAM-1 and 16 MB is stored in RAM-2.

Example 15.4

What is available is a 1K× 8 chip of the type shown in Fig. 15.25. This chip, as shown in the diagram, gets activated only when select input $\overline{CS1}$ is LOW and select input CS2 is HIGH. Show how two such ROMs can be connected to get 2K × 8 ROM without using any additional logic.

Solution

- Figure 15.26 shows the arrangement.
- The address bit AB_{10} is low for the first 1024 address inputs (from 00000000000 to 01111111111) and ROM-1 is selected.
- For the remaining 1024 address inputs (from 10000000000 to 11111111111), the AB_{10} bit is HIGH, thus enabling ROM-2.

Example 15.5

Figure 15.27 shows an arrangement of four memory chips, each 16 ×4 RAM with an active LOW chip select input. Determine the total capacity and the word size. Which RAMs will put data on the data bus when the address input is 00001101. Also, determine the address input range for which RAM-1 and RAM-2 will be active.

Solution

- For address inputs $(00000000)_2$ to $(00001111)_2$, RAM-1 and RAM-2 are selected.
- RAM-1 stores four higher bits and RAM-2 stores four lower bits of data words corresponding to the 16 address inputs mentioned above.

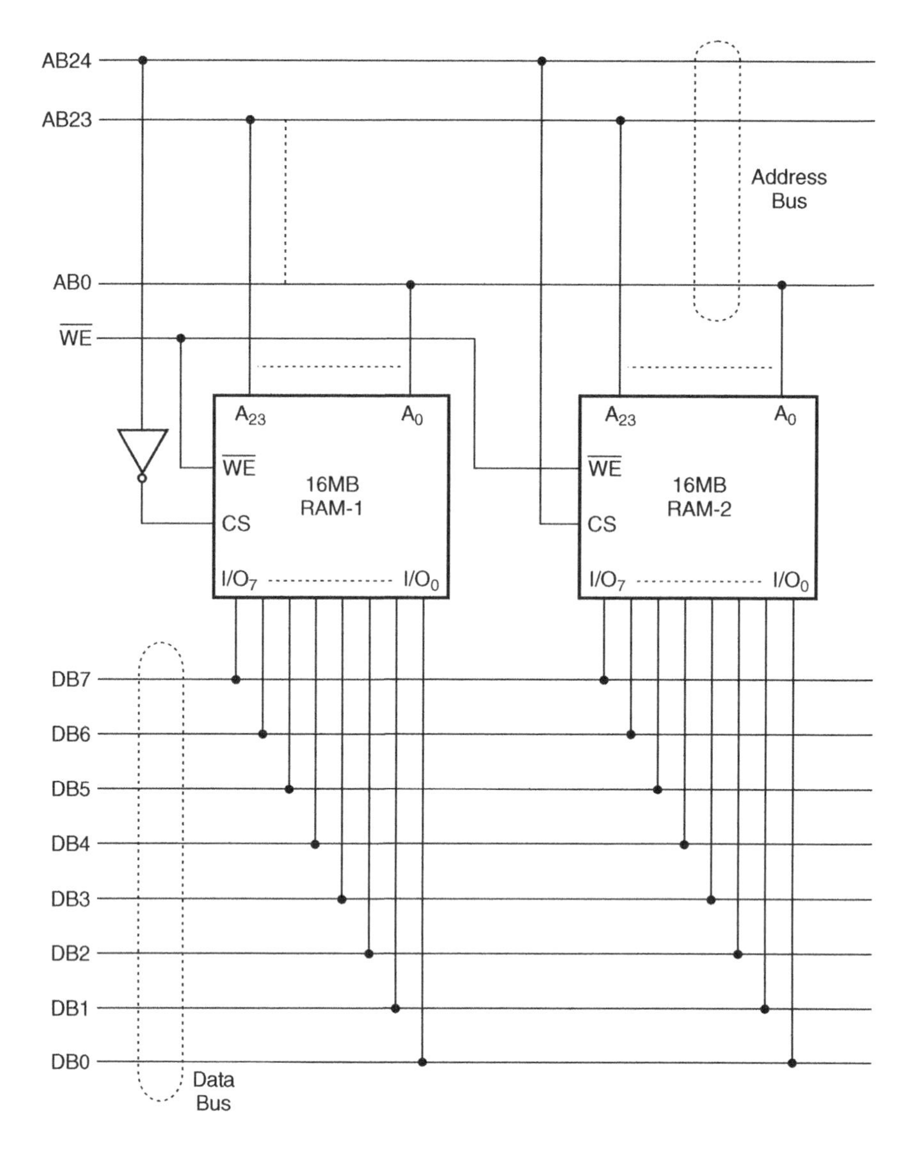

Figure 15.24 Solution to example 15.3.

- This gives us a capacity of 16×8.
- Now, for address inputs $(00010000)_2$ to $(00011111)_2$, RAM-3 and RAM-4 are selected.
- Similarly, RAM-3 and RAM-4 respectively store four upper bits and four lower bits of data words corresponding to these address inputs.
- This again gives a capacity of 16×8.
- Thus, the overall capacity is 32×8.
- The word size is 8.

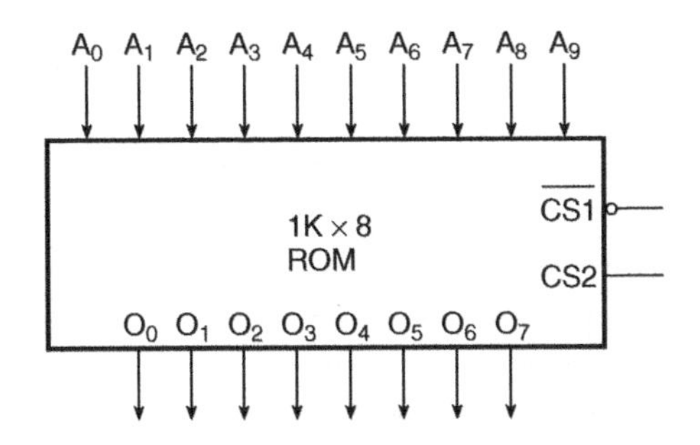

Figure 15.25 Example 15.4.

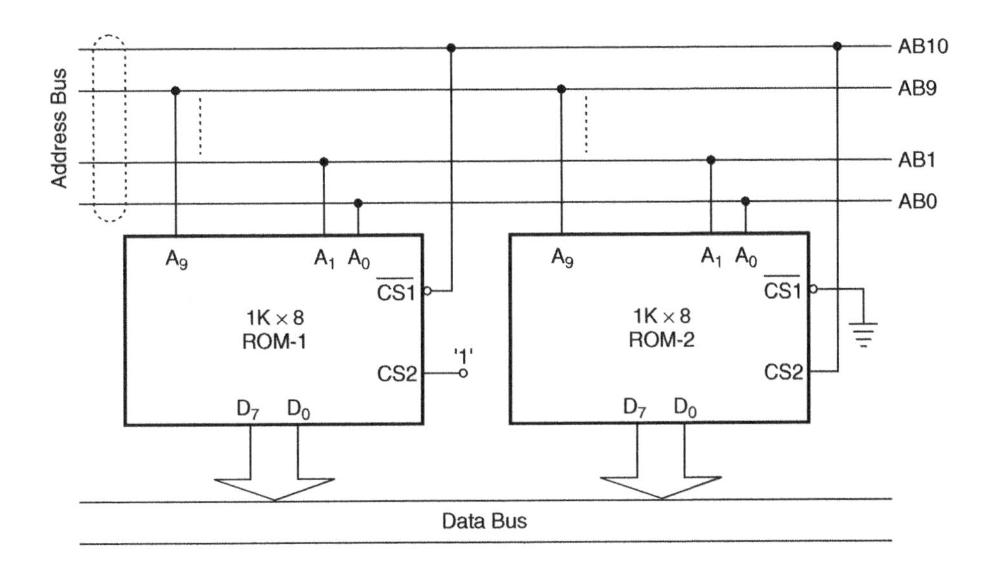

Figure 15.26 Solution to example 15.4.

- For an address input 00001101, RAM-1 and RAM-2 will be selected.
- The address input range for which RAM-1 and RAM-2 are active is $(00000000)_2$ to $(00001111)_2$.

15.8 Input and Output Ports

Input and output ports were briefly introduced in the earlier part of the chapter in Section 15.1.3. As outlined earlier, these are categorized as serial and parallel ports. The commonly used serial and parallel ports are described in the following paragraphs.

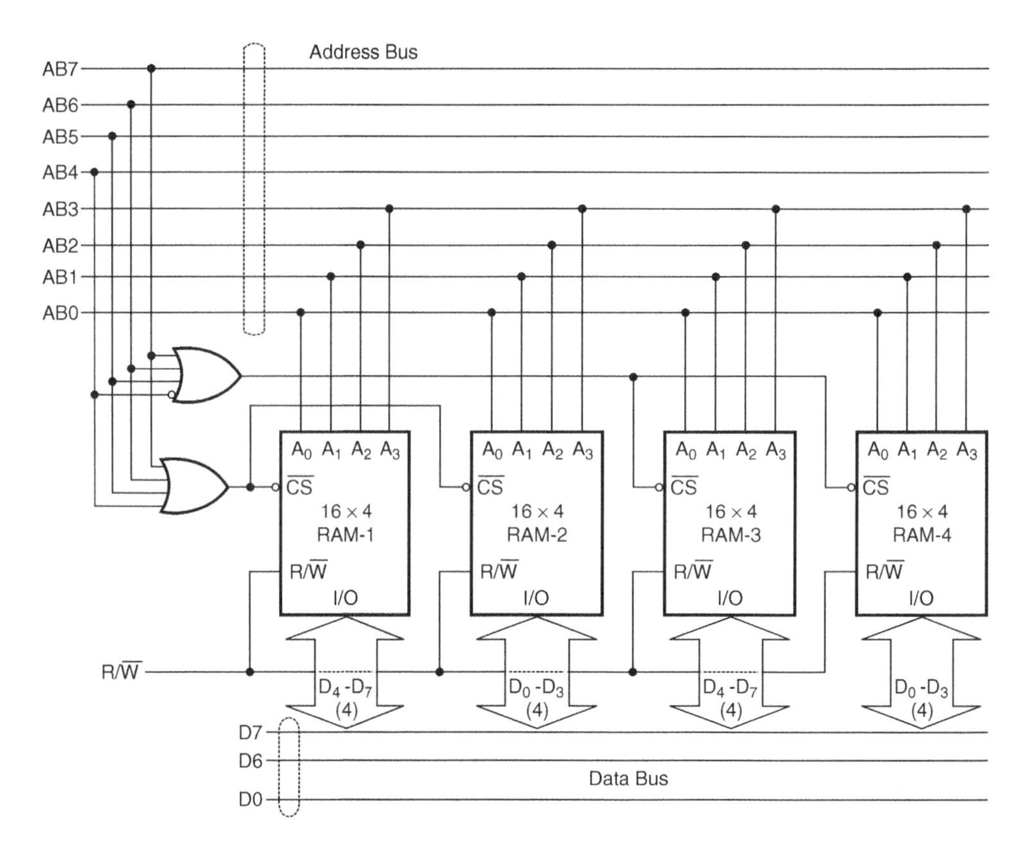

Figure 15.27 Example 15.5.

15.8.1 Serial Ports

A serial port is a physical communication interface through which the information transfer takes place one bit at a time. Serial ports are used to connect mouse, keyboard and modems to the computer. Some of the commonly used serial standards include the RS-232C port, PS/2, FireWire and USB.

15.8.1.1 RS-232C Port

RS-232 is one of the oldest and most well-known standards for serial interfaces approved by the Electronic Industries Association (EIA). It was developed to interface data terminal equipment (DTE) with data communication equipment (DCE). RS-232C, a variant of the RS-232 standard, is the most relevant for the computer world. RS-232C is mostly used to connect modem and other communication devices to the computer. In this case the computer is referred to as the DTE and the attached device as the DCE.

The RS-232C standard specifies 25 communication lines between the DTE and the DCE. Hence, the standard RS-232C connector is a 25-pin connector (DB-25). For personal computer applications,

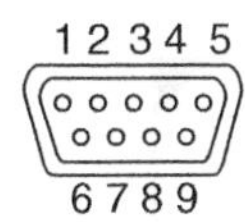

1. DCD (Data Carrier Detect)
2. RD (Recieve Data)
3. TD (Transmit Data)
4. DTR (Data Terminal Ready)
5. GND (Ground)
6. DSR (Data Set Ready)
7. RTS (Request To Send)
8. CTS (Clear To Send)
9. RI (Ring Indicator)

Figure 15.28 DE-9 connector.

not all the 25 pins are required. Hence, most personal computers have a nine-pin connector (DE-9). Figure 15.28 shows the DE-9 connector along with its pin assignments.

The maximum specified cable length for the RS-232C interface is 50 ft for a data transmission rate of 20 kbaud. As the cable length increases, the transmission rate decreases. The RS-422 and RS-423 standards have higher transmission speeds than RS-232C. They also support larger cable lengths. However, RS-232C remains the most commonly used serial port.

15.8.1.2 FireWire

FireWire is the name of the interface specified by the IEEE standard 1394. This high-speed serial bus standard is used for interfacing graphics and video peripherals such as digital cameras and camcoders to the computer. FireWire can be used to connect up to 63 devices in a cyclic topology. It supports both plug-and-play and hot swapping. It is available in two versions, namely FireWire 400 and FireWire 800. FireWire 400 hardware is available in six-pin and four-pin connectors and can support data rates of 100, 200 and 400 Mbits/s. The four-pin connector is used mostly in consumer electronic goods and the six-pin connector is used in computers.

FireWire 800 is based on the IEEE 1394b standard and supports a data rate of 786.432 Mbits/s. It has a nine-wire connection.

15.8.1.3 Universal Serial Bus (USB)

The USB port was introduced in the year 1997 and is used to connect printers, mouse, scanners, digital cameras and external storage devices to the computer. Different versions of the USB standard include 0.9, 1.0, 1.1 and 2.0, with USB 2.0 being the latest. Another variant of the USB standard is the radio spectrum based USB implementation, known as Wireless USB.

A USB port can be used to connect 127 devices. It supports two data rates of 1.5 Mbits/s (low speed) and 12 Mbits/s (full speed). Most of the USB 2.0 devices also support data rates of 480 Mbits/s (Hi speed). USB is a four-wire connection and is available in two standard types referred to as type A and type B. Miniature versions of the USB connector are also available, namely Mini-A and Mini-B. Figure 15.29 shows different types of USB connector, along with their pin details.

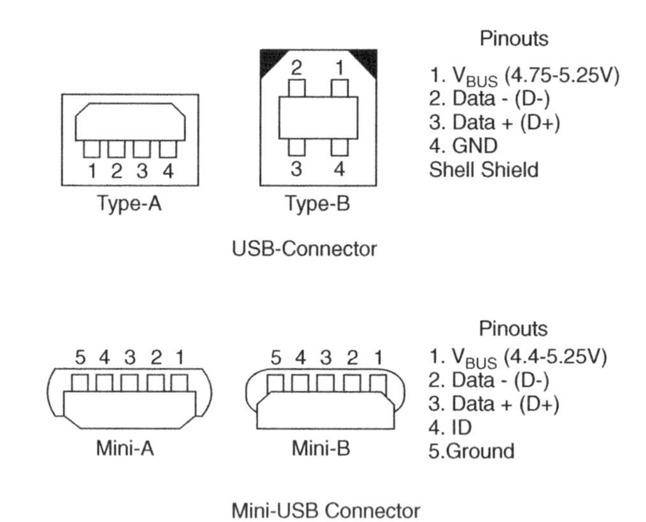

Figure 15.29 USB connector.

15.8.1.4 PS/2 Connector

PS/2 connectors are used for connecting the keyboard and mouse to a personal computer. The PS/2 mouse and PS/2 keyboard connectors are similar to each other, except for the fact that the PS/2 keyboard connector has an open-collector output. PS/2 mouse and keyboard connectors have replaced the DE-9 and five-pin DIN connectors respectively. Figure 15.30 shows the PS/2 connector with the pin details.

15.8.2 Parallel Ports

Parallel ports send multiple bits at the same time over a set of wires. They are used to connect printers, scanners, CD burners, external hard drives, etc., to the computer. Commonly used standard parallel ports include IEEE-488, the small computer system interface (SCSI) and IEEE 1284.

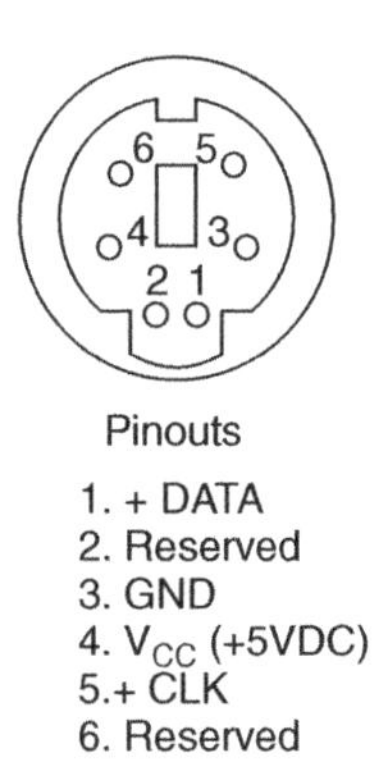

Pinouts
1. + DATA
2. Reserved
3. GND
4. V_{CC} (+5VDC)
5.+ CLK
6. Reserved

Figure 15.30 PS/2 connector.

15.8.2.1 IEEE-488

IEEE-488 is a short-range parallel bus standard widely used in test and measurement applications. It is also referred to as a general-purpose interface bus (GPIB). The IEEE 488 standard specifies a 24-wire connection for transferring eight data bits simultaneously. Other connections include eight control signals and eight ground lines. The maximum data rate is 1 MB/s in the original standard and about 8 MB/s with the modified standard (HS-488). Figure 15.31 shows the pin connections and pin details.

15.8.2.2 Small Computer System Interface (SCSI)

SCSI is a widely used standard for interfacing personal computers and peripherals. SCSI is a standard given by the American National Standards Institute (ANSI). There are several variations of this standard, and one variant may not be compatible with another. Some of the SCSI versions include SCSI-1, SCSI-2, Wide SCSI, Fast SCSI, Fast Wide SCSI, Ultra SCSI, SCSI-3, Ultra SCSI-2 and Wide Ultra SCSI-2. Description of all these interfaces is beyond the scope of this book.

15.8.2.3 IEEE-1284

IEEE 1284 is a standard that defines bidirectional parallel communications between computers and other devices. It supports a maximum data rate of 4 MB/s. It supports three types of connector: DB-25 (type A) for the host connection, Centronics 36-pin (type B) for the printer or device connection and Mini Centronics 36-pin (type C), a smaller alternative for the device connection. IEEE 1284-I

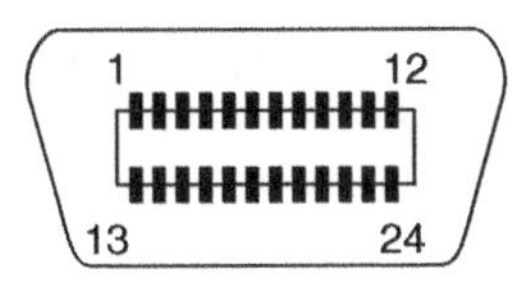

1EEE 488 Connector

Pinouts

1. Data 1/O1
2. Data 1/O2
3. Data 1/O3
4. Data 1/O4
5. EOI (End or Identity)
6. DAV (Data Valid)
7. NRFD (Not Ready for Data)
8. NDAC (Not Data Accepted)
9. IFC (Interface Clear)
10. SRQ (Service Request)
11. ATN (Attention)
12. Sheild Ground
13. Data 1/O5
14. Data 1/O6
15. Data 1/O7
16. Data 1/O8
17. REN (Remote Enable)
18. P/O Twisted Pair with 6
19. P/O Twisted Pair with 7
20. P/O Twisted Pair with 8
21. P/O Twisted Pair with 9
22. P/O Twisted Pair with 10
23. P/O Twisted Pair with 11
24. Signal Ground

Figure 15.31 Pin connections and details of the IEEE-488 connector.

devices use IEEE 1284-A and IEEE 1284-B connectors, while IEEE 1284-II devices use IEEE-1284-C connectors. The type C connector is not very popular.

15.8.3 Internal Buses

Input/output ports are used to connect the computer to external devices. Input and output standards described in the previous sections are referred to as external bus standards. In addition to these external buses, computers also have internal buses that carry address, data and control signals between the CPU, cache memory, SRAM, DRAM, disk drives, expansion slots and other internal devices. Internal buses are of three types, namely the local bus, the PCI bus and the ISA bus.

15.8.3.1 Local Bus

This bus connects the microprocessor to the cache memory, main memory, coprocessor and PCI bus controller. It includes the data bus, the address bus and the control bus. It is also referred to as the primary bus. This bus has high throughput rates, which is not possible with buses using expansion slots.

15.8.3.2 PCI Bus

The peripheral control interconnect (PCI) bus is used for interfacing the microprocessor with external devices such as hard disks, sound cards, etc., via expansion slots. It has a VESA local bus as the standard expansion bus. Variants of the PCI bus include PCI 2.2, PCI 2.3, PCI 3.0, PCI-X, PCI-X 2.0, Mini PCI, Cardbus, Compact PCI and PC/104-Plus. The PCI bus will be superseded by the PCI Express bus. PCI originally had 32 bits and operated at 33 MHz. Various variants have different bits and data transfer rates.

15.8.3.3 ISA Bus

The industry-standard architecture (ISA) bus is a computer standard bus for IBM-compatible computers. It is available in eight-bit and 16-bit versions. The VESA local bus was designed to solve the bandwidth problem of the ISA bus. It worked alongside the ISA bus where it acted as a high-speed conduit for memory-mapped I/O and DMA, while the ISA bus handled interrupts and port-mapped I/O. Both these buses have been replaced by the PCI bus.

15.9 Input/Output Devices

Input/output devices are human–machine interface devices connected to the computer. Input devices are used for entering data into the computer. They convert the raw data to be processed into a computer-understandable format. Output devices convert the processed data back into a user-understandable format. This section briefly describes the commonly used input/output devices.

15.9.1 Input Devices

As mentioned before, input devices convert the raw data to be processed into a computer-understandable format. Input devices can be broadly classified into various types, depending upon the type of input data they handle. Commonly used input devices include keyboard devices, pointing devices, image and video input devices and audio input devices.

15.9.1.1 Keyboard Devices

Keyboards are designed for the input of text and characters and also to control the operation of a computer. Keyboards have an arrangement of keys where each press of a key corresponds to some action. Keyboards are available in different types and sizes. Keyboard and pointing devices are also referred to as data entry input devices.

15.9.1.2 Pointing Devices

These include the computer mouse, trackball, joystick, touch screen, light pen and so on. The mouse is a handheld device whose motion is translated into the motion of a pointer on the display. It is one of the most popular input devices used with microcomputers. A joystick consists of a handheld stick that pivots about one end and transmits its angle information to the computer. Touch screens are input devices that sense the touch event and send processing signals to the computer. Touch screens are available in various types including resistive, surface wave, capacitive, infrared, strain gauge, optical imaging and so on. Light pens are devices that transmit their coordinates to the machine when placed against the CRT screen of the machine. Hence, they allow the user to point to displayed objects on the screen or to draw on the screen, similarly to a touch screen but with greater position accuracy.

15.9.1.3 Image and Video Input Devices

These devices, as the name suggests, take some image or video as the input and convert it into a format understandable by the computer. These include magnetic ink character recognition (MICR), optical mark recognition (OMR), optical character recognition (OCR), scanners, digital cameras and so on. MICR devices are used to detect the printed characters with magnetically charged ink and convert them into digital data. They are widely used in the banking industry for the processing of cheques. An OMR device senses the presence or absence of a mark but not the shape of the character. It is a very popular input device for surveys, census compilations and other similar applications. OCR devices are used for translating images of text or handwritten data into a machine-editable text or for translating pictures or characters into a standard encoding scheme (ASCII or Unicode).

A scanner is a device that analyses an image such as a photograph, printed text, etc., of an object and converts it to a digital image. OCR, OMR and image scanners are also referred to as data automation input devices. A digital camera is an electronic device used to capture and store photographs electronically instead of using photographic film.

15.9.2 Output Devices

Output devices convert the processed data back into a user-understandable format. Like an input device, an output device, too, acts as a human–machine interface. Printers, plotters and displays are the commonly used output devices. Computer output microfilm (COM) is another form of computer output where huge amounts of data can be outputted and stored in a very small size.

15.9.2.1 Printers

A printer is a device that produces a hard copy of the documents stored in electronic form, usually on a physical print medium such as paper. Printers can be broadly classified as 'impact printers' and 'nonimpact printers'. An 'impact printer' is one where the characters are formed by physically striking the type-device against an inked ribbon. Dot-matrix printers, daisy wheel printers, ball printers and drum and chain printers belong to this category. The dot-matrix printer is the most popular in this category. The 'dot matrix' is the basis of the printing mechanism in dot-matrix printers. The dot matrix is formed by arranging a number of small rods in a specified number of rows and columns. The number of rows and the number of columns in the dot matrix may vary from printer to printer. In order to print a character, the corresponding configuration of rods are stricken. The larger the number of dots in the dot matrix, the better is the printer quality.

Impact printers have been largely replaced by nonimpact printers. In this case, there is no physical contact with the paper. The characters are formed by using heat (in thermal printers), laser beam (in laser printers), ink spray (in inkjet printers), photography (in xerographic printers) and so on. Thermal printers are low-cost serial printers that use a number of small heating elements to construct each character from a dot-matrix print head. They use a special kind of heat-sensitive paper that turns black when heated. An inkjet printer sprays small droplets of ink rapidly from tiny nozzles onto the surface of the paper to form characters. A laser printer consists of a toner and a light-sensitive drum and works in a similar manner to a photocopier machine, except that, instead of working photographically from a printed document, the laser printer uses a laser beam to create the image.

15.9.2.2 Plotters

A plotter is a printer-like device used for producing hard-copy outputs of maps, charts, drawings and other forms of graphics. It is a vector graphics printing device that operates by moving a pen over the surface of the paper. Different types of plotter include pen plotters, electrostatic plotters and dot-matrix plotters. There are two types of pen plotter, namely the flat-bed plotter and the drum plotter. In the case of flat-bed plotters the pens move and the paper is stationary, whereas in the case of drum plotters the pens are stable and the paper is moved on a drum.

The electrostatic plotter works like a nonimpact-type electrostatic printer. It electrostatically charges the surface of a special kind of paper at the desired points and then passes the paper through a toner containing ink particles of opposite charge. The ink adheres to the paper surface only at charged points. The dot-matrix plotter works on the same principle as the impact-type dot-matrix printer.

15.9.2.3 Displays

Displays are devices used to display images on the screen in accordance with the signals generated by the computer. Displays are of various types including cathode ray tube (CRT) displays, liquid crystal displays (LCDs), plasma displays and organic light-emitting diode (OLED) displays. The CRT is a vacuum tube employing a focused beam of electrons from the cathode to hit the luminescent screen. The LCD is a display device made up of a number of colour or monochrome pixels arrayed in front of a light source or reflector. Each pixel comprises a liquid crystal molecule.

The plasma display is a flat-panel display where visible light is created by a phosphorus screen excited by discharged inert gases. The OLED is a special type of LED in which the emissive layer comprises a thin film of organic compounds.

15.9.2.4 Computer Terminals

Computer terminal in general refers to the entire range of devices that are connected to a computer and can be used to enter data into the computer system and receive the processed data as output. A computer terminal is used both as an input and as an output device. Typically, it consists of a keyboard and a CRT. Based on the capabilities and performance features, terminals are classified as dumb, smart and intelligent terminals. Depending upon the type of data the terminals are capable of displaying, they are classified as alphanumeric and graphic terminals. Detailed description of the various types of terminal is beyond the scope of this book.

15.10 Secondary Storage or Auxiliary Storage

Secondary storage devices are used for the mass nonvolatile storage of data and programs. It is often not practical to build a very large-sized primary memory to meet all the storage requirements of the system as it will increase the size and cost. That is where secondary storage is useful. Usually, it is located physically outside the machine. Although it is not an essential component in theoretical terms, the secondary storage is almost indispensable if one wants to exploit the full potential of a computer. Secondary storage devices are also referred to as auxiliary storage devices.

Owing to its semiconductor nature, the primary storage can be accessed much faster than any of the storage media used for secondary storage. The secondary storage on the other hand is economical as far as cost per unit data stored is concerned and has an unlimited storage capacity. It is also safe from getting tampered with by any unauthorized persons. Commonly used secondary storage devices include magnetic, magneto-optical and optical storage devices. Another emerging secondary storage device is the USB flash drive.

15.10.1 Magnetic Storage Devices

Magnetic storage devices include magnetic hard disks, floppy disks and magnetic tapes.

15.10.1.1 Magnetic Hard Disks

Hard disks are nonvolatile random access secondary data storage devices, i.e. the desired data item can be accessed directly without actually going through or referring to other data items. They store the data on the magnetic surface of hard disk platters. Platters are made of aluminium alloy or a mixture of glass and ceramic covered with a magnetic coating. Figure 15.32 shows the internal structure of a typical hard disk. As can be seen from the figure, there are a few (two or more) platters stacked on top of each other on a common shaft. The shaft rotates these platters at speeds of several thousand rpm. Each platter is organized into tracks and sectors (Fig. 15.33), both having a physical address used by the operating system to look for the stored data. Tracks are concentric circles used to store data. Each track is further subdivided into sectors so that the total number of sectors per side of the magnetic disk is the product of the number of tracks per side and the number of sectors per track. And if it is a

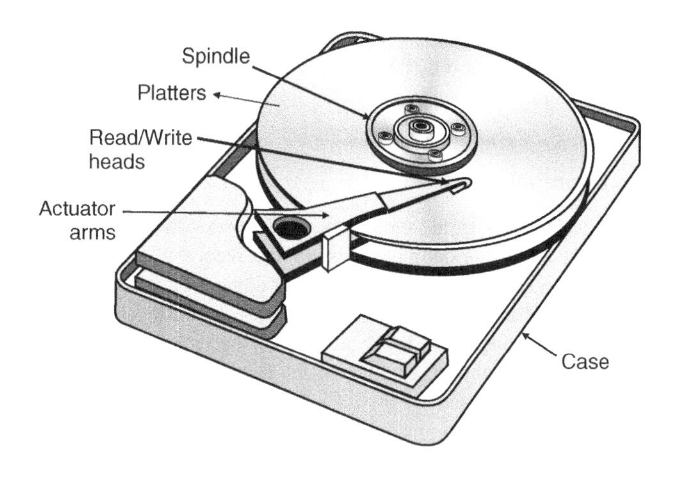

Figure 15.32 Internal structure of a typical hard disk.

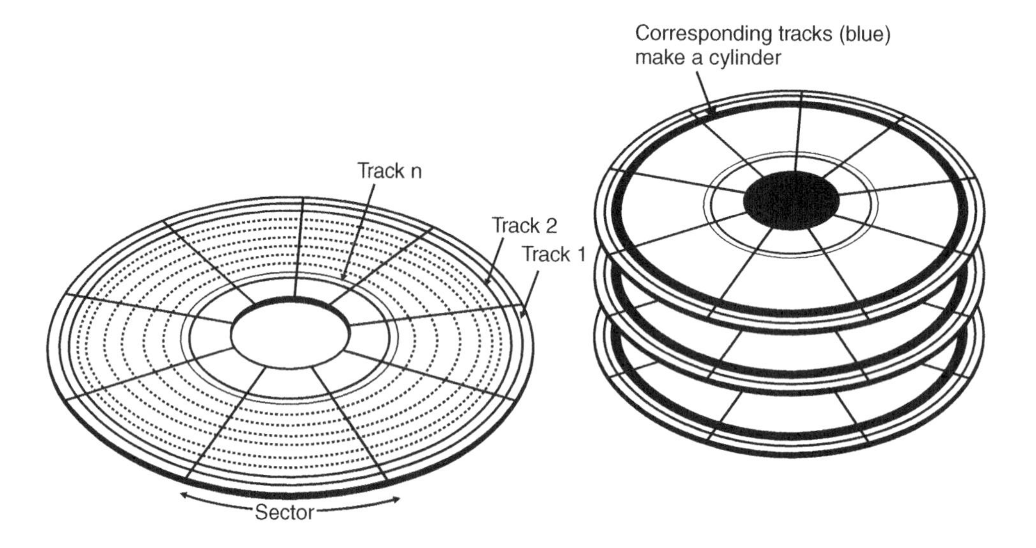

Figure 15.33 Tracks and sectors in a hard disk.

double-sided disk, the total number of sectors gets further multiplied by 2. From known values of the total number of sectors and the number of bytes stored per sector, the storage capacity of the disk in bytes can then be computed.

There is a read/write head on one or both sides of the disk, depending upon whether it is a single-sided or a double-sided disk. The head does not physically touch the disk surface; it floats over the surface and is close enough to detect the magnetized data. The direction or polarization of the magnetic domains on the disk surface is controlled by the direction of the magnetic field produced by the write head according to the direction of the current pulse in the winding. This magnetizes a small spot on the disk surface in the direction of the magnetic field. A magnetized spot of one polarity represents a binary '1', and that of the other polarity represents a binary '0'.

One of the most important parameters defining the performance of the hard disk is the size of the disk. Disks are available in various sizes ranging from 20 GB to as large as 80 GB. Other parameters defining the hard disk performance include seek time and latency time. *Seek time* is defined as the average time required by the read/write head to move to the desired track. *Latency time* is defined as the time taken by the desired sector to spin under the head once the head is positioned over the desired track.

15.10.1.2 Floppy Disks

Floppy disks are removable disks made of flexible polyester material with magnetic coating on both sides. Important parts of a floppy disk are shown in Fig. 15.34. Floppy disks are also organized in the form of tracks and sectors similar to a hard disk. A floppy disk drive unit is required to read data from or write data into a floppy disk. A read/write head that forms a part of the drive unit does this job. During a read or write operation, the disk rotates to the appropriate position and the head makes a physical contact with the disk to do the desired operation.

Earlier floppy disks were available in 5.25 inch size with a storage capability of 360 kB. They were known as double-sided double-density (DSDD) floppy disks. They have been superseded by 3.5 inch floppy disks having a storage capability of 1.44 MB. Floppy disks are fast being replaced by CD disks and USB drives.

15.10.1.3 Magnetic Tapes

Magnetic tapes are sequential access secondary storage devices used for storing backup data from mass storage devices. In sequential access storage devices, in order to access a particular data item, one has to pass through all the data items stored prior to it. The magnetic tapes are run on machines called tape

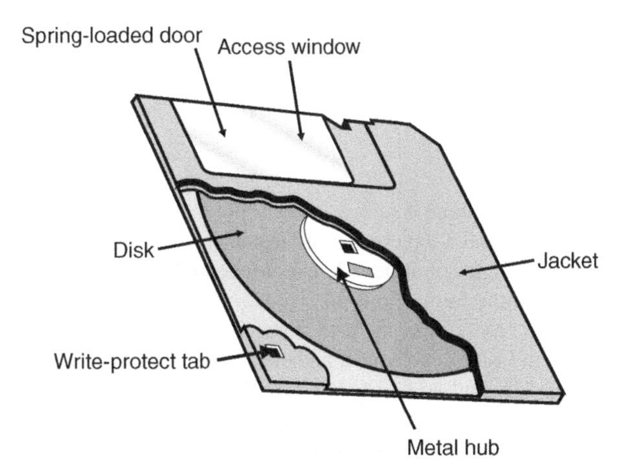

Figure 15.34 Important parts of a floppy disk.

drive units. The data on the tape are represented by tiny magnetized spots, with the presence of a spot representing a binary digit '1' and its absence representing a binary digit '0'. A simple and commonly used method of representing data on tapes is by using EBCDIC code. Magnetic tape is available in the form of reels, cassettes and cartridges. Reels are the most popular type.

15.10.2 Magneto-Optical Storage Devices

Magneto-optical storage devices use a combination of magnetic and optical technologies for data storage. The magnetic coating used in the case of these devices requires heat to alter the magnetic polarization, making them extremely stable at ambient temperatures. For the data write operation, a laser beam having sufficient power is focused onto a tiny spot on the disk. This raises the temperature of the spot. Then the magnetic field generated by the write head changes the polarization of the magnetic particles of that spot, depending upon whether a '1' or a '0' needs to be stored.

For the read operation, a laser beam with less power is used. It makes use of the 'Kerr effect', where the polarity of the reflected beam is altered depending upon the polarization of the magnetic particles of the spot.

15.10.3 Optical Storage Devices

One of the most significant developments in the field of storage media has been that of optical storage devices. Having arrived on the scene in the form of CD-Audio (Compact Disk-Audio) in the early 1980s, since then optical disks have undergone tremendous technological development. These are available in various forms, namely CD-ROM (Compact Disk Read Only Memory), WORM disks (Write Once Read Many), CD-R (Compact Disk Read), CD-RW (Compact Disk Read/Write) and DVD-ROM (Digital Versatile Disk Read Only Memory).

An optical disk differs from a conventional hard disk (solid magnetic disk) in the method by which information is stored and retrieved. While hard disks use a magnetic head to read and write data, in the case of an optical disk this is done with a laser beam. The high storage density of optical disks primarily results from the ability of the coherent laser beam to be focused onto a very tiny spot. The main advantages of optical disks include their vast storage capacity, immunity to illegal copying and their easy removability. Also, they do not transfer viruses from one user to the next.

15.10.3.1 CD-ROM

A CD-ROM is a disk comprising three coatings, namely polycarbonate plastic on the bottom, a thin aluminium sheet for reflectivity and a top coating of lacquer for protection. It can store up to 660 MB of data. It is formatted into a single spiral track having sequential sectors. CD-ROMs are prerecorded at the factory and store data in the form of pits and lands.

These are classified by the access time and data transfer rate. The performance of CD-ROM disks is enhanced by spinning them faster to achieve a higher transfer rate and faster access time. These are rated as 2X, 4X, 6X, 16X, 24X and so on. A 16X CD-ROM drive will be 16 times faster than the original drives. The spinning rate of the drive is the number of revolutions per minute. Its seek time is the time the drive takes to locate a track where desired data are stored. The time for which the drive has to wait for data to rotate under it is the latency. The sum of seek time and latency is the access time.

The read operation (Fig. 15.35) is performed by using a low-power laser beam. The laser beam is focused onto pits and lands. Laser light reflected from a pit is 180°out of phase with the light reflected from land. This light is detected by a photodiode followed by a processing circuitry. As the disk rotates, a series of pits and lands are sensed and the data stored in them is read.

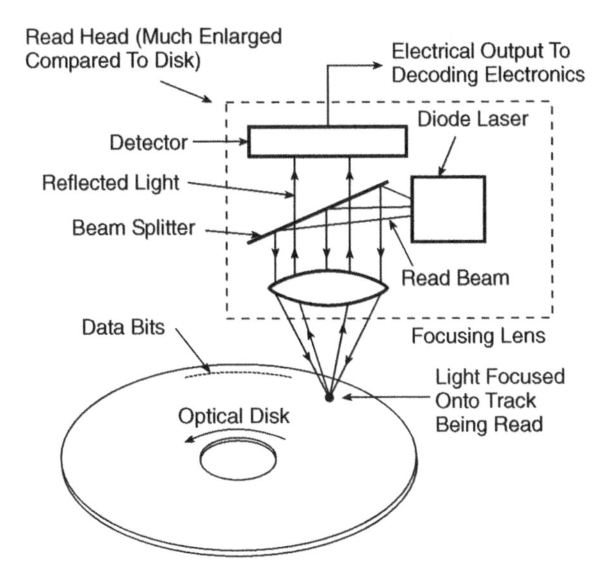

Figure 15.35 Use of a laser beam for CD READ operation.

15.10.3.2 WORM

This is a type of optical storage device where the data can be written once by the user, cannot be erased later but can be read many times. A low-power laser beam is used to burn microscopic pits on the disk surface. Burned surfaces represent a binary '1' and unburned areas represent a binary '0'.

15.10.3.3 CD-R

This is a type of WORM that allows multiple write sessions to different areas of the disk. In this case a laser is used to write data in the form of microscopic pits in an organic dye layer. The presence and absence of a bump indicate a '1' and '0' respectively.

15.10.3.4 CD-RW

In this case, data can be recorded, erased, rewritten and read many times. Recording of data is done by changing the state of the material from a well-structured crystalline state to a less ordered amorphous state.

15.10.3.5 DVD-ROM

Originally the term DVD was an abbreviation of Digital Video Disk, but today it is used for referring to Digital Versatile Disks. It has a much higher storage density than a CD-ROM. This is because the pit size is smaller in the case of DVD-ROMs.

CD-ROMS are single-side storage devices, whereas DVD-ROMs are available in single-sided as well as double-sided formats. As against the 660 MB storage capacity of a CD-ROM, a single-sided DVD of the same size offers 4.7 GB in a single layer. A double-layer or double-sided DVD would offer 9.4 GB of storage capacity, and a double-sided, double-layer DVD would have up to 17 GB, which is about 30 times the storage capacity available on a CD-ROM. DVD-R and DVD-RAM are the counterparts of CD-R and CD-RW.

15.10.4 USB Flash Drive

USB flash drives are plug-and-play flash-memory data storage devices integrated with the USB interface. They are lightweight, rewritable, erasable devices with storage capacities ranging from 8 MB to 64 GB.

Review Questions

1. With the help of a block schematic, describe the role of various elements in a computer system.
2. Explain the difference between:

   (a) a *sequential access memory* and a *random access memory*;
   (b) a *memory write* operation and a *memory read* operation;
   (c) EEPROM and UVEPROM;
   (d) synchronous SRAM and asynchronous SRAM.

3. Explain in brief the concept of cache memory.
4. With the help of a diagram, describe the functioning of different parts of a typical SRAM.
5. Compare the performance features of an SRAM and a DRAM. What is DRAM refreshing? Which type of RAM would you expect in battery-operated equipment?
6. Why do we need to have secondary storage devices when the computer already has a primary storage? Distinguish between magnetic tape and magnetic disk as a secondary storage device.
7. Briefly describe the following:

   (a) a serial port and a parallel port;
   (b) the internal bus system of a computer;
   (c) auxiliary storage devices.

8. What are the commonly used input and output ports in a computer system? Briefly describe the applications of each one of them.

Problems

1. A certain memory is specified as 16K × 8. Determine (a) the number of bits in each word, (b) the number of words being stored and (c) the number of memory cells.

   (a) 8; (b) 16 384; (c) 131 072

2. A certain memory is specified as 32K × 8. Determine (a) the number of address input lines, (b) the number of data input lines, (c) the number of data output lines and (d) the type of decoder.

(a) 15; (b) 8; (c) 8; (d) 1-of-15 decoder

3. It is desired to construct a 64K × 16 RAM from an available RAM chip specified as 16K × 8. Determine the number of RAM chips required for the same.

8

4. The following data refer to a hard disk: number of tracks per side = 600; number of sides = 2; number of bytes per sector = 512; storage capacity in bytes = 21 504 000. Determine the number of sectors per track for this hard disk.

35

Further Reading

1. Tocci, R. J. and Ambrosio, F. J. (2002) *Microprocessors and Microcomputers: Hardware and Software*, Prentice-Hall, NJ, USA.
2. Rafiquzzaman, M. (1990) *Microprocessors and Microcomputer-based System Design*, CRC Press, FL, USA.
3. Keeth, B. and Baker, J. (2000) *DRAM Circuit Design: A Tutorial* (IEEE Press Series on Microelectronic Systems), John Wiley & Sons–IEEE Press, New York, USA.
4. Prince, B. (1999) *High Performance Memories: New Architecture DRAMs and SRAMs – Evolution and Function*, John Wiley & Sons, Ltd, Chichester, UK.
5. Axelson, J. (1997) *Parallel Port Complete: Programming, Interfacing and Using the PC's Parallel Port*, Lakeview Research, Madison, WI, USA.
6. Axelson, J. (1998) *Serial Port Complete*, Lakeview Research, Madison, WI, USA.

16

Troubleshooting Digital Circuits and Test Equipment

This chapter looks at two interrelated aspects of digital circuit troubleshooting, namely *troubleshooting* and the *test equipment*. The chapter is divided into two parts: the first part discusses troubleshooting guidelines for a variety of digital devices and circuits; the second part deals with test and measuring equipment. The chapter begins with general guidelines to troubleshooting digital circuits and then moves on to discuss techniques for troubleshooting specific digital building blocks such as logic gates, flip-flops, counters, registers, arithmetic circuits, memory devices and so on. In the second part of the chapter, some of the more commonly used test and measuring equipment is discussed at length. The test instruments covered here are not necessarily ones that are required by a troubleshooter during the course of fault finding. They also include instruments that are the result of advances in digital technology and have a digital-dominated internal hardware. In fact, this constitutes one of the most important areas where digital technology has so strongly manifested itself. Some of these instruments, such as the digital multimeter (DMM), the logic probe and the digital storage oscilloscope (DSO), are the essential tools of any digital circuit troubleshooter. The chapter is adequately illustrated with a large number of case studies related to digital circuit troubleshooting.

16.1 General Troubleshooting Guidelines

Irrespective of the type and complexity of the digital circuit to be troubleshot, the following three-step procedure should be followed:

1. Fault detection or identification.
2. Fault isolation.
3. Remedial measures.

Digital Electronics: Principles, Devices and Applications Anil Kumar Maini

Fault detection means knowing the nature of the fault, which could be done by comparing the actual or present performance of the circuit with the ideal or desired performance. Complete knowledge about the nature of the fault often gives an idea about the nature of tests and measurements to be performed to isolate the fault. It is therefore important that the nature of the fault is properly understood and appreciated in terms of the functions performed by various parts of the overall digital circuit or system.

Fault isolation means performing tests and making measurements with the available diagnostic tools to know precisely where the fault lies. This could be in the form of a faulty component or a shorted or open track and so on. The level of documentation that is available plays an important role in deciding about the type of measurements to be made to isolate the fault. Comprehensive documentation helps in significantly reducing the time period required to actually latch on to the faulty component or area. Again, the faults could either be internal to the components and devices, digital integrated circuits, for instance, or external to the components. These two types of fault are discussed in the following paragraphs.

Remedial measures follow the fault isolation. This could mean repairing of tracks or replacement of one or more components.

16.1.1 Faults Internal to Digital Integrated Circuits

Digital circuits and systems are dominated by the use of digital integrated circuits (ICs). The number of discrete devices is usually much smaller than the number of ICs used. Therefore, the knowledge of typical faults that can occur in digital ICs is central to fault isolation in digital systems. The most commonly observed defects or failures in digital ICs are as follows:

1. Shorting of input or output pins to V_{CC} or ground terminals or shorting of tracks.
2. Open circuiting of input or output pins.
3. Shorting of two pins other than ground and V_{CC} pins.
4. Failure of the internal circuitry of the IC.

16.1.1.1 Internal Shorting of Input or Output Pins to GND or V_{CC}

This is one of the commonly observed faults internal to digital ICs. Shorting of one or more of the input or output pins internally to GND puts a permanent LOW on the pin(s). This could have several manifestations depending upon the nature of the IC and also upon the nature of the component driving these pins. Some of these manifestations are given in the following examples:

1. If an input pin that is internally shorted to GND is being driven from an output pin of another IC, that particular output pin will face a permanent ground and will be affected accordingly. A pulsating signal, if originally present at that pin, will vanish.
2. If the shorted input terminal happens to be that of a NAND gate, the output of the gate will permanently go to the logic HIGH state and will not respond to any changes on the other input.
3. If the shorted input pin is the PRESET input of a presettable, clearable *J–K* flip-flop with active LOW PRESET and CLEAR inputs, the output of this particular flip-flop will always be in the logic HIGH state irrespective of the status of the *J* and *K* inputs.
4. Shorting of the output pin to GND puts a permanent logic LOW on that pin, and this particular output does not respond to changes on the corresponding input pins.

Shorting of input or output pins to V_{CC} puts a permanent HIGH on those pins. If it is the output pin, it again fails to respond to any changes on the corresponding input pins, and, if it is the input pin, it affects the output response of the IC depending upon the nature of the IC. The following examples illustrate this point further:

1. If it is the input of the NAND gate, a permanent HIGH on the input permanently transforms it into an inverter circuit, which means that the NAND gate no longer performs its intended function.
2. If it happens to be the input terminal of an OR gate, it drives its corresponding output to a permanent logic HIGH state.

16.1.1.2 Open Circuiting of Input or Output Pins

Open circuiting of input and output pins occurs for reasons internal to the IC when the fine wire that connects the IC pin to the relevant location on the chip breaks. The effects of open circuiting can be serious too. For instance, an open on the input or output pin makes it a floating terminal, and, if the IC belongs to the TTL logic family, it will be treated as logic HIGH. It could even lead to overheating and subsequent damage to the IC. An open on the input pin also prohibits any genuine changes on the pin from reaching the input on the chip, with the result that the output fails to respond to those changes. Similarly, an open on the output pin affects the response of the subsequent IC to whose input this particular output is connected.

16.1.1.3 Shorting of Two Pins Other than GND and V_{CC} Pins

This fault forces the affected pins to have the same logic status at all times. For obvious reasons, the output responds incorrectly. Such a situation also leads to shorting of the two pins from where these affected (internally shorted) pins are being fed. The ultimate effect on the performance depends upon the nature of the ICs involved.

16.1.1.4 Failure of the Internal Circuitry of the IC

Failure of the internal circuitry could be anything from damage to a certain active device to increase in the resistance value of a certain on-chip resistor. Bearing in mind the complexity of the internal circuitry of the present-day digital ICs, there could be numerous possibilities. However, the occurrence of such a fault is not very common.

16.1.2 Faults External to Digital Integrated Circuits

The commonly observed faults external to digital ICs include the following:

1. Open circuits.
2. Short circuits.
3. Power supply faults.

16.1.2.1 Open Circuit

An open circuit could be caused by any of a large number of factors, such as a broken track (usually a hairline crack that is very difficult to notice with the naked eye), a dry solder leading to a loose or intermittent connection, a bent or broken pin on the IC, which disallows the signal from reaching that pin, and even a faulty IC socket, where the IC pin does not make a good contact with the socket. Any of the above-mentioned fault conditions would produce a break in the signal path. Such a fault condition can be easily located by switching off the power to the circuit and then establishing the continuity in the suspected areas with the help of a multimeter.

16.1.2.2 Short Circuit

A short circuit could be caused by an improperly etched PCB leading to unetched copper between tracks, solder bridges tending to short two points that are close to each other, such as adjacent pins of an IC, and other similar factors reflecting poor-quality PCB making, wiring and soldering techniques. Such a fault could also be easily located with the help of a multimeter by switching off the power to the circuit.

16.1.2.3 Faulty Power Supply

The third commonly observed fault external to the ICs results from a faulty power supply. There are in fact two commonly observed conditions that generally lead to an apparent power supply fault. One of them could be a catastrophic failure of the power supply that feeds DC voltages to the V_{CC} or V_{DD} pins. The result could be either a complete absence of or a reduction in these DC voltages. The other possible condition could be the overloading of the power supply, which means that the power supply is being asked to deliver a current that is greater than it is designed for. Such a condition is usually due to a fault internal to the IC. In some cases, the fault could be external to the IC too. In such cases it would be good practice to check the power supply and ground status of all the digital ICs being used. An overloading caused by some kind of fault internal to the IC often leads to an increased ripple on the power supply line. Having confirmed such a situation, it would again be good practice firstly to rule out any possibility of a short or a very low resistance path external to the ICs. After that, the ICs could be removed one at a time until the situation is corrected. The IC whose removal restores normalcy is the one that has developed an internal fault. The next obvious step is that of replacing the faulty IC with a fresh one. Sometimes, more than one IC develops internal faults so as to load the power supply. In that case it is necessary to replace all of them to restore normal functioning.

The general guidelines outlined above are applicable to troubleshooting digital circuits using digital ICs of different complexities, from logic gates to counters, registers and arithmetic building blocks. Application of these guidelines to some simple case studies related to troubleshooting of combinational circuits is presented in the following examples.

Example 16.1

Refer to the simple combinational circuit of Fig. 16.1. The logic status of the different input and output pins of the ICs used in this circuit, as observed with the help of a logic probe, is as follows: pin 1 of IC-1 is LOW; pin 2 of IC-1 is pulsing; pin 3 of IC-1 is LOW; pin 4 of IC-1 is HIGH; pin 5 of IC-1 is pulsing; pin 6 of IC-1 is pulsing; pin 1 of IC-2 is indeterminate; pin 2 of IC-2 is pulsing; pin 3 of IC-2 is indeterminate. What in your opinion is the most probable cause of this faulty condition? Give justification wherever required. The ICs used here belong to the 74HC logic family.

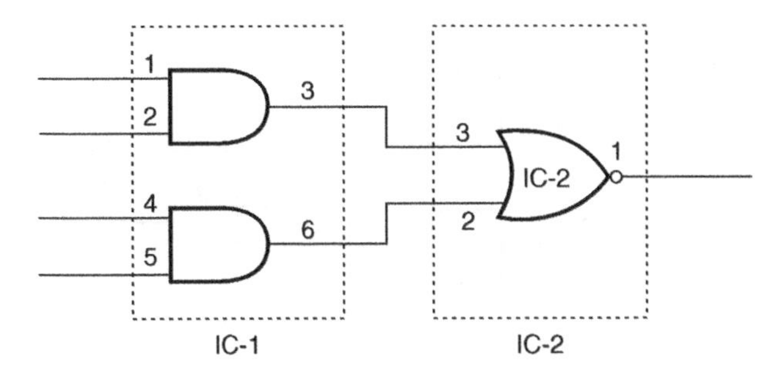

Figure 16.1 Combinational circuit (example 16.1).

Solution

At the outset, the functional status of each one of the building blocks used in this combinational logic circuit is looked at. The upper AND gate is disabled as one of its inputs is observed to have a logic LOW, with the result that its output should be a logic LOW. This is confirmed by the logic probe measurement at pin 3 of this IC. The lower AND gate is enabled as one of its inputs is in the logic HIGH state. Therefore, the output of this gate should be the same as the other input of this gate, which is a pulsed waveform. The output of this gate is a pulsed one, as confirmed by the logic probe measurement at pin 6 of IC-1.

Pin 6 of IC-1 is connected to pin 2 of IC-2. Pin 2 of IC-2 is one of the inputs of a two-input NOR gate. Pin 2 of IC-2 shows the presence of a pulsed waveform, which confirms that it is being properly fed from pin 6 of IC-1. Now, pin 3 of IC-1 is in the logic LOW state, and this is connected to pin 3 of IC-2. Therefore, pin 3 of IC-2 should have shown a logic LOW status. This is, however, not the case, as demonstrated by logic probe measurement. The indeterminate state at pin 3 of IC-2 also manifests itself at pin 1 of IC-2, which is understandable when CMOS ICs are being dealt with.

The indeterminate status of pin 3 of IC-2 only indicates that there is an open circuit somewhere in the path from pin 3 of IC-1 to pin 3 of IC-2. This can be verified with the help of a logic probe and tracing the path and identifying the spot where the genuine logic LOW status changes to an undesired indeterminate status. Remember that CMOS ICs treat floating inputs as indeterminate states.

Example 16.2

Figure 16.2(a) shows the implementation of a two-input multiplexer that is supposed to have the functional table of Fig. 16.2(b). Instead, it is behaving like the functional table of Fig. 16.2(c). The ICs used are from the TTL family. The observations made at different pins of the three ICs used in the circuit are listed in Table 16.1. What is the most probable cause of this faulty behaviour?

Solution

If we look at the logic status of various pins of IC-1, IC-2 and IC-3 for $S = 0$, we find that the inverter in IC-1 is not working properly. Its output should have been $\overline{S}$ and not logic '0'. The two AND gates in IC-2 and the OR gate in IC-3 are functioning as per their respective truth tables. Even the inverter seems to be doing its job when the input is a logic '1'. Such behaviour of the inverter is possible only when the input to this inverter is always a logic '1', irrespective of the logic status of S.

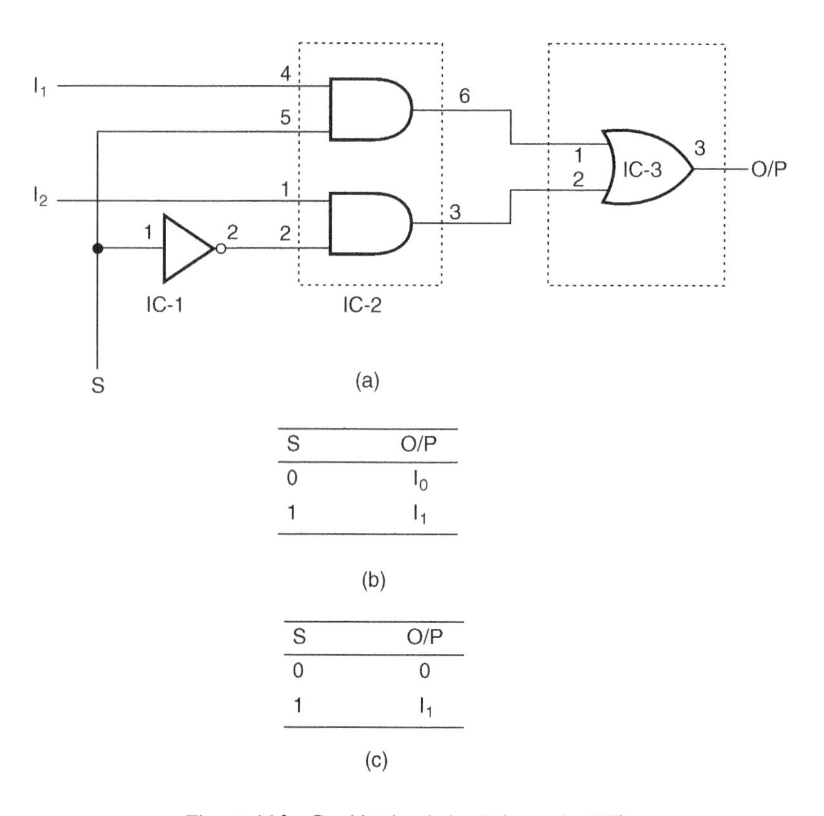

S	O/P
0	I_0
1	I_1

(b)

S	O/P
0	0
1	I_1

(c)

Figure 16.2 Combinational circuit (example 16.2).

Table 16.1 Example 16.2.

Pin/IC	$S = 0$	$S = 1$
Pin 1 (IC-1)	0	1
Pin 2 (IC-1)	0	2
Pin 1 (IC-2)	I_2	I_2
Pin 2 (IC-2)	0	0
Pin 3 (IC-2)	0	0
Pin 4 (IC-2)	I_1	I_1
Pin 5 (IC-2)	0	1
Pin 6 (IC-2)	0	I_1
Pin 1 (IC-3)	0	I_1
Pin 2 (IC-3)	0	0
Pin 3 (IC-3)	0	I_1

Probable reasons for such behaviour are as follows:

1. Pin 2 of IC-1 is internally shorted to GND.
2. Pin 2 of IC-2 is internally shorted to GND.
3. Pin 1 of IC-1 is internally open, which means that it is floating and is therefore treated as a logic HIGH input as the IC belongs to the TTL family.

The first two reasons can be ruled out one by one by checking the continuity between pin 2 of IC-1 and GND and also between pin 2 of IC-2 and GND. If the meter shows no continuity in the two cases, these reasons are ruled out. In such a case, the third reason seems to be the most probable cause.

16.2 Troubleshooting Sequential Logic Circuits

The troubleshooting guidelines for combinational circuits that have been outlined and illustrated in the previous pages with the help of troubleshooting exercises are equally valid in the case of sequential logic circuits such as flip-flops, counters, registers, etc. Faults such as open and short circuits affect all categories of digital building blocks, including both combinational and sequential circuits. However, the effects of open and short circuits in the case of sequential logic devices can be far more serious and difficult to analyse than they would be in the case of logic gates and other combinational building blocks. This is due to the memory characteristics of flip-flops, on account of which the output of a sequential device or circuit depends not only on the present inputs but also on the past inputs. A noise pulse, if large enough in amplitude and duration, could induce a change in the logic status at the output of a logic gate. However, the logic gate would get back to its original status after the noise pulse has vanished. On the other hand, the same noise pulse induced state transition in the case of a flip-flop is permanent.

Let us take the case of a floating input due to an internal or external open circuit. A floating input is highly prone to picking up noise. The most susceptible inputs from the viewpoint of noise pick-up in the case of flip-flops are the CLOCK, PRESET and CLEAR inputs, as these inputs, if activated by noise pick-up due to an internal or external open circuit, can cause the flip-flop to behave erratically. The other possible fault condition is a short circuit at one or more of the inputs of the flip-flop. Again, the symptoms in the case of flip-flops would be different from those in the case of combinational circuits.

Yet another condition that is particularly troublesome in the case of clocked sequential circuits arises from what is known as *clock skew*. Clock skew is basically the difference in the time of arrival of the clock signal at the clock inputs of various sequential devices such as flip-flops comprising a complex synchronous sequential circuit. This time delay, if more than the propagation delay associated with each of the individual devices, could cause serious problems. This can be best explained with the help of a simple illustration. Refer to Fig. 16.3(a). It shows a simple sequential circuit comprising a cascade arrangement of two D flips-flops. The outputs Q_l and Q_2 are initially in the logic '0' state. With the occurrence of LOW-to-HIGH transition of the first clock pulse, Q_l should go to the logic '1' state, whereas Q_2 should stay in the logic '0' state. However, if for some reason the clock signal reaching the clock input of FF-2 is delayed from the clock input of FF-l input by more than the propagation delay associated with the individual flip-flops, the Q_2 output would also go to the logic '1' state. This is obvious from the waveforms shown in Fig. 16.3(b). The dotted block in Fig. 16.3(a) represents the clock signal delay. The reasons for this clock skew could be long connecting lines, parasitic capacitance at clock inputs and so on. Since these undesired parameters change with temperature and other circuit conditions, the behaviour of affected devices is usually erratic and unpredictable.

Example 16.3

Refer to the flip-flop circuit of Fig. 16.4. The D input to the flip-flop is tied to GND. The Q output of the flip-flop is expected to go to the logic '0' state with the application of a clock pulse. However, it does not do so, as shown by the observations recorded by the troubleshooter: pin 1 of IC-1 is in the logic '1' state; pin 2 of IC-1 is in the logic '1' state; pin 3 of IC-1 is in the logic '0' state; pin 2 of IC-2 is in the logic '0' state; pin 3 of IC-2 is pulsing; pin 5 of IC-2 is in the logic '1' state. List various possible causes of occurrence of this fault. Isolate them one by one to arrive at the actual fault.

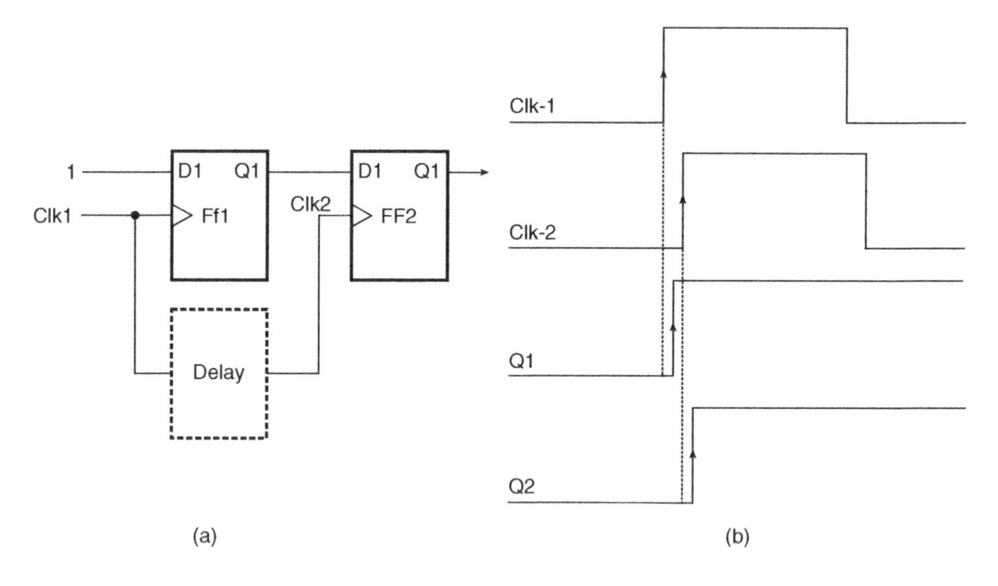

Figure 16.3 Clock skew problem.

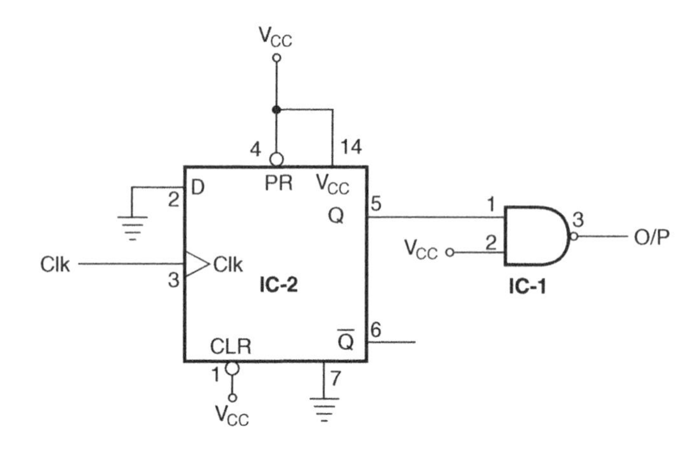

Figure 16.4 Sequential circuit (example 16.3).

Solution

Various possible causes that can lead to the above fault conditions are as follows:

1. Pin 1 of IC-1 is either externally or internally shorted to V_{CC}.
2. Pin 5 of IC-2 is either externally or internally shorted to V_{CC}.
3. Pin 4 of IC-2 is either externally or internally shorted to GND.
4. IC-2 has some kind of internal failure, which stops it from responding to inputs.

5. Pin 6 of IC-2 is externally or internally shorted to GND.

A continuity check can be used to rule out one by one the first, second, third and fifth causes. Remember that, although pin 6 is not used in the circuit, if it is shorted to GND it will force the Q output permanently to go to the logic '1' state owing to the cross-coupling arrangement in the internal structure of the flip-flop. Incidentally, pin 6 in IC-2 happens to be very close to the GND pin, which is pin 7. Even a solder bridge between pin 6 and pin 7 could lead to this. What is important to note is that a troubleshooter may tend to ignore IC pins that are not used, but even those unused pins in the IC can cause faulty conditions. Once the continuity check rules all these possibilities out, the IC can be replaced.

Example 16.4

Figure 16.5 shows a cascaded arrangement of three D flip-flops belonging to the TTL family of digital ICs. The circuit shown here is only a small part of a complex digital circuit. Each of the three flip-flops has a clock input-to-output propagation delay of 15 ns. The expected and observed outputs of the flip-flops for the first few clock cycles are listed in Table 16.2. Although the circuit shown here is that of a three-bit shift counter, the observed outputs are nowhere near to what they should have been in the case of a three-bit shift counter. Identify the possible cause for the observed outputs being different from the expected ones. All flip-flops are observed to be in the logic '0' state just before application of

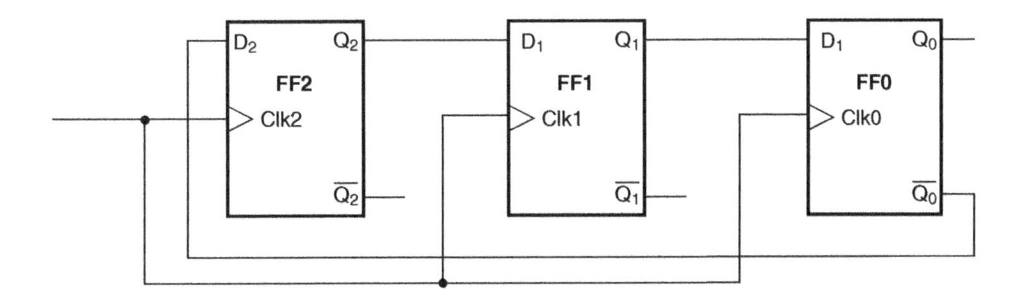

Figure 16.5 Three-bit shift counter (example 16.4).

Table 16.2 Example 16.4.

Clock pulse	Expected output			Actual output		
	Q_2	Q_1	Q_0	Q_2	Q_1	Q_0
0	0	0	0	0	0	0
1	1	0	0	1	1	1
2	1	1	0	0	0	0
3	1	1	1	1	1	1
4	0	1	1	0	0	0
5	0	0	1	1	1	1
6	0	0	0	0	0	0
7	1	0	0	1	1	1

the clock signal. The clock signals appearing at the input terminals of the three flip-flops, when seen individually, are observed to be clean and free of any noise content.

Solution

Initially, $Q_2 = Q_l = Q_0 = 0$ and $D_2 = l$ as D_2 is fed from $\overline{Q_0}$. Therefore, with the occurrence of the first clock pulse, Q_2 is expected to go to the logic '1' state. Since $D_l = D_0 = 0$, Q_l and Q_0 are expected to remain in the logic '0' state. However, Q_2, Q_l and Q_0 are observed to make a transition to the logic '1' state. Now this could have been possible if $D_2 = D_1 = D_0 = 1$, which is not the case. This would be remotely possible if there were an external or an internal open at all the D inputs, making them floating inputs. Since the ICs used here are TTL ICs, these floating inputs would be treated as logic HIGH. All this seems to be valid for only the first clock pulse, because, if this were true, the three outputs would subsequently stay in the logic '1' state. Here, all outputs are observed to be toggling. Whether there is any internal or external open or short can be verified with a continuity check using a multimeter.

There is another possibility. As we know, clock skew is a problem that quite often bothers flip-flop timing. Whether or not the fault could possibly be due to the clock skew problem will now be examined. This is not an arbitrary choice. If the statement of the problem is carefully read, it is stated there that the given circuit is only a part of a bigger circuit, and also that clock signals have been observed only individually at the relevant inputs of different flip-flops. It is therefore quite possible that the clock signals at the clock inputs of different flip-flops are not synchronous. If the clock inputs

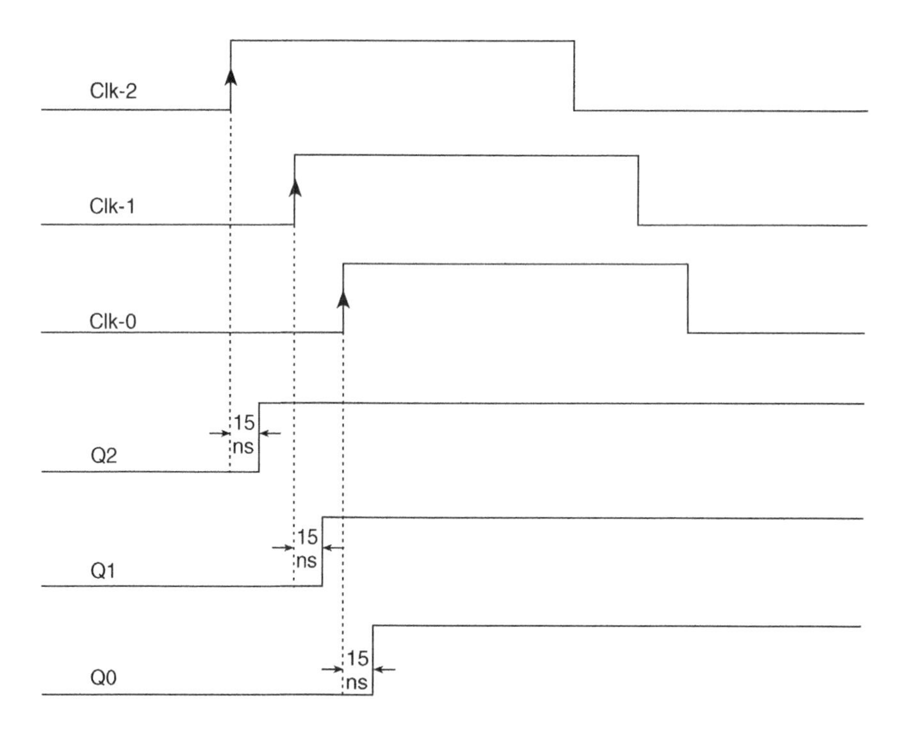

Figure 16.6 Waveforms (troubleshooting exercise 16.4).

to flip-flops FF-2 and FF-l are examined simultaneously and it is discovered that the clock input to FF-l is delayed from the clock input to FF-2 by more than 15 ns, FF-l will make a transition to the logic '1' state with the first clock transition. Similarly, if the clock input to FF-0 is delayed by more than 15 ns from that to FF-l or by more than 30 ns from that to FF-2, even FF-0 is going to make a transition to the logic '1' state with the first relevant transition of the clock signal. And what is more important is that all other observed outputs for subsequent clock pulses, as shown in Table 16.2, are also valid under these circumstances. The waveforms shown in Fig. 16.6 illustrate how this clock delay can cause a fault condition. Thus, this seems to be the most probable reason for the present fault condition.

16.3 Troubleshooting Arithmetic Circuits

The arithmetic circuits also fall into the category of combinational circuits. Therefore, the troubleshooting tips are similar to those described at length in the previous pages. It would be worth reiterating again that knowledge of the internal structure and functional aspects of the ICs used helps a lot in identifying the reasons for a fault. The following troubleshooting exercise illustrates the point.

Example 16.5

Figure 16.7 shows a four-bit binary adder–subtractor circuit configured around a four-bit parallel binary adder (type number 7483) and a quad two-input EX-OR gate (type number 7486). The arrangement works as an adder when the ADD/SUB input is in the logic '0' state, and as a subtractor when ADD/SUB is in the logic '1' state. The circuit has developed a fault. It is functioning satisfactorily as a subtractor. However, when it is used as an adder, it is observed that the SUM output is not $A+B$ but $A+B+l$ instead. What do you think is the probable reason for this behaviour?

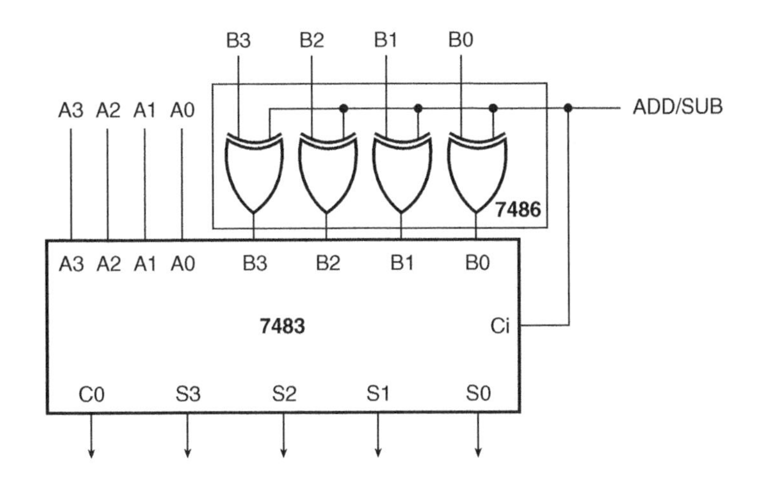

Figure 16.7 Adder–subtractor circuit (example 16.5).

Solution

- Since the circuit is functioning properly as a subtractor (when ADD/SUB = 1), this implies that:

  (a) IC 7486 is functioning properly;
  (b) the C_i input is in the logic '1' state.

- The present fault would occur only if the C_i input were in the logic '1' state, even when ADD/SUB = 0. This is possible in the case of either of the following two situations:

  (a) There is an external open between the ADD/SUB input and the C_i input. This would make C_i a floating input, which would be treated as a logic '1' in a TTL IC.
  (b) There is an internal open at the C_i input, which would have the same end result.

- The fault can be isolated with the help of a continuity check.

16.4 Troubleshooting Memory Devices

This section outlines the general procedure used for testing and troubleshooting memory devices. As will be seen in the paragraphs to follow, troubleshooting memory devices is far more complex than troubleshooting other digital building blocks. The procedure outlined earlier for digital building blocks such as logic gates, flip-flops, counters, registers, arithmetic circuits, etc., is not valid for testing memory devices such as RAM and ROM. One thing that is valid is that it is equally important fully to understand the operation of the system before attempting to troubleshoot. You must also remember that there is a lot of digital circuitry outside the memory device that is a part of the overall memory system. This may include a decoder circuit and some combinational logic.

16.4.1 Troubleshooting RAM Devices

The most common method of testing a RAM system involves writing known the pattern of 0s and 1s to each of the memory locations and then reading them back to see whether the location stored the pattern correctly. This way, both READ and WRITE operations are checked. One of the most commonly used patterns is the 'checkerboard pattern' where all memory locations are tested with a 01010101 pattern and then with a 10101010 pattern. There are many more patterns that can be used to check various failure modes in RAM devices. No check, however, guarantees 100 % accuracy. A chip that passes a checkerboard test may fail in another test. But if the chip fails in the checkerboard test, it is certainly not good.

RAM check is performed automatically. In the majority of computers and microprocessor-based systems, every time the system is powered, the CPU runs a memory-test program that is stored in the ROM. The operator can also execute this memory-test routine on request. The system displays some message after the test is over. After that, remedial action can be initiated.

16.4.2 Troubleshooting ROM Devices

ROM devices cannot be checked by writing and reading back known patterns of 0s and 1s, as was done in the case of RAM devices. ROM is a 'read only memory' device and its testing should basically involve reading the contents of each location of the ROM and then comparing them with what it is actually supposed to contain.

ROM testing is done with the help of a special instrument that can be used to read the data stored in each location of the ROM. It cannot be tested, like a RAM, by writing some pattern of 0s and 1s and then reading them back. One of the methods is to read data in each location and produce a listing of those data for the user to compare with what the ROM is actually supposed to store. But, of course, the process becomes highly cumbersome for large-capacity ROM chips.

Another approach is to have a reference ROM plugged into the test instrument along with the test ROM. The instrument reads data in each of the locations on the test ROM and then compares them with the data stored on the corresponding locations of the reference ROM.

Yet another method is to use a CHECKSUM. Checksum is a code that is stored in the last one or two locations of the ROM. It is derived from the addition of different data words stored in different locations of the ROM under test. For instance, if the data words stored in the first three locations are 11001001, 10001110 and 11001100, then the checksum up to this point will be 00100011. When the test instrument reads data in the test ROM, it creates its own checksum. It compares the checksum with the one already stored in the test ROM. If the two match, the ROM may be considered to be a good one. We have used the word 'may' because even wrong data can possibly lead to a correct checksum. However, if the checksums do not match, it is definitely a faulty ROM.

16.5 Test and Measuring Equipment

As outlined at the beginning of the chapter, the test and measuring instruments discussed in this part of the chapter are not only the ones that a digital system troubleshooter or analyser has generally to make use of; some of the instruments described here have an internal hardware dominated by digital technology and its advances. The test equipment covered at length in the following pages include the digital multimeter, digital oscilloscope, logic probe, logic analyser, frequency counter, synthesized function generator and arbitrary waveform generator. Computerized instrumentation and equipment–computer interface standards are discussed towards the end of the chapter.

16.6 Digital Multimeter

In a *digital multimeter*, the analogue quantity to be measured (current, voltage, resistance) is firstly transformed into an equivalent voltage if the parameter to be measured is current or resistance. The transformed analogue voltage is then digitized using an A/D converter (ADC). To be more precise, the analogue voltage is converted into a pulse train whose frequency depends upon the magnitude of the voltage. The pulses are counted over a known gating period in a counter. The counter outputs are decoded and displayed. The displayed count represents the magnitude of the parameter under measurement.

In another approach that is also in common use the input analogue voltage is compared with a ramp from a ramp generator. The comparator generates a gating pulse whose width equals the time interval between the ramp amplitude rising from zero to the analogue voltage under measurement. The counter in the ADC counts clock pulses of a known frequency over this gating interval, and the counter count is decoded and displayed. Thus, while in the former method there is a voltage-to-frequency (V/F) conversion and the equivalent frequency representing the analogue voltage is counted over a fixed gating interval, in the latter method a fixed frequency is counted over a variable gating interval, with the gating interval being proportional to the analogue voltage. Different techniques of analogue-to-digital conversion have been discussed in detail in Chapter 12 on data conversion circuits.

16.6.1 Advantages of Using a Digital Multimeter

The digital multimeter has the advantages of offering unambiguous display with no allowance for any human error, improved accuracy (±0.1% as against ±3% in analogue meters) and improved resolution (+0.1% as against 1% in analogue meters). Other advantages include easy incorporation of features such as *autoranging*, automatic polarity and diode/transistor test and so on. The cost advantage that used to exist in favour of analogue meters has narrowed down to a small amount with advances in IC technology. Digital multimeters are fast replacing analogue meters even for routine measurements. However, analogue meters are relatively immune to noise and are preferred in an electrically noisy environment.

16.6.2 Inside the Digital Meter

Figure 16.8 shows the schematic arrangement of a typical digital meter. The signal scalar at the input is basically an attenuator/amplifier block and is partly used for range selection function. In autoranging meters, the input signal level is sensed on application of the input signal, and the signal scalar gain is selected accordingly. The signal conditioner generates a DC voltage proportional to the input signal. The ADC employed is usually the integrating-type ADC, single slope or dual slope, with the latter being the preferred one because of its higher accuracy, insensitivity to changes in integrator parameters and low cost. All the building blocks depicted in Fig. 16.8, except for the display, are available on a single chip. ICL 7106/7107 is an example.

16.6.3 Significance of the Half-Digit

Digital multimeters (DMMs) invariably have a display that has an additional half-digit. We have $3\frac{1}{2}$-, $4\frac{1}{2}$- and $5\frac{1}{2}$-digit digital multimeters rather than 3-, 4- and 5-digit multimeters. While the usually so-called full digits can display all digits from 0 to 9, a half-digit can display either a '0' or a '1'. The

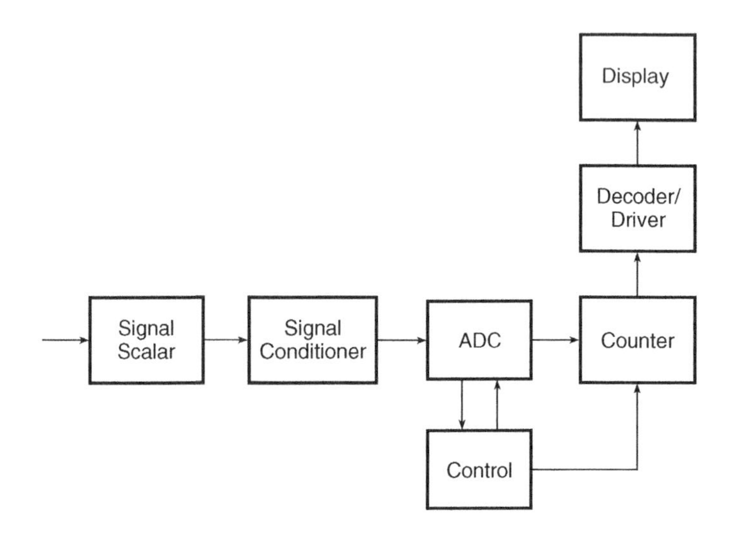

Figure 16.8 Block schematic of a digital meter.

addition of a half-digit in the MSB position of the display preserves the resolution of the multimeter up to a higher range. For instance, a three-digit multimeter has a resolution of 0.1 V up to 99.9 V. A $3^1/_2$-digit meter with practically no additional hardware would give you a resolution of 0.1 V up to 199.9 V. This increase in resolution range comes with the addition of one additional seven-segment display and no change in hardware complexity. The display resolution is also sometimes expressed in terms of counts. The $3^1/_2$-digit DMM has a 2000 count resolution. DMMs with a 4000 count resolution, referred to as $3^3/_4$-digit meters, are also commercially available. These meters will also have four seven-segment displays but have some additional hardware.

Digital multimeters are made in a large variety of sizes, shapes and performance specifications, ranging from pen-type $3^1/_2$-digit DMMs to $7^1/_2$-digit high-resolution benchtop versions. Handheld versions are available, typically up to $4^1/_2$-digit resolution. The majority of them have an in-built diode test, transistor test and continuity check features. Some of them even offer L-C measurement and frequency measurement without any significant change in price. Figure 16.9 shows a photograph of one such multimeter (the Fluke 115 multimeter). It has a 6000 count display and an in-built continuity check, diode test, frequency measurement, capacitance measurement, etc., in addition to conventional functions. Figure 16.10 shows a photograph of a high-end benchtop version of a digital multimeter (Fluke 8845A).

Example 16.6

The specification sheet of a certain $3^1/_2$-digit digital multimeter lists its display to be a 4000 count display. Determine the resolution offered by the multimeter for the following measurements:

(a) the maximum DC voltage that can be measured with a resolution of 0.1 V;
(b) the maximum resistance value that can be measured with a resolution of 1 Ω.
(c) the maximum DC current that can be measured with a resolution of 10 μA.

Figure 16.9 Handheld digital multimeter. Reproduced with permission of Fluke Corporation.

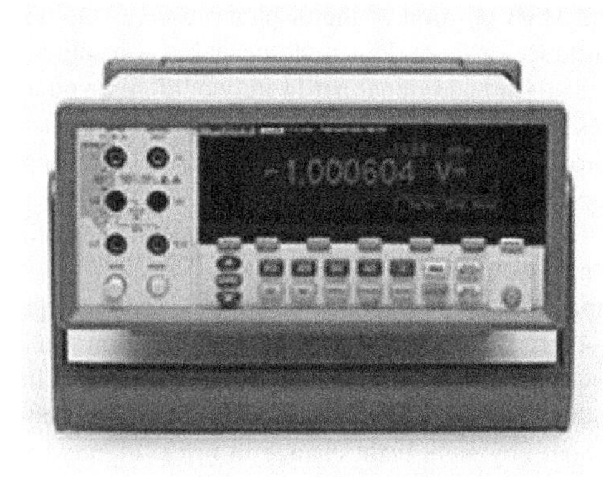

Figure 16.10 Benchtop digital multimeter. Reproduced with permission of Fluke Corporation.

Solution

(a) 399.9 V;
(b) 3999 Ω;
(c) 39.99 mA.

16.7 Oscilloscope

After the multimeter, the oscilloscope is the most commonly used item of electronic test equipment. Be it the electronics industry or a research laboratory, the oscilloscope is an indispensable test and measurement tool for an electronics engineer or technician. Most of us regard the oscilloscope as an item of equipment that is used to see pulsed or repetitive waveforms. However, very few of us are familiar with the actual use of the multiplicity of front-panel controls on the oscilloscope and the potential that lies behind the operation of each one of these controls.

With the arrival of the digital storage oscilloscope (DSO), the functional potential of oscilloscopes has greatly increased. The digital storage oscilloscope enjoys a number of advantages over its analogue counterpart.

16.7.1 Importance of Specifications and Front-Panel Controls

It is very important to have a clear understanding of the performance specifications of oscilloscopes. The specification sheet supplied by the manufacturer contains scores of specifications. Each one of them is important in its own right and should not be ignored. Although some of them explain only the broad features of the equipment and do not play a significant role as far as measurements are concerned, these are important when it is required to choose one for a given application. In fact, the performance specifications of an oscilloscope and the operational features of its front-panel controls cannot be considered in isolation. One complements the other. Not only does the correct interpretation of specifications help in the selection of the right equipment for an intended application, their appreciation is almost a prerequisite to a proper understanding of the functional potential of front-panel controls.

16.7.2 Types of Oscilloscope

Technology is often the single most important criterion forming the basis of oscilloscope classification. Different types of oscilloscope include analogue oscilloscopes, CRT storage type analogue oscilloscopes, digital storage oscilloscopes and sampling oscilloscopes. Digital storage oscilloscopes and sampling oscilloscopes are often clubbed together under digital oscilloscopes. Analogue oscilloscopes are briefly described in the following paragraphs. This is followed up by a detailed description of digital oscilloscopes.

16.8 Analogue Oscilloscopes

The analogue oscilloscope displays the signal directly and enables us to see the waveform shape in real time. The signal update rate in an analogue oscilloscope is the fastest possible as there is only the beam retrace timing and the trigger rearm between two successive sweeps. Consequently, an analogue oscilloscope has a much higher probability of capturing the desired event than any other type of oscilloscope. Analogue oscilloscopes find wide application for viewing both repetitive and single-shot events up to a bandwidth of about 500 MHz. Analogue oscilloscopes do not give a desirable display when viewing very low-frequency repetitive signals or single-shot events. In such cases, the display is nothing but a bright dot moving slowly across the screen to trace the waveform. Such waveforms are not at all convenient to analyse and need some kind of photographic memory.

16.9 CRT Storage Type Analogue Oscilloscopes

A CRT storage type analogue oscilloscope overcomes this problem by using a special type of CRT. In one such type, the phosphor dots have higher persistence. As a result, the moving dot leaves behind a visible trail as it sweeps across the screen, even at much lower sweep speeds. There are two main types of storage mode currently in use for these oscilloscopes: the bistable storage mode, which is capable of storing signals for many hours, and the more popular variable-persistence storage mode, which can store signals for a maximum of 10 min. The majority of commercially available CRT storage oscilloscopes have the option of both the above-mentioned storage modes.

The CRT storage type oscilloscope is an excellent choice for slowly changing signals. As the writing rate is faster than that of the conventional analogue oscilloscopes, it is extremely good for viewing fast transient events. It can be used to store both repetitive and single-shot signals having a bandwidth of up to 500 MHz or so. Oscilloscope type 7934 from Tektronix, for instance, has a bandwidth of 500 MHz and a maximum writing speed of 4000 cm/ms. Even handheld versions of these scopes with a reasonably good writing speed are available. Analogue storage oscilloscope technology is fast being replaced by digital storage oscilloscope technology owing to the far superior performance features of the latter.

16.10 Digital Oscilloscopes

In a digital oscilloscope, the signal to be viewed is firstly digitized inside the scope using a fast A/D converter. The digitized signal is stored in a high-speed semiconductor memory to be subsequently retrieved from the memory and displayed on the oscilloscope screen. There are two digitizing techniques, namely real-time sampling and equivalent-time sampling. The digital storage oscilloscopes (DSOs) use real-time sampling, as shown in Fig. 16.11, so that they can capture both repetitive and

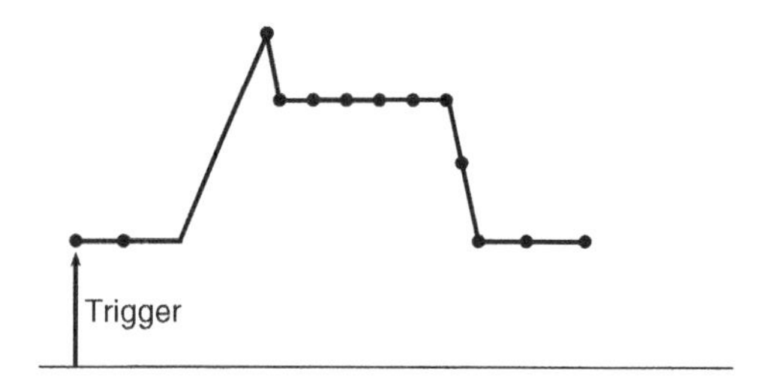

Figure 16.11 Real-time sampling.

single-shot signals. In digital storage oscilloscopes, the digitizer samples the entire input waveform with a single trigger. Sampling oscilloscopes use equivalent-time sampling and are limited to capturing repetitive signals. Some digital storage oscilloscopes also use equivalent-time sampling to extend their useful frequency range for capturing repetitive signals. The equivalent-time sampling technique is thus applicable to only stable repetitive signals and can be implemented in at least three different ways, namely sequential single-sample, sequential sweep and random interleaved sampling (RIS).

In the *sequential single-sample technique* (Fig. 16.12), the digitizer acquires a single sample with the first trigger pulse. It then waits for the second trigger, and, on receipt of the second trigger, a time delay equal to the reciprocal of the desired sampling rate is executed and then the second sample is acquired. The trigger-to-acquisition delay is incremented by the desired intersample period Δt for each subsequent acquisition. The resulting capture has thus an equivalent sample rate of $1/\Delta t$. Clearly, this method is slow, as N trigger cycles would be needed to gather N samples, and the scopes using this type of digitizing technique cannot provide real-time operation.

In *sequential sweep equivalent-time sampling* (Fig. 16.13), a sweep of samples spanning the desired display time range is acquired for each trigger. Here, N samples are acquired in M trigger cycles, where $N = kM$. On receipt of each trigger, k sequential samples are acquired at sample rate f_s. These are stored in every Mth location of the acquisition memory allocated for N samples. k samples of the first sweep are acquired directly on receipt of the trigger. Subsequent sweeps have an increasing delay between trigger receipt and sweep initiation, with the delay increment being equal to $1/Mf_s$ with reference to trigger detection in order to give an apparent sample rate of Mf_s.

The *random interleaved sampling* (RIS) technique uses a memory distribution scheme that is philosophically similar to that of sequential sweep equivalent-time sampling, with the difference that the samples are random with respect to the trigger. Sampling in this case occurs on both sides of

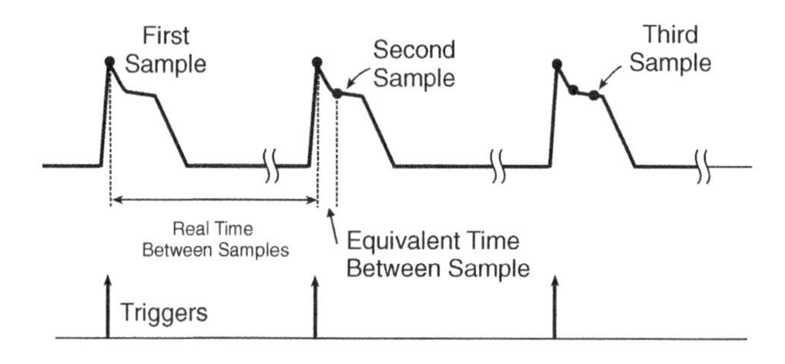

Figure 16.12 Sequential single-sample technique.

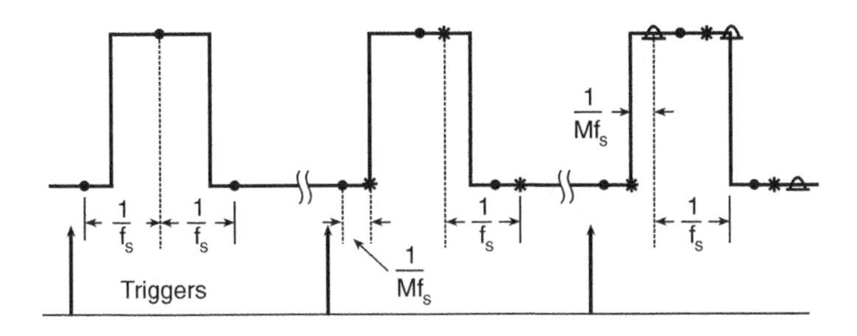

Figure 16.13 Sequential sweep equivalent-time sampling.

trigger points, which gives this technique a 'pretrigger view' capability not available in the first two equivalent-time sampling techniques, as both methods gather signals only following the receipt of a trigger.

If we wanted to view a 1 GHz signal, the sweep speed requirement would be enormous. Even if we were successful in achieving this high speed, the beam would be almost invisible. We have often noticed that, as the time-base setting is made faster, we are forced to adjust the intensity control to maintain an acceptable intensity level setting. Another major problem in designing a real-time oscilloscope for viewing very high-frequency signals (in the GHz range) is the difficulty in building such a high bandwidth in the vertical amplifier. A sampling oscilloscope using any of the equivalent-time sampling techniques outlined above is an answer to all these problems. In such scopes it is not imperative to take a sample or a group of samples from each cycle of the signal to be viewed. The next adjacent sample or group of samples may be 10 000 cycles away. As a result, the bandwidth of the vertical amplifier can afford to be much lower than the frequency of the signal.

Another type of sampling oscilloscope, although not very common in use, is the analogue sampling oscilloscope, where a conventional sample/hold circuit consisting of an electronic switch and a capacitor is used for signal acquisition (Fig. 16.14). It can be used to view high-frequency repetitive signals in nonstorage mode, unlike the digital sampling scopes where the signal is sampled digitally and then stored in semiconductor memory for subsequent retrieval. It can also be used for viewing high-frequency repetitive signals in storage mode, although not in real time (Fig. 16.15).

Digital storage oscilloscopes are also available in a large variety of sizes, shapes, performance features and specifications. Battery-operated, handheld digital storage oscilloscopes with a bandwidth as high as 200 MHz are common (Fig. 16.16). The digital phosphor oscilloscope (DPO) is a big step forward in DSO technology. It captures, stores, displays and analyses, in real time, three dimensions of signal information, i.e. amplitude, time and distribution of amplitude over time. This third dimension

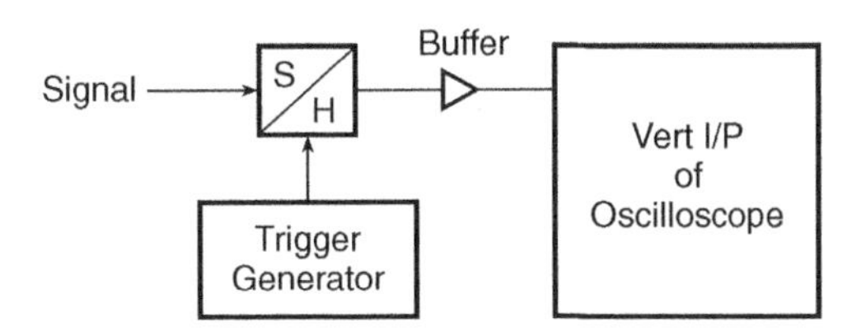

Figure 16.14 Analogue sampling oscilloscope (nonstorage mode). Reproduced with permission of Fluke Corporation.

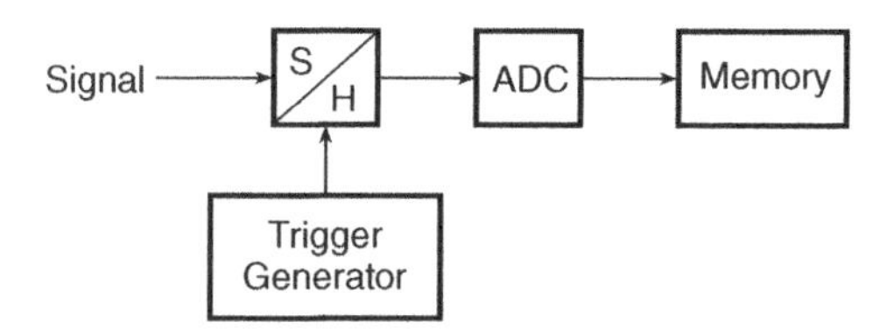

Figure 16.15 Analogue sampling oscilloscope (storage mode). Reproduced with permission of Fluke Corporation.

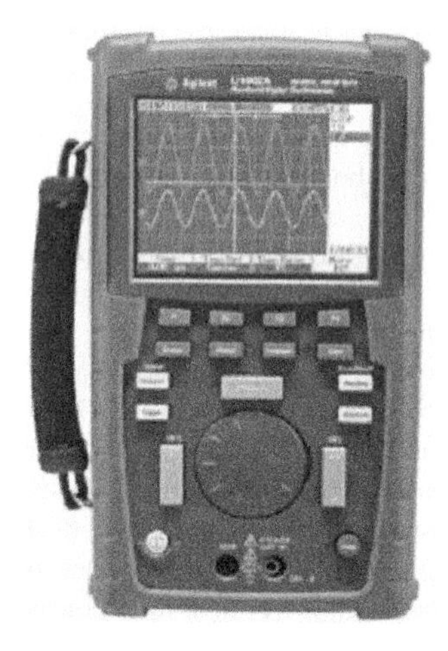

Figure 16.16 Handheld digital storage oscilloscopes. Reproduced with permission of Fluke Corporation.

offers the advantage of interpretation of signal dynamics, including instantaneous changes and the frequency of occurrence displayed in the form of quantitative intensity information.

16.11 Analogue Versus Digital Oscilloscopes

Almost all oscilloscopes available today use one or a combination of the technologies discussed above. Each technology has its own benefits and shortcomings. While signal manipulation and its consequent benefits are the strong point of the digital technology, extremely fast update rates coupled with low cost is a feature associated with analogue scopes. In fact, many state-of-the-art oscilloscopes are not simply analogue or digital. They offer advantages of both technologies.

16.12 Oscilloscope Specifications

Although oscilloscopes are characterized by scores of performance specifications, not all of them are important. Important specifications of analogue and digital oscilloscopes are briefly described in the following paragraphs.

16.12.1 Analogue Oscilloscopes

Key specifications include bandwidth (or rise time), vertical sensitivity and accuracy. Other features such as triggering capabilities, display modes, sweep speeds, etc., are secondary in nature.

16.12.1.1 Bandwidth and Rise Time

The bandwidth and rise time specifications of an oscilloscope are related to one another. Each can be calculated from the other. Bandwidth (in MHz) = 350/rise time (in ns).

Bandwidth is the most important specification of any oscilloscope. It gives us a fairly good indication of the signal frequency range that can be viewed on the oscilloscope with an acceptable accuracy. If we try to view a signal with a bandwidth equal to the bandwidth of the oscilloscope, the measurement error may be as large as 40 % (Fig. 16.17). As a rule, the oscilloscope bandwidth should be 3–5 times the highest frequency one is likely to encounter in order to keep the measurement error to less than 5 %.

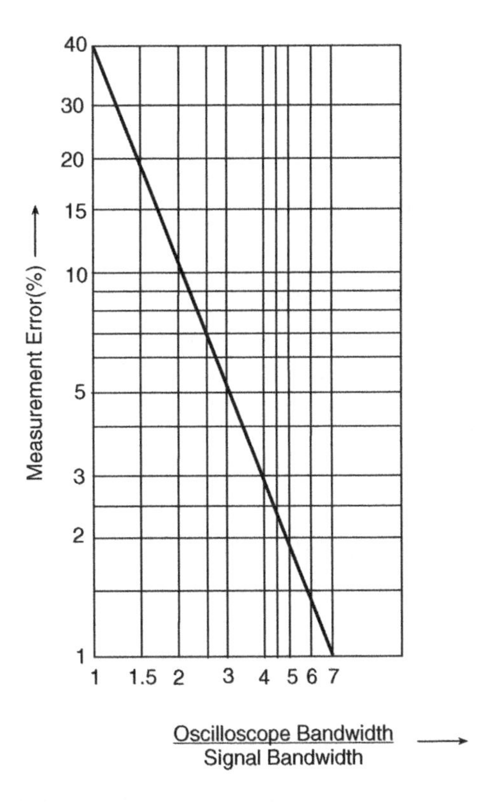

Figure 16.17 Measurement error as a function of oscilloscope bandwidth.

16.12.1.2 Vertical sensitivity

The vertical sensitivity specification tells us about the minimum signal amplitude that can fill the oscilloscope screen in the vertical direction. A 5 mV/div sensitivity is quite common. Oscilloscopes with a sensitivity specification of 1 mV/div are also available. Sensitivity and bandwidth are often trade-offs. Although a higher bandwidth enables us to capture high-frequency signals, there is a good possibility of unwanted high-frequency noise being captured if the oscilloscope has a higher sensitivity too. That is why most of the high-sensitivity scopes have bandwidth limit controls to enable a clear view of low-level signals of moderate frequencies. It is also important that the oscilloscope we choose has an adequate V/div range to make possible a full-screen or near-full-screen display for a wide range of signal amplitudes.

16.12.1.3 Accuracy

The accuracy specification indicates the degree to which our measurement conforms to a true and accepted standard value. An accuracy of $\pm 1 - 3\%$ is typical. Almost all oscilloscopes are provided with a $\times 5$ magnification in the V/div selector switch. This alters the nominal vertical deflection scale from say 5 mV/div–5 V/div to 1 mV/div–1 V/div. It may be mentioned here that the accuracy suffers with the magnifier pull. Most of the manufacturers list accuracy specifications separately for the two cases for the oscilloscopes manufactured by them.

16.12.2 Analogue Storage Oscilloscope

With the CRT storage-type oscilloscope, the stored *writing speed* is usually the main criterion for choosing the instrument. The speed of a CRT storage scope depends on the speed of the input signal (signal frequency) and the size of the trace it draws.

16.12.3 Digital Storage Oscilloscope

Just like an analogue scope, the specification sheet of a digital oscilloscope contains scores of specifications that at first sight may appear quite confusing. A closer look at these specifications, particularly the decisive ones, will make one appreciate the performance capabilities of digital oscilloscopes. The real strength of a digital oscilloscope lies in the following specifications: bandwidth, sampling rate, vertical resolution, accuracy and acquisition memory.

16.12.3.1 Bandwidth and Sampling Rate

The *bandwidth* is an important specification of digital oscilloscopes, just as it is for analogue oscilloscopes. The bandwidth, which is primarily determined by the frequency response of input amplifiers and filters, must exceed the bandwidth of the signal if the sharp edges and peaks are to be accurately recorded.

The *sampling rate* is another vital digital scope specification. In fact, the sampling rate determines the true usable bandwidth of the scope. While the bandwidth is associated with the analogue front end of the scope (amplifiers, filters, etc.) and is specified in Hz, the sampling rate is associated with the digitizing process and, if it is not adequate, degrades the bandwidth. A clear understanding of sample rate specification is thus important when it comes to establishing the adequacy of a particular sample rate to achieve a given bandwidth. Digital oscilloscope specification sheets often contain two sample rates, one for single-shot events and the other for repetitive signals. In some cases, both repetitive and single-shot events are sampled at the same rate, although the bandwidth capability of the oscilloscope for the two cases is different. It is lower in the case of single-shot events.

Theoretically, the Nyquist criterion holds true, and this criterion states that at least two samples must be taken for each cycle of the highest input frequency. In other words, the highest input frequency (also called the Nyquist frequency) cannot exceed half the sample rate. Given this condition, a sinx/x interpolation algorithm can exactly reproduce a digitized signal. An interpolation algorithm is the mathematical function used by an oscilloscope to join two successive sample points while reconstructing the signal. The sinx/x interpolation has a tendency to amplify noise in the signal, particularly when each cycle is sampled only twice. With (sinx/x) interpolation, four samples per cycle are found to be quite adequate. The additional sample points effectively enhance the signal-to-noise ratio for sinx/x interpolation. With straight-line interpolation, at least ten samples are required per cycle for good results.

For repetitive signals, however, even a smaller sample rate does the job, as explained in the case of sampling oscilloscopes. Thus, it becomes important to look into the sample rate specification together with the interpolation algorithm used. For instance, in a digital storage oscilloscope with a single-shot sample rate of 400 MS/s (where MS stands for megasamples), using the sinx/x interpolation technique can give us a single-shot bandwidth of 100 MHz, while the same sample rate will provide a bandwidth of only 40 MHz if a straight-line interpolation algorithm is used instead. Thus, the single-shot bandwidth capability of a digital storage oscilloscope must always be gauged by its single-shot sample rate. The sample rate in samples per second should be at least twice the highest frequency component or 4 times the highest frequency component for good results, or anywhere between 2 and 4, assuming sinx/x interpolation. For repetitive signals, if it is not a real-time DSO, the sampling rate could be smaller.

16.12.3.2 Memory Length

Memory length is a vital digital oscilloscope specification and should not be considered to be an insignificant one. Not only does it affect the sample rate and consequently the single-shot bandwidth, longer memories also have many more peripheral benefits. The sample rate as quoted by the manufacturer always refers to the maximum digitizing rate attainable in single-shot mode. Interestingly, the quoted sample rate figure does not hold true for the entire range of time-base settings. For a given memory length, the attainable sample rate is observed to decrease as the time base is made slower. Some manufacturers offer record length, which is nothing but the size of the memory used while displaying the signal. Suppose a particular DSO has a memory length of 1K and a quoted sample rate specification of 100 MS/s. In the limit when the record length equals the memory length, we can store approximately 1000 samples. At the given sample rate, the displayed waveform will cover a time span of 10 ms, i.e. a time-base setting of 1 ms/div, if the waveform is to cover the full screen in the horizontal direction. If the time-base setting is changed to 10 ms/div, the effective sample rate would be limited to only 10 MS/s, thus reducing the single-shot bandwidth. The only method to maintain the sample rate at the quoted value for a larger time-base setting range is to have a longer acquisition memory. The effect of memory length on single-shot bandwidth as a function of time-base setting is expressed by

$$\text{Sample rate} = \text{memorylength}/(10 \times \text{time} - \text{base setting})$$

The '10' is the total number of divisions in the horizontal direction.

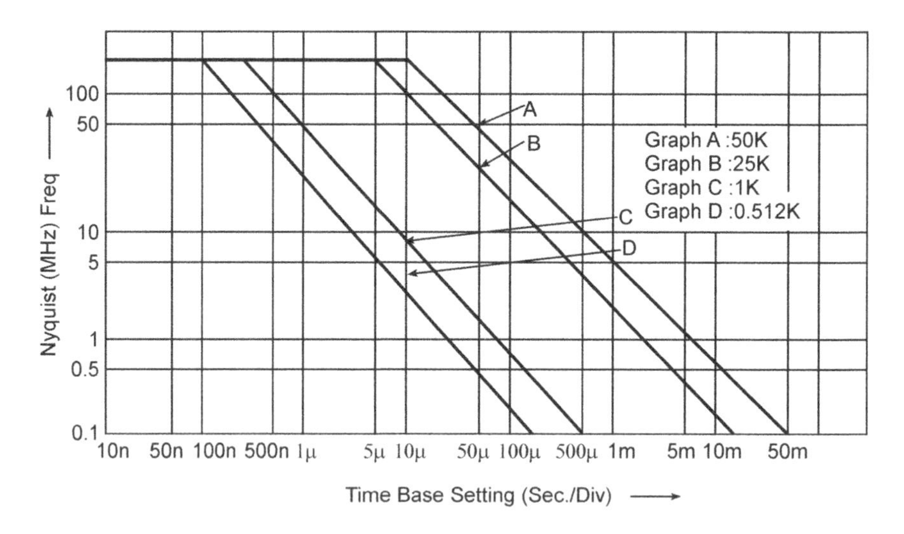

Figure 16.18 Effective sampling rate – time-base setting graph.

Figure 16.18 shows the changes in sample rate as a function of time-base setting for digital oscilloscopes of different memory lengths. Given two oscilloscopes with identical sample rate and single-shot bandwidth specifications, the one with the longer acquisition memory has a decisive edge. Hence, it must be accorded due importance when choosing one to meet your requirements. For a given time resolution, a longer memory enables events of longer duration to be recorded. For instance, a DSO with a 1K memory can record a 1 s transient with a time resolution of 1 ms, whereas a DSO with a 10K memory can record a 10 s long event with the same time resolution. In other words, for the same transient duration, longer memories give enhanced time resolution. Long memories also help in acquiring hard-to-catch signals and also minimize signal reconstruction distortion.

16.12.3.3 Vertical Accuracy and Resolution

The accuracy specification tells us how closely the measurement matches the actual value. The accuracy of a DSO is affected by various sources of error, including gain and offset errors, differential nonlinearity, quantization error and so on. The quantization error indirectly indicates vertical resolution, i.e. uncertainty associated with any reading or the ability of the oscilloscope to see small changes in amplitude measurements. Choosing a scope with fewer than eight bits of resolution is not recommended. Resolution specification must not be considered in isolation from accuracy specification. For instance, more than eight bits of resolution is meaningless when the overall accuracy itself is $\pm 1\%$. An eight-bit resolution gives a $\pm 0.4\%$ uncertainty, which is fairly acceptable if the overall accuracy is $\pm 1\%$, as can be seen from Table 16.3. Also, digital oscilloscopes with more than seven bits of resolution can resolve signal details better than visual measurements made with analogue oscilloscopes.

To sum up our discussion on the available oscilloscope types and the selection criteria for choosing the right one, it can be said that both analogue and digital oscilloscopes have their advantages and shortcomings. The suitability of a particular type must always be viewed in terms of intended application. Although digital oscilloscopes can perform many functions that analogue versions cannot, analogue oscilloscope technology, too, has reached high performance standards. It is important to

Table 16.3 Uncertainty of an oscilloscope as a function of the number of bits.

Number of bits	Uncertainty (%)
6	1.6
7	0.8
8	0.4
9	0.2
10	0.1
11	0.05
12	0.02

understand the critical specifications of each type and then decide whether it fits an intended application. The key specifications to look for in analogue scopes are bandwidth, vertical sensitivity and accuracy, whereas the strength of a digital oscilloscope must be ascertained from its bandwidth, sample rate, vertical resolution, accuracy and memory length.

16.13 Oscilloscope Probes

The oscilloscope probe acts as a kind of interface between the circuit under test and the oscilloscope input. The signal to be viewed on the oscilloscope screen is fed to the vertical input (designated as the *Y* input) of the oscilloscope. An appropriate probe ensures that the circuit under test is not loaded by the input impedance of the oscilloscope vertical amplifier. This input impedance is usually 1 MΩ, in parallel with a capacitance of 10–50 pF. The most commonly used general-purpose probes are the 1X, 10X and 100X probes. These probes respectively provide attenuation by factors of 1 (i.e. no attenuation), 10 and 100. That is, if we are measuring a 10 V signal with a 10X probe, the signal actually being fed to the oscilloscope input will be 1 V. 10X and 100X probes are quite useful for measuring high-amplitude signals. Another significant advantage of using these probes is that the capacitive loading on the circuit under test is drastically reduced.

Refer to the internal circuit of the 10X probe as shown in Fig. 16.19. The RC time constant of the probe equals the input RC time constant of the oscilloscope. Since the resistance of the probe is 9 times the input resistive component of the oscilloscope, in order to provide attenuation by a factor of 10, the probe capacitance has got to be smaller than the input capacitance of the scope by the same amount. As a result, the circuit under test with a 10X probe will never see a capacitance of more than 5 pF.

16.13.1 Probe Compensation

The probe is compensated when its RC time constant equals the RC time constant of the oscilloscope input. With this, what we see on the screen of the scope is what we are trying to measure independent of the frequency of the input signal. If the probe is not properly compensated, the signal will be attenuated more than the attenuation factor of the probe at higher frequencies owing to reduction in the effective input impedance of the vertical input of the scope.

To check for probe compensation, the probe can be used to see the calibration signal (the CAL position on the front panel) available on the oscilloscope. If the probe is properly compensated, the

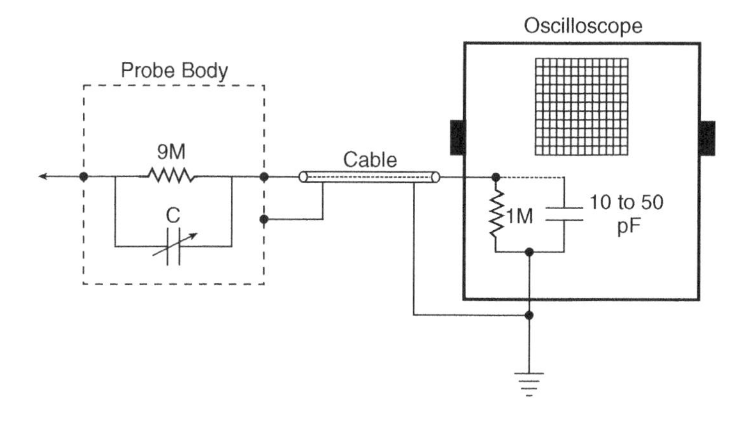

Figure 16.19 Internal circuit of 10X probe.

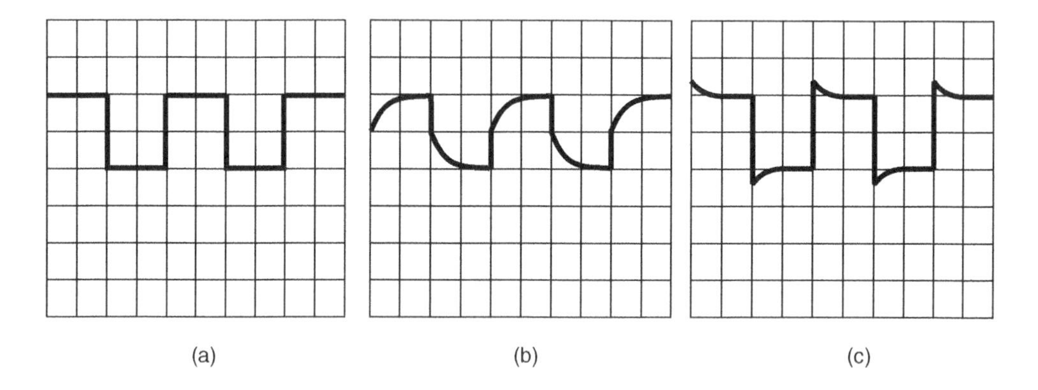

Figure 16.20 Probe compensation.

CAL signal will appear in perfect rectangular shape [Fig. 16.20(a)] with no rounding-off of edges [Fig. 16.20(b)] or any spikes on fast transitions [Fig. 16.20(c)]. Rounding-off of edges indicates too little a probe capacitance, while spikes indicate too large a probe capacitance. The probe capacitance can be adjusted by turning a screw or rotating the probe barrel after loosening the locking nut (in some probes) to get a perfect calibration signal.

16.14 Frequency Counter

The most basic function of a frequency counter is the measurement of an unknown frequency. Modern frequency counters, however, offer much more than just frequency measurement. Other related parameters such as the *time period,* which is the reciprocal of frequency, the *time interval* between two events and the *totalize count*, which is nothing but the cumulative count over a known period, are other functions that are available with present-day frequency counters. These instruments, offering a variety of measurement options, are usually referred to as *universal counters.*

16.14.1 Universal Counters – Functional Modes

The functions available with modern universal counters, other than measurement of an unknown frequency, are time interval measurement, period, time interval average, totalize, frequency ratio A/B, phase A relative to B and pulse width.

16.14.1.1 Time Interval Measurement

This mode measures the time that elapses between the occurrence of two events. One of the events, called the start signal, is usually fed into one of the channels, while the other, called the stop signal, feeds the second channel. The resolution of measurement is typically 10 ns or better. A typical application of this measurement mode is in determination of the propagation delay in logic circuits. Variations of this mode can be used to measure pulse width and rise/fall times.

16.14.1.2 Time Interval Average

This mode can be used to improve the measurement resolution in the time interval measurement mode for a given clock frequency. The resolution improves as the square root of the number of measurements. That is, an average of 100 measurements would give a 10-fold improvement in resolution.

16.14.1.3 Period

In this mode, the time period of the input signal is measured by counting clock pulses between two successive leading or trailing edges of the input signal. Again, the period average function can be used to improve upon the measurement resolution for a given clock. For instance, if the measurement were done for 100 periods instead of one period for a given clock frequency, the measurement resolution would also improve by a factor of 100.

16.14.1.4 Totalize

The totalize mode gives a cumulative count of events over a known time period.

16.14.1.5 Frequency Ratio A/B

This gives the ratio of the frequencies of signals fed to the A and B channels. This feature can be used to test the performance of prescalers and frequency multipliers.

16.14.1.6 Phase A Relative to B

This compares the phase delay between signals with similar frequencies.

16.14.2 Basic Counter Architecture

Figure 16.21 shows the architecture of a frequency counter when it is being used in the frequency measurement mode. The oscillator section, comprising a crystal-based oscillator and a frequency divider

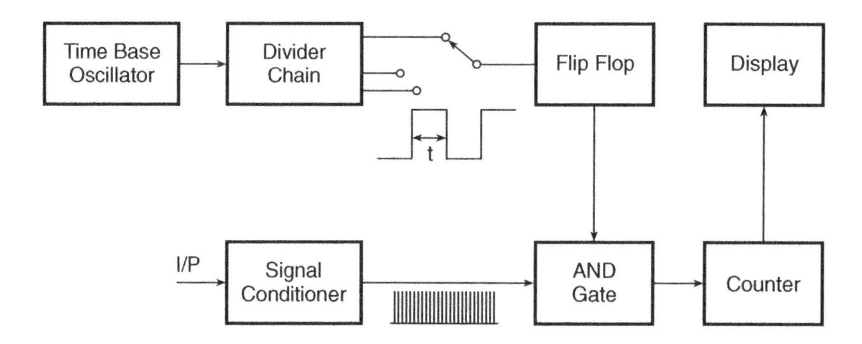

Figure 16.21 Counter architecture. Reproduced with permission of Fluke Corporation.

chain, generates the clock pulses. The clock pulses are used to trigger a flip-flop whose output serves to enable or disable the AND gate. When the AND gate is enabled, the input signal, after passing through the signal conditioning section comprising level shifting amplifiers, comparators, etc., reaches the counter. In the simplest case, if the AND gate is enabled for 1 s (which is the case when the flip-flop clock input is 1 Hz), then the counter count will represent the signal frequency. The measurement resolution in this case would be 1 Hz. The measurement resolution can be improved by enabling the AND gate for a longer time. For instance, a 0.1 Hz clock at the flip-flop input would give a 10 s gate time and a consequent 0.1 Hz resolution. Similarly, a shorter measurement for a gate time of 0.1 s (corresponding to a clock of 10 Hz) gives a measurement resolution of 10 Hz.

The same building blocks, when slightly rearranged as shown in Fig. 16.22, can be used to measure the time period. Enabling and disabling of the AND gate are now determined by the frequency of the input signal and not by the clock frequency. The number stored in the counter here is proportional to the number of clock pulses that reach the counter during the period of the input signal. The same set-up can be used for time interval (TI) measurement by having two input signal channels, with one enabling the AND gate by, say, setting the flip-flop and the other disabling the same by resetting the flip-flop.

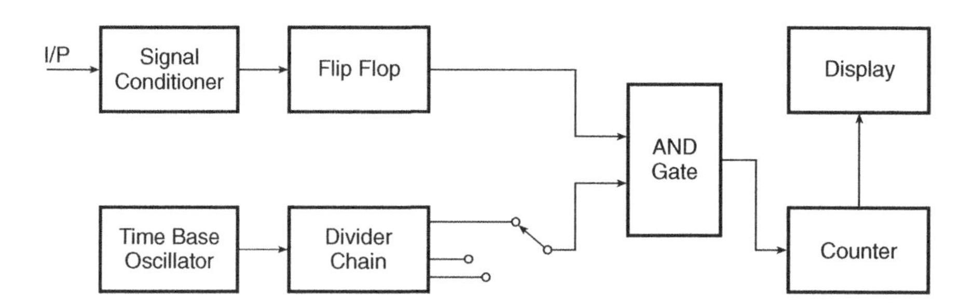

Figure 16.22 Time period measurement using a frequency counter.

16.14.3 Reciprocal Counters

The reciprocal counter overcomes some of the limitations of the basic counter architecture described in the previous paragraphs. Most important of all is its inadequate resolution, particularly when measuring low frequencies. The basic counter had a resolution of 1 Hz for a gate time of 1 s and the resolution could be enhanced only by increasing the gate time. If the gate time cannot be increased in a certain application, the resolution is restricted to 1 Hz. The basic counter measures frequency in terms of event count. Depending upon the gate time, which is 1 s or any other decade value such as 10, 100, etc., the decimal point appropriately placed in the count gives frequency. What is important to note here is that computation of frequency involves computation of the event count only. The frequency, which is given by the event count divided by the time taken, is calculable from the event count itself if the time is 1 s, 10 s, 100 s, etc.

In a reciprocal counter, both events as well as time are computed and the ratio of the two gives the frequency. The advent of the reciprocal counter was made possible owing to the availability of digital logic that could perform arithmetic division economically and with precision. Figure 16.23 shows the reciprocal counter hardware. The processor is the heart of the counter hardware and controls almost every other building block. The synchronizing and routing logic block routes the A and B channel inputs and the time-base signal to the *event* and *time* counters. The routing is determined by the measurement function. The computations are done in the processor block.

As a matter of comparison, let us see how the two counters having an internal clock of 10 MHz would respond to measurement of a signal frequency of 50.38752 Hz. The basic counter will display 50 Hz, assuming a gate time of 1 s as the event count will be 50. The reciprocal counter will also have an event count of 50 but it will also measure time with a resolution of 100 ns (for a 10 MHz clock), equal to 0.9923328 s. The measured frequency will therefore be 50.38752 Hz. The frequency resolution offered for a 10 MHz clock is seven digits, equal to 0.000005 Hz in the present case for a 1 s gate. The resolution could be further enhanced by increasing the clock frequency. Since clock frequencies of up to 500 MHz are practical, a reciprocal counter would give a resolution of 2 ns for a 1 s gate time.

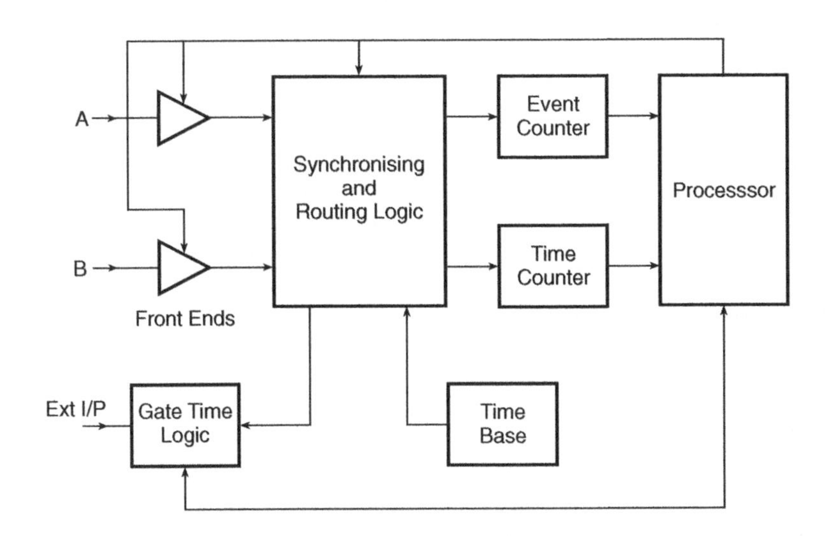

Figure 16.23 Reciprocal counter architecture.

The resolution of reciprocal counters can be further enhanced by using a technique called interpolation. It is possible to achieve a nine-digit resolution with a 10 MHz clock using interpolation techniques that otherwise would require a clock of 1 GHz. This is particularly important when we are looking for a given resolution in shorter gate times. The details of interpolation techniques are beyond the scope of this text.

16.14.4 Continuous-Count Counters

The counter architectures discussed in the previous paragraphs had a counter that counted for a known period equal to the gate time. These counters have a dead time when the gate is disabled. Such counters could miss vital information that could be important to the measurement. The continuous-count counter architecture is based on the fact that, if different measurements of a certain parameter of a signal were not disjoint and the relationship that they had were made use of, the measurement resolution could be significantly enhanced by applying what we call curve-fitting algorithms. These counters have all the attributes of reciprocal counters, with the additional ability of reading the event, the time and the counter without having to disable the gates.

16.14.5 Counter Specifications

The data sheets and manuals of universal counters contain detailed specifications of the instrument. The important ones include *sensitivity*, *bandwidth*, *resolution*, *accuracy* and *throughput*.

16.14.5.1 Sensitivity

This refers to the smallest signal that the instrument can measure and is usually expressed as mV (RMS) or peak-to-peak. A sensitivity of 10–20 mV (rms) is typical. In the majority of measurement situations, sensitivity is not the issue.

16.14.5.2 Bandwidth

The bandwidth of the counter is its front-end bandwidth and is not necessarily the same as the maximum frequency that the counter is capable of measuring. Measuring a signal frequency higher than the instrument's bandwidth only reduces its sensitivity specification and requires a larger minimum input signal. However, the bandwidth does affect the measurement accuracy in the case of some parameters. Rise time is one such parameter. Thus, it is always preferable to choose a counter with as high a bandwidth as possible. Bandwidth is not explicitly mentioned in the specifications. However, it can be estimated by looking at variation in sensitivity across the frequency range of the instrument.

16.14.5.3 Resolution

Resolution refers to the minimum resolvable frequency increment (in the case of frequency measurement) and time increment (in the case of time interval measurement). The resolution is usually very close to the least significant digit and is often ±1 count or LSD. Noise in the input signal, noise in the front end and input signal slew rate are some of the factors that affect resolution.

16.14.5.4 Accuracy

Accuracy is related to resolution but is not the same as resolution. Factors such as time-base (or clock) accuracy and trigger accuracy must be considered along with the resolution specification to determine the ultimate accuracy of frequency measurement. Time-base error affects measurement accuracy as follows:

$$\text{Frequency accuracy} = \text{resolution} \pm \text{time base error} \times \text{frequency}$$

Trigger level accuracy is the precision with which the trigger level can be set. If there is an error in the trigger level setting, the trigger timing is changed, thus affecting measurement accuracy.

16.14.5.5 Throughput

Throughput is related to resolution. For instance, increasing the gate time of a certain frequency measurement increases the measurement resolution by the same factor, but it slows down the throughput by almost the same amount. Other factors affecting the throughput are more related to the speed of the microprocessor and the interface system. Two factors to be watched here are the number of measurements the counter can deliver through the interface and the speed with which the counter can switch between different functions or set-ups. If short gate times are being used and/or measurements are being switched between different functions repeatedly, these factors become important.

16.14.6 Microwave Counters

The counter architectures discussed in the preceding paragraphs (conventional, reciprocal, continuous count) are usually good enough up to 500 MHz or so. Counters meant for carrying out measurements at RF frequencies beyond 500 MHz and microwave frequencies employ a different architecture. There are two types of architecture in use for building microwave counters. One uses a prescaler while the other is based on down-conversion.

Prescaler counters use a prescaler placed between the front end and the gating circuitry of the counter. In fact, prescalers are available inside the counters as an optional channel to extend the frequency range of measurement. Extension up to 3 GHz is typically available with a prescaler. Prescalers are not used with pulsed microwave counters owing to their tendency to self-oscillation. When used with a basic counter, a prescaler causes degradation of resolution. This is because the frequency resolution of a basic counter is dependent upon the contents of the event counter and, owing to the location of the prescaler before the gating circuitry, its contents cannot be read. The resolution is not affected when the same is used in a reciprocal counter.

In a microwave counter based on down-conversion architecture, the input signal frequency is down-converted to produce an intermediate frequency (IF). The IF, which is the difference between the input signal frequency and the local oscillator (LO) frequency, is then counted. The actual frequency is then computed from LO + IF. Covering a frequency range of tens of GHz for an LO is an expensive proposition. The solution is to use a relatively lower-frequency LO (approximately 200 MHz). The LO drives a step recovery diode that produces a sharp pulse with usable harmonics up to the desired range. This pulse drives a sampler which samples points of the input signal. The resulting IF is low-pass filtered and counted. The actual input frequency is then given by $N \times \text{LO} + \text{IF}$, where N is the harmonic of the LO that goes through the mixing operation. One of the methods for determining N is to measure the IF at two slightly different LO frequencies. N is then given by (IF1–IF2) / (LO2–LO1). However, all this is the instrument's headache and may take several tens of milliseconds only. Figure 16.24

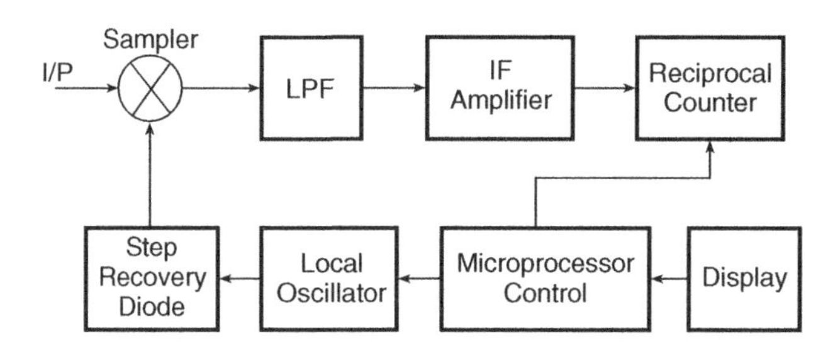

Figure 16.24 Microwave counter architecture.

shows the architecture of a microwave counter. Pulse microwave counters use similar architecture with additional gating circuitry to position the gate comfortably within the pulse.

16.15 Frequency Synthesizers and Synthesized Function/Signal Generators

Frequency synthesizers generate sinusoidal signals of extremely high frequency stability and exceptional output level accuracy. Frequency synthesizers and similar instruments such as synthesized function/signal generators are used to provide test signals for characterization of devices, subsystems and systems. Synthesized function generators, in addition to providing spectrally pure and accurate CW sinusoidal signals, also provide other waveforms such as ramp, triangle, square and pulse. Synthesized signal generators, in addition to providing spectrally pure and accurate CW signals, also have modulation capability and can be used to generate AM, FM, PM and pulse-modulated signals. There is another class of synthesized function generators called synthesized arbitrary waveform generators. The majority of synthesized function generators have a limited arbitrary waveform generation capability built into them. However, these are available as individual instruments also. All the above-mentioned instruments have one thing in common, that is, the synthesis of a signal that lends ultrahigh frequency stability and amplitude accuracy to the generated waveform. They therefore have more or less similar architecture for a given technique used for frequency synthesis.

16.15.1 Direct Frequency Synthesis

The frequency synthesizer in its basic form uses a reference oscillator, which is an ultrastable crystal oscillator, and other signal-processing circuits to multiply the oscillator frequency by a fraction M/N (where M and N are integers) in order to generate the desired output frequency. One such arrangement is shown in Fig. 16.25. It comprises an assortment of frequency multipliers and dividers, mixers and band-pass filters (BPFs). The diagram shows the use of this architecture to generate 17 MHz. In this arrangement, if the BPF has a pass band centred around 3 MHz, the output will be 3 MHz as the mixer produces both sum and difference components. This method of frequency synthesis has several disadvantages, not least that the technique is highly hardware intensive and therefore expensive. Another disadvantage is loss of phase continuity while switching frequencies, with the result that this technique has not found favour with designers.

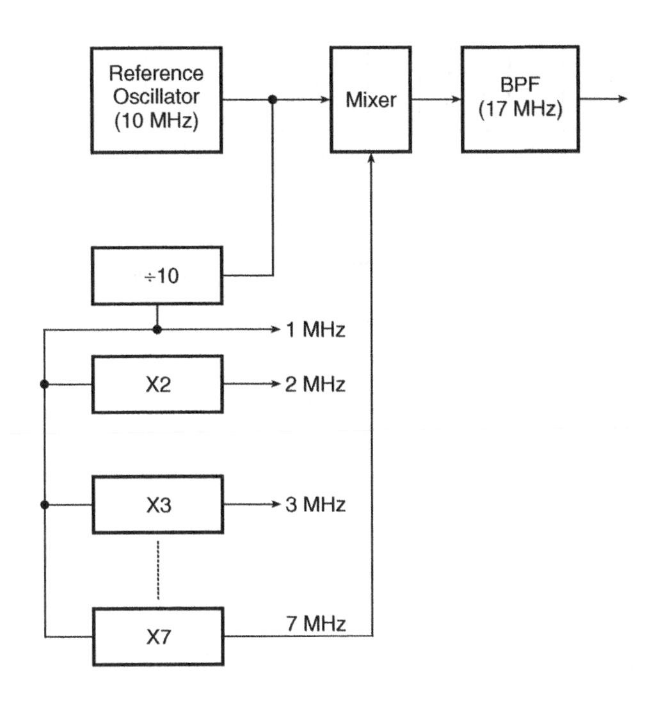

Figure 16.25 Frequency synthesizer architecture – direct frequency synthesis.

16.15.2 Indirect Synthesis

In indirect synthesis, the output is not directly derived from the quartz crystal based reference oscillator. Instead, the reference oscillator is used in a phase-locked loop wired as a frequency multiplier to generate an output frequency that is M/N times the reference oscillator frequency. The output is taken from the VCO of the phase-locked loop. Figure 16.26 shows the basic arrangement. If we insert a divide-by-N circuit between the reference oscillator and the phase detector signal input and a divide-by-M circuit between the VCO output and the phase detector VCO input, then the loop will lock with the VCO output as $f_{ref} \times (M/N)$. The frequency resolution of this architecture is f_{ref}/N, where f_{ref} is the frequency of the reference oscillator. The loop frequency switching speed is of the order of 10 times the period of reference frequency input to the loop phase detector. That is, if we desired a frequency resolution of 1 Hz, the switching time would be of the order of 10 s, which is highly unacceptable. Another disadvantage of this architecture is that frequency multiplier loops also multiply noise at the phase detector, which manifests itself in the form of noise sidebands at the VCO output. This restricts the maximum multiplication factor to a few thousands in this arrangement, which limits the resolution. If a finer resolution is needed, sequences of multiplication, division and addition are used that involve more than one phase-locked loop. One such arrangement is shown in Fig. 16.27. The synthesizer output in this case is given by $f_{ref} \times [m/(N_1 \times N_2) + 1]$.

This technique can be extended to get any desired resolution. Since the multiplication numbers are low and the loop frequency is high, the output will have low noise sidebands. Also, the synthesizer is capable of fast frequency switching. Another popular method of indirect synthesis is fractional N synthesis, where a single PLL is made to lock to the noninteger multiple of the loop reference. This

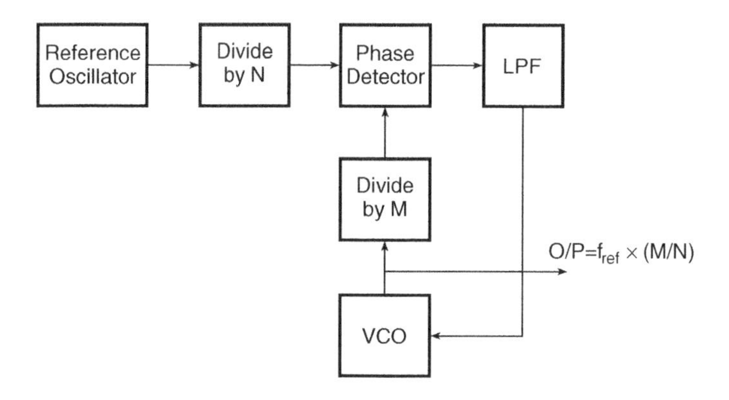

Figure 16.26 Frequency synthesizer architecture – indirect synthesis.

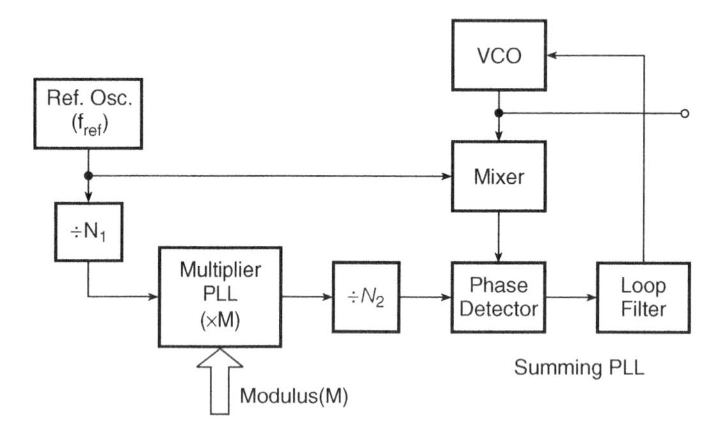

Figure 16.27 Indirect synthesis using more than one PLL.

technique can be used to achieve a frequency resolution of microhertz order at switching speeds of the order of a millisecond or so. Figure 16.28 shows the basic architecture. The configuration functions as follows.

The integer part of the desired multiplier is supplied to the digital divider placed between the VCO output and the phase detector in the form of its dividing factor. The fractional part is supplied to the accumulator. The accumulator is clocked by the reference source derived from the crystal oscillator. The quantum of fractional input is added to the accumulator contents every clock cycle. The VCO output is $N \times F$ times the reference input when the loop is locked. The circuit functions in such a way that the contents of the accumulator predict the expected phase detector output resulting from the frequency difference of the two phase detector input signals. The D/A converter is then so scaled and polarized that its output waveform cancels the phase detector output waveform. The two waveforms are added in the analogue adder, sampled and filtered to provide the oscillator control voltage. Also, to keep the phase detector output within its linear range, whenever the phase difference between the

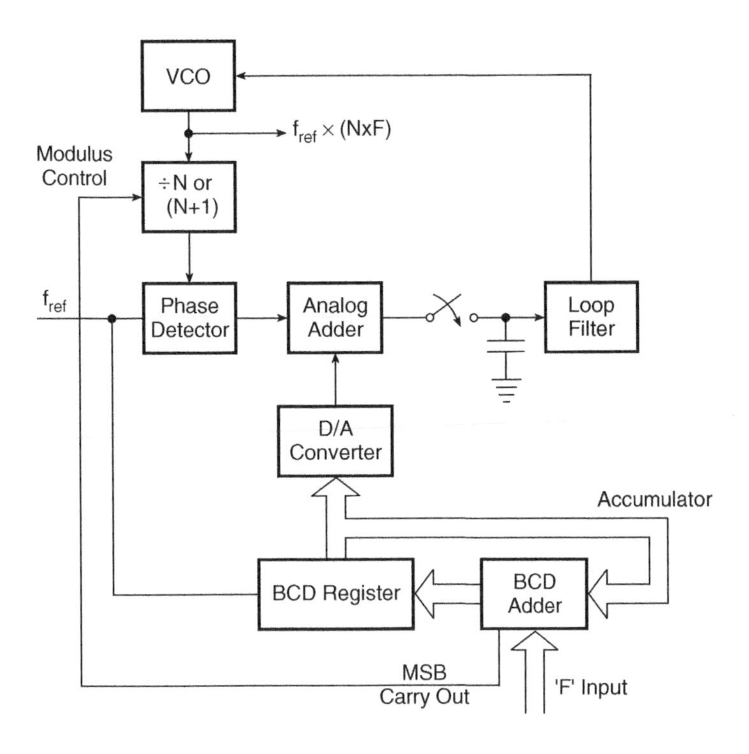

Figure 16.28 Fractional N synthesis.

two inputs to the phase detector tends to become 360° , which is the maximum the phase detector can tolerate without going out of range, the phase of the divider output (which is ahead of the reference input in phase) is retarded by 360° by either changing the divider modulus to $N+1$ momentarily or by any other means. In the architecture shown, the modulus is changed to accomplish this on receiving a command from the BCD adder at the time of accumulator overflow.

16.15.3 Sampled Sine Synthesis (Direct Digital Synthesis)

This method of frequency synthesis is based on generating the waveform of desired frequency by first producing the samples as they would look if the desired waveform were sampled or digitized according to the Nyquist sampling theorem, and then interpolating among these samples to construct the waveform. As the frequency is the rate of change in phase, this information is made use of to generate samples. The sine of different phase values is stored in a memory, which is addressed by phase increment information stored in an accumulator. Figure 16.29 shows a simplified block schematic representation of direct digital synthesis. When the accumulator is clocked at a fixed frequency, the contents of the accumulator jump by the phase increment whose digital equivalent information is stored in the phase increment register (PIR). By changing the contents of the PIR, the output frequency can be changed. The rate at which the look-up table in the memory is addressed is given by the clock frequency and phase increment during one clock period as given by the PIR contents. For instance, if the contents of the PIR represented a phase angle of 36°, then the digital samples present at the output

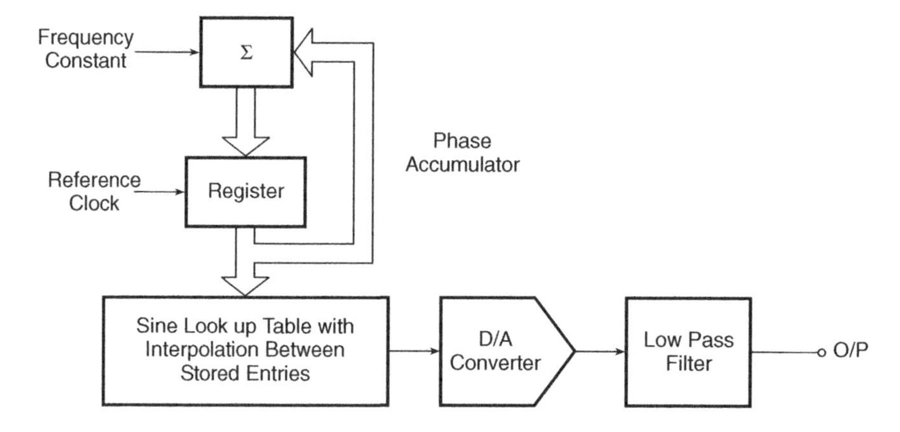

Figure 16.29 Direct digital synthesis.

of the memory would correspond to phase differences of 36, 72, 108, 144, 180, 216, 252, 288, 324 and 360° to complete one cycle of output waveform. The 10 samples will be produced in 10 clock cycles. Therefore, the output frequency will be one-tenth of the clock frequency. In general, the output frequency is given by

$$[\phi/2\pi] \times f_{clock} \tag{16.1}$$

where ϕ is the phase increment in radians.

The digital samples are converted into their analogue counterparts in a D/A converter and then interpolated to construct the waveform. The interpolator here is a low-pass filter. Relevant waveforms are shown in Fig. 16.30.

This method of synthesis derives its accuracy from the fact that both the phase increment information and the time in which the phase increment occurs can be computed to a very high degree of accuracy. With the frequency being equal to the rate of change in phase, the resulting waveform is highly

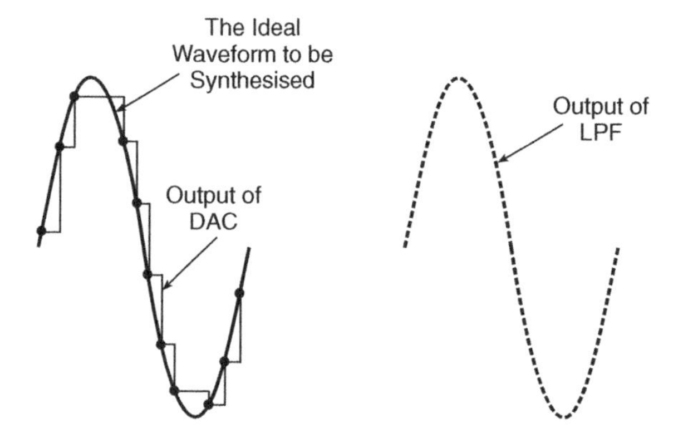

Figure 16.30 Direct digital synthesis – relevant waveforms.

stable. The most important feature of this technique, however, is its capability to provide instantaneous switching. This is possible because the size of the angle increments between two consecutive table look-ups may be changed instantaneously. The limitations of this technique are the *quantization noise* and *aliasing* inherent in any sampled data system. Another serious disadvantage is the presence of spurious components owing to imperfections and inaccuracies in the D/A converter. The highest frequency that can be synthesized is limited by the maximum speed of the available digital logic. The usable frequency range of the direct digital synthesis output may be extended by a variety of techniques. However, depending upon the technique used, some of the advantages of this technique may be lost. As in the case of more conventional synthesizers, the output of a direct digital synthesizer may be doubled, mixed with other fixed sources or used as a reference inside a PLL.

16.15.4 Important Specifications

Frequency range, resolution, frequency switching speed and signal purity are the important synthesizer specifications.

16.15.4.1 Frequency Range and Resolution

While considering the *frequency range,* it is important to note whether the claimed frequency range is being covered in a single band or a series of contiguous bands. This aspect is significant from the viewpoint of noise performance, which may be different in different bands in cases where the frequency range is covered in more than one band. This often leads to a larger transient when the frequency switching involves switchover of the band also. Frequency resolution is usually the same throughout the range. It is typically 0.1 Hz, although a resolution as fine as 1 mHz is also available in some specific instruments.

16.15.4.2 Frequency Switching Speed

The *frequency switching speed* is a measure of the time required by the source to stabilize at a new frequency after a change is initiated. In the PLL-based synthesizers it depends upon the transient response characteristics of the loop. The switching time is typically several hundreds of microseconds to tens of milliseconds in PLL-based synthesizers and a few microseconds in instruments using the direct digital synthesis technique.

16.15.4.3 Signal Purity

The *signal purity* tells how well the output signal approximates the ideal single spectral line. Phase noise is one parameter that affects signal purity. This refers to the sidebands that result from phase modulation of the carrier by noise. It is specified as the total sideband power (in decibels) with respect to the carrier. The presence of spurious signals resulting from undesired coupling between different circuits within the instrument and distortion products in the signal mixers also spoils signal purity.

16.15.5 Synthesized Function Generators

Synthesized function generators are function generators with the frequency precision of a frequency synthesizer. The hardware of a synthesized function generator is similar to that of a frequency synthesizer with additional circuitry to produce pulse, ramp, triangle and square functions. These instruments with additional modulation capability are referred to as *synthesized signal generators*.

Direct digital synthesis described in the earlier pages of this chapter is almost invariably used in synthesized function/signal generator design. Advances in digital technology have made these synthesized function/signal generators truly versatile. Synthesized sine wave output up to 30 MHz and other functions such as pulse, ramp, triangle, etc., up to 100 kHz, all with a resolution of 1 μHz, are available in contemporary synthesized function generators.

Figure 16.31 shows one such synthesized function generator (Fluke 271 DDS function generator) that employs direct digital synthesis for achieving a high level of stability. It offers sine, square, triangle and ramp outputs of up to 10 MHz.

16.15.6 Arbitrary Waveform Generator

The *arbitrary waveform generator* (AWG) is a signal source that is used to generate user-specified custom analogue waveforms. Using a custom stimulus waveform and measuring the response waveform provides realistic characterization of the device or system under test. The contemporary AWG allows generation of almost any conceivable waveform.

Direct digital synthesis again is the heart of an arbitrary waveform generator. Figure 16.32 shows the hardware. It looks very similar to the one shown in Fig. 16.29. The sequential amplitude values of the waveform to be generated are stored in the RAM. The size of the RAM decides the number of samples that can be stored, which in turn decides the maximum number of samples into which one period of the desired waveform can be divided. These sample values can be entered into the RAM from the keyboard. Once the sample values are loaded into the RAM, they can be stepped through at a repetition rate governed by the *frequency word* input to the phase accumulator in the same way as explained in the case of a frequency synthesizer. The complexity of the waveform that can be synthesized by this process is limited by the size of the RAM. As a rule of thumb, a minimum of about 3–4 samples per cycle of the highest frequency in the waveform should be used. This is intended to eliminate aliasing. Figure 16.33 shows a typical arbitrary waveform possible in a typical arbitrary waveform generator.

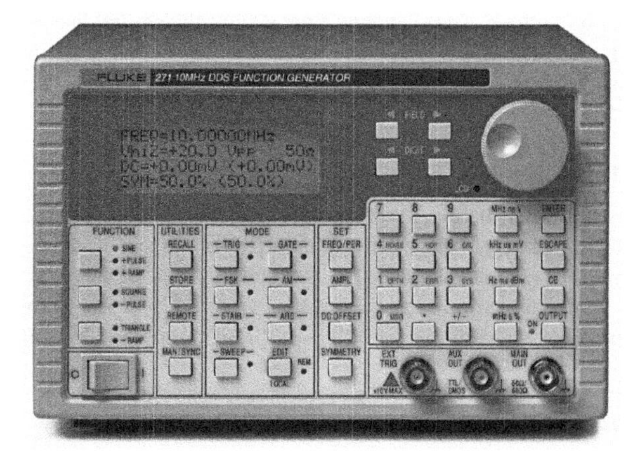

Figure 16.31 Synthesized function generator. Reproduced with permission of Fluke Corporation.

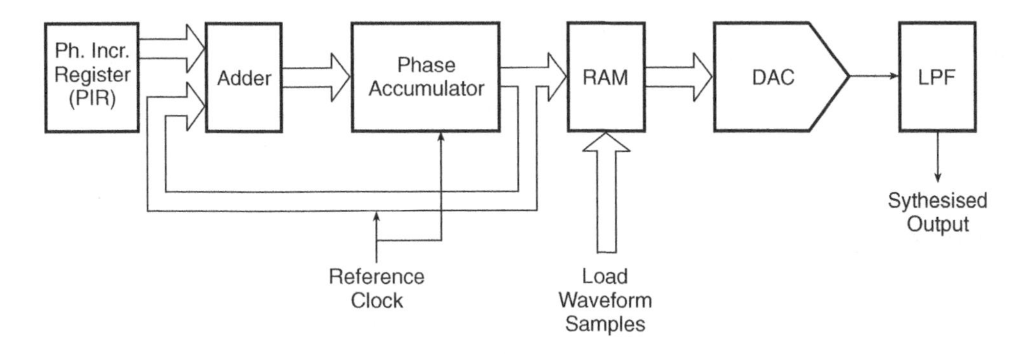

Figure 16.32 Arbitrary waveform generator architecture.

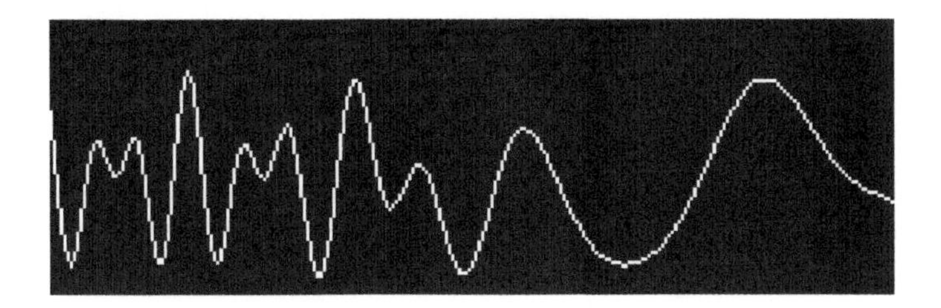

Figure 16.33 Arbitrary waveform generator – typical waveform.

16.16 Logic Probe

The *logic probe* is the most basic tool used for troubleshooting of digital circuits. It is a small, handheld pen-like test instrument with a metallic tip on one end (Fig. 16.34). The instrument can be used to ascertain the logic status of various points of interest such as the pins of digital integrated circuits in a digital circuit. The logic status is indicated by a glowing LED. There may typically be three LEDs

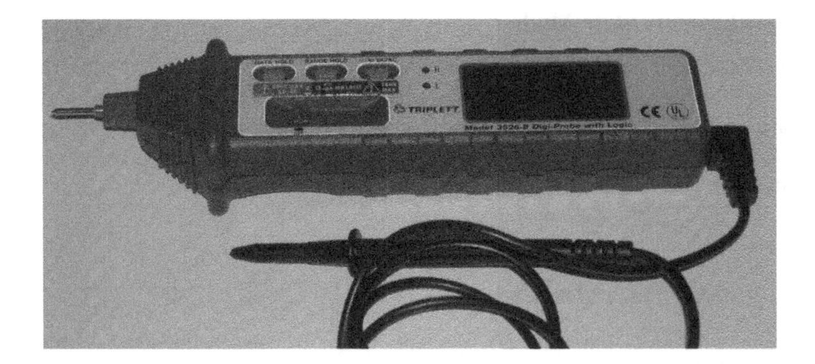

Figure 16.34 Logic probe. Reproduced with permission of Phoronix/Michael J. Larabel.

to indicate logic HIGH, logic LOW and Hi-Z states. Depending upon the actual logic status where the probe is touched, the corresponding LED comes on. The probe can be used to determine open and short circuits. Also, the probe has settings for different logic families to accommodate different acceptable voltage level ranges for logic LOW and HIGH status for different logic families.

16.17 Logic Analyser

The *logic analyser* is used for performance analysis and fault diagnosis of digital systems. Logic analysers have become a very relevant and indispensable diagnostic tool in the present-day instrumentation scenario, with the whole gamut of electronic instruments being centred on microprocessor/microcomputer-based digital architecture. In addition, most logic analysers can be configured to format their outputs as a sequence of microprocessor instructions, which makes them useful for debugging software too.

16.17.1 Operational Modes

The logic analyser works in one of two modes of operation, namely the asynchronous timing mode and the synchronous state mode. A brief description of each of these two modes is given in the following paragraphs.

16.17.1.1 Asynchronous Timing Mode

In this mode of operation, the signals being probed are recorded either as logic '0' or logic '1'. The logic analyser provides the time base referred to as the 'internal clock. The time base determines when data values are clocked into the memory of the analyser. On screen, the asynchronous mode display looks similar to an oscilloscope display except for the number of channels that can be displayed, which is much larger in the case of a logic analyser.

16.17.1.2 Synchronous State Mode

In this mode of operation, samples of signals are stored in the memory on a clock edge, referred to as the external clock, supplied by the system under investigation. The logic analyser samples new data values or states only when directed by the clock signal. On a given clock edge, the logic states of various signals constitute a group. The logic analyser display in this mode shows progression of states represented by these groups.

16.17.2 Logic Analyser Architecture

Figure 16.35 shows the block schematic arrangement of a logic analyser. Important constituents of all logic analysers include probes, memory, trigger generator, clock generator, storage qualifier and user interface.

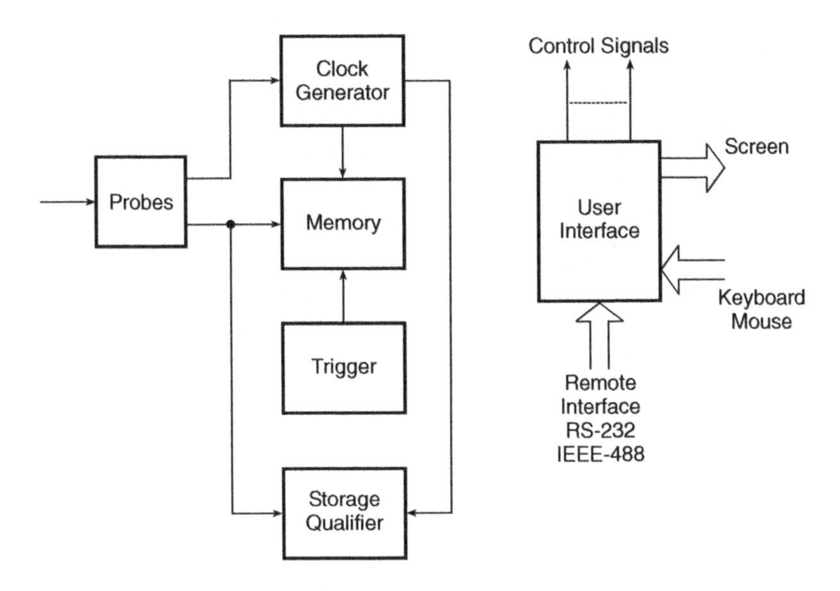

Figure 16.35 Logic analyser architecture.

16.17.2.1 Probes

Probes are used to provide physical connection to the circuit under test without causing any loading effects, so that the logic signal of interest is not unduly affected and its timing integrity is maintained. The probes usually operate as voltage dividers. By this, the comparators inside the probe are asked to handle the lowest possible voltage slew rate that enables higher-speed signals to be captured. These comparators have an adjustable threshold to make the probes compatible with different logic families as different families have different voltage thresholds. The comparators transform the input signals into logic ls and 0s.

16.17.2.2 Memory

The memory stores the sampled logic values. Addresses for given samples are supplied internally. In a typical measurement using a logic analyser, the user is interested in observing the logic signals around some event called the measurement trigger and the samples have a timing relationship with this trigger event. These samples are placed in the memory, depending upon the instantaneous value of the internally supplied address.

16.17.2.3 Trigger

Logic analysers have both a combinational (or word-recognized) trigger mode and an external trigger mode. In the combinational trigger mode, the trigger circuitry compares the incoming data with a word programmed by the user from the front panel. A trigger signal is generated when the incoming data match with the programmed word. Data are being sampled and stored in the memory by either an internal or an external clock. On the occurrence of a trigger, the stored data samples are displayed on the screen.

16.17.2.4 Clock Generator

As stated earlier, the clock is either internal or external, depending upon whether the selected operational mode is the asynchronous timing mode or the synchronous state mode. The two modes were described in Section 16.17.1. Again, in the timing mode there are two commonly used approaches. Some logic analysers offer both approaches.

In the first approach, called the continuous storage mode, the clock is generated at the selected rate irrespective of the activity occurring on the input signals. The logic status of the input signal is stored in the memory on every clock cycle [Fig. 16.36(a)]. In the second approach, called the transitional timing mode, the input signals are again sampled at the selected rate, but the clock generator circuitry allows the samples to be stored in the memory only if one or more signals change their logic status. Thus, the memory storage locations are used only if inputs change, leading to more efficient use of memory. For each sample, however, a time marker is recorded, as shown in Fig. 16.36(b).This approach offers a distinct advantage when long time records of infrequent or bursts of finely timed events are to be recorded.

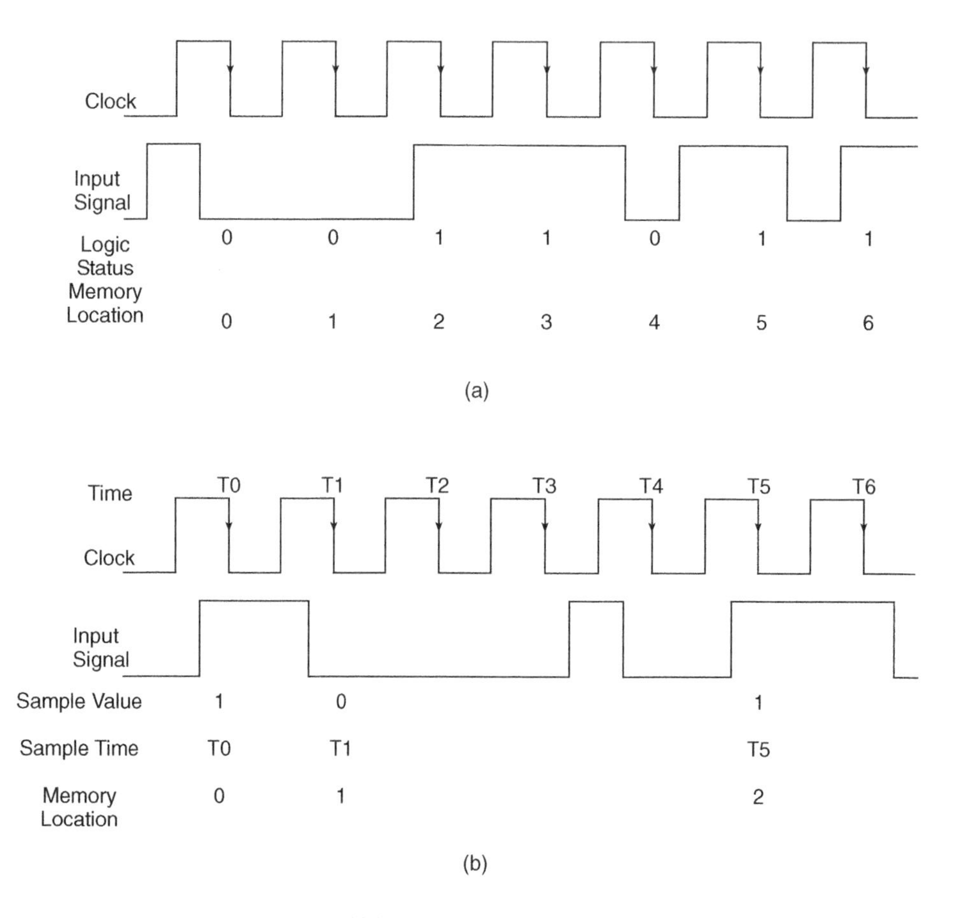

Figure 16.36 Logic analyser – relevant waveforms.

16.17.2.5 Storage Qualifier

The function of the storage qualifier is to determine which data samples are clocked into the memory. The storage qualifier block looks at the sampled data and tests them against a criterion. If the criterion is met, the clocked sample is stored in the memory. This feature is particularly useful in troubleshooting microprocessor architectures. For instance, if the circuit under test is a microprocessor bus, this function can be used to separate bus cycles to a specific I/O port from cycles to all other ports or from instruction cycles.

16.17.2.6 User Interface

Benchtop logic analysers typically use a dedicated keyboard and CRT display. Graphical user interfaces (GUIs) similar to those available on personal computers are also available with many products. Also, interfaces such as RS-232, IEEE-488 or local area network (LAN) enable the use of the instrument from a personal computer or a workstation. Remote interfaces are important in manufacturing applications. LAN interfaces have emerged as critical links in research and development activities where these instruments can be tied to project databases.

16.17.3 Key Specifications

Some of the important specifications of logic analysers include sample rate, set-up and hold times, probe loading, memory depth and channel count. Trigger resources, the availability of preprocessors/inverse assemblers, nonvolatile storage and the ability of the logic analyser to store time value along with captured data are the other key features.

16.17.3.1 Sample rate

The *sample rate* in the timing mode determines the minimum resolvable time interval. Since the relationship of the sample clock and the input signal transition is random, two edges of the same signal can be measured to an accuracy of two sample periods. Measuring a transition on one signal with respect to a transition on another signal can also be done with an accuracy of two sample periods plus whatever skew exists between the channels. In the state mode, the sample rate determines the maximum clock rate that can be measured in the target state machine.

16.17.3.2 Set-up and Hold Times

The *set-up and hold time* specification in the case of logic analysers is similar to that in the case of flip-flops, registers and memory devices. Like these devices, a logic analyser also needs stable data for a specified time before the clock becomes active. This specified time is the set-up time. The *hold time* is the time interval for which the data must be held after the active transition of the clock to enable data capture. The hold time is typically zero for logic analysers.

16.17.3.3 Probe Loading

It is desired that the target system not be perturbed by probe loading. Logic analysers with a sampling rate of equal to or less than 500 MHz have probe specifications of typically 100K and 6–8 pF. Analysers having a sample rate greater than 1 GHz usually come with SPICE models for their probes so as to enable the users to know the true impact of probe loading on signal integrity.

16.17.3.4 Memory depth

The *memory depth* determines the maximum time window that can be captured in the timing mode or the total number of states or bus cycles that can be captured in the state mode. Most of the logic analysers offer 4K to 1M samples of memory.

16.17.3.5 Channel count

Channel count is the number of available input channels. Together with maximum rate, channel count determines the cost of instrument.

16.18 Computer–Instrument Interface Standards

Quite often, in a complex measurement situation, more than one instrument is required to measure a parameter. In another situation, the system may require a large number of parameters to be measured simultaneously, with each parameter being measured by a dedicated instrument. In such measurement situations, the management of different instruments becomes very crucial. This has found a solution in automated measurement set-ups where various instruments are controlled by a computer. Another reason for instruments being placed into such automated measurement set-ups is to achieve capabilities that the individual instruments do not have. If there were a single instrument that did all the measurements the user required, automated them and compiled all the data in the required format, probably there would be no need for an integrated system. The probability of a single system doing all this is extremely remote when there are a large number of different measurements to be made. Yet another reason for having a computer-controlled instrument system is that it enables the user to make measurements faster and free of any human error.

In an integrated measurement set-up there has to be transfer of data back and forth between different instruments and also between individual instruments and computer. Different interface standards have evolved to allow transfer of data. The IEEE-488 interface is the most commonly used one for the instrument–computer interface. This and some of the other popular interface standards are briefly discussed in the following paragraphs.

16.18.1 IEEE-488 Interface

The IEEE interface has evolved from the Hewlett-Packard interface bus (HP-IB), also called the general-purpose interface bus (GP-IB). Presently, it is the standard interface bus used internationally for interconnecting programmable instruments in an automated measurement set-up.

Figures 16.37(a) and (b) show the general interface and bus structure of IEEE-488/HP-IB. Figure 16.37(a) shows the interconnection of different types of programmable device such as talkers, listeners, controllers, etc. A listener is an instrument that can only receive data from other instruments. A printer is an example of a listener-type instrument. A talker such as a frequency counter is capable of transmitting data to the other instruments connected to the bus. There are some instruments that may perform both the functions. In the listening mode, they receive instructions to carry out certain

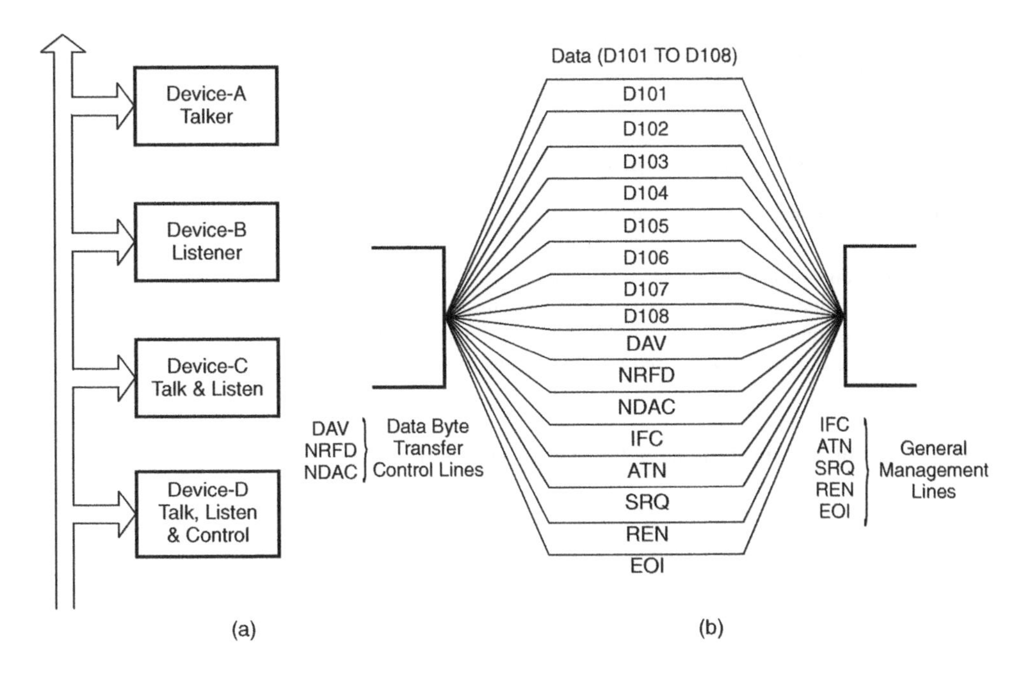

Figure 16.37 IEEE-488 interface standard.

measurements, and in the talking mode they transmit the results of measurements. A controller is supposed to manage the interface bus.

The interface bus has 16 lines and uses a 24-pin connector. A maximum of 15 devices/instruments can be connected to this interface bus in parallel. A typical data rate is 250–500 kbytes/s over the full transmission path which ranges from 2 to 20 m. The maximum data rate is 1 Mbyte/s. There are eight lines dedicated for data transfer (D-101 to D-108) in bit parallel format. There are three lines for data byte transfer control (DAV, NRFD and NDAC) and five lines for general interface management (IFC, ATN, SRQ, REN and EOI). Different lines in the interface bus carry addresses, program data, measurement data, universal commands and status bytes to and from the devices interconnected in the system. The data lines are held at +5 V for logic '0' and pulled to ground for logic '1'.

The other popular instrument interface buses that allow interconnection of stand-alone instruments and computers are the VXI-bus, the PCI bus and the MXI-bus. These interface buses are more relevant to the fast-growing concept of virtual instrumentation and therefore are discussed in the next section on virtual instruments.

16.19 Virtual Instrumentation

Advances in software development and rapid increase in the functional capabilities available on the PC platform have changed the traditional instrumentation scenario. The scene is fast changing from the box-like conventional stand-alone instruments to printed circuit cards offering various instrument functions. These cards are inserted either into a card cage, called the mainframe, or into a PC slot. These acquire the measurement data which are then processed in the computer and subsequently displayed on the monitor in a format as required by the user. Such an instrumentation concept is commonly referred to as *virtual instrumentation.*

16.19.1 Use of Virtual Instruments

There are four types of virtual instrumentation set-up:

1. A set of instruments used as a virtual instrument.
2. A software graphical panel used as a virtual instrument.
3. Graphical programming techniques used as a virtual instrument.
4. Reconfigurable building blocks used as a virtual instrument.

16.19.1.1 Set of Instruments as a Virtual Instrument

In complex measurement situations, usually more than one instrument is required to do the intended measurement. An instrumentation set-up that is used to qualify various subsystems and systems for electromagnetic compatibility (EMC) is an example. In such a set-up, as shown in Fig. 16.38, the computer receives measurement data from all the stand-alone instruments, works on the data and then displays the measurement results. Another similar set-up that has been customized to perform a certain test on a certain specific product, however, would not be classified as a virtual instrument.

16.19.1.2 Software Graphical Panel as a Virtual Instrument

In this type of virtual instrumentation set-up, the instrumentation hardware is controlled by a personal computer from a keyboard or a mouse. The PC screen is used to display the measurement results (Fig. 16.39). The instrumentation hardware could be a traditional box-like instrument or a PC card offering the desired measurement function. The computer control of the instrument is through an interface bus such as IEEE-488.

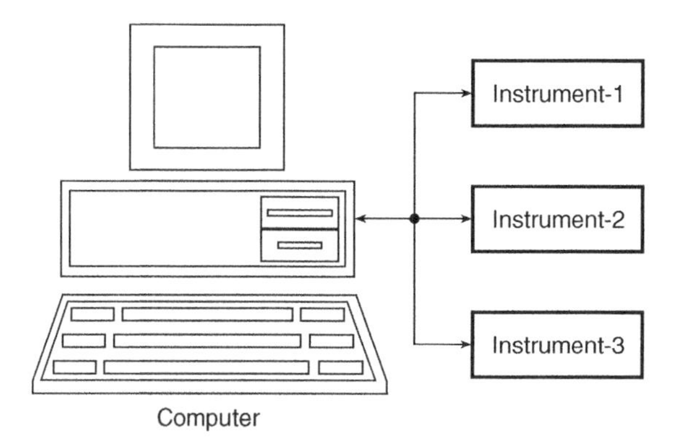

Figure 16.38 Set of instruments as a virtual instrument. Reproduced with permission of Fluke Corporation.

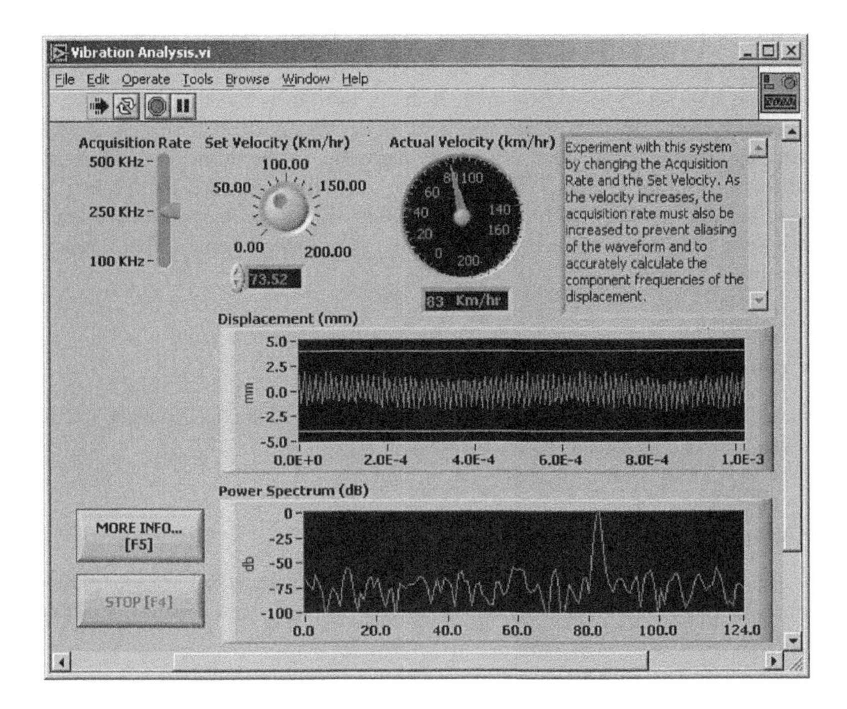

Figure 16.39 Software graphical panel as a virtual instrument. Photo courtesy of National Instruments Corporation.

16.19.1.3 Graphical Programming Technique as a Virtual Instrument

In a typical computer-controlled instrument set-up, the software to do the job is written using a textual programming language such as C, BASIC, Pascal and FORTRAN. Owing to the constant increase in computer power and instrument capabilities, the development of software that makes full use of the instrumentation setup has become a tedious and time-consuming job if it is done using one of the available textual programming languages. There has been a distinct trend to move away from the conventional programming languages and to move towards graphical programming languages. A graphical programming equivalent of a program is a set of interrelated icons (graphical objects) joined by lines and arrows. The use of a graphical programming language leads to a drastic reduction in programming time, sometimes by a factor as large as 10.

Having written a graphical program for a certain test, all icons appear on the screen with programmed interactions. It may be mentioned here that with graphical language the instrument control as well as the program flow and execution are determined graphically. A graphical programming product lists the interface buses and instruments that are supported by it. Graphical programming languages are typically used where one wants to decrease the effort needed to develop a software for instrument systems. However, they require substantial computing power, and the size of these programs can reduce the speed of application in some cases.

16.19.1.4 Reconfigurable Building Blocks as a Virtual Instrument

If one looks into the building blocks of various instruments, one is sure to find a lot of commonality. Building blocks such as front ends, A/D converters, D/A converters, DSP modules, memory modules, etc., are the commonly used ones. One or more of these building blocks are invariably found in voltmeters, oscilloscopes, spectrum analysers, waveform analysers, counters, signal generators and so on. In an instrumentation set-up comprising more than one instrument function there is therefore likely to be lot of redundant hardware.

A fast-emerging concept is to have instrument hardware in the form of building blocks that can be configured from a graphical user interface (GUI) to emulate the desired instrument function. These building blocks could be reconfigured at will to become voltmeters, oscilloscopes, spectrum analysers, waveform recorders and so on. A graphical panel would represent each virtual instrument.

16.19.2 Components of a Virtual Instrument

The basic components of a virtual instrument as shown in Fig. 16.40 are the computer and display, the software, the bus structure and the instrument hardware.

16.19.2.1 Computer and Display

The majority of virtual instruments are built around personal computers or workstations with high-resolution monitors. The chosen computer should meet the system requirements as dictated by the software packages.

16.19.2.2 Software

The software is the brain of any virtual instrument set-up. The software uniquely defines the functional capabilities of the instrument set-up, and in most cases it is designed to run industry-standard operating systems for personal computers and workstations.

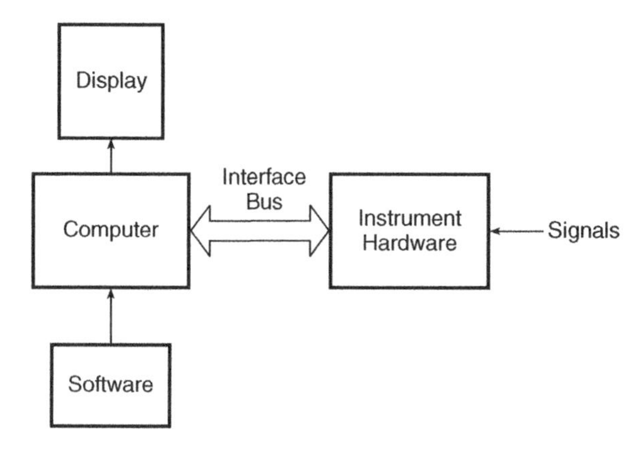

Figure 16.40 Components of a virtual instrument.

16.19.2.3 Interface Bus Structure

Commonly used interface bus structures for a computer–instrument interface are the IEEE-488, described in Section 16.18.1, the PC-bus and the VXI-bus. The other two are described here.

In a PC-bus virtual instrument set-up, the instrument function available on a printed circuit card (Fig. 16.41) is inserted directly into a vacant slot in the personal computer. Since these cards are plugged directly into the computer back plane and contain no embedded command interpreter as found in IEEE-488 instruments, these cards are invariably delivered with driver software so that they can be operated from the computer. PC-bus instruments offer a low-cost solution to building a data acquisition system. Owing to the limited printed circuit space and close proximity to sources of electromagnetic interference, PC-bus instruments offer a lower performance level than their IEEE-488 counterparts.

VXI-bus instruments are plug-in instruments that are inserted into specially designed card cages called mainframes (Fig. 16.42). The mainframe contains power supplies, air cooling, etc., that are common to all the modules. VXI-instruments combine the advantages of computer back-plane buses and IEEE-488. A VXI-bus instrument has high-speed communication as offered by computer back-plane buses (such as the VME-bus) and a high-quality EMC environment that allows high-performance instrumentation similar to that found in IEEE-488 instruments.

One of the methods to communicate with VXI instruments is via IEEE-488, as shown in Fig. 16.43. In this case, an IEEE-488 to VXI-bus converter module is plugged into the VXI-bus mainframe. The mainframe then interfaces with the IEEE-488 interface card in the computer using the standard interface cable. The set-up is easy to program, but the overall speed is limited by the IEEE-488 data transfer rate.

Another technique is to use a higher-speed interface bus between the hardware mainframe and the computer. One such bus is the MXI-bus, which is basically an implementation of the VXI-bus on a flexible cable. In this case, the VXI-MXI converter is plugged into the mainframe and an MXI-interface

Figure 16.41 Instrument function on a PC card. Photo courtesy of National Instruments Corporation.

Figure 16.42 VXI-bus instruments. Photo courtesy of National Instruments Corporation.

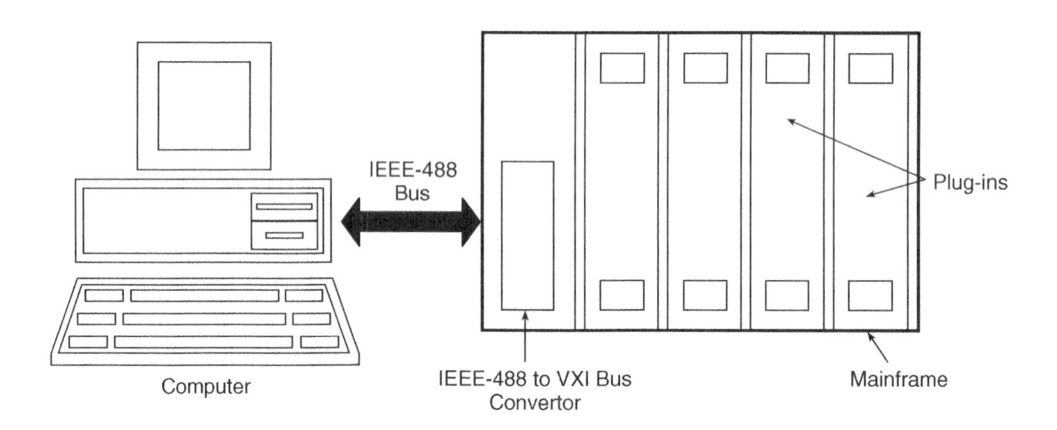

Figure 16.43 VXI instrument–PC interface using IEEE-488.

card with software is installed in the computer. This set-up allows the use of off-the-shelf PCs to communicate with VXI instruments at speeds much faster than IEEE-488 instruments.

Yet another approach is to insert a powerful VXI-bus computer in the hardware mainframe to take full advantage of the VXI-bus instruments. The disadvantage of such a set-up is that, owing to the low volume requirement of VXI computers, these may not be able to match the industry standard personal computers on the price performance criteria. The set-up is shown in Fig. 16.44.

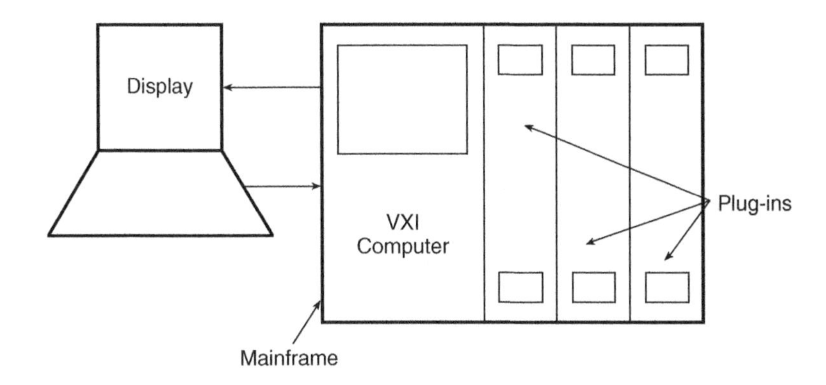

Figure 16.44 VXI-bus computer inside the hardware mainframe.

16.19.2.4 Instrument Hardware

The instrument hardware comprises of sensors and other hardware components that acquire the data and condition it to a level and form so that it can be processed in the computer to extract the desired results.

Review Questions

1. Briefly outline the different steps involved in the troubleshooting of digital circuits. In what possible ways can an internal open or short circuit in a digital IC manifest itself?
2. Why is the troubleshooting of sequential logic circuits a more cumbersome task than in the case of combinational logic? Explain with the help of a suitable illustration.
3. Briefly describe commonly used methods of diagnosing faulty ROM and RAM devices.
4. Distinguish between an analogue storage oscilloscope and a digital storage oscilloscope. Briefly describe the major performance specifications of analogue and digital scopes.
5. With reference to a digital storage oscilloscope, briefly explain the following:

   (a) How does the effective sampling rate depend upon the acquisition memory?
   (b) What do you understand by real-time sampling and equivalent-time sampling?
   (c) What is the difference between bandwidth and sampling rate?

6. Briefly describe the counter architecture when it is used in:

   (a) frequency measurement mode;
   (b) time interval measurement mode.

7. What are reciprocal counters? How does a reciprocal counter provide a much higher resolution even when the frequency of the signal is very low?
8. Briefly describe the following with respect to frequency counters:

   (a) bandwidth;
   (b) resolution;

(c) accuracy;
(d) throughput.

9. Write short notes on:

   (a) sampled sine synthesis;
   (b) virtual instrumentation.

10. Briefly describe various test and measurement functions that can be performed by a logic analyser. Distinguish between asynchronous and synchronous modes of operation of a logic analyser.

Problems

1. Figure 16.45 shows a *D* flip-flop wired around a *J–K* flip-flop that belongs to the TTL family of devices. The *D* input in this circuit has been permanently tied to V_{CC}. The logic probe observations at the *J* and *K* inputs respectively show logic HIGH and logic LOW status, as expected. The *Q* output of this circuit is supposed to go to logic HIGH status with the first LOW-to-HIGH transition of the clock input. However, the *Q* output is observed to be a pulsed waveform with the frequency of the signal being one-half of the clock frequency. What is the most probable cause of this unexpected behaviour of the circuit?

   The K input of the J-K flip-flop is internally open. The K input is therefore floating and behaves as if it were in the logic HIGH state. This converts it into a toggle flip-flop

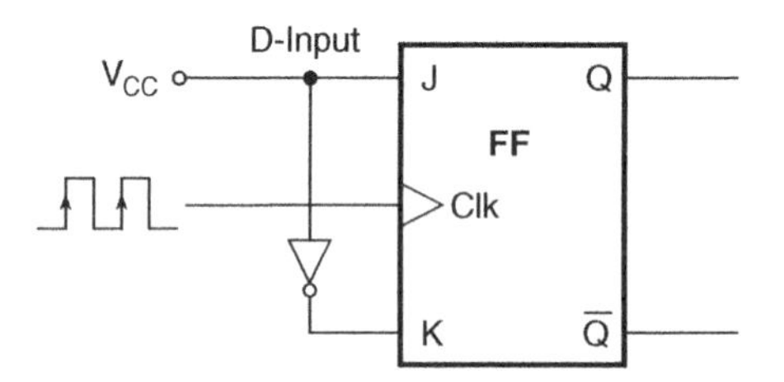

Figure 16.45 *D* flip-flop (problem 1).

2. Figure 16.46 shows the block schematic arrangement of a three-bit ring counter configured around three *D* flip-flops. The expected and actual outputs of the flip-flops for the first few clock cycles are listed in Table 16.4. Each of the flip-flops has a propagation delay of 15 ns. Identify the possible cause of observed outputs being different from the expected outputs. The clock signals appearing at the clock input terminals of the flip-flops when seen individually are observed to be clean and free of any noise content. Flip-flops FF-1 and FF-0 are initially cleared to the logic '0' state. The *Q* output of FF-2 is initially in the logic '1' state.

   The fault is possibly due to the clock skew problem. The clock input to FF-1 is delayed from the clock input to FF-2 by a time period that is greater than 15 ns. Also, the clock input to FF-0 is delayed from the clock input to FF-1 by a time period that is greater than 15 ns

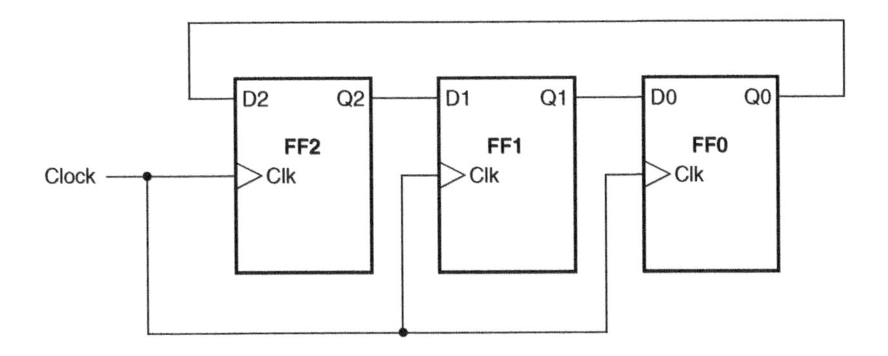

Figure 16.46 Three-bit ring counter (problem 2). Photo Courtesy of National Instruments Corporation.

Table 16.4 Problem 2.

Clock pulse	Expected output			Actual output		
	Q_2	Q_1	Q_0	Q_2	Q_1	Q_0
0	1	0	0	1	0	0
1	0	1	0	0	0	0
2	0	0	1	0	0	0
3	1	0	0	0	0	0
4	0	1	0	0	0	0
5	0	0	1	0	0	0
6	1	0	0	0	0	0

3. A digital storage oscilloscope is specified to have a sample rate of 400 MS/s and an acquisition memory of 20K. (a) Determine the slowest possible time-base setting for which the specified sample rate is achievable. (b) If the time-base setting were 1 ms per division, what sampling rate would be achievable in this case?

   (a) 5μs/div; (b) 2 MS/s

4. A transient of 100 ms is to be captured on a digital storage oscilloscope on full screen in the horizontal direction. If the transient is to be recorded at a sampling rate of 100 kS/s, what should the minimum size of the acquisition memory be?

   10K

Further Reading

1. Tomal, D. and Widmer, N. S. (2004) *Electronic Troubleshooting*, McGraw-Hill, USA.
2. Coombs Jr, C. F. (1999) *Electronics Instrument handbook*, McGraw-Hill Inc., USA.
3. Webster, J. G. (1999) *The Measurement, Instrumentation and Sensors Handbook*, CRC Press (in cooperation with IEEE Press), FL, USA.
4. Whitaker, J. C. (1996) *The Electronics Handbook*, CRC Press (in cooperation with IEEE Press), FL, USA.

Index

Digital Electronics: Principles, Devices and Applications Anil Kumar Maini

Printed and bound by CPI Group (UK) Ltd, Croydon, CR0 4YY

16/04/2025

14658384-0005